Biology *for the* Informed Citizen

WITH PHYSIOLOGY

Biology *for the* Informed Citizen

WITH PHYSIOLOGY

Donna M. Bozzone
Saint Michael's College

Douglas S. Green
Saint Michael's College

New York   Oxford
OXFORD UNIVERSITY PRESS

Oxford University Press is a department of the University of Oxford.
It furthers the University's objective of excellence in research,
scholarship, and education by publishing worldwide.

Oxford   New York
Auckland   Cape Town   Dar es Salaam   Hong Kong   Karachi
Kuala Lumpur   Madrid   Melbourne   Mexico City   Nairobi
New Delhi   Shanghai   Taipei   Toronto

With offices in
Argentina   Austria   Brazil   Chile   Czech Republic   France   Greece
Guatemala   Hungary   Italy   Japan   Poland   Portugal   Singapore
South Korea   Switzerland   Thailand   Turkey   Ukraine   Vietnam

Copyright © 2014 by Oxford University Press

For titles covered by Section 112 of the US Higher Education Opportunity
Act, please visit www.oup.com/us/he for the latest information about
pricing and alternate formats.

Published in the United States of America by
Oxford University Press
198 Madison Avenue, New York, NY 10016
http://www.oup.com

Oxford is a registered trademark of Oxford University Press.

CIP data is on file at the Library of Congress
ISBN: 978-0-19-538199-3

Printing number: 9 8 7 6 5 4 3 2

Printed in the United States of America
on acid-free paper

For all of our teachers, especially Allison and Samantha

Contents in Brief

Contents

UNIT 2. REPRODUCTION, INHERITANCE, AND EVOLUTION

CHAPTER 5 Cancer 127
How Can It Be Prevented, Diagnosed, and Treated?

case study Xeroderma pigmentosum 128

CHAPTER 6. Reproduction 155
What Kind of Baby Is It?

case study The Fastest Woman on Earth 156

CHAPTER 7 Plants, Agriculture, and Genetic
Engineering 185
*Can We Create Better Plants
and Animals?*

case study   Golden Rice 186

CHAPTER 8 Health Care and the Human
Genome 217
*How Will We Use Our New Medical
and Genetic Skills?*

case study   Carrie Buck and the American
Eugenics Movement 218

CHAPTER 9 Evolution 245
How Do Species Arise and Adapt?

case study Lactose Intolerance and the Geographic Variation of Human Traits 246

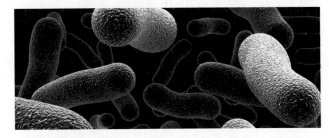

CHAPTER 10 The Evolution of Disease 283
Why Do We Get Sick?

case study Deadly Malaria 284

UNIT 3. PHYSIOLOGY: THE BODY IN HEALTH AND DISEASE

CHAPTER 11 Homeostasis 313
*Why Is It Important That the Body
Maintain Its Internal Balance?*

case study   Max Gilpin 314

CHAPTER 12 Circulation and Respiration 343
*What If Your Body Doesn't Get the
Oxygen It Needs?*

case study   Blood Doping at the Tour de
France 344

CHAPTER 13 The Nervous System 377
Does Your Brain Determine Who You Are?

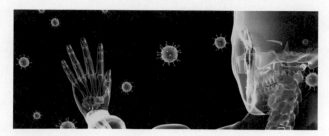

CHAPTER 14 Infectious Disease and the Immune System 411

How Are Invaders Repelled, Evaded, or Killed?

CHAPTER 15 Nutrition, Activity, and Wellness 439

How Can We Live a Healthy Lifestyle?

UNIT 4. INTERACTING WITH NATURE

CHAPTER 18 Human Population Growth 531
*How Many People Can a Single
Planet Hold?*

case study   A Story About Bacteria 532

Preface

We wrote *Biology for the Informed Citizen* because we love biology and are convinced that everyone should have a basic understanding of biology to function as a fully engaged, contributing member of society. As we researched topics, designed our approach, and wrote, we were thinking of you—students majoring in fields other than biology. You possess perspectives, interests, dispositions, and expectations that differ somewhat from those of most students majoring in biology, and that is what makes it so much fun and so rewarding to teach you.

One of the challenges in trying to foster an understanding and appreciation of the importance of biology is that our educational system and society tend to compartmentalize science rather than seeing it as a central aspect of modern life. In this era of deep specialization, we are even more in need of conversation, communication, and understanding among specialists and one another. In reality, the integration of knowledge—not simply within biology but also among sciences, social sciences, humanities, and the arts in general— is essential for confronting and finding solutions to the challenges we all face. You have the potential to play an important role in meeting these challenges and helping to find solutions precisely because your particular interests allow you to see biology from different perspectives. And the biology you will learn will enrich your understanding of and strengthen the connections among the things you already know.

Biology for the Informed Citizen presents biology in the context of important cultural and social issues you are likely to encounter now and in the future. In writing this book, we chose to address biology in a way that will help you learn what you need to know about biology to make informed decisions in your life; become effective, engaged citizens; and understand, at least in principle, the new opportunities and challenges modern biology provides. Although you may be interested in studying biology for its own sake, we recognize that you may be most interested in the consequences of biology: what it says about your health, disease, and the environment, for example.

Although our motivation for writing this book was to teach you, along with the guidance of your course instructor, the major concepts of biology, evolution, and the process of science so that you can apply your knowledge as informed consumers and users of scientific information, we also benefited in some unexpected ways. We became more informed scientists, teachers, and parents. We hope that you have as much pleasure reading and learning as we did in creating this book for you.

Sadly, Doug Green, my husband and coauthor of this book, passed away before its completion. However, I am delighted to see the project come to fruition.

Donna Bozzone
Saint Michael's College
Colchester, Vermont

xix

Approach: Cases, Concepts, and Consequences

Our goal for *Biology for the Informed Citizen* was to write a book that, more than any other non-science-majors biology book, helps students **connect the concepts of biology to the consequences of biology**—the consequences that students can and should see in every facet of their lives, if only they are trained to identify them. This text teaches the concepts of biology, evolution, and the process of science so that students can apply their knowledge as informed consumers and users of scientific information.

In order to help students become biologically and scientifically literate, we wove two major themes into every chapter: the process of science and the theory of evolution. Our rationale is that if students are going to learn and then apply what they have learned, they need to know **not only "what we know" but also "how we know what we know."** Therefore, each chapter includes stories of real scientists who had interests and curiosities—not altogether different from some of the students reading this text. We hope these stories will motivate students to think critically in their daily lives. Throughout this book we also emphasize the theory of evolution—the most central of all biological concepts—to help students see the big picture underlying the magnificent diversity and awe-inspiring mechanisms of the living world.

Features

Because students come to their biology courses, and this text, with a rich set of interests, we included features to help students make connections between their present knowledge and the biology they are learning.

Case Studies

Each chapter opens with a rich case study that highlights an issue or challenge with biological significance and focuses on the consequences of biology. These cases motivate the material in each chapter and demonstrate ways in which an understanding of biology can be used to make informed decisions about important issues. Examples of cases we introduce include "Sickle Cell Disease, Malaria, and Human Evolution" (Chapter 4); "The 'Infidelity Gene'" (Chapter 1), and "Dying for the 'Perfect Body'" (Chapter 15), which address how genes influence our health, personal relationships, and body image, respectively. In the remaining sections of the chapter, we weave in the biology needed for a fuller understanding of the issue or challenge. As a result, students learn specific biological concepts in a context that shows why they are important and enables students to make connections between biology and other fields of study.

case study

Sickle Cell Disease, Malaria, and Human Evolution

FIGURE 4.1 Malaria is caused by the parasite *Plasmodium falciparum*, which is carried by the *Anopheles* mosquito. When one of these mosquitoes bites an individual, the parasitic cells take up residence in the blood, destroying red blood cells, the cells that carry oxygen through the body.

malaria :: a disease caused by a parasite carried by the *Anopheles* mosquito; it is often fatal

protozoan :: a simple, single-celled organism

red blood cell :: a cell that carries oxygen through the blood

Tony Allison grew up in Kenya. His father, a farmer, had relocated the family from England in 1919. As a boy, Tony went on long excursions with professional naturalists to observe and help collect birds for the Natural History Museum in London. He also visited the archeological excavation site of Louis Leakey, the preeminent anthropologist of his time, and became intrigued with human evolution and the relationships among the various tribes he saw in Kenya.

During one of the Allisons' holidays on the beaches of Malindi in Kenya, Tony contracted malaria; he was only 10 years old. **Malaria** is a terrible disease and often fatal. It is caused by a **protozoan**, a simple single-celled organism, and carried by a specific type of mosquito. When one of these mosquitoes bites a person, it injects this protozoan, *P. falciparum*, into the individual (**FIGURE 4.1**). These parasitic cells take up residence in the blood and destroy **red blood cells**, the cells that carry oxygen through the blood. There is currently no vaccine against malaria. Tony's experience with malaria led him to switch gears: rather than becoming a naturalist, he decided to become a physician.

After earning his undergraduate degree in South Africa, Tony moved to England to finish his medical training at Oxford University. Although he enjoyed medical school, he had not lost his keen interest in human evolution. Tony was convinced that there had to be a way to measure human evolutionary relationships more precisely than anthropologists did by looking only at bones.

In 1949, when Oxford University sent a group of scientists to Kenya to survey and study plants and animals all over the country, Tony jumped at

Process of Science

Biology in particular, and science in general, represent one way of asking questions and evaluating the answers; it is not the only way. Still, the specific manner in which scientists go about learning about the natural world is both powerful and successful. And as a way of thinking, it is practical for many questions, not just scientific ones. Thus, we highlight the process of science in the *How Do We Know?* essays featured in each chapter. These essays help students move beyond memorizing facts to get them thinking critically about how we know what we know.

All too often, people do not appreciate the human dimension of biology. Everything ever discovered or solved is literally the result of a person or many people thinking that the question being pursued was the *most* interesting and important thing to be found. They simply *had* to work on this problem—it was like an itch that had to be scratched. *Scientist Spotlight* essays in each chapter provide biographical information and historical context about the real individuals whose scientific discoveries have made tremendous impacts on all of our lives.

Real-World Applications

Biology does not exist as a disconnected field of study. In fact, to understand biology well, one needs to be conversant with the ways that biology connects to the larger culture. The inverse is also true: to understand our culture fully, one needs to be familiar with biology. More specifically, biological research, ideas, and knowledge intersect with global issues, ethics, and social responsibility. *Life Application* essays in each chapter present real-world examples illustrating how biological knowledge has been used to help individuals and society at large make informed decisions on a range of issues.

Advances in scientific research directly affect us in our day-to-day lives. A great deal of what is known in biology and continues to be studied depends on the development and implementation of specific methods and techniques. *Technology Connection* essays in each chapter provide students with information on specific methods or techniques that biologists use to answer questions. These essays show how the tools of scientific research are being used to shape the world in which we live.

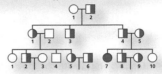

Pedagogy

Every chapter in *Biology for the Informed Citizen* includes carefully crafted tools to help students learn and reinforce biological concepts.

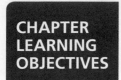

CHAPTER LEARNING OBJECTIVES

After reading this chapter, you should be able to answer the following questions:

4.1 What Is Sickle Cell Disease?
Specify the cause and consequences of sickle cell disease and the relationship betwe...

4.2 C...
Prev... Explain...
sickle...
molec...

4.1 What Is Sickle Cell Disease?
Specify the cause and consequences of sickle cell disease and the relationship between sickle cell disease and malaria.

4.3 Where Is Our Genetic Information Stored?
Describe the study that disproved the concept of pangenesis.

4.4 How Did Mendel Discover the Rules of Inheritance?
Outline how Mendel conducted his experiments related to inheritance and the resulting set of rules.

4.5 How Much Do Mendel's Rules Explain?
Identify the three reasons that Mendel's rules fail to explain inheritance completely.

Chapter Learning Objectives at the start of each chapter (based on Bloom's taxonomy) correspond to the main headings and provide a framework for the key concepts to help students focus on what is most important.

Question-Based Chapter Titles and Section Headings model the spirit of inquiry at the heart of the scientific process.

Simple and Clear Illustrations in each chapter help students visualize important concepts. The art program uses a consistent format to help guide students through complex processes. For example, Chapter 2 introduces the steps of the scientific method, and the figures that highlight scientific experiments throughout the book help to reinforce these steps—blue: observations and facts; purple: hypotheses; pink: predictions; light green: hypothesis testing; and dark green: evaluation and/or results. Brief *figure captions* provide a running summary of the chapter, reinforcing main points discussed in the text.

FIGURE 4.11 In the 1880s, August Weismann performed an experimental test of pangenesis using mice. Weismann concluded that the information to construct the tails of the mice did not reside in the tail itself.

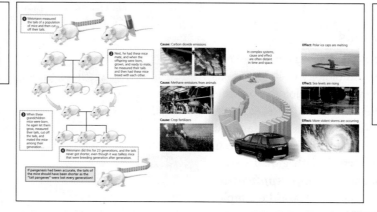

FIGURE 2.7 Every event or outcome in nature has a cause or source. A scientist can learn about causes by observing the effects that occur; note that in this figure all of the causes on the left contribute to the effects on the right.

A *Marginal Glossary* defines key terms in the margins of the pages on which the terms appear so students can easily find definitions and explanations when preparing for exams.

allele :: a different form of a gene

Each chapter concludes with a *Biology in Perspective* section that places the chapter concepts in larger context.

Bulleted *Chapter Summaries* at the end of each chapter are organized around the chapter learning objectives and highlight and reinforce the main concepts.

Chapter Summary

4.1 What Is Sickle Cell Disease?
Specify the cause and consequences of sickle cell disease and the relationship between sickle cell disease and malaria.
- Sickle cell disease is a severe inherited genetic disorder.
- In sickle cell disease, red blood cells that are normally disc shaped and flexible become rigid and pointy shaped.
- Sickled cells get stuck in blood vessels, which can lead to blood clots and other serious physical consequences, including kidney failure, paralysis, and heart failure.
- The inheritance of the gene associated with sickle cell disease also confers resistance against malaria.

4.2 Could Molecular Medicine Prevent Sickle Cell Disease?
Explain how Linus Pauling's research into sickle cell disease ushered in the age of molecular medicine.
- Linus Pauling and his colleagues examined two types of hemoglobin (HbA and HbS).

4.4 How Did Mendel Discover the Rules of Inheritance?
Outline how Mendel conducted his experiments related to inheritance and the resulting set of rules.
- Mendel chose his experimental organism carefully: garden peas, which are simple to cultivate and whose mating can be controlled manually.
- Mendel focused on one trait at a time and followed the cross for more than one generation, collecting quantitative data and keeping detailed records.
- Mendel's research revealed some basic rules of inheritance.
  o Genes can come in more than one form, or allele.
  o Alleles of a particular gene sort individually into gametes during meiosis.
  o Certain traits are dominant, while others are recessive.

4.5 How Much Do Mendel's Rules Explain?
Identify the three reasons that Mendel's rules fail to explain inheritance completely.

Basic multiple choice and short-answer *Review Questions* at the end of each chapter ask students to recall core information presented in the chapter. Answers to the multiple choice questions appear at the end of the book.

The Thinking Citizen advanced questions at the end of each chapter ask students to think critically and analytically about the main chapter concepts.

The Informed Citizen advanced questions at the end of each chapter ask students to apply biological concepts to relevant cultural and social issues.

Review Questions

1. What evidence did Dr. Allison use to understand the relationship between malaria and sickle cell disease?
2. What happens to the blood cells in an individual with sickle cell disease? What can trigger a sickling incident?
3. If an individual inherits an HbS protein and an HbA protein, which of the following is likely to be true?
   a. The individual will likely develop full-blown sickle cell disease.

6. What were the principal discoveries that Mendel made from his monohybrid crosses?
   a. Genes can come in more than one form; some traits are dominant and others are recessive; pangenesis is a proven theory.
   b. Genes can come in more than one form; the alleles of a particular gene sort individually into gametes during meiosis; some traits are dominant and others are recessive.

The Informed Citizen 125

The Thinking Citizen

1. How could strenuous physical activity and/or high altitude potentially lead to organ failure in a person with sickle cell disease?
2. Predict the outcomes of matings between the following if the theory of pangenesis were correct:

of the sickle cell trait if malaria were to be eliminated entirely? Why?
5. Imagine you checked the news, online or print, and you read that the "gene for alcoholism" had been isolated. Is this claim likely to be correct? Why or why not?

The Informed Citizen

1. Tony Allison supported his theory about the relationship between malaria and the selective advantage of the sickle cell trait by doing three studies that involved human subjects. Is it ethical to do research on humans? If so, what safeguards are necessary? If not, why not? How would you learn about human structure and function?
2. In one of the studies that Allison joined, a pharmaceutical company was testing its drug to see whether it was an ef-

are unequivocally associated with sickle cell disease, cystic fibrosis, Huntington's chorea, or muscular dystrophy. Should all individuals be screened? If so, what should be done with this information? Should it be off-limits to insurance companies and employers, or do they have a right to know to minimize their own financial risks? Is it acceptable to refuse screening even though an individual might add to the healthcare burden of society? Is it acceptable to force

Organization

Biology for the Informed Citizen covers the **foundational concepts** that comprise a standard non-science-majors biology course but does so on a "need-to-know" basis, placing biological topics within the context of important cultural and social issues, but **without excessive detail**. We thought carefully about which topics to include and which to omit, with the goal of providing the needed biological coverage in a framework that we hope students will enjoy reading! This book is organized into four units. (*Biology for the Informed Citizen* is also available "*without Physiology*": omitting five chapters on homeostasis; circulation and respiration; the nervous system; infectious disease and the immune system; and nutrition, activity, and wellness.)

Unit 1: The Scientific Study of Life

In the first unit, we introduce the main themes of the text. In **Chapter 1**, we address the relevance of biology to informed citizenship, how biology affects our view of nature, ourselves, and our society, and we also introduce the main features of life. Chapter 1 emphasizes the centrality of evolution in biology, an emphasis we maintain throughout the text. **Chapter 2** addresses the importance of science to the study of biology and medicine. We discuss the scientific method in theory and by example, and make several points about the philosophy of science. Finally, we compare science with other ways of knowing and contrast science and pseudoscience.

Unit 2: Reproduction, Inheritance, and Evolution

In the second unit, we explore the molecular, cellular, and evolutionary basis of life. The first three chapters provide foundational information about cells, genes, and inheritance. In **Chapter 3**, we use human development, from fertilization to childbirth, to introduce our discussion of cell biology. **Chapter 4** explores how inheritance works and how the molecules DNA, RNA, and protein contribute to the production of the physical and functional characteristics of organisms, including those seen in sickle cell disease. In **Chapter 5**, we address cancer and use it to motivate our discussion of how cell division is regulated.

In the next three chapters, we emphasize topics related to informed citizenship. These topics include fertility and genetic screening, cloning and genetic engineering, and gene therapy and stem cells. **Chapter 6** examines reproduction. We think that understanding the biology of reproduction is essential for both responsible family planning and stewardship of our natural resources. Our discussion of genetically engineering plants in **Chapter 7** provides an opportunity to introduce plant structure and photosynthesis. **Chapter 8** addresses biological determinism and the extent to which genes explain complex human characteristics.

The unit concludes with a closer examination of evolution, addressing the evolutionary basis of medicine and health. In **Chapter 9**, we complete the story of what evolution is and how it works, pointing out interesting and puzzling aspects of the natural world and showing how they can be accounted for by evolution operating over long periods of time. In **Chapter 10**, we examine why it is that we get sick, how diseases evolve, and what evolution can tell us about how diseases can be prevented and controlled.

Unit 3: Physiology: The Body in Health and Disease

The premise of the third unit is that making good, individual decisions about health is an act of informed citizenship. In this unit, we examine physiology from three perspectives: how the body monitors and regulates its internal state, how the functioning parts of the body communicate to coordinate their activities, and how physiology mediates the body's response to the outside world.

We interweave these themes throughout the unit. In **Chapter 11**, we introduce homeostasis, illustrating the concept with thermoregulation and a discussion of water balance in the urinary system. **Chapter 12** introduces circulation and respiration, **Chapter 13** examines the nervous system, and **Chapter 14** addresses the immune system. In **Chapter 15**, we examine the digestive and musculoskeletal systems and discuss diet and dietary supplements, steroids, and eating disorders to arrive at suggestions for a healthy lifestyle.

This approach allows us to cover the major organ systems while emphasizing the fact that organ systems work together to produce good health and a properly functioning body.

Unit 4: Interacting with Nature

The final unit looks at ecology. In **Chapter 16**, we examine how individuals of different species interact with one another and with nature and how those interactions drive an ecosystem's function. **Chapter 17** discusses biodiversity and the impact humans have on biodiversity and the impact of biodiversity on us. Finally, in **Chapter 18**, we examine human population growth and its consequences. Advances in agriculture, medicine, engineering, and technology have supported healthy population growth for most of our history, but the continued growth of the human population is unsustainable. Because we can recognize the problem, however, we can also discover ways to solve it.

Learning Package

Oxford University Press offers instructors and students a comprehensive ancillary package, designed to help students become fully informed citizens and to assist instructors in meeting this objective. The following resources are available for qualified adopters of *Biology for the Informed Citizen*.

For Students

Companion Website at www.oup.com/us/bozzone

The FREE and OPEN companion website offers a number of study tools, including online quizzes (over 1000 questions) and a curated guide to relevant animations, videos, podcasts, and more.

Biology for the Informed Citizen Dashboard Online Homework

Dashboard delivers quality content, tools, and assessments to track student progress in an intuitive, web-based learning environment. Assessments created by Everett Weber, Brandi King, and Heather Passmore (Murray State University) are designed to accompany *Biology for the Informed Citizen* and automatically graded so instructors can easily check students' progress as they complete their assignments. The color-coded gradebook illustrates at a glance where students are succeeding and where they can improve so on-the-fly instructors can adapt lectures to student needs. For students, this means quality content and instant feedback. Dashboard features a streamlined interface that connects instructors and students with the functions they perform most, and simplifies the learning experience by putting student progress first. All Dashboard content is engineered to work on mobile devices, including the iOS platform. Our goal is to create a platform that is simple, informative, and mobile. For more information about Dashboard please visit www.oup.com/dashboard.

Study Guide

Authored by Sharon Gilman (Coastal Carolina University), who is also the author of the test item file (see below for more info), the Study Guide provides students with brief summaries and step-by-step analyses of each chapter, additional review questions, and thoughtful advice and study tips. (ISBN 9780199958016)

For Instructors

In addition to Dashboard described above, qualified adopters of *Biology for the Informed Citizen* can access the following teaching tools for immediate download at the companion website (**www.oup.com/us/bozzone**):

Text Image Library

All text images are available in PowerPoint and jpeg formats. Instructors who adopt the text gain access to every illustration, photo, figure caption, and table from the text in high-resolution electronic format, and some multipart figures are optimized by being broken down into their constituent parts for clear projection in large lecture halls. The image library includes:

- **Art slides** in PowerPoint and jpeg formats with figures exactly as they appear in the text

- **Unlabeled art slides,** in which text labels are turned off
- **Lecture note slides** with outlines for each chapter that can be edited, which makes preparing lectures faster and easier than ever

Instructor's Resource Manual with Video and Animation Guide

The Instructor's Resource Manual is a collection of materials designed to help instructors build and implement a course around *Biology for the Informed Citizen*. The manual includes several kinds of supplemental instructional aids ranging from a list of chapter learning outcomes, to full chapter outlines and summaries, to question prompts for in-class discussions and activities. At the heart of the manual is a *curated and annotated guide to high-quality and freely available animations, movie clips, videotaped lectures, podcasts, and presentations* of core concepts covered in the text, all vetted and collected in one place for convenient access. The Video and Animation Guide for each chapter includes a web link to a customized YouTube playlist, which includes several relevant videos that highlight, illustrate, and expand on the concepts covered in the text.

Test Item File

Written by Sharon Gilman (also author of the Study Guide), the test item file includes over 1200 multiple choice, true/false, and short-answer questions in editable Microsoft Word format. Questions are organized by chapter section number and learning objectives, and each item is identified according to Bloom's taxonomic categories of knowledge, comprehension, and application.

Computerized Test Bank

Using the test authoring and management tool Diploma, the computerized test bank that accompanies this text is designed for both novice and advanced users. Diploma enables instructors to create and edit questions, create randomized quizzes and tests with an easy-to-use drag-and-drop tool, publish quizzes and tests to online courses, and print quizzes and tests for paper-based assessments. Available on the Ancillary Resource Center (ARC).

Ancillary Resource Center (ARC)

The Ancillary Resource Center (ARC) is a convenient, instructor-focused single destination for resources to accompany your text. Accessed online through individual user accounts, the ARC provides instructors access to up-to-date ancillaries at any time while guaranteeing the security of grade-significant resources. In addition, it allows OUP to keep instructors informed when new content becomes available. The ARC for *Biology for the Informed Citizen* includes the Test Item File, the Computerized Test Bank, the Text Image Library, and the Instructor's Resource Manual with Video and Animation Guide. For more information about ARC please visit www.oup-arc.com.

Ebook

Available through CourseSmart.

Acknowledgments

There are many people who deserve heartfelt thanks for their support and help throughout the creation of this book. These generous individuals fall into four categories: family, friends, the wonderful people at Oxford University Press, and the many individuals who reviewed chapters and artwork throughout many stages of this project. In the family category, we are grateful for the encouragement, support, and love of our daughters, Samantha and Allison. Their interests, good questions, perspectives, and suggestions made this a better book. Also, they never lost patience—at least not noticeably—with their distracted parents. We would also like to thank Bill and Janet Bozzone for their moral support and good dinners, too.

All of our friends in the biology department at Saint Michael's deserve a thank you for good discussions and for making it a lovely thing to come to work. Denise Martin, Declan McCabe, and Doug Facey went above and beyond the call of duty in ways that are known to them and difficult to describe. Along with those three, my friends and colleagues in our education department Valerie Bang-Jensen and Mary Beth Doyle kept me going. We are also appreciative of the support that Saint Michael's College provided (so much so that the College is included in the friend category), especially a sabbatical leave for Doug in fall 2009 and one for Donna in spring 2010.

The team at Oxford University Press was extraordinary. Jason Noe, senior editor, in his cheerful, tenacious way kept this project moving forward with the right combination of encouragement (prodding) and keeping hands off. He is terrific. John Haber, our initial development editor, was instrumental in helping us to say what we were trying to convey in a more lucid and pleasing way. When he left OUP, we felt a bit panicked, but fortunately our fears turned out to be utterly unfounded. Lisa Sussman, senior development editor, stepped in, and she has been fantastic—a gifted editor and talented writer in her own right, as well as indispensable for the development of the *Biology for the Informed Citizen* art program. We would like to thank the editorial assistants who helped us over the years of developing the text, including Melissa Rubes, Katie Naughton, Caitlin Kleinschmidt, and Andrew Heaton. We would also like to thank Patrick Lynch, editorial director; John Challice, vice president and publisher; Jason Kramer and David Jurman, marketing managers; Christine Naulty and Meghan Daris, marketing assistants; Frank Mortimer, director of marketing; Jolene Howard, market development; and Bill Marting, national sales manager. We would also like to thank the exceptional and dedicated production team who helped take this book from ideas and drafts to a final, published reality, including Lisa Grzan, production manager; Barbara Mathieu, senior production editor; Michele Laseau, art director; and the team at Precision Graphics.

Reviewers, Class Testers, and Focus Group Participants

We are especially grateful to the extraordinary group of over 200 dedicated colleagues teaching non-science majors who provided thoughtful commentary as reviewers, class testers, and focus group participants as we developed the text's manuscript, illustrations, and supplements program. We started this journey with the goal of providing instructors with a set of tools that could help them reach out to their students. We wanted students to see how by becoming scientifically literate, they could improve their lives and those of their fellow citizens. Your comments and suggestions were invaluable in that effort and in helping us to refine the final version of the first edition.

Reviewers

Over the course of development, we extensively reviewed *Biology for the Informed Citizen* at over 145 colleges and universities, with approximately 175 reviewers. We read each review and incorporated feedback wherever we could in order to develop this first edition so it would be the best option for you and your students. We extend our heartfelt appreciation to the following reviewers:

Manuscript Reviewers

Sylvester Allred	Northern Arizona University
Tara Devi S. Ashok	University of Massachusetts Boston
Yael Avissar	Rhode Island College
Ellen Baker	Santa Monica College
Andrew Baldwin	Mesa Community College
Roberta Batorsky	Middlesex County College
Emma Benenati	Northern Arizona University
Morgan Benowitz-Fredericks	Bucknell University
Brenda Bourns	Seattle University
Mark Buchheim	The University of Tulsa
Sara Carlson	The University of Akron
Aaron Cassill	The University of Texas at San Antonio
Deborah Cato	Wheaton College
Michelle Cawthorn	Georgia Southern University
Thomas T. Chen	Santa Monica College
Thomas F. Chubb	Villanova University
John L. Clark	University of Alabama
Michael F. Cohen	Sonoma State University
Claudia Cooperman	University of South Florida
James W. Cosgrove	Montgomery College
Helen Cronenberger	The University of Texas at San Antonio
Michael S. Dann	Pennsylvania State University
Paula Raelynn Deaton	Sam Houston State University
Buffany DeBoer	Wayne State College
Leif D. Deyrup	University of the Cumberlands
Hartmut Doebel	The George Washington University
Paul Farnsworth	University of New Mexico
Michele Finn	Monroe Community College
Susan W. Fisher	Ohio State University
Brandon Lee Foster	Wake Technical Community College
Lori Frear	Wake Technical Community College
Wendy Jean Garrison	University of Mississippi
Vaughn M. Gehle	Southwest Minnesota State University
Sharon L. Gilman	Coastal Carolina University
Mary Gobbett	University of Indianapolis
Brandon Groff	Eastern Michigan University
Laine Gurley	Harper College
Kristy Halverson	The University of Southern Mississippi

Janelle Hare	Morehead State University
Mesha Hunte-Brown	Drexel University
Allison Hunter	Worcester Polytechnic Institute
Evelyn F. Jackson	University of Mississippi
Arnold Karpoff	University of Louisville
Christopher J. Kirkhoff	McNeese State University
Peter Kourtev	Central Michigan University
Jeff Kovatch	Marshall University
Ellen Shepherd Lamb	The University of North Carolina at Greensboro
Ann S. Lumsden	Florida State University
Molly MacLean	University of Maine
Lisa Maranto	Prince George's Community College
Karen McCort	Eastern New Mexico University, Ruidoso Branch Community College
Diane L. Melroy	University of North Carolina Wilmington
Tim Metz	Campbell University
Scott M. Moody	Ohio University
Brenda Moore	Truman State University
Cynthia E. Morgan	Austin Peay State University
Mario Mota	University of Central Florida
Ann Murkowski	North Seattle Community College
Rajkumar Nathaniel	Nicholls State University
Fran Norflus	Clayton State University
Paul Eugene Olson	University of Central Oklahoma
David Oppenheimer	University of Florida
Wiline Pangle	Central Michigan University
Mark A. Paulissen	Northeastern State University
Ashley Ramer	The University of Akron
Terri S. Richardson	California State University, Northridge
Darryl Ritter	Northwest Florida State College
Pamela Sandstrom	University of Nevada, Reno
Georgianna Saunders	Missouri State University
Malcolm Schug	The University of North Carolina at Greensboro
Caryn Self-Sullivan	Georgia Southern University
Alice Sessions	Austin Community College
Justin Shaffer	North Carolina Agricultural and Technical State University
Edward A. Shalett	Mesa Community College
Jack Shurley	Idaho State University
Ayodotun Sodipe	Texas Southern University
Bethany B. Stone	University of Missouri
Mark Thogerson	Grand Valley State University
Jeffrey Thomas	Queens University of Charlotte
William Unsell	University of Central Oklahoma
Sheela Vemu	Northern Illinois University
Janet Vigna	Grand Valley State University
Timothy S. Wakefield	John Brown University
Kristen L.W. Walton	Missouri Western State University
Everett Weber	Murray State University
Aimee K. Wurst	Lincoln University

Illustration Reviewers

Sylvester Allred	Northern Arizona University
Megan Anduri	California State University, Fullerton
Yael Avissar	Rhode Island College
Emma Benenati	Northern Arizona University

Gayle Birchfield	Austin Peay State University
Lisa G. Bryant	Arkansas State University Beebe
Mark Buchheim	The University of Tulsa
Sara Carlson	The University of Akron
Aaron Cassill	The University of Texas at San Antonio
Deborah Cato	Wheaton College
Hartmut Doebel	The George Washington University
Lori Frear	Wake Technical Community College
Wendy Jean Garrison	University of Mississippi
Vaughn M. Gehle	Southwest Minnesota State University
Renaud Geslain	College of Charleston
Brandon Groff	Eastern Michigan University
Arnold Karpoff	University of Louisville
Janet E. Kübler	California State University, Northridge
Diane L. Melroy	University of North Carolina Wilmington
Krista Peppers	University of Central Arkansas
Malcolm Schug	The University of North Carolina at Greensboro
Justin Shaffer	North Carolina Agricultural and Technical State University
Jack Shurley	Idaho State University
Ayodotun Sodipe	Texas Southern University
Bethany B. Stone	University of Missouri
Jeffrey Thomas	Queens University of Charlotte
Kristen L.W. Walton	Missouri Western State University

Supplements Program Reviewers

Kim Atwood	Cumberland University
Yael Avissar	Rhode Island College
Roberta Batorsky	Middlesex County College
Erin Baumgartner	Western Oregon University
Sara Carlson	The University of Akron
Deborah Cato	Wheaton College
Buffany DeBoer	Wayne State College
Hartmut Doebel	The George Washington University
Lori Frear	Wake Technical Community College
Seth Jones	University of Kentucky
Peter Kourtev	Central Michigan University
Brenda Leady	University of Toledo
Rajkumar Nathaniel	Nicholls State University
Justin Shaffer	North Carolina Agricultural and Technical State University
Jeffrey Thomas	Queens University of Charlotte
Jennifer Wiatrowski	Pasco-Hernando Community College
Dwina W. Willis	Freed-Hardeman University

Market Development Reviewers

We would also like to thank the extensive list of faculty at over eighty schools who participated in our first-edition market development review program for sharing their insights and suggestions.

Fernando Agudelo-Silva	College of Marin
Justin Anderson	Radford University
Kim Atwood	Cumberland University
Melissa Barlett	Mohawk Valley Community College
Katrinka Bartush	University of North Texas

Tonya Bates	University of North Carolina at Charlotte
Jane Beers	John Brown University
Tiffany Bensen	University of Mississippi
Donna Bivans	Pitt Community College
Michelle Boone	Miami University
Steven Brumbaugh	Green River Community College
Rebecca Bryan	Metropolitan State University of Denver
Lisa Bryant	Arkansas State University Beebe
Stylianos Chatzimanolis	University of Tennessee at Chattanooga
Peter Chen	College of DuPage
Catherine Chia	University of Nebraska Lincoln
Claudia Cooperman	University of South Florida
Jacquelyn Duke	Baylor University
Kari Eamma	Tarrant County College Northeast
Deborah Gelman	Pace University
Sandra Gibbons	Moraine Valley Community College
Paul Gier	Huntingdon College
Mary Gobbett	University of Indianapolis
Andrew Goliszek	North Carolina A&T State University
Laine Gurley	Harper College
Judy Haber	California State University Fresno
James Harper	Sam Houston State University
Olivia Harriott	Fairfield University
Timothy Henkel	Valdosta State University
Kelley Hodges	Gulf Coast State College
Jill Holliday	University of Florida
Sue Hum-Musser	Western Illinois University
Carl Johansson	Fresno City College
Scott Johnson	Wake Technical Community College
Staci Johnson	Southern Wesleyan University
Claudia Jolls	East Carolina University
Barbara Juncosa	Citrus College
Brenda Leady	University of Toledo
Sarah Leupen	University Of Maryland Baltimore City
David Luther	George Mason University
Paul Luyster	Tarrant County College
James Malcolm	University of Redlands
Christiane Meyer Healey	University of Massachusetts Amherst
Liza Mohanty	Olive-Harvey College
Jamie Moon	University of North Florida
Rajkumar Nathaniel	Nicholls State
Dana Newton	College of The Albemarle
Trent Nguyen	Tarrant County College
Fran Norflus	Clayton State University
Brandi Norman	University of North Carolina at Pembroke
Katrina Olsen	University Of Wisconsin Oshkosh
Mark Pilgrim	Lander University
Nicola Plowes	Mesa Community College
Mary Poffenroth	San Jose State University
Gerald Posner	Broward College
Vanessa Quinn	Purdue University North Central
Erin Rempala	San Diego City College
Angel Rodriguez	Broward College
Rob Ruliffson	Minneapolis Community and Technical College
Lynn Rumfelt	Gordon State College
Ann Rushing	Baylor University

Michael Rutledge	Middle Tennessee State University
Anna Schmidt	University of Wisconsin Platteville
Erik Scully	Towson University
Jyotsna Sharma	University of Texas at San Antonio
Heidi Sleister	Drake University
Marc Smith	Sinclair Community College
Ayodotun Sodipe	Texas Southern University
Kathryn Spilios	Boston University
Sonja Stampfler	Kellogg Community College
Louise Steele	Kent State University
Zuzana Swigonova	University of Pittsburgh
Pamela Thinesen	Century College
Rani Vajravelu	University of Central Florida
Sheela Vemu	Northern Illinois University
Paul Verrell	Washington State University
Stephen Wagener	Western Connecticut State University
Lisa Weasel	Portland State University
Jennifer Wiatrowski	Pasco-Hernando Community College
Dwina Willis	Freed-Hardeman University
Kimberly Zahn	Thomas Nelson Community College
Martin Zahn	Thomas Nelson Community College
Ted Zerucha	Appalachian State University

Class Testers

Over the course of development, we extensively class-tested *Biology for the Informed Citizen* at eighteen schools with eighteen instructors and 400 student participants. We thank all of them sincerely for their involvement, comments, and enthusiasm, which helped shape the final version of the first edition.

Michael Anders	Tarrant County College
Gayle Birchfield	Austin Peay State University
William Blaker	Furman University
Karen Bledsoe	Western Oregon University
Brenda Bourns	Seattle University
Yavuz Cakir	Benedict College
Sarah Cooper	Arcadia University
James Courtright	Marquette University
Kenneth Filchak	University of Notre Dame
Lori Frear	Wake Technical Community College
Andrew Goliszek	North Carolina A&T State University
MaryKate Holden	Greensboro College
Staci Johnson	Southern Wesleyan University
Jennifer Landin	North Carolina State University
Jonas Okeagu	Fayetteville State University
Brian Shmaefsky	Lone Star College Kingwood
Ayodotun Sodipe	Texas Southern University
Aaron Sullivan	Houghton College

Focus Group Participants

We would also like to thank those individuals who participated in our focus group activities at the NABT (National Association of Biology Teachers) meetings:

Dallas, Texas, 2012:

| Erin Baumgartner | Western Oregon University |
| Claudia Womack Cash | Tarrant County College Northeast Campus |

Sehoya Cotner	University of Minnesota
Kathy Gallucci	Elon University
Kristy Halverson	University of Southern Mississippi
Mickey Laney-Jarvis	Rogue Community College
Sunita Rangarajan	Collin College
David Tanner	University of North Texas at Dallas
Lisa Turnbull	Lane Community College

Atlanta, Georgia, 2013:

Pieter deHart	Virginia Military Institute
Maria Isabel Fernandez	Georgia Gwinnett College
Kristy Halverson	University of Southern Mississippi
Jennifer Kneafsey	Tulsa Community College
Melissa Masse	Tulsa Community College
Fran Norflus	Clayton State University
Theus Rogers	Georgia Gwinnett College
Danilo Sanchez	San Joaquin Delta College
Brian Shmaefsky	Lone Star College Kingwood

Biology *for the* Informed Citizen

WITH PHYSIOLOGY

CHAPTER LEARNING OBJECTIVES

After reading this chapter, you should be able to answer the following questions:

1.1 How Does Biology Affect Your Life?

Explain how understanding biology can help you make informed decisions as a citizen.

1.2 What Are the Features of Life?

Use the four principal features of life to discriminate living things from nonliving things.

1.3 How Do Organisms Function?

Describe the relationship among atoms, macromolecules, and cells, and explain how enzymes maintain cell function.

1.4 How Do Organisms Reproduce?

Describe the relationship among DNA, genes, and alleles, and compare how genes are inherited in sexual and asexual reproduction.

1.5 How Does Life Evolve?

Create a scenario describing how a population of organisms might adapt to a new environment.

1.6 What Patterns of Diversity Are Found in Nature?

Explain the connection among common ancestors, the diversity of environments on Earth, and the scientific classification system.

The natural world is full of wonders, mysteries, and surprises.

CHAPTER 1

The Nature of Biology and Evolution

Why Does Biology Matter to You?

Becoming an Informed Citizen . . .

Biological research advances at an unprecedented rate. Breakthroughs change our society and our lives, providing us with many opportunities. However, they also raise moral, ethical, and personal challenges. To understand and evaluate these challenges, you must understand the science behind them.

utterflies are beautiful—but have you ever wondered why they are so brightly colored? Eating an apple or a cookie can give you the jolt you need to get going. How does your body use the energy stored in the foods that you eat so that you can function? People say you look a lot like your mother but not so much like your father. Why? The blind mole rat never sees a thing. Not only is it blind, it lives its entire life underground. But why does it still have eyes—and why are they completely covered with skin? And why are there over 1 million species of beetles but only four kinds of great apes, the group to which we belong?

The natural world is full of puzzles, mysteries, and surprises. Nature has inspired poets, artists, scientists, and explorers. Biology is your key to understanding nature. This book seeks to satisfy your curiosity about biology and the natural world.

But curiosity is only one reason for studying biology. Biological research advances at an unprecedented rate. Breakthroughs change our lives and our society. They provide us with many opportunities, but they also raise moral, ethical, and personal challenges. If you want to understand and evaluate these challenges, you will need to understand the science behind them. This book introduces you to concepts that will help you understand how biology plays an increasingly important role in your life.

To illustrate this role, consider John and Mary Smith: a married couple who suddenly realized they needed to know more biology.

case study

The "Infidelity Gene"

John and Mary Smith met in college and married soon after graduation. They were lucky enough to find good jobs in the same area, bought a condo, and settled into a comfortable life.

Eventually John and Mary decided to start a family. They heard about DNA testing: a new family planning procedure using DNA samples to screen for possible genetic differences that could affect their child's health and development. This sounded like a wise precaution, and they ordered the tests. When the results came back, John and Mary were relieved to learn that they each had a clean bill of genetic health as far as their future children were concerned. There was one personal glitch, however: the tests revealed that John had the "infidelity gene" (**FIGURE 1.1**).

Mary had never heard of an infidelity gene. In fact, Mary wasn't entirely sure what a gene was, and she had no idea how—or whether—a gene could cause her husband to be unfaithful. Mary consulted her doctor and searched online to find out more about this gene, but what she read was confusing. The following is a sample of her findings.

The term "infidelity gene" is the popular name for a gene called *AVPR1a*. (The names of human genes are always capitalized and italicized.) *AVPR1a* controls the production of a protein known as the *arginine vasopressin receptor*. This protein is found on the membranes of nerve cells in the brain and is responsible for transporting the hormone *arginine vasopressin* from the

FIGURE 1.1 The term "infidelity gene" is the popular name for an allele of the *AVPR1a* gene, which controls production of a protein responsible for transporting the hormone arginine vasopressin from the bloodstream to the nerve cell. This particular allele produces less protein and so transports less hormone, which some scientists believe results in less interest in a mate.

bloodstream to the nerve cell. Several versions of the *AVPR1a* gene are found in humans. Different versions of the same gene are called **alleles**. John's particular allele of *AVPR1a* made less protein than other alleles, which means that less of the hormone made it to John's brain cells. Some scientists conclude that less hormone means less interest in a mate, which could lead to infidelity.

Mary understood very little of what she learned about *AVPR1a*. How do genes control protein production? What are hormones? Why do hormones need receptors? What do hormones do inside your brain? The one thing Mary did understand, however, was that scientific studies showed that males with this particular allele of *AVPR1a* are more likely to commit adultery. Was their marriage in trouble?

allele :: different version of the same gene

1.1 How Does Biology Affect Your Life?

For most people, Mary's story will raise more questions than answers—just as it did for her. Can genes really control your behavior this way? (For further discussion on this topic, see *Technology Connection: Identifying the "Infidelity Gene."*) If you plan a family, should you get screened for possible genetic problems? What will you do if you learn of any? Do other genes affect our behavior

Technology Connection

Identifying the "Infidelity Gene"

Our case study is hypothetical, but the infidelity gene is real—though whether it affects humans is an open question.

The *AVPR1a* gene controls the transport of arginine vasopressin, a hormone, from the blood to the brain. Sexual activity and other social interactions between individuals release this hormone in the brain, which helps to make you feel good. Scientists believe this hormone helps to strengthen the pair bond between two individuals. If this is true, then individuals with alleles of *AVPR1a* that transport less of the hormone to brain cells may experience a weaker pair bond. Such individuals may therefore be more likely to seek sexual gratification outside the pair bond.

The original scientific work on *AVPR1a* was done on several species of *voles*, a mouselike organism, not on humans. The prairie vole is *monogamous*: one male and one female form a pair bond and mate only with each other. The meadow vole and the montane vole are *promiscuous*: both sexes mate several times with different partners, and no pair bond forms. When researchers examined the *AVPR1a* gene in these species, they discovered that the monogamous species and the promiscuous species had different versions (called alleles) of the gene. The popular press picked up on this result and labeled the allele found in the promiscuous species the "infidelity gene."

Researchers are now beginning to study humans to see whether versions of *AVPR1a* alleles have the same effect in us as they do in voles. In 2008, researchers studied 1000 couples in Sweden who had been together at least five years. Couples who reported being in a less intimate or less stable relationship were also more likely to have an allele of *AVPR1a* that transports less of the hormone to the brain cells. In fact, this allele is similar to the one found in promiscuous species of voles.

A second study examined game-playing behavior instead of potential infidelity. Scientists invited 200 subjects to play certain games. In this case, the scientists found that the

Prairie voles are monogamous; one male and one female form a pair bond and mate only with each other.

Meadow voles are promiscuous; both sexes mate several times with different partners and no pair bond forms.

The original scientific work on *AVPR1a* was done on several species of *voles*, a mouselike organism.

"infidelity" allele was associated with increased levels of selfish, merciless game playing. This result has led the popular media to give the allele a second name, the "ruthless gene." Neither name is a scientific term, and, to add to the confusion, both should be identified as an allele of a gene, not the gene itself.

A quick Internet search will yield thousands of hits for the infidelity gene or the ruthless gene, and even the prestigious scientific publication *Nature* has succumbed to calling these alleles by their popular names. And there really is a company that will test your DNA (or that of your significant other) to determine whether you have the allele—only US$99 as of this writing!

Despite the popular attention, these studies are at best preliminary. A relatively small number of people have been tested, and the studies have not been repeated. As we will see in Chapter 2, the scientific method relies on studies that can be repeated in order to see how widely applicable the results are. We will also see that an *association* between an allele and a behavior, even if it clearly exists, does not prove that the allele *causes* the behavior. Whether a test for this particular allele of the *AVPR1a* gene provides useful information is still an open question. ::

and health? Why do we have the genes we do—and why would anyone have genes that appear to do more harm than good?

Advances in biology have given us new abilities and a lot more information about our lives. But these advances require us to learn more if we want to use them effectively. One of the main goals of this book is to help you understand biology better and enable you to make better decisions about how to use your new biological skills. In later chapters, we will discuss in greater detail genes, hormones, receptors, and other questions raised here. For now, try putting yourself in John and Mary's place. What do you know, what do you need to know, and what actions would you take?

During your life, you'll face many questions about your personal health: diet, exercise, injuries, diseases and ways to treat them, and eventually (and sadly) questions about aging, cancer, cardiovascular problems, and end-of-life choices. Scientists and medical researchers will continue to discover new information that will affect the answers to all these questions. Of course, no one can keep up with the discoveries in all these areas. But luckily you won't have to go it alone. For all these decisions, you'll be working with doctors and counselors who can advise you about your choices. Even so, understanding the biology behind the choices will help you make decisions that are right for you.

Biology affects more than your personal life, however. Every year science and biology issues raise questions that affect your public life, life in your community, in your country, and on the planet. The Internet provides a wealth of information that is just a click away, a resource undreamed of a decade or two ago. You can consult newspapers, television broadcasts, blogs, and many other sources. But ultimately you need to evaluate what you learn and make responsible decisions that contribute positively to your life and the lives of others.

Suppose, for example, that you serve on a jury, and DNA evidence plays a large role in the case. Your understanding of the biology of DNA could affect whether you send a person to prison for life. Suppose you sit on the school board for your local district, and a citizen's group petitions to have intelligent design taught alongside evolution in science class. Your decision will influence the education of hundreds or even thousands of children (**FIGURE 1.2**). Is there a right to life, or a right to die? Should we support stem cell research or genetic

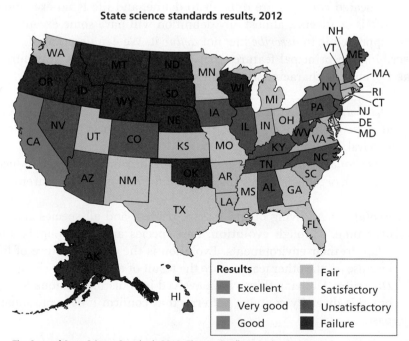

State science standards results, 2012

Results	
Excellent	Fair
Very good	Satisfactory
Good	Unsatisfactory
	Failure

The State of State Science Standards 2012, Thomas Fordham Institute:
http://www.edexcellencemedia.net/publications/2012/2012-State-of-State-Science-Standards/2012-State-of-State-Science-Standards-FINAL.pdf

Science writers display signs in advance of a hearing in 2008 by the Texas State Board of Education. The Board was preparing to hear testimony on school standards that would encourage teaching alternative explanations for evolution.

FIGURE 1.2 A 2012 analysis of science standards in education gave a majority of states mediocre to awful grades. One of the criteria used to determine the grades was whether or not state standards undermined evolution.

FIGURE 1.3 The diversity of life is astounding. Organisms come in an amazing array of shapes, sizes, and complexity.

evolution :: theory that explains how all living organisms are related and how existing populations adapt to their environments and new species arise

engineering, gay marriage, or DNA databanks? Should we allow the hunting of top predators, such as wolves, or allow farmers to plant genetically modified crops?

Your understanding of biology can influence how you exercise your most fundamental right and responsibility as a citizen: your vote. You might not vote on these issues directly, but you certainly will vote for legislators who will take positions on them and who will control funding for different avenues of scientific research. Good citizens take an active role in making decisions that affect their community, and making good decisions requires that you be informed about the issues.

The issues you will confront are complex and multifaceted—involving politics, economics, ethics, and other disciplines. As biological research advances, more and more of these complex issues will involve biology as well. We have written this book to help you understand the biology that underlies these issues. We hope that your understanding will lead you to exercise good citizenship by making informed decisions about the biological issues that affect you and your community.

The social consequences of biology are important, but biology is fascinating on its own as well. Our planet is home to an astonishing collection of living organisms (**FIGURE 1.3**). Life comes in every size, shape, and color, and the ways that organisms go about making their living are equally diverse. In fact, diversity is one of the main features of life. In spite of this diversity, living organisms have many features in common. Understanding the diversity and common features of life is in fact the subject of the theory of evolution, as we will see in this chapter, and it is the first step to understanding biology.

1.2 What Are the Features of Life?

Most complicated concepts are difficult to define, and life is no exception. Because life is so diverse, almost any definition will have some exceptions. A better approach is to *describe* life, not *define* it. We can describe life best by describing the principal features that all living things have in common (**FIGURE 1.4**) and the characteristics of life itself that arise as a result:

- *Metabolism:* All living organisms take in energy and matter (i.e., food) and use them in a controlled way to promote their responsiveness, survival, and reproduction.
- *Inheritance and reproduction:* All living organisms can reproduce, and each organism inherits biological information from its parents or parent.
- *Evolution:* All living organisms are related, and all species change over time. Through **evolution**, new species arise and populations adapt to their environments. Evolution is the central feature of life because *all* the other features are the result of evolutionary change.
- *Diversity:* All living things have evolved into distinct, unique forms, but this diversity contains patterns that confirm life's evolutionary past.

This book examines each of life's main features in detail. In the remaining sections of this chapter, we will lay the groundwork by introducing each feature and explaining why it is important for understanding biology.

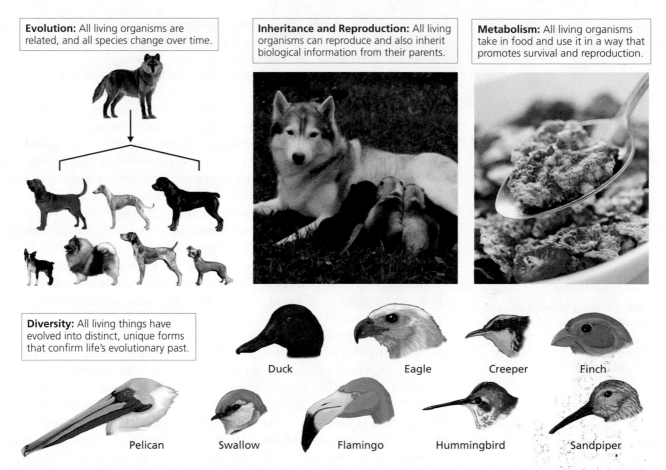

FIGURE 1.4 All living organisms have four principal features in common: metabolism, inheritance and reproduction, evolution, and diversity.

1.3 How Do Organisms Function?

Although John and Mary may not have thought of it in this way, John's infidelity gene is part of his biochemistry: to understand it, we have to apply the basics of chemistry to biology. The gene is composed of DNA, which itself is made from other subunit molecules called nucleotides. These subunits are made from even smaller parts: atoms. One of life's wonders is that atoms can be arranged to make complicated structures that contribute to complex characteristics and behaviors—maybe even affecting whether a person is faithful to his or her partner. To understand how organisms are put together, function, and behave, it is essential to understand certain fundamental concepts of chemistry and biochemistry.

Atoms, Chemical Bonding, and Molecules

Every physical thing in existence is composed of **atoms**. There are more than 90 unique kinds of atoms, one for each element. An **element** is a substance composed of only one type of atom. Atoms themselves are composed of three types of subatomic particles: **protons**, **neutrons**, and **electrons**. All atoms have the same general structure: a **nucleus**, or central sphere made of protons and neutrons, and electrons that orbit around the nucleus (**FIGURE 1.5**). Protons are positively charged, electrons have a negative charge, and neutrons

atom :: the building block of every physical thing in existence; there is one type of atom for each element

element :: a substance composed of only one type of atom

proton :: positively charged subatomic particle inside the nucleus of an atom

neutron :: neutral subatomic particle inside the nucleus of an atom

electron :: negatively charged subatomic particle that orbits the nucleus of an atom

nucleus :: central sphere of an atom made of protons and neutrons, with electrons orbiting

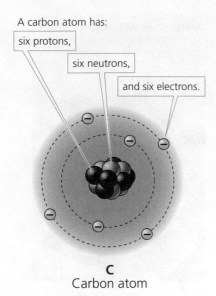

C
Carbon atom

FIGURE 1.5 Every physical thing in existence is composed of atoms. All atoms have the same general structure: electrons orbiting around a nucleus or central sphere made of protons and neutrons.

orbital :: the outermost shell of an atom that can accommodate a specific number of electrons

chemical bonding :: sharing of electrons in the outer shell of more than one atom

molecule :: formed through the chemical bonding of atoms

have no electrical charge. Each type of element has atoms of a unique size and with a specific number of protons. For example, a carbon atom has six protons and six neutrons in the nucleus and six electrons orbiting.

The chemistry of living things deals with a relatively small subset of elements—only around 25. And four elements account for close to 96% of all atoms in organisms: carbon, oxygen, hydrogen, and nitrogen. Yet this subset of the elements can make a variety of structures, including us.

The personality of an atom—whether it is reactive or inert—depends on the number and arrangement of electrons it has orbiting its nucleus. The outermost **orbital**, or shell, of an atom can accommodate a specific number of electrons. When the shell is filled or complete, the atom is nonreactive. No electrons need to be gained or lost to complete the outer shell. In contrast, if an atom needs just one electron to complete its outer shell, or if an atom has only one electron in its outer shell, it will be highly reactive. Atoms like sodium and chlorine are each one electron away from a completed shell: sodium needs to get rid of one, and chlorine needs to gain one (**FIGURE 1.6**). Both of these atoms react easily with one another to make sodium chloride, or common table salt.

Many atoms have partially filled outer shells. They can't get rid of enough electrons or gain enough to complete their outer shells in this way. Instead, they share electrons. The atoms get close enough together so that the electron orbitals of each overlap. The electrons in the outer shell of each atom orbit *both* atoms. As a result, each atom has a complete shell at least some of the time. This sharing of electrons is referred to as **chemical bonding** (**FIGURE 1.7**). Chemical bonding combines atoms to make **molecules**. For example, carbon and oxygen can chemically bond to each other to make carbon dioxide, the gas each of us exhales.

Macromolecules

Living cells (the fundamental units of life) and organisms are comprised of mostly water and carbon-based **macromolecules**, or large molecules made

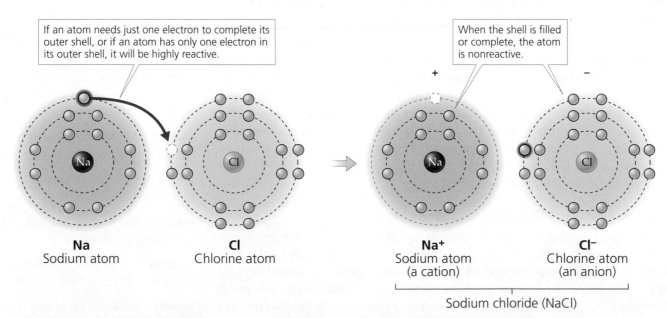

FIGURE 1.6 Atoms like sodium and chlorine are each one electron away from a completed shell, making them highly reactive. When they react with one another, they make sodium chloride, or common table salt.

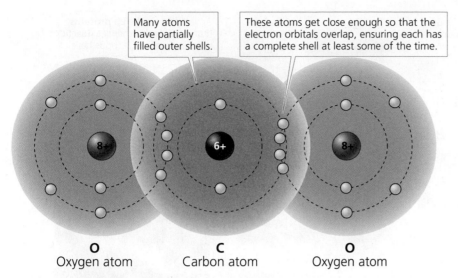

Many atoms have partially filled outer shells.

These atoms get close enough so that the electron orbitals overlap, ensuring each has a complete shell at least some of the time.

O
Oxygen atom

C
Carbon atom

O
Oxygen atom

FIGURE 1.7 Chemical bonding combines atoms to make molecules. Carbon and oxygen can chemically bond to each other to make carbon dioxide.

up of many atoms. These macromolecules are the nucleic acids, proteins, carbohydrates, and lipids (**FIGURE 1.8**). Each type of macromolecule is made of specific types of molecular subunits. For instance, amino acids are hooked together to make proteins. Because there are 20 different types of amino acids in cells that can be combined in any order, a bewildering array of proteins can be made. Let's consider each of these macromolecules, the subunits that comprise them, and examples of biological functions.

macromolecule :: large molecule made up of many atoms

- *Nucleic acids:* As we saw in Mary and John's case, DNA, one type of nucleic acid, is the physical material of which the infidelity gene and other hereditary instructions are made. Nucleic acids are made from subunits called nucleotides.
- *Proteins:* Amino acids make up proteins that serve functional roles like regulating the chemical reactions in cells. For example, John's infidelity gene controls the production of the arginine receptor protein. Proteins also have structural functions: they make the many parts of cells and organisms.
- *Carbohydrates:* Sugars, such as glucose, are joined together to make carbohydrates like cellulose, the structural material of plant cell walls. Some carbohydrates, like starch, act as a food source.
- *Lipids:* Subunits called *fatty acids* and *glycerol* join together to make lipids, or fats. Besides being an excellent energy source and the molecule animals use to store excess food, fats insulate the body and serve as shock absorbers for internal organs. Lipids are also an essential structural component of cell membranes.

Cells

Cells, the smallest units of life, are discrete worlds that maintain the conditions needed to support life (**FIGURE 1.9**). The boundary of the cell, the cell membrane, regulates precisely what can enter and leave. Because cells and organisms contain so much water, the chemical reactions related to life actually take place in a unique aqueous, or watery, environment. Anything inside

cell :: the smallest unit of life

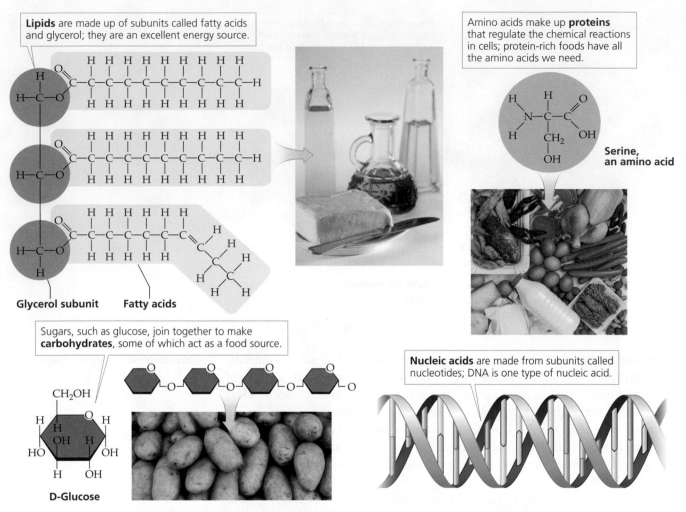

Lipids are made up of subunits called fatty acids and glycerol; they are an excellent energy source.

Glycerol subunit   **Fatty acids**

Amino acids make up **proteins** that regulate the chemical reactions in cells; protein-rich foods have all the amino acids we need.

Serine, an amino acid

Sugars, such as glucose, join together to make **carbohydrates**, some of which act as a food source.

D-Glucose

Nucleic acids are made from subunits called nucleotides; DNA is one type of nucleic acid.

FIGURE 1.8 Living cells and organisms are comprised of mostly water and carbon-based macromolecules. These macromolecules are the nucleic acids, proteins, carbohydrates, and lipids.

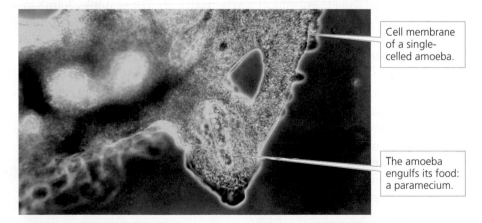

Cell membrane of a single-celled amoeba.

The amoeba engulfs its food: a paramecium.

FIGURE 1.9 Cells are the units of all life. The cell membrane regulates precisely what may enter and leave, and most water-soluble molecules on either side cannot pass across the membrane unassisted.

Scientist Spotlight

Carol W. Greider (1961–)

Carol Greider was doing laundry in the early morning before getting her two young daughters up and ready for school when the phone rang unexpectedly. She picked it up. "Hello?" A voice at the other end had astonishing news: "Congratulations, you have been awarded the 2009 Nobel Prize in Physiology or Medicine!" Carol Greider decided to finish the laundry later. Not your stereotypical scientist.

Born in 1961 in California, Greider began doing research as an undergraduate at the University of California at Santa Barbara. She loved it and decided to become a professional scientist. Although a gifted student with excellent grades, research experience, and letters of recommendation, Greider had earned such bad GRE scores (as a result of dyslexia that had previously been unknown to her) that almost no graduate schools would consider her for admission. Fortunately, the University of California at Berkeley interviewed and accepted her. In fact, it was in the lab of Elizabeth Blackburn at UC Berkeley that Greider began the project that led ultimately to a Nobel Prize for them both.

Greider and Blackburn's big breakthrough was the discovery of a molecule called *telomerase*. Telomerase is like a cell's fountain of youth. It helps prevent some aging effects and allows a cell to divide over and over again. This is particularly important in multicellular organisms. In a healthy body, cells divide to replace others that are damaged or worn out. As the organism ages, its cells become less effective at dividing and replacing other cells, which accounts for some of the problems aging produces.

Telomerase works by placing a protective cap, or telomere, on the ends of chromosomes inside the cell. This cap protects chromosomes during the cell's life and during reproduction. The chromosomes contain the cell's genetic information in the

Carol Greider won the 2009 Nobel Prize in Physiology or Medicine as a result of her work with the molecule telomerase.

form of DNA. If the chromosomes become damaged, the cell may die or become unable to reproduce.

In the research Greider has done since the discovery of telomerase, she has shown that as cells age, telomeres shorten. Once telomeres get too short, they lose their protective power. The chromosomes are damaged and the cell dies. Greider has also discovered that cancer cells have overactive telomerase and telomeres that are too long. As a result, cancer cells are immortal. Evidently, telomere length regulates the life span of a cell, and telomerase is the molecule responsible for that length.

Greider's research has propelled her to turn these lab results into strategies for treating patients. She is driven by her intense curiosity and the challenge of, in her words, "diving into the unknown." ::

the cell that can dissolve in water, such as salts and many small molecules, will therefore exist in a solution state. In contrast, molecules such as fats, which do not dissolve in water, will remain separate.

Cell membranes enclose cells so successfully in part because most water-soluble molecules on either side cannot pass across the membrane unassisted. If that were to happen, the cells would die. (To learn more about the relationship between cell aging and cancer, see *Scientist Spotlight: Carol W. Greider*).

As we will discuss in later chapters, there are thus two chemical worlds in cells and organisms, a water-soluble one and a fat-soluble one. And that has important consequences for a range of functions—including cell communication, fertilization, digestion, water balance, and even thinking.

Chemical Reactions and Enzymes

Two broad categories of chemical reactions take place in cells and organisms: molecules are taken apart to produce smaller components, or molecules are joined together to make larger structures. Whether a reaction occurs at all depends on whether there is energy available. Some reactions require a large input of energy because the products of the reactions are structurally more complex than the starting materials. In contrast, some reactions occur without a net input of energy.

Reactions inside organisms depend upon balancing energy input and output. Biosynthetic reactions need a net input of energy. For instance, DNA is more structurally complex than the nucleotides that form it. It costs energy to make this macromolecule. Breakdown reactions, like the digestion of starch from a potato, yield energy ultimately but the reaction needs a kick start; enzymes serve this purpose.

enzyme :: protein that makes it easier for chemical reactions to occur

Most **enzymes** are proteins that make it easier for chemical reactions to occur. They lower the degree to which "start-up" energy is needed. Enzymes do this in a variety of ways such as orienting reacting molecules so they can bind efficiently or by actually bending molecules so they are easier to break into parts (**FIGURE 1.10**). Enzymes play an especially important role in regulating all chemical reactions in cells and organisms, including those responsible for digesting food, growing new cells, or replicating DNA, for example.

Energy Extraction and Use

metabolism :: all of the chemical reactions in cells that sustain life

The responsiveness, survival, and reproduction of all organisms depend on **metabolism**—the chemical reactions in cells that sustain life. The ultimate source of energy that supports life is the sun. Plants are able to capture light energy from the sun and convert it into chemical bond energy in sugars and related molecules. Organisms extract energy from the food they eat. The chemical bond energy present in food molecules is converted to a form that can be used for any of a variety of functions, including protein production, walking, or mating. The reactions of metabolism are carefully regulated by the actions of specific enzymes.

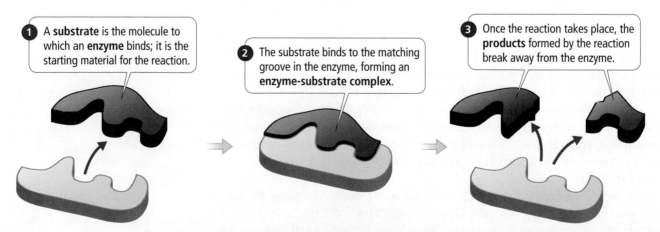

1. A **substrate** is the molecule to which an **enzyme** binds; it is the starting material for the reaction.

2. The substrate binds to the matching groove in the enzyme, forming an **enzyme-substrate complex**.

3. Once the reaction takes place, the **products** formed by the reaction break away from the enzyme.

FIGURE 1.10 Enzymes are proteins that play an important role in regulating all chemical reactions in cells and organisms. They help reacting molecules to bind efficiently or to bend, so they are easier to break into parts, as shown in this example.

1.4 How Do Organisms Reproduce?

In our case study, John learned that he had the infidelity gene—or, more accurately, the infidelity allele. We've discussed some of the consequences of having this allele. But we haven't discussed *why* John had it in the first place. Part of the answer is that John inherited the allele from his father or mother. This fact might have helped John and Mary evaluate their concerns about the allele. If John's parents had a happy marriage—in spite of having the infidelity allele—the prospects for John and Mary look better. At least they could conclude that the allele doesn't necessarily *cause* infidelity. What else can we learn about life by studying inheritance and reproduction?

Inheritance

Inheritance is the way life perpetuates itself from one generation to the next. Organisms like us, for example, inherit from two sources: the mother and father. Each parent has the same set of genes (if we ignore sex differences). But each parent has a different combination of alleles for these genes. This means that you inherit the same set of genes as all humans, but your particular combination of alleles, some from the father and some from the mother, is unique (unless you have an identical twin or other identical multiple-birth sibling). You probably resemble your parents and your siblings, but each individual varies genetically from others (**FIGURE 1.11**).

Inheritance involves passing *information* from one generation to the next. For us, only a microscopic egg and sperm cell are required. But these tiny entities contain all the genetic information needed to make a human (**FIGURE 1.12**). These are the instructions a new organism needs to grow,

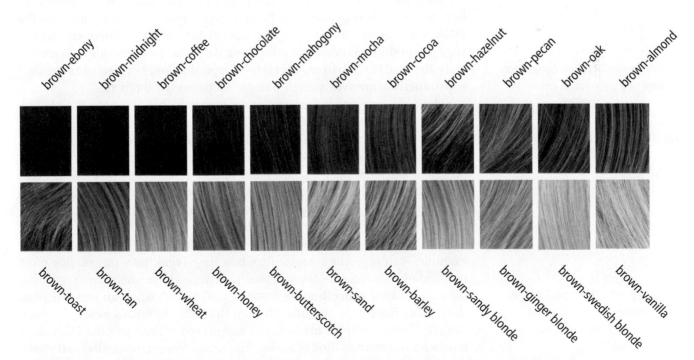

FIGURE 1.11 Every human being inherits the same set of genes as all other humans, but the particular combination of alleles is unique (unless you have an identical twin or other identical multiple-birth sibling). For example, individuals may have different combinations of alleles for hair color, resulting in different shades.

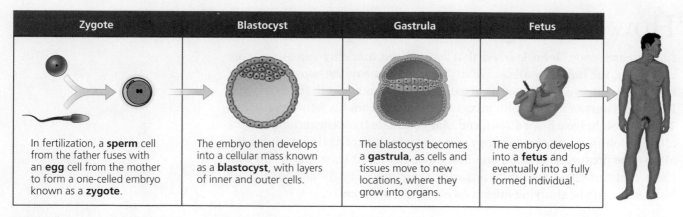

Zygote	Blastocyst	Gastrula	Fetus
In fertilization, a **sperm** cell from the father fuses with an **egg** cell from the mother to form a one-celled embryo known as a **zygote**.	The embryo then develops into a cellular mass known as a **blastocyst**, with layers of inner and outer cells.	The blastocyst becomes a **gastrula**, as cells and tissues move to new locations, where they grow into organs.	The embryo develops into a **fetus** and eventually into a fully formed individual.

FIGURE 1.12 Only a microscopic egg and sperm cell are required to pass on the genetic information needed to make a human. This information includes the instructions a new organism needs to grow, survive, and eventually reproduce.

survive, and eventually reproduce. An individual ("within" a generation) uses inherited information to direct the basic processes of living. This information, also passed on to offspring ("between" generations), allows them to do the same.

Reproduction

Inheritance works the way it does because of the way organisms reproduce. The steps in reproduction make new individuals and ensure that each inherits the full complement of genetic information. We know now that genetic information is stored in DNA. We also know that the chemical properties of DNA allow it to be used both to store information and to pass it on to the next generation. In nature this happens in two ways.

The simplest form of reproduction occurs in individual cells, like the cells in your body or in organisms like bacteria that have only a single cell. First the DNA is copied, so that two complete sets exist. Then the cell divides, and one copy goes into each daughter cell. These daughter cells are (almost) genetically identical to the parent. This type of reproduction, referred to as **asexual reproduction**, involves a single parent—so no sex (**FIGURE 1.13**).

The DNA copying mechanism is very accurate—but not perfect. Although the chances are much less than one in a million, sometimes errors occur when DNA is copied. These mistakes are called **mutations**. Because of mutations, any cell may have slight genetic differences from its parent. When a mutation occurs in a gene, it makes a new allele of that gene. That's why alleles exist, like the allele of *AVPR1a* that may be related to infidelity. While there is a *very* slight chance that John's copy of *AVPR1a* was a mutation, and that he didn't inherit it from his parents, this is unlikely.

A few multicellular organisms—flatworms, sea stars, and plants, for example—can reproduce asexually when they break into pieces. But most multicellular organisms, even those that reproduce asexually, reproduce sexually. **Sexual reproduction** (**FIGURE 1.14**) occurs when an egg cell from the mother fuses with a sperm cell from the father to form a new individual. Sexually formed individuals don't just have a copy of their parents' DNA: they have a unique combination of alleles. The actual traits a new individual shows depend on how the alleles in this particular set interact.

But why reproduce at all? Unfortunately, bodies wear out, the chances of fatal accidents accumulate, and over long time spans, habitats change too

asexual reproduction :: reproduction that involves a single parent and no sex

mutation :: a physical change in the DNA

sexual reproduction :: reproduction that occurs when an egg cell from a female fuses with a sperm cell from a male to form a new individual

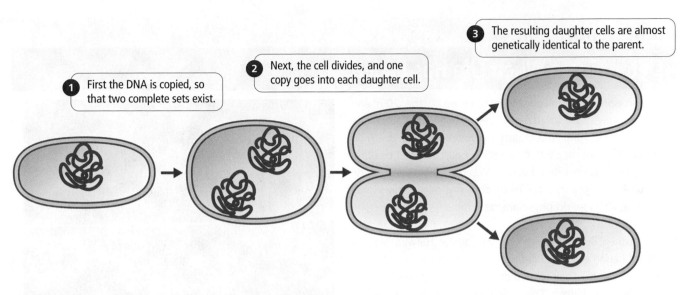

① First the DNA is copied, so that two complete sets exist.

② Next, the cell divides, and one copy goes into each daughter cell.

③ The resulting daughter cells are almost genetically identical to the parent.

FIGURE 1.13 Asexual reproduction involves a single parent—no sex. It occurs in the cells in your body and in single-celled organisms like bacteria.

much for a single individual to tolerate. Ironically, a universal feature of life is death (see *Life Application: Determining When Life Has Ended*, p. 18).

The fact that all individuals die makes reproduction and inheritance a very significant part of life. From an evolutionary perspective, reproduction is the single most important event in life. Two facts about these processes are particularly noteworthy:

- Living organisms are part of an unbroken chain of ancestors and descendants, extending back into time. Reproduction divides this

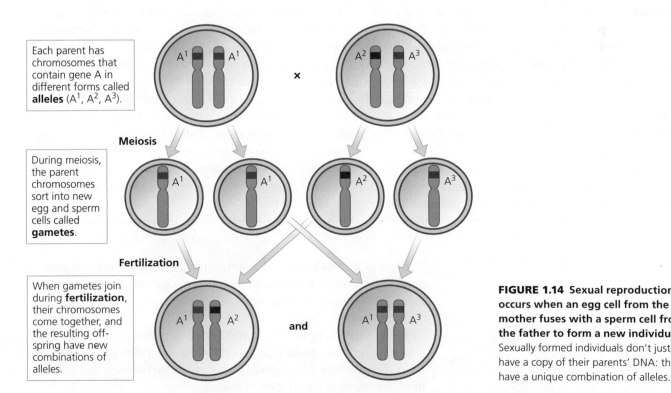

Each parent has chromosomes that contain gene A in different forms called **alleles** (A¹, A², A³).

Meiosis

During meiosis, the parent chromosomes sort into new egg and sperm cells called **gametes**.

Fertilization

When gametes join during **fertilization**, their chromosomes come together, and the resulting off-spring have new combinations of alleles.

FIGURE 1.14 Sexual reproduction occurs when an egg cell from the mother fuses with a sperm cell from the father to form a new individual. Sexually formed individuals don't just have a copy of their parents' DNA: they have a unique combination of alleles.

Determining When Life Has Ended

The amazing medical advances achieved in the 20th and 21st centuries have saved many lives. Yet they have also made it more challenging to establish a definitive time of death. In the not-too-distant past, if a person's heart stopped beating, he or she was considered dead. With our abilities to provide medical life support, a still heart does not necessarily signify death at all. How do physicians determine whether a person has died?

In the 1960s, a group of physicians at Harvard Medical School began efforts to define clear criteria for neurological, or brain, death. They refined and formalized guidelines, and by 1980 the Uniform Determination of Death Act (UDDA) was drafted as a model of legislation for individual states to pass. Supported by the President's Commission for the Study of Ethical Problems in Medicine and Biomedical and Behavioral Research, the UDDA states: "An individual who has sustained either (1) irreversible cessation of circulatory and respiratory functions, or (2) irreversible cessation of all functions of the entire brain, including the brain stem, is dead."

Based on state guidelines, hospitals have strict protocols for determining brain death. First, only appropriately trained physicians who have met certain professional standards may designate brain death. Clinical evaluations precede this determination to verify that three essential features are present: coma, absence of brainstem reflexes, and apnea. A coma is a deep state of unconsciousness from which a person cannot be aroused. The brainstem is the most posterior part of the brain and is responsible for controlling certain critical but not neurologically complex functions. A person with a dead brainstem has no cough reflex, no gag reflex, no corneal reflex, no normal eye movement, and no pupil response—even when a physician tries to elicit these behaviors. Apnea means that a person has stopped breathing independently.

In the event that a patient appears to be brain dead, physicians perform a complex battery of tests twice, with several hours separating the two tests, to make sure that the

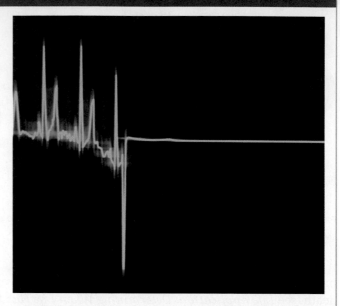

In the event that a patient appears to be brain dead, physicians perform a complex battery of tests twice. In some circumstances, the brain death diagnosis will be verified by an electroencephalogram (EEG).

individual has an irreversible loss of brain function. Precautions are also taken to rule out the possibility of some potentially reversible cause for a lack of brain function (a drug overdose, for example). In some circumstances, the diagnosis of brain death will be verified by a brain scan, to see whether there is any blood flow, or by an electroencephalogram (EEG)—a method of detecting electrical activity in the brain or determining whether it has "flatlined."

If a person meets the criteria for brain death, and if he or she is an organ donor, the certification of brain death generally requires the signatures of two physicians. Neither can be directly involved in performing the transplant operation. This extra step serves to allay concerns that a determination of death may have been influenced by a need for healthy organs for transplantation. ::

chain into generations, and inheritance guarantees some degree of genetic continuity. Ever since life got started, new life has come only from existing life, not from nonliving sources (see *How Do We Know? Spontaneous Generation*).

•   The combination of mutation and sexual shuffling of alleles guarantees that individuals in one generation differ from each other and from the previous generation. This genetic variation is the basis for life's most central feature: evolution.

Spontaneous Generation

If you have ever had the unpleasant experience of coming across a dead animal in the woods or on the side of the road, you probably noticed that the carcass was covered with maggots. Perhaps you even knew that these creatures were simply fly *larvae*, the term for young insects. Most of us would find this observation unsettling or even repulsive. Yet no one would conclude that the dead animal had "turned into" maggots. People in ancient times, however, believed that maggots arose "spontaneously" and directly from an animal's remains. We know today that the only way new individuals arise is from their parents, through reproduction. But how was this result established scientifically?

The first person to tackle whether life could arise spontaneously was the Italian natural philosopher Francesco Redi. In 1668, Redi developed the hypothesis that maggots appear in rotting meat not as a result of "spontaneous generation," but because maggots develop from eggs laid by flies. To test this hypothesis, Redi set up a simple demonstration: he placed meat in glass jars. One set of jars was left open. A second set was sealed shut. The third set was covered with gauze. The results were conclusive. Maggots appeared only on the meat that was in the open jars. Flies entered the jars, landed on the meat, and laid eggs. No maggots were observed in the meat in the sealed jars or the jars covered by gauze.

Although Redi's experiment put to rest the idea that animals could arise spontaneously, many still held on to the notion that microorganisms such as bacteria could do so because they were such small and simple forms of life. It wasn't until almost 200 years after Redi's work that Louis Pasteur, a French microbiologist and chemist, performed the critical experiment that demonstrated that *all* forms of life come only from the reproduction of other living organisms. Pasteur was motivated by a contest sponsored by the French Academy of Sciences (FAS) in 1859. The FAS offered a prize for the best experiment that would either prove or disprove spontaneous generation.

Pasteur cultured bacteria in S-shaped glass flasks. He poured culture broth into these "swan-neck" flasks and then boiled the broth to sterilize it; no bacteria remained. The flasks were left open to the air, but because of the S-shaped neck, dust particles could not fall into the flask. As a consequence, even though air could enter, no bacteria did. The broth remained sterile. When Pasteur broke off the neck, so that

the broth was directly under the opening of the flask, a teeming population of bacteria materialized within a day. Although Pasteur was not the only scientist to address whether microorganisms could generate spontaneously, he was certainly the person who did so most decisively. It should come as no surprise that Pasteur won the contest by a unanimous vote! ::

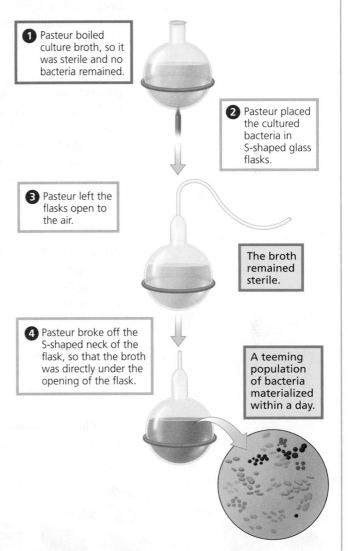

1 Pasteur boiled culture broth, so it was sterile and no bacteria remained.

2 Pasteur placed the cultured bacteria in S-shaped glass flasks.

3 Pasteur left the flasks open to the air.

The broth remained sterile.

4 Pasteur broke off the S-shaped neck of the flask, so that the broth was directly under the opening of the flask.

A teeming population of bacteria materialized within a day.

Louis Pasteur, a French microbiologist and chemist, performed the critical experiment that demonstrated that *all* forms of life come only from the reproduction of other living organisms.

1.5 How Does Life Evolve?

Let's return to the question of why John had the infidelity allele. Noting that John inherited it from one of his parents simply pushes the question back a generation—why did one of *them* have it?

Humans have about 25,000 genes, and many (perhaps most) of them have several alleles. Some alleles are very common. For example, alleles for blue eyes and blond hair are common in Scandinavia, while alleles for black hair and dark eyes are common in Africa. The chances of having a particular allele depend on how common it is in a population—that is, how many people in the group have it. In any human population, a certain number of people will have the infidelity allele of *AVPR1a*. Unfortunately for John, one or both of his parents fell into this category.

Over time the frequency of some alleles increases, while the frequency of other alleles decreases. If the frequency of an allele goes up, more people will have it, and if it decreases, fewer people will have it. Any change in allele frequency, for any reason, reflects *a genetic change in the population* (**FIGURE 1.15**). This is the modern biological definition of evolution. Later in this book, in Chapter 9, we will look thoroughly at how Charles Darwin made his discovery and the evidence for evolution. And, again and again, especially in Chapters 3–8, we will see how genes are part of our inheritance— and what humans are doing with that knowledge. Let's start, though, with an overview of Darwin's theory, to see why evolution serves as a unifying theme of biology.

Darwin's Theory

When Darwin proposed his theory of evolution in 1859, he did not know about DNA, genes, or alleles. But he did know about inheritance, and he discovered the significance of the fact that individuals vary. Here are the facts Darwin worked with:

- Individuals in a population vary from one another.
- Variations are inherited, at least in part.
- In every population, many or most of the offspring die before reaching maturity.

What conclusion would you draw from these easily verified facts? Darwin asked this question: *which* offspring would die? Darwin's discovery was

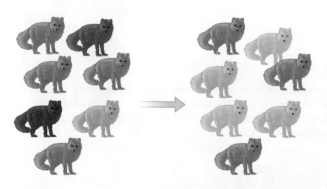

White arctic foxes can blend in better with their winter environment, so over time, alleles for darker-colored foxes decreased, while alleles for white-colored foxes increased.

FIGURE 1.15 Over time, the frequency of some alleles increases, while the frequency of other alleles decreases. Any change in allele frequency, for any reason, reflects a genetic change in the population.

simple and yet profound. Some of the variations, Darwin reasoned, would help certain offspring survive better than others. Offspring that were a little bigger, or a little faster, or that were able to hide or find food better—these offspring had a better chance of surviving than others that lacked these variations.

The survivors were more likely to reproduce, and when they did, their offspring would inherit these variations. Variations that promoted survival would become more common, while variations that hindered survival would become less common. The next generation would not only be different from the previous one, it would be different in a specific way. It would have more individuals with variations that helped them survive the challenges of the particular habitat in which they lived.

Darwin referred to this process as "descent with modification" (**FIGURE 1.16**). He realized that a species could change over time, which was a revolutionary idea in the 1800s although other scientists were reaching the same conclusion. Darwin used the term **natural selection** to describe the *mechanism* by which these changes occurred—the ways that conditions in the natural environment "selected" which individuals would survive and reproduce.

natural selection :: the term Darwin used to describe the mechanism by which species change over time; conditions in the natural world select which individuals will survive and reproduce

Extending Darwin's Theory Through Time

Very often in science, ideas that seem simple can have a major impact on our thinking. If we open the door to the possibility that species can change, where does this concept lead?

One of the first realizations was that as species change, they become better adapted to their environment. In Darwin's time, geologists were beginning

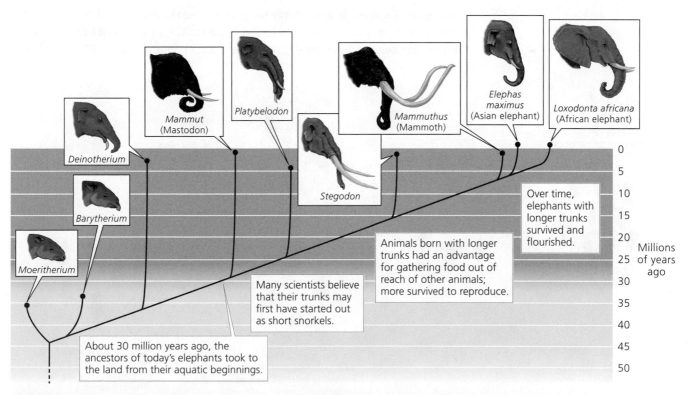

FIGURE 1.16 "Descent with modification" was the name Darwin gave to the idea that each generation would vary from the previous one in a specific way. Variations that promoted survival would become more common, while variations that hindered survival would become less common.

to show that conditions on Earth changed over time—almost as surprising as species changing. Clearly these processes would have to go hand in hand. If the environment changed, species would either change to keep up or become extinct.

A second realization was that if species changed enough, they might even change into a different species. A single species could have populations living in many different and widespread areas. If conditions in these areas changed independently of one another, as the different populations adapted to their local conditions they would become increasingly different. Given sufficient time, Darwin thought, they would separate into two distinct species where originally there had been only one (**FIGURE 1.17**). The idea that a new species could be created by natural means was earthshaking and difficult for many to accept.

But the most far-reaching conclusion of Darwin's theory emerges if we run this process *backward* in time. Instead of species changing, becoming different, and splitting into more species, we see species becoming more similar and merging together. Viewed from this perspective, two species are related because they descended from a common ancestor—just as you and your cousins are related through a common grandparent.

Because all forms of life are part of a continuous chain of reproduction, the logical conclusion of Darwin's theory is that *all* species are related, all part of an ever-expanding tree of life. In the century and a half since Darwin, biologists have learned a great deal about evolution. They have extended and modified Darwin's ideas, and found some new ones. But Darwin's original ideas about evolution remain largely in place today.

Evolution Is the Unifying Theme of Biology

The goal of science is to discover the reasons for the events and situations that occur in the natural world. Evolution implies that all forms of life—including ours—are related to some degree; this is a fact that can be difficult to see at

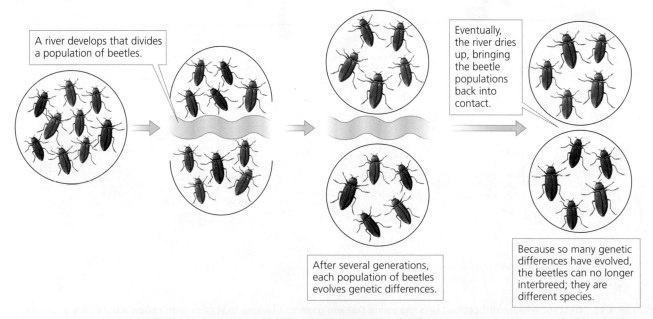

A river develops that divides a population of beetles.

Eventually, the river dries up, bringing the beetle populations back into contact.

After several generations, each population of beetles evolves genetic differences.

Because so many genetic differences have evolved, the beetles can no longer interbreed; they are different species.

FIGURE 1.17 Darwin came to the realization that if species changed enough, they might even change into a different species.
The idea that a new species could be created by natural means was earthshaking and difficult for many to accept.

first. But this idea explains so many facts of nature that biologists see evolution as the central theme that ties together all of biology.

Our own animal group provides a good example. We belong to the *vertebrates*: animals with internal skeletons made of bones (mostly) and with well-defined heads that contain our brain and sensory organs. Vertebrates are a diverse group (**FIGURE 1.18**). Whales, dinosaurs, and elephants are among the largest animals that ever lived. The dwarf goby fish, one of the smallest vertebrates, is less than a centimeter long. And a flea toad fits comfortably on your thumbnail. Some vertebrates are fierce predators; others eat plants. Cheetahs can run at over 70 miles per hour, and falcons dive at over 200 miles per hour. On the other hand, tree sloths typically move less than a foot per minute, and they may not move at all for long periods of time. Vertebrates live in the deepest oceans, on the highest mountains, and in the desert, the rainforest, prairies, and marshes—just about everywhere from the equator to the poles.

Since vertebrates show such differences, you might expect their bodies to be different too. But all vertebrates have the same body plan: a skull attached to a vertebral column, which supports ribs and ends in a tail. Vertebrates all have similar bone structure (**FIGURE 1.19**). Bones vary in size and shape and are sometimes lost, but for the most part the set of bones is the same. Vertebrates have many of the same muscles that move these bones. All vertebrates have the same organs in roughly the same places, make similar systems of blood vessels, and have the same organization in their nervous systems.

If you examine them in detail, you learn that all vertebrates are variations on a theme rather than new designs for a particular way of life. We see this pattern over and over in nature. And this is precisely the pattern that evolution

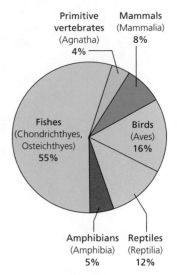

FIGURE 1.18 Vertebrates are a diverse group comprised of mammals, birds, reptiles, amphibians, and fish. They reside just about everywhere, and they are also some of the largest and smallest and the fastest and slowest animals that ever lived.

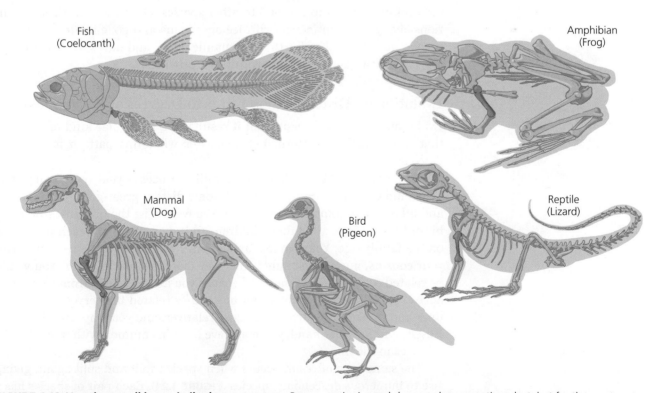

FIGURE 1.19 Vertebrates all have similar bone structure. Bones vary in size and shape and are sometimes lost, but for the most part, the set of bones is the same; the femur is highlighted in each illustration.

predicts. Vertebrates inherited their similarities from common ancestors and evolved differences as they adapted to environments. It is difficult to explain this pattern of diversity and similarity any other way. The eminent scientist Theodosius Dobzhansky put it succinctly in a 1973 paper: "Nothing in biology makes sense except in the light of evolution."

1.6 What Patterns of Diversity Are Found in Nature?

diversity :: the variety of life forms in the natural world

Scientists have described about 1.3 million species. But we have a long way to go: by some estimates, there may be 30 million living species and another 50 *billion* that have gone extinct. Life is diverse, and **diversity** is one of the most obvious facts of life. All you have to do is look around you and count the number of life forms you see to appreciate natural diversity. In Chapter 17, we will look at the challenge of preserving the great diversity of life for future generations.

Life's diversity presents biologists with two tasks. First, we must explain why this diversity exists. Second, we need to organize this diversity in a meaningful way, one that reflects its underlying causes.

Life Is Diverse

New species arise when older species adapt to new and different living conditions. The diversity of life goes hand in hand with the diversity of living conditions our planet has to offer (**FIGURE 1.20**). Every region on Earth presents a distinct set of physical conditions, like temperature, moisture, and amount of sunlight. Within each region, there are numerous ways to get food and shelter, and many ways to respond to other species. Over time, all these conditions change, as continents drift, ice ages come and go, climate varies, and species go extinct. The Earth is a dynamic place, and new species arise that can take advantage of the opportunities it provides.

Evolutionary Diversification Leads to Degrees of Relatedness

Evolution results in diversity, but it results in a particular kind of diversity that has a distinct pattern. Let's consider what this pattern is and why it occurs.

It's helpful to start with a more familiar process—your own family tree. You could trace your ancestry back to some distant great-great-grandparent and fill out your complete family tree. You would list the first ancestor, all of his or her children, all of their children, and so on until you reach your place on the family tree. Your parents and siblings would be on the tree, as would your cousins, uncles, and aunts, and ever more-distant relatives. You would be related to everyone on the tree because you all descended from a common ancestor. However, you would not be *equally* related to everyone. Your siblings and parents are particularly close relatives, and you may resemble them closely. On the other hand, you may have little in common with more distant relatives in the tree.

The same type of result occurs when species split and split again, giving rise to numerous descendant species (**FIGURE 1.21**). Each pair of species has a common ancestor. But if that common ancestor lived long ago, the species

FIGURE 1.20 The diversity of life goes hand in hand with the diversity of living conditions on the planet. Over time, these conditions change, leading some species to go extinct, while others arise that can take advantage of opportunities.

may not be closely related. There may be a long chain of species leading back to the common ancestor, and there will have been a lot of time for the species to diverge and become different.

On the other hand, two species may have a common ancestor that lived very recently—more like a mother species than a great-great-great-grandmother species. These species may have much in common, such as similar genetics, appearance, and behaviors. In fact, you might observe a species in the process of splitting, in which the various subgroups are so similar it's hard to tell them apart.

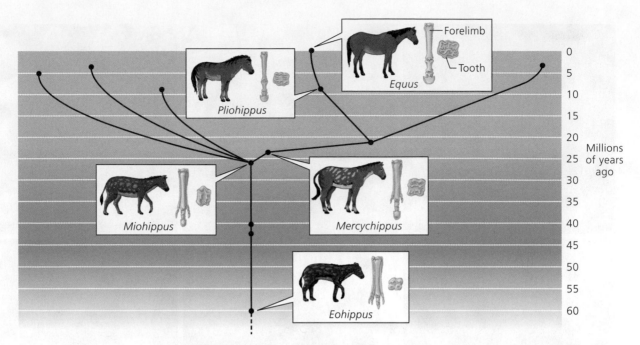

FIGURE 1.21 Each pair of species has a common ancestor. If that common ancestor lived long ago, however, there will have been a lot of time for the species to diverge and become different.

Our understanding of evolution allows us to make a strong prediction about diversity in nature. If new species arise when older species split, we should see patterns of relatedness: groups of species that are closely related (like a family), that are part of larger but less closely related groups (like an extended family), and that are part of still larger and less closely related groups (perhaps like a village founded long ago by only a few couples). This type of pattern is called a hierarchy. If life's diversity can be organized into a hierarchy of relatedness, this is good evidence that evolution produces new species in the way Darwin envisioned: older species split in two as they adapt to new and different conditions.

Organizing Hierarchies in the Diversity of Life

Species lie at one end of life's hierarchy. Members of a species are genetically so similar that they can interbreed successfully. Let's consider dogs as an example. Dogs vary dramatically, the result of our own "artificial" selection. But even a Great Dane and a Chihuahua can produce offspring (granted, they may need a little help). All dogs are in the same species.

It's easy to see that dogs are part of a larger group that includes wolves and coyotes. A still larger grouping includes foxes. Dogs are carnivores, which puts them in an even larger group that includes cats and seals. All of these species are mammals, and all mammals are vertebrates. Finally, all vertebrates are animals. Natural diversity consists of a hierarchical organization of groups within groups within groups.

People have undoubtedly observed this hierarchy of similarity and diversity for ages. The hierarchy was formally organized in the early 1700s by Carolus Linnaeus. He defined a hierarchy with seven levels, some of which may be familiar to you:

kingdom
  phylum (plural "phyla")
    class
      order
        family
          genus (plural "genera")
            species (both singular and plural)

A kingdom consists of one or more phyla, a phylum contains one or more classes, and so forth down to species. Each level contains fewer and more closely related groups than the one above it. Linnaeus used his hierarchy to *classify* every species known at the time. **Classification** means assigning every species to a genus, each genus to a family, continuing on to kingdom (**FIGURE 1.22**).

One consequence of Linnaeus' work is that each species has a unique **scientific name**—its particular classification from kingdom to species. These names for the specific levels are in Latin and designed to be descriptive of the grouping. The Latin word for dog is *canis*, for example, and *Canidae* is the name of the dog family. It's cumbersome to use the complete classification to name a species, so by convention a species' scientific name includes its genus and species names only. Both are italicized since the words are Latin; the genus is capitalized, and the species name is not.

The most important consequence of Linnaeus's work, though he did not realize it, is that a species' classification is its family, or, more specifically, its

classification :: the assignment of every species to a genus, family, order, class, phylum, and kingdom

scientific name :: the unique classification of each species from kingdom to species

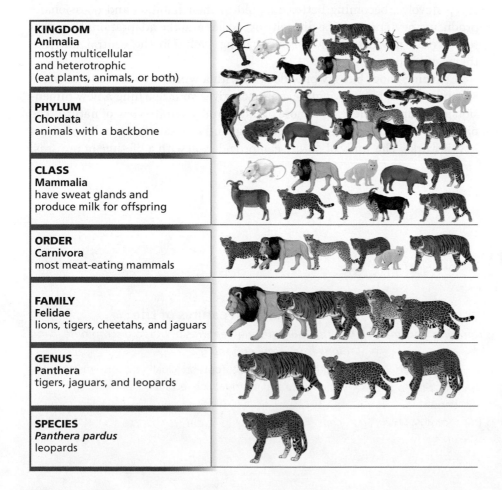

KINGDOM **Animalia** mostly multicellular and heterotrophic (eat plants, animals, or both)	
PHYLUM **Chordata** animals with a backbone	
CLASS **Mammalia** have sweat glands and produce milk for offspring	
ORDER **Carnivora** most meat-eating mammals	
FAMILY **Felidae** lions, tigers, cheetahs, and jaguars	
GENUS **Panthera** tigers, jaguars, and leopards	
SPECIES ***Panthera pardus*** leopards	

FIGURE 1.22 Carolus Linnaeus formally organized the hierarchy of life in the early 1700s. He defined a hierarchy with seven levels: kingdom, phylum, class, order, family, genus, and species.

evolutionary tree. Tracing a species back through its evolutionary tree allows you to see what species it is related to, and how close the relationship is, just as you could do with relatives in your own family tree. Knowing these relationships helps you understand the biology of species better. For example, you know that if a species is in class *Mammalia*, it gives birth to live young, feeds them milk, has hair, and maintains a constant body temperature. The simple fact that species can be classified and organized this way is strong evidence for evolution and the origin of species.

Biology in Perspective

Why does biology matter to you? Knowing more about biology will help you make better personal decisions regarding your health and that of your family. It might even save you from the kind of worry and stress our case study couple, John and Mary, experienced. Understanding biology can also make you a better citizen. Citizenship requires making informed decisions on issues that affect your community, country, and planet. While these issues are complex and multidisciplinary, biology will play an increasingly important role as scientists make advances in biological and medical research.

Biology is the study of life. Life on Earth is astonishingly diverse, but all living things have several features in common. All living things have a metabolism and use molecules and energy in a controlled way to support their life. All living things store genetic information and pass it on to their descendants when they reproduce. And although individuals don't evolve, over time species evolve, becoming better adapted to their habitats and occasionally producing new species. The diversity of species and the adaptations they possess are all products of evolution, making evolution the central unifying theme of biology.

Learning about biology provides you with a window to the natural world around you. Adding a scientific approach gives your learning a new dimension and helps you develop a more detailed and accurate view of nature. The natural world is complex and fascinating, and it is satisfying to learn about how it works. The study of nature can provide you with a lifetime of rewards.

Chapter Summary

1.1 How Does Biology Affect Your Life?
Explain how understanding biology can help you make informed decisions as a citizen.

- Understanding biology can help you make informed decisions about health care, aging, cancer, disease, voting, and legislation, among many other things.
- New technologies offer opportunities but may also present risks.

1.2 What Are the Features of Life?
Use the four principal features of life to discriminate living things from nonliving things.

- Living things exhibit four principal features: metabolism, inheritance and reproduction, evolution, and diversity.

1.3 How Do Organisms Function?

Describe the relationship among atoms, macromolecules, and cells, and explain how enzymes maintain cell function.

- All substances are composed of atoms; each atom has a spherical nucleus, composed of neutrons (no charge) and protons (positive charge), orbited by negatively charged electrons.
- Elements are pure chemical substances in which all of the atoms contain the same number of protons.
- Living things are mostly composed of four elements: carbon, oxygen, hydrogen, and nitrogen.
- The reactivity of an atom is determined by the number of electrons in its outer shell; when atoms share electrons, they form chemical bonds.
- Living cells are composed mostly of water and macromolecules; these macromolecules—nucleic acids, proteins, carbohydrates, and lipids—are the smallest discrete units that maintain the conditions required to support life.
- Most chemical reactions require energy.
- Metabolism is the chemical reactions in cells that sustain life.

1.4 How Do Organisms Reproduce?

Describe the relationship among DNA, genes, and alleles, and compare how genes are inherited in sexual and asexual reproduction.

- Organisms inherit information that codes for traits on DNA from their parents; alleles are different versions of a gene.
- Asexual reproduction involves a single parent; the offspring have the same DNA as the parent.
- Most multicellular organisms reproduce sexually by combining the DNA from two parents. Eggs from the mother and sperm from the father carry the genetic information required for a new organism, and the exchange of information results in a new combination of alleles.
- Mistakes made copying DNA are called mutations; mutations and sexual reproduction create the variation required for evolution.

1.5 How Does Life Evolve?

Create a scenario describing how a population of organisms might adapt to a new environment.

- The modern biological definition of evolution is a change in allele frequency in a population.
- Natural selection is a process by which population allele frequencies change to those that are better adapted to the environment.
- Because all organisms require a living ancestor, every species is related to each other through common ancestors.
- All forms of life on Earth are related in some way; this is a unifying theme of biology.

1.6 What Patterns of Diversity Are Found in Nature?

Explain the connection among common ancestors, the diversity of environments on Earth, and the scientific classification system.

- The Earth is dynamic in space and time. Through evolution, species adapt to their specific environments.
- Species that evolved in the recent past have a more recent common ancestor. As a result, they have more common features.
- Biologists use a pattern of relatedness, based on a hierarchy of relatedness similar to a family tree, to classify organisms.
- The most important aspect of the classification system is that the tree represents the evolutionary history of organisms.

Key Terms

allele 5
asexual reproduction 16
atom 9
cell 11
chemical bonding 10
classification 27
diversity 24
electron 9

element 9
enzyme 14
evolution 8
macromolecule 10
metabolism 14
molecule 10
mutation 16
natural selection 21

neutron 9
nucleus 9
orbital 10
proton 9
scientific name 27
sexual reproduction 16

Review Questions

1. What does the "infidelity gene" (*AVPR1a*) do in your body? How does this affect your brain?

2. What is an allele?

   a. A different gene

   b. A gene that encodes for hair color

   c. A gene that encodes for eye color

   d. A similar gene

   e. A different version of the same gene

3. List the four major features of life.

   a. Metabolism, inheritance and reproduction, evolution, and diversity

   b. Metabolism, evolution, diversity, and decomposition

   c. Metabolism, creationism, inheritance and reproduction, and diversity

   d. Inheritance and reproduction, evolution, diversity, and death

   e. Inheritance, evolution, diversity, and death

4. How do atoms bond to make molecules?

5. What are enzymes, and what is their function?

   a. Enzymes are proteins that make it more difficult for chemical reactions to occur.

   b. Enzymes are lipids that make it easier for chemical reactions to occur.

   c. Enzymes are proteins that make it easier for chemical reactions to occur.

   d. Enzymes are molecules that break down other molecules.

   e. Enzymes are carbohydrates that make it easier for chemical reactions to occur.

6. What is the difference between asexual and sexual reproduction?

   a. Asexual reproduction occurs only in the cells of the human body; sexual reproduction occurs between two parents of different organisms.

   b. Asexual reproduction involves a single parent; sexual reproduction occurs when an egg cell from the mother fuses with a sperm cell from the father to form a new individual.

   c. Asexual reproduction involves two parents; sexual reproduction involves one parent.

   d. Asexual reproduction involves a single parent; sexual reproduction occurs only when a human egg cell from the mother fuses with a human sperm cell from the father to form a new individual.

   e. Asexual reproduction always involves a single parent; sexual reproduction may or may not involve two parents in forming a new individual.

7. List two reasons why children vary genetically from their parents.

8. What did Darwin mean by "descent with modification"? Is descent with modification an observable fact of nature?

9. Briefly describe how natural selection works.

10. What are the seven levels in the Linnaean system of classification? How is this classification system used to define scientific names?

The Thinking Citizen

1. What are some of the ways that nature inspires you or has a positive impact on your life?

2. What are some specific examples of issues or questions you might encounter in your life for which knowledge of biology might be helpful?

3. When cells divide, they copy their DNA and pass one complete copy to each daughter cell. Would this process work to make egg or sperm cells? Explain why or why not.

4. Would natural selection work if the younger generation were genetically identical to the older generation? That is, are mutation and (for organisms that reproduce sexually) the combining of DNA from both parents required for natural selection to work? .

The Informed Citizen

1. (a) Use the Internet to search for the "infidelity gene" (or "ruthless gene"). Select several sites and try to evaluate each for clarity, accuracy, and usefulness. (b) Search for a commercial lab that will test your DNA for the presence of the infidelity gene. This process is called direct-to-consumer (DTC) marketing. In this case, DTC marketing gives people genetic information without the benefit of a trained genetic counselor or other medical professional. Do you think most people will be able to understand the test results on their own? Should the government regulate DTC marketing of genetic screening procedures? Should it be eliminated in favor of medically supervised tests?

2. Do you think your genes influence your behavior? Do they control your behavior? If a person has the "infidelity gene" and the hormone imbalance it produces, is fidelity still a choice this person can make? Why or why not?

3. Suppose a public hearing is convened in your town to discuss, and eventually vote on, the bulleted scenarios. For each of the scenarios, answer these questions: (a) What information would you need to draw your conclusion or to form a viewpoint? (b) From where or whom would you seek this information? (c) What specific questions would you need to have answered to make an informed choice or decision? (d) How would you vote?

   - A biotechnology company has plans to build a facility in your community that will provide genetically modified organisms to local farmers.

   - A group of parents submit a petition to your local school board asking that intelligent design be taught in biology courses as an alternative explanation to natural selection.

   - The local university seeks state funding to build a new research facility where scientists will study human genetics, research animal cloning, and perform stem cell research.

   - Your town proposes to add fluoride to the drinking water.

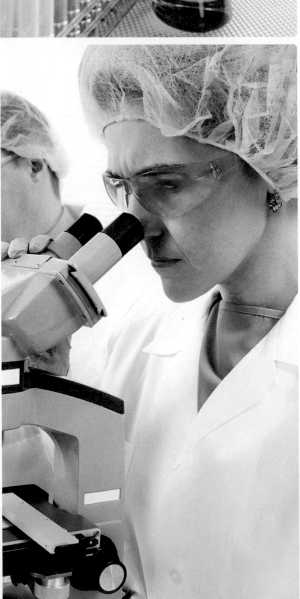
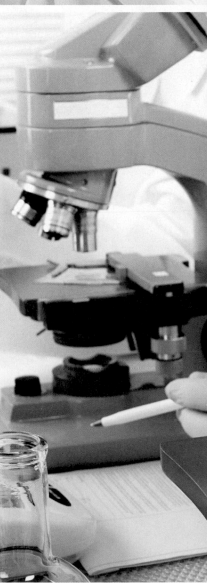

CHAPTER LEARNING OBJECTIVES

After reading this chapter, you should be able to answer the following questions:

2.1 How Would a Scientist Investigate Childbed Fever?
Outline the steps that Ignaz Semmelweis took to find the cause of childbed fever.

2.2 How Does Science Work?
Determine whether a given statement is an observation, hypothesis, or theory, and use the steps of the scientific method to design and evaluate a simple experiment.

2.3 What Assumptions Does Science Make About Nature?
Use the assumptions of science to determine whether a statement is scientific.

2.4 What Are the Principal Features of Science?
Explain the principal features of science.

2.5 How Does Science Differ from Other Ways of Knowing?
Distinguish science from other ways of knowing.

2.6 How Does Science Differ from Pseudoscience and Quackery?
Determine whether a claim is likely science, pseudoscience, or quackery.

Most scientists try to solve a particular problem, answer a specific question, or determine how a certain process works. The work of a scientist can increase our knowledge of the world, and it can help save lives and improve our quality of life.

CHAPTER 2

The Nature of Science

How Do We Know How the World Works?

Becoming an Informed Citizen ...

There is a lot we do not understand about the world. What we do know comes from the fact that the world appears to operate in a regular, repeatable way. Science exploits the world's regularity in particular ways that allow us to better understand how it works.

lbert Einstein once noted that *"the most incomprehensible thing about the world is that it is comprehensible."* There is much that we do not understand about the world and the way it works. But it is surprising that we understand as much as we do. Our knowledge comes from the fact that the world appears to follow a set of rules. Science exploits the world's regularity in ways that enable us to discover these rules. As we learn more about the rules, we come to understand the world better. This is the principal aim of science.

Most scientists try to solve a particular problem, answer a specific question, or determine how a certain process works. Science includes a number of tools, procedures, and working assumptions that facilitate these endeavors. Sometimes a scientist is motivated simply by curiosity, but often there are important consequences to a scientist's work that impact medicine, industry, and society. A few scientists, such as Isaac Newton, Charles Darwin, and Albert Einstein, developed comprehensive theories that explain a great deal about how our world works. The work of a scientist can increase our knowledge of the world, and it can help save lives and improve our quality of life.

Consider how scientific reasoning helped a doctor discover a cause and treatment for a tragic disease, childbed fever.

case study

The Mysterious Case of Childbed Fever

The pregnant young woman slowly climbed the stairs of the most modern hospital in Austria, the *Allgemeines Krankenhaus*, or General Hospital, of Vienna (**FIGURE 2.1**). Her destination was the maternity ward. It was 1846, and she was in labor. At the top of the stairs, there was a doorway opening into

FIGURE 2.1 During the 1840s, Ignaz Semmelweis investigated the cause of childbed fever at the Allgemeines Krankenhaus (General Hospital) in Vienna.

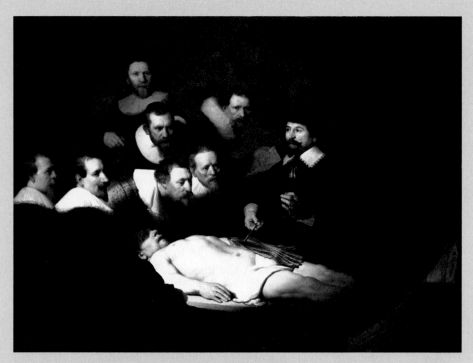

FIGURE 2.2 **Autopsies are thorough examinations conducted by doctors to determine the cause of death.** This famous painting by Rembrandt, *The Anatomy Lesson of Dr. Tulp* (1632), features an autopsy of a male patient.

a reception area. On the right was a hallway leading to Division 1, where doctors and medical students delivered babies. The hallway on the left led to Division 2, where midwives and their students attended the laboring mothers. As the young woman was led to the right, she hesitated, remembering that other women had told her to avoid the doctors, since there were rumors that many of the women who went to the right never came back. She had heard stories about women in labor who, upon being sent to Division 1, would panic and beg to be discharged from the hospital and sent home.

The young woman delivered a baby boy, and at first it seemed to her that her future was a bright one. However, a few hours after her baby was born, she started to feel a pain in her lower abdomen. The pain was much worse when any of the sore areas were touched. She became nauseous, vomiting frequently, and her pain worsened. Suffering with a high fever and a racing pulse, her abdomen filled with gas, causing even greater pain. She continued to worsen in all regards, and just three days after her baby's birth she died.

Upon opening the body of the deceased woman during the autopsy (**FIGURE 2.2**), the doctors and the medical students who were there to observe encountered a terrible, foul stench in her abdomen. Inside, they observed that her uterus wall was swollen, and pus and putrid flesh were evident in the body cavity. The skin and flesh were filled with gaseous bubbles and a foul-smelling fluid. The doctors determined that the woman had died of childbed fever, a disease that claimed the lives of one out of every 10 women who delivered babies in Division 1 of the hospital.

Why was giving birth so dangerous in the 1840s, and what has changed? We will find out shortly, while learning more about how the world works, thanks to the science of biology. This chapter will introduce you to the nature of science.

How Would a Scientist Investigate Childbed Fever?

Ignaz Semmelweis was a Hungarian, born in 1818 in the city now called Budapest, the fifth of 10 children in a prosperous family. He earned his medical degree in obstetrics in 1844. For two years, he worked on the medical staff in Division 1 of the Allgemeines Krankenhaus before becoming head of the division in 1846. In that year alone, 459 women died of childbed fever in his division, and Semmelweis became obsessed with discovering the cause of the disease and a way to prevent it.

Looking for Clues

One of the first things Semmelweis did was to compile records of births and childbed-fever–related deaths at the hospital. His results for the years 1841–1846 are shown in **TABLE 2.1**.

These records provided Semmelweis with his first clues about the causes of childbed fever. He discovered that death rates were over 2.5 times higher in Division 1. During this period, doctors and their medical students (all male) treated the women in Division 1, while midwives and the midwife students (all female) treated the women in Division 2 (although occasionally doctors worked in Division 2 as well). In addition to these data from the hospital, Semmelweis had anecdotal information: that women who delivered babies outside the hospital, either at home or on their way to the hospital, hardly ever contracted childbed fever.

Possible Causes

Whatever the cause of childbed fever, it appeared to involve practices at the hospital itself rather than some epidemic among the public. Semmelweis quickly set about investigating the cause by narrowing down the possibilities.

Was it miasma? One of the first possibilities Semmelweis pursued was that childbed fever was caused by **miasma**: harmful, toxic vapor supposedly exhaled by sick people or exuded by garbage or sewers into the air around them. According to this widely held belief, breathing in miasma could cause people to become sick, which indeed happened regularly to people surrounded by other sick people or by garbage and sewers.

miasma :: harmful, toxic vapor supposedly exhaled by sick people or exuded by garbage or sewers into the air around them

Table 2.1 :: Maternal Births and Deaths by Division, 1841–1846

	DIVISION 1			DIVISION 2		
Year	Births	Deaths	Rate (%)	Births	Deaths	Rate (%)
1841	3036	237	7.8	2442	86	3.5
1842	3287	518	15.8	2659	202	7.6
1843	3060	274	9.0	2739	164	6.0
1844	3157	260	8.2	2956	68	2.3
1845	3492	241	6.9	3241	66	2.0
1846	4010	459	11.4	3754	105	2.8
TOTALS	20,042	1989	9.9	17,791	691	3.9

Sources: I. Semmelweis, *The Etiology, Concept, and Prophylaxis of Childbed Fever*, 1861; and adapted from D. Gillies, *Hempelian and Kuhnian Approaches in the Philosophy of Medicine: The Semmelweis Case*, n.d.

Semmelweis predicted that if miasma were the cause, he could reduce the death rate by increasing the ventilation and air circulation in the divisions. He took measures to keep windows open and found other ways to introduce fresh air. However, none of these measures reduced the incidence of childbed fever, which led Semmelweis to conclude that miasma was not the cause. Even if miasma had been implicated as a cause, Semmelweis realized that this would not explain the differences in death rate between the two divisions, since both were in the same building and even shared a common entryway.

Was it the birthing position? Semmelweis considered another possibility based on the method doctors used to deliver babies. In Division 1, doctors had the mother lie on her back during labor, while in Division 2, midwives had the mother lie on her side. If lying on one's back was somehow a cause of childbed fever, Semmelweis predicted that changing the birthing position would reduce the death rate. He instructed doctors to deliver babies while the mother laid on her side, the same practice used in Division 2. Unfortunately, changing the birthing position had no effect on the incidence of childbed fever.

Were the male medical students being too rough on the patients? One of the main differences between the two divisions was the presence of male medical students in Division 1 and the presence of female midwife students in Division 2. Thus, Semmelweis considered whether the male medical students were too rough when they examined and treated pregnant women. This idea led Semmelweis to reduce the number of medical students by half, from around 40 to 20, and to reduce the number of examinations that each patient underwent. Unfortunately, this practice also failed to reduce the death rate significantly.

What else could it be? As 1846 came to an end, Semmelweis became despondent. He had tested many different possibilities, but the results of his tests indicated that none of them caused childbed fever. In March 1847, Semmelweis tried to renew himself by taking a vacation to Venice. When he returned on March 20, he learned of the death of his friend and fellow doctor Jakob Kolletschka. Kolletschka's death was tragic, but it provided the final clue Semmelweis needed to determine the cause of childbed fever.

"Cadaverous Particles"

Jakob Kolletschka had been stuck in the finger by a medical student's scalpel during an autopsy on a woman who had died of childbed fever. Within a few days, Kolletschka died of a massive infection. When his body was autopsied, doctors found pus, putrefied tissue, and many of the other abnormalities typically found in the cadavers (dead bodies) of women who died from childbed fever. Semmelweis concluded that Kolletschka in fact died from childbed fever, even though he clearly was not pregnant. This led Semmelweis to the conclusion that the corpse of a victim of childbed fever contained "cadaverous particles" that caused the disease (**FIGURE 2.3**). Today, we would call these particles "germs," or more specifically **microorganisms** (like bacteria). If these particles entered the body of a healthy person, that person would catch the disease as well. In the case of Kolletschka, the cadaverous particles contaminated the scalpel during the woman's autopsy and entered his body when the scalpel cut his finger.

microorganism :: a living thing that cannot be seen without the use of a microscope, such as bacteria

Semmelweis's concept of cadaverous particles explained many aspects of the childbed fever mystery. Only doctors and medical students performed autopsies, most of them on women who died of childbed fever. Typically, doctors and their medical students performed autopsies in the morning and then went on to Division 1 to examine patients and deliver babies. Sterilization of equipment and latex gloves were not part of the medical repertoire in 1847.

Ignaz
Semmelweis

FIGURE 2.3 Semmelweis discovered that childbed fever could be greatly reduced by washing hands and instruments thoroughly to eliminate "cadaverous particles" (what today we know as microorganisms). This 1966 painting by Robert Thom, *Semmelweis—Defender of Motherhood*, depicts medical professionals washing their hands.

FIGURE 2.4 Semmelweis required all medical personnel to wash their hands before entering the ward to examine women or deliver babies. The process outlined here should take about the time it takes to sing "Happy Birthday" twice.

Doctors and students washed their hands after performing autopsies, but often their hands still smelled from the corpse, suggesting that their hands could be contaminated with cadaverous particles. These particles could be transferred to patients during an examination or when delivering a baby, leading to infection of the mother. Since the midwives in Division 2 did not perform autopsies, patients in this area would be infected at a much lower rate.

This idea relating to cadaverous particles led to a simple prediction. If cadaverous particles on the hands of doctors and medical students caused childbed fever, then more rigorous hand cleaning would reduce the incidence of the disease. Semmelweis placed a bowl of diluted chlorine bleach at the door to Division 1. He required all medical personnel to wash their hands and instruments with it (**FIGURE 2.4**), using a brush to clean under fingernails, before entering the ward to examine women or deliver babies.

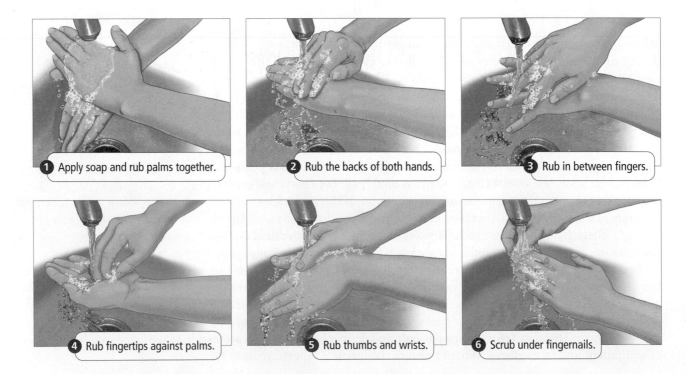

1 Apply soap and rub palms together.

2 Rub the backs of both hands.

3 Rub in between fingers.

4 Rub fingertips against palms.

5 Rub thumbs and wrists.

6 Scrub under fingernails.

Table 2.2 :: Maternal Births and Deaths by Division, 1847–1858

Year	DIVISION 1			DIVISION 2		
	Births	Deaths	Rate (%)	Births	Deaths	Rate (%)
1847	3490	176	5.0	3306	32	1.0
1848	3556	45	1.3	3319	43	1.3
1849	3858	103	2.7	3371	87	2.6
1850	3745	74	2.0	3261	54	1.7
1851	4194	75	1.8	3395	121	3.6
1852	4471	181	4.0	3360	192	5.7
1853	4221	94	2.2	3480	67	1.9
1854	4393	400	9.1	3396	210	6.2
1855	3659	198	5.4	2938	174	5.9
1856	3925	156	4.0	3070	125	4.1
1857	4220	124	2.9	3795	83	2.2
1858	4203	86	2.0	4179	60	1.4
TOTALS	**47,935**	**1712**	**3.6**	**40,870**	**1248**	**3.1**

Sources: I. Semmelweis, *The Etiology, Concept, and Prophylaxis of Childbed Fever*, 1861; and adapted from D. Gillies, *Hempelian and Kuhnian Approaches in the Philosophy of Medicine: The Semmelweis Case*, n.d.

This procedure yielded spectacular success. Within months, the mortality in Division 1 dropped to the level seen in Division 2, approximately 3%. Semmelweis kept childbed-fever–related death records for the next 12 years. The rates varied slightly over time but remained consistently low, with no major differences between the two divisions (**TABLE 2.2**).

With over 150 years of scientific and educational advances since Semmelweis's time, it is easy to understand why his approach was successful. His cadaverous particles were actually bacteria, which cause a lethal infection when they enter a person's bloodstream (*see Life Application: Childbed Fever*).

Life Application

Childbed Fever

Outbreaks of childbed fever occurred regularly in the cities of Europe and America for almost two centuries, between 1700 and 1900. During this time, the practice of medicine advanced dramatically, and many hospitals were built in the larger cities. But more hospitals and more doctors attending births just made the problem worse. There was considerable evidence linking childbed fever to hospitals. For example, in London between the years of 1831 and 1843, 10 mothers per 10,000 died of childbed fever with home delivery, while 600 per 10,000 died in hospital maternity wards. There was also evidence of the link between doctors and contagion. Many doctors across the globe pursued investigations similar to those of Semmelweis and reached similar conclusions. In the 1840s Boston physician Oliver Wendell Holmes Sr. published a paper arguing that doctors should wash their hands and sterilize their instruments to reduce the transmission of the disease.

Unfortunately, science doesn't always work out the way it should. The results reported by Semmelweis, Holmes, and others were attacked by a number of prominent physicians. Despite the evidence, these physicians did not accept the idea that childbed fever was contagious. In fact, they felt it was preposterous that well-trained doctors could be responsible for transmitting disease. When hospitals initiated hygienic practices, they saved the lives of many women. Yet it was not until the early 1900s that the majority of doctors accepted

(*Continued*)

Childbed Fever (*Continued*)

that childbed fever and other diseases are contagious and caused by germs. As the germ theory took hold, modern hygienic and sterilization practices became the norm.

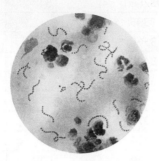

Childbed fever is known to be caused by the bacterium *Streptococcus pyogenes*, shown here. This is the same bacterium that causes strep throat and other more serious diseases such as toxic shock and flesh-eating disease.

Today, childbed fever, more technically referred to as *puerperal sepsis*, is known to be caused by a bacterium: *Streptococcus pyogenes*. This is the same bacterium that causes strep throat, scarlet fever, rheumatic fever, and a variety of skin infections. *S. pyogenes* normally lives on the skin and inside the nasal passages. Typically, up to 20% of the population is infected with the bacterium, but as long as it lives outside of our bodies, it does no harm. Once it infects the bloodstream, however, the bacterium changes its genetic expression in ways that cause the diseases just mentioned.

Childbed fever still occurs occasionally, but with early detection it is easily treated with antibiotics, and death is rare. It is intriguing that the bacterium is still present, although outbreaks of the disease are gone. Some biologists hypothesize that the bacterium has evolved into a less virulent form, since for most parasites killing the host is a bad idea. ::

Chlorine bleach kills bacterial cells, and washing hands in bleach drastically reduces the risk of transmitting bacteria from doctor to patient. In 1847, however, none of this was known. Doctors at the time were well aware that microorganisms existed, but most did not believe that microorganisms could cause disease. As a result, many doctors did not adopt Semmelweis's practices, even though they clearly were successful in reducing death from childbed fever. It was not until about 25 years later that research by scientists like Louis Pasteur and Robert Koch would establish the link between microorganisms and disease, now known as the "germ theory" of disease (see *Scientist Spotlight: Robert Koch*).

2.2 How Does Science Work?

scientific method :: set of procedures scientists use in their investigations; includes four steps: observations and facts, hypotheses and predictions, testing, and evaluation and interpretation of results

Semmelweis found a treatment for childbed fever because he took a scientific approach to the problem. In this section, we examine the **scientific method**: the set of procedures scientists use in their investigations (**FIGURE 2.5**, p. 42). Depending on the particular question or problem being explored, scientists may not always follow all of these steps in the order described, but they will emphasize certain components of the scientific method, as appropriate.

Observations and Facts

observation :: what you can see, hear, smell, taste, or feel physically

fact :: something that has actual existence

A scientific investigation begins with **observations** (what you can see, hear, smell, taste, or feel physically) and **facts** (things you know to be true) about the natural world. These relate to some phenomenon, process, or pattern that we see in nature. Scientists are curious. They want to explain the observations and facts they see. They want to discover the cause, reason, or mechanism that explains these observations and facts. Semmelweis observed two facts at the outset of his investigation: (1) deaths due to childbed fever had higher incidence in the hospital than outside, and (2) within the hospital, the incidence

Robert Koch (1843–1910)

Robert Koch was born December 11, 1843, in the Upper Harz Mountains of Germany. Even as a boy, he was smart, determined, and persistent. He was also methodical. By the time Koch was five years old, he had actually taught himself to read by looking at newspapers. No one had helped him in this endeavor. These personal characteristics, already evident in childhood, were ultimately critical for the successes that Koch achieved as a scientist.

Robert Koch was instrumental in the development of the germ theory of disease.

Interested in nature and biology from childhood, Koch attended the University of Göttingen, earning an M.D. in 1866. He began private practice in 1867 and in 1870 volunteered for service in the Franco-Prussian War. After the war, in 1872, Koch served as the district medical officer in a rural area of Prussia that is now part of Poland. Although he was a skillful and effective physician, Koch's restless intellect was not satisfied with the practice of medicine alone. He set up a laboratory for research in his four-room house. It was here that he labored for the next four years, alone, with a microscope his wife gave him as a gift and other equipment he made himself. During this period, Koch had no access to libraries and no scientific colleagues with whom he could exchange ideas.

Koch first took an interest in *anthrax*, a deadly scourge for farm animals and farmers. Although the bacterium associated with anthrax had been discovered earlier, Koch wanted to show scientifically that this bacterium was indeed the specific *cause* of this disease. Koch injected mice, using slivers of wood that he made, with blood from the spleens of animals that had died from anthrax. For a comparison, or control, he also injected another group of mice with blood from spleens of healthy animals. Koch carefully observed both populations of mice, and to his delight, the results were unequivocal. The mice inoculated with the anthrax bacteria died of the disease, whereas the control mice remained healthy.

Koch continued his study of anthrax bacteria and developed pure cultures of them for microscopic study. He photographed the bacteria and the structures they formed. Anthrax bacteria make these structures, called spores, when growth conditions are poor. Koch showed that spores could germinate to create new bacterial cells. These observations solved a mystery associated with anthrax: how could animals get the disease even if they had not been exposed to other sick animals? The answer is that anthrax spores can remain dormant in the soil for many years until there is an opportunity to infect a new host. The ability of anthrax to form long-lasting, deadly spores is one reason for concern about its use as a biological weapon.

Koch's discoveries about the ability of anthrax to form long-lasting deadly spores led to concern about the use of anthrax in biological weapons. In 2001, an anthrax scare shut down the U.S. Capitol Complex, which houses the Senate and House of Representatives.

By 1880 Koch had established his own professional research lab, where he developed many new techniques for growing and observing bacteria. Koch also traveled all over the world studying the infectious origins of many diseases including tuberculosis, cholera, plague, malaria, and typhus. The recipient of many awards, Koch won the 1905 Nobel Prize in Physiology or Medicine for his work on the germ theory of disease. ::

Steps of Scientific Method

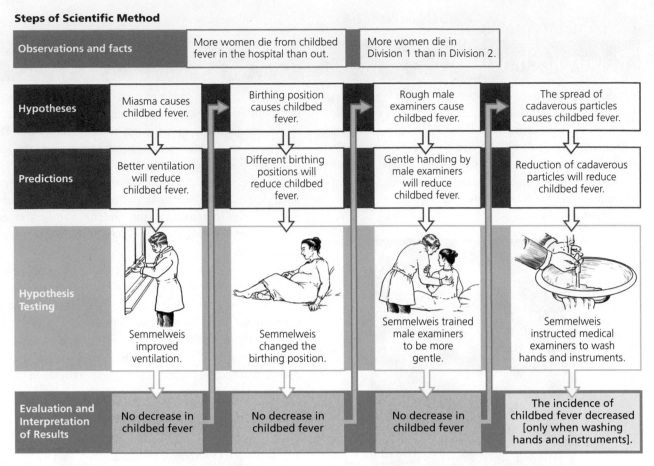

FIGURE 2.5 **Semmelweis used the steps of the scientific method to investigate the cause of childbed fever.**

was higher in Division 1 than in Division 2. His investigation focused on finding the cause of these two facts.

Hypotheses and Predictions

hypothesis :: possible cause or mechanism that could explain observations and facts

The second step in a scientific investigation is to develop a **hypothesis**. A hypothesis is a possible cause or mechanism that could explain the observations and facts. Semmelweis developed a number of hypotheses for the cause of childbed fever: miasma, birthing position, or rough treatment by male examiners. Scientists often develop several hypotheses during the course of their investigation if their initial hypothesis turns out to be incorrect.

prediction :: educated speculation about what an outcome will be

Every scientific hypothesis leads to a **prediction**, or an educated speculation about what an outcome will be. If you understand how something works, you can predict how it will change or work differently under different conditions. Semmelweis came up with a different prediction for each of his hypotheses. For example, if his hypothesis that miasma causes childbed fever were correct, the prediction would be that ventilation will reduce the incidence of the disease. All scientific predictions have this form: *if* the hypothesis is correct, *then* a certain set of conditions should yield a certain set of results.

Testing

test :: a procedure that sets up the conditions the predictions require

The third step is to **test** the hypothesis by designing a procedure to set up the conditions the predictions require. If you look back at the form predictions take, you can see how the test should be done. If the predicted results occur, your hypothesis may be correct. If they don't, your hypothesis is probably wrong. When Semmelweis tested his hypotheses, most of them turned out to be wrong. He added ventilation, he changed the birthing position, and he reduced examinations by male students—in each case, his hypothesis predicted that the incidence of childbed fever would go down. But it didn't. His hypotheses were not supported by the results. This is a common, and very frustrating, part of science. But Semmelweis's final hypothesis, that cadaverous particles transmit the disease, led to a prediction that came true when it was tested: washing hands and instruments to remove the particles reduced the number of cases. It took a while, but careful testing allowed Semmelweis to discover the cause of childbed fever, which led in turn to a way to prevent it.

There are several ways to test a hypothesis, but a **controlled experiment**—in which a scientist keeps all variables (possible factors that could affect the outcome of the test) the same except for the one under investigation—is one of the most common. Generally, there are many extraneous factors that can affect the outcome of a test. Suppose, for example, you developed a hypothesis that predicts a specific type of fertilizer will increase plant growth. If you grow your plants outside in a garden, factors like rainfall, clouds, and cold temperatures will affect your results and may overpower any effects the fertilizer might have. In an experiment, the scientist controls all the factors except the one under investigation. Plants could be grown in a greenhouse under constant temperature and in groups that all received equal amounts of light and moisture but varied in how much fertilizer they got. Because the amount of fertilizer is the only difference among the plant groups, the scientist can conclude that any corresponding differences in growth are due to the fertilizer.

controlled experiment :: a test or manipulation in which a scientist keeps all variables (possible factors that could affect the outcome of the test) the same except for the one under investigation

Evaluation and Interpretation of Results

The fourth and final step is to evaluate and interpret the test results. We have already discussed this somewhat. If the predicted results occur, this provides support for the hypothesis. If they don't, this is good evidence that the hypothesis is wrong. But given the results, what does the scientist do next?

In general, the first thing a scientist does is repeat the experiment. If the predicted results do not occur, it could be because some mistake was made performing the experiment. Repeating the experiment can determine whether this is the case. But if repeated failures convince the scientist that the hypothesis is incorrect, it's back to the drawing board. Like Semmelweis, the scientist comes up with a new hypothesis, a new set of predictions, and a newly designed test. There is no guarantee, but persistence often pays off with a successful test that supports the new hypothesis.

If the predicted results occur, the scientist wants to know whether they will occur consistently. If repeated experiments all yield the same results, the scientist has more confidence in the hypothesis.

Even if the test results consistently support the hypothesis, the scientist still is not done (as you might expect, science is never "done"). Most hypotheses lead to more than one set of predictions. To make the case stronger, a scientist will stick with the hypothesis but design other experiments to test these other predictions. Eventually, the scientist can accumulate a good deal of evidence, all of which may support the hypothesis. The accumulation of evidence helps the scientist to convince others that the hypothesis is true, perhaps in a published paper or a presentation at a conference.

But in science, no matter how much evidence you have to support a hypothesis, *the hypothesis is never proven with absolute certainty.* Proof like that happens only in mathematics, not in the natural world. Even if a large number of experiments support a hypothesis, you can't be certain that the next one will do so as well. And nature is so complex, with so many factors that affect one another, that even if you get your expected results, you can't be certain you have the correct cause. Some other factor you have not considered could be responsible for the results (see *How Do We Know? Hypothesis Testing and Scientific Proof*).

Despite these difficulties, it would be wrong to think that science can never lead to valid conclusions. Just take a moment to consider all the things that science has made possible: computers, improved medical care and increased longevity, and robotic probes that explore the surface of Mars, to name just a few. The fact that scientific ideas are not proven means that eventually they may be replaced with even more accurate ideas. But until then, the practical results of current scientific inquiry provide us with many benefits.

How Do We Know?

Hypothesis Testing and Scientific Proof

Science is based on a type of logic known as inductive reasoning—the search for general truths from specific observations. For example, suppose that the Republicans I know are conservatives. If I then generalize to say that all Republicans are conservatives, I have used inductive reasoning: *if* these Republicans are conservatives, *then* all Republicans are conservatives.

In this example, my observations (*if* . . .) are the *premises*, which are followed (*then* . . .) by the *conclusion*. If the premises are false, then the conclusion is false, too. It is possible that one of these Republicans is a libertarian or that he or she subscribes to some other ideology or set of beliefs. But even if all the premises are true, the conclusion can still be false. The conclusion makes a prediction, and further observations may prove that prediction wrong. After all, no matter how many conservative Republicans you observe, it takes only one liberal Republican that you did not observe to falsify the statement.

Even if a predicted result does occur, this does not prove that a hypothesis is true. There may be a number of different hypotheses that make the same prediction. For example, in the past, many people believed that the sun orbited the Earth. Let's make that our hypothesis: the sun orbits the Earth once a day, clockwise as seen from above the North Pole. Our hypothesis predicts that the sun will rise in the east and set in the west every day. And it does, but our hypothesis is still false. We neglected to consider another hypothesis that makes the same prediction: the Earth rotates about its axis once a day, counterclockwise, and this rotation is responsible for the rising and setting of the sun.

A

B

Even if a predicted result occurs, such as the rising and setting of the sun every day, this does not prove that a hypothesis is correct. Many people used to believe that the sun orbited the Earth (A), when in fact another hypothesis, the Earth rotates round its axis once a day (B), makes the same prediction.

Even though you can't prove inductive statements about nature, they are still useful and meaningful. Just like science, induction begins with a series of empirical observations. The world is a complex place, so you can never observe *everything* about it, or even a part of it. Since your observations are incomplete, any conclusions you draw from them could be incorrect. But the nice property of induction is that as you add more observations, you increase the chances that your conclusion is correct. In science, this means that repeated experiments and related work by other scientists provides more information about nature and brings us ever closer to understanding how nature truly works.

It is important to understand inductive reasoning if you are to understand the nature of scientific "proof." Even if a hypothesis (or theory) has support from prior research, it will still be false if it fails to make correct predictions in the future. A truly scientific hypothesis is falsifiable: it makes predictions that could fail repeated tests. It is wrong, however, to conclude that science cannot prove anything. If a hypothesis makes correct predictions under repeated testing, a scientist can conclude that the results support the hypothesis. We can then treat this hypothesis as proven until contradictory evidence arises. ::

Scientific Theories

You may be familiar with a number of scientists who are famous because they developed a scientific theory. Isaac Newton's theory of gravity, Charles Darwin's theory of evolution by natural selection, and Albert Einstein's theory of relativity are good examples of scientific theories, and these scientists are famous because of them. But what exactly is a scientific theory?

Theories are similar to hypotheses in some ways, but they provide a bigger picture of how some aspect of nature works, and they may weave together supporting evidence from several scientific fields. Both theories and hypotheses explain why certain events occur in the natural world. Both make predictions that can be tested. Hypotheses, however, focus on explaining smaller and more specific events. Theories are broader in scope and explain a wider variety of events. You can think of successful hypotheses as representing individual pieces in the puzzle of nature. Theories put these pieces together (**FIGURE 2.6**).

Darwin's theory of natural selection, which we encountered in Chapter 1, is a good example. There have been thousands of studies done to determine if natural selection can account for some specific trait or behavior in a given species: the peacock's tail or the lion's mane, or the behavior of a prairie dog that risks death by giving alarm calls instead of hiding when a predator approaches. Each of these studies tests a hypothesis about natural selection in a specific species. Darwin's theory of natural selection provides a general explanation of how adaptations in *all* organisms evolved. Every individual scientific study that tests some detail of natural selection has the potential to support Darwin's theory, as do results from a number of other scientific fields including geology, paleontology, and cosmology.

In science, a theory is as close to a proof as one can get. Like hypotheses, however, *theories are only supported by the evidence, not proven.* Given enough time, new theories can supersede earlier ones, though this can take a long time. Newton's theory of gravity lasted four centuries before Einstein's theory of relativity replaced it.

Scientific theories tie together results of numerous studies, often from several different disciplines. Theories also provide a framework for further investigations as scientists pursue new questions in new contexts to test them. Over time, a successful theory increases our understanding of the fundamental processes of nature and the rules by which nature operates. The idea that nature operates according to rules that we can discover and understand is relatively recent. As we see next, this idea is based on a number of assumptions that are critical to how science works.

theory :: an idea, supported by evidence, which provides a bigger picture than a hypothesis of how some aspect of nature works; it may weave together supporting evidence from several scientific fields

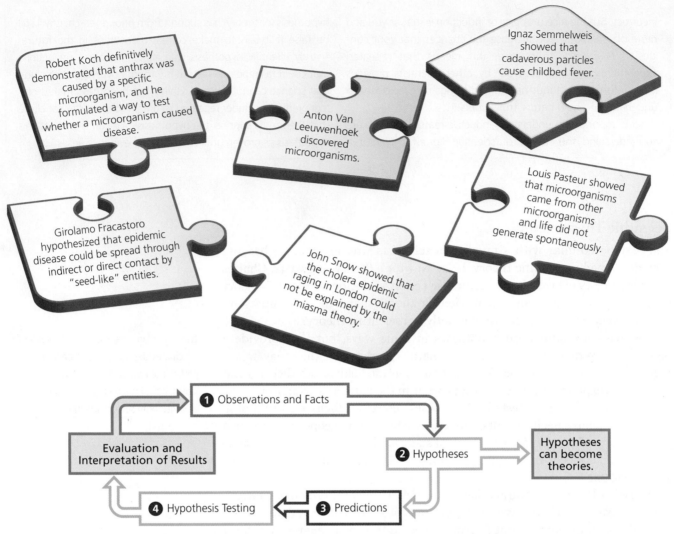

FIGURE 2.6 **While hypotheses represent individual pieces in the puzzle of nature, theories put these pieces together.** All of these hypotheses and observations made by different scientists contributed to the germ theory of disease, which states that infectious disease is caused by microorganisms.

2.3 What Assumptions Does Science Make About Nature?

Science is built on a number of assumptions about the nature of the world. These assumptions have come from years of research and experiments by scientists all over the globe. The assumptions are based on four related principles, or rules, that have been shown to hold true over hundreds of years of study: cause and effect, consistency, repeatability, and materialism.

Cause and Effect

cause and effect :: every event or outcome in nature has a source; if a scientist sets up the correct conditions, the results can be predicted in advance

First and foremost, science assumes that the rules of **cause and effect** (**FIGURE 2.7**) hold for all natural phenomena. Every event or outcome in nature has a cause or source, and since cause and effect are related, a scientist can learn about causes by observing the effects that occur. As we saw previously, cause and effect is the key to testing a hypothesis. If a scientist sets up the correct causes (conditions), the effects (results) can be predicted in advance.

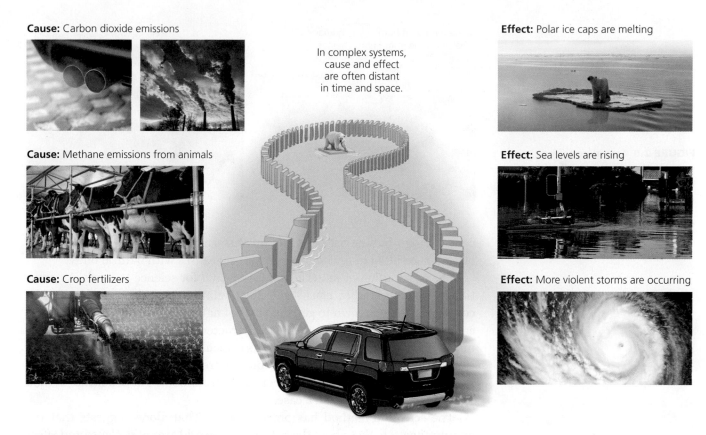

Cause: Carbon dioxide emissions

Cause: Methane emissions from animals

Cause: Crop fertilizers

In complex systems, cause and effect are often distant in time and space.

Effect: Polar ice caps are melting

Effect: Sea levels are rising

Effect: More violent storms are occurring

FIGURE 2.7 Every event or outcome in nature has a cause or source. A scientist can learn about causes by observing the effects that occur; note that in this figure all of the causes on the left contribute to the effects on the right.

Consistency and Repeatability

Science also assumes that events are consistent and repeatable. That is, if the same conditions (causes) are set up, the same results (effects) will occur. This assumption is important because it leads to the idea that there are general rules in nature, and that these rules will operate in a similar manner over and over again. It is also a reason why scientists repeat their experiments and try to repeat those of other scientists. Experimental results that can be regularly repeated become more strongly established and are more likely to be true. On the other hand, results that cannot be repeated by other scientists become suspect and may likely be the result of experimental error.

Experimental repetition has another consequence: it makes science *self-correcting*. Individual scientists can make mistakes in how they perform an experiment or interpret its results. Through the collective action of many scientists repeating these experiments, mistaken results may be weeded out.

Materialism

Another assumption that science makes is that effects in the natural world all have natural causes, rather than supernatural ones, a position known as **materialism** (**FIGURE 2.8**). Notice that all the hypotheses Semmelweis developed were based on natural causes. He did not hypothesize that childbed fever was caused by demonic possession or by divine retribution for the mother's sins. Those hypotheses may have seemed reasonable to people two or three

materialism :: the idea that effects in the natural world all have natural causes, rather than supernatural ones

The miracle of human life: human babies come from human beings.

Storks don't deliver babies.

FIGURE 2.8 According to the concept of materialism, all effects in the natural world have natural causes, rather than supernatural ones.

centuries ago, but the rise of science in Western culture over the past centuries often has resulted in replacing supernatural causes with natural ones.

The assumption of materialism is *required* for science to work. As mentioned, science studies cause-and-effect relationships that are repeatable, consistent, and conform to general rules of nature. Events with supernatural causes need not have any of these characteristics; they can appear quite unpredictable. Imagine Isaac Newton's difficulty in studying gravity if sometimes the apple fell straight down from the tree, while other times it floated, rose, or shot off horizontally, all seemingly at random. The scientific method would not be useful in such a case.

The scientific method has proven useful. That alone suggests that the assumptions it makes about the nature of the world are valid. Cause and effect do hold, and events have natural causes and are generally repeatable. Many scientists are so convinced by science's success that they adopt a completely materialistic view: *all* events, they believe, have natural causes, and the supernatural does not exist. Here, science can come into conflict with other approaches to knowledge, particularly religion. However, other scientists think instead of the supernatural as a limit to science: it is something that science simply cannot study. Either point of view has more to do with philosophy than science.

2.4 What Are the Principal Features of Science?

As we have seen, the scientific method gives scientists a procedure to follow when they perform their investigations. Science also makes a number of assumptions about how nature works. Many ways of learning use a similar logic, but a few other features combine to make science unique. Science is based on empirical evidence, it can be tested, and it can be applied more generally to other situations. Semmelweis relied on these features when he investigated the cause of childbed fever (**TABLE 2.3**).

Empirical Evidence

empirical evidence :: information that one gets from direct observation, from experience, or from the results of experiments and other tests of hypotheses

Science is based on evidence rather than on opinion or belief. **Empirical evidence** includes information that one gets from direct observation, from experience, or from the results of experiments and other tests of hypotheses. Empirical evidence in the form of background information helps a scientist develop a hypothesis, a possible mechanism to explain a particular fact or

Table 2.3 :: Principal Features of Science, as Illustrated by Semmelweis's Studies

Feature	Semmelweis's Studies
Empirical evidence	Observed that spread of cadaverous particles was associated with childbed fever
Testability	Tabulated death rate before and after the initiation of hand washing
Generality	Determined that findings were relevant to all maternity cases, not just those in the Allgemeines Krankenhaus

observation. The results of a scientific investigation produce more empirical evidence that other scientists can use.

Most important, every scientific hypothesis and theory is evaluated strictly by how well it explains the existing empirical evidence. No matter how popular an idea is, how logical, or how well it conforms to ideology or belief, if it does not fit in with the available evidence, it is wrong. Semmelweis probably expected that childbed fever was caused by miasma, since this was the prevailing view of disease at the time. But the evidence did not support miasma as a cause—the evidence prevailed, and the miasma hypothesis failed.

Testability

A commitment to **testability**—or a procedure for determining the evidence in support of a hypothesis—is the central tenet of science. When scientists hypothesize a cause for a particular fact or observation, they want to know if they are right. In science, that means their hypothesis is supported by evidence and makes a successful prediction. Science is self-correcting because tests like these can be repeated. Controlled experiments are the main form of testing, but in cases where experiments are not feasible, other types of observations can provide a test.

testability :: a procedure for determining the evidence in support of a hypothesis

Regardless, for a hypothesis to be testable, it must be **falsifiable**; it must be able to be proved wrong. As we have seen, a successful test or experiment provides only *support* for a hypothesis, not proof that it is right. But an unsuccessful test, one in which the predicted results do not occur, gives conclusive evidence that the hypothesis is false (provided the experiment was performed properly). False hypotheses eventually are identified and discarded, leaving only those hypotheses that are supported by empirical results. Semmelweis developed several hypotheses about the cause of childbed fever and tested each one. Falsifying his hypotheses—showing the hypotheses were incorrect—though certainly frustrating, helped him rule out many possible causes and eventually led him to find the true cause of the disease.

falsifiable :: able to be proved wrong

Generality

Most scientific investigations are narrow in scope and apply to a particular event, process, or situation. But scientists are interested in the **generality** of their work, how widely it applies to situations other than the specific ones they tested. When scientists present or publish their work, for example, they are careful to point out how their work relates to the work of other scientists. Scientific ideas are tested under a variety of conditions, at different times, and at different places. Over time, scientists develop an understanding of the general principles that underlie these individual results. These general principles are embodied in scientific theories that explain how the world works and the rules by which it operates. Discovering these principles of nature is one of the

generality :: how widely a scientific investigation applies to situations other than the specific ones scientists tested

Throat Cultures

Although childbed fever is no longer the terrible problem it once was, the bacterium that causes it, *Streptococcus pyogenes*, still infects humans. Strep throat, for example, is due to one type of *S. pyogenes*. This illness is characterized by a very sore throat, red pharynx, and fever. If you have an unusually sore throat—so painful that it can be difficult to swallow—it is important to get tested for strep. While most untreated strep infections eventually clear up, rheumatic fever develops in 3% of the cases. This can lead to heart and kidney problems, as well as arthritis.

How is strep throat detected? First, the healthcare provider will examine the patient's throat for redness, swelling, or patches of pus. He or she will gently touch the patient's throat, particularly the tonsils and pharynx, with a sterile cotton swab. With the rapid strep test, the swab is dipped in a test tube containing chemicals that will extract molecules from the bacteria if it is present. Next, an antibody that binds to molecules specific to *S. pyogenes* is added, along with a third substance that will change color if it detects the binding of antibodies to *S. pyogenes* molecules. The rapid test takes only minutes to get results. It is about 80% sensitive, meaning

In a positive throat culture for strep throat, the presence of hemolysis of blood agar can be detected. *S. pyogenes* causes the destruction of red blood cells in the agar, resulting in a clearing of the red color.

the test will fail to detect 20% of the positive cases. It is 95% accurate, which means it will read positive or negative incorrectly only 5% of the time. If the rapid test gives a positive result, antibiotic treatment begins immediately so that *S. pyogenes* will be eliminated from the patient.

Swab samples are also taken from the patient's throat so that a culture test can be performed. With this technique, the sample from the patient is cultured on a specific type of microbiological medium called blood agar. *S. pyogenes* is *hemolytic*, meaning it will destroy the red blood cells in its vicinity, thus forming a white halo around the bacterial colony. It can take a couple of days to get results from the culture test. Nevertheless, healthcare providers will always run it to make sure they catch the cases of strep throat that were undetected by the rapid test. The culture test, although more time consuming, is highly sensitive (~95%) and accurate (95%). Even though most cases of strep throat do not produce serious health problems, the infection is so contagious that it is essential to detect and treat it, so that people with compromised immune systems or weakened health are not exposed and placed at risk. ::

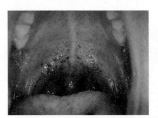
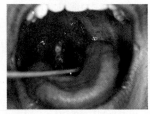

A   **B**

Strep throat is detected through an exam in which the patient's throat is examined for redness, swelling, or patches of pus (A). Then a cotton swab is used to touch the patient's tonsils and pharynx (B).

major goals of science. The results Semmelweis found could be applied beyond the divisions of the Allgemeines Krankenhaus. Wherever his procedures were applied they saved lives, and today we understand the general truth that microorganisms often cause disease. (See *Technology Connection: Throat Cultures* for a discussion of one method that scientists use to test for infectious microorganisms today.)

In this chapter, we have discussed a number of features of science, but these features are not unique to science. Many other disciplines rely on one or more of them, and you have likely used some of them in your own work. Taken collectively, however, this particular combination of features distinguishes science from other ways of acquiring knowledge and allows it to stand alone as a separate human enterprise.

How Does Science Differ from Other Ways of Knowing?

The search for truth or the answers to questions is not unique to science. We humans seek to understand ourselves and the world we inhabit. As you have seen, science explores particular questions about the natural world with a specific set of practices. And science, as a means to learn about the natural world, has been breathtakingly successful.

Does this mean that science is superior to other ways of knowing? Not at all. In fact, there are some questions that science and its tools are not equipped to help us answer. Other areas of knowledge attempt to address other types of questions, and they do so using methods appropriate for their subject area. For example, we can analyze a poem for its aesthetic value, explore what it means to live a good life, or seek to understand God. These are the tasks of literary analysis, philosophy, and theology, respectively.

In some cases, the distinction between science and another field is easy to describe. For example, science and theology could hardly be more different. Scientists never invoke the supernatural or refer to miracles to explain how the natural world works. Divine revelation is never a source of scientific understanding regarding the natural world. In contrast, many theologians do believe in a spiritual, nonmaterial world and in revealed knowledge. Some interpret this distinction as a conflict between science and religion. It does not need to be. Science and religion usually seek answers to completely different questions. Scientists are not going to answer questions about metaphysical or spiritual domains. Neither is religious scripture the place to go to learn about the material world (**FIGURE 2.9**).

Science cannot address the question of whether there is…

A divine being

an afterlife

or reincarnation

Religion cannot explain why…

A tumor forms
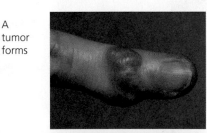

ice caps melt

male pattern baldness exists
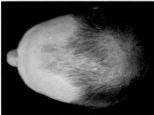

FIGURE 2.9 Scientists are not going to answer questions about metaphysical or spiritual domains. Neither is religious scripture the place to go to learn about the material world.

It can be difficult to draw a clear line between science and some fields. For instance, cognitive psychologists may have a great deal in common with biologists who study the features of the brain. Similarly, historians who try to piece together long-ago events share this approach with evolutionary biologists who research fossils. This rich cross-fertilization among disciplines should come as no surprise. In fact, collaboration among different specialists is a key for tackling big problems like the health of our planet, human well-being, and our collective futures.

2.6 How Does Science Differ from Pseudoscience and Quackery?

pseudoscience :: fake science

quackery :: promoting the use and/or purchase of remedies even when there is no scientific evidence or plausible rationale for their effectiveness

Being able to recognize the distinctions that separate science from other areas of knowledge is interesting and valuable. However, knowing the difference between science and **pseudoscience** (fake science) or **quackery**—promoting the use and/or purchase of remedies even when there is no scientific evidence or plausible rationale for their effectiveness—is essential for making informed decisions about many aspects of life. Indeed, lives can be negatively affected when people choose to follow the "latest" health remedy advertised on television or when they organize their lives according to the predictions of astrological charts or tarot readings.

Pseudoscience attempts to look like actual science so that its assertions might appear valid. However, unlike science, pseudoscience begins with a claim and looks only for things that support it. Controlled experiments are never done. In fact, direct tests of any kind, even if possible, are generally avoided. Pseudoscience is indifferent to facts and tries to persuade with

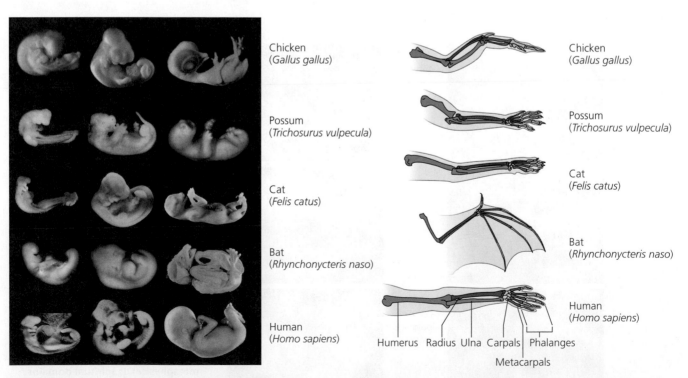

FIGURE 2.10 Evolution predicts that species change and that new species have formed during billions of years of life on Earth. The geological record, the fossil record, evidence from embryos, and molecular studies all support this theory.

FIGURE 2.11 Creationists claim that biblical scripture supports the notion that dinosaurs and humans inhabited the Earth at the same time, despite evidence to the contrary. In the landmark federal trial *Kitzmiller v. Dover* (2005), the federal district court of Harrisburg, Pennsylvania, found that intelligent design was a form of creationism (religious theory) and therefore could not be taught in American public schools.

appeals to emotion, sentiment, or distrust of established knowledge. Unlike real science, pseudoscience does not progress; nothing is revised or learned.

Let's consider creationism as an example of pseudoscience. Creationism is the idea that a divine being created all species in a single event. That may be a valid theological notion, but it is not a scientific hypothesis. Creationism does not allow for controlled studies to test it, and predictions from it have never been observed. All too often, evidence that refutes creationism as a scientific event is ignored. For example, evolution predicts that species change and that new species have formed during billions of years of life on Earth. The geological record, the fossil record, evidence from embryos, and molecular studies all confirm just that (**FIGURE 2.10**). Moreover, controlled studies in evolution involving selection and selective breeding over many years have produced stickleback fish that can withstand colder temperatures, tame foxes, and mice with increased activity levels.

Creationists disregard this mountain of information. Sometimes, they make up "facts." In contrast to claims made by some creationists, evidence from the fossil record clearly shows that dinosaurs and humans did *not* inhabit the Earth together at the same time (**FIGURE 2.11**). Finally, creationism is a static idea, not an inquiry into nature and a way of learning more. There is no research agenda, no revision of ideas, and no progress. Nothing new is learned.

As bad as pseudoscience is, quackery is even worse. Quackery takes advantage of people when they are sick and vulnerable. The rising cost of health care and the inability of many to get access to medical help have made this a golden age for quacks. We see the claims in the media daily: a vitamin or nutritional supplement supposedly will cure fatigue; a botanical will cure depression; copper bracelets will reduce arthritis pain (**FIGURE 2.12**); a hormone

FIGURE 2.12 Quackery takes advantage of people when they are sick and vulnerable. Are copper bracelets really the best "remedy" for arthritis compared with anti-inflammatory medications?

pill promotes weight loss; a shark cartilage extract will fight cancer . . . the list goes on and on.

How can we avoid being tricked? Your best protection is to know more about biology and the nature of science. By the end of this book, you will have plenty of practice. You should be skeptical about claims that appear too good to be true. You should also trust your health to knowledgeable caregivers who rely on tested and verified treatments.

Biology in Perspective

There are a number of ways to learn about how the world works. However, the best way we have discovered so far to learn about the world of nature is through science. Put simply, science works. The scientific method has been used successfully to find causes, treatments, and cures of diseases, and to find ways to improve our health and increase our longevity. Throughout this text, you will encounter many other examples of scientific success.

Science assumes that the events in the world occur through a chain of cause and effect, that the causes are natural rather than supernatural, and that as a result the world operates in a consistent and repeatable fashion. The validity of scientific ideas depends on the evidence that supports them and on how well they explain events in the natural world. Ultimately scientists search for theories that integrate broad areas of science and articulate fundamental natural laws. While science is not the only way that humans gain knowledge, it is the most successful approach for understanding the natural world.

Taken by itself, the scientific method is simply a procedure for learning about nature. The procedure is neither good nor bad, ethical or unethical. However, science is practiced by human beings, and decisions about how (or whether) to use the knowledge gained from scientific investigations are made by humans. This is why it is important for citizens to understand how science works and use their knowledge to ensure that science is used in ways that enrich our lives. It is also essential to use this knowledge to recognize and reject pseudoscientific claims and quack remedies.

Chapter Summary

2.1 How Would a Scientist Investigate Childbed Fever?

Outline the steps that Ignaz Semmelweis took to find the cause of childbed fever.

- Semmelweis identified several potential causes for the deaths of women following childbirth, including miasma, birthing position, rough handling by students, and "cadaverous particles" unintentionally carried by doctors from autopsies to the maternity ward.

- Semmelweis tested each potential cause by changing conditions one at a time; the only change that had any effect on deaths was having doctors wash hands and instruments in diluted bleach prior to entering the doctors' wing.

- Procedures instituted by Semmelweis reduced women's death rate in the doctors' wing to the level of the midwives' wing.

2.2 How Does Science Work?

Determine whether a given statement is an observation, hypothesis, or theory, and use the steps of the scientific method to design and evaluate a simple experiment.

- The scientific method involves making observations and compiling facts about the natural world, constructing hypotheses, testing the predictions made by hypotheses, and evaluating evidence to determine whether the hypotheses are supported.
- One method of testing hypotheses is through controlled experiments.
- Scientific hypotheses and theories are both explanations of the natural world that make testable predictions. Both hypotheses and theories are only supported by evidence, never proven.

2.3 What Assumptions Does Science Make About Nature?

Use the assumptions of science to determine whether a statement is scientific.

- Science makes some important assumptions about the nature of the world.
  - o There is a link between cause and effect.
  - o Events are consistent and repeatable.
  - o Effects in the world have natural causes rather than supernatural ones.
- While some scientists believe the assumptions of science reflect reality, other scientists see the assumptions as limiting the types of questions that can be answered through science.

2.4 What Are the Principal Features of Science?

Explain the principal features of science.

- Scientific hypotheses and theories are evaluated strictly by how well they explain existing empirical evidence.

- Testability is a central tenet of science; in order for scientific hypotheses and theories to be testable, they must be falsifiable.
- Although most scientific investigations are narrow in scope, they gain value when results can be applied beyond the particular experiment in which they were conducted.

2.5 How Does Science Differ from Other Ways of Knowing?

Distinguish science from other ways of knowing.

- Science uses a particular set of methods, assumptions, and features to answer questions. Other disciplines use different sets of practices.
- Science can be strongly contrasted with theology. Acceptable methods specific to theology include invocation of the supernatural, divine inspiration, and reliance on scriptural evidence.
- Many fields, such as history and psychology, can incorporate scientific methods into their discipline.

2.6 How Does Science Differ from Pseudoscience and Quackery?

Determine whether a claim is likely science, pseudoscience, or quackery.

- Pseudoscience looks like science, but it begins with a claim and then selects only evidence that supports the claim.
- Creationism is an example of pseudoscience because it disregards evidence contrary to its predictions, it makes up "facts," and it refuses to change with new evidence.
- Quackery uses pseudoscience to support unfounded claims of medical cures and treatments.

Key Terms

cause and effect 46
controlled experiment 43
empirical evidence 48
fact 40
falsifiable 49
generality 49

hypothesis 42
materialism 47
miasma 36
microorganism 37
observation 40
prediction 42

pseudoscience 52
quackery 52
scientific method 40
test 42
testability 49
theory 45

Review Questions

1. What causes did Semmelweis think could be responsible for the spread of childbed fever when he first began to pursue possibilities?

   a. Miasma, sunlight, delivery method

   b. Bacteria, sunlight, delivery method

   c. Miasma, delivery method, rough treatment of patients

   d. Bacteria, delivery method, rough treatment of patients

   e. Miasma, bacteria, sunlight

2. What was Semmelweis's evidence for his cadaverous particle hypothesis?

3. What are the major steps in the scientific method?

   a. Observations and facts, hypotheses and predictions, hypothesis testing, evaluation and interpretation of results

   b. Observations and facts, hypothesis testing, evaluation and interpretation of results, repeatability

   c. Hypotheses and predictions, hypothesis testing, evaluation and interpretation of results, consistency

   d. Observations and facts, hypothesis testing, cause and effect, evaluation and interpretation of results

   e. Observations and facts, hypothesis and predictions, materialism, hypothesis testing

4. What is the relationship (or difference) between a hypothesis and a prediction?

5. What is a controlled experiment?

   a. An experiment in which a scientist tests all possible factors at the same time

   b. An experiment in which four possible factors are under investigation

   c. An experiment conducted in a temperature-controlled environment

   d. An experiment in which a scientist keeps all possible factors that could affect the outcome of the test the same except for the one under investigation

   e. An experiment that is consistent and repeatable

6. List two reasons why a scientist might repeat an experiment.

7. What are the differences between a hypothesis and a theory?

8. What is materialism?

9. What are three main features of science that help to distinguish science from other forms of knowledge?

10. What does it mean for a hypothesis to be falsifiable?

    a. The hypothesis is material.

    b. The hypothesis is false.

    c. The hypothesis does not match the prediction.

    d. The hypothesis is right.

    e. The hypothesis must be able to be proved wrong.

11. What is inductive reasoning?

12. Which of the following is NOT true of pseudoscience?

    a. It is fake science.

    b. It relies on controlled experiments.

    c. It begins with a claim and looks only for things that will support it.

    d. Controlled experiments are never done, and direct tests of any kind are generally avoided.

    e. It tries to persuade with appeals to emotion, sentiment, or distrust of established knowledge.

13. What is quackery? Describe an example of quackery that you have seen advertised in the media.

The Thinking Citizen

1. Why do scientists reject supernatural explanations as a way to explain natural phenomena?

2. Suppose scientists developed a new drug that they claimed cured the common cold. What tests and experiments would need to be done to support or refute their claim?

3. Two scientists are planning to study prey capture by wolves in Yellowstone National Park. The first scientist develops this hypothesis: the smartest wolves will capture the most prey. The second scientist's hypothesis is this: the fastest wolves will capture the most prey. Which, if either, of these is a valid scientific hypothesis? One of the four "features of science" discussed in the text will help you evaluate these hypotheses for their scientific validity.

4. How would you describe the relationship shown in **FIGURE 2.13** between global average temperature and number of pirates in the world (e.g., "As the number of pirates goes down, _____")? Is this a cause-and-effect relationship? If so, explain the cause and the effect it produces. If not, how do you explain the fact that these two variables appear to be correlated?

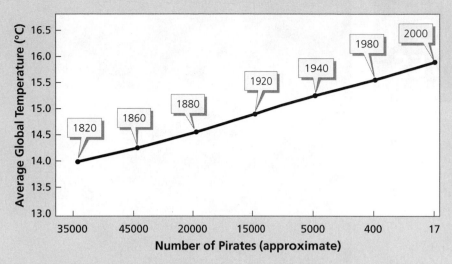

FIGURE 2.13 This figure depicts the global average temperature versus the number of pirates. These data should not be taken too seriously!

The Informed Citizen

1. Science is expensive. Many (though not all) scientific investigations done in the United States are funded by the government—that is, through taxpayers' dollars. Scientists apply to government organizations like the National Institutes of Health (NIH) and the National Science Foundation (NSF) for money to perform their research. Given that taxpayers pay for this research, is it reasonable for these organizations to prioritize funding, allocating most for research projects that will be of direct, immediate benefit to people (sometimes called *applied research*)? Are there ways to justify funding research designed primarily to simply discover more new things, such as the behavior of a marine organism or how a plant attracts its pollinators (sometimes referred to as *basic research*)?

2. Describe an example (real or hypothetical) of scientific research that you think is unethical. Is there a distinction between *unethical research* and *unethical uses* to which the research is applied? What sorts of guidelines would you suggest to distinguish ethical research from unethical? Who do you think is in the best position to develop such guidelines, and what are your suggestions for how such guidelines could be enforced?

CHAPTER LEARNING OBJECTIVES

After reading this chapter, you should be able to answer the following questions:

3.1 What Are the Units of Life?
Explain the consequences of cell theory for embryogenesis.

3.2 What Cell Structures Play a Role in Embryo Development?
Describe the cell structures that play a role in embryo development.

3.3 How Do Eggs and Sperm Form?
Show the possible chromosomes in gametes produced through meiosis in a given cell.

3.4 What Happens in Fertilization?
Describe the process of fertilization.

3.5 How Does an Embryo Form and Ultimately Become a Fetus?
Outline the stages of mitosis and the ways that cells become specialized.

3.6 What Are the Key Events of Pregnancy?
Differentiate fetal from embryonic development, and identify key points in development during pregnancy.

3.7 What Happens in Labor and Delivery?
Summarize the stages of labor and important events that are likely to occur in each stage.

3.8 How Do Twins Form?
Distinguish the development of identical from fraternal twins.

3.9 What Can Conjoined Twins Tell Us About Biology and Ourselves?
Assess the value of conjoined twin research and what conjoined twins can tell us about ourselves.

How do we come to be living, breathing, thinking, and feeling individuals with distinct limbs, organs, and the character that makes us who we are? That is the extraordinary riddle of human development.

CHAPTER 3

Human Development

How Do Cells Make a Person?

Becoming an Informed Citizen . . .

Examining human development reminds us of the diversity of life, including human life. Although we understand a great deal about what makes a body, we know much less about what makes a person. And while biology certainly plays a part, there is so much more.

human development :: the process by which an individual grows and matures from a single cell embryo inside a mother's womb to a baby that can survive on its own

Ow do we come to be living, breathing, thinking, and feeling individuals? How does an embryo become us—with distinct arms, legs, organs, and the individual character that makes us who we are? That is the extraordinary riddle of **human development**—*the process by which an individual grows and matures from a single cell embryo inside a mother's womb to a baby that can survive on its own.*

The answers to this puzzle begin even before the embryo forms, with the most basic unit of life, the cell. One biologist, Albert Claude, calls the cell "the mansion of our birth," not just our first home, but all of "our acquired wealth." As we see in this chapter, a single cell already includes the genetic heritage that we bring into the world from our parents. But to develop into humans, cells must communicate with one another and reproduce. We shall see how that small population of cells forms first an embryo, then the more mature stage called a fetus, and, ultimately, a baby. Each part in the journey relies on cell communication, cell movement, and the ability to regulate how genetic information is used.

No two people develop in exactly the same ways—not even identical twins. How can development have such a tremendous impact on a person's life? Consider some *unexpected* development in the story of the Hensel twins.

case study

Unusually Close Sisters

Abigail and Brittany Hensel entered the world on March 7, 1990. The words out of the obstetrician's mouth said it all: "They've got one body and two heads." Still under sedation, the mother, Patty, heard the word "Siamese" and asked, "I had cats?" But no, she had had what people often called Siamese twins, two babies with their bodies connected even after birth.

conjoined twin :: identical twins whose bodies are physically attached at some location

Abigail and Brittany are **conjoined twins**—their bodies are physically attached at some location (**FIGURE 3.1**). Each has her own head, neck, nervous system, heart, esophagus, and gall bladder. Yet they are *parapagus*, or joined at the side, and share many organs. Together they have three kidneys, four lungs (two of which are joined), one small intestine, one large intestine, one pelvis, one bladder, one set of reproductive organs, one liver, and one ribcage. Their spines are joined together at the pelvis, and they have just two legs and two arms. (A third undeveloped arm was removed from their chest when they were infants.)

Even more astonishing, Abigail and Brittany can live much of a normal life. The sisters learned to walk by the age of 15 months, a remarkable achievement given that each girl controls just one side of her body. To do something as ordinary as learning to walk, Abigail and Brittany had to agree on which direction to go; if not, they went in circles. In fact, they are a model of cooperation: they play piano (Abigail the right hand and Brittany the left). They swim, ride horses, play basketball, type, and respond to e-mails—all without having to speak. Each girl passed her driving test and is licensed to drive.

FIGURE 3.1 Abigail and Brittany Hensel are conjoined twins. Each has her own head, neck, nervous system, heart, esophagus, and gall bladder, but they are physically connected at the side and share many organs.

Abigail controls the pedals, radio, and defogger, while Brittany is in charge of the lights and turn signals. They coordinate steering.

Although they share more than other twins do, Abigail and Brittany are quite different in personality, likes and dislikes, and ambitions. Their tastes in food and clothes are different, they style their hair differently, and they experience separately the need to sleep. In school, Abigail and Brittany were good at different subjects. They buy tickets for two seats at the movies and celebrate their birthdays on separate days.

What makes Abigail and Brittany so different? How do humans develop, and what does that variation tell us about the essence of being an individual? At what point during human development does someone become a person? As we will see in this chapter, conjoined twins are only one of the many potential variations in human development.

What Are the Units of Life?

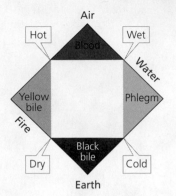

FIGURE 3.2 Before the invention of microscopes, the humoral theory held that a healthy body required four humors: blood, phlegm, yellow bile, and black bile. When these fluids were not in balance, illness resulted.

cell :: the most basic unit of life

embryogenesis :: The development of cells into embryos and bodies that are able to perform the tasks necessary for life

cell theory :: the realization that all living organisms are made of cells; it was first proposed in 1839 by Matthias Schleiden and Theodor Schwann

We start with the most basic unit of life, the **cell**. From cell structure and function, we can then understand how embryos develop to produce a body, like yours or that of the Hensel sisters, able to perform the tasks necessary for life. That process of development is called **embryogenesis**. Scientists who study it are embryologists.

Before the invention of microscopes in the 17th century, scientists had limited understanding of the substance of life. According to one early idea, a healthy body required four humors: blood, phlegm, yellow bile, and black bile (**FIGURE 3.2**). When these fluids were not in balance, illness resulted. Blood-letting, or the removal of blood, was used well into the 19th century to treat fever and other ailments. Not only did this fail to work, it was potentially harmful. Repeated bleedings of George Washington in response to a headache probably killed him.

Surgical procedures and autopsies revealed the existence of organs (such as the heart) and organ systems (such as the one for blood circulation), and what scientists and physicians described as "tissues and fibers." Later it became apparent that human tissues and fibers are composed of an even smaller unit, the cell.

Of course, that discovery relied on an important tool of science, the microscope. Robert Hooke, in 1663, first observed well-defined chambers in slices of cork (**FIGURE 3.3**). Anton van Leeuwenhoek enthusiastically explored these cells through the early 1700s (see *Scientist Spotlight: Anton van Leeuwenhoek*). As microscopes improved dramatically in the 19th century, it turned out that *all* organisms, plant and animal alike, are composed of cells. There are even single-celled organisms, such as bacteria.

The realization that all living organisms are made of cells is called the **cell theory**. The theory, first proposed in 1839 by Matthias Schleiden and Theodor Schwann, reinforces the common ancestry of life. Knowledge about cells also opens the door to understanding how our bodies function in health and disease.

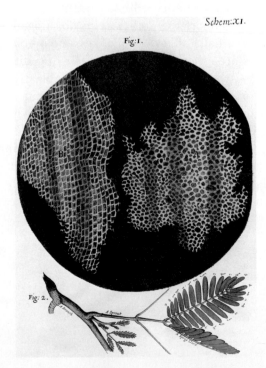

FIGURE 3.3 In 1663, Robert Hooke first observed well-defined chambers (cells) in slices of cork wood. This is a color-enhanced version of the illustration Hooke provided in his book *Micrographia*, published in 1667.

Scientist Spotlight

Anton van Leeuwenhoek (1632–1723)

Anton van Leeuwenhoek started and finished his life as a tradesman. He was born in Holland in 1632, was apprenticed to a linen drapers shop in 1648, and went into business for himself as a fabric merchant in 1654. In 1668, however, he learned how to grind lenses and make simple microscopes, and he began to look at *everything*. Some historians speculate that he was inspired by Robert Hooke's popular book *Micrographia*, which had illustrations of magnified objects.

In all, van Leeuwenhoek made over 500 microscopes. They opened up the previously unknown world of the very small. What did he examine and observe? To name a few examples, he scraped teeth, collected saliva, and observed the bacteria present. He examined pond water and saw tiny invertebrates, protozoa, and single-celled green algae. After cutting himself shaving, he examined the blood on his razor and saw red blood cells. Van Leeuwenhoek wrote to the Royal Society of London, describing his discoveries. Despite his long, rambling letters and lack of formal education, the society recognized the merit of his work. He was elected as a full member in 1680, and he continued sending descriptions of his observations until the last days of his life.

Van Leeuwenhoek risked social censure with perhaps his most important discovery—sperm cells in animals. At first, he did not want to study the question of what is in human semen; he was too concerned about its propriety. Ultimately, he observed his own semen—"acquired not by sinfully defiling himself," he testified, but "by hastening from the marriage bed to the microscope." Later he showed that sperm are present in the semen of other animals. He also observed large numbers of sperm in the uterus and fallopian tubes of recently mated dogs and rabbits. These observations were essential to our understanding of the cellular basis of reproduction.

Anton van Leeuwenhoek discovered sperm cells in animals using microscopes of his own construction.

This Dutch biologist realized how unusual his drive and curiosity were:

> Some go to make money out of science, or to get a reputation in the learned world. But in lens-grinding and discovering things hidden from our sight, these count for naught. And I am satisfied too that not one man in a thousand is capable of such study, because it needs much time . . . and you must always keep thinking about these things if you are to get any results. And over and above all, most men are not curious to know: nay, some even make no bones about saying, what does it matter whether we know this or not? (translation from "Leeuwenhoek, Anton van [1632–1723]," in *Encyclopedia of World Biography* [Gale, 1998])

Anton van Leeuwenhoek exemplifies the intense desire to observe, learn, and understand that characterizes all good scientists. ::

3.2 What Cell Structures Play a Role in Embryo Development?

Each of us developed from a single cell (**FIGURE 3.4**), as did twins like Abigail and Brittany. As we saw in Chapter 1, our bodies are composed of *eukaryotic* cells, with an enclosed nucleus and specialized structures called **organelles**. (Prokaryotic cells, such as bacteria, do not possess organelles, and we consider them in greater detail in Chapter 10.) Organelles play a role in the structure, function, and development of all multicellular organisms, including humans. The common features of all eukaryotic cells are striking evidence that all species share a common ancestor. All eukaryotic cells include a cell membrane, nucleus, mitochondria, endomembrane system, and cytoskeleton.

organelle :: specialized structure in eukaryotic cells

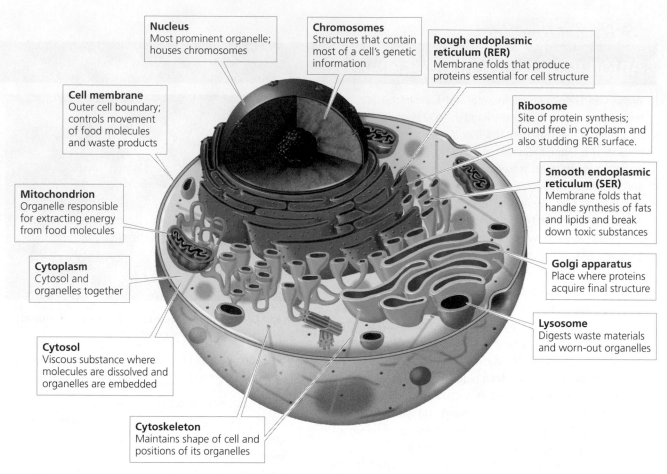

Nucleus
Most prominent organelle; houses chromosomes

Chromosomes
Structures that contain most of a cell's genetic information

Rough endoplasmic reticulum (RER)
Membrane folds that produce proteins essential for cell structure

Cell membrane
Outer cell boundary; controls movement of food molecules and waste products

Ribosome
Site of protein synthesis; found free in cytoplasm and also studding RER surface.

Smooth endoplasmic reticulum (SER)
Membrane folds that handle synthesis of fats and lipids and break down toxic substances

Mitochondrion
Organelle responsible for extracting energy from food molecules

Golgi apparatus
Place where proteins acquire final structure

Cytoplasm
Cytosol and organelles together

Lysosome
Digests waste materials and worn-out organelles

Cytosol
Viscous substance where molecules are dissolved and organelles are embedded

Cytoskeleton
Maintains shape of cell and positions of its organelles

FIGURE 3.4 The most basic unit of life is the cell. Shown here is an animal cell.

Cell Membrane

cell membrane :: outer boundary of the cell body, where the cell encounters its environment

cytosol :: a viscous substance in a cell where small molecules are dissolved and organelles are embedded

cytoplasm :: in a cell it is the cytosol and organelles together

The **cell membrane**, or outer boundary of the cell body, is where the cell encounters its environment. It regulates what enters and leaves the cell. For example, the membrane controls the movement of food molecules into the cell and the removal of waste products. The cell membrane also maintains the environment *inside* the cell—essential for the many thousands of chemical reactions that occur in cells. The membrane also encloses a viscous substance called the **cytosol**. Here, small molecules are dissolved and organelles are embedded. The cytosol and organelles together are called the **cytoplasm**.

The cell membrane plays many roles in human development: it can detect chemical signals from other cells, starting with communication between a human egg and sperm cell. It can also stick to other cells to make multicellular structures like tissues and organs.

Nucleus

nucleus :: the most prominent organelle inside the cell; it houses the chromosomes that contain the genetic information

chromosome :: structure that contains most of the genetic information of the cell; housed in the nucleus

The **nucleus** is the most prominent organelle inside the cell. The nucleus houses the **chromosomes**, the structures that contain most of the genetic information of the cell. These genetic instructions are written in the language of the molecule DNA. The instructions are organized into genes. The DNA of genes directs the production of molecules that are important for a cell's structure, function, and even identity. During embryogenesis, gene expression—which genes are turned off or on—is carefully regulated so that the correct types of cells form where they should.

Mitochondria

Mitochondria are the organelles responsible for extracting energy from food molecules. Muscle cells have many mitochondria, which is no surprise given the energy demands on muscles. Mitochondria are also abundant in the cells of developing embryos, because biological development takes energy.

mitochondrion :: organelle responsible for extracting energy from food molecules

Endomembrane System

Like a factory, the **endomembrane system** of the cell produces important molecules and delivers them to key locations within and outside the cell. In human development, it provides the molecules essential to such organs as brain, lung, liver, or skin. It also collects, packages, and removes waste materials. Otherwise, those materials would lead to improper brain development and other defects. The endomembrane system consists of three structures—the endoplasmic reticulum, the Golgi apparatus, and lysosomes.

endomembrane system :: produces important molecules and delivers them to key locations within and outside the cell; consists of the endoplasmic reticulum, the Golgi apparatus, and lysosomes

Located near the nucleus, the **endoplasmic reticulum (ER)** is a series of membrane folds. The *smooth endoplasmic reticulum* (SER) handles the synthesis of fats and lipids including those that compose the cell membrane. The SER also breaks down and eliminates some toxic substances. Liver cells are especially rich in SER, since this organ detoxifies the blood.

endoplasmic reticulum (ER) :: a series of membrane folds located near the nucleus; smooth endoplasmic reticulum handles the synthesis of fats and lipids, and rough endoplasmic reticulum produces proteins

The *rough endoplasmic reticulum* (RER) produces proteins, molecules essential for cell structure and regulation. The RER gets its name from the **ribosomes** that stud its surface. The ribosomes thread the protein into the interior of the RER as they are making it. Once inside, some proteins are packaged in bubble-like vesicles, which then travel to other destinations in the cell, delivering their protein cargo.

ribosome :: studs the surface of the rough endoplasmic reticulum (RER) and threads the protein into the interior of the RER as it is being made

In the **Golgi apparatus**, proteins acquire their final structure. Some proteins, for example, become digestive enzymes, molecules capable of breaking down other molecules. Vesicles also leave the Golgi apparatus and travel to the cell membrane. There, they release their contents to the space outside the cell. Cells in the respiratory system, for example, secrete mucus in this way.

Golgi apparatus :: the place where many proteins acquire their final structure

Lysosomes digest waste materials and worn-out organelles. For example, when mitochondria age and no longer function, they are marked for destruction and delivered to lysosomes. The lysosomes also recycle the molecules so that they can be reused. Similarly, red blood cells, which live for no more than three months, are destroyed by certain cells of the immune system. These immune cells engulf the old red blood cells and digest them in lysosomes present within these immune cells.

lysosome :: digests waste materials and worn-out organelles and recycles the molecules so they can be reused

Cytoskeleton

The **cytoskeleton (FIGURE 3.5)** maintains the shape of the cell and the positions of its organelles. Its fine fibers also allow for cell motion. The cytoskeleton has three major components: *microtubules, microfilaments*, and *intermediate filaments*. Microtubules are hollow rods that determine cell shape and help with cell movement. Flagella, the motile tails on swimming sperm cells, are made largely of microtubules. Microfilaments are thin, strong, and versatile and also aid in cell movement. A ring of microfilaments, along with the help of some other proteins, pinches the cell membrane so that cells may divide in two. Cell division happens a great deal in human development. Intermediate filaments strengthen cells and areas of the embryo, such as the covering of the cell nucleus.

cytoskeleton :: maintains the shape of the cell and the positions of the organelles

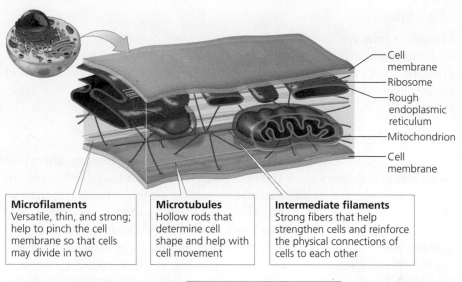

Microfilaments
Versatile, thin, and strong; help to pinch the cell membrane so that cells may divide in two

Microtubules
Hollow rods that determine cell shape and help with cell movement

Intermediate filaments
Strong fibers that help strengthen cells and reinforce the physical connections of cells to each other

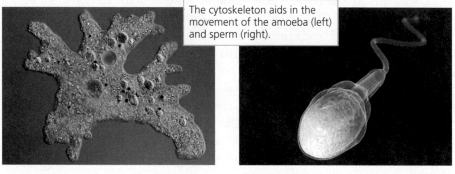

The cytoskeleton aids in the movement of the amoeba (left) and sperm (right).

FIGURE 3.5 The cytoskeleton maintains the shape of the cell and also aids in cell movement. It has three major components: microtubules, microfilaments, and intermediate filaments.

gamete :: sex cell

meiosis :: a special type of cell division that produces eggs or sperm

homologue :: one in a pair of chromosomes

testis :: male sex organ where sperm are produced

ovary :: female sex organ where eggs are produced

diploid cell :: cell with two complete sets of chromosomes

haploid cell :: a cell with just one set of chromosomes

interphase :: the stage in cell division when cells prepare materials that will be necessary for meiosis or mitosis to occur

3.3 How Do Eggs and Sperm Form?

All organisms that undergo sexual reproduction make **gametes**, or sex cells. In many species, including humans, males make sperm and females make eggs (**FIGURE 3.6**).

A special type of cell division, called **meiosis**, produces eggs or sperm. In humans, body cells contain 46 chromosomes organized into two sets of 23 (**FIGURE 3.7**). Pairs of chromosomes are called **homologues**. Meiosis halves the number of chromosomes. At the same time, each gamete receives a complete set of genetic instructions to pass on to the next generation.

Sperm cells form in male sex organs, the **testes**, egg cells in female sex organs, the **ovaries**. The process involves the repackaging of chromosomes so that DNA can move without being physically damaged. Critical events along the way make it appear as if meiosis is a series of discrete steps, or *phases*. In fact, phases progress smoothly from one to the next.

Meiosis entails two rounds of cell division (**FIGURE 3.8**). It starts with a single **diploid cell**, a cell with two complete sets of chromosomes. It ends with four **haploid cells**, each with just one set of chromosomes.

Meiosis I: The First Round of Cell Division (**FIGURE 3.9**)

*   *Interphase:* Cells in **interphase** are hard at work preparing materials that will be necessary for meiosis to occur. Although sometimes called a "resting stage," it is a very busy time. The chromosomes replicate, all organelles continue to function, and the cell increases in size. At this stage, the chromosomes are loosely packaged and

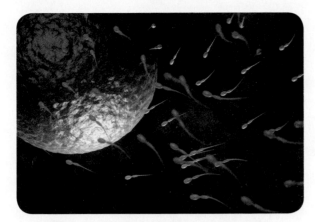

FIGURE 3.6 All organisms that undergo sexual reproduction make gametes, or sex cells. In many species, including humans, males make sperm and females make eggs.

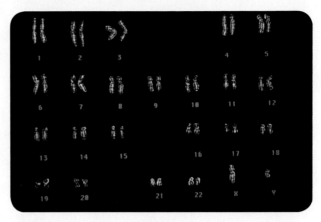

FIGURE 3.7 In humans, body cells contain 46 chromosomes, which are organized into two sets of 23. Pairs of chromosomes are called homologues.

individual chromosomes are hard to see even with a powerful light microscope.

- *Prophase I:* During **prophase I**, the chromosomes become shorter and thicker, and they line up with their homologues. The membrane surrounding the nucleus comes apart, leaving the chromosomes in the cytoplasm.

Meanwhile, the rest of the cell is busy. The microtubules of the cytoskeleton disassemble and form the **spindle**, a structure that will later help to distribute the chromosomes into new cells. In its completed form, it looks like a starburst of fibers, emanating from the two ends or poles of the cell. These fibers extend into the center of the cell, where they can "catch" a chromosome. This continues until they have snagged each of the paired chromosomes.

- *Metaphase I:* In **metaphase I**, specific proteins associated with the fibers of the mitotic spindle pull the chromosomes toward both poles.

prophase I :: in meiosis, the stage in the first round of cell division when the chromosomes become shorter and thicker and line up with their homologues

spindle :: formed from the microtubules of the cytoskeleton, this is a structure that helps to distribute chromosomes to new cells

metaphase I :: in meiosis, the stage in the first round of cell division when paired chromosomes line up in single file down the middle of the cell

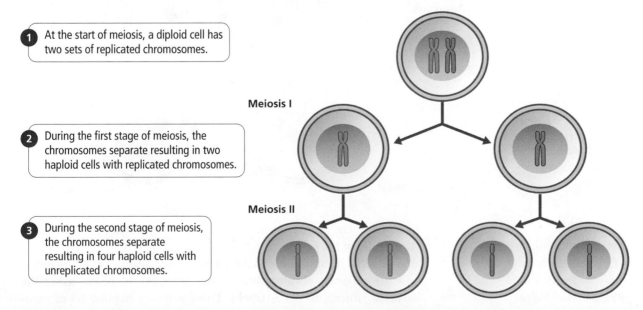

1 At the start of meiosis, a diploid cell has two sets of replicated chromosomes.

Meiosis I

2 During the first stage of meiosis, the chromosomes separate resulting in two haploid cells with replicated chromosomes.

Meiosis II

3 During the second stage of meiosis, the chromosomes separate resulting in four haploid cells with unreplicated chromosomes.

FIGURE 3.8 Meiosis, a special type of cell division, produces eggs or sperm. It starts with a single diploid cell, a cell with two complete sets of chromosomes, and ends with four haploid cells, each with just one set of chromosomes.

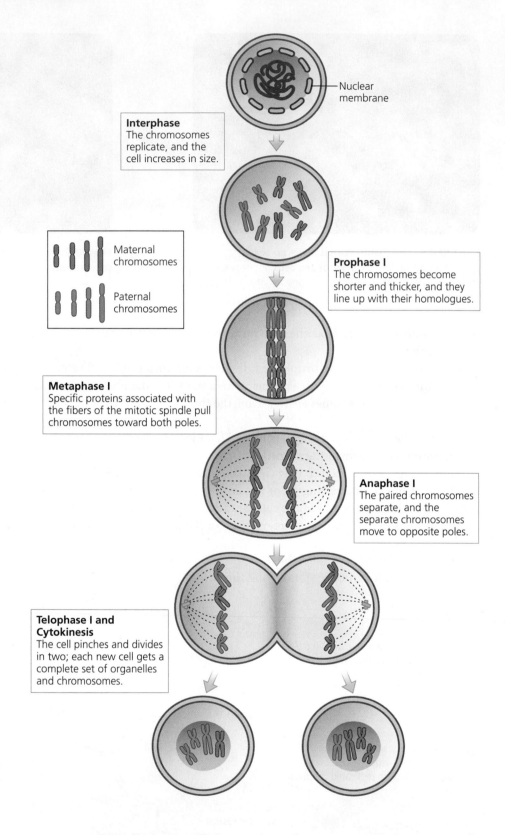

Interphase
The chromosomes replicate, and the cell increases in size.

Maternal chromosomes

Paternal chromosomes

Prophase I
The chromosomes become shorter and thicker, and they line up with their homologues.

Metaphase I
Specific proteins associated with the fibers of the mitotic spindle pull chromosomes toward both poles.

Anaphase I
The paired chromosomes separate, and the separate chromosomes move to opposite poles.

Telophase I and Cytokinesis
The cell pinches and divides in two; each new cell gets a complete set of organelles and chromosomes.

Nuclear membrane

FIGURE 3.9 Meiosis I is the first round of cell division; it ends with two haploid cells. The chromosomes are still in their replicated form, with two chromatids each.

anaphase I :: in meiosis, the stage in the first round of cell division when paired chromosomes separate and move to opposite poles, leaving a haploid set of chromosomes on each side of the cell

As a result, the paired chromosomes line up in single file, right down the middle of the cell.

- *Anaphase I:* In **anaphase I**, the paired chromosomes now separate, and the separate chromosomes move to opposite poles, using the shrinking microtubules as tracks. There is now a haploid set of chromosomes on each side of the cell.

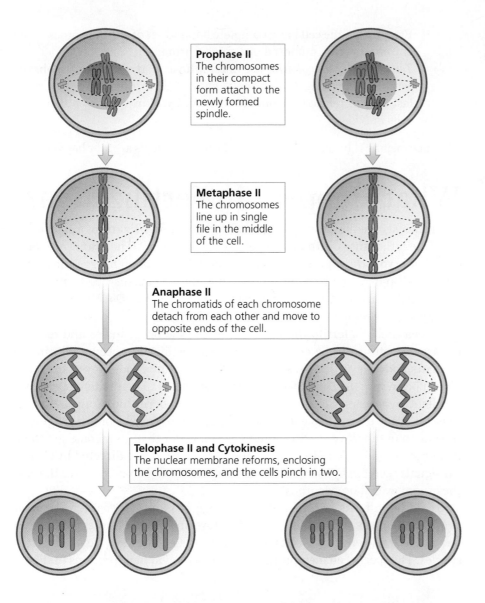

FIGURE 3.10 Meiosis II is the second round of cell division; the two attached sister chromatids separate. Meiosis II ends with four haploid cells, and each chromosome has one chromatid.

Prophase II
The chromosomes in their compact form attach to the newly formed spindle.

Metaphase II
The chromosomes line up in single file in the middle of the cell.

Anaphase II
The chromatids of each chromosome detach from each other and move to opposite ends of the cell.

Telophase II and Cytokinesis
The nuclear membrane reforms, enclosing the chromosomes, and the cells pinch in two.

- *Telophase I and cytokinesis:* In **telophase I and cytokinesis,** the cell at last pinches and divides in two. Each new cell gets a complete set of organelles and chromosomes.

Meiosis I ends with two haploid cells. Meiosis is not over yet, though, because each of these chromosomes has *two* copies of its genetic information called **chromatids**. At the end of meiosis I, each chromosome is composed of two identical chromatids attached to each other. The task of meiosis II is to separate them.

Meiosis II: The Second Round of Cell Division (FIGURE 3.10)

- *Prophase II:* Each stage of meiosis II occurs in both cells produced in meiosis I. At the start of **prophase II,** the chromosomes, still in their compact form, attach to the newly formed spindle.
- *Metaphase II:* In **metaphase II,** chromosomes are captured by one or more microtubules from each side. They line up in single file in the middle of the cell.
- *Anaphase II:* In **anaphase II,** the two chromatids of each chromosome detach from each other and move to opposite ends of the cell.

telophase I and cytokinesis :: in meiosis, the stage in the first round of cell division when the cell pinches and divides in two; each new cell has a complete set of organelles and chromosomes

chromatid :: one of two copies of the DNA and proteins that make up a replicated chromosome; replicated chromosomes are composed of two chromatids

prophase II :: in meiosis, the stage in the second round of cell division when chromosomes, still in their compact form, attach to the newly formed spindle

metaphase II :: in meiosis, the stage in the second round of cell division when chromosomes line up in single file in the middle of the cell

anaphase II :: in meiosis, the stage in the second round of cell division when the two chromatids of each chromosome detach from each other and move to opposite ends of the cell; each half of the cell has a complete haploid set of chromosomes

telophase II :: in meiosis, the stage in the second round of cell division when the nuclear membrane reforms, enclosing the chromosomes

cytokinesis :: the stage in cell division when the cells are pinched in two

Each half of the cell has a complete haploid set of chromosomes. Each chromosome is composed of a single chromatid.

- *Telophase II and cytokinesis*: In **telophase II**, the nuclear membrane reforms, enclosing the chromosomes. The chromosomes return at last to their relaxed, decondensed form. **Cytokinesis** now pinches the cells in two.

Since meiosis II began with two cells, the final products are four haploid cells.

3.4 What Happens in Fertilization?

fertilization :: the process of a sperm cell from the father being fused with an egg cell from the mother

zygote :: a one-celled embryo produced as a result of fertilization

After meiosis, we now have the egg and sperm. In **fertilization**, a sperm cell from the father fuses with an egg cell from the mother. The brand-new one-celled embryo is the **zygote**. For fertilization to occur, though, eggs and sperm must find and recognize each other. They must then fuse to form a new diploid cell, with two full sets of chromosomes (see *How Do We Know? Eggs and Sperm Are Both Needed for Fertilization*).

In sea urchins, fertilization occurs outside the body: males and females release gametes into the ocean. In humans, fertilization occurs within the woman's body. Nevertheless, sperm still face an arduous journey to the fallopian tubes where they will meet the egg (**FIGURE 3.11**). Fortunately, sperm are excellent swimmers. The female helps transport sperm too. Contractions of muscles in the uterus make it possible for sperm to travel from the vagina to the fallopian tubes within 30 minutes, much faster than they could get there just by swimming. Once in the fallopian tubes, sperm are directed by chemical signals secreted by the egg. When the sperm find themselves near the egg, they swim even more vigorously.

In addition to its cell membrane, the egg has another coating, the *zona pellucida* (**FIGURE 3.12**). This protective covering shields the egg cell from damage and also ensures that it will not be fertilized by sperm from a different species. The sperm cell membrane and *zona pellucida* have proteins that bind

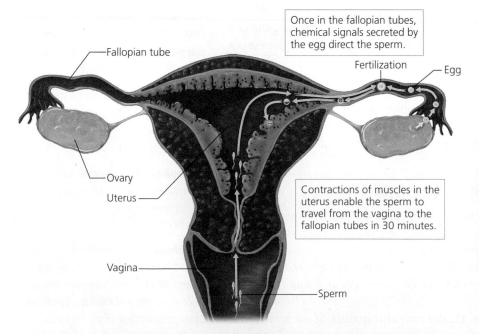

Once in the fallopian tubes, chemical signals secreted by the egg direct the sperm.

Fallopian tube

Fertilization

Egg

Ovary

Uterus

Contractions of muscles in the uterus enable the sperm to travel from the vagina to the fallopian tubes in 30 minutes.

Vagina

Sperm

FIGURE 3.11 In humans, fertilization occurs within the woman's body. Sperm face an arduous journey to the fallopian tubes, where the egg may be fertilized.

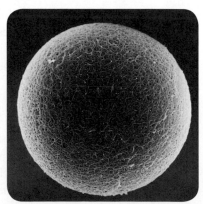

FIGURE 3.12 The *zona pellucida* is a protective covering that shields the egg cell from damage. Shown here is a scanning electron micrograph (SEM) of the surface of the *zona pellucida*.

Eggs and Sperm Are Both Needed for Fertilization

It takes a male and a female for human or animal reproduction, and both contribute to the development of the newborn. But exactly what do they contribute? For centuries, no one knew. Ideas ranged from physical material to energy—or more mysterious forces.

In 1667, Anton van Leeuwenhoek discovered sperm cells in the semen of humans and other animals. In fertilization, he proposed, sperm enter the egg. Unfortunately, he could not provide evidence for his hypothesis. Mammalian eggs were not even discovered until 1797.

Already, however, scientists were out to test van Leeuwenhoek's hypothesis. In 1784, the Italian scientist Lazzaro Spallanzani placed little trousers on male frogs and let them try mating with female frogs. Despite their best efforts, no offspring developed. Spallanzani also noticed drops of semen on the frog pants. When he combined these drops with frog eggs, fertilization took place.

In 1824, the French scientists J. L. Prevost and J. B. Dumas demonstrated that sperm cells, not just semen, were necessary to trigger development in amphibian eggs. Peering into a microscope in 1854, the British scientist George Newport finally watched frog sperm enter frog eggs.

In 1876 the German scientist Oskar Hertwig looked within these cells. He demonstrated that as the egg and sperm fuse in sea urchin, their nuclei fuse as well. The new nucleus indeed has genetic information from each parent. ::

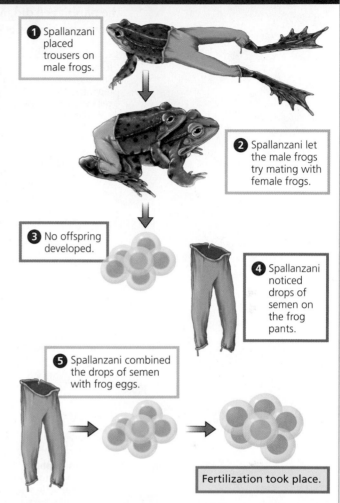

1 Spallanzani placed trousers on male frogs.

2 Spallanzani let the male frogs try mating with female frogs.

3 No offspring developed.

4 Spallanzani noticed drops of semen on the frog pants.

5 Spallanzani combined the drops of semen with frog eggs.

Fertilization took place.

In 1784, the Italian scientist Lazzaro Spallanzani placed little trousers on male frogs and let them try mating with female frogs. Despite their best efforts, no offspring developed.

together and attach the two cells to each other. Once bound, the sperm releases enzymes that digest a hole in the *zona pellucida*, and another set of proteins triggers cell fusion. The egg cell now springs into action releasing enzymes to remove all sperm-binding proteins from the *zona pellucida*. This ensures that only one sperm fuses with an egg. The zygote starts development with exactly one set of chromosomes from each parent—none missing and none extra. If fertilization is not successful, the egg is absorbed by the body.

3.5

How Does an Embryo Form and Ultimately Become a Fetus?

After fertilization, embryogenesis begins. It starts with **cleavage**, the division into new cells that will form the earliest stage of development, or the **embryo**. Cleavage involves **mitosis**, a second type of cell division, which is somewhat

cleavage :: the start of embryogenesis when the cells divide into new cells

embryo :: the earliest stage of development

mitosis :: a type of cell division during which all of the chromosomes in a cell are copied or replicated, and each new cell receives genetic information identical to the parent cell

different from meiosis. During mitosis (**FIGURE 3.13**), all of the chromosomes in a cell are copied or replicated, and each new cell receives genetic information identical to the parent cell. In humans, cells have 46 chromosomes, meaning two sets of 23. After mitosis, each cell has 46 chromosomes, just as the parent cell did. Mitosis is therefore less complex than meiosis: it involves only one round of cell division.

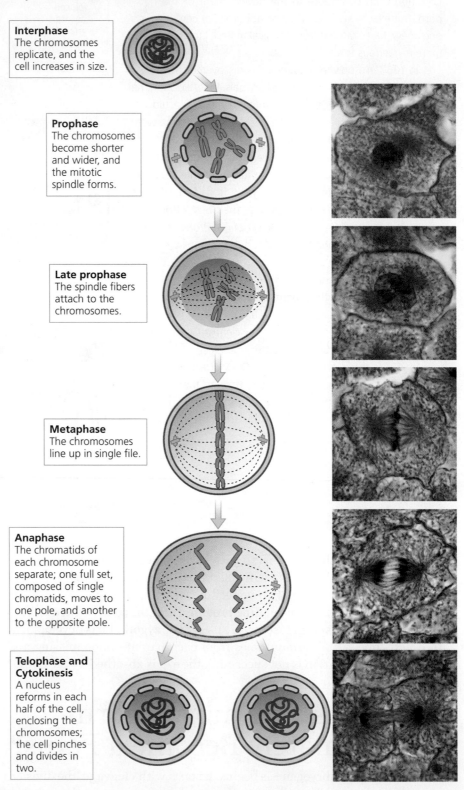

Interphase
The chromosomes replicate, and the cell increases in size.

Prophase
The chromosomes become shorter and wider, and the mitotic spindle forms.

Late prophase
The spindle fibers attach to the chromosomes.

Metaphase
The chromosomes line up in single file.

Anaphase
The chromatids of each chromosome separate; one full set, composed of single chromatids, moves to one pole, and another to the opposite pole.

Telophase and Cytokinesis
A nucleus reforms in each half of the cell, enclosing the chromosomes; the cell pinches and divides in two.

FIGURE 3.13 During mitosis, all of the chromosomes in a cell are copied or replicated, and each new cell receives genetic information identical to the parent cell.

Mitosis

Mitosis governs all cell division except for making an egg or sperm. You'll recognize many of the steps from similar stages in meiosis:

- Mitosis begins with *interphase*. Here the chromosomes replicate, and the cell makes other materials necessary for cell division. The cell increases in size.
- In *prophase*, the chromosomes become shorter, wider, and easier to see with a microscope. The nuclear membrane disappears, and the mitotic spindle forms. As in meiosis, microtubules extend from opposite poles of the cell and bind to chromosomes. Each chromosome is composed of two chromatids.
- In *metaphase*, the chromosomes line up in single file, as specific proteins associated with the microtubules pull on them from each pole.
- In *anaphase*, the chromatids of each chromosome separate. One full set of chromosomes, composed of single chromatids, moves to one pole, and one set moves to the opposite pole. As a consequence, each half cell has a full complement of genetic instructions.
- During *telophase and cytokinesis*, a nucleus reforms in each half of the cell, enclosing the chromosomes. The chromosomes resume their relaxed form, and the cell pinches and divides in two. In humans, the newly formed cells again possess two sets of 23 chromosomes.

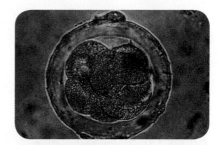

FIGURE 3.14 In humans and other mammals, the first couple of cleavages create a cluster of cells, none of which has specialized.

In humans and other mammals, the first couple of cleavages create a cluster of cells, none of which has specialized (**FIGURE 3.14**). By the time the embryo reaches the eight-cell stage, the cells compact together into a tight group. At the 16-cell stage, there are two populations of cells present: the outer ones will help to make the **placenta**, the organ that connects the developing fetus to the mother's uterine wall, and other structures that are needed to support the development of the embryo and fetus. The inner cell population will eventually form all of the cells, tissues, and organs of the embryo and fetus. At this point in development, the inner cells are **embryonic stem cells**: they will produce all of the types of cells in the body.

placenta :: an organ that connects a developing fetus to the mother's uterine wall and provides for an exchange of nutrients and waste elimination

embryonic stem cell :: cell that can produce any type of cell in the body

Gastrulation and Organ Formation

During **gastrulation** (**FIGURE 3.15**), cells and tissues move to new locations, where they will grow into organs. For example, the cells that give rise to the skin remain on the exterior surface of the embryo, while the cells that will ultimately form the heart move into the interior.

Organ formation involves many of the same types of cell and tissue rearrangement seen in gastrulation. For example, the brain and spinal cord of the central nervous system start their development as a flat sheet of cells. This tissue rolls up to form a **neural tube**. The anterior, or front part, of the tube expands, ultimately making the brain. The posterior portion, or rear part, becomes the spinal cord (**FIGURE 3.16**).

As with all aspects of development, mistakes can have serious consequences. In the case of the central nervous system, failure to close the anterior region of the neural tube will likely result in a fetus that has no brain. Survival is not possible. If the posterior region of the neural tube does not close, spina bifida occurs (**FIGURE 3.17**). The health of an individual with spina bifida depends on the location and size of the opening.

gastrulation :: a stage in embryonic development in which cells and tissues move to new locations, where they will grow into organs

neural tube :: the tissue that starts from a flat sheet of cells and rolls up, ultimately becoming the brain and spinal cord

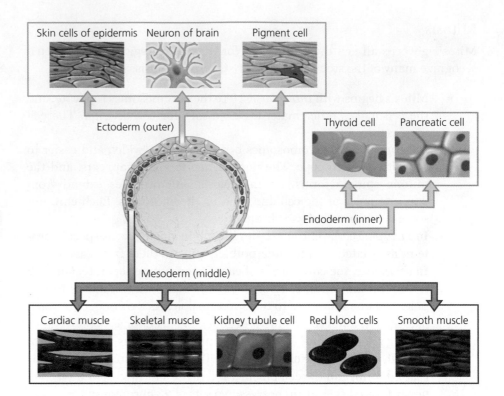

FIGURE 3.15 During gastrulation, cells and tissues move to new locations, where they will grow into organs.

As for Abigail and Brittany, or other conjoined twins, it is not entirely clear how errors in setting up the organization of their bodies led to the particular patterns of organ formation observed in them. Ideas about how conjoining occurs are discussed later.

Differentiation

For an embryo to become a fetus, its cells must take on new roles on the way to becoming human organs like the heart and lungs. Hans Spemann, an embryologist, captured the wonder of development this way: "We are standing and walking with parts of our body which could have been used for thinking had they developed in another part of the embryo."

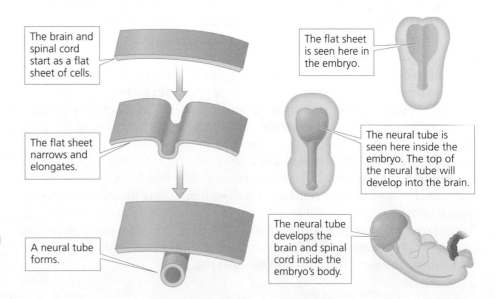

The brain and spinal cord start as a flat sheet of cells.

The flat sheet is seen here in the embryo.

The flat sheet narrows and elongates.

The neural tube is seen here inside the embryo. The top of the neural tube will develop into the brain.

A neural tube forms.

The neural tube develops the brain and spinal cord inside the embryo's body.

FIGURE 3.16 Organ formation involves many of the same types of cell and tissue rearrangement seen in gastrulation. For example, the brain and spinal cord of the central nervous system start their development as a flat sheet of cells.

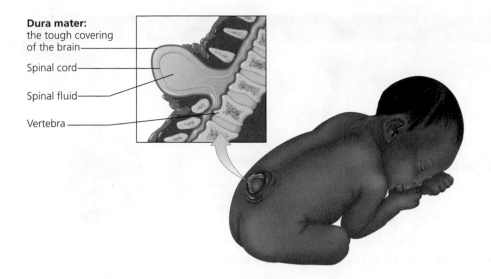

Dura mater: the tough covering of the brain

Spinal cord

Spinal fluid

Vertebra

FIGURE 3.17 Mistakes in the development of the central nervous system can have serious consequences. If the posterior region of the neural tube does not close, *spina bifida* occurs; in serious cases, part of the spinal cord may protrude through the bones.

The zygote, the very first embryonic cell, can produce any type of cell in the body. Cells in an adult body no longer have that unlimited potential. A human adult, in fact, has more than 200 types of specialized cells. The change from cells that have unlimited potential to cells that are specialized is called cell **differentiation**. In the process, cells take on a specific structure, set of behaviors, and function.

Cell differentiation (**FIGURE 3.18**) takes place through a series of commitments. Think of a marriage. Marriage does not generally happen moments after a couple meets. Instead, it comes at the end of a series of choices and an increasing level of commitment. When two individuals meet, there needs to be some "chemistry" between them. They might go out a few times to see whether the relationship has a future. If all goes well, they agree to see one another exclusively. Eventually they may decide to become engaged. Finally, they make the big commitment, marriage. They are fully "differentiated."

Returning to cells, let's consider the differentiation of the muscle cells in the heart (**FIGURE 3.19**). Again, we start with the zygote, which has unlimited potential: it has made no commitment to become any one type of cell. During cleavage, inner and outer cell populations arise, and muscle cells originate from the inner cells. During gastrulation, three major types of embryonic cells emerge: *ectoderm*, *mesoderm*, and *endoderm*. These three types are not yet fully specialized, but each will give rise only to specific cells. Muscle cells derive from the mesoderm alone. The mesoderm produces several different

differentiation :: the change from cells that have unlimited potential to cells that are specialized

FIGURE 3.18 Cell differentiation takes place through a series of commitments that is in some ways analogous to courtship and marriage.

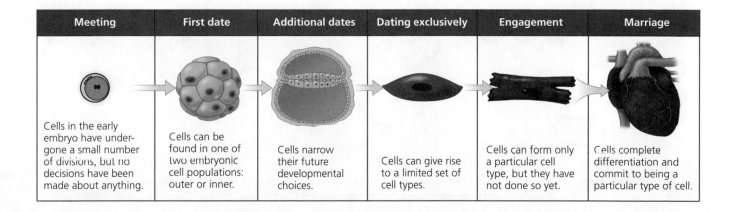

Meeting	First date	Additional dates	Dating exclusively	Engagement	Marriage
Cells in the early embryo have undergone a small number of divisions, but no decisions have been made about anything.	Cells can be found in one of two embryonic cell populations: outer or inner.	Cells narrow their future developmental choices.	Cells can give rise to a limited set of cell types.	Cells can form only a particular cell type, but they have not done so yet.	Cells complete differentiation and commit to being a particular type of cell.

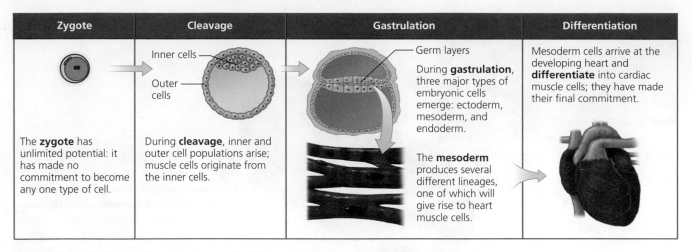

Zygote	Cleavage	Gastrulation	Differentiation
The **zygote** has unlimited potential: it has made no commitment to become any one type of cell.	Inner cells / Outer cells During **cleavage**, inner and outer cell populations arise; muscle cells originate from the inner cells.	Germ layers During **gastrulation**, three major types of embryonic cells emerge: ectoderm, mesoderm, and endoderm. The **mesoderm** produces several different lineages, one of which will give rise to heart muscle cells.	Mesoderm cells arrive at the developing heart and **differentiate** into cardiac muscle cells; they have made their final commitment.

FIGURE 3.19 This figure illustrates how cells differentiate into cardiac muscles of the heart.

mesoderm lineages, one of which will give rise to heart, or cardiac, muscle cells. Finally, these mesoderm cells arrive at their final destination (the developing heart) and differentiate into cardiac muscle cells. They have made their final commitment.

Gene Expression

What guides the process? What ensures that a healthy individual gets the right complement of healthy organs? First, in complex organisms like humans, cell differentiation depends on cell communication. Chemical signals between neighboring cells tell them when and where to differentiate. This "conversation" enables the construction of elaborate, precisely functioning organs like the eye.

Second, cell differentiation depends on using just the right genetic information, a process called **gene expression** (**FIGURE 3.20**). All cells contain the

gene expression :: the process during which the information encoded in genes is used

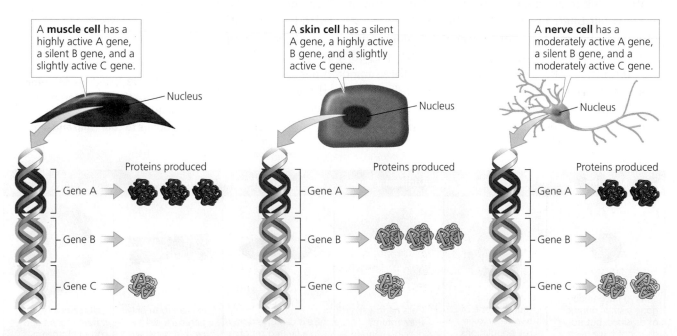

FIGURE 3.20 Cell differentiation depends on cell communication and gene expression. Chemical signals between neighboring cells tell them when and where to differentiate, and different cells use different combinations of the same genes to specialize or differentiate.

same genes, but different genes are used in different cells. Some genes are uniquely expressed in brain cells and a different set in liver cells. If you think of all your genes as a fully loaded MP3 player, then all of your cells possess all of the songs. Gene expression is like listening to just the songs you need.

Cell differentiation results in highly specialized cells—with their own shapes, internal structure, and behavior. For example, the white blood cells of the immune system are amorphous blobs; these cells crawl around the body, ever vigilant for infection. In contrast, muscle cells are long and thin, connected to other muscle cells in parallel. They can thus coordinate their individual contractions so that the entire muscle contracts.

3.6 What Are the Key Events of Pregnancy?

Now we know how we get from an egg and sperm to an embryo and the early steps of making organs. Let's take a closer look at pregnancy, starting with conception and ending with labor and delivery. All this has to happen to produce a healthy newborn.

Embryonic Development: Conception to Eight Weeks (FIGURE 3.21)

It might seem that becoming pregnant is the easiest thing in the world. After all, a great deal of effort goes into avoiding unwanted pregnancy. In fact, the odds of conceiving during a given cycle of fertility are only 15–25%, depending on age, health, and other factors. The ability to conceive decreases starting at age 30. A couple who is trying to conceive but has not been successful will not be considered infertile until at least close to a year of effort.

During the first week after conception, cleavage produces the cells that will form the embryo and chorion, a component of the placenta. In week 2, the embryo, still only the size of the period at the end of this sentence, implants in the wall of the **uterus**—also known as the womb. Gastrulation occurs, and cell specialization is underway: ectoderm, mesoderm, and endoderm are formed.

uterus :: the womb of a mother

The third week includes **neurulation**, or formation of the earliest stages of the central nervous system, and the beginning of heart development. In week 4, heart development continues and limb buds, or swellings that will become limbs, can be seen. The embryo is barely larger than a letter on this page. In week 5, a discernible nose, eyes, and ears appear. By weeks 6 and 7, fingers and toes are evident, and the embryo has a skeleton made of cartilage. By the end of week 8, the embryo is approximately 1.5 inches long, bone is beginning to replace cartilage, and all organ systems are developing. At this point, embryonic development is finished and fetal development begins.

neurulation :: formation of the earliest stages of the central nervous system

Fetal Development: Three Months to Nine Months (FIGURE 3.22)

During the rest of development, organ systems are refined, and the **fetus**—the stage of development that follows the embryo—grows dramatically. During the third month, the gender of the fetus can be discerned and fingernails form. Also, the placenta is fully developed by the 10th week. In the fourth month, the fetus is approximately 6 inches long and weighs 6 ounces. The skeleton is visible using **ultrasound** imaging (see *Technology Connection: Ultrasound*, p. 79). Eyelashes, eyebrows, and hair on the head are all present. If the mother has had a previous pregnancy, she can feel and identify fetal movement.

fetus :: the stage of development that follows the embryo

ultrasound :: a device that relies on high-frequency sound waves to send back images of internal organs

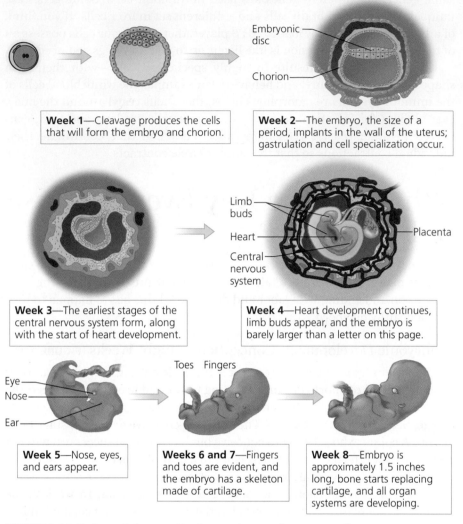

Week 1—Cleavage produces the cells that will form the embryo and chorion.

Week 2—The embryo, the size of a period, implants in the wall of the uterus; gastrulation and cell specialization occur.

Embryonic disc

Chorion

Week 3—The earliest stages of the central nervous system form, along with the start of heart development.

Week 4—Heart development continues, limb buds appear, and the embryo is barely larger than a letter on this page.

Limb buds

Heart

Central nervous system

Placenta

Week 5—Nose, eyes, and ears appear.

Weeks 6 and 7—Fingers and toes are evident, and the embryo has a skeleton made of cartilage.

Week 8—Embryo is approximately 1.5 inches long, bone starts replacing cartilage, and all organ systems are developing.

Eye

Nose

Ear

Toes   Fingers

FIGURE 3.21 Embryonic human development occurs from conception to eight weeks.

Month 3 (12–15 weeks)	Month 4 (16–19 weeks)	Month 5 (20–23 weeks)	Months 6 and 7 (24–31 weeks)	Months 8 and 9 (32–39 weeks)
The gender of the fetus can be discerned, and fingernails form.	The fetus is approximately 6 inches long and weighs 6 ounces; the skeleton is visible using ultrasound imaging, and eyelashes, eyebrows, and hair are present.	A stethoscope placed on the mother's abdomen can detect a fetal heartbeat, and a mother can feel fetal movement.	Organ systems continue to mature, and the eyes open.	The fetus increases in size, lungs mature, and body hair disappears; a typical newborn weighs around 7 pounds and is approximately 20 inches in length.

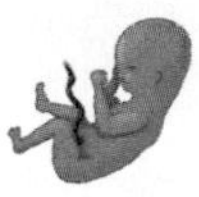

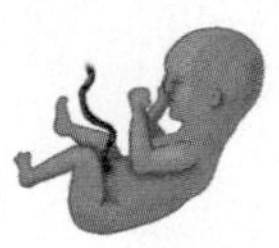

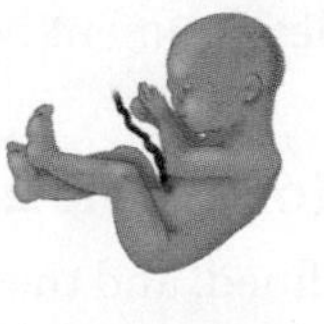

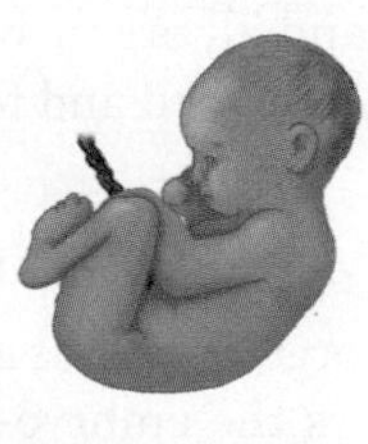

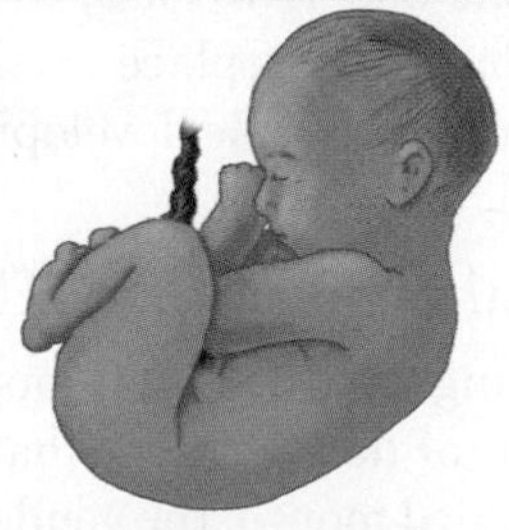

FIGURE 3.22 Human fetal development occurs from three months to nine months.

Technology Connection

Ultrasound

Not long ago, a "baby book" opened with a picture of the newborn infant. In many families, that place of honor is now occupied by a black-and-white, slightly fuzzy image of a tiny fetus. Ultrasound imaging has become a routine part of pregnancy care in developed countries.

Just as bats use sound to "visualize" their world and fishermen use fish finders, an ultrasound examination relies on high-frequency sound waves. As the waves echo back from an object, they are sent through a recording device called a transducer and transformed into an image on a computer. By measuring the waves, we can determine the shape of an object, its distance, its size, and whether it is liquid, solid, or both. Ultrasound imaging operates in real time. You see a live image.

During pregnancy, ultrasound is used to monitor fetal growth, development, and the health of the placenta. It can also identify many developmental problems 20 weeks into pregnancy—or even sooner. It can detect anencephaly (the lack of a brain), dwarfism, spina bifida (failure to close the neural tube), cleft palate, and heart abnormalities, for example.

Ultrasound also has other medical uses. It can image such organs as the heart and its major blood vessels, liver, breasts, spleen, gallbladder, pancreas, kidneys, bladder, uterus, ovaries, and scrotum. Ultrasound may even guide biopsies, or tissue samples drawn by a needle. It can assess damage after a heart

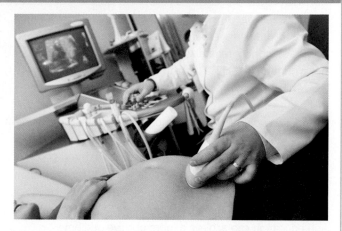

Ultrasound imaging has become a routine part of pregnancy care in developed countries.

attack, verify the presence of blood clots, measure the size of blood vessels to see whether there is a buildup of fatty deposits, and visualize tumors. A modification, called Doppler ultrasound, can measure blood flowing through the heart.

Because it is versatile, noninvasive, and safe, the use of ultrasound is growing. New breakthroughs are leading to the next generation of ultrasound scanners. These include 3D ultrasound, in which a three-dimensional image is generated from successive two-dimensional images, and 4D ultrasound—or 3D images in real time. ::

In the fifth month, a stethoscope placed on the mother's abdomen can detect a fetal heartbeat. Even a first-time mother can feel fetal movement. The sixth and seventh months are marked by a continual maturing of organ systems. The body of the fetus is covered by a layer of fine hair called lanugo. By the seventh month, the fetus is 12 inches long and weighs around 3 pounds. The eyes open, and if the fetus is male, the testes will have descended into the scrotum.

By this point, the fetus is positioned head down (**FIGURE 3.23**), the orientation appropriate for delivery. A fetus born after 7 months has roughly a 90% chance to survive. The eighth and ninth months are characterized by increases in size, maturation of the lungs, disappearance of body hair, and deposits of fat for insulation. A typical newborn weighs around 7 pounds or so and is approximately 20 inches in length.

Because human embryos and fetuses develop inside their mothers, they are protected from such things as heat, cold, rain, and wind. Yet that location makes them vulnerable to other assaults. Whatever the mother puts in her body also winds up in her offspring. If she drinks alcohol during pregnancy, the consequences for her baby can be grave and include the

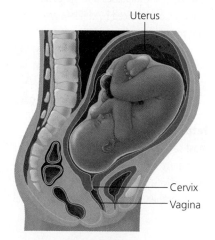

FIGURE 3.23 By the end of seven months, the fetus is positioned head down, the orientation appropriate for delivery. A fetus born after 7 months has roughly a 90% chance of survival.

Life Application

Fetal Alcohol Syndrome

Fetal alcohol spectrum disorders (FASDs) encompass a range of health problems experienced by children who were exposed to alcohol prenatally. The most severe form, fetal alcohol syndrome (FAS), is a lifelong condition.

Children with FAS may suffer from a variety of symptoms—including poor growth, abnormal facial features, hyperactivity, poor reasoning skills, vision problems, hearing problems, and mental disabilities. FAS occurs in from 0.2 to 1.5 out of every 1000 births in the United States, while the less severe form is present in up to 6 out of every 1000. These children are at a higher risk for psychiatric problems and criminal behavior. Although there is no treatment, a supportive home and education can achieve a more positive outcome.

FASD is completely preventable: women should not drink alcohol during pregnancy. There is no known safe amount. Even occasional drinking can lead to FASD.

While prevention is quite simple in principle, it can be challenging in practice. Many women have no idea that they are pregnant soon after conception. If a woman is sexually active and drinks alcohol, she should use an effective contraceptive to prevent pregnancy. If a woman is trying to start a family, she should not drink. If she learns that she is pregnant and drinks, she should stop—the sooner, the better.

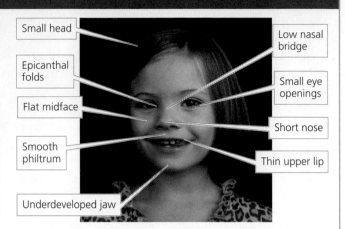

Children with fetal alcohol syndrome (FAS) may suffer from a variety of symptoms—including poor growth, abnormal facial features, hyperactivity, poor reasoning skills, vision problems, hearing problems, and mental disabilities.

Prenatal exposure to alcohol is not the only problem. Exposure to other drugs—including cocaine (crack or powder), heroin or other opiates, barbiturates, and PCP—can have terrible consequences. ::

---

fetal alcohol spectrum disorder (FASD) :: encompasses a range of health problems experienced by children who were exposed to alcohol prenatally

risk of **fetal alcohol spectrum disorders** (**FASDs**), the most severe of which is **fetal alcohol syndrome** (see *Life Application: Fetal Alcohol Syndrome*).

 What Happens in Labor and Delivery?

fetal alcohol syndrome (FAS) :: the most severe FASD; symptoms may include poor growth, abnormal facial features, hyperactivity, poor reasoning skills, vision problems, hearing problems, and intellectual disabilities

Normally the fetus triggers labor once the lungs are fully developed. A surfactant is a substance that coats the tissues lining the inside of the lungs so that they don't stick together. Without adequate surfactant, the sticky tissues make breathing very difficult. In fact, immature lungs are one of the most serious health challenges facing infants born prematurely. When a fetus is ready to be born, it secretes surfactant, which causes an immune response in certain fetal cells. These cells migrate to the uterus, where they secrete a chemical signal that causes the uterus to contract.

Light uterine contractions occur throughout pregnancy. The contractions that mark labor are different. At the onset of labor, contractions occur every 15–20 minutes, each lasting around a minute. As delivery nears, contractions occur every 1–2 minutes, each lasting a minute.

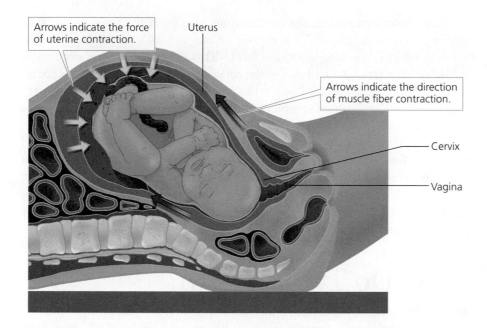

Arrows indicate the force of uterine contraction.

Uterus

Arrows indicate the direction of muscle fiber contraction.

Cervix

Vagina

FIGURE 3.24 During labor, the cervix stretches and causes a uterine contraction, which stimulates release of the hormone oxytocin, which in turn stimulates more uterine contraction and pushes the fetus downward.

Although the timing of events varies, most labors involve the same scenario: the lower end of the uterus, or **cervix**, stretches and causes a uterine contraction (**FIGURE 3.24**). This contraction stimulates release of the hormone oxytocin, which in turn stimulates more uterine contraction. As these contractions push the fetus downward, pressure from the fetus makes the cervix stretch more. This cycle repeats until eventually the cervix dilates (or expands) to give an opening 4 inches (10 centimeters) in diameter.

Labor and delivery are divided into three stages (**FIGURE 3.25**). In stage 1, uterine contractions pull the lower part of the uterus up toward the fetus's

cervix :: the lower end of the uterus

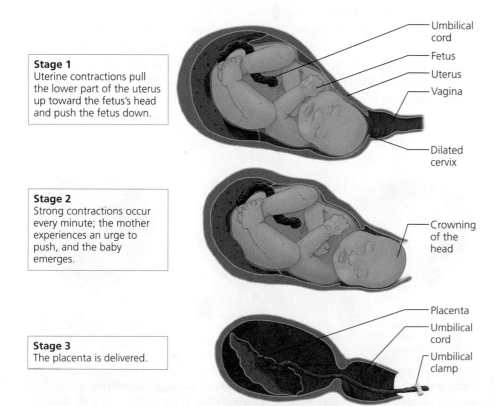

Stage 1
Uterine contractions pull the lower part of the uterus up toward the fetus's head and push the fetus down.

Umbilical cord

Fetus

Uterus

Vagina

Dilated cervix

Stage 2
Strong contractions occur every minute; the mother experiences an urge to push, and the baby emerges.

Crowning of the head

Stage 3
The placenta is delivered.

Placenta

Umbilical cord

Umbilical clamp

FIGURE 3.25 Labor and delivery are divided into three stages.

amnion :: sac in which the embryo is suspended

head. As described earlier, the contractions also push the fetus down, thus helping cervical dilation. The **amnion**, or sac in which the embryo is suspended, breaks, releasing amniotic fluid. (This event is often referred to as breaking water.) Stage 1 ends when the cervix is fully dilated. Stage 2 is sometimes called "hard labor." Contractions are very strong, occurring every minute or so. The mother experiences an almost irresistible urge to push. The baby emerges. Finally, in stage 3 the placenta is delivered.

Unlike some animals, human newborns are essentially helpless and remain so for quite some time. Scientists hypothesize that humans are born less fully developed than many other mammals because of the size of a human head. If the fetus grew much further, its head would be too big to deliver without surgery. Even now, some women have babies too big to deliver vaginally and have to resort to a Caesarian, or surgical, delivery.

3.8 How Do Twins Form?

Abigail and Brittany are twins, although of a very unusual kind. Even so, most pregnancies produce only one baby. Twins are born in a little more than 1% of pregnancies, unless the woman has been treated with hormones for infertility; the odds of twins then increase to 3%.

Identical twins come from a single fertilized egg that splits into two embryos sometime during very early development (**FIGURE 3.26**). About a third of twins are identical. The types of identical twins can be divided into three categories.

- In 33% of identical twins, the embryos have completely separate chorions (a component that forms in the placenta). Because the tissue that develops this support structure for the embryo itself forms at five days post-conception, it is clear that twinning occurs prior to five days after conception.
- In 66% of identical twins, the embryos share a chorion but have separate amnions, the fluid-filled sac that surrounds and cushions the embryo. Twinning in these cases occurs after five days—the embryonic tissue that will form the chorion is present—but before nine days post-conception, the time when the amnion forming tissue would have developed.

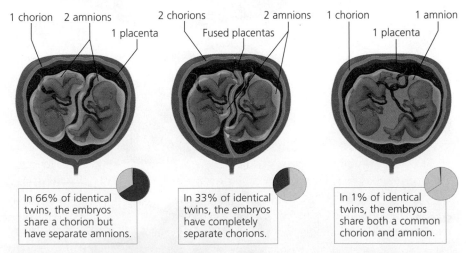

1 chorion   2 amnions   1 placenta

2 chorions   Fused placentas   2 amnions

1 chorion   1 amnion   1 placenta

In 66% of identical twins, the embryos share a chorion but have separate amnions.

In 33% of identical twins, the embryos have completely separate chorions.

In 1% of identical twins, the embryos share both a common chorion and amnion.

FIGURE 3.26 Identical twins come from a single fertilized egg that splits into two embryos sometime during very early development. About a third of twins are identical.

- In 1% of identical twins, the embryos share *both* a common chorion and a common amnion. Separation of the twins must happen later than nine days after conception. When twins separate this late during embryogenesis, they are at risk for being conjoined.

Fraternal twins, in contrast, occur when the mother releases two eggs, and each is fertilized by a separate sperm. Although they develop in the same uterus, fraternal twins are no more closely related genetically than siblings born to the same parents. A set of fraternal twins can consist of two boys, two girls, or one boy and one girl.

3.9 What Can Conjoined Twins Tell Us About Biology and Ourselves?

Amazingly, conjoined twins form in 1 out of every 50,000 pregnancies—and their births are energetically reported. Given that the world population is more than 6 billion, we would expect there to be at least 60,000 sets of conjoined twins alive on Earth today. Why does conjoining seem so rare?

Not all pregnancies are successful. Especially when the embryo or fetus has serious genetic or physical defects, the pregnancy may abort or end spontaneously. This event is called a miscarriage. Approximately 40–60% of all conceptions result in miscarriages. Although 1 out of every 50,000 pregnancies involves conjoined twins, only 1 out of every 200,000 births produce conjoined individuals. In a majority of these cases, the fetus dies, thus resulting in a stillbirth. Even when conjoined twins are delivered alive, their chances of survival are lower than 25%. For some types of conjoining, survival is less than 5%.

Abigail and Brittany are especially rare. They are joined side by side, sharing many organs and structures. About three-fourths of conjoined twins are joined at the chest, abdomen, or both. Joining occurs at the buttocks in 18% of twins and at the ischium (a part of the hip) in 6%. Finally, head-to-head joining happens in 2% of conjoined twins. The angle of joinings may also vary, as may the sizes of each twin—anything from slightly different sizes to a situation in which one of the twins is "parasitic," or not fully formed and often not completely conscious.

Explanations for Conjoining

Conjoining has had many explanations, few taken seriously today. A 16th-century French physician, Ambroise Paré, offered several: God was showing his wrath or his power; the pregnant woman looked at a terrible sight; her uterus was too tight; she sat the wrong way during pregnancy. Explanations like this do not count as scientific hypotheses. Why? It is difficult to think of ways to test them.

The true causes of conjoining are unknown, and the few clues so far don't lead to a clear avenue of inquiry. Conjoining is much more common in females: 75% of conjoined twins are sisters. It also occurs more often in India and Africa, less often in China and the United States. Some scientists argue that the early embryos in conjoined twins fail to separate properly. Yet an increasing number of scientists think that this "improper fission" cannot be the whole story. How can it explain which parts are attached and at what angle? One possibility is the fusion of embryos. Twin embryos develop in very close proximity, within a single amnion. Perhaps they bump into each other and stick together.

Whatever their cause, conjoined twins show how tightly the body regulates a growing embryo. Conjoined twins are the result of the best effort to follow a normal developmental program. Although their body plan sure looks different from one we are used to seeing, conjoined twins are beautiful in their own ways.

What Conjoining May Tell Us About Biology

Scientists often learn more from studying rare events than they do from everyday occurrences. For example, research on cancer cells tells us an enormous amount about the structure and function of normal cells. Similarly, conjoined twins can help us understand the mechanisms by which embryos form a body, because development has followed an unexpected path.

Conjoining is not unique to humans; it can occur in other organisms, too. It can also be produced experimentally in many organisms, so that we can study it more closely. Hans Spemann's lab first produced conjoined embryos in the 1920s in amphibians—creatures that can live both on land and in water, like frogs and toads (**FIGURE 3.27**). Spemann, a pioneer in experimental embryology, had assigned a research problem to a graduate student, Hilde Mangold: how do tissues interact and communicate to direct embryogenesis? Mangold removed small pieces of tissue from one set of embryos and transplanted them to another set of embryos. To Spemann's amazement, when one piece of tissue was transferred, it actually directed the cells of the other embryo to form a second body axis. The result was tadpoles, or newborn frogs, joined at the abdomen.

This basic experiment has been repeated thousands of times in many different organisms. Scientists have learned about the molecular signals, cell interactions, and changes in gene expression that accompany both conjoining and normal development. Spemann received the Nobel Prize in Physiology or Medicine in 1935 for his contributions to experimental embryology. Sadly, Mangold did not share the award, because she had died in a tragic accident, and Nobel Prizes are never awarded after death.

What Conjoining May Tell Us About Ourselves

As much as conjoining reveals about biology, it tells us even more about ourselves as people. Conjoining, like other physical differences, forces us to confront how we define "normal." It also forces us to ask how we treat people who are different from ourselves. Throughout history, the answer has been influenced by culture, religious views, and individual experience.

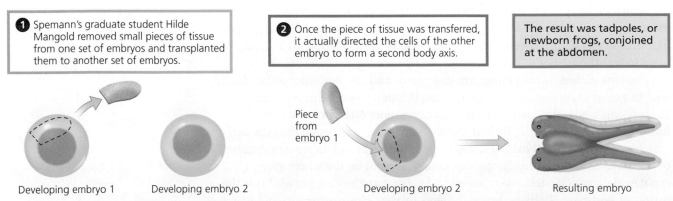

1 Spemann's graduate student Hilde Mangold removed small pieces of tissue from one set of embryos and transplanted them to another set of embryos.

2 Once the piece of tissue was transferred, it actually directed the cells of the other embryo to form a second body axis.

The result was tadpoles, or newborn frogs, conjoined at the abdomen.

Developing embryo 1   Developing embryo 2

Piece from embryo 1

Developing embryo 2   Resulting embryo

FIGURE 3.27 Conjoining is not unique to humans; it can also be produced experimentally in many organisms. Hans Spemann's lab first produced conjoined embryos in the 1920s in amphibians.

In ancient times, conjoined twins might be worshipped as gods, banished, or killed. In one of the earliest reported cases, Eliza and Mary Chulkhurst were born in England in 1100, joined at the hip. When one of the sisters died, the other refused to attempt separation. She died shortly thereafter. Upon their deaths, the Chulkhurst sisters, also known as the Biddenden maids, left 20 acres of land to the poor. Ever since, cakes imprinted with their images are baked and distributed to the poor at Easter.

The Biddenden maids have been held in high regard for almost 900 years. Much more often, conjoined twins have been treated as freaks and displayed in sideshows or circuses. In some cases, the experience was profoundly humiliating, and in others, the twins managed to eventually take charge of their situation and have successful careers. There are probably few if any examples from the historical record in which the twins did not suffer exploitation for at least part of their lives. The seemingly inevitable display of conjoined twins for some manner of entertainment was prevalent from at least the 15th century into the 20th century.

Perhaps the most famous conjoined twins in history, Chang and Eng Bunker (**FIGURE 3.28**), were born in Siam (now Thailand) in 1811. Billed as the "Siamese Twins," the Bunkers enjoyed a lucrative career in show business, touring around the world with an act that included acrobatics and other physical feats. They eventually settled in North Carolina, owned and ran a store, and married two sisters. Amazingly, they fathered 21 children between the two of them. (None were twins.) Many of their numerous descendants continue to live in the area of North Carolina where the brothers made their home.

Violet and Daisy Hilton, born in 1908 in England, were joined at the buttocks. The twins were sold to their single mother's boss, who saw the girls as a way to make money. The sisters were forced into show business and held against their will for almost 20 years. They finally escaped and sued for their independence. The Hilton sisters were then able to chart their own professional course and performed: singing, dancing, and playing saxophone and violin. They also appeared in some motion pictures, including *Freaks* and *Chained for Life*. They did well for quite a while, but after 20 or so years of touring and performing, interest in them decreased. After their agent abandoned them in Charlotte, North Carolina, the sisters got a job in a grocery store, made themselves a home in Charlotte, and generally fit into the community.

Finally, Clara and Altagracia Rodriguez were born in 1973, joined at the abdomen. By then most people's notions of conjoined twins had shifted from oddities to individuals. Medical science, too, had developed, so that surgical separation was now feasible in many cases. The Rodriguez sisters were separated successfully in 1974 by a team of 23 doctors headed by Dr. C. Everett Koop (**FIGURE 3.29**).

When conjoined twins are born now, efforts are generally undertaken to separate the individuals. However, some conjoined twins share a critical organ, such as the heart, and separation would kill one twin or leave them both profoundly disabled. Such would be the case for the Hensel twins, for example.

Conjoining and other types of physical difference challenge the concept of normal. Abigail and Brittany Hensel are living full lives with interests, ambitions, friends, and loved ones. Chang and Eng Bunker probably experienced fulfilling lives as well. In contrast, the Hilton sisters had difficult journeys. Some of the greatest challenges faced by those with a "disability" are the ways that others treat them.

Conjoining brings into sharp relief questions about what it means to be a person. What makes *anyone* an individual? Although identical twins,

FIGURE 3.28 Perhaps the most famous conjoined twins in history, Chang and Eng Bunker, were born in 1811. Billed as the "Siamese Twins," the Bunkers enjoyed a lucrative career in show business before settling in North Carolina, marrying two sisters, and fathering 21 children between the two of them.

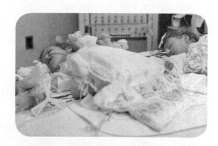

FIGURE 3.29 Clara and Altagracia Rodriguez were born in 1973, joined at the abdomen. The sisters were separated successfully in 1974 by a team of 23 doctors headed by Dr. C. Everett Koop.

conjoined or not, possess identical genetic material, each is clearly an individual. As we saw at the start of this chapter, Abigail and Brittany are quite different from each other in many ways. Without a doubt, there is more to our development as individuals than our genes alone.

With respect to the question of when personhood is established, no one knows for certain. Since gastrulation does not occur until 14 days, and twinning can take place at least up until then, it is hard to argue that we are defined as individuals before this time. Another way to consider this question is to ask: when does human life begin? This question is surprisingly difficult to answer in a biological sense or even in a religious manner, since different faith traditions have different views. We will return to a full consideration of this issue in Chapter 8.

Biology in Perspective

An understanding of human embryogenesis and development opens a door to the informed and intelligent consideration of many of the challenges and choices we face as a society with regard to reproduction. Learning about human development also reveals the fascinating biological story of how each of us came to be.

Examining human development reminds us of the diversity of life, and it challenges us to confront our views about what is normal. Even though we are all the same species, the bodies we have are so very variable. A look around any room filled with people will demonstrate how different we are from one another. Conjoined twins are just one dramatic example. They force us to ask how we engage with others who are different from us. The answer says a lot about each of us as individuals and all of us as a society. Although we understand a great deal about what makes a body, we know much less about what makes a person. Biology certainly plays a part, but there is so much more.

Chapter Summary

3.1 What Are the Units of Life?
Explain the consequences of cell theory for embryogenesis.
- Cell theory states that cells are the most basic unit of life, all living things are composed of cells, and all living things come from existing cells.
- Embryogenesis is the process of development from single cells to whole organisms.

3.2 What Cell Structures Play a Role in Embryo Development?
Describe the cell structures that play a role in embryo development.
- Humans are composed of eukaryotic cells with membrane-enclosed nuclei and organelles.

- Cell membranes are the outer boundary of cells that regulate what enters and leaves cells and detect chemical signals.
- The nucleus houses the chromosomes and most of the genetic information for the cell as DNA.
- Mitochondria extract energy from food molecules.
- The endomembrane system consists of three membrane-bound structures: the endoplasmic reticulum (ER), including the smooth endoplasmic reticulum (SER) and rough endoplasmic reticulum (RER); the Golgi apparatus; and lysosomes.
- The cytoskeleton maintains the shape of the cell and is responsible for cell movement and organelle placement.

3.3 How Do Eggs and Sperm Form?

Show the possible chromosomes in gametes produced through meiosis in a given cell.

- Almost all human cells are diploid with 23 chromosome pairs (46 chromosomes total).
- Human eggs and sperm (gametes) are haploid; they have only one set of 23 chromosomes and are formed through meiosis.
- Meiosis entails two rounds of cell division and results in four haploid cells.
- Meiosis I includes interphase, prophase I, metaphase I, anaphase I, and telophase I and cytokinesis; it ends with two haploid cells. The chromosomes are each composed of two chromatids.
- Meiosis II includes prophase II, metaphase II, anaphase II, and telophase II and cytokinesis; it results in four haploid cells. The chromosomes are each composed of one chromatid.

3.4 What Happens in Fertilization?

Describe the process of fertilization.

- Fertilization occurs when a sperm fuses with an egg; the one-celled embryo is called a zygote.
- In humans, sperm must swim to the fallopian tubes to reach a fertile egg; they are aided by contractions of the uterus and chemical signals secreted by the egg.
- A coating outside the cell membrane of the egg (*zona pellucida*) contains proteins that protect the egg; sperm with the correct protein bind to the egg cell and release enzymes that digest a hole to the egg membrane; the egg removes the sperm-binding proteins from the *zona pellucida* so that only one sperm will fuse with the egg.

3.5 How Does an Embryo Form and Ultimately Become a Fetus?

Outline the stages of mitosis and the ways that cells become specialized.

- Embryogenesis begins with cleavage.
- Mitosis involves a single division resulting in two cells with exactly the same chromosomes as the mother cell; the phases are similar to meiosis: interphase, prophase, metaphase, anaphase, and telophase and cytokinesis.
- When the zygote divides into the 16-cell stage, the cells separate into two groups: the outer cells make the chorion, and the inner cells become the embryo.
- During gastrulation, cells in the embryo move into three layers, which will give rise to specific cell types.
- The neural tube forms following gastrulation; the anterior part becomes the brain and the posterior becomes the spinal cord.
- Cell differentiation is controlled by chemical signals between cells.

3.6 What Are the Key Events of Pregnancy?

Differentiate fetal from embryonic development, and identify key points in development during pregnancy.

- During the first week of conception, cleavage produces cells that will form the embryo and the chorion.
- In week 2, the embryo implants in the wall of the uterus and gastrulation occurs.
- In week 3, the earliest stages of the nervous system and heart begin.
- Weeks 4–7 involve further differentiation and development of some external features such as ears, nose, fingers, and toes.
- At week 8, fetal development begins: organ systems are refined and the fetus grows dramatically.
- In the third month, sex can be determined.
- In the fifth month, the heartbeat can be heard with a stethoscope.
- By the seventh month, the fetus is positioned head down, appropriate for delivery.
- In the eighth and ninth months, size increases and the lungs mature.

3.7 What Happens in Labor and Delivery?

Summarize the stages of labor and important events that are likely to occur in each stage.

- When a fetus's lungs are fully developed, they secrete surfactant, which also causes the uterus to contract.
- At onset of labor, the contractions occur every 15–20 minutes; near delivery, contractions occur every 1–2 minutes.
- Stretching of the cervix releases oxytocin, which causes contractions that push the fetus downward, further stretching the cervix; this repeats until the cervix dilates to a 4-inch diameter.
- Labor and delivery consist of three phases: in stage 1, the fetus is pushed down and the amnion breaks, releasing amniotic fluid; during stage 2, strong contractions occur and the mother experiences an almost irresistible urge to push; in stage 3, the placenta is delivered.
- If there are complications caught early in a labor, the doctor may opt for surgical delivery.

3.8 How Do Twins Form?

Distinguish the development of identical from fraternal twins.

- Identical twins come from a single fertilized egg that splits into two embryos early in development.
  - 33% of identical twins have separate chorions.
  - 66% of identical twins share a chorion but have separate amnions.
  - 1% of identical twins share both a chorion and an amnion; this category of twins is at risk of becoming conjoined.
- Fraternal twins occur when two separate eggs are fertilized by separate sperm. Fraternal twins are no more closely related than regular siblings.

3.9 What Can Conjoined Twins Tell Us About Biology and Ourselves?

Assess the value of conjoined twin research and what conjoined twins can tell us about ourselves.

- The survival rate of conjoined twins is very low.
- Conjoined twins can be produced experimentally in many organisms; these experiments help us understand the process of development.
- Conjoined twins were previously often exploited as oddities, but now conjoined twins are more likely to be treated as individuals with individual personalities.
- Although it is now possible to surgically separate many conjoined twins, some conjoined twins cannot be separated without causing significant harm to one or both individuals.

Key Terms

amnion 82
anaphase I 68
anaphase II 69
cell 62
cell membrane 64
cell theory 62
cervix 81
chromatid 69
chromosome 64
cleavage 71
conjoined twin 60
cytokinesis 70
cytoplasm 64
cytoskeleton 65
cytosol 64
differentiation 75
diploid cell 66
embryo 71
embryogenesis 62

embryonic stem cell 73
endomembrane system 65
endoplasmic reticulum (ER) 65
fertilization 70
fetal alcohol syndrome (FAS) 80
fetal alcohol spectrum disorder (FASD) 80
fetus 77
gamete 66
gastrulation 73
gene expression 76
Golgi apparatus 65
haploid cell 66
homologue 66
human development 60
interphase 66
lysosome 65
meiosis 66
metaphase I 67
metaphase II 69

mitochondrion 65
mitosis 71
neural tube 73
neurulation 77
nucleus 64
organelle 63
ovary 66
placenta 73
prophase I 67
prophase II 69
ribosome 65
spindle 67
telophase I and cytokinesis 69
telophase II 70
testis 66
ultrasound 77
uterus 77
zygote 70

Review Questions

1. What is the cell theory?
   a. The idea that bloodletting can be used to maintain balance within cells
   b. The realization that all living organisms are made up of cells
   c. The idea that cells have specialized structures called organelles
   d. The realization that all cells have a nucleus
   e. The idea that all cells must divide through the process of mitosis

2. Identify the principal function of each organelle or cell structure: (a) nucleus; (b) mitochondria; (c) rough endoplasmic reticulum; (d) smooth endoplasmic reticulum; (e) Golgi apparatus; (f) lysosome; (g) cell membrane.

3. Give an example of the function of each component of the cytoskeleton: (a) microfilaments; (b) microtubules; (c) intermediate filaments.

4. Starting with a diploid cell that has six chromosomes (two chromatids per chromosome), diagram meiosis I and meiosis II.

5. Starting with a diploid cell that has six replicated chromosomes (meaning two chromatids per chromosome), diagram mitosis.

6. What protects the egg from being fertilized by sperm of another species?
   a. Fallopian tubes
   b. Meiosis
   c. Gametes
   d. Homologues
   e. *Zona pellucida*

7. What are the main events of the first two months of human development?

8. What are the main events of fetal development, and when does each occur?

9. How is labor triggered by the fetus?

   a. When a fetus is ready to be born, the cervix will shrink for a time, causing a contraction that stimulates the release of the hormone oxytocin.

   b. When a fetus is ready to be born, it will secrete phlegm, which will block the lungs and send a signal to the uterus to contract.

   c. When a fetus is ready to be born, it will kick the wall of the uterus, causing a contraction.

   d. When a fetus is ready to be born, it will secrete surfactant, which causes certain fetal cells to migrate to the uterus, where they secrete a chemical signal that causes the uterus to contract.

   e. When a fetus is ready to be born, neurulation will occur, causing contractions.

10. Describe the three stages of labor and delivery.

11. Describe the difference between the formation of fraternal and identical twins.

   a. Identical twins arise from a single fertilized egg, which splits into two embryos sometime during very early development; fraternal twins arise when the mother secretes two eggs, and each egg is fertilized by a separate sperm.

   b. Identical twins arise when the mother secretes two eggs, and each egg is fertilized by a separate sperm; fraternal twins arise from a single fertilized egg, which splits into two embryos sometime during very early development.

   c. Identical and fraternal twins both arise from a single fertilized egg, but in the case of fraternal twins, the egg is fertilized by separate sperm.

   d. Identical and fraternal twins both arise when the mother secretes two eggs, but in the case of fraternal twins, the eggs are fertilized by the same sperm.

   e. There is no significant difference between the development of identical and fraternal twins.

12. Why is there a greater risk of conjoined twins forming if twins separate later than nine days after conception?

13. What are two ideas regarding how conjoined twins form?

The Thinking Citizen

1. Why is it necessary to produce gametes by meiosis rather than mitosis? What would the consequence be if gametes were instead formed by mitosis?

2. Given the high rate of spontaneous miscarriage (40–60% of all conceptions), why do you think we have not declared this situation a public health crisis?

3. Human and other mammalian embryos are attached to the mother's circulatory system and obtain nutrients and oxygen delivered from the placenta. Birds do not develop in their mothers' bodies. Where do bird embryos get their nutrients and oxygen?

4. Even though identical twins have identical genes, they are not the same person. Why is this so? What is responsible for their individualities?

The Informed Citizen

1. Many thoughtful people disagree about when human life begins. What information would you need in order to figure out when an embryo or fetus becomes a person?

2. People do not generally have funeral ceremonies to mourn miscarried embryos or fetuses. Should we? Why or why not?

3. Many infants born dramatically premature have serious health problems for the rest of their lives. Do you think that heroic medical efforts should always be made to save these infants, no matter the cost? Why or why not?

4. If a baby is born with a physical difference, it is generally the parents who decide whether to allow surgery. Is it always appropriate for parents to make this choice? Why or why not? Can you think of any examples in which no decision should be made about surgery until the child is old enough to make his or her own choice?

5. What are your thoughts about reality television programs, movies, or live shows that are focused on people with physical differences?

At its most basic, a gene is the instruction to make a specific protein. The notion that there is a gene "for" eye color misses the mark; rather, we should think of genes as *used for* or *used in* eye development.

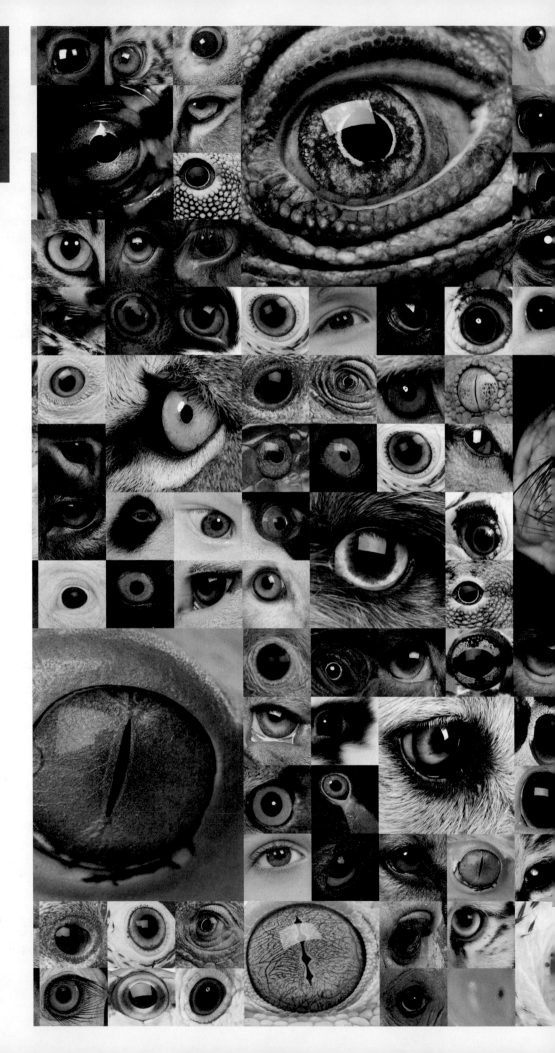

CHAPTER 4

Inheritance, Genes, and Physical Characteristics

Does Disease Have a Genetic Basis?

Becoming an Informed Citizen ...

We all want to know where we came from and why we look and act the way we do. Science is making great strides in understanding genes and even manipulating what they can do, but those advances also have real consequences for individuals and society.

e all want to know where we came from. Why do we look and act the way we do? How much of who we are can be traced to our genes?

Science is making tremendous advances in understanding **genes**—and even manipulating what they do. In this chapter, we explore how inheritance works. We see how the genetic information passed down from your parents plays a role in your unique development. To do so, we must learn the basic rules of inheritance. We need to know how the molecules DNA, RNA, and protein produce what is coded in our genetic instructions.

The function of a gene is widely misunderstood. At its most basic, a gene is simply the instructions to make a specific protein. The notion that there is a gene "for" eye color, or height, or intelligence, or aggression, or sexual orientation, or any other characteristic misses the mark entirely. Instead of a gene *for*, we should think of genes as *used for* or *used in*: for example, we can speak of the gene used in eye development. And many factors influence that development—including the environment. Genes act within a community of other genes as well, not as solitary actors.

In light of all this complexity, a single gene can have far-ranging consequences. Take the case of two devastating diseases—sickle cell disease and malaria.

gene :: hereditary unit consisting of DNA

case study

Sickle Cell Disease, Malaria, and Human Evolution

FIGURE 4.1 Malaria is caused by the parasite *Plasmodium falciparum,* which is carried by the *Anopheles* mosquito. When one of these mosquitoes bites an individual, the parasitic cells take up residence in the blood, destroying red blood cells, the cells that carry oxygen through the body.

malaria :: a disease caused by a parasite carried by the *Anopheles* mosquito; it is often fatal

protozoan :: a simple, single-celled organism

red blood cell :: a cell that carries oxygen through the blood

Tony Allison grew up in Kenya. His father, a farmer, had relocated the family from England in 1919. As a boy, Tony went on long excursions with professional naturalists to observe and help collect birds for the Natural History Museum in London. He also visited the archeological excavation site of Louis Leakey, the preeminent anthropologist of his time, and became intrigued with human evolution and the relationships among the various tribes he saw in Kenya.

During one of the Allisons' holidays on the beaches of Malindi in Kenya, Tony contracted malaria; he was only 10 years old. **Malaria** is a terrible disease and often fatal. It is caused by a **protozoan**, a simple single-celled organism, and carried by a specific type of mosquito. When one of these mosquitoes bites a person, it injects this protozoan, *P. falciparum*, into the individual (**FIGURE 4.1**). These parasitic cells take up residence in the blood and destroy **red blood cells**, the cells that carry oxygen through the blood. There is currently no vaccine against malaria. Tony's experience with malaria led him to switch gears: rather than becoming a naturalist, he decided to become a physician.

After earning his undergraduate degree in South Africa, Tony moved to England to finish his medical training at Oxford University. Although he enjoyed medical school, he had not lost his keen interest in human evolution. Tony was convinced that there had to be a way to measure human evolutionary relationships more precisely than anthropologists did by looking only at bones.

In 1949, when Oxford University sent a group of scientists to Kenya to survey and study plants and animals all over the country, Tony jumped at the opportunity to revisit the land of his childhood and to study human populations. He accompanied the expedition team on their travels throughout

Kenya, and at each location he collected blood samples from individuals of each of the various tribes, including the Kikuyu, Luo, and Masai tribes. He tested the samples for red blood cell membrane proteins to see whether there might be a pattern that would reveal genetic relationships. Tony also tested for the presence of **sickle cells** (**FIGURE 4.2**), a type of abnormal red blood cell. Prior to embarking on the trip, Tony had learned from other scientists that sickle cells, common in some areas of Africa and inherited, produced **sickle cell disease**—a painful and, in the early twentieth century, generally fatal disorder.

To Tony's disappointment, the analysis of red blood cell membrane proteins failed to reveal a pattern. In contrast, the sickle cell results presented an unexpected surprise. Tony observed the sickle cell trait in more than 20% of the individuals in tribes living along the coast or near Lake Victoria. Yet fewer than 1% of people in tribes living in dry land areas or at high altitudes exhibited the trait. Tony wondered why there was such a big difference in its prevalence and why it seemed to depend on where a person lived.

Sometimes a good idea comes like a bolt out of the blue. Tony knew that the density of mosquito populations varied in different locales, as did the incidence of malaria. In lowlands and wet areas, the mosquitoes were abundant, malaria was prevalent, and the incidence of the sickle cell trait was high. In the dry highlands, mosquitoes were few and far between, malaria was not a problem, and the sickle cell trait was rare. Was there a relationship? Tony hypothesized that possession of the sickle cell trait provided some resistance to infection with *P. falciparum*, the malaria parasite, but he could not test the idea because he had to return to England to finish medical school.

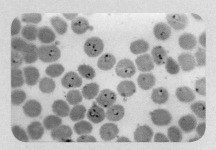

FIGURE 4.2 Normal red blood cells appear disc shaped, while sickle cells appear in the shape of a sickle blade.

sickle cell :: a type of abnormal red blood cell that is shaped like a sickle

sickle cell disease :: an inherited genetic disorder in which red blood cells sickle easily, leading to potentially serious physical consequences

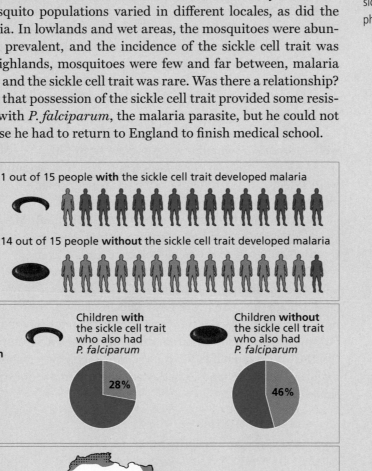

Study #1
30 individuals were inoculated with *P. falciparum* for a drug trial

1 out of 15 people **with** the sickle cell trait developed malaria

14 out of 15 people **without** the sickle cell trait developed malaria

People who developed malaria

Study #2
Allison tested the blood of 290 children

Children **with** the sickle cell trait who also had *P. falciparum*

28%

Children **without** the sickle cell trait who also had *P. falciparum*

46%

Study #3
Allison analyzed blood samples taken from 5000 people representing more than 30 different tribes from various locations throughout Africa

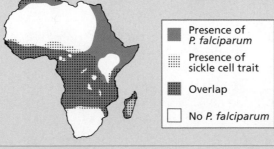

Presence of *P. falciparum*

Presence of sickle cell trait

Overlap

No *P. falciparum*

FIGURE 4.3 In 1953, Tony Allison designed three studies to see whether the inheritance of the gene associated with sickle cell disease would confer resistance against malaria.

Distribution of malaria

Distribution of sickle cell trait

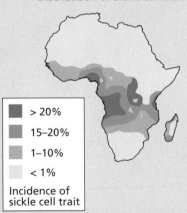

■	> 20%
■	15–20%
■	1–10%
■	< 1%

Incidence of
sickle cell trait

FIGURE 4.4 In Tony Allison's final study, he discovered that the sickle cell trait was rare or absent in locations where malaria was absent and common in locations where malaria was a serious problem.

Tony completed medical school in 1952 and received the financial and technical support he needed to return to Africa in 1953. As Dr. Allison, his goals were straightforward: he wanted to see whether the inheritance of the gene that is associated with sickle cell disease would indeed confer resistance against malaria.

Dr. Allison designed three studies to address his research question (**FIGURE 4.3**). In the first study, he piggybacked onto a drug trial being conducted by a pharmaceutical company. Scientists injected *P. falciparum* into healthy volunteers. When subjects developed malaria, the scientists administered their test drug to see whether it was effective. Allison had access to all 30 of the individuals in the study. He took blood samples from all and looked for a relationship between those possessing the sickle cell trait and those who developed malaria. The results were dramatic. Fourteen out of 15 infected subjects who did not have the sickle cell trait developed malaria. Only 1 out of 15 subjects who had the sickle cell trait did so.

Allison was excited about these results, but he knew that adults living in mosquito-infested areas probably had been exposed to *P. falciparum* earlier in their lives. The results may have been due to immune responses remaining from previous infections. To address this possibility, Allison undertook a second study. He offered free medical checkups to children at the local farmers' markets. He administered physical exams and collected a small amount of blood from each child. He examined the blood samples for two things: the presence or absence of the sickle cell trait and whether *P. falciparum* was in the blood.

Eventually Allison collected data for 290 children. Once again, the results were dramatic. *P. falciparum* was present in only 28% of children who had the sickle cell trait. In contrast, 46% of children who did not have the trait were infected and the number of parasites present was greater than that observed in the children with the sickle cell trait.

In Allison's final study, he traveled throughout Africa collecting blood samples from 5000 individuals representing more than 30 tribes. He observed that the frequency of the sickle cell trait ranged from 0% to 40%. In all cases, the sickle cell trait was rare or absent in locations where malaria was absent and common in locations where malaria was a serious problem (**FIGURE 4.4**).

Through his three studies, Allison demonstrated unequivocally that possession of the sickle cell trait afforded some protection against malaria. However, many questions remained. How does a person develop sickle cell disease? What accounts for the relationship between sickle cell disease and malaria? For our purposes, let's start with a simpler question: what exactly is sickle cell disease?

4.1 What Is Sickle Cell Disease?

In 1910, Dr. James Herrick of Chicago took a drop of blood from a dental student and examined it with a microscope. The student suffered from **anemia** and was deficient in **hemoglobin**, which is the protein that carries oxygen in red blood cells. Symptoms of anemia may include yellowing of eyes, fatigue, dizziness, shortness of breath, and chest pain, to name just a few. Herrick saw something quite unusual: instead of being disc shaped, the red blood cells were shaped like commas or sickles (curved blades generally used for harvesting crops). The cause of the sickling was not known. In 1917, Dr. Victor Emmel placed blood from an individual with sickle cell disease onto a slide and sealed it from the air. He observed that the red blood cells sickled as the amount of oxygen available to them decreased.

By 1923, it was clear to scientists that sickle cell disease was genetic. A person could inherit the disorder from parents. And in 1949, scientists realized that there were two forms of the disease. In the severe type, red blood cells sickle easily; this is sickle cell disease (or *sickle cell anemia*). In the less severe version, red blood cells do not sickle unless they are completely without oxygen. Someone with this less severe form is referred to as "being a carrier" or "possessing the sickle cell trait."

anemia :: a disease in which a person is deficient in hemoglobin

hemoglobin :: the protein that carries oxygen in red blood cells

How Sickling Happens

Sickle cell disease is the most commonly inherited blood disorder in the United States. Approximately 72,000 Americans have this disease. One out of every 500 African Americans is affected.

While normal red blood cells are flexible, sickle cells are rigid and can get stuck in narrow blood vessels, thus forming blood clots (**FIGURE 4.5**). Cells, tissues, and organs "downstream" from a blood clot will be deprived of oxygen. If the lack of oxygen persists, tissue and organ damage are certain. The range of physical consequences is quite broad and potentially serious: muscle cramps, intense pain, anemia, swollen joints, kidney failure, and heart failure are some examples (**FIGURE 4.6**). A sickling incident can be triggered by many events such as physical exertion or other health issues such as infections, heart disease, or respiratory ailments.

FIGURE 4.5 Normal blood cells are flexible, but sickle cells are rigid and can get stuck in narrow blood vessels, forming blood clots. The range of physical consequences is quite broad and potentially serious.

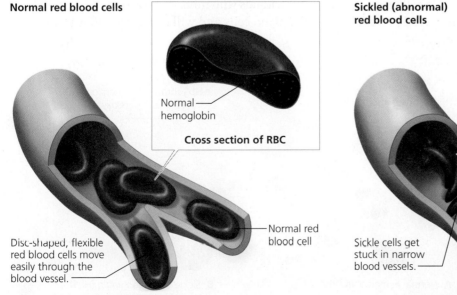

Normal red blood cells

Normal hemoglobin

Cross section of RBC

Normal red blood cell

Disc-shaped, flexible red blood cells move easily through the blood vessel.

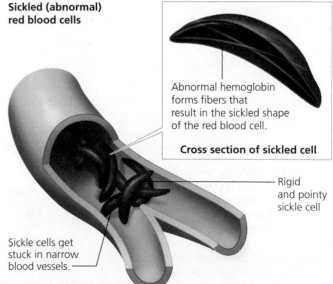

Sickled (abnormal) red blood cells

Abnormal hemoglobin forms fibers that result in the sickled shape of the red blood cell.

Cross section of sickled cell

Rigid and pointy sickle cell

Sickle cells get stuck in narrow blood vessels.

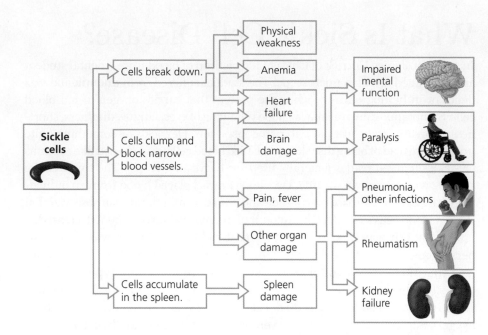

FIGURE 4.6 Sickled cells can lead to a range of serious physical consequences caused by lack of oxygen, including physical weakness, intense pain, anemia, heart failure, kidney failure, impaired mental function, and paralysis.

HbA :: normal hemoglobin that carries oxygen molecules throughout the body

HbS :: abnormal hemoglobin; although it can carry oxygen, it forms long fibers once it has given up its oxygen, causing red blood cells to sickle

The principal biochemical defect of sickle cell disease is in the molecule hemoglobin. Located inside red blood cells, normal hemoglobin (**HbA**) carries oxygen molecules throughout the circulatory system, delivering them to cells and tissues within the body. Sickle cell hemoglobin (**HbS**) is abnormal. Although it can carry oxygen, unlike HbA, HbS forms long fibers once it has given up its oxygen, causing the red blood cells to become sickle shaped; deoxygenated HbA does not form fibers (**FIGURE 4.7**).

Sickle Cell Disease and Inheritance

The inheritance pattern of sickle cell anemia is well understood (**FIGURE 4.8**). If a person inherits genes that encode for HbA protein, he or she will not have sickle cell disease. Inheritance of one normal gene and an altered one that directs the synthesis or production of HbS protein means that the person will be a carrier of the disorder. Such a person will have *both* HbA and HbS proteins in his or her red blood cells, but he or she will not develop full-blown sickle cell disease. Finally, individuals who inherit two altered genes will have only HbS protein in their red blood cells and will definitely suffer from sickle cell disease.

FIGURE 4.7 Both normal hemoglobin (HbA) (A) and sickle cell hemoglobin (HbS) (B) carry oxygen molecules throughout the circulatory system, delivering them to cells and tissues within the body. However, when oxygen is low, HbS forms long fibers because of the attraction of the HbS molecules to one another, causing cells to become sickle shaped.

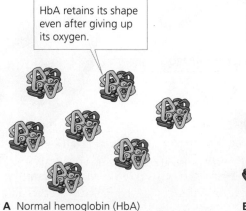

HbA retains its shape even after giving up its oxygen.

HbS forms long fibers once it has given up its oxygen.

A Normal hemoglobin (HbA)

B Sickle cell hemoglobin (HbS)

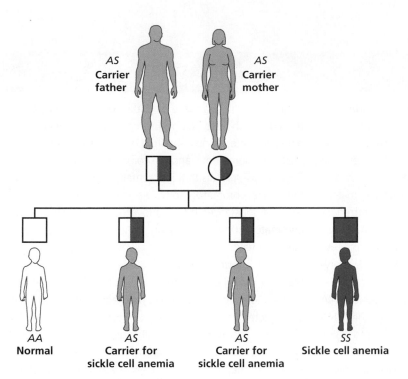

FIGURE 4.8 Only individuals who inherit genes that encode for HbS protein from both a mother and a father will suffer from sickle cell disease. In the figure, *A* stands for the HbA allele, and *S* stands for the HbS allele.

Individuals will also differ sharply in their susceptibility to malaria. People who have only HbA protein are vulnerable to infection, while those with both HbA and HbS proteins are resistant. Having only HbS provides malaria protection but guarantees sickle cell disease. Evidently, individuals who have the gene that encodes HbA protein *and* the gene that directs the synthesis of HbS protein have the best-case scenario possible in a malarial area: they have resistance to a deadly disease, malaria, but don't develop a different potentially fatal problem, sickle cell disease.

How does the sickle cell trait confer protection against malaria? Scientists have shown that sickled cells are more difficult for *P. falciparum* to enter. Even parasites that manage to get into the red blood cells while they are still shaped normally will be killed by a sickling event.

Even though the inheritance, biochemistry, and molecular biology of sickle cell disease are well understood, we have no effective treatments. Physicians can only prevent symptoms from developing and relieve them when they occur. Public health efforts to screen for carriers and children with sickle cell disease have been only partially successful. The problem remains serious: at least 8% of all African Americans are carriers of the sickle cell trait.

4.2 Could Molecular Medicine Prevent Sickle Cell Disease?

Research to understand the biochemical basis of sickle cell disease began almost accidentally. In 1945, Linus Pauling (**FIGURE 4.9**), one of the most important chemists of the 20th century, was visiting Harvard University to deliver a lecture. While there, he met with other scientists to share research findings and questions. In his meeting with professor William Castle, Pauling learned about sickle cell disease and how red blood cells sickle in the absence of oxygen.

FIGURE 4.9 Linus Pauling revealed that sickle cell disease was due to a mutated form of hemoglobin.

Technology Connection

Electrophoresis

"Who done it?" On television and in the movies, DNA tests often hold the answer. You know the scenario. DNA is collected from the saliva on a cigarette butt carelessly left behind at the crime scene. If the police have a suspect and there is enough evidence, a court order can require the accused to provide a DNA sample. But how do we know whether the samples match?

One of the first steps is to chop up the DNA into manageable fragments. Chemicals called enzymes can cut a DNA molecule at specific nucleotide sequences. For example, the enzyme EcoR1 cuts the molecule between A and G everywhere the sequence GATTC appears. Because each of us has unique DNA sequences, the enzymes will produce DNA sequences of different lengths. By comparing the DNA fragments from different samples, scientists can determine whether the DNA came from the same person.

Electrophoresis supplies the comparison. First developed in the 1930s by Arne Tiselius, this technique is used widely for forensic investigation. It also identifies genes for paternity determination or disease screening and diagnosis.

Electrophoresis works much like the chromatography you might have done in elementary school. Back then, you might have taken a piece of paper towel or filter paper and dotted it with food coloring. When you dip the edge of the paper in water, it absorbs fluid, which travels up the paper past the colored dots, and the colors separate. The green dot, for example, reveals that it is actually a mixture of yellow and blue. Like chromatography, electrophoresis separates mixtures of substances by the movement of molecules. Unlike chromatography, it relies on their movement through an electrical field.

With electrophoresis, a mixture of DNA fragments is placed in a well cut into the gel, a solid matrix similar in consistency to gelatin. The gel, submerged in a conductive liquid, is subjected to an electric field. Because DNA is negatively charged when in solution, it will migrate toward the positively charged end of the field. Small fragments migrate the fastest, large ones the slowest. Eventually, the mixture of DNA pieces separate into a ladder of fragments that can be dyed and thus visualized as stripes called bands. By comparing the bands produced by different DNA mixtures, the forensic scientist can tell whether the DNA on the cigarette did indeed come from the suspect. ::

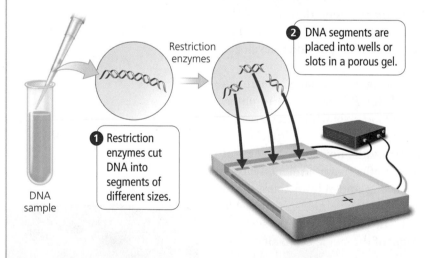

① Restriction enzymes cut DNA into segments of different sizes.

DNA sample

Restriction enzymes

② DNA segments are placed into wells or slots in a porous gel.

③ When an electric current passes through the chamber, DNA fragments move toward the positively charged end.

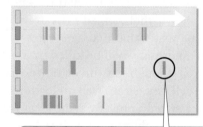

④ Smaller DNA segments migrate the furthest and the fastest; large ones migrate the least far and the slowest.

Gel electrophoresis is used to separate mixtures of molecules such as proteins, RNA, and DNA. It is one of the most important tools in biomedical, biotechnological, conservation biological, and forensic science laboratories.

Pauling seized on this information and set out to study the HbA and HbS proteins. What exactly was different between the two molecules? He wanted to show that disease could be studied—and even prevented or cured—at the biochemical level. Pauling's big idea was that molecular medicine was the path to the future.

electrophoresis :: a widely used method for separating mixtures of molecules, including proteins, DNA, and RNA

Pauling and his coworkers examined the HbA and HbS proteins using a technique known as **electrophoresis** (see *Technology Connection: Electrophoresis*). They found that the two types of hemoglobin were different chemically: HbA and

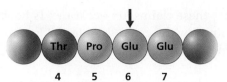
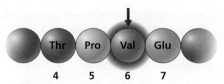

| 4 | 5 | 6 | 7 |
Normal amino acid sequence with glutamic acid in position 6

| 4 | 5 | 6 | 7 |
Valine replaces glutamic acid in position 6 of amino acid sequence

FIGURE 4.10 In HbS, the amino acid that is in position 6 on the chain is valine instead of glutamic acid (as it is in HbA). A change in just one amino acid produces an abnormal protein that causes the disease.

HbS had a different distribution and localization of electrical charges on their surfaces. Also, they demonstrated that the red blood cells of individuals who were carriers contained an approximately equal mix of HbA and HbS. Finally, Pauling and his colleagues revealed that the sickling that occurs in low oxygen is because HbS forms long filaments under those conditions, thus deforming the cells.

Although biochemical information was important, Pauling's insights about the results were game changing. First, he realized that the type of hemoglobin present in a person was related to the genetics of the disease. A person with two normal hemoglobin genes showed no evidence of the disease at all. Inheritance of one normal gene and one sickle cell gene produces a carrier. An individual who suffers from sickle cell disease has inherited two sickle cell genes, one from each parent.

Second, he demonstrated that one could trace the cause of a disease to an alteration in the molecular structure of a protein. Finally, Pauling argued that since sickle cell disease is inherited, genes must determine the structure of proteins. This was years before anyone knew what genes did or what they were made of. Pauling ushered in the age of **molecular medicine**, although it took decades for other scientists to catch up to this idea that understanding the way genes, proteins, and other molecules work could be the key to diagnosing and treating disease.

What precisely was different between the HbA and HbS molecules? In 1957, Vernon Ingram tackled this problem using a more sensitive technique he had developed. He examined HbA and HbS structures and showed that they differed in only one tiny place.

Proteins are composed of subunits called **amino acids**. The most basic level of protein structure is simply the types and order of amino acids hooked together like snap beads. In HbS, the amino acid that is in position 6 on the chain is the wrong one. Instead of glutamic acid, as is present in HbA, HbS has valine in that spot (**FIGURE 4.10**). Evidently, a change in a gene results in the alteration in amino acid sequence, thus producing an abnormal protein that causes a disease. And as if these connections are not sufficiently amazing, here is one more of a different type: the source of the HbS protein for Ingram's study was none other than Dr. Tony Allison!

molecular medicine :: a field of medicine that aims to understand the relationship between health and genes, proteins, and other molecules, in order to diagnose and treat disease

amino acid :: a subunit of protein

4.3 Where Is Our Genetic Information Stored?

Newspapers, magazines, television, and the Internet bombard us with stories about inheritance. We hear that genes are related to everything from the risk of disease to intelligence, musical talent, athleticism, and even sexual orientation. How can we assess whether these claims are accurate? Is how we look, act, and live primarily determined by the genes we inherit? For an answer, let's first consider how genetic information is passed from one generation to the next.

Humans have been interested in how heredity works since at least the dawn of agriculture. Plant breeding for crop improvement and animal breeding to increase milk yield, strength, or meat production were practiced long before anyone had a clue about the mechanisms responsible for these successes. People developed significant practical knowledge regarding breeding, and for most, this was all a person needed to know. However, as is true in all time periods and places, there were some individuals who were seized by the need to know *how* a thing works. They wanted to understand the exact mechanisms of inheritance.

One early idea that persisted for more than 2000 years was **pangenesis**. According to this notion, each part of the body produced a characteristic

pangenesis :: an idea that persisted for more than 2000 years and that held that each part of the body produced a characteristic seed that traveled to reproductive organs; this idea was debunked by August Weismann in the 1880s

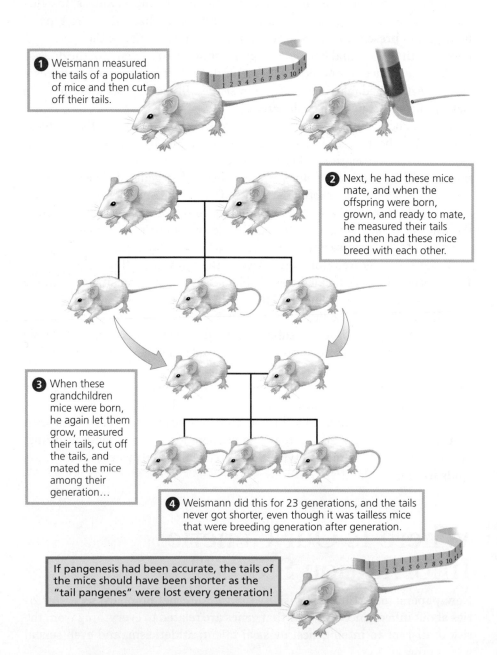

① Weismann measured the tails of a population of mice and then cut off their tails.

② Next, he had these mice mate, and when the offspring were born, grown, and ready to mate, he measured their tails and then had these mice breed with each other.

③ When these grandchildren mice were born, he again let them grow, measured their tails, cut off the tails, and mated the mice among their generation…

④ Weismann did this for 23 generations, and the tails never got shorter, even though it was tailless mice that were breeding generation after generation.

If pangenesis had been accurate, the tails of the mice should have been shorter as the "tail pangenes" were lost every generation!

FIGURE 4.11 In the 1880s, August Weismann performed an experimental test of pangenesis using mice. Weismann concluded that the information to construct the tails of the mice did not reside in the tail itself.

"seed" that traveled to reproductive organs. When mating occurred, the "seeds" were released by males and females and thus combined. As a consequence, the seeds from the parents formed the respective parts of the offspring. For example, the nose seeds, arm seeds, and leg seeds of the parents mixed, and the result was a blending of characteristics. It is easy to understand why pangenesis seemed like a realistic mechanism. If you look at offspring, human or other, they often resemble a blend of their parents' characteristics. Also, there was not a better explanation available yet.

Finally in the 1880s, August Weismann performed an experimental test of pangenesis (**FIGURE 4.11**). He took a population of mice and removed their tails (with their hypothetical seeds). He then bred offspring from these mice and measured whether the tails were shorter.

When Weismann measured the tails, he found they were just as long as the tails of the parents. Unfortunately for the mice, Weismann was a patient and methodical man. He repeated this experiment for seven years. Mice of this 23rd generation had normal-sized tails. Weismann concluded that the information to construct the tail does not reside in the tail itself. So where is it? The answer is the nucleus of each cell, in a molecule called DNA. The rest of this chapter tells the story.

4.4 How Did Mendel Discover the Rules of Inheritance?

The path to answering the questions of where genetic information resides and how inheritance works is a long one. In fact, it began 20 years before Weismann's experiment. Gregor Mendel (**FIGURE 4.12**), an Augustinian monk from the Brunn Monastery (in what is now the Czech Republic) undertook a series of carefully planned experiments. In doing so, he revealed the basic rules of inheritance.

Thanks to one of those accidents of circumstance, Mendel happened to be assigned to a monastery with a long tradition of scientific pursuits. It was equally fortunate that he was well trained in science and mathematics. The religious order's original plan was for Mendel to be a science teacher. Luckily for science, Mendel turned his formidable mind to the problem of inheritance.

FIGURE 4.12 Gregor Mendel was an Augustinian monk who conducted a series of carefully planned experiments that revealed the basic rules of inheritance.

Mendel's Experiments

Mendel made some important decisions as he designed his experiments (**FIGURE 4.13**):

1. He chose garden peas as the organism to study. Peas are simple to cultivate, and a person can easily control the matings by manually applying pollen to the female parts of the flowers. Mendel's initial genetic experiments were done with mice, but the church hierarchy pulled the plug on this work. Apparently all of that mouse sex was deemed inappropriate. It appears that the fathers did not understand that plants mate too.

2. Mendel focused on one **trait**, or characteristic, at a time, rather than just recording the overall appearance of the offspring of a **cross** (mating). For example, he looked at traits that were discrete and easy

trait :: characteristic

cross :: mating

1 **Choose experimental organism carefully.** Mendel chose garden peas, which are simple to cultivate and whose matings may be controlled by manually applying pollen to the female parts of another flower using a paintbrush.

2 **Focus on one trait at a time.** Rather than just recording the overall appearance of the offspring of a cross, Mendel focused on one trait (such as flower color, pod color, or stem length) at a time.

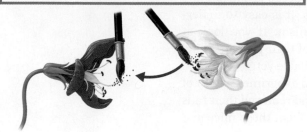

Flower color

Pod color

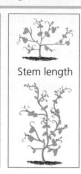
Stem length

3 **Follow the cross for more than one generation.** Mendel followed the crosses beyond the first set of offspring.

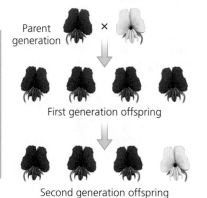
Parent generation ×

First generation offspring

Second generation offspring

4 **Collect quantitative data.** Mendel counted the number of progeny possessing a trait and kept detailed records.

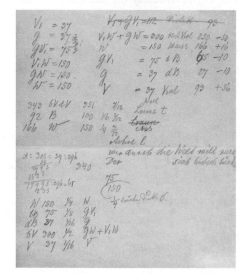

FIGURE 4.13 **Mendel followed four steps that led to his success.**

1 Mendel manually pollinated purple and white flowers (P generation).

2 In the first generation (F$_1$), all of the plants had purple flowers.

3 Mendel crossed the purple flowers using either self- or cross-pollination.

4 In the second generation (F$_2$), 705 plants had purple flowers, and 224 plants had white flowers.

Mendel concluded that purple was dominant to white and that flower color must be regulated by two hereditary factors.

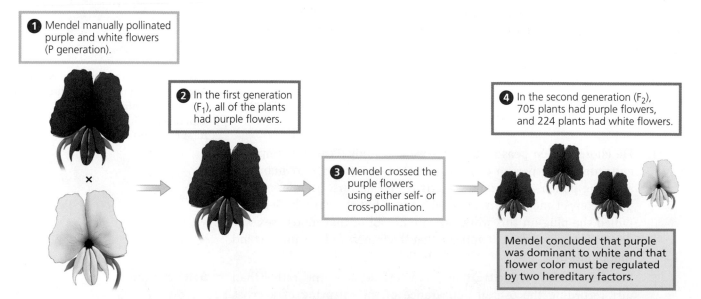

FIGURE 4.14 **Mendel crossed pea plants that had purple flowers with pea plants that had white flowers.** In the first generation of offspring, all of the flowers were purple, but in the second generation, Mendel noted a mix of purple flowers and white flowers.

to identify, such as yellow peas versus green peas or white flowers versus purple flowers. In humans, examples of discrete traits include attached or unattached earlobes, left- or right-handedness, and red blood cells with HbA or HbS protein.

3. Mendel followed the crosses for more than one generation. This made it possible to see whether a characteristic that disappeared in the children of a cross would reappear in the grandchildren.

4. Finally, Mendel collected numerical data: he counted the number of progeny possessing a particular trait.

Mendel began his experimental crosses with pure-breeding parental strains that always produced offspring just like themselves; self-pollinating plants with purple flowers had offspring that had purple flowers and plants with white flowers had offspring with white flowers. In one of Mendel's experimental crosses, he mated pure-breeding plants with purple flowers with other pure-breeding plants that had white flowers (**FIGURE 4.14**). After these flowers formed seeds (peas), the seeds were planted. The plants that grew produced flowers. In this first cross, all of the flowers were purple. Next, Mendel crossed these purple flowers among themselves, planted the resulting peas, and observed the flowers produced by these "grandchildren" plants. He reported that 705 plants produced purple flowers and 224 had white.

A lot of information is packed into this experiment. First, Mendel observed that the "white" trait disappeared from view in the progeny of the first cross but reappeared in the next cross. This result showed that purple was dominant to white. However, the white reappeared. This demonstrated that whatever was responsible for the production of this color still existed.

Second, Mendel figured out that flower color must be regulated by two hereditary factors. He reasoned that one factor is received from each parent, because each must contribute hereditary instructions. He knew that the offspring in the first cross had a purple factor; after all, the flowers were purple. Because white showed up in the grandchildren, Mendel argued that a white factor must exist in the children. People just could not see the expression of its information. Therefore, there must be *two* factors in the children: one white and one purple. When the children mated, the purple and white factors from each made very specific combinations in their offspring.

Gametes and Monohybrids

We now know that these factors sort into different gametes, or sex cells (see Chapter 3). In the cross of the children, the gametes from the mother had *either* a purple *or* a white factor. Since the grandchildren are each produced by the union of one male gamete and one female gamete, only the following combinations of factors are possible: purple, purple; purple, white; and white, white (**FIGURE 4.15**).

Mendel knew nothing about cells, cell division, chromosomes, or meiosis, but he did understand statistics. He realized that if you had two independent factors sort into pairs, the following ratios would occur: 1 purple, purple; 2 purple, white; and 1 white, white. Mendel observed that plants with a purple factor and a white factor produced flowers that looked identical to those produced by plants with two purple factors. If we consider only how the flowers look, Mendel's hypothesis predicted that the ratio of purple to white flowers in the grandchildren should be 3:1. Mendel's results were 705 purple to 224 white flowers, or a ratio of 3.15:1, amazingly close.

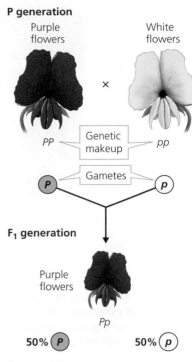

P generation

Purple flowers × White flowers

PP — Genetic makeup — pp

P — Gametes — p

F₁ generation

Purple flowers

Pp

50% P 50% p

F₂ generation

Sperm from F₁ (Pp) plant

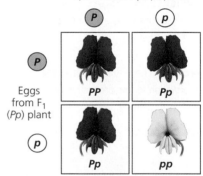

Eggs from F₁ (Pp) plant

	P	p
P	PP	Pp
p	Pp	pp

3 purple flowers : 1 white flower

FIGURE 4.15 This figure depicts Mendel's genetic cross examining flower color, with symbols representing the inherited factors *P* (purple) and *p* (white).

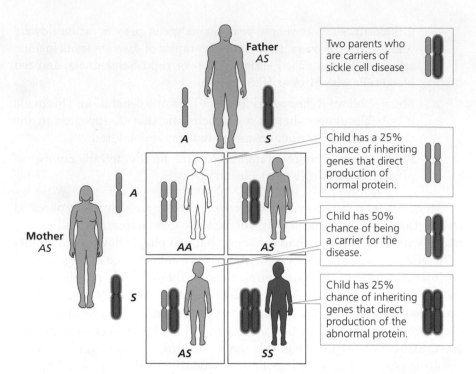

Two parents who are carriers of sickle cell disease

Father
AS

A   *S*

Mother
AS

A

S

	AA	AS
	AS	SS

Child has a 25% chance of inheriting genes that direct production of normal protein.

Child has 50% chance of being a carrier for the disease.

Child has 25% chance of inheriting genes that direct production of the abnormal protein.

FIGURE 4.16 Mendel's rules apply to all organisms that mate and with respect to all monohybrid crosses, including sickle cell disease.

monohybrid :: a type of cross in which scientists pay attention to one trait at a time

factor :: a gene that possesses the information for the production of a specific protein

The type of cross in which we pay attention to one trait at a time is called **monohybrid**. Mendel repeated his analysis of monohybrid crosses in peas for six more traits: seed shape, seed color, pod shape, pod color, stem height, and flower position. He established that each trait is encoded by two factors, one from each parent. While Mendel did not understand the mechanism, other scientists have shown that the **factors** are genes that possess the information for the production of a specific protein. Meiosis sorts these genes or factors into gametes. When gametes fuse upon fertilization, genes from the mother and father are brought together again.

Although we have considered the details of the monohybrid cross and inheritance with respect to peas, the rules apply to all organisms that mate. For example, sickle cell disease follows the same pattern (although we can't undertake a controlled mating experiment with people, even volunteers) (**FIGURE 4.16**). A person who has sickle cell disease has two genes, each of which has the information for the synthesis of HbS protein. Someone with only normal hemoglobin has two genes that encode for HbA protein. And if an individual has one gene that encodes for HbA and another for HbS, he or she is a carrier of the disease.

Two parents who produce normal hemoglobin will have children that do so too. Two parents with sickle cell disease will have children who suffer from this disorder. The children of parents who are both carriers have a 25% chance of making HbA protein only, a 25% chance of making only HbS and having sickle cell disease, and a 50% chance of making HbA and HbS and being carriers. Mendel's rules predict just that.

Mendel's Rules

Mendel's research on monohybrid crosses revealed much of the basic information about inheritance:

allele :: a different form of a gene

1. Genes can come in more than one form, or **allele**. For example, the pea flower color gene can be either a purple or white allele. Similarly, the gene that encodes for hemoglobin can be the normal or sickle cell allele, which encodes for HbA or HbS protein, respectively.

2. The alleles of a particular gene sort individually into gametes during meiosis. In other words, a pea plant with a purple allele and a white allele would produce gametes each possessing one of these alleles. A person who is a carrier of sickle cell disease will make gametes, 50% of which have the normal hemoglobin allele and 50% the sickle cell allele.

3. Finally, pea flower color is a trait that exhibits a dominant/recessive relationship between the purple and white alleles. Purple is the **dominant** trait, or the one seen when both alleles appear together. When present together, the two different alleles produce a flower just as purple as one developed by a plant with two purple alleles. White, which is then not seen, is the **recessive** trait.

dominant :: a trait that is seen even when one allele is present

recessive :: a trait that is not seen when only one allele is present

 # How Much Do Mendel's Rules Explain?

Do Mendel's rules explain inheritance completely? Not exactly. Suppose that we compare the physical outcomes for individuals with two normal alleles, two sickle cell alleles, or one normal and one sickle cell allele. We find that there is a lot more to the story than Mendel might have predicted. First, alleles can interact, and so can genes. Second, genes do not always affect just one characteristic. Finally, gene expression depends on the environment.

Alleles Can Interact, and So Can Genes

Sickle cell disease inheritance is more complicated than a simple dominant/recessive relationship between alleles. One copy of the sickle cell allele cannot totally dominate the effect of the normal HbA allele. Instead, the alleles interact (**FIGURE 4.17**). That is why a carrier has a mild form of the disease. Carriers do not have completely normal red blood cells, but they are not afflicted with sickle cell disease. As you saw earlier, carriers make both HbA and HbS proteins, and each red blood cell contains both types of hemoglobin. Red blood cells of carriers don't sickle unless the oxygen levels are very low. Interactions like these are actually much more common than dominant or recessive alleles.

Genes can interact with each other in other ways, too. For example, several different alleles influence whether you are blond, brunette, black haired, or red

Red blood cell	Hemoglobin	Sensitivity to oxygen concentration	Resistance to malaria
Normal	HbA	Not sensitive	Not resistant
Carrier	HbA and HbS	Intermediate sensitivity	Resistant
Sickled	HbS	Very sensitive	Resistant

FIGURE 4.17 There is a predictable relationship between the type of hemoglobin present in a person's red blood cells and whether he or she will experience sickling or resistance to malaria.

FIGURE 4.18 Genes can interact with one another in different ways. If an individual has the genes to make a certain set of skin pigments but also one gene that says "make no pigment at all," the skin color directions will be irrelevant; the individual will end up with albinism.

cystic fibrosis :: a disease in which a thick, abnormal mucus accumulates in lungs and the digestive system

haired. To make things more complicated, another gene is associated with baldness. Suppose someone has two "no hair" alleles. He might also have two blond alleles, but he would still have a bald head. In another example, if a person has a "no skin pigmentation" allele, it will not matter if she also has alleles for light or dark skin. She will end up with albinism, a condition in which individuals have no pigmentation in their skin, hair, or eyes (**FIGURE 4.18**). Sickle cell disease depends on other genes as well. Genes can influence the degree to which cells are sticky and form blood clots, even during a sickling event.

Genes May Affect More Than One Characteristic

Most genes also affect more than one characteristic. The sickle cell gene illustrates this point well. It also makes red blood cells less susceptible to infection by *P. falciparum*.

A similar situation is seen in **cystic fibrosis** (CF), a disease in which a thick, abnormal mucus accumulates in lungs and the digestive system. The CF allele encodes for an abnormal chloride channel in cell membranes. Interestingly, carriers of the CF allele are resistant to cholera and other serious infectious intestinal diseases.

Because a single gene can affect multiple traits, it could be wrong to remove genes even to cure disease. You could solve one problem but cause an even worse one. The consequences are unpredictable and potentially grave.

Gene Expression Depends on the Environment

Finally, how genes contribute to physical characteristics can depend on the environment. Flowers of *Hydrangea* vary in color depending on the acidity of the soil in which they grow (**FIGURE 4.19**). Some animal embryos develop differently depending on whether they are exposed to predators during their development. For example, in the absence of predators, carp are a slender fish. In contrast, exposure to predators during development results in a fatter, hunchbacked shape—no doubt more difficult to eat. Our development, too, is influenced by exposure to environmental factors, especially *in utero* (in the womb).

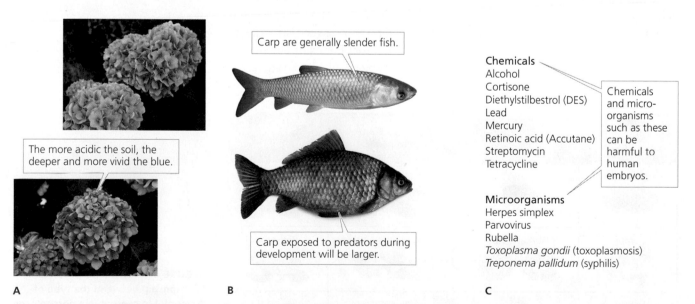

Carp are generally slender fish.

The more acidic the soil, the deeper and more vivid the blue.

Carp exposed to predators during development will be larger.

Chemicals
Alcohol
Cortisone
Diethylstilbestrol (DES)
Lead
Mercury
Retinoic acid (Accutane)
Streptomycin
Tetracycline

Chemicals and microorganisms such as these can be harmful to human embryos.

Microorganisms
Herpes simplex
Parvovirus
Rubella
Toxoplasma gondii (toxoplasmosis)
Treponema pallidum (syphilis)

A   B   C

FIGURE 4.19 Gene expression can depend on the environment. Hydrangea vary in color due to the acidity of the soil (A), carp vary in size depending on whether or not they are exposed to predators during their development (B), and humans can be affected by exposure to pathogens and chemicals *in utero* (C).

What Are Genes Made Of?

Suppose you went to a genetic counselor to find out whether you are at risk for passing on a genetic disorder to your children. The counselor needs some basic information:

- What genes are on your chromosomes, and how are the alleles for a particular gene sorted into gametes during meiosis?
- What gametes could each parent contribute during fertilization?

The resulting probability of having children with specific alleles can then be calculated in many cases using a technique known as **pedigree analysis.** (See *How Do We Know? Pedigree Analysis.*)

pedigree analysis :: a family history study; it shows a pattern of inheritance

How Do We Know?

Pedigree Analysis

In 1866 the physician Paul Broca wondered whether the breast cancer from which his wife suffered was hereditary. He observed a cluster of cases in his wife's own family: 10 out of 24 women had breast cancer. Broca mapped a family history, or pedigree, to see if he could figure out the pattern of inheritance. The idea that some traits "run in the family" is a very old one. Broca saw that they also run in patterns.

Today, pedigree analysis helps genetic counselors determine the risk of developing or passing on traits, such as disease. First, a counselor interviews the individual about the occurrence of the disease in other family members. The counselor then constructs a chart and applies the rules of inheritance.

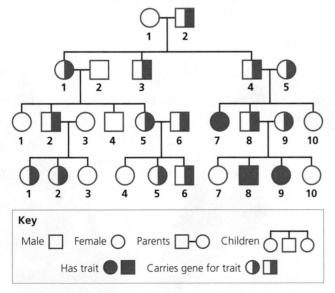

The pattern of inheritance of sickle cell disease in a family can be seen in a pedigree analysis.

For a disease like sickle cell disease or cystic fibrosis, in which a single gene is responsible, successful interpretation is likely. In other cases, such as ovarian cancer, disease depends on more than one gene as well as the environment. Still, a pedigree will point to the risks of developing the disease relative to someone with a different genetic background. It can also reveal unexpected patterns of inheritance across generations. ::

For instance, if a man who is a carrier of the sickle cell allele and a woman with normal hemoglobin genes have children, there is a 50% chance that these children will be carriers of the sickle cell trait, making both HbA and HbS proteins, and a 50% likelihood that they will make only HbA protein.

Sometimes, however, a person wants to be tested to see whether he or she carries a specific allele of a gene—one that does not show an obvious pattern of inheritance. In that case, these methods won't suffice. The techniques to identify specific genes for medical or forensic reasons all rely on knowledge about the molecular basis of genes. Thanks to our understanding of DNA, genes can be identified, manipulated, and even changed. To identify genes, unravel the causes and mechanisms of many diseases, and design drugs to treat these ailments, we need to know the basics of DNA structure. We need to know what genes are made of.

DNA (**deoxyribonucleic acid**) is a type of nucleic acid located in the cell nucleus. DNA is the physical material of which genes are made. Chromosomes are composed principally of DNA (**FIGURE 4.20**). DNA holds our genetic information.

DNA molecules are composed of subunits called **nucleotides**. There are four different nucleotides in DNA: adenine (A), thymine (T), guanine (G), and cytosine (C). Each has similar components: a sugar (deoxyribose), a phosphate group, and a nitrogen-rich ring structure called a *base*. To form the DNA molecule, the nucleotides hook together in a specific manner: the sugar of one nucleotide connects to the phosphate of another nucleotide. In turn, its sugar group attaches to a phosphate of another nucleotide and so on until

DNA (deoxyribonucleic acid) :: a type of nucleic acid located in the cell nucleus; it is the physical material of which genes are made

nucleotide :: a subunit of a DNA molecule

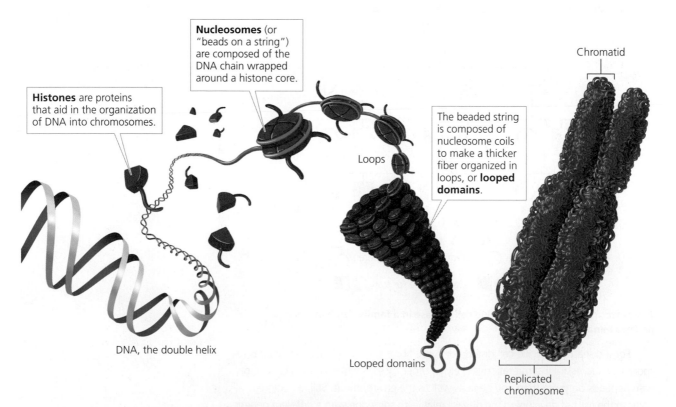

Nucleosomes (or "beads on a string") are composed of the DNA chain wrapped around a histone core.

Histones are proteins that aid in the organization of DNA into chromosomes.

The beaded string is composed of nucleosome coils to make a thicker fiber organized in loops, or **looped domains**.

Loops

Chromatid

DNA, the double helix

Looped domains

Replicated chromosome

FIGURE 4.20 DNA holds our genetic information. Chromosomes are composed of DNA.

a long strand of DNA is made. As a consequence, DNA has a sugar-phosphate backbone with bases sticking out from one side (**FIGURE 4.21**).

This is only the first aspect of DNA structure. The final form is a double helix: two single strands of DNA bound to each other in a shape like a twisting staircase. The attachment of the strands occurs thanks to the complementary base-pairing interactions between pairs of nucleotides. The base pair combinations are specific; A and T bind to each other and G and C to each other. No other combination works.

Let's put all of this information together for a look at DNA structure as a whole. The nucleotides of a particular DNA sequence bind to each other, forming a sugar-phosphate backbone. The bases stick out from this linear structure—in, for example, the sequence AATTGG. The DNA strand that will bind to this sequence is 100% predictable: TTAACC. Once scientists worked out these basic rules, they tackled the question of DNA function. Eventually, this fundamental understanding of DNA structure and complementary base pairing led to the development of the molecular techniques that make possible genetic testing, gene manipulations, biotechnology, drug development, and forensic science. (For a discussion of the discovery of DNA's double-helix structure, see *Scientist Spotlight: Rosalind Franklin*.)

4.7 How Does DNA Function?

By 1928, scientists had demonstrated the inheritance of genes but did not know much at all about the physical basis of genes. It took scientists two steps to figure it out. First, Frederick Griffith found that a substance in bacteria could transform how other bacteria function to cause disease. Second, Oswald Avery, Colin MacLeod, and Maclyn McCarty showed that the mysterious substance was in fact DNA.

Transformation

Griffith found a clue to DNA's function while studying a species of bacteria that caused pneumonia. These bacteria came in two forms. One, referred to as smooth, grew on petri dishes as glistening, moist colonies. The other, rough, produced colonies that were drier looking and not smooth at all. Griffith knew that the smooth bacteria caused pneumonia, whereas the rough type did not. Griffith was interested in what was different between the two types of bacteria so that one caused a disease and the other was harmless.

Griffith tried a lot of manipulations of the two types of bacteria (**FIGURE 4.22**, p. 111). He would then infect mice to see whether they would develop pneumonia from the experimentally altered bacteria. The results from one of the tests were truly amazing. Griffith killed a population of smooth bacteria by heating them. He combined the dead smooth bacteria with living rough bacteria. Next, he infected mice with rough bacteria alone, dead smooth bacteria alone, or the mixture of dead smooth and live rough bacteria. Griffith monitored the health of the mice carefully. Only one group got sick—the mice infected with the mixture of dead smooth and live rough bacteria. The other two test populations remained completely healthy.

Griffith isolated bacteria from the sick mice and observed that they were all smooth. Somehow the physical characteristics of the formerly rough bacteria had been altered by the dead smooth bacteria. Griffith referred to this process as **transformation** because the bacteria had changed their

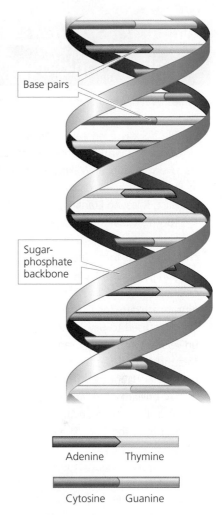

FIGURE 4.21 DNA has a sugar-phosphate backbone with bases sticking out from one side. The final form is a double helix with two single strands of DNA bound to each other in a shape like a twisting staircase.

Base pairs

Sugar-phosphate backbone

Adenine   Thymine

Cytosine   Guanine

transformation :: a term that Frederick Griffith used to describe the process by which dead smooth bacteria altered rough bacteria in experiments; the transforming substance was later found to be DNA

Scientist Spotlight

Rosalind Franklin (1920–1958)

James Watson and Francis Crick are most closely associated with discovering the structure of DNA. One strand of the double helix is even called Watson, and the other is Crick. However, the work of other scientists was essential to their discovery. Maurice Wilkins shared the 1962 Nobel Prize in Physiology or Medicine with Watson and Crick. Another scientist, Rosalind Franklin, received little recognition.

Born in London to a wealthy family, Rosalind Franklin was interested in getting answers to questions, even as a child. She was fortunate to attend one of the few schools in London that taught chemistry and physics to girls, and she decided at age 15 to become a scientist. Franklin passed the entrance exam for Cambridge University in 1938, but her father refused to pay. He was against *any* woman attending university. Fortunately, Franklin's aunt stepped in and offered to fund her niece's education. Franklin's mother also supported her daughter's ambition. Finally, her father relented.

A brilliant student, Franklin completed her degree in 1941 and earned a Ph.D. in 1945. Her scientific talent was evident from the start. She published several research papers before she was 26 years old. Franklin spent 1947–1950 working in a research laboratory in Paris.

While in Paris, Franklin became expert in a technique called X-ray crystallography. In 1951, she was offered a job as a research associate in the laboratory of John Randall at King's College, London. She was assigned to research the structure of DNA.

With X-ray crystallography, a molecule is bombarded with X-rays. The pattern produced, called a diffraction pattern, gives clues to the positions of the atoms. Imagine shining a flashlight on a kitchen chair and figuring out its structure from its shadow.

Maurice Wilkins was another research associate in Randall's lab. Even though Wilkins and Franklin were peers, he initially misunderstood their relationship and thought she was his assistant. To make things worse, women in England were not permitted in campus dining halls, senior common rooms, or many pubs. Feeling isolated, Franklin threw herself into her work.

 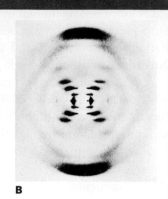

A   **B**

An expert in X-ray crystallography, Rosalind Franklin (A) produced a beautiful image of crystallized DNA (B) that was essential for figuring out the structure of this molecule.

She soon discovered that crystals of DNA take two forms. When these forms mix together, it is impossible to get a clear diffraction image. It is like shining the flashlight on a chair that has a table in front of it. However, she then figured out how to separate the forms, and she achieved excellent pictures of the A form. The scientist J. D. Bernal described them as "the most beautiful photographs of any substance ever taken." Finally, her analysis demonstrated that DNA was double stranded. The molecule had a sugar-phosphate backbone on the outside and bases paired on the inside. But how were they paired?

The answer came from Watson and Crick. Unbeknown to Franklin, Wilkins showed them her photograph of DNA. They took what they had learned from their own research, combined it with Franklin's pivotal information, and quickly described DNA's structure.

Franklin never knew exactly what had happened. Her accompanying research paper did not make the huge splash enjoyed by Watson and Crick. Sadly, she did not live to enjoy the accolades that her work deserved. She died of ovarian cancer in 1958 at the age of 37. Franklin was, however, recognized in her day as a first-class scientist with wide-ranging insights into chemical structures. She kept working in the lab until weeks before her death. ::

characteristics dramatically. Moreover, this change was heritable. Once transformed, these "new" smooth bacteria had offspring that were also smooth. Griffith, however, had no idea what this "transforming substance" was, just that it survived heating. Nevertheless, he realized that he had discovered a physical material that could change bacteria in a way that could be inherited.

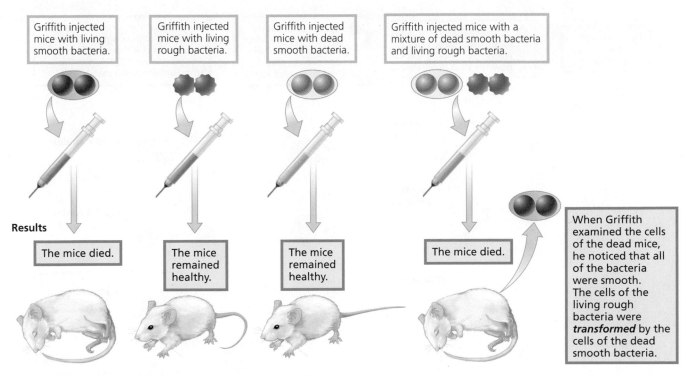

FIGURE 4.22 In 1928, Frederick Griffith found a clue to DNA's function, which he called transformation, while studying a species of bacteria that caused pneumonia.

The "Transforming Substance"

The next step was to figure out the identity of this transforming material. Sadly, Griffith was unable to complete this work because he was killed, while working in his lab in London, in a bombing raid during World War II. In fact, the next piece of the puzzle was not filled in until the mid-1940s. Like so many basic research questions, this one was put aside because of World War II, a time when scientists focused their attentions on military service or projects related to the war.

Finally in 1944, Avery, MacLeod, and McCarty demonstrated that DNA was responsible for the transformation. These scientists performed a series of experiments modeled after Griffith's (**FIGURE 4.23**). They killed smooth bacteria. They then used enzymes to eliminate carbohydrate, protein, fat, RNA, or DNA from the virulent filtrate (the liquid that is left after you filter something). Next, they combined each filtrate from which they had removed *one* type of molecule with the rough bacteria. Only the mixture containing DNA combined with rough bacteria resulted in the appearance of smooth bacteria. DNA was the transforming substance.

4.8 What Processes Must DNA Accomplish?

Long before the work of Griffith or Avery, MacLeod, and McCarty, it was clear to scientists that the molecule of inheritance has to accomplish three functions:

1. It must replicate itself.
2. It must be able to change, or *mutate*.
3. It must produce the proteins responsible for physical changes.

The rest of this chapter tackles the details of these three processes.

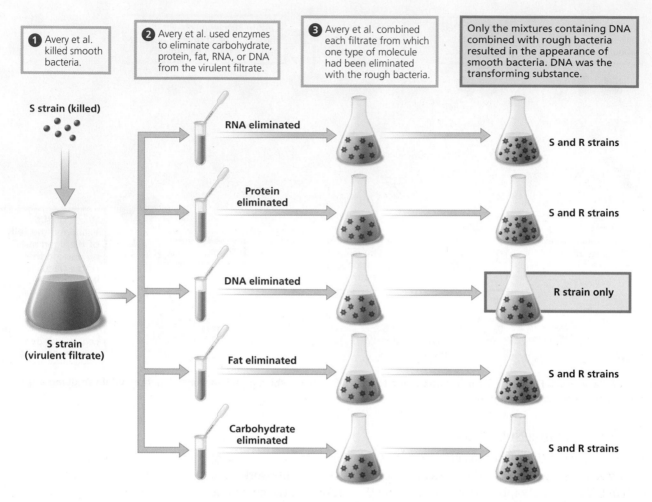

① Avery et al. killed smooth bacteria.

② Avery et al. used enzymes to eliminate carbohydrate, protein, fat, RNA, or DNA from the virulent filtrate.

③ Avery et al. combined each filtrate from which one type of molecule had been eliminated with the rough bacteria.

Only the mixtures containing DNA combined with rough bacteria resulted in the appearance of smooth bacteria. DNA was the transforming substance.

S strain (killed)

S strain (virulent filtrate)

RNA eliminated — S and R strains

Protein eliminated — S and R strains

DNA eliminated — R strain only

Fat eliminated — S and R strains

Carbohydrate eliminated — S and R strains

FIGURE 4.23 In 1944, Oswald Avery, Colin MacLeod, and Maclyn McCarty demonstrated that DNA was the substance responsible for the transformation discovered by Griffith.

Replication

replication :: the process by which the double helix of DNA unzips and new strands form

The first characteristic of DNA is that it can replicate. **Replication** is the process by which the double helix of DNA unzips and new strands form. During mitosis, identical cells form, chromosomes replicate, and a complete set sorts into each daughter cell (see Chapter 3). Similarly, during meiosis, chromosomes replicate, and two rounds of cell division parcel out single complete sets of chromosomes to individual gametes. Both mitosis and meiosis must involve DNA replication.

When James Watson and Francis Crick discovered the molecular structure of DNA in 1953, they suggested a hypothesis for how replication might work. Because of complementary base-pairing rules (A binds to T, G to C), the information for accurate replication is built right into the molecule.

To see why, imagine that the double strands of the DNA molecule separate from each other—like a zipper opening. To form two new molecules, each separate strand must now link to other nucleotides to form a *complementary strand*. As they separate, each strand reveals a sequence of nucleotides, and each nucleotide can pair with one and only one other nucleotide. Correct base-pairing thus ensures that the newly made strands will have the correct sequences (**FIGURE 4.24**).

No one in 1953 knew *how* the information in DNA led to the development of physical traits. Even so, scientists reasoned that copying of genetic instructions

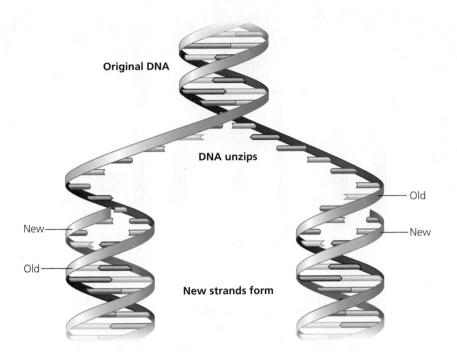

Original DNA

DNA unzips

Old

New

New

Old

New strands form

FIGURE 4.24 One important characteristic of DNA is its ability to replicate. During DNA replication, the double helix unzips and new strands form.

had to be accurate for inheritance to work. Over more than a decade, scientists collected evidence that supports Watson and Crick's idea about DNA replication.

Mutation

The second characteristic of DNA is that it can mutate. A **mutation** is a change in DNA. Genes for a given trait come in different forms, as we have seen for the gene that codes for normal and sickle cell hemoglobin. Genes can be altered or mutated in a way that is inherited. DNA must be able to do this, too.

If DNA sequences and genes remained identical one generation after the next, never changing, so would life. There would be none of the great diversity of life we see all around us. Instead, all living things would be genetically identical. But we *do* see diversity—in millions of species. We see variation even within a species and between parents and children. Why?

The answer is that although DNA replication is accurate, it is not perfect. Sometimes an organism's DNA is damaged. As a consequence, there is a small probability that genes can become altered, or *mutate*. In addition to the point mutation present in the sickle cell allele in which a nucleotide in the DNA is replaced by an incorrect nucleotide (a substitution point mutation), there are four more types of mutations: deletion, duplication, inversion, and transloca- tion. A deletion removes a segment from the chromosome, a duplication repeats a segment, an inversion reverses a segment within a chromosome, and a trans- location moves a segment from one chromosome to another (**FIGURE 4.25**). As a result, the nucleotide sequences of the chromosome are altered.

Most of the time, no physical effect can be discerned as a result of muta- tions. At other times, however, the change can be lethal, as with certain rare cancer-causing genes. In still other cases, the change in DNA, or mutation, produces an allele that confers a selective advantage to those who possess it.

For example, a mutation in the gene *CCR5* prevents binding of the human immunodeficiency virus (HIV) to immune cells. Individuals who possess this allele are resistant to acquired immune deficiency syndrome (AIDS). And, as we have discussed, the sickle cell allele provides protection against malaria.

mutation :: a change in DNA; five examples of mutations are point, deletion, duplication, inversion, and translocation

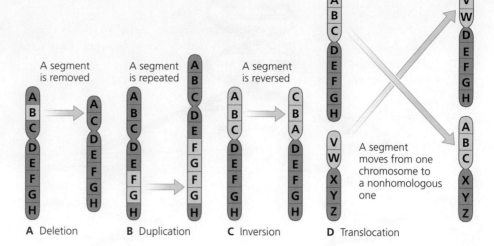

FIGURE 4.25 A second characteristic of DNA is its ability to mutate or change. In addition to the type of mutation seen in sickle cell disease, there are other types of mutations that can involve chromosome changes, including deletion (A), duplication (B), inversion (C), and translocation (D).

Organisms have developed ways to protect their DNA or even repair it. However, DNA mutations are sometimes due to faulty DNA repair. Exposure to ultraviolet light from the sun, tanning beds, or radiation, certain industrial chemicals like benzene, and activities like smoking tobacco can all damage DNA. The damage can be relatively minor, such as a change in or loss of one nucleotide. It can also be major, involving the deletion of a large sequence of DNA. In some cases, the cell can recognize relatively small mutations and enzymes can repair the damage by placing the correct nucleotides into position (**FIGURE 4.26**). If the repair is not done accurately, the

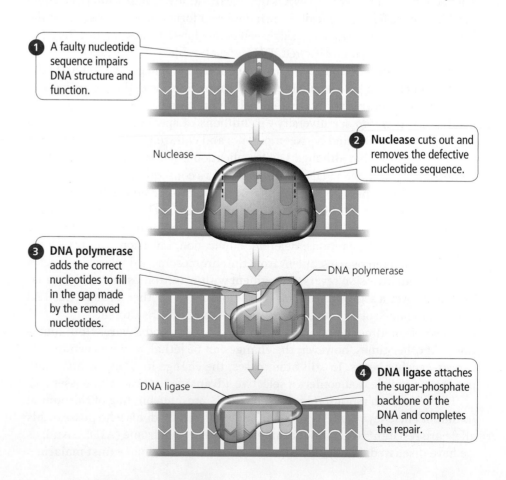

FIGURE 4.26 In some cases, enzymes can repair the damage of small mutations by placing the correct nucleotides into position.

mutation will persist. In other circumstances, the defect is too large, and repair is impossible.

Sometimes mutations originate during replication. Accurate DNA replication relies on enzymes that "proofread" the newly formed strand to make sure that the sequences are correct. Some mutations are the consequences of errors in this mechanism. If the wrong nucleotide is added during replication and the mistake goes undetected, the incorrect sequence will become a component of the DNA molecule for generations to come.

Protein Production

The last characteristic of DNA is that it can direct the production of specific proteins. Linus Pauling's sickle cell research demonstrated that the normal hemoglobin allele and the sickle cell allele encode for different proteins. As we saw in Section 4.4, individuals with two normal alleles made HbA protein, those with two sickle cell alleles produced HbS protein, and carriers made both types of hemoglobin. The connection between genes and proteins was clear. What remained unknown in 1949 was *how* the information in DNA directs the synthesis of a specific protein.

As was true for deciphering the details of DNA structure, replication, and mutation, it took a legion of scientists from many labs in many countries over a decade to work out the details. They discovered how DNA fulfills the third requirement of genetic material, along with replicating and mutating: how it is involved in the production of proteins and the development of physical characteristics. Encoding for proteins has three aspects—the *genetic code* itself, its *transcription* into RNA, and finally its *translation* into proteins (**FIGURE 4.27**). In this way, the sequence of nucleotides is ultimately translated into a specific sequence of amino acids connected together to make a protein.

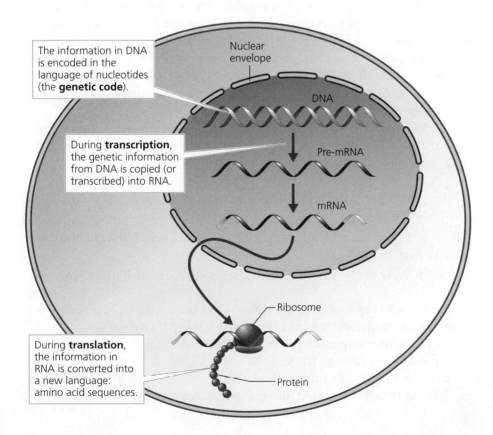

The information in DNA is encoded in the language of nucleotides (the **genetic code**).

During **transcription**, the genetic information from DNA is copied (or transcribed) into RNA.

During **translation**, the information in RNA is converted into a new language: amino acid sequences.

Nuclear envelope

DNA

Pre-mRNA

mRNA

Ribosome

Protein

FIGURE 4.27 A third characteristic of DNA is its ability to direct protein synthesis. Encoding for proteins involves three steps: the reading of the genetic code, the transcription of the genetic code into RNA, and the translation of RNA into proteins.

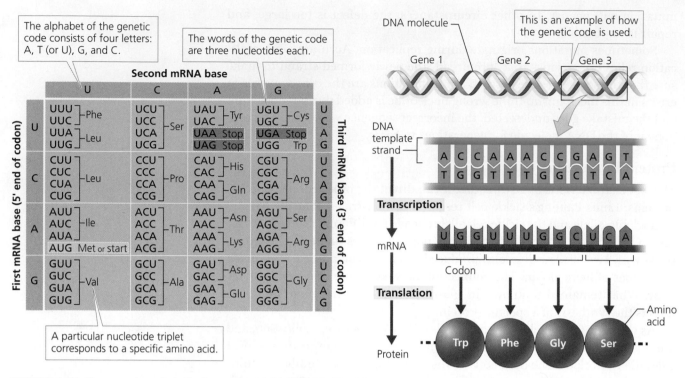

FIGURE 4.28 The genetic code is the language of nucleotides. The genetic code encodes the information in DNA.

genetic code :: the information encoded in DNA; it is the language of nucleotides

The Genetic Code The information encoded in DNA is in a language of nucleotides called the **genetic code** (**FIGURE 4.28**). The alphabet of the genetic code consists of four letters: A, T, G, and C. The "words" are three nucleotides each. A particular nucleotide triplet corresponds to a specific amino acid that will be incorporated into a protein. Only four nucleotides can be present in any sequence of three. Therefore, the maximum number of triplets possible is $4 \times 4 \times 4$, or 64. The proteins present in organisms are composed of combinations of 20 different types of amino acid. As a result, the genetic code has more than enough information potential to direct protein synthesis.

Transcription DNA is not translated directly into protein. Other intermediary molecules, RNAs, are required. RNA is structurally similar to DNA, with three exceptions:

1. The sugar in the nucleotides is ribose, not deoxyribose.
2. The nucleotide uracil (U) is used instead of T, but U and A are complementary base pairs (just as A and T are complementary base pairs in DNA).
3. RNA is single stranded rather than double stranded like DNA.

Three major types of RNA are essential for the synthesis of protein: messenger RNA, transfer RNA, and ribosomal RNA. All types of RNA are produced by **transcription**, or the process of copying or transcribing genetic information from DNA to RNA.

transcription :: the process of copying or transcribing genetic information from DNA to RNA

DNA never leaves its safe location in the nucleus. Consequently, its information has to be transported into the cytoplasm, where proteins are made.

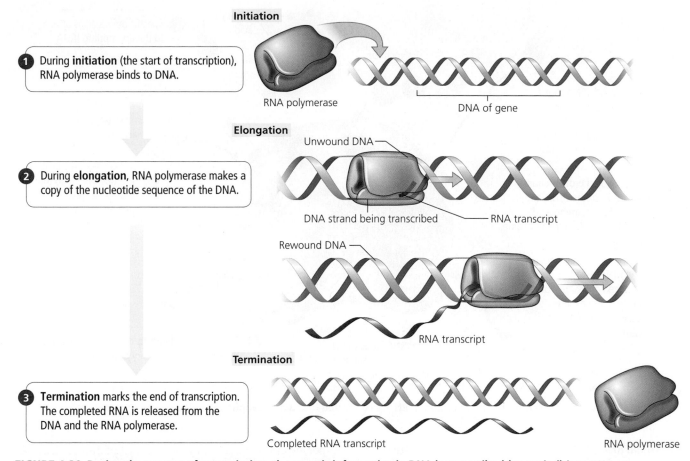

Initiation

① During **initiation** (the start of transcription), RNA polymerase binds to DNA.

RNA polymerase

DNA of gene

Elongation

② During **elongation**, RNA polymerase makes a copy of the nucleotide sequence of the DNA.

Unwound DNA

DNA strand being transcribed

RNA transcript

Rewound DNA

RNA transcript

Termination

③ **Termination** marks the end of transcription. The completed RNA is released from the DNA and the RNA polymerase.

Completed RNA transcript

RNA polymerase

FIGURE 4.29 During the process of transcription, the genetic information in DNA is transcribed (or copied) into RNA.

If you think of the DNA as the blueprints for a building, RNA is the photocopy of those instructions that you can take to the construction site.

Transcription begins in the nucleus when the double-stranded structure of a specific gene opens, so that single-stranded structures are revealed (**FIGURE 4.29**). Much as in DNA replication, the single-stranded regions of DNA bind nucleotides according to complementary base-pairing rules: A and U join together, and G joins to C. Once the entire DNA sequence has been transcribed, the RNA detaches and moves into the cytoplasm. The DNA restores its double-stranded structure.

Translation The information present in DNA is in "nucleotide language." This information is transcribed into RNA, another molecule that is in nucleotide language. To move the information from the RNA to protein entails **translation** into a new language: amino acid sequences.

Once transcription has been completed, all three types of RNA are located in the cytoplasm, and each plays its role in the synthesis of protein (**FIGURE 4.30**):

- **Messenger RNA (mRNA)** sequences are composed of nucleotide triplets. These **codons** identify precisely which amino acids should be joined together and in what order.
- **Transfer RNA (tRNA)** is a carrier molecule. At one end it has a triplet nucleotide sequence, called the **anticodon**. The other end has a binding

translation :: the process by which the information in RNA is converted to a new language: amino acid sequences

messenger RNA (mRNA) :: codons that identify precisely which amino acids should be joined together and in what order

codon :: a specific mRNA sequence of three nucleotides

transfer RNA (tRNA) :: a carrier molecule; at one end is the anticodon, and at the other is a binding site to which a specific amino acid is attached

anticodon :: a triplet nucleotide sequence at one end of tRNA

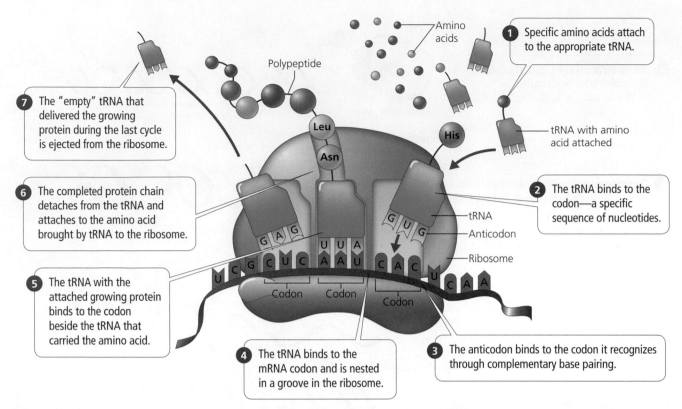

7 The "empty" tRNA that delivered the growing protein during the last cycle is ejected from the ribosome.

6 The completed protein chain detaches from the tRNA and attaches to the amino acid brought by tRNA to the ribosome.

5 The tRNA with the attached growing protein binds to the codon beside the tRNA that carried the amino acid.

1 Specific amino acids attach to the appropriate tRNA.

tRNA with amino acid attached

2 The tRNA binds to the codon—a specific sequence of nucleotides.

tRNA
Anticodon
Ribosome

4 The tRNA binds to the mRNA codon and is nested in a groove in the ribosome.

3 The anticodon binds to the codon it recognizes through complementary base pairing.

Amino acids
Polypeptide
Leu
Asn
His

FIGURE 4.30 During translation, the information in RNA is converted into a new language: amino acid sequences.

ribosomal RNA (rRNA) :: forms ribosomes, clamp-like structures that hold mRNA and tRNA in the correct orientation so that the amino acids can be connected together

site to which a specific amino acid is attached. A host of different tRNAs each carry the specific amino acid requested by the mRNA codon.

• Finally, **ribosomal RNAs** (rRNA) form ribosomes, clamp-like structures that hold mRNA and tRNA in the correct orientation so that the amino acids can be connected together.

The overall picture that emerges is incredible. DNA, the repository of genetic information, resides in the nucleus. It possesses information for the production of all proteins in the cell. Transcription is the first step for moving the information present in DNA on its journey to the information built into protein. Finally, translation is efficient and successful because ribosomes, composed of various rRNAs, hold the growing protein in the correct arrangement.

Let's return to hemoglobin to see how this process works. The gene that encodes for the synthesis of hemoglobin opens. The nucleotide sequence that encodes for which amino acids should be hooked together is transcribed into mRNA. The anticodon of tRNAs bear amino acids, and these bind one by one to the codons of mRNA. They thus deliver the appropriate amino acids in the correct order. Because of the work of the ribosomes, the entire mRNA sequence is translated to make HbA protein. In the case of the sickle cell alleles, translation produces HbS protein, not HbA.

As mentioned earlier, these two types of hemoglobin differ only by one amino acid. The alleles themselves differ by only one nucleotide in one triplet of the DNA and hence in just one mRNA codon. Yet the physical consequences are profound, because this tiny difference is responsible for meaningful changes in the structure of the hemoglobin protein. The structural differences between HbA and HbS proteins have huge physical consequences: they may mean the difference between having sickle cell disease or not.

4.9 Why Is Protein Structure So Important?

When amino acids join together to produce a protein, the final structure depends on the identities of each amino acid. Because of the differences in their size, shape, and electrical charge, a string of amino acids will fold into the shape that is the most chemically stable. If the wrong amino acid, with different chemical properties, is inserted during protein synthesis, the entire protein may fold incorrectly (**FIGURE 4.31**). As a result, the shape of the protein will be altered, dramatically changing its function. The information present in DNA and RNA is in the language of nucleotides. In contrast, shape is the language of protein structure.

Suppose you ball your hand into a fist. Too belligerent? Suppose instead that you open it fully to reveal your empty palm, close it except for sticking up your thumb, or close it except for pointing one finger. Each shape has a different meaning. The closed fist might mean anger or a threat, the open palm is friendly, the thumbs-up might mean cheerful agreement, and your index finger could be pointing to something of interest. (We won't even ask about your middle finger.) Shape is one way that your hand can convey information.

In the case of hemoglobin, the shape of the molecule allows it to fulfill its primary function: binding oxygen and delivering it to cells and tissues in the

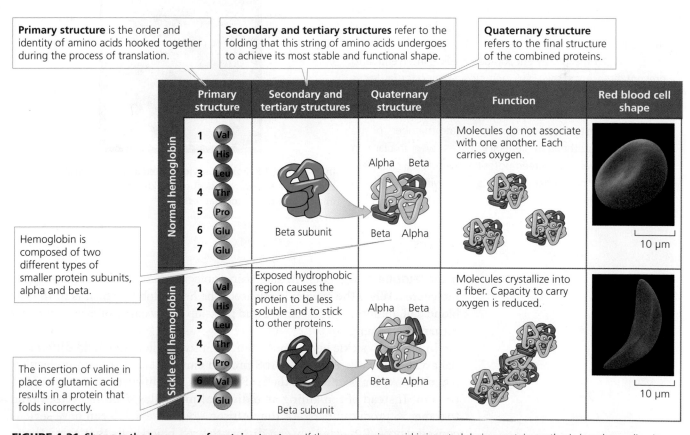

Primary structure is the order and identity of amino acids hooked together during the process of translation.

Secondary and tertiary structures refer to the folding that this string of amino acids undergoes to achieve its most stable and functional shape.

Quaternary structure refers to the final structure of the combined proteins.

Hemoglobin is composed of two different types of smaller protein subunits, alpha and beta.

The insertion of valine in place of glutamic acid results in a protein that folds incorrectly.

FIGURE 4.31 Shape is the language of protein structure. If the wrong amino acid is inserted during protein synthesis (as when valine is substituted for glutamic acid in sickle cell disease), the protein may fold incorrectly, changing the shape of the protein and, as a result, its function.

Life Application

The Effectiveness of Genetic Screening

Scientists and physicians have the ability to screen for carriers of many genetic diseases. Yet the outcomes of screening programs are quite variable. If we measure success as a reduction in the number of infants born with a particular disorder, screening for Tay Sachs disease has been spectacularly effective, whereas screening for sickle cell disease is disappointing so far. The difference in how well the screening programs have worked shows that it is important to educate patients. It is important, too, for healthcare professionals to be culturally sensitive and aware.

Tay Sachs disease afflicts children who are born with two copies of a mutant allele. The normal allele encodes for a protein called *hexosaminidase A*. This protein breaks down a type of fatty molecule called GM2, which is present in the brain and nervous system. A child with Tay Sachs disease is missing normal hexosaminidase A, and as a consequence GM2 builds up, thus destroying brain and nerve cells. Symptoms begin to appear by the age of 6 months. From then until death, usually before age 5, the nervous system deteriorates.

While not exclusive to a specific ethnic group, the Tay Sachs allele is especially prevalent in people of Ashkenazi ancestry (Eastern European Jews). One out of every 27 people from this community is a carrier of the Tay Sachs trait. Because Jewish teaching explicitly encourages medical intervention to prevent disease and save lives, genetic screening for Tay Sachs is consistent with religious beliefs. Add to this the active engagement of community and religious leaders by physicians and the strenuous efforts to educate members of the community, and the ingredients for success were in place. In fact, the campaign to screen potential parents led to a 90% reduction in the number of infants born with Tay Sachs in North America.

In contrast, the screening program for sickle cell disease reads like a script for how *not* to do something properly. Testing of African Americans began in the early 1970s—but without consent, without community outreach, and without adequate public education. When people learned that they had the sickle cell trait, they were rarely told they were *not* going to be sick. Nor did they learn the chances that their children would develop the disease. As if this were not bad enough, individuals who tested positive were discriminated against and unable to get life or health insurance. Until 1981, the U.S. Air Force Academy refused entrance to any applicant who carried the sickle cell trait. Even after these discriminatory practices were stopped, it was too late to build the trust necessary for this program to succeed. Instead, the program's legacy was an enhanced suspicion that medical genetics has racist intentions. ::

In 2004, the U.S. Postal Service issued a commemorative stamp to raise awareness about sickle cell disease and the importance of testing babies.

body (**FIGURE 4.32**). Each hemoglobin molecule can bind to four oxygen molecules. When the oxygen is released, the hemoglobin, which is in the red blood cell, maintains its structure and picks up a new cargo of oxygen when it returns to the lungs.

In contrast, sickle cell hemoglobin, with its single amino acid difference, behaves differently. Although HbS protein can bind four oxygen molecules, when the oxygen is unloaded the HbS has a slightly different shape from HbA protein. Instead of remaining as individual molecules, HbS molecules bind together to produce fibers. As Tony Allison saw, these fibers can have serious consequences. Because they can distort the red blood cells that carry them, cell sickling results. Depending on where the sickled cells form, blood flow can be blocked anywhere in the body. Tissues downstream from a blood clot will be deprived of oxygen and nutrients. If the clot is not dissolved, tissues will die. When blood flow to vital organs is impaired, the clots can be fatal. And this entire disease scenario was triggered by just one error in protein structure.

The importance of protein structure and shape is not unique to sickle disease and hemoglobin. Most cell components and regulatory molecules are proteins. Whether it is cell communication, cell division, development, cell adhesion, regulation of gene expression, or reception of hormone signals, proteins are the main players. It is not hard to imagine what could result from the presence of abnormal proteins: cancer, birth defects, autoimmune diseases, diabetes—just about anything. (See *Life Application: The Effectiveness of Genetic Screening* for a discussion of the potential cultural complexities of genetic testing.)

Biology in Perspective

Sickle cell disease reveals a great deal about how a genetic mutation can be inherited and result in a disease. It also demonstrates how even a complex syndrome can be caused by the malfunctioning of a single protein.

Sickle cell disease starts with a mutation in DNA. The RNA that is transcribed from this mutated gene is translated into hemoglobin S, a protein that differs from normal hemoglobin A by only one amino acid. Because HbS attains a final structure and set of surface electrical charges different from those of HbA, it forms fibers in red blood cells when oxygen levels are low. The HbS fibers in turn deform red blood cells into a sickle shape. Sickled cells do not behave like normal red blood cells: they get stuck in narrow blood vessels, thus producing clots. Depending on the size and location of the clot, symptoms experienced by an individual with sickle cell disease range from muscle pain to organ damage to death. All of this results from one incorrectly placed amino acid.

It is not correct, however, to think of the sickle cell allele as all bad. Individuals who have one HbS allele and one HbA allele might experience mild sickle cell symptoms under certain conditions. But they are also resistant to developing malaria, another potentially lethal disease.

The sickle cell allele illustrates a general reality about genes and inheritance. Genes do not function in isolation. The path between the information present in DNA and its arrival in protein and the ultimate journey from there to the development of physical characteristics is not so straightforward. Alleles interact with each other, as do genes. A single gene can affect numerous characteristics. The entire genetic background of an individual matters. Finally, the environment plays a role in how genes behave, too.

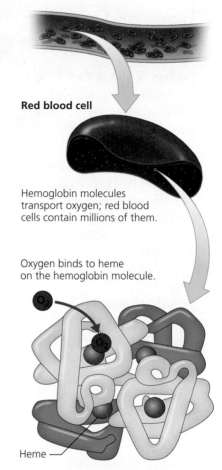

Red blood cell

Hemoglobin molecules transport oxygen; red blood cells contain millions of them.

Oxygen binds to heme on the hemoglobin molecule.

O_2

O_2

Heme

Hemoglobin molecule

FIGURE 4.32 The shape of the hemoglobin molecule allows it to fulfill its primary function: binding to oxygen and delivering it to cells and tissues in the body.

Chapter Summary

4.1 What Is Sickle Cell Disease?

Specify the cause and consequences of sickle cell disease and the relationship between sickle cell disease and malaria.

- Sickle cell disease is a severe inherited genetic disorder.
- In sickle cell disease, red blood cells that are normally disc shaped and flexible become rigid and pointy shaped.
- Sickled cells get stuck in blood vessels, which can lead to blood clots and other serious physical consequences, including kidney failure, paralysis, and heart failure.
- The inheritance of the gene associated with sickle cell disease also confers resistance against malaria.

4.2 Could Molecular Medicine Prevent Sickle Cell Disease?

Explain how Linus Pauling's research into sickle cell disease ushered in the age of molecular medicine.

- Linus Pauling and his colleagues examined two types of hemoglobin (HbA and HbS).
- Pauling demonstrated that the cause of sickle cell disease could be traced to an alteration in the molecular structure of a protein.
- Pauling concluded that since sickle cell disease is inherited, genes must determine the structure of proteins.
- It took decades for other scientists to catch up to this idea that understanding the way genes, proteins, and other molecules work could be the key to diagnosing and treating disease.

4.3 Where Is Our Genetic Information Stored?

Describe the study that disproved the concept of pangenesis.

- In the 1880s, August Weismann performed an experimental test of pangenesis.
- Weismann took a population of mice and removed their tails and then bred offspring from these mice and measured whether the tails were shorter.
- The tails of the offspring were just as long as the tails of the parents.
- Weismann repeated this experiment for seven years.
- Weismann concluded that the information to construct the tail does not reside in the tail itself.
- Later scientists discovered that this information resides in the nucleus of each cell in a molecule called DNA.

4.4 How Did Mendel Discover the Rules of Inheritance?

Outline how Mendel conducted his experiments related to inheritance and the resulting set of rules.

- Mendel chose his experimental organism carefully: garden peas, which are simple to cultivate and whose mating can be controlled manually.
- Mendel focused on one trait at a time and followed the cross for more than one generation, collecting quantitative data and keeping detailed records.
- Mendel's research revealed some basic rules of inheritance.
  - o Genes can come in more than one form, or allele.
  - o Alleles of a particular gene sort individually into gametes during meiosis.
  - o Certain traits are dominant, while others are recessive.

4.5 How Much Do Mendel's Rules Explain?

Identify the three reasons that Mendel's rules fail to explain inheritance completely.

- Alleles can interact, and so can genes.
- Most genes affect more than one characteristic.
- How genes contribute to physical characteristics can depend on the environment.

4.6 What Are Genes Made Of?

Detail the structure of DNA.

- DNA is the physical material of which genes are made.
- DNA molecules are composed of nucleotides, and there are four different types: adenine (A), thymine (T), guanine (G), and cytosine (C).
- Each nucleotide has a sugar, a phosphate group, and a nitrogen-rich ring structure called a base.
- To form the DNA molecule, the nucleotides hook together in a specific manner.
- As a consequence, DNA has a sugar-phosphate backbone with bases sticking out from one side.
- The final form is a double helix: two single strands of DNA bound to each other in a shape like a twisting staircase.

4.7 How Does DNA Function?

Describe how scientists uncovered the physical basis of genes.

- Frederick Griffith found a clue to DNA's function while studying a species of bacteria that caused pneumonia.
  - He killed a population of smooth bacteria.
  - He combined the dead smooth bacteria with living rough bacteria and infected mice with either rough bacteria alone, dead smooth bacteria alone, or a combination of the two.
  - Only the mice infected with the combined mixture got sick.
  - After isolating the bacteria from the sick mice, he observed that they were all smooth; he had discovered a physical material that could change bacteria in a way that could be inherited, but he had no idea what the "transforming substance" was.
- Scientists Oswald Avery, Colin MacLeod, and Maclyn McCarty showed that the mysterious substance was in fact DNA.
  - They killed smooth bacteria.
  - They used enzymes to eliminate carbohydrate, protein, fat, RNA, or DNA from the virulent filtrate.
  - They combined each filtrate from which they had removed *one* type of molecule with the rough bacteria.
  - Only the mixture containing DNA combined with rough bacteria resulted in the appearance of smooth bacteria; DNA was the transforming substance.

4.8 What Processes Must DNA Accomplish?

Diagram the processes of DNA replication, mutation, and protein production.

- DNA must accomplish three functions: replication, mutation, and protein production.
- During replication, the double helix of DNA unzips and new strands form.
- During mutation, a change in the DNA occurs; five examples are point mutation, deletion, duplication, inversion, and translocation.
- During transcription, the genetic information is transcribed or copied from DNA to RNA.
- During translation, the information in RNA is converted to amino acid sequences.

4.9 Why Is Protein Structure So Important?

Explain why a protein's structure is so important to its function.

- Protein structure determines a protein's function; a change in protein structure can dramatically alter a protein's function.
- Most cell components and regulatory molecules are proteins.
- Abnormal proteins can result in cancer, birth defects, autoimmune diseases, diabetes—just about anything.

Key Terms

allele 104
amino acid 99
anemia 95
anticodon 117
codon 117
cross 101
cystic fibrosis 106
DNA (deoxyribonucleic acid) 108
dominant 105
electrophoresis 98
factor 104
gene 92

genetic code 116
HbA 96
HbS 96
hemoglobin 95
malaria 92
messenger RNA (mRNA) 117
molecular medicine 99
monohybrid 104
mutation 113
nucleotide 108
pangenesis 100
pedigree analysis 107

protozoan 92
recessive 105
red blood cell 92
replication 112
ribosomal RNA (rRNA) 118
sickle cell 93
sickle cell disease 93
trait 101
transcription 116
transfer RNA (tRNA) 117
transformation 109
translation 117

Review Questions

1. What evidence did Dr. Allison use to understand the relationship between malaria and sickle cell disease?

2. What happens to the blood cells in an individual with sickle cell disease? What can trigger a sickling incident?

3. If an individual inherits an HbS protein and an HbA protein, which of the following is likely to be true?

   a. The individual will likely develop full-blown sickle cell disease.

   b. The individual will likely have no resistance to malaria, if exposed.

   c. The individual will likely have some resistance to malaria, if exposed, and will also be unlikely to develop full-blown sickle cell disease.

   d. The individual will likely have no resistance to malaria, if exposed, and will also be unlikely to develop full-blown sickle cell disease.

   e. The individual will likely develop malaria, if exposed, and will also be likely to develop full-blown sickle cell disease.

4. What did Linus Pauling and his colleagues learn about sickle cell disease?

   a. An individual with two normal hemoglobin genes would be unlikely to exhibit sickle cell disease.

   b. An individual with two normal hemoglobin genes would be likely to exhibit sickle cell disease under certain conditions.

   c. An individual who suffers from sickle cell disease has inherited one normal and one sickle cell gene.

   d. Sickle cell disease is not an inherited disease.

   e. There is no such thing as a carrier when it comes to sickle cell disease.

5. What is pangenesis?

   a. The idea that genetic information is stored in DNA.

   b. The idea that certain traits are dominant and others are recessive.

   c. The idea that genes can come in more than one allele.

   d. The idea that each part of the body produces a characteristic seed that is released by males and females during mating, resulting in offspring with blended traits.

   e. The idea that alleles sort individually into gametes during meiosis.

6. What were the principal discoveries that Mendel made from his monohybrid crosses?

   a. Genes can come in more than one form; some traits are dominant and others are recessive; pangenesis is a proven theory.

   b. Genes can come in more than one form; the alleles of a particular gene sort individually into gametes during meiosis; some traits are dominant and others are recessive.

   c. Genes can come in more than one form; pangenesis is a proven theory; the alleles of a particular gene sort individually into gametes during meiosis.

   d. There is no such thing as dominant and recessive traits; genes can come in more than one form; the alleles of a particular gene sort individually into gametes during meiosis.

   e. Pea flower color is a trait that exhibits a dominant/recessive relationship between the purple and white alleles; gene expression depends on the environment; genes always affect more than one trait.

7. What were the reasons that Mendel was successful whereas others who wanted to understand the rules of inheritance were not?

8. What is an allelic interaction?

9. What is a gene interaction?

10. What is an example of the environment influencing the expression of genes?

11. Describe or sketch the general features of the structure of DNA.

12. For the DNA sequence GCATTTGCCAACTGA, please answer the following:

    a. What is the complementary DNA sequence?

    b. What is the mRNA sequence that would be transcribed from it?

    c. What are the specific anticodons that would bind to the codons of the mRNA?

13. Describe the evidence that demonstrates that genes are made of DNA.

14. Describe or diagram the pathway of information flow starting with DNA and ending with a protein.

15. Explain the relationship between protein structure and function.

16. Suppose a man who carries the sickle cell trait marries a woman who also carries the trait. What is the likelihood that their children will have sickle cell disease? What are the chances of the children being carriers?

The Thinking Citizen

1. How could strenuous physical activity and/or high altitude potentially lead to organ failure in a person with sickle cell disease?

2. Predict the outcomes of matings between the following if the theory of pangenesis were correct:
   - A black dog and a white dog
   - A man missing his left ear with a woman with an extra toe on each foot
   - A gerbil with long hair and a gerbil with short hair
   - A pea plant with red flowers and a pea plant with white flowers
   - An Olympic swimmer and a completely out-of-shape couch potato

3. What observations can you make that refute the theory of pangenesis?

4. Why does the incidence of the sickle cell trait in people in a geographic area vary in relationship to the incidence of malaria? What do you predict would occur to the incidence of the sickle cell trait if malaria were to be eliminated entirely? Why?

5. Imagine you checked the news, online or print, and you read that the "gene for alcoholism" had been isolated. Is this claim likely to be correct? Why or why not?

6. Assume that a specific gene is associated with alcoholism. Would a person who has two copies of this gene definitely be an alcoholic? Why or why not?

7. Suppose Mendel wanted to demonstrate that the offspring of a white-flowered and a purple-flowered parent actually has a "hidden" white allele even though it looks purple. What cross could demonstrate that the white allele is indeed present?

8. Even after Avery and his colleagues showed DNA's role in heredity, many scientists were skeptical. DNA is so simple, skeptics said. Its double helix has just four possible base pairs. Surely protein would be a more likely component of our genes. What would you say in reply?

The Informed Citizen

1. Tony Allison supported his theory about the relationship between malaria and the selective advantage of the sickle cell trait by doing three studies that involved human subjects. Is it ethical to do research on humans? If so, what safeguards are necessary? If not, why not? How would you learn about human structure and function?

2. In one of the studies that Allison joined, a pharmaceutical company was testing its drug to see whether it was an effective treatment for malaria. Is it ethical to infect people with a disease to see whether you can cure it? If not, how can you test whether a drug does indeed treat a disease successfully and safely? Is it even ethical to test drugs on people? If so, what safeguards are needed? If not, how should drugs be tested safely in humans?

3. The rights and responsibilities of individuals and the public sometimes seem to come in conflict. Consider genes that are unequivocally associated with sickle cell disease, cystic fibrosis, Huntington's chorea, or muscular dystrophy. Should all individuals be screened? If so, what should be done with this information? Should it be off-limits to insurance companies and employers, or do they have a right to know to minimize their own financial risks? Is it acceptable to refuse screening even though an individual might add to the healthcare burden of society? Is it acceptable to force people to be screened against their will, or would that violate a right to privacy?

4. Suppose it is possible to remove or permanently disable the gene associated with a disease. Should it be done? What are the potential risks and benefits? Does the patient get to decide whether to have the treatment, or is it a legal obligation for the good of society?

CHAPTER LEARNING OBJECTIVES

After reading this chapter, you should be able to answer the following questions:

5.1 How Does Cancer Make You Sick?
Use appropriate terminology to describe cancer and the reasons why cancer makes you ill.

5.2 How Do Cancer Cells Differ from Normal Cells?
Contrast the behavior of normal and cancerous cells.

5.3 What Is the Life Cycle of a Cell?
Use a signal transduction pathway to explain how cells become cancerous.

5.4 In What Ways Is Cancer a Genetic Disorder?
Describe the role that mutations play in cancer development.

5.5 What Risk Factors Are Associated with Cancer?
Identify actions you can take to reduce your risk of cancer.

5.6 How Is Cancer Diagnosed?
Explain how cancers are diagnosed.

5.7 How Is Cancer Treated?
Contrast the three methods used to treat cancer.

5.8 How Can Cancer Be Prevented?
Enumerate steps to decrease cancer risk.

Every day approximately 1500 people in the United States are diagnosed with cancer. Some cancers and causes are well understood, while others are less so.

CHAPTER 5

Cancer

How Can It Be Prevented, Diagnosed, and Treated?

Becoming an Informed Citizen ...

Cancer is a disease that is likely to touch each of us in some way at some point in our lives. All cancers are the consequence of alterations or mutations of genes responsible for regulating cell division. But what is cancer exactly? What causes it? Can it be prevented? Are there effective treatments? These are just some of the questions we investigate.

dward R. Murrow was one of journalism's most important figures. He was a pioneer in television journalism. He was well respected for his integrity and commitment to the truth. His face became well known as television developed into an integral source of information for millions.

Murrow's ever-present cigarette was just as familiar. Reminiscing about his 1965 death, the broadcast journalist Alexander Kendrick remarked, "On CBS Radio, the news of Ed Murrow's death, reportedly from lung cancer, was followed by a cigarette commercial."

You might not be surprised about the placement of that commercial. Perhaps you think that knowledge about the dangers of cigarette smoking is relatively recent. Only it isn't. The connection between tobacco and cancer was suspected centuries ago. Moreover, clear evidence of the relationship was published in 1950. Why, then, did a person as intelligent and informed as Edward R. Murrow continue to engage in a behavior that increased his chances of developing cancer, a disease that was often a death sentence in his day?

A disease of uncontrolled cell proliferation, **cancer** impacts the lives of most Americans because they themselves have the disease or a friend or family member does. Every day approximately 1500 people in the United States are diagnosed with cancer. Close to 40% of all Americans will be diagnosed with cancer, and half of these people will die from the disease. Hundreds of millions of dollars are spent on cancer research and treatments each year.

In this chapter, we'll discuss what cancer is and what causes it. We'll also outline the tools of prevention and treatment. Far from being a modern phenomenon, descriptions of cancer date back to the ancient Egyptian, Greek, and Roman empires. Some cancer causes are well understood, while others are less so. And some cancers may be caused or exacerbated by certain behaviors, while others may result from the genes we inherit. Let's take a look at one such genetic disorder that leads to a high incidence of skin cancer.

cancer :: a disease of uncontrolled cell proliferation

case study

Xeroderma pigmentosum

cyst :: a fluid-filled sac

melanoma :: an invasive and deadly skin cancer

***Xeroderma pigmentosum* (XP)** :: a rare disorder in which individuals are extremely sensitive to ultraviolet (UV) light and more easily develop skin cancers if they are exposed to sunlight or other sources of UV light

Thirteen-year-old Jeff Markway finally got around to showing his mother the golf ball–sized lump on his right arm. He demonstrated for her how the lump would move up and down when he wiggled his right thumb. Not surprisingly, Jeff's mother was alarmed by what she saw and rushed her son to the doctor. At first their doctor was not concerned because he thought the lump was simply a **cyst**, a fluid-filled hollow, and not a serious problem at all. When the doctor attempted to drain the cyst, however, instead of finding fluid in the lump, the doctor observed that it was filled with cells. The lump was not a cyst at all; it was cancer. In fact, the tumor formed because Jeff had **melanoma**, a particularly invasive and deadly skin cancer (**FIGURE 5.1**). How could a 13-year-old develop melanoma—or any cancer, for that matter?

Jeff and his family were referred to Dr. Henry Lynch, a physician and scientist whose specialty was the study of genetic factors important in cancer. Jeff was diagnosed with ***Xeroderma pigmentosum* (XP)**, a rare disorder in which people are extremely sensitive to ultraviolet (UV) light and develop skin cancers more easily if they are exposed to sunlight or other sources of UV

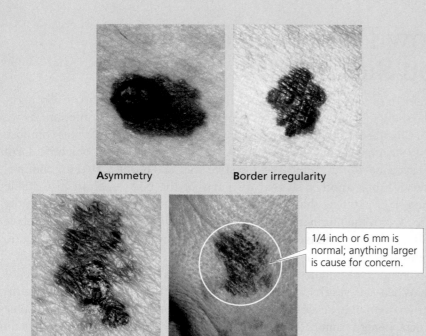

Asymmetry   Border irregularity

Color   Diameter

1/4 inch or 6 mm is normal; anything larger is cause for concern.

FIGURE 5.1 Melanoma is a particularly invasive and deadly skin cancer. The ABCD characteristics of a mole that should prompt a medical examination include (A) asymmetry, (B) border irregularity, (C) changes in color, and (D) diameter (large size).

light. Lynch examined Jeff's entire family and found that his sisters Patrice and Kathleen and his twin brothers Phil and Gregg all had XP. In fact, of the seven siblings of the Markway family, five suffered from the disease. XP affects only 1 out of every 250,000 people, and yet most of the Markway children had it. Clearly there is a genetic component to this disorder.

Individuals with XP who are under the age of 20 years have a 2000-fold higher incidence of skin cancer than those without XP. Also, the median age of skin cancer for individuals with XP is 8 years, while the median for individuals without XP is 58 years. Evidently, XP increases the sensitivity of skin to sunlight and speeds up the process whereby UV light exposure causes skin cancer.

Dr. James Cleaver learned of the Markway family's extreme sensitivity to UV light from an article he read in the newspaper. Interested in the effects of UV radiation on DNA, Cleaver eagerly contacted Dr. Lynch to see whether it would be possible to study their cells. They willingly agreed.

Cleaver's study unlocked the secret of exactly what went wrong in XP. Normal cells exposed to UV radiation experience DNA damage, but the cells are able to repair it. In XP cells, the UV damage to DNA is permanent (**FIGURE 5.2**). The cells are incapable of doing DNA repair. Consequently, the UV-exposed XP cells accumulate DNA damage, some of which renders the cells more likely to become cancerous. Affected individuals must avoid sunlight and UV light completely. Many go outdoors only at night.

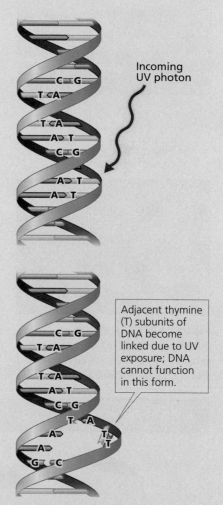

Incoming UV photon

Adjacent thymine (T) subunits of DNA become linked due to UV exposure; DNA cannot function in this form.

FIGURE 5.2 Cells exposed to ultraviolet (UV) light experience DNA damage. Normal cells repair the damage, but in *Xeroderma pigmentosum* (XP) cells, the UV damage is permanent, and the cells are more likely to become cancerous.

5.1 How Does Cancer Make You Sick?

tumor :: a lump that results from the production of extra cells

benign :: harmless

Some growths, such as warts, are benign.

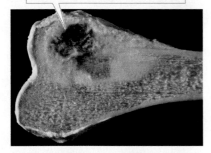

Some growths, such as the bone tumor osteosarcoma, continue to increase in size and may even become malignant, spreading to other places in the body.

FIGURE 5.4 Sometimes the balance between cell birth and death is not properly regulated, which results in the production of extra cells.

In order to function smoothly, the cells of the body live as a carefully regulated society in which each type of cell has its own place and function. As we saw in Chapter 1, cells assemble into tissues such as tendons. Tissues compose organs, such as the heart, and organs with a shared overall objective form organ systems—for example, the digestive system (**FIGURE 5.3**). Cells communicate with one another so that all parts of the body operate in a stable manner. For instance, the amount of cell proliferation is generally carefully balanced with the number of cells that die. In this way, the body can replace many worn-out or damaged cells, thus avoiding problems with tissue, organ, or organ system function.

Sometimes the balance between cell birth and death is not properly regulated. When this results in the production of extra cells, a lump, or **tumor**, might form. In many cases, these growths are **benign**, or harmless, as is true for warts. Benign tumors do not spread from their sites of origin and often do not continue growing. In other situations, the growths are **malignant**, meaning cell division continues in an uncontrolled manner, resulting in cancer (**FIGURE 5.4**).

There are three reasons that the production of extra cells creates such serious health consequences for the body, including death:

1. *Tumors can grow quite large.* In doing so, they impair the function of the organ from which they originate. They may even press down on blood vessels, blocking circulation and reducing or eliminating blood flow to organs. Organs starved of oxygen and nutrients because of impaired circulation can die.

2. *Cancer can become invasive.* It can break out of its place of origin, damaging more than one organ in the process. Individual cancer cells may leave a tumor, enter the bloodstream, and produce secondary tumors at new sites, a process called **metastasis** (**FIGURE 5.5**). This is what people mean when they say that a cancer has spread.

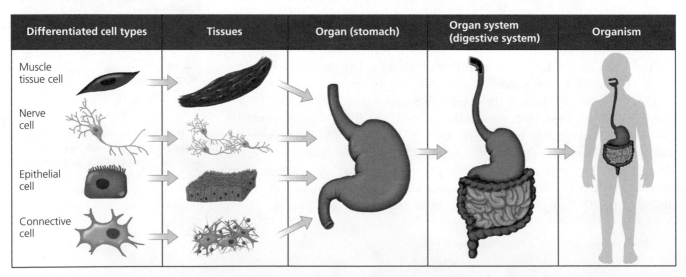

Differentiated cell types	Tissues	Organ (stomach)	Organ system (digestive system)	Organism
Muscle tissue cell				
Nerve cell				
Epithelial cell				
Connective cell				

FIGURE 5.3 In order to function smoothly, the body is organized in a structural hierarchy. Cells assemble into tissues, tissues make organs, organs form organ systems, and organ systems comprise an organism.

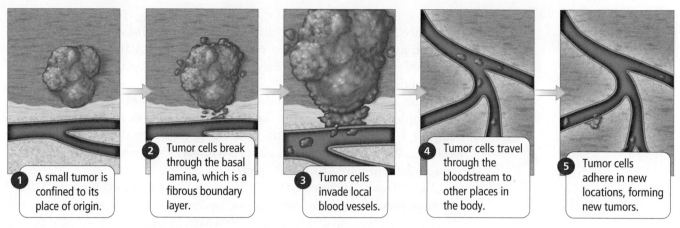

1. A small tumor is confined to its place of origin.

2. Tumor cells break through the basal lamina, which is a fibrous boundary layer.

3. Tumor cells invade local blood vessels.

4. Tumor cells travel through the bloodstream to other places in the body.

5. Tumor cells adhere in new locations, forming new tumors.

FIGURE 5.5 In a process known as metastasis, individual cancer cells break away from a tumor and enter the bloodstream, producing secondary tumors at new sites.

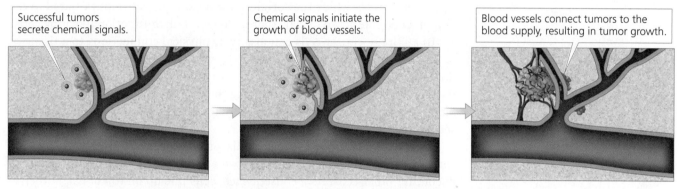

Successful tumors secrete chemical signals.

Chemical signals initiate the growth of blood vessels.

Blood vessels connect tumors to the blood supply, resulting in tumor growth.

FIGURE 5.6 In angiogenesis, successful tumors secrete chemical signals that cause blood vessels to sprout from the circulatory system. These blood vessels connect tumors to the blood supply and the oxygen and food molecules it carries.

3. *Tumors can deprive other cells of oxygen.* In **angiogenesis** (**FIGURE 5.6**), successful tumors secrete chemical signals that cause blood vessels to sprout from the circulatory system—the organ system responsible for transporting nutrients and other materials throughout the body—grow toward the tumor, and connect the tumor to the blood supply of the body. This event results in the normal, healthy cells of the body receiving less oxygen and food molecules. Eventually, tissues and organs become physically stressed, seriously impairing normal function.

Cancer is a complex disease: it is really a set of diseases all unified by the characteristic of uncontrolled cell proliferation. In all, there are more than 100 types of cancer. At first glance, this number might provoke a sense of hopelessness, but the different types of cancer result from similar causes. All cancers involve some breakdown in the normal pathways of cell cycle regulation.

malignant :: cell growth that continues in an uncontrolled manner, resulting in cancer

metastasis :: a process by which individual cancer cells leave a tumor, enter the bloodstream, and produce secondary tumors at new sites

angiogenesis :: a process by which successful tumors secrete chemical signals that cause blood vessels to sprout from the circulatory system

5.2 How Do Cancer Cells Differ from Normal Cells?

In order to understand what goes wrong with cancer cells so that cell proliferation is no longer controlled, it makes sense to compare the behaviors and characteristics of normal cells to those of cancer cells. Cancer turns out to involve a breakdown in cell communication.

Normal cells communicate with one another using chemical signals that influence many cell behaviors: division, migration, differentiation or specialization, and even programmed cell death. Also, normal cells have specific morphologies, or shapes, and they attach to other cells in an orderly manner to form tissues and organs.

When grown outside the body in a culture dish, normal cells will reach a certain density and then stop dividing. When a cell encounters another in culture, both cease moving and dividing so that they don't pile up on one another. Also, normal cells in culture will divide only a certain number of times; the cells have a finite life span.

In contrast, cancer cells do not behave in an organized manner. Often, because they are damaged, they ignore the chemical signals produced by cells that tell them to stop dividing, to differentiate, or to die. In fact, some cancer cells act as if they are receiving signals telling them to divide, even when signal molecules are absent. Cancer cells also lose the characteristics unique to the tissue from which they originate. For example, a tumor initiated in the lung may be composed of cells that no longer look or function like lung cells.

In addition, cancer cells can lose their physical connections to a tissue or organ. They may then detach and secrete enzymes called proteases that break down other cells. These enzymes can burn a hole in tissue, exposing another tissue or organ to cancer, or permitting the cancer cells to escape and join the bloodstream. The cancer cells will circulate throughout the body, forming secondary tumors in new locations.

In culture conditions, cancer cells proliferate even after they reach a certain density. They continue to move around and divide even when they encounter other cells. As a consequence, cells pile up on top of each other in culture. Finally, for many cancer cells, there is no limit to the number of times they can divide. The cells are immortal.

5.3 What Is the Life Cycle of a Cell?

Even though there are more than 100 types of cancer, there are only a few kinds of defects in the cell communication mechanisms that control cell division. Signal molecules, receptors that bind these signals, and even genes can be abnormal. To understand exactly what goes wrong, consider first how cells *normally* regulate cell division.

The Molecules That Regulate Cell Division

growth factor :: a signal molecule that enables normal cells to divide

growth inhibitor :: a signal molecule that prevents cell division

First, many normal cells require signal molecules called **growth factors** in order to grow. They also respond to another type of signal molecule, a **growth inhibitor**, which prevents cell division. Cancer cells manage to grow in the absence of growth factors, they make a steady supply of growth factors themselves, or they may be insensitive to growth inhibitors.

signal transducer :: a molecule inside the cell that relays the information once a signal molecule binds to a receptor

Second, the growth factor, or signal that triggers cell division in a normal cell, does so by binding to a receptor present in the cell membrane. This binding of the signal molecule to the receptor activates the **signal transducer**, a molecule inside the cell that relays the information acquired when the signal molecule is bound to the receptor. This information passes from molecule to molecule, like a bucket brigade, until ultimately molecules called transcription factors are activated (**FIGURE 5.7**). This last step is essential because

transcription factor :: a molecule that regulates which genes are active in a cell

transcription factors regulate which genes are active in a cell. The proteins

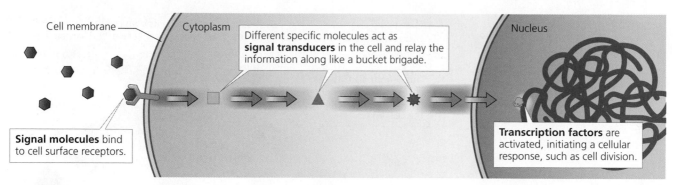

FIGURE 5.7 Cells communicate to regulate cell division. Different specific molecules act as signal transducers in the cell and relay the information along like a bucket brigade.

produced due to the activity of specific genes are directly responsible for triggering cell division.

In cancer cells, the receptors themselves can be permanently activated even when no signals are present. Similarly, the signal transducers can be stuck in an activated state even when there is no receptor at all. In addition, transcription factors may be produced and genes regulated even in the absence of all of the preceding steps of the communication pathway. Cancer cells have the capacity to bypass the regulatory steps of this control pathway. Different types of cancers do so in different ways.

Finally, normal cells eventually become damaged from age, wear and tear, or accidents. Sometimes the defects can be repaired. For example, cells have enzymes that can repair many types of DNA damage. However, when the defects are too severe to repair, especially when the DNA is seriously damaged, normal cells will undergo programmed cell death, or **apoptosis**. Cancer cells may lack the ability to repair DNA, as seen in *Xeroderma pigmentosum*, or they may ignore the signal to die no matter how abnormal their DNA might be (**FIGURE 5.8**). As a consequence, cancer cells continue to divide, accumulate more and more errors, and grow increasingly abnormal.

apoptosis :: cell death

The Cell Cycle

Cell division plays an important role in the lives of all organisms. Considering humans as an example, new cells are made regularly to replace worn-out blood cells, skin, and the lining of the intestines, and in males, to produce sperm. As we discussed in Chapter 3, cell division is also an essential component of embryogenesis, the growth of babies into adults. It is essential, too, to wound healing throughout one's life.

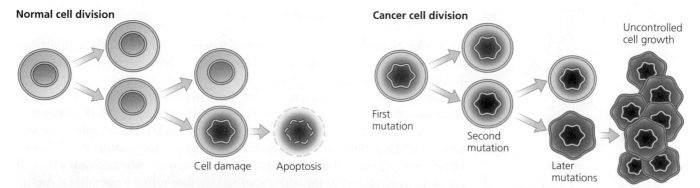

FIGURE 5.8 Normal cells eventually become damaged from age, wear and tear, or accidents. When the defects are too severe to repair, normal cells will undergo apoptosis, or programmed cell death, while cancer cells will continue to divide and grow increasingly abnormal.

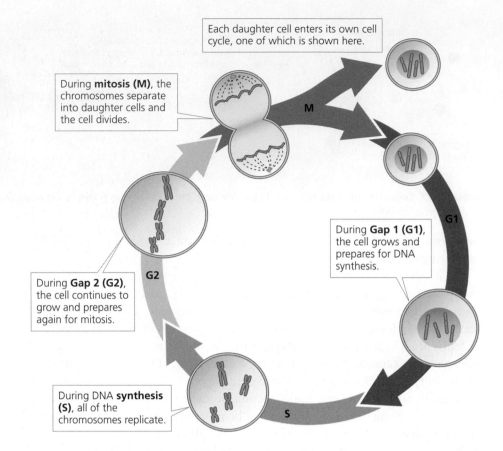

Each daughter cell enters its own cell cycle, one of which is shown here.

During **mitosis (M)**, the chromosomes separate into daughter cells and the cell divides.

During **Gap 1 (G1)**, the cell grows and prepares for DNA synthesis.

During **Gap 2 (G2)**, the cell continues to grow and prepares again for mitosis.

During DNA **synthesis (S)**, all of the chromosomes replicate.

FIGURE 5.9 **The cell cycle consists of four phases that prepare for and carry out cell division: mitosis (M), Gap 1 (G1), DNA synthesis (S), and Gap 2 (G2).**

cell cycle :: four distinct phases of cell life: mitosis (the chromosomes separate into daughter cells), Gap 1 (the cell grows), DNA synthesis (the chromosomes replicate), and Gap 2 (the cell grows and prepares again for mitosis)

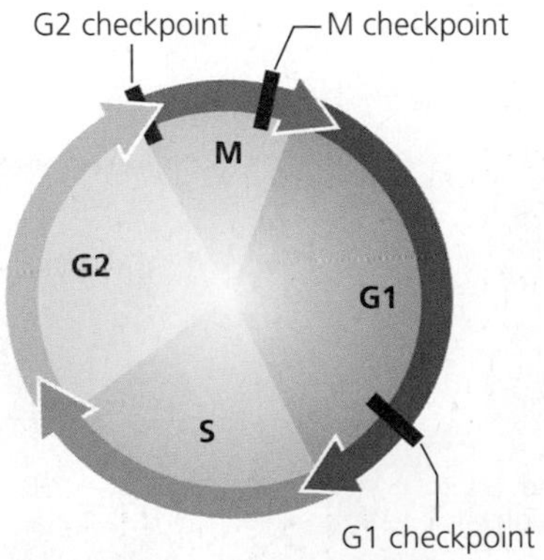

FIGURE 5.10 **Under normal conditions, a series of checkpoints carefully regulate the timing and degree of cell proliferation.**

Clearly, cell division is a critical aspect of the entire life cycle. Learning how cells regulate proliferation is key to understanding what can go wrong when cells transform into cancer.

The **cell cycle** begins when the cell originates from the division of another cell and ends when it divides (**FIGURE 5.9**). The cycle comprises four distinct phases:

1. During mitosis (M), the chromosomes separate into the daughter cells and the cell divides (Chapter 3).

2. In Gap 1 (G1), the cell grows and prepares for the DNA synthesis.

3. During DNA synthesis (S), all of the chromosomes replicate (Chapter 4).

4. In Gap 2 (G2), the cell continues to grow and prepare again for mitosis.

G1, S, and G2 constitute interphase.

Under normal conditions, a series of "checkpoints" carefully regulate the timing and amount of cell proliferation within the cell cycle (**FIGURE 5.10**). Each of these checkpoints provides the cell with an opportunity for quality control. For example, there is a checkpoint late in G1 during which the cell assesses whether it has adequate raw materials to complete S. If it does not, cell division will be prevented. If the cell were to begin DNA replication and run out of the molecules needed to complete the job, the resulting cells would be severely defective. Similarly, another checkpoint in mitosis ensures that all chromosomes attach to the mitotic spindle before mitosis completes. Again, loss of a chromosome, such that daughter cells did not get the full set, would dramatically disable a cell.

Many cancer cells ignore the checkpoint controls. They go marching through the cell cycle even if conditions inside the cells are far from normal.

5.4 In What Ways Is Cancer a Genetic Disorder?

Compared to normal cells, all cancer cells have altered DNA, and as a result of cell division they pass the altered DNA on to their offspring. In this sense, cancer is certainly a genetic disease. The alteration or mutation of the DNA means that genes don't function normally. In some cases, the sequence of nucleotides in the DNA is changed so that a particular protein might be over-produced, made with an abnormal structure, or not produced at all. If the protein was one important in the communication pathway that regulates cell division, the mutation may be the first step toward transforming the cell into a cancerous one. Similarly, if the mutated gene is altered in some way in which its expression can no longer be properly controlled, and if the gene is part of the regulatory network controlling the cell cycle, the mutant cell may be on its way to becoming cancerous.

If the DNA damage is too severe, or if the genes that encode for the synthesis of DNA repair enzymes are themselves impaired, DNA repair won't occur. The mutated DNA will be passed on to daughter cells upon cell division. Eventually the inability to repair DNA damage will result in the accumulation of mutations and an increased likelihood of cancer. Mutations of certain types of genes make a cell vulnerable to transformation to the cancerous state.

Oncogenes and Tumor Suppressor Genes

There is a complex network of controls that regulate whether or not a cell divides. At each step along the pathways comprising the network, there are proteins that say "yes" or "no" to cell division. These proteins influence whether or not a cell will divide. One of the most amazing set of discoveries in the late 20th and early 21st centuries is that cells have genes that encode proteins that behave like a car accelerator and urge cells to go ahead and divide. These **proto-oncogenes** normally control cell division quite precisely. There are also **tumor suppressor** genes that act in a manner analogous to car brakes, telling the cell, "no, don't divide."

Proto-oncogenes can be altered to form **oncogenes** that are like an accelerator stuck "on" (**FIGURE 5.11**). Oncogenes tell the cell to divide. Similarly, tumor suppressor genes can be altered so they no longer function, disabling the "brakes" on cell division; this situation also encourages cell division. Cancer cells result from an accumulation of several mutations. In all cases, at least one tumor suppressor must be disabled and several oncogenes activated. For a discussion of experiments showing that genes may be inserted into normal cells to make them cancerous, see *How Do We Know? Cancer-Causing Genes from Malfunctioning Normal Genes*, p. 137.

The proto-oncogenes and tumor suppressor genes balance each other under normal circumstances but fail to do so in cancer cells. For example, in certain types of **leukemia**, a type of blood cell cancer, the tumor suppressor gene *NF-1* is mutated. The function of the normal *NF-1* gene is to inhibit

proto-oncogene :: a gene that encodes for a protein that regulates cell division

tumor suppressor :: a gene that encodes for a protein that tells a cell not to divide

oncogene :: a gene that tells the cell to divide in the absence of normal instructions to do so

leukemia :: a type of blood cell cancer in which the tumor suppressor gene is mutated

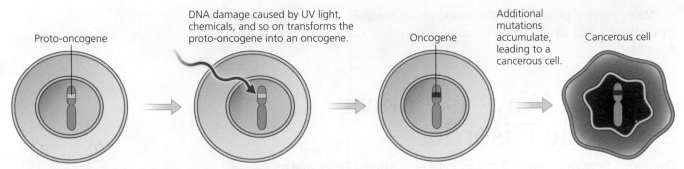

Proto-oncogene

DNA damage caused by UV light, chemicals, and so on transforms the proto-oncogene into an oncogene.

Oncogene

Additional mutations accumulate, leading to a cancerous cell.

Cancerous cell

FIGURE 5.11 Proto-oncogenes are genes that encode proteins to urge cells to divide. DNA damage can transform a proto-oncogene into an oncogene—one that tells cells to divide in the absence of normal instructions to do so, leading to cancerous cells.

a protein encoded by a proto-oncogene call *ras*. The normal *ras* gene activates cell division. When *NF-1* is mutated, it is unable to stop *ras* function. As a consequence, uncontrolled cell proliferation occurs.

There have been at least 100 oncogenes discovered so far, as well as a few dozen tumor suppressor genes. Certain types are found commonly in many cancers. Assessing the specific types of oncogenes present in tumors and determining which particular tumor suppressors are faulty provide valuable information for diagnosis, predicting disease outcomes, and in some cases treatment. For example, in approximately 30% of breast cancers, the onco-gene *erbB-2* overproduces a growth factor receptor, the erbB-2 protein, on breast cell membranes. When too much erbB-2 protein is present, cell proliferation is out of control. The drug **Herceptin** binds to erbB-2 protein and stops it from triggering cell division. Herceptin is quite effective in regaining control of the division of breast cells. As we learn more about oncogenes and tumor suppressors, it is likely that we will develop new drugs and treatments in our arsenal to fight cancer.

Herceptin :: a drug that is used to treat a certain type of breast cancer by binding to the erbB-2 protein and stopping it from triggering cell division

Chromosomal Abnormalities

In addition to the relatively small genetic changes that occur when DNA mutates (Chapter 4), large-scale alterations in the inherited instructions can occur when chromosomes, or parts of chromosomes, break, rearrange their structures, stick to other chromosomes, or get lost (**FIGURE 5.12**). Just as was

FIGURE 5.12 Chromosome abnormalities may also lead to cancer. Pieces of chromosomes may be lost, inappropriately duplicated, turned upside down, or moved to another chromosome entirely.

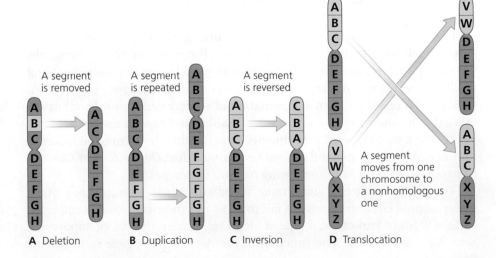

A segment is removed

A segment is repeated

A segment is reversed

A segment moves from one chromosome to a nonhomologous one

A Deletion   **B** Duplication   **C** Inversion   **D** Translocation

How Do We Know?

Cancer-Causing Genes from Malfunctioning Normal Genes

By the late 1970s, scientists had learned that certain viruses are tumorigenic, or capable of causing cancer. They also discovered that these viruses carried specific genes responsible for transforming the cells into a cancerous state and that these cancer genes could be found in normal cells too. Scientists wondered whether cancer was due to stray cellular genes that were introduced into cells when they were infected by tumorigenic viruses *or* if the cancers could be attributed to improperly functioning but previously normal genes.

Dr. Chiaho Shih and Dr. Robert Weinberg did an experiment to address this question. They took normal cells and cultured them in the laboratory. They treated these cells with a carcinogen, or cancer-causing substance, isolated from coal tar. As a consequence of the treatment, the cells transformed and became cancerous. Next Shih and Weinberg isolated the DNA from the cancer cells and inserted it into a second set of normal cells. They reasoned that if the cells became cancerous,

it would provide evidence that our cells contain genes that can become abnormal and cause cancer; no virus is needed. Indeed, that is exactly what happened. Some of the recipient cells became cancerous because of the added DNA.

Dr. Weinberg next wondered whether this result was an unusual event that would happen only if he used cells he had transformed in culture as the source of cancer genes. Would the experiment work if the source of potential cancer genes was an actual tumor isolated from a patient? To answer this question, Weinberg, along with his colleague Dr. Geoffrey Cooper, isolated DNA from a human bladder cancer. They inserted this DNA into normal cells in culture. As was the case in the Shih and Weinberg experiment, the added cancer DNA, this time from a human body, was able to transform cells in culture to a cancerous state. These experiments showed dramatically that cellular DNA can be altered in such a way that cancer can result. ::

Shih and Weinberg experiment

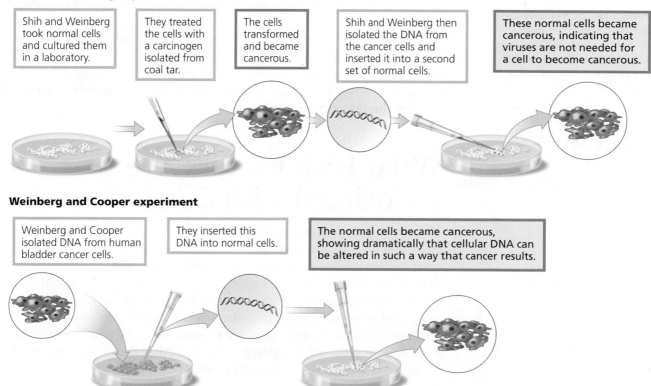

| Shih and Weinberg took normal cells and cultured them in a laboratory. | They treated the cells with a carcinogen isolated from coal tar. | The cells transformed and became cancerous. | Shih and Weinberg then isolated the DNA from the cancer cells and inserted it into a second set of normal cells. | These normal cells became cancerous, indicating that viruses are not needed for a cell to become cancerous. |

Weinberg and Cooper experiment

| Weinberg and Cooper isolated DNA from human bladder cancer cells. | They inserted this DNA into normal cells. | The normal cells became cancerous, showing dramatically that cellular DNA can be altered in such a way that cancer results. |

Dr. Chiaho Shih and Dr. Robert Weinberg conducted an experiment to address whether cancer was due to stray cellular genes introduced by a tumorigenic virus *or* whether it was due to previously normal genes functioning improperly. Then, Weinberg and Dr. Geoffrey Cooper isolated DNA from a human bladder to see whether the experiment would work if the tumor were isolated from a patient.

Before translocation

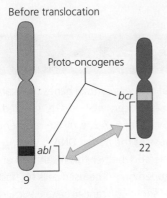

After translocation

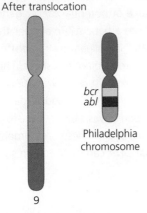

FIGURE 5.13 The Philadelphia chromosome, which is associated with chronic myelogenous leukemia (CML), forms from a translocation event between two normal chromosomes. The chromosome rearrangement repositions two proto-oncogenes, *abl* and *bcr*, next to each other in such a way that both are permanently stuck "on."

the case for mutation, if chromosomal abnormalities turn proto-oncogenes into oncogenes or disable tumor suppressors, then the regulation of cell proliferation may be compromised and cancer can result.

The first case in which scientists were able to demonstrate that a specific chromosomal alteration was responsible for a particular cancer was the link revealed between the *Philadelphia chromosome* and *chronic myelogenous leukemia* (CML). In 1960, two scientists, Dr. Peter Nowell and Dr. David Hungerford, examined the chromosomes of CML patients and compared them to the chromosomes of healthy individuals. They discovered that all CML patients and no healthy patients had an unusual, small chromosome in their cells. Nowell and Hungerford named it the Philadelphia chromosome in honor of the city in which it was discovered.

In the 1970s, Dr. Janet Rowley examined the Philadelphia chromosome more closely, using methods that had not been available to Nowell and Hungerford when they did their study. She found that two chromosomes (numbers 9 and 22) exchanged pieces with each other. One of the products of this exchange was the Philadelphia chromosome.

Years after this discovery, in 1984, Dr. Gerard Grosveld used molecular biological techniques to show that the chromosome rearrangement that produced the Philadelphia chromosome actually repositioned two proto-oncogenes, *abl* and *bcr*, next to each other in such a way that both were permanently stuck "on" (**FIGURE 5.13**). As a result, cell division occurred without control.

The Philadelphia chromosome is not a unique incident. At least 200 specific chromosomal abnormalities have been consistently linked to particular types of cancer.

It is clear that all cancers share one common feature: DNA mutates in some way so that the network of cell division controls no longer operates effectively. How do these DNA changes originate? Are some people more susceptible to random mutation, or are there external causes of cancer that work by altering or damaging DNA?

5.5 What Risk Factors Are Associated with Cancer?

The causes of cancer are quite complex and sensitive to a person's environment and lifestyle. In fact, the majority of cancers, at least two-thirds, are influenced by factors we can control such as smoking, poor diet, excessive alcohol use, lack of exercise, and excessive exposure to UV radiation. We have less control (and sometimes no control) over other risk factors, including the infections we get, carcinogens that we encounter in the workplace, pollution, the genes we inherit from our parents, our age, and our wealth. All of these factors may affect cancer development: they damage or alter the DNA of the genes that regulate cell proliferation.

Smoking

Although the relationship between tobacco use and cancer was suspected for hundreds of years, it wasn't until 1948 that a young medical student, Ernest Wynder, decided to undertake a careful study of the subject. The idea came to Wynder after he had observed an autopsy of a man who had died of lung cancer. The man's widow had said that her husband had smoked two packs of cigarettes a day, and Wynder wondered whether there was a connection.

Because Wynder was a student, he did not have the resources to do a research project by himself. He therefore approached Dr. Evarts Graham, a physician and professor at Washington University's medical school. Dr. Graham, a heavy smoker himself, did not see the need for a study about smoking and lung cancer. Nevertheless, Graham was supportive of Wynder's project, and he approved the research. Wynder's well-designed and -implemented study showed that cigarette smokers were 40 times more likely to develop lung cancer than people who did not smoke. When Dr. Graham saw Wynder's data, he was so convinced about the link between smoking and cancer that he quit smoking. Sadly, Graham died of lung cancer six years later.

The risk of cancer and other diseases that result from cigarette smoking is quite high. Smoking is responsible for at least 30% of all cancer deaths. Approximately 80–90% of all lung cancers can be attributed to smoking (**FIGURE 5.14**). Even more striking, tobacco is responsible for as many as 4 million deaths per year worldwide. Of these tobacco-related deaths, 40% are due to lung cancer.

Even though public health education has helped to reduce cigarette smoking in the United States, tobacco use is still common in China, India, Eastern Europe, and Africa—all heavily populated regions of the world. Scientists estimate that unless smoking patterns change significantly, tobacco will kill 10 million people per year by 2030.

Diet and Exercise

Whereas smoking accounts for at least 30% of all cancer deaths in the United States, another 35% can be attributed to diet. In the United States, the problem comes from eating too much red meat, too much saturated fat, too few fruits and vegetables, and an inadequate amount of fiber. The link between diet and cancer is easy to see if one looks at the health of immigrants. For example, the incidence of breast cancer in American women is relatively high (12%), whereas in Japanese women (living in Japan), it is very low (less than half the incidence seen in American women). However, within one generation

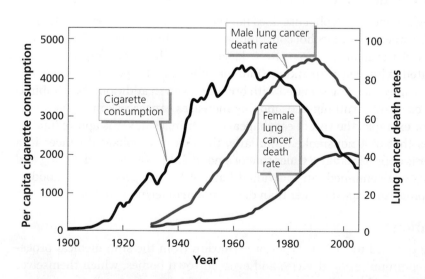

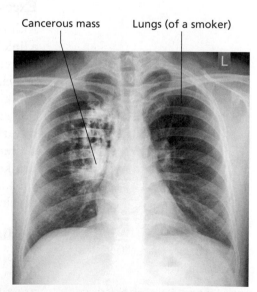

FIGURE 5.14 Smoking is responsible for 80–90% of lung cancers. The risk of developing lung cancer increases with the amount smoked.

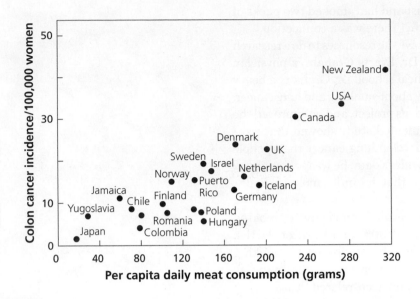

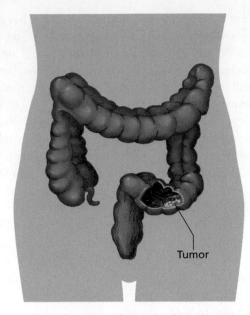

FIGURE 5.15 Consumption of meats and fats is linked to an increased risk of colon and rectal cancers. The graph relates the incidence of colon cancer in women to daily meat consumption.

of migrating to the United States and adopting an American diet, the incidence of breast cancer in Japanese women is just as high as in American women whose families have been in the United States for many generations.

Similarly, compared to countries in which diet is rich in fruits, vegetables, and fiber, the incidence of cancer of the colon and rectum are 10 times higher in areas where the consumption of meat and fats is significant: the United States, Western Europe, and New Zealand (**FIGURE 5.15**). Fortunately, studies have revealed that switching to a diet richer in fiber, fruits, and vegetables will lower cancer risk dramatically.

With respect to exercise, the most critical time period is in childhood. Excessive weight gain and the buildup of body fat increases the risk of eventually developing breast or prostate cancer. In fact, a lack of exercise and extra body fat in childhood are responsible for 5% of prostate and breast cancers in adults.

Excessive Alcohol Use

Although moderate alcohol use does not seem to increase cancer risk, there is a link between heavy alcohol consumption and certain cancers. For example, cancers of the mouth, throat, liver, larynx (voice box), and esophagus are associated with heavy drinking (four or more drinks per day). Excessive drinking may also be associated with breast cancer. In addition, the combination of excessive drinking and smoking increases cancer risk in a synergistic way. For example, the risk of esophageal cancer in a smoker is approximately 6 times that of a nonsmoker. Similarly, the risk of esophageal cancer in a drinker is approximately 6 times that of a nondrinker. But when drinking and smoking are combined, an individual has a 40 times greater likelihood of developing esophageal cancer than does a nondrinker/nonsmoker.

Radiation

radiation :: energy transmitted as waves or subatomic particles

X-ray :: a type of relatively high energy radiation

We are exposed to various types of **radiation** from the sun, medical procedures, communication devices, and even our own bodies, which themselves are a source. Although excessive exposure to high-energy radiation such as **X-rays** can indeed cause certain cancers, especially leukemia, the overall risk

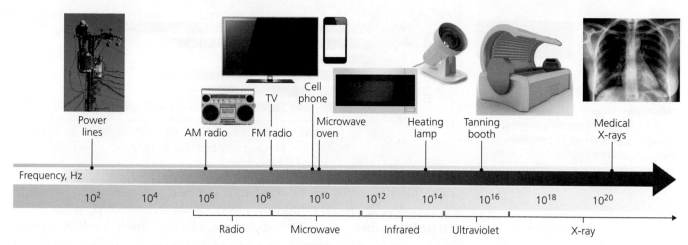

FIGURE 5.16 Ultraviolet irradiation and X-rays are the most dangerous types of radiation with respect to cancer causation. Exposure to other forms of electromagnetic radiation—such as that emitted by electrical power lines, radios, televisions, cell phones, and microwave ovens—do not appear to increase cancer risk to a measurable degree.

is actually quite low. Exposure to other forms of electromagnetic radiation such as that emitted by radios, phones, cell phones, wireless systems, television, and electricity does not appear to increase cancer risk to a measurable degree (**FIGURE 5.16**). The most dangerous type of radiation with respect to cancer causation is ultraviolet irradiation.

Responsible for more than 90% of all skin cancers, the UV rays present in sunlight and tanning beds are certainly an avoidable risk. Unfortunately, the incidence of skin cancer is on the rise. In 1930, people started collecting data about the incidence of the most deadly form of skin cancer, melanoma. The number of cases of melanoma has risen dramatically since 1930. The current incidences, corrected for population size, are 10 times higher than those of 50 years ago. Melanoma is also a leading type of cancer among young adults. Melanoma, an aggressive, metastatic cancer, is diagnosed in 50,000 people each year. Of these, 7000 will die of the disease. In fact, one of Jeff Markway's brothers succumbed to melanoma.

Infection

The relative risk of cancer-causing infections varies a great deal throughout the world depending upon the relative degree of affluence and poverty. Although viruses are the most common type of infectious agents responsible for cancer, at least one type of bacteria (*Helicobacter pylori*) is associated with stomach cancer, and there are flatworms responsible for bladder and liver cancers.

Let's explore a specific example that demonstrates the relationship between infection and cancer—the **human papillomavirus (HPV)** (**FIGURE 5.17**). Responsible for most of the world's cancers of the genitals and anus, HPV is a sexually transmitted virus. The risk of being infected is directly related to the age at which a person becomes sexually active and also to the number of sexual partners. The earlier sexual activity begins and the greater number of partners, the greater the likelihood of infection. Sexually inexperienced individuals exhibit the lowest incidence of HPV infection.

The prevalence of HPV infection is higher in developing nations (15%) than in developed ones (7%). HPV is responsible for 80–90% of cervical cancers worldwide. Although rare in the United States, cervical cancer is the second most common cancer affecting women in the world. The relative rarity of cervical cancer in the United States is due to the wide use of the **Pap test**,

human papillomavirus (HPV) :: a sexually transmitted virus that is responsible for most of the world's cancers of the genitals and anus

Pap test :: a method used to detect cervical cancer by collecting and examining cervical cells

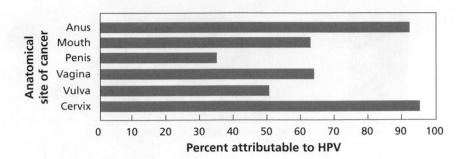

FIGURE 5.17 HPV is a sexually transmitted virus linked with several types of cancers, including cervical cancer and cancers of the genitals, anus, and mouth.

in which cervical cells are collected and examined to screen women for early signs of cervical cancer. In 2006, a vaccine, Gardasil, which prevents infection from certain types of HPV, was approved for use. (See *Scientist Spotlight: Peyton Rous* for an account of the discovery of tumor-forming viruses.)

Workplace Carcinogens and Pollution

The first description of workplace cancer dates back to the 1700s, when the physician Percival Pott noted the greater incidence of scrotal cancer among chimney sweeps compared to men and boys in other occupations. The World Health Organization's International Agency for Research in Cancer (IARC) has identified more than 40 specific materials in the workplace associated with increases in particular cancers (**TABLE 5.1**). Exposure to cancer-causing materials in the workplace is responsible for 5% of all cancer deaths in the United States. In 1950, before safety improvements were put into place, the cancer death incidence due to workplace exposures was 10%. Because these safety

Scientist Spotlight

Peyton Rous (1879–1970)

Although he did not appreciate it fully at the time, Peyton Rous ushered in an exciting era of cancer research with his discovery of tumor-forming viruses in 1909. Eventually he came to understand the importance of his work, as did everyone else; Rous was finally awarded a Nobel Prize in 1966.

Born in Texas in 1879, Peyton Rous grew up in Baltimore, where he earned his college and medical degrees from Johns Hopkins University. Rous decided to devote himself to medical research rather than clinical practice, and after a brief time at the University of Michigan, he joined the Rockefeller Institute in 1909. Rous remained at the Rockefeller until his death in 1970.

Rous's discovery of a tumorigenic virus came about in an unexpected way. In 1909, a farmer from Long Island made the journey into New York City in order to see Dr. Rous. The farmer brought his prize Plymouth hen with him, hoping that Rous could help: the chicken had a tumor in her right breast muscle. Rous knew he could not cure the chicken, but he somehow managed to convince the farmer to leave the bird for further study.

Peyton Rous ushered in an exciting era of cancer research with his discovery of tumor-forming viruses in 1909 and eventually received the Nobel Prize in 1966 at the age of 87.

Rous did two critical experiments with this bird. First, he removed the tumor and transplanted it to another chicken (from that same farmer's flock). The chicken that had received the transplant developed cancer. Evidently, cancer could be transferred from one bird to another. Second, Rous ground up the tumor tissue, collected the liquid that was squeezed out of the tumor, filtered the extract to make sure no cells or solid debris remained, and injected the filtrate into another chicken. Amazingly, the injected chicken developed cancer. Apparently some infectious agent, too small to be seen with the microscopes available in 1909–1911, caused cancer. We now know that this agent is a virus, named the Rous sarcoma virus in honor of its discoverer.

Rous abandoned his cancer work in 1915. He turned his attention to other biomedical questions, because he was unable to produce tumors in mammals by exposing them to tumorigenic viruses. As a consequence, Rous incorrectly thought the phenomenon he observed was specific to chickens and therefore not of general significance. Although he returned to his cancer research in 1934, he did so in a scientific community that was not ready to believe that viruses could cause cancer.

Rous was 87 years old by the time he was awarded the Nobel Prize. Because the Nobel Prize is given only to living recipients and never posthumously, it is fortunate that Rous lived a long life. Still, it was a long wait for a well-deserved honor. It must have been an awkward but happy moment in the family when Rous's son-in-law, Alan Hodgkin, won a Nobel Prize before him, in 1963, for his discoveries about how nerve cells work. One can presume that when Rous won the 1966 prize, it brought great relief to everyone! ::

Table 5.1 :: A Sampling of Materials and Occupations Associated with Increases in Particular Cancers

Substance or Mixture	Occupation or Industry in Which the Substance Is Found	Site(s) of Cancer
Ionizing radiation and sources, including X-rays, γ rays, neutrons, and radon gas	Radiologists; technologists; nuclear workers; underground miners; plutonium workers; cleanup workers following nuclear accidents; aircraft crew	Bone, leukemia, lung, liver, and thyroid
Asbestos	Mining and milling; insulating; shipyard workers; sheet-metal workers; asbestos cement industry	Lung
Talc containing asbestiform fibers	Manufacture of pottery, paper, paint, and cosmetics	Lung
Wood dust	Logging and sawmill workers; pulp and paper and paperboard industry; woodworking trades	Nasal cavities and paranasal sinuses
Arsenic and arsenic compounds	Nonferrous metal smelting; use of arsenic-containing pesticides; wool fiber production; mining of ores containing arsenic	Skin, lung, and liver
Cadmium and cadmium compounds	Cadmium-smelter workers; battery production workers; dye and pigment production	Lung
Selected nickel compounds, including combinations of nickel oxides and sulfides in the nickel-refining industry	Nickel refining and smelting; welding	Lung and nasal cavity and sinuses
Benzene	Solvents in the shoe production industry; chemical, pharmaceutical, and rubber industries; printing industry; gasoline additive	Leukemia
Coal tars and pitches	Production of refined chemicals and coal tar products; coal gasification; aluminum production; road paving and construction	Skin
Mineral oils, untreated and mildly treated	Used as lubricant by metal workers, machinists, and engineers; printing industry; used in cosmetics and medicinal and pharmaceutical preparations	Skin, bladder, lung, and nasal sinuses
Soots	Chimney sweeps; heating-unit service personnel; brick masons; building demolition workers; insulators; firefighters; work involving burning of organic materials	Skin, lung, and esophagus

Source: Adapted from http://www.ncbi.nlm.nih.gov/pmc/articles/PMC1247606/table/t3-ehp0112-001447/. Environ Health Perspect. 2004 November; 112(15): 1447–1459. Published online 2004 July 15. doi: 10.1289/ehp.7047

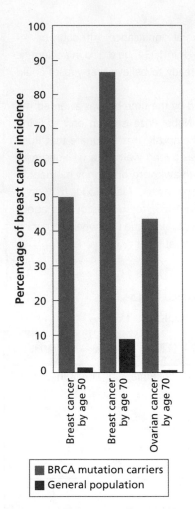

FIGURE 5.18 Mutations in the *BRCA-1* and *BRCA-2* genes dramatically increase the chances of developing breast or ovarian cancer.

improvements are generally found in developed nations, the cancer incidence due to workplace exposure will continue to be high in developing nations and even grow because of the rapid industrialization that is presently under way.

As for the environment outside the workplace, scientists estimate that 2% of all cancers can be attributed to air, water, and/or soil pollution.

Inheritance

Most cancers, probably more than 90%, display no family history at all. Cancers appear sporadically in a family lineage. In approximately 5–10% of cancers, including some breast cancers, an inherited mutation might be the culprit.

As mentioned in Chapter 4, Paul Broca observed that many members of his wife's family had breast cancer. He mapped his wife's family on a pedigree chart and showed that out of 24 women, 10, including Broca's wife, had breast cancer. This cluster of cases in her family was strong evidence that an inherited factor was responsible.

More than 100 years later, scientists have discovered two genes that increase dramatically the chances of developing breast cancer. Mutations in these two genes, *BRCA-1* and *BRCA-2*, account for 90% of all hereditary breast cancers. A woman with mutant *BRCA* genes faces an 84% probability that she will develop breast cancer by the age of 70 (**FIGURE 5.18**). Scientists have also learned what the normal *BRCA* genes are supposed to do: they are responsible for DNA repair, and one type also acts as a tumor suppressor.

Age

Put simply, the likelihood of cancer increases as you age (**FIGURE 5.19**). Although there are childhood cancers, as well as some cancers that occur in young people, the majority of cancers appear later in life. Because cancer results from an accumulation of errors and damage to DNA, it takes time for enough mistakes to happen so that the cell passes a tipping point and becomes fully malignant and invasive.

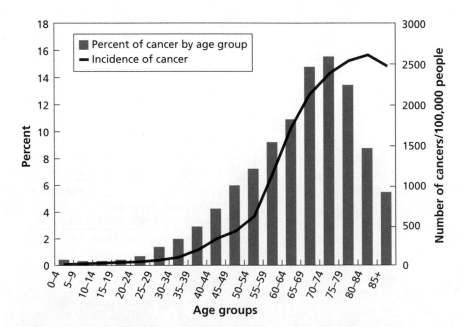

FIGURE 5.19 The likelihood of developing cancer increases with age. The bars show the percentages of all cancers in the population appearing in each age group, and the line shows the incidence of cancer in each age group.

Poverty

Living in poverty increases a person's risk of cancer. This is true for poor people living in wealthy countries and also those living in less prosperous ones. The incidence of all types of cancer except those of breast, prostate, colon, and rectum are higher in the poor.

The reasons for this association between poverty and cancer are many and complex. First, several factors that are linked with cancer are more prevalent in the poor: smoking, poor nutrition, exposure to carcinogens in the workplace, and exposure to infectious agents (sometimes due to lack of clean water). Second, poor people don't have easy access to health care. Cancer screening is not generally available. Problems are diagnosed too late, if at all, and medical treatment can be difficult to obtain and too expensive. This disparity between the health states of poor people and those who are more affluent is a serious concern that needs to be addressed for economic, political, and moral reasons. Similarly, workplace and environmental exposures to carcinogens are societal problems for which all stakeholders—individuals, businesses, society, and government—share a responsibility to act.

5.6 How Is Cancer Diagnosed?

Physicians have many effective tools for diagnosing cancer, but one of the most important sources of information is the patient. You are the most informed person about your body. You know if something is "different" or "does not feel right." Most of the time, the cough that seems to persist or the ache that won't go away is due to something that is not serious at all. Nevertheless, regular checkups and physician visits when you have a problem or concern are important ways of keeping healthy. **TABLE 5.2** includes the American Cancer Society's Guidelines for the early detection of cancer.

Some bodily events should prompt a trip to your doctor:

- A lump in your breast, testis, or elsewhere
- Changes in a wart or mole, especially in size or color
- Changes in bowel or bladder habits
- Persistent indigestion or heartburn; difficulty swallowing
- Unusual bleeding

Again, most of the time these symptoms don't indicate cancer, but sometimes they do.

Physicians use important tools to screen people for early signs of cancer. For example, women 40 years and older are encouraged to have yearly **mammograms**, a form of X-ray, to detect possible breast cancers (**FIGURE 5.20**). Women are also screened for cervical cancer by the Pap test to make sure they are normal. The relative rarity of cervical cancer in the United States is thanks to the wide use of the Pap test to screen women for early signs of cervical cancer. Both men and women can be screened for cancer of the colon or rectum using **colonoscopy**, a procedure wherein the interior of the large intestine, rectum, and anus are examined to determine whether any abnormal growths are present. Physicians also examine patients manually to see whether any lumps are present. For example, the prostate gland and testes in men are checked this way. If any of these screening methods reveals something that warrants closer attention, additional tests will be done.

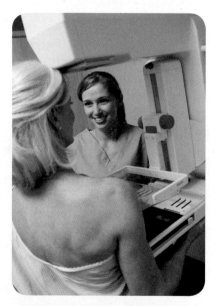

FIGURE 5.20 Mammograms are a form of X-ray that can be used to detect breast cancers even at relatively early stages.

mammogram :: a form of X-ray used to detect possible breast cancer

colonoscopy :: a procedure used to examine the large Intestine, rectum, and anus to screen men and women for colon or rectal cancer

Table 5.2 :: American Cancer Society 2013 Guidelines for the Early Detection of Cancer (Most Adults)

Cancer	Guidelines
Breast	• Yearly mammogram for women starting at age 40 • Clinical breast exam for women once every 3 years (ages 20–30) and annually (40 years and over) • Optional breast self-exam for women starting at age 20

Colorectal	• Screening should begin at age 50 for both men and women. Options include:
	  o Flexible sigmoidoscopy every 5 years
	  o Colonoscopy every 10 years
	  o Double-contrast barium enema every 5 years
	  o Virtual colonoscopy every 5 years
	  o Fecal occult blood test every year

Cervical	• Screening should begin at age 21 • Pap test once every 3 years (women ages 21–29) • Pap test plus an HPV test every 5 years (women ages 30–65)
Prostate	• Starting at about age 50, men should talk with a doctor about the pros and cons of testing to determine whether it is the right choice • If men decide to be tested, they should have the PSA blood test with or without a rectal exam—how often depends on their PSA level
Cancer-related checkups	• Women and men age 20 and older should have periodic health exams • Health exams should include health counseling and, depending on age and gender, exams for cancers of the thyroid, oral cavity, skin, lymph nodes, testes, and ovaries

Source: http://www.cancer.org/healthy/findcancerearly/cancerscreeningguidelines/american-cancer-society-guidelines-for-the-early-detection-of-cancer.

computerized tomography (CT) scan :: an advanced type of X-ray procedure that provides a more detailed look at internal structures by sending thin X-ray beams at different angles

magnetic resonance imagery (MRI) scan :: a medical imaging technique that uses a magnetic field and radio waves to visualize the inside of a body

The details of the next round of tests would be determined by the physician's specific concerns. Such tests could include blood sampling to look directly at the blood cells themselves or to measure specific proteins associated with particular cancers. For example, the amount of PSA, a protein found in high amounts in individuals with prostate cancer, could be determined. Also, visualization methods like ultrasound, **computerized tomography (CT) scans** (see *Technology Connection: Computerized Tomography [CT] Scans*), or **magnetic resonance imagery (MRI) scans** (**FIGURE 5.21**) provide a good look at the internal structures of your body, including the

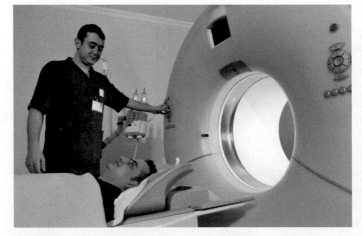
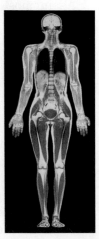

FIGURE 5.21 Magnetic resonance imaging (MRI) (A) produces clear, detailed images of the internal structures of the body (B).

A

B

Technology Connection

Computerized Tomography (CT) Scans

Using a standard X-ray machine, we can see inside the body because the rays can penetrate skin and other soft tissues. Other parts of the body are more opaque and do not permit the X-rays to pass through so easily; the X-rays are absorbed. These opaque tissues are structures that show up on the film as dark areas. For example, bones can be clearly viewed in this way. X-rays can also be used to view tumors, which in many cases are denser than the surrounding tissues and are able to absorb more X-rays. As a consequence, they will appear darker than the surrounding normal tissue.

Computerized tomography (CT) scanning is a more advanced type of X-ray procedure. CT scans provide a better and more detailed look at internal structures, including tumors, because rather than sending a single straight beam of X-rays through the tissue, CT scanners use a computer-controlled system to send thin X-ray beams at different angles, resulting in three-dimensional images. As with the

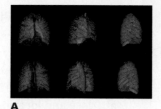 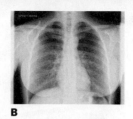

A   **B**

CT scans use a computer-controlled system to send thin X-ray beams at different angles, resulting in three-dimensional images (A) that are better and more detailed than standard X-rays (B).

standard X-ray method, tumors will absorb more X-rays and appear darker than normal tissue. However, in this case, a tumor will appear in three dimensions rather than as a flat spot. CT scanning is one of several visualization techniques that have revolutionized medical diagnosis in general and for cancer in particular. ::

details of your organs. Finally, your physician may take a biopsy, a procedure in which cells are collected and examined to see whether they are cancerous (**FIGURE 5.22**).

Some cancer screens are designed so that *all* potential cancers will be detected. As a consequence, there will be some "false positives." For example, 10% of women who have mammograms will be called back for additional examination. The vast majority of the time, this second look reveals nothing worrisome. The mammogram is designed in this way so that there will be no "false negatives"—women who are told there is no cancer when there is.

5.7

How Is Cancer Treated?

If the careful and extensive testing done by the physician reveals cancer, the next step is to make decisions about treatment. Three principal methods are used to treat cancer: surgery, radiation therapy, and chemotherapy. These treatments are often used in combination.

Surgery

Surgery is performed on more than 50% of all cancer patients. In some cases, it is relatively straightforward—the removal of a skin cancer, for example. In other situations, it is a daunting task, as in the removal of a cancerous lung. Surgery is also sometimes used to perform a biopsy.

Surgery is most successful when the tumor is localized and has not spread, or metastasized. The physician will remove the tumor and have

In a needle biopsy, cells are removed with a syringe.

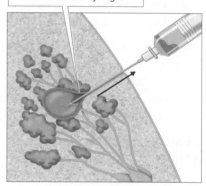

In a surgical biopsy, a larger piece of tissue is removed with a scalpel or tiny scissors.

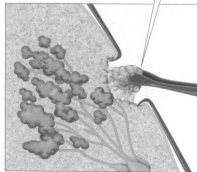

FIGURE 5.22 In a biopsy, cells or tissue are removed and examined to see whether they are cancerous.

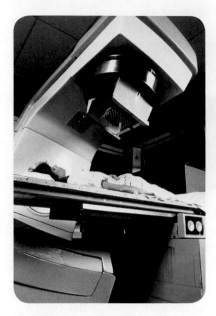

FIGURE 5.23 Radiation therapy is used to kill cancer cells that cannot be removed surgically or to shrink a tumor prior to surgery.

pathologist :: a doctor who specializes in the study of disease

radiation therapy :: radiation delivered either externally or internally to kill cancer cells

chemotherapy :: a treatment method using drugs that impair the ability of cells to replicate; these medications may be taken orally or injected directly into the bloodstream

immunotherapy :: treatment that directly targets a cancer or boosts the immune system to make other cancer therapies more effective

angiogenesis inhibitor :: a drug that stops the growth of blood vessels, inhibiting the ability of tumors to attach to the circulatory system and grow

pathologists, doctors who specialize in the study of disease, examine it in order to see how serious the cancer is by evaluating the degree to which the chromosomes and cellular structures appear abnormal. This information will provide guidance regarding other possible treatments.

Radiation Therapy

The objective of **radiation therapy** is to kill cancer cells that cannot be removed surgically or in some cases to shrink a tumor prior to surgery. Radiation is delivered to the cancer in one of two ways. For some cancers, the radiation exposure is external (**FIGURE 5.23**). After determining the appropriate dose and location, the tumor is bombarded with a focused beam of X-rays in order to kill the cells. In other situations, the radiation exposure projects from an internal source. In this circumstance, a localized source of radiation, namely, a tiny "seed" containing radioactive material, is placed near or directly in the tumor. Again, the objective is to kill the cancer cells.

Chemotherapy

Chemotherapy is generally used to treat cancers that have spread from the primary tumor. It is also sometimes used when there is a concern that the tumor may have metastasized even though the spreading is not clearly observable. Chemotherapy uses drugs that impair the ability of cells to replicate. These drugs may be taken orally or injected directly into the bloodstream (**FIGURE 5.24**). As with radiation therapy, the goal of chemotherapy is to kill cancer cells.

Chemotherapy sometimes produces challenging side effects. Because the drugs target all cells undergoing mitosis, populations of cells in the body that normally divide are especially vulnerable. For example, in a healthy person, blood cells, hair follicle cells, and the cells lining the intestines continually renew themselves. Chemotherapy drugs can impact these cell populations, causing anemia and a weakened immune system, hair loss, and nausea.

Cancer Treatments on the Horizon

Researchers continue to develop new potential treatments for cancer. Two promising avenues are immunotherapy and drugs to inhibit angiogenesis.

In **immunotherapy**, the patient's immune system is used to fight cancer. Some immunotherapies act to fight cancers directly, while others focus on boosting the immune system so that therapies can work better. In some cases, antibodies, or molecules produced by immune system cells, are made in the lab and administered to patients to target specific cancers, such as certain types of breast cancers.

Angiogenesis inhibitors are drugs that stop the growth of blood vessels. When used to treat cancer, they inhibit the ability of tumors to attach to the circulatory system and obtain the nutrients and oxygen necessary for growth. There are some angiogenesis inhibitors being used to treat a small number of specific cancers at the present time, with many more inhibitors undergoing clinical trials on patients.

Why Cancer Treatments Sometimes Fail

Even the most aggressive cancer treatments are not always successful. Cancer cells are not uniform or unchanging in their characteristics. In a given tumor, each new round of cell division can produce cells that are increasingly abnormal and ever more invasive.

Ironically, the process of natural selection can shift the cancer cell population so that the cells resistant to chemotherapy increase in number to the point where the drugs are no longer effective. When this occurs, the patient might be able to try a different battery of drugs, but eventually cancer cells may develop resistance to this new therapy. Ultimately, some cancers will recur and will be unresponsive to any treatment. At that point, the medical goal is to keep the patient pain free and to help with end-of-life issues.

5.8 How Can Cancer Be Prevented?

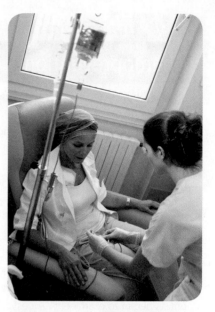

FIGURE 5.24 Chemotherapy kills cancer cells by using drugs that impair the ability of cells to replicate.

Jeff Markway suffered from a genetic disorder that dramatically increased his chances of developing melanoma. Except for avoiding UV light from the sun or other sources, he really had no other means of prevention. In contrast to Jeff's situation, many cancers are preventable for most people. As you have read, at least 65% of all cancers in the United States are due to smoking, a poor diet, and lack of exercise. Excessive exposure to sunlight is the principal cause of skin cancer. And in the United States, the most common way to become infected with a cancer-causing pathogen is through promiscuous sexual activity. Clearly there are ways that we can modify our behaviors to lower our individual risk for cancer.

Is there actually evidence that cancer can be prevented? After all, if a person has to make potentially difficult lifestyle changes, it would be comforting to know that the effort will pay off. Let's revisit smoking to see a case that demonstrates the effectiveness of behavior modification for the prevention of cancer.

The first publication detailing the link between cigarette smoking and lung cancer was published in the United States in 1950. A few months later, a paper detailed a similar study done in Great Britain. The British study verified the association between smoking and lung cancer that had been described in the U.S. publication by Wynder. In 1955, British men under the age of 55 had the highest lung cancer incidence in the world. After the reported relationship between lung cancer and smoking finally captured the attention of the government, things changed. First, because cigarette tar is a potent carcinogen, British cigarettes were redesigned to lower tar levels. Second, the government launched a public health education campaign that succeeded in reducing the number of people who smoked. Thanks to these efforts, and the willingness of people to change long-ingrained habits, the lung cancer incidence in British men under the age of 55 dropped by more than 65%. Prevention *does* work.

So what can you do to decrease your cancer risk? You know enough to figure it out, but **TABLE 5.3** provides some suggestions. To learn how the foods we eat can help prevent cancer, an idea referred to as **chemoprevention**, see *Life Application: Chemoprevention*.

chemoprevention :: the action of natural or manufactured materials used to prevent or halt cancer development

Table 5.3 :: Simple Steps to Reduce Your Chances of Cancer

- Don't smoke.
- Eat a diet rich in fruits, vegetables, and fiber, and lower in saturated fat.
- Avoid excessive exposure to UV light either from the sun or a tanning bed.
- Exercise and maintain a healthy weight.
- Don't engage in risky sexual behavior.
- Avoid excessive alcohol—especially if you smoke, because these two factors act synergistically to increase the incidence of certain cancers.
- Don't expose yourself to known carcinogens at home or work without protective clothing and proper precautions.
- Get regular checkups.

Life Application

Chemoprevention

It seems that new diets are constantly appearing in the media, including magazines and television. They have spawned a new wave of cookbooks describing different diets: Atkins, South Beach, anti-inflammatory, Mediterranean, and Paleolithic, to name a few. All of these food plans claim to provide health benefits. Is it possible that what we eat can indeed influence our health in a disease-specific manner?

We know that consumption of some foods increases our chances of developing cancer (note the correlation between animal fats and colon cancer), and that reducing or avoiding them will decrease our risk. But are there particular things we could eat that would decrease our risk even more? In the 1950s, some scientists began to examine this question. They wondered whether we could use natural or manufactured materials to prevent or halt cancer development. This idea is called chemoprevention, and research in this area has produced some promising results.

Studies have shown that a diet rich in fruits, vegetables, and fiber and low in saturated fat decreases the risk of developing certain cancers, such as those of the colon and rectum. However, scientists want to learn which phytochemicals, components derived mostly from plants, are responsible for this beneficial effect. One group of researchers studied dithiolethione, a phytochemical found in cruciferous vegetables such as broccoli, cabbage, and cauliflower. Controlled experimental studies have shown that dithiolethione inhibits the development of tumors in the lung, colon, mammary gland, and bladder when tested in laboratory animals. Other studies suggest that dithiolethione interferes with cancer formation because it activates liver enzymes that in turn detoxify carcinogens.

R=CH₃; ADT
R=H; ACS-1

This is an example of dithiolethione found in these vegetables.

Research has shown that dithiolethione, a phytochemical found in cruciferous vegetables such as broccoli, cabbage, and cauliflower, inhibits the development of tumors in the lung, colon, mammary gland, and bladder when tested in laboratory animals.

Chemoprevention research has blossomed in recent years. Many materials have been tested in laboratory experiments, and in some cases, human population studies are under way as well. It is likely that as we learn more, we may be able not only to decrease the likelihood of developing cancer, but also to take active steps to preserve health even more effectively. ::

Scientists estimate that if we paid more attention to cancer prevention instead of focusing mostly on treatment, at least 10 times more lives could be saved. Prevention of cancer also spares people fear, physical pain, and the financial burden of the disease. The realization that most cancers are preventable does not mean that someone who gets cancer is to blame. Instead, this knowledge should empower each of us to make good choices that will enhance our health.

Even if we were to eliminate all external causes of cancer, scientists estimate that we would still see 20–25% of the cancers that occur now. Some cancers are inevitable. This disease is complex and is the result of interactions between factors external to the body and events that occur in our cells. And chance plays an important role too.

 # Biology in Perspective

Cancer is a disease of uncontrolled cell proliferation. All cancers are the consequence of alterations or mutations of genes responsible for regulating cell division. Scientists have learned a great deal about how cancers form and how they might be prevented. There are several known factors that increase a person's risk of developing cancer. Some of them are avoidable, but not all, and some factors must be addressed by society at large, including the government.

Early diagnosis dramatically increases cancer survival rates. There are many methods for diagnosing cancer, including physical examination, certain blood tests, and visualization tools such as ultrasound, CT scanning, and MRI. Cancer treatments include surgery, radiation therapy, and chemotherapy, or a combination of these methods. As is true of cancer risk itself, poverty influences the chances that an individual will be able to get medical care for early detection and whether all possible tests and treatments will be available. Taking care of everyone in need of medical care requires more than what an individual can do for him- or herself.

Chapter Summary

5.1 How Does Cancer Make You Sick?

Use appropriate terminology to describe cancer and the reasons why cancer makes you ill.

- Malignant tumors can spread cancer; benign tumors are unlikely to be dangerous.
- Cells continue to divide uncontrolled in malignant tumors, but if cells stop uncontrolled cell division they form benign tumors.
- Tumors can press on other organs, nerves, and blood vessels; metastasize and form secondary tumors; and deprive other cells of oxygen through angiogenesis.

5.2 How Do Cancer Cells Differ from Normal Cells?

Contrast the behavior of normal and cancerous cells.

- Cells communicate to control cell division through chemical signals.
- Cancerous cells no longer respond properly to chemical signals and often lose their original function.

5.3 What Is the Life Cycle of a Cell?

Use a signal transduction pathway to explain how cells become cancerous.

- The cell cycle has four steps (M, G1, S, and G2). Chemical signals regulate cell division at checkpoints in the cell cycle.
- Chemical signals regulate gene activity through signal transcription pathways. Receptors sense chemical signals and transfer the signals to signal transducers that pass the signals to transcription factors that turn genes on or off.
- Mutations to any part of a signal transduction pathway can change cell division and prevent apoptosis.

5.4 In What Ways Is Cancer a Genetic Disorder?

Describe the role that mutations play in cancer development.

- Proto-oncogenes are normal genes that turn on cell division, while tumor suppressor genes stop cell division.
- Mutated proto-oncogenes or tumor suppressor genes can cause uncontrolled cell division and may ultimately become oncogenes that cause cancer.

5.5 What Risk Factors Are Associated with Cancer?

Identify actions you can take to reduce your risk of cancer.

- Any compound or activity that increases the risk of mutations has the potential to cause cancer.
- Risk due to infectious agents can be reduced through vaccines (e.g., Gardasil) and behavior.
- Some cancer risks, such as smoking and alcohol consumption, can work synergistically to create risk many times greater than either would cause on its own.
- Protective equipment, safety protocols, and manufacturing practices can reduce workplace exposure to carcinogens.
- Cancer risk increases with age, with certain inherited genes, and with poverty because of the likelihood of increased exposure to risk factors.

5.6 How Is Cancer Diagnosed?

Explain how cancers are diagnosed.

- Some bodily events should prompt a trip to a doctor, including the following: lumps in your body, changes in warts and moles, changes in bowel or bladder habits, persistent indigestion, difficulty swallowing, and unusual bleeding.
- Regular cancer screening can help catch cancers early.

- Once a cancer is suspected, further testing is usually required. Tests may include chemical methods (e.g., PSA) or visualization methods (e.g., ultrasound, CT scans, and MRI).
- Ultimately, doctors usually take a biopsy, which is sent to a pathologist to determine whether it is cancerous.
- Most cancer tests are sensitive to reduce false negatives but increase false positives.

5.7 How Is Cancer Treated?
Contrast the three methods used to treat cancer.

- Cancer is treated through three primary methods: surgery, radiation therapy, and chemotherapy; cancer treatment usually involves multiple methods.

- Cancers can become resistant to chemotherapy through the process of natural selection.
- Two promising avenues of cancer treatment on the horizon are immunotherapy and angiogenesis inhibitors.

5.8 How Can Cancer Be Prevented?
Enumerate steps to decrease the risk of cancer.

- You can take steps to reduce or eliminate cancer risk factors.
- Cancer prevention could reduce cancer-related deaths 10-fold.

Key Terms

angiogenesis 131
angiogenesis inhibitor 148
apoptosis 133
benign 130
cancer 128
cell cycle 134
chemoprevention 149
chemotherapy 148
colonoscopy 145
computerized tomography (CT) scan 146
cyst 128
growth factor 132

growth inhibitor 132
Herceptin 136
human papillomavirus (HPV) 141
immunotherapy 148
leukemia 135
magnetic resonance imagery (MRI) scan 146
malignant 130
mammogram 145
melanoma 128
metastasis 130
oncogene 135

Pap test 141
pathologist 148
proto-oncogene 135
radiation 140
radiation therapy 148
signal transducer 132
transcription factor 132
tumor 130
tumor suppressor 135
Xeroderma pigmentosum (XP) 128
X-ray 140

Review Questions

1. Explain the three primary ways that cancer makes someone sick.

2. Which of the following is NOT a reason why the production of extra cells creates serious health consequences for the body?
   a. Tumors can grow quite large.
   b. Tumors can be benign.
   c. Cancer can become invasive.
   d. Individual cancer cells can metastasize.
   e. Tumors can deprive other cells of oxygen.

3. Describe how cancer is a breakdown in cell communication and community.

4. Diagram the cell cycle. Where are potential checkpoints located?

5. What is the relationship between proto-oncogenes and oncogenes?

6. How do oncogenes contribute to tumor formation?
   a. Oncogenes tell the cell to divide; they are like an accelerator stuck in the "on" position.
   b. Oncogenes tell the cell to stop dividing; they act like a car brake.
   c. Oncogenes are similar to tumor suppressor genes; they tell the cells to stop dividing.
   d. Oncogenes encode proteins to urge cells to divide.
   e. Oncogenes do not contribute to tumor formation.

7. What is a tumor suppressor, and what role does it play in tumor formation?

8. The p53 tumor suppressor gene regulates whether the cell cycle can proceed between G1 and S phase, and G2 and M phase. It stops the cell cycle to allow for DNA repair, but if the damage is irreparable, it will trigger apoptosis. Why is it important to stop at those points in the cell cycle?

9. All of the following are risk factors associated with cancer EXCEPT which one?

   a. Lack of exercise

   b. Poor diet

   c. Smoking

   d. Young age

   e. Excessive exposure to UV radiation

10. Why is cancer more prevalent in the elderly and the poor?

11. Why is accurate DNA repair necessary to prevent cancer?

12. What cancer warning signs should signal that you need to see the doctor?

13. What are some of the routine tests that can be done to detect cancer?

14. Distinguish between false positive and false negative test results.

15. An oncologist might use _____ to reduce the metastasized cancer spreading throughout a person before removing or destroying the primary tumor through _____ or _____.

    a. radiation therapy; surgery; phytochemicals

    b. radiation therapy; surgery; colonoscopy

    c. surgery; chemotherapy; radiation therapy

    d. surgery; radiation therapy; chemoprevention

    e. radiation therapy; surgery; chemotherapy

16. What are some ways to prevent cancer?

The Thinking Citizen

1. Suppose you are responsible for developing cancer treatments. Explain how you would target your research based on what you know of signal transduction pathways related to cancer.

2. Why do people persist in dangerous behaviors that are known to cause cancer?

3. Explain how an individual eating all the right foods, exercising, and not exposed to carcinogens could develop cancer and another person who smoked a pack a day for 40 years might not have cancer.

4. What are specific choices you can make to reduce your chances of developing or dying from cancer?

5. Pick one of the occupations or workplaces listed in **TABLE 5.1** (on pg. 143). Describe safety precautions you would expect to be in place at that location. Explain how the precautions would help reduce the likelihood of cancer.

6. Do you think it is possible to achieve a cancer-free human population (0% incidence)? Why or why not?

7. An important area of research concerns whether angiogenesis inhibitors can stop tumor growth. Why would the inhibition of angiogenesis be a potential way to kill or limit the growth of tumors?

The Informed Citizen

1. Prevention is a proven method for reducing cancer incidence. How would you go about trying to communicate *effectively* with middle school and high school students about the choices they need to make for a more healthy life?

2. What is society's responsibility for individuals who develop cancer because of bad decisions they have made in their lives?

3. Many people seek alternative and complementary medical therapies for their cancer. Should insurance companies cover the costs of treatments that people *believe* to be effective but for which no evidence exists to support this belief? Why or why not?

4. Compare the likely cancer risks between two locations in your region with different socioeconomic levels. What are the sources of those different risks? How could you help decrease the cancer risks in each community? Why do you suppose people would move to a community that has a potential environmental source of carcinogens?

5. Suppose you were in charge of the part of the federal budget targeted to fund projects concerned with cancer. What proportions of the funds would you allocate for prevention and for treatment? How would you make your decision?

6. Despite all that people know about preventing harmful UV radiation and the availability of sunscreen, skin cancer is still on the rise. Why is that? What other factors might be responsible?

7. You have just been told that you have cancer based upon a preliminary test. The false positive rate is 10%, and the false negative rate is 1%. Further testing is painful but has a lower rate of false positives. The cancer treatment is painful, dangerous, and disfiguring. Do you get additional tests, go straight into treatment, or wait another year or two to get tested again? What additional information do you want? How would that information help you choose a course of action?

CHAPTER LEARNING OBJECTIVES

After reading this chapter, you should be able to answer the following questions:

6.1 How Do Males and Females Form?
Describe how chromosome and hormone instructions interact to determine sex.

6.2 What Happens If the Hormonal Signals Are Missing or Misread?
Predict the outcome of a change in hormone production or response.

6.3 How Do Men Produce Sperm?
Outline the steps of sperm production in males.

6.4 How Do Women Produce Eggs?
Outline the steps of egg production in females.

6.5 How Can Pregnancy Be Prevented?
Compare methods used to prevent pregnancy.

6.6 What Causes Infertility, and How Can It Be Treated?
Predict causes of infertility and propose potential treatments.

6.7 How Can We Tell If a Fetus or Baby Is Healthy?
Explain the purpose of and methods for testing fetal and newborn health.

6.8 What Tests Are on the Horizon?
Evaluate the ramifications of further refinements in genetic tests and treatments.

Reproduction is all about communication. It involves communication between two parents, such as this courtship display between two blue-footed boobies, but also communication between individual cells and between the embryo and the mother's body.

CHAPTER 6

Reproduction

What Kind of Baby Is It?

Becoming an Informed Citizen …

Sex and the urge to have offspring are biological drives. Understanding the biology of reproduction is, therefore, essential for understanding what makes us human and for preserving our natural resources.

*P*roducing offspring is the most important thing we do as a species. Although individuals may be childless, as a species we are diligently making babies—so many, in fact, that we risk overpopulating the world (Chapter 18). Understanding the biology of reproduction is thus essential for both responsible family planning and stewardship of our natural resources.

Reproduction is all about communication. Of course, human reproduction involves two people expressing their hopes and desires. But communication also takes place between individual cells. Embryos listen to chemical signals, too, to determine what sex organs to develop—male or female. Other communications regulate sperm and egg formation, and sperm and eggs must read each other's signals for successful fertilization to occur. The embryo and uterus communicate as well, so that the embryo can implant in the mother's body and stay there safely for an entire pregnancy. The fetus even signals the start of labor, with dramatic responses from the mother's body (Chapter 3).

Reproduction is at the core of who we are as individuals. One might define a man as a person capable of fathering a child and a woman as someone capable of giving birth. But is it really that simple? Do humans fall neatly into two categories? Does it even matter? It mattered a great deal to the fastest woman on Earth.

case study

The Fastest Woman on Earth

Running through the countryside of Limpopo on dirt roads, Caster Semenya (**FIGURE 6.1**) dreamed of being a champion. In 2008, Semenya took her first step onto the world stage of athletics, winning a gold medal at the Commonwealth Youth Games. She followed this victory with another gold medal at the 2008 African Junior Athletics Championship. In fact, Semenya's performance was so spectacular that she qualified for the 2009 World Championship in Berlin. Only 18 years old, Semenya appeared to be on top of her game—a world-class athlete with a dazzlingly bright future.

FIGURE 6.1 Caster Semenya was subjected to gender testing and accusations that she was not really a woman because she looked masculine and performed so much better than her competitors in track and field events.

But then the rumors and accusations began to surface. Some doubted that Semenya was actually a woman. She looked masculine, they contended, and performed so much better than her competitors that she *must* be male. Semenya was subjected to two rounds of gender testing, the first in Pretoria. The second was in Berlin, a day before her race. Here she was, a teenager from a tiny rural village in South Africa, being examined in ways she did not understand, without her family there for support. Her parents never even knew the tests were being done.

Successfully passing the tests did not stop the rumors. In many ways, the situation actually got worse. After Semenya ran successfully in the semifinal round, a TV reporter outside of the stadium hollered at her: "I hear you were born a man!" Although visibly distressed, she managed to keep her focus and win the world championship.

It turns out that the International Association of Athletics Foundations (IAAF) had broken confidentiality rules and leaked the rumor. Worse, Australia's *Daily Telegraph* claimed that test results showed that Semenya has external female genitals but no ovaries or uterus. The tabloid also claimed that she had undescended testes and testosterone levels three times higher than those of an average woman (but much lower than a man's). While we don't know if these reports are accurate (the tests are supposed to be private), we do know that Semenya was raised as a girl and sees herself as female. All of a sudden, at the age of 18, she was being told that she is not who she thinks she is.

Caster Semenya returned to the University of Pretoria, her athletic future uncertain. The IAAF allowed her to keep her gold medal because she did not commit fraud by competing as a woman. However, she did not run competitively with women in international track and field competitions for almost a year as she waited for the IAAF's decision regarding her eligibility. Finally, Semenya got some good news when, in 2010, she was declared free to compete professionally with other female athletes. She won a silver medal in the 800 meters at both the 2011 world championships and the 2012 Summer Olympics in London. Semenya also carried South Africa's flag in the opening ceremonies at those Olympics.

The IAAF had a huge problem. It wanted to answer a very difficult if not impossible question: what is the ultimate difference between males and females? Is it simply the ability to father a child or give birth once a person is sexually mature? If so, what about those who are unable to have children? Is there something that can be measured or tested that can show unequivocally that someone is a man or a woman? Or might there be still other outcomes to sexual development? Could someone be human and yet neither a man *nor* a woman?

6.1 How Do Males and Females Form?

disorder of sexual development (DSD) :: a condition sometimes noted at birth in which the genitalia are atypical

When a baby is born, one of the first questions everyone asks is whether it is a boy or a girl. It is not always easy to answer. Some scientists estimate that as much as 1.7% of the population has a **disorder of sexual development** (DSD). For some, that means development is profoundly different from the standard definition of female and male. Others could go through life never knowing they *had* a DSD. Why is there so much variability? To answer, we need to know how males and females form.

The Stages of Sex Determination

What determines sex? As we saw in Chapter 3, it all begins with *gametes*, the egg and sperm that unite to form a new organism, and it depends on the species (**FIGURE 6.2**). In some turtles and lizards, for example, the sex of the offspring is determined by the temperature at which eggs are incubated. Equal numbers of males and females will hatch at some temperatures, but if the environment is cooler or warmer, the sex ratio is skewed dramatically. In the marine worm *Bonellia*, newly hatched young move randomly, before settling on a solid surface. Exactly where the hatchling attaches determines its sex. If the young attaches to the front of a female body, it will develop as a male. If instead it attaches somewhere else, female development will be the outcome.

For some species of fish, sex is not even permanent. Born a male, the clownfish, *Amphiprion ocellaris*, changes into a female later on in life (**FIGURE 6.3**). In contrast, the black sea bass, *Centropristis striata*, starts its life as a male and can become female later. As these examples show, the development of male or female anatomy is not just complex: it may also have several stages, each with its own plan.

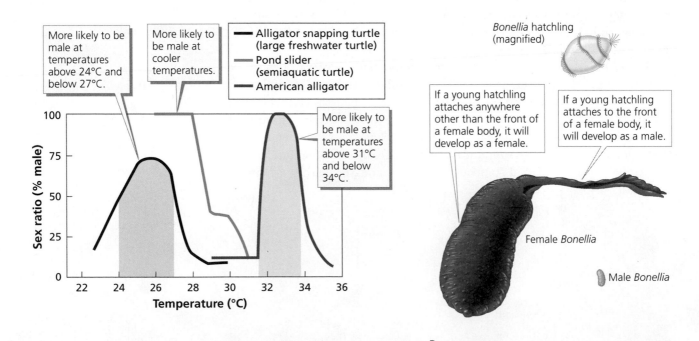

A

More likely to be male at temperatures above 24°C and below 27°C.

More likely to be male at cooler temperatures.

— Alligator snapping turtle (large freshwater turtle)
— Pond slider (semiaquatic turtle)
— American alligator

More likely to be male at temperatures above 31°C and below 34°C.

Sex ratio (% male)

Temperature (°C)

B

Bonellia hatchling (magnified)

If a young hatchling attaches anywhere other than the front of a female body, it will develop as a female.

If a young hatchling attaches to the front of a female body, it will develop as a male.

Female *Bonellia*

Male *Bonellia*

FIGURE 6.2 The mechanism of sexual determination differs by organism. In some reptiles, the sex of the offspring is determined by the temperature at which eggs are incubated (A), while in the marine worm *Bonellia,* the location at which a hatchling attaches determines its sex (B).

In humans and other mammals, the first set of instructions is found on the sex chromosomes (**FIGURE 6.4**). Humans have two types of sex chromosomes, X and Y, and normal cells have pairs of chromosomes. Males are XY and females XX. When gametes are produced for reproduction, males make sperm that have an X chromosome or a Y chromosome. Females make eggs that have an X chromosome. When the egg and sperm fuse together, the resulting zygote will get X from the mother and X or Y from the father.

So far this seems simple enough: XX means that the embryo will develop into a girl, and XY means that a boy will form. Girls have female sex organs, or **ovaries**. Boys have male sex organs, or **testes**. However, chromosomal sex is just the first piece of information in sexual development. It is responsible for **primary sex determination**: whether ovaries or testes form.

Secondary sex determination governs the development of other sex-related body characteristics (**FIGURE 6.5**). Men have external **genitals**, or sex organs. In addition to their external genitals (the vulva), women also have breasts, a **uterus** (or womb), and a vagina. Men and women also differ in the structure of the pelvic bone, voice tone, and the locations of body fat and hair. Chemical signals called **hormones** are responsible for these and other secondary sex characteristics. Let's take a look at the interplay of genes and hormones when sexual development goes "according to plan."

Chromosome Instructions

Until an embryo is six weeks old, it is not possible to tell whether it is male or female by its appearance; it is still bipotential for sex, meaning it has not yet made a physical commitment to be male or female. However, structures that will eventually become ovaries or testes are already present in the lower abdomen, awaiting instructions from specific genes (**FIGURE 6.6**).

Even these instructions are more than just XX in females versus XY in males. Ordinarily, the Y chromosome has the *SRY* gene, which is required for

FIGURE 6.3 For some species of fish, sex is not permanent. Born a male, the clownfish, *Amphiprion ocellaris*, changes into a female later on in life.

ovary :: female sex organ

testis :: male sex organ

primary sex determination :: the step in sexual development that determines whether ovaries or testes form

secondary sex determination :: the step in sexual development that governs the development of sex-related body characteristics such as external genitals, structure of pelvic bone, voice tone, and locations of body fat and hair

genital :: external sex organ

uterus :: the womb; the hollow, muscular organ located between the bladder and the rectum of females

hormone :: chemical signal responsible for physiological or developmental responses, including secondary sex characteristics

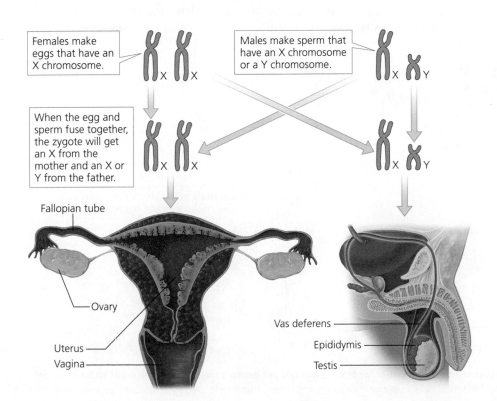

Females make eggs that have an X chromosome.

Males make sperm that have an X chromosome or a Y chromosome.

When the egg and sperm fuse together, the zygote will get an X from the mother and an X or Y from the father.

Fallopian tube
Ovary
Uterus
Vagina

Vas deferens
Epididymis
Testis

FIGURE 6.4 In humans and other mammals, primary sex determination comes from the sex chromosomes. They are responsible for whether ovaries or testes form.

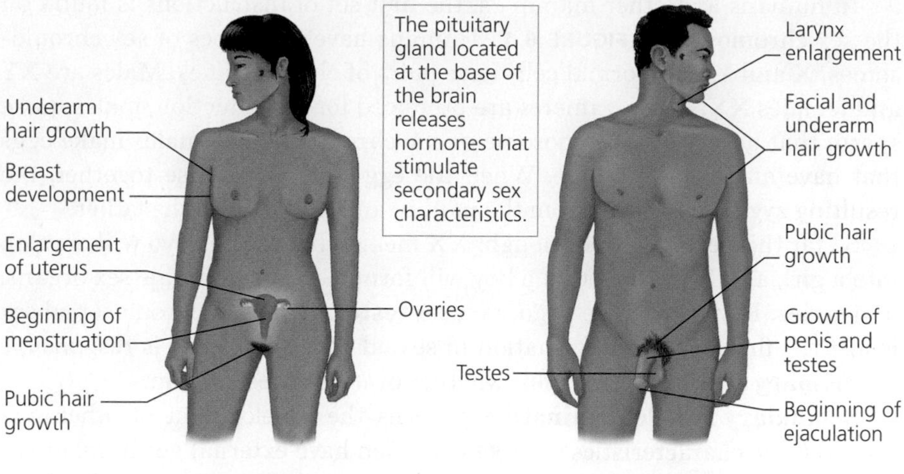

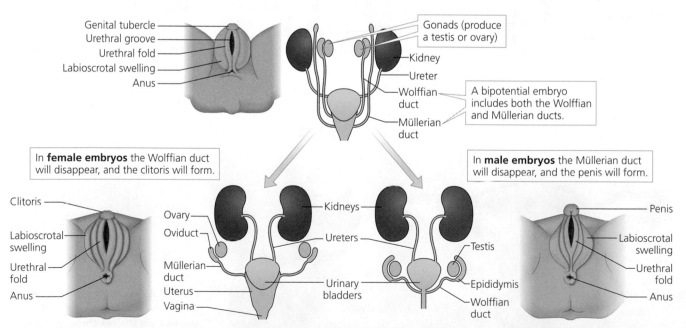

FIGURE 6.5 Secondary sex determination governs the development of other sex-related body characteristics.

Underarm hair growth

Breast development

Enlargement of uterus

Beginning of menstruation

Pubic hair growth

The pituitary gland located at the base of the brain releases hormones that stimulate secondary sex characteristics.

Ovaries

Larynx enlargement

Facial and underarm hair growth

Pubic hair growth

Growth of penis and testes

Testes

Beginning of ejaculation

the development of testes. If a mutation disables SRY, even though the embryo is XY, ovaries develop. If an XX embryo has the SRY gene because of a genetic error, testes develop. In normal females, the X chromosomes both have a gene called *DAX1*. An individual who is missing an X chromosome will be female but won't develop normal ovaries, because there is not enough DAX1.

Secondary sex characteristics develop in several stages—in the embryo, in fetal development, and at puberty. Until at least the seventh week of development, embryos, whether male or female, have the precursors of the internal features of both sexes. These features amount to a set of tubes and ducts—the Wolffian ducts in males and the Müllerian ducts in females. In XX embryos, the Wolffian ducts will disappear, while the Müllerian ducts will disappear in males.

How do genetic instructions handle all this internal sculpting? Cells in the testes and ovaries secrete the necessary hormones.

testosterone :: a hormone that promotes the development of the Wolffian ducts, the prostate gland, and the penis and regulates the descent of testes into the scrotum

Hormone Instructions

In XY embryos, the testes secrete two critical hormones. Anti-Müllerian hormone triggers the demise of Müllerian ducts. **Testosterone** promotes the

Genital tubercle
Urethral groove
Urethral fold
Labioscrotal swelling
Anus

Gonads (produce a testis or ovary)

Kidney

Ureter

Wolffian duct

Müllerian duct

A bipotential embryo includes both the Wolffian and Müllerian ducts.

In **female embryos** the Wolffian duct will disappear, and the clitoris will form.

In **male embryos** the Müllerian duct will disappear, and the penis will form.

Clitoris

Labioscrotal swelling

Urethral fold

Anus

Ovary
Oviduct

Müllerian duct
Uterus
Vagina

Kidneys

Ureters

Urinary bladders

Testis

Epididymis

Wolffian duct

Penis

Labioscrotal swelling

Urethral fold

Anus

FIGURE 6.6 Until an embryo is six weeks old, it is bipotential for sex—it has not yet made a physical commitment to be male or female. However, internal structures that will eventually become ovaries or testes and external structures that will eventually become the clitoris or penis are already present.

development of Wolffian ducts, the prostate gland, and the penis. Testosterone also inhibits breast formation and regulates the descent of testes into a pouch called the **scrotum**. In XX embryos, ovaries secrete **estrogen**, which promotes the development of the Müllerian ducts, oviducts (the passageway from the outside of the body to the ovaries), the uterus, and the upper end of the vagina. Ovaries also secrete some testosterone, just not as much as testes do.

External genitals do not start out looking male or female either (Figure 6.6). At six weeks, the embryo has a small bud between its legs. Eventually, it will form either a penis or clitoris. By nine weeks, there are two swellings on either side of a urogenital groove. By 14 weeks, this groove has disappeared in male embryos, and the scrotum has formed from the swellings. In females, the groove becomes the opening of the vagina, and the swelling becomes other folds of the female genitals (called the labia majora and labia minora). Amazingly, all these changes are regulated by hormones. That is why most babies are easy to identify as male or female at birth.

The final changes come with puberty, thanks to a dramatic increase in sex hormones. These changes make the differences between males and females even more obvious—most of the time.

scrotum :: a pouch that holds the testes

estrogen :: a hormone secreted by the ovaries that promotes the development of the Müllerian ducts, oviducts, the uterus, and the upper end of the vagina

6.2 What Happens If the Hormonal Signals Are Missing or Misread?

In almost 2% of men and women, however, events play out differently. In fact, there are more than 40 different types of disorders of sexual development (DSD). Two dramatic examples are androgen-insensitive syndrome in women and pseudohermaphroditism in men.

Androgen-Insensitivity Syndrome

In **androgen-insensitivity syndrome** (**AIS**), individuals who are XY do not develop as males because their cells are deaf to the instructions saying "make male parts." More technically, they lack receptors to bind to, or stick to, testosterone (**FIGURE 6.7**). Because AIS individuals are XY, the SRY gene functions correctly, and they develop testes. The testes make testosterone, which

androgen-insensitivity syndrome (AIS) :: a disorder of sexual development in which an individual who is XY does not develop as a male because the individual's cells lack receptors to bind to testosterone

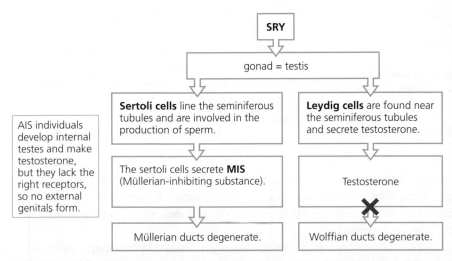

Eden Atwood is a jazz singer with AIS who lectures frequently about the disorder.

FIGURE 6.7 Individuals with androgen-insensitivity syndrome (AIS) lack the ability to respond to testosterone signals. Even though they are XY, they develop female physical traits.

Life Application

Gender Testing in Sports

Athletic competition brings out the best in some people—and the worst in others. In the 1930s, Stella Walsh and Helen Stephens dominated women's track and field. During the 1936 Olympic Games, the press reported rumors that both women were actually men. Walsh herself accused Stephens of being male. The Olympic Committee examined Stephens and confirmed that she had female genitals.

Well into the 1960s, the press fueled claims about gender identity. Male athletes from Eastern Europe were said to have bound their genitals in order to compete as women.

In the 1930s, Stella Walsh (left) and Helen Stephens (right) dominated women's track and field. During the 1936 Olympic Games, the press reported rumors that both women were actually men.

The reports worked up a nationalistic fervor. *There is no way*, people cried, *that those Communists are going to get away with cheating!* To compete in the 1966 European Athletics Championship, women had to agree to a humiliating examination. They had to see a gynecologist and also stand naked in front of a committee to have their external genitals checked. Yet of 243 women examined, none had abnormalities.

Chromosome testing to verify gender began with the 1968 Olympics. Unfortunately, these tests introduced new problems, for not everyone with a Y chromosome develops as a male.

In 1991, testing for the SRY gene was the "new and improved" way to verify gender. Results were difficult to interpret. At the 1992 Barcelona Olympics, some 2000 women were evaluated, and 15 were positive for SRY. (Further information about these athletes has remained confidential.) In 1996, at the Atlanta Olympics, 3000 were tested, and eight were positive for SRY. Of these, seven had AIS and one had another DSD that prevented testosterone from being active. All eight were allowed to compete.

Gender testing of athletes has never revealed anyone deliberately misrepresenting her gender. However, many women have had their lives turned upside down. Maria Patino, a hurdler from Spain, went to the World University Games in Japan. Patino has AIS (she is XY) and forgot to bring the letter from her doctor verifying her as female. Patino was required to take a gender verification test, and the results leaked out. She was stripped of all past titles, her scholarship was taken away, and her boyfriend left her. Patino was eventually reinstated in 1988 because she was able to show that her body could not use testosterone.

There is no evidence that women athletes with DSDs have an edge. What drove the enthusiasm for gender testing, at least in part, was fear—fear that men would masquerade as women to win competitions. In 1999, the Olympics Committee finally abandoned compulsory gender testing. Now, just as it has always been for male athletes, female athletes will be spared the ordeal of proving who they are. ::

causes Müllerian ducts to degenerate. However, without the right testosterone receptors, no external male genitals form. Rather, the testes remain internal. And since the Müllerian ducts are destroyed, no internal female body parts form either.

AIS embryos are exposed to estrogen from the mother's ovaries and the embryo's internal testes, too. This estrogen is responsible for breast development and the female pattern of hair and fat. As a result, a person with AIS will look and be female, even though she is XY and has developed testes but no ovaries. In many cases, women who have AIS were completely unaware of it until they tried to find out why they did not menstruate or why they were unable to become pregnant.

Sometimes AIS can be partial. Whispers of testosterone signals get through. When this occurs, a woman may be XY and possess external female genitals, but she may also have some male characteristics—such as greater muscle mass, little breast development, and a less female-shaped pelvis. Could Caster Semenya have partial AIS? We don't know.

Pseudohermaphroditism

Sexual development can also take an unexpected path in males. In **pseudo-hermaphroditism**, for example, an XY individual develops testes but is missing an enzyme that enables testosterone to send the right signals. External genitals appear female, but the Müllerian duct degenerates, and no female internal ducts form.

Then, at puberty, the adrenal gland produces testosterone, and everyone is in for a big surprise. What seemed to be a clitoris enlarges to reveal itself as a penis. The voice deepens, and muscles enlarge. Breasts do not develop, and menstruation does not begin. Having started life as a girl, this individual develops into a man. Although male pseudohermaphroditism is rare in the general population, it is less unusual in an isolated village in the Dominican Republic. There, 2% of the population that is born looking female develop into males after puberty. People in this village call the condition *guevedoce*, or "penis at age 12."

So far, we have referred to individuals with DSD as male or female. But should we instead leave open how to designate anyone in that 2% of the population? Does it even make sense to say that someone changes sex at puberty? How else do men and women develop differently? See *Life Application: Gender Testing in Sports* to learn how the question of whether and how to designate male and female has consequences beyond the individual.

pseudohermaphroditism :: a disorder of sexual development in which an individual who is XY develops testes but is missing an enzyme that lets testosterone send the right signals until puberty; having started life as a girl, such an individual develops into a man

How Do Men Produce Sperm?

Males are sperm factories. They make sperm continuously, from the time they reach puberty practically until death. (Sperm quality degrades as a man ages.) The production of sperm takes about 9–10 weeks. It involves chemical signals and communication, cell division, and cell differentiation (Chapter 3).

The Testes

Sperm are produced in the testes. Measuring approximately 5 cm in length, the testes are located in the pouch called the scrotum (**FIGURE 6.8**). Small muscles hold the testes in place. These muscles also regulate the position of the testes relative to the body. If the air is cold, testes are brought closer. When it is warm, the muscles relax and the testes hang lower. This "yo-yo" keeps the sperm healthy. Sperm are very sensitive to temperature. If they get too warm or too cold, they can't function.

Testes are sensitive body parts. One good hit from a baseball, a foot, or anything else produces significant pain. Why, then, are testes external to the body, where they are vulnerable to injury or accident? As we'll see in

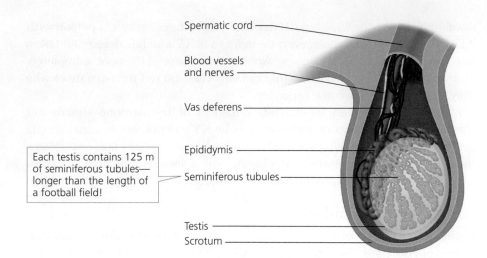

Each testis contains 125 m of seminiferous tubules—longer than the length of a football field!

Spermatic cord
Blood vessels and nerves
Vas deferens
Epididymis
Seminiferous tubules
Testis
Scrotum

FIGURE 6.8 Measuring approximately 5 cm in length, the testes are located in the pouch called the scrotum.

stem cell :: self-renewing cell

epididymis :: coiled tube sitting atop each testis, whose cells secrete chemical signals that help sperm to complete differentiation

Chapter 9, if the testes were inside the body, where the temperature is higher, they could not do their job.

The testes consist of around 30 wedge-shaped subunits. Each of these contains two or three tightly packed coils, called the seminiferous tubules. Astonishingly, each of the testes contains 125 m of seminiferous tubules—longer than the length of a football field! Lining the walls of the seminiferous tubules is a population of self-renewing cells, or **stem cells**. When these cells undergo mitosis (Chapter 3), on average one of the daughter cells replenishes the stem cell population. The other, called a primary spermatocyte, begins the long journey to becoming a sperm cell.

The primary spermatocyte undergoes meiosis I, producing secondary spermatocytes (Chapter 3). The secondary spermatocytes complete meiosis II. Finally, the spermatids made by this division differentiate into sperm (**FIGURE 6.9**).

Sperm and Semen

When sperm leave the testes, they are not completely mature. If these sperm were placed in a culture dish with a fertile egg, nothing would happen. The sperm don't have information yet about what to do. They get their next set of instructions when they move into another coiled tube, the **epididymis**. Cells

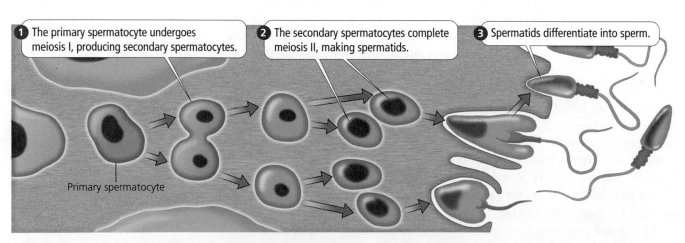

❶ The primary spermatocyte undergoes meiosis I, producing secondary spermatocytes.

❷ The secondary spermatocytes complete meiosis II, making spermatids.

❸ Spermatids differentiate into sperm.

Primary spermatocyte

FIGURE 6.9 When stem cells lining the seminiferous tubules undergo mitosis, one of the daughter cells, the primary spermatocyte, begins the long journey to becoming a sperm cell.

of the epididymis secrete chemical signals that help sperm to complete differentiation. Sitting atop each testis, the epididymis would be approximately 6 m long if straightened out—another marvel of packaging. Sperm are stored in the last stretch of the epididymis and remain there until a male is sexually aroused. The sperm then speed into yet another tube (the vas deferens), past ejaculatory ducts, and into the urethra, which runs through the penis. Finally, in ejaculation, the sperm are propelled outside the body.

Sperm are swimming cells. They need some type of liquid to get anywhere. Several glands secrete materials into the mix of fluids that bathe the sperm. Besides serving as the passageway for sperm, the urethra also carries urine. Even before ejaculation, two bulbourethral glands secrete a clear fluid into the urethra to neutralize any urine residue that might be present. The **prostate gland** secretes a milky white fluid that neutralizes the acidity of the vagina so that sperm will survive. Finally, two seminal vesicles secrete the sugar fructose, which serves as an energy source for sperm, and **prostaglandin**, a chemical signal that causes muscle contractions in the female reproductive tract. These contractions help move sperm closer to the egg. The final combination of sperm and all of these secretions is called **semen**.

prostate gland :: secretes a milky white fluid that neutralizes the acidity of the vagina

prostaglandin :: a chemical signal that causes muscle contractions in the female reproductive tract

semen :: the final combination of sperm and secretions by the prostate gland, seminal vesicles, and bulbourethral glands

Hormones and Sperm Production

Sperm have only one job to do: fuse with an egg and together make a zygote. Inadequate sperm production could lead to infertility. Making too many sperm could lead to problems with sperm quality. Fortunately, the entire process is carefully regulated by hormones.

The hormone that starts the whole chain of events is secreted by a part of the brain called the **hypothalamus** (**FIGURE 6.10**). When gonadotropin-releasing hormone (GnRH) is released by the hypothalamus, it stimulates the release of other hormones by the **pituitary gland**, also located in the brain. The pituitary secretes follicle-stimulating hormone (FSH) and luteinizing hormone (LH). In females, FSH and LH play a role in egg formation. In males,

hypothalamus :: a part of the brain that secretes the gonadotropin-releasing hormone (GnRH)

pituitary gland :: a part of the brain that secretes follicle-stimulating hormone (FSH) and luteinizing hormone (LH)

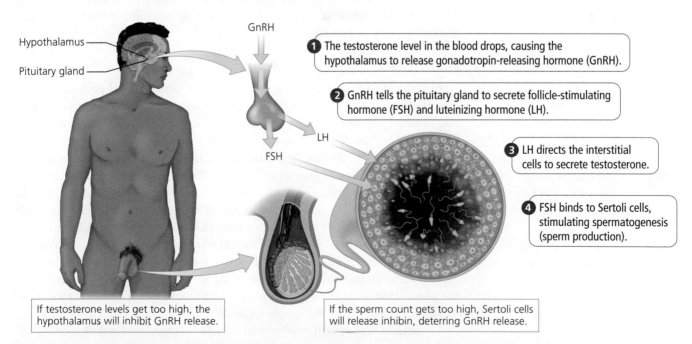

Hypothalamus

Pituitary gland

GnRH

LH

FSH

1 The testosterone level in the blood drops, causing the hypothalamus to release gonadotropin-releasing hormone (GnRH).

2 GnRH tells the pituitary gland to secrete follicle-stimulating hormone (FSH) and luteinizing hormone (LH).

3 LH directs the interstitial cells to secrete testosterone.

4 FSH binds to Sertoli cells, stimulating spermatogenesis (sperm production).

If testosterone levels get too high, the hypothalamus will inhibit GnRH release.

If the sperm count gets too high, Sertoli cells will release inhibin, deterring GnRH release.

FIGURE 6.10 Sperm production is regulated by a series of hormones. The hypothalamus releases GnRH and stimulates the pituitary gland, which in turn releases FSH and LH, stimulating testosterone production by the testes.

FSH stimulates sperm production, while LH triggers testosterone production by the interstitial cells of the testes. Interstitial cells are located in the tissues between seminiferous tubules.

The hypothalamus is the main player that monitors this hormone cocktail. If the testosterone level in the blood drops below a certain threshold, the hypothalamus secretes GnRH. Next, GnRH tells the pituitary to release LH and FSH. The LH directs the interstitial cells to secrete testosterone. FSH binds to Sertoli cells, which line the seminiferous tubules, stimulating spermatogenesis, the production of sperm cells.

If testosterone levels get too high, the hypothalamus can also shut the whole thing down by inhibiting GnRH release. And if the sperm count is too high, Sertoli cells release another signal, called inhibin, which deters GnRH release. Inhibin also acts directly on the pituitary to stop FSH release.

The careful dance between cells and hormonal signals makes the production of sperm sensitive to chemicals that resemble sex hormones, such as certain drugs or pollutants. Molecules that disrupt normal hormone signals can have serious consequences for sexual development, secondary sex characteristics, and fertility. Examples include anabolic steroids, which are sometimes abused for body building, and atrazine, a pesticide.

6.4 How Do Women Produce Eggs?

oogenesis :: the production of fertile eggs

Women have a much more involved task than men when it comes to gamete formation and reproducing. Even the production of fertile eggs, or **oogenesis**, is more complicated than the production of sperm. In addition to releasing eggs, a woman's body also needs to prepare for pregnancy. Whereas men develop sperm continuously, women are born with all the immature eggs, or **primary oocytes**, they will ever have.

primary oocyte :: immature egg

Men are able to ejaculate sperm multiple times per day, but women release one egg per month except during pregnancy. Behind that monthly pattern is the interaction between the brain and the female reproductive system—again orchestrated by hormones.

The Ovaries

Women have their gamete-making organs, or ovaries, tucked safely inside their bodies (**FIGURE 6.11**). In addition to maturing and releasing eggs, ovaries also produce hormones that are important for the development of secondary sex characteristics. The two ovaries alternate producing an egg each month. When an egg is propelled from the ovary, it enters a small tube called an **oviduct**. Each ovary is positioned next to its own oviduct. Motile hairlike structures line the inside of the oviduct and help to sweep the eggs into the uterus. Approximately the size and shape of an upside-down pear, the thick-walled, muscular uterus is the place where the fetus develops (Chapter 3). The **cervix** comprises the lower part of the uterus and the opening to the vagina.

oviduct :: a small tube down which an egg travels when it is propelled from the ovary

cervix :: the lower part of the uterus and the opening to the vagina

Oogenesis begins before a girl is even born. A female fetus develops approximately 2 million potential egg cells. By the time a baby girl is born, she has about 700,000, although by puberty only 350,000 remain. Fewer than 500 are ever released for possible fertilization.

Oogenesis, like sperm production, entails meiosis—but not at all continuously. Each oocyte starts meiosis I in the ovary and pauses, for years, until puberty. Think about it: a woman who is 35 years old has had oocytes stuck in

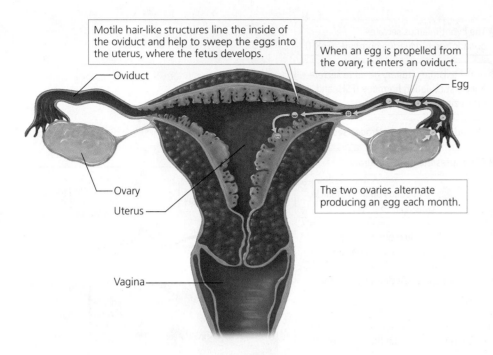

Motile hair-like structures line the inside of the oviduct and help to sweep the eggs into the uterus, where the fetus develops.

When an egg is propelled from the ovary, it enters an oviduct.

Oviduct

Egg

Ovary

Uterus

The two ovaries alternate producing an egg each month.

Vagina

FIGURE 6.11 Ovaries are the site of egg production in a woman's body. They mature and release eggs.

meiosis I for more than 35 years. No wonder the quality of eggs diminishes as women age.

Once puberty occurs, one oocyte completes meiosis I each month. One of these cells, the secondary oocyte, keeps most of the cytoplasm. The other tiny cell, called a polar body, does not develop further. The secondary oocyte starts meiosis II but stops and, about 12 hours later, bursts from the ovary. This event is **ovulation**. Some women feel ovulation as a sharp pain in the lower abdomen. If it fuses with a sperm, the secondary oocyte will complete meiosis II, producing a mature egg.

ovulation :: the release of an egg from the ovary

However, a woman's reproductive system is responsible for more than simply producing eggs. It must also support pregnancy should fertilization occur. Each month, an ovary releases an egg for possible fertilization, and the uterus prepares for pregnancy. These events are tightly coordinated by a sequence of hormones.

The Ovarian Cycle

At the start of an ovarian cycle, the hypothalamus secretes GnRH (**FIGURE 6.12**). As a consequence, the pituitary releases FSH and LH. (Remember that the hypothalamus, GnRH, LH, and FSH all play roles in forming sperm, too.) Each primary oocyte is surrounded by a single layer of cells; together, they are called a **follicle**. In women, FSH causes the follicle to grow. FSH and LH also stimulate estrogen secretion. Fluid builds up in the follicle, causing it to swell—until it bursts out of the ovary, releasing the egg.

follicle :: a primary oocyte surrounded by a single layer of cells

After ovulation, the body has no idea whether the egg has been fertilized. It begins the preparation for pregnancy in any case. First, the ruptured follicle develops into a structure called the corpus luteum. The corpus luteum secretes estrogen and yet another hormone, **progesterone**, which plays a role in the development of the lining of the uterus needed for a potential pregnancy. In fact, the estrogen and progesterone levels in the blood surge so high that no more follicles ripen as long as the corpus luteum lasts. It's as if the body is saying, "Let's hold off getting another egg ready until we see what

progesterone :: a hormone secreted by the corpus luteum that is used in the development of the endometrium or for preventing the development of a second follicle during one menstrual cycle

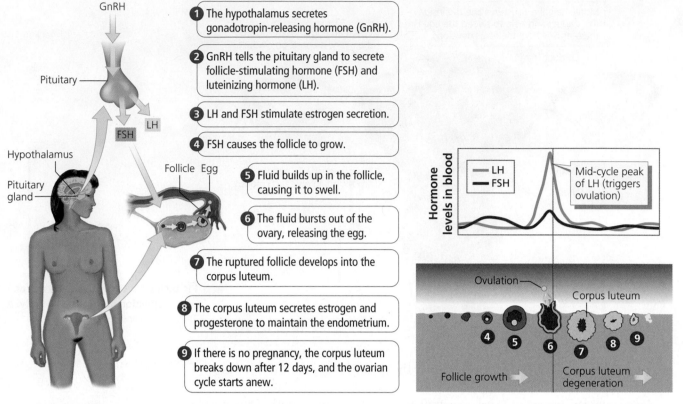

FIGURE 6.12 Egg production is regulated by a series of hormones. At the start of the ovarian cycle, the hypothalamus releases GnRH and stimulates the pituitary gland, which in turn releases FSH and LH, stimulating estrogen production and causing the follicle to grow until it bursts, releasing the egg.

menstruation :: the shedding of the lining of the uterus that supports fetal development

endometrium :: the lining of the uterus produced during the menstrual cycle to support the development of an embryo if a pregnancy occurs

happens to the one we just launched." But this holding pattern is temporary. If there is no pregnancy, the corpus luteum breaks down after 12 days, and the ovarian cycle starts anew.

The Uterine Cycle

If there is fertilization, it is up to the uterus to make sure that there is a place for the embryo to develop. Every month, if a woman is not already pregnant, the uterus prepares its inner surface for pregnancy. This uterine cycle (**FIGURE 6.13**) has three phases, starting with **menstruation**, the shedding of the lining of the uterus that supports fetal development. Once the old lining, called the **endometrium**, is gone, a new one is produced—stimulated by increasing estrogen levels. The body still has no clue whether a pregnancy or even intercourse is going to happen in the next couple of weeks. Yet the uterus gets ready just in case.

The second phase, ovulation, occurs roughly halfway through the 28-day cycle. The estrogen now prompts the cervix to secrete a thin, clear mucus that is easy for sperm to swim in.

You saw that the corpus luteum can survive around 12 days, even if there is no pregnancy. It also has an important job in the luteal phase—the third and final phase of the uterine cycle. It secretes estrogen and progesterone, which maintains the endometrium. If there is no pregnancy, then the corpus luteum degenerates, estrogen and progesterone levels plummet, and the endometrium breaks down. The menstrual fluid that flows from the body is a combination of blood and endometrial tissue.

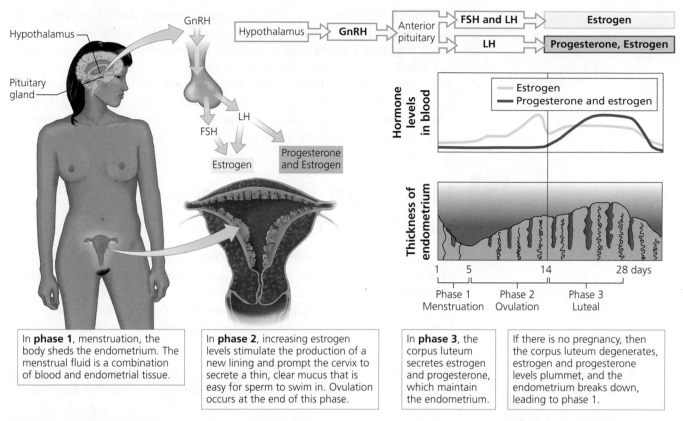

In **phase 1**, menstruation, the body sheds the endometrium. The menstrual fluid is a combination of blood and endometrial tissue.

In **phase 2**, increasing estrogen levels stimulate the production of a new lining and prompt the cervix to secrete a thin, clear mucus that is easy for sperm to swim in. Ovulation occurs at the end of this phase.

In **phase 3**, the corpus luteum secretes estrogen and progesterone, which maintain the endometrium.

If there is no pregnancy, then the corpus luteum degenerates, estrogen and progesterone levels plummet, and the endometrium breaks down, leading to phase 1.

FIGURE 6.13 Every month, if a woman is not already pregnant, the uterus prepares its inner surface for pregnancy. This uterine cycle has three phases.

Hormones and Pregnancy

Once a woman ovulates, the egg has only a brief time to encounter sperm. The journey from the ovary to the oviduct, uterus, vagina, and uterus takes about three days. However, fertilization leading to a successful pregnancy almost always occurs in the oviduct, where swimming sperm meet the egg.

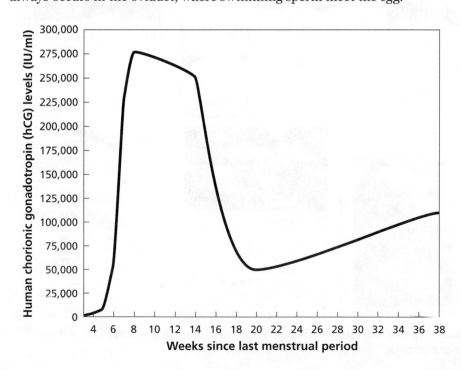

FIGURE 6.14 If fertilization occurs, the embryo travels down the oviduct and into the uterus, and the endometrium secretes human chorionic gonadotropin (hCG) hormone. During the early phases of pregnancy, hCG levels rise dramatically.

When fertilization does occur, the embryo travels down the oviduct and into the uterus. The embryo burrows into the wall of the uterus, a process called implantation. As a consequence, the endometrium does not break down and now secretes another hormone, human chorionic gonadotropin (hCG) (**FIGURE 6.14**). This hormone allows the corpus luteum to survive for another six to seven weeks. By the time the corpus luteum finally stops secreting estrogen and progesterone, the placenta makes enough progesterone to maintain the uterine lining on its own.

How Can Pregnancy Be Prevented?

The key to preventing pregnancy is keeping eggs and sperm apart. Of the roughly 300 million sperm ejaculated during sexual intercourse, only about 200 reach an oviduct. The female reproductive tract does its best to help the sperm succeed on their journey (see *How Do We Know? The Female Reproductive Tract Helps Sperm Find an Egg*). Even so, some of these sperm

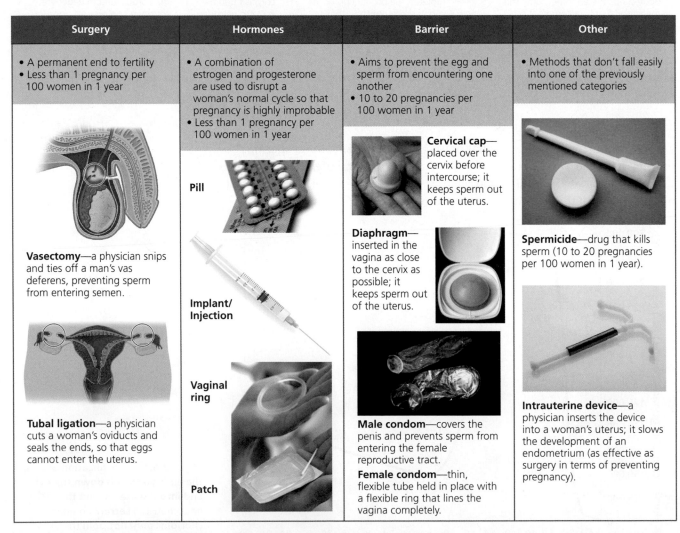

Surgery	Hormones	Barrier	Other
• A permanent end to fertility • Less than 1 pregnancy per 100 women in 1 year	• A combination of estrogen and progesterone are used to disrupt a woman's normal cycle so that pregnancy is highly improbable • Less than 1 pregnancy per 100 women in 1 year	• Aims to prevent the egg and sperm from encountering one another • 10 to 20 pregnancies per 100 women in 1 year	• Methods that don't fall easily into one of the previously mentioned categories

Vasectomy—a physician snips and ties off a man's vas deferens, preventing sperm from entering semen.

Tubal ligation—a physician cuts a woman's oviducts and seals the ends, so that eggs cannot enter the uterus.

Pill

Implant/ Injection

Vaginal ring

Patch

Cervical cap—placed over the cervix before intercourse; it keeps sperm out of the uterus.

Diaphragm—inserted in the vagina as close to the cervix as possible; it keeps sperm out of the uterus.

Male condom—covers the penis and prevents sperm from entering the female reproductive tract.

Female condom—thin, flexible tube held in place with a flexible ring that lines the vagina completely.

Spermicide—drug that kills sperm (10 to 20 pregnancies per 100 women in 1 year).

Intrauterine device—a physician inserts the device into a woman's uterus; it slows the development of an endometrium (as effective as surgery in terms of preventing pregnancy).

FIGURE 6.15 Birth control methods can be divided into four categories: surgery, hormones, barrier, and other. Each of these methods has various rates of success in preventing pregnancy.

The Female Reproductive Tract Helps Sperm Find an Egg

Fertilization takes teamwork—and not just the kind you might be thinking of. Sperm face a long and difficult swim to find a fertile egg. They could not do it without assistance. Fortunately, the female reproductive system has ways to help.

After ejaculation into the vagina, sperm speed toward the cervix—and their first fresh challenge. Cervical mucus selects for sperm that swim well. Before ovulation, the molecules in the mucus are arranged like a mesh, capturing poor swimmers. When a woman has a fertile egg available, the same fibers line up side by side to make a smooth path for

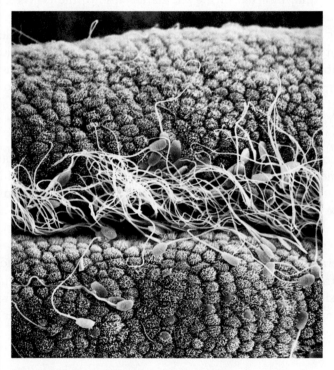

Scientists have found that the sperm that stick to the oviduct are healthier and swim better than those that remained unattached.

sperm. Contractions in the muscles of the uterus help move the sperm along even more quickly.

Once sperm enter the oviduct, there are only a few thousand left from a population that started with hundreds of millions. Luckily for the survivors, the oviduct is ready and waiting. It contains a reservoir for sperm, where sperm can bind to the oviduct's inner surface.

Sperm that stick to the oviduct have been found to be healthier and swim better than those that remained unattached, according to scientists who cultured specimens from male volunteers with pieces of the inner surface of oviducts. Other studies in cows, pigs, and rabbits have confirmed that sperm collect in the oviduct, binding to specific molecules. When ovulation occurs, the sperm are set free—if only in small numbers—to locate the egg.

Dr. Joanna Ellington, a veterinarian and sperm physiologist, found an unusual way to demonstrate that a sperm reservoir exists in humans, too. After she and her husband decided they did not wish to have any more children, she scheduled a tubal ligation for herself. However, Ellington told her surgeon that she and her husband were going to have intercourse first. She asked to have pieces of her oviducts removed during surgery, so that she could see whether sperm were attached to the inner surface. They were.

Scientists have found that fertilization can occur as late as five days after intercourse. Sperm can "relax" in the oviduct all that time, waiting for ovulation to occur. Once an egg is released, the sperm are directed to it by chemical signals secreted by the egg. Sperm can also sense the higher temperature in its vicinity.

All this assistance is not just altruism on the part of the female reproductive tract. The path from the vagina to the egg is like an obstacle course, where each step helps select for healthy, strong sperm. They are the winners, and the egg is first prize. ::

---

enter the oviduct near the ovary that did *not* ovulate. They have no chance at finding a healthy, fertile egg. As discussed in Chapter 3, it is a wonder that babies are ever conceived. Yet they are—enough that many people want to plan whether and when to bring a new person into the world. The birth control, or **contraceptive**, industry earns billions of dollars worldwide each year. Birth control methods can be divided into four categories: surgery, hormones, barrier, and other (**FIGURE 6.15**).

contraceptive :: birth control

Surgery

Surgery can provide essentially a permanent end to fertility. Men can undergo a **vasectomy**. In this procedure, a physician snips and ties off the vas deferens,

vasectomy :: a surgical procedure in which the vas deferens is snipped and tied off to prevent sperm from entering semen

tubal ligation :: a surgical procedure in which a woman's oviducts are cut and the ends are sealed so that the eggs cannot get into the uterus

preventing sperm from entering semen. Women can have a **tubal ligation**. Here, a physician cuts the oviducts and seals the ends so that eggs cannot enter the uterus.

Hormones

Hormones can be used to disrupt a woman's normal cycle so that pregnancy is impossible either because follicle development is inhibited or because endometrium formation is prevented. A combination of estrogen and progesterone can be delivered in a pill (sometimes daily), an implant, an injection, a vaginal ring, or a patch.

Barrier

cervical cap :: a barrier contraceptive that keeps sperm out of the uterus

diaphragm :: a barrier contraceptive placed in the vagina as close to the cervix as possible; it bars the way to the uterus

condom :: a barrier contraceptive that covers the penis and prevents sperm from entering the female reproductive tract

After abstinence, barrier methods are the oldest form of birth control. Modern barriers aim to prevent the egg and sperm from encountering one another. Placed over the cervix before intercourse, the **cervical cap** keeps sperm out of the uterus. A **diaphragm**, inserted in the vagina as close to the cervix as possible, similarly bars the way to the uterus. Male **condoms** take a different strategy: by covering the penis, they prevent sperm from entering the female reproductive tract at all. Finally, female condoms—thin, flexible tubes held in place with a flexible ring—line the vagina completely. Besides preventing contact between eggs and sperm, condoms also help prevent the spread of sexually transmitted diseases, including HIV/AIDS.

Other

spermicide :: drug that kills sperm

intrauterine device (IUD) :: a contraceptive inserted into a woman's uterus to slow the development of an endometrium

Finally, there are some birth control methods that don't fall easily into one of the previously mentioned categories. **Spermicides** are drugs that kill sperm. They are most effective when used together with a barrier method. An **intrauterine device** (**IUD**), which is inserted into a woman's uterus by a physician, slows the development of an endometrium. If an embryo did form, it would not be able to implant in the uterus.

The world desperately needs reliable, affordable, culturally sensitive, and easily available contraceptives. Not only do many couples want to plan their families in a responsible manner, but the population is growing fast. One new approach, a vaccine against hCG, has been successful in limited clinical trials. Future vaccines for women could target sperm.

6.6 What Causes Infertility, and How Can It Be Treated?

According to the American Medical Association (AMA), approximately 15% of all couples are infertile. **Infertility** is defined as a failure to conceive and become pregnant after a year of regular, unprotected intercourse. In 40% of the cases, the cause can be traced to the man, in 40% to the woman, and in 20% to both.

infertility :: failure to conceive and become pregnant after a year of regular, unprotected intercourse

Causes of Infertility

A common reason for male infertility is a low sperm count. Sometimes the problem is not the overall number of sperm, but a shortage of healthy ones (**FIGURE 6.16**). Scientists do not completely understand why some men fail to make enough normal sperm. But there is a clear relationship between sperm counts and smoking, alcohol, or drug use. In addition, injury to the testes can

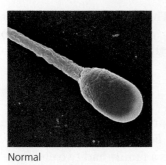
Normal

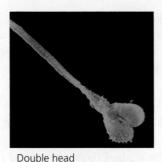
Double head

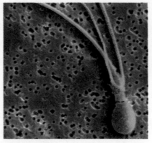
Multiple tails

Malformed head

FIGURE 6.16 One reason for male infertility is abnormal sperm.

damage the delicate network of ducts, making it difficult for sperm to enter the vas deferens.

Sperm counts are declining in men in developed, industrialized countries (**FIGURE 6.17**). Reasons proposed by scientists include a sedentary job or lifestyle (which elevates the temperature of the testes, and sperm cannot tolerate high temperatures), or exposure to harmful pollutants.

Infertility in women may also have several causes. Sometimes the hormonal cycles don't operate properly. Consequently, ovulation fails to take place or the uterine lining is not prepared properly. Another possibility is **endometriosis**, a painful condition in which uterine tissue grows elsewhere in the body—usually in the oviducts and in the ovaries. The tissue responds to hormones just like it would in the uterine lining. It grows and is shed with each cycle, but it is not easily eliminated from the body. This can lead to inflammation and disruption of ovulation. Sometimes female infertility is caused by blocked oviducts due to **pelvic inflammatory disease** (**PID**). PID can be the result of a sexually transmitted disease or even an infection caused by an IUD.

Disorders of sexual development (DSD) can cause infertility, too. For example, XY women who have complete androgen-insensitivity syndrome

endometriosis :: a painful condition in which uterine tissue grows elsewhere in the body, leading to inflammation and disruption of ovulation

pelvic inflammatory disease (PID) :: infertility caused by blocked oviducts that may be a consequence of a sexually transmitted disease or infection caused by an IUD

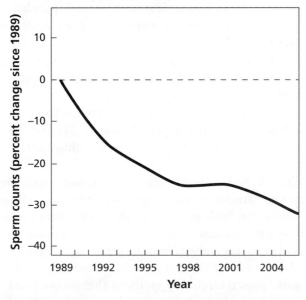

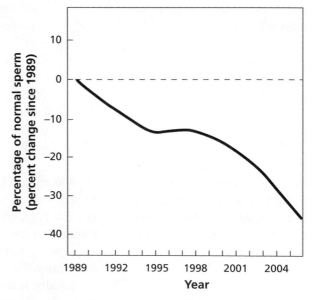

Source of Data: M. Rolland, J. Le Moal, V. Wagner, D. Royere, J. De Mouzon, Decline in semen concentration and morphology in a sample of 26,609 men close to general population between 1989 and 2005 in France. *Human Reproduction 2012*; **28**: 462-470.

FIGURE 6.17 Sperm counts and overall quality of sperm have been declining in men in developed, industrialized countries. In France, for example, both the total number of sperm and the percentage of normal sperm have declined in 35-year-old men.

(AIS) are sterile; they lack ovaries and a uterus. This returns us to our earlier question: what defines male or female? Must a male have the potential to father a child? Must a female be able to give birth? If there are solutions to infertility, could they change who we are?

Infertility Treatments

The billions of dollars spent each year to prevent pregnancy are rivaled by the billions devoted to making pregnancies happen. For most of human history, an infertile couple who wished to stay together as a family had two choices—remain childless or adopt a child. In 1978, however, all that changed with the first "test tube" baby. Now a healthy woman in her 30s and a mother herself, Louise Brown led the vanguard of a revolution in reproductive medicine, **assisted reproductive technology (ART).**

ART is used to bring healthy sperm and eggs together so that fertilization is possible. Its ultimate goals are the implantation of an embryo, a successful pregnancy, and the birth of a healthy baby.

assisted reproductive technology (ART) :: reproductive procedure that is used to bring healthy sperm and eggs together so that fertilization is possible

artificial insemination :: sperm from the male partner are placed in the female partner's vagina

- With **artificial insemination**, sperm from the male partner are placed in the female partner's vagina. This method can be helpful for couples in which the man has a low sperm count, because sperm can be collected and concentrated. If the sperm concentration is still too low, sperm from another donor can be used instead.

intrauterine insemination :: sperm are placed directly into the uterus

- **Intrauterine insemination** is a variation of artificial insemination. Here sperm are placed directly into the uterus. The idea is that sperm with motility problems won't have to swim as far to meet an egg. Also, if the woman's vagina is a hostile environment for her partner's sperm, depositing sperm directly into the uterus can help.

in vitro fertilization (IVF) :: immature oocytes are extracted and placed in a petri dish in a solution that permits them to mature, and they are combined with sperm from a male partner and then transferred into the uterus

- With **in vitro fertilization (IVF)** (**FIGURE 6.18**), the method used to conceive Louise Brown, a physician isolates one or more eggs from the female partner. The oocytes are placed in a petri dish in a nutrient-rich solution. Sperm, generally collected by the male partner, are then combined with the oocytes. In two to four days, a small number of embryos, each at approximately the eight-cell stage, are transferred into the uterus. One or more may implant, resulting in a pregnancy.
- Gamete intrafallopian transfer (GIFT) is a variation of IVF. The egg and sperm are placed into the oviduct together, or the one-celled zygote is placed there.
- Intracytoplasmic sperm injection (ICSI) is yet one more variation. With this method, a physician injects a single sperm into an egg. If an embryo results, it will be placed in the uterus no later than the eight-cell stage.

surrogacy :: an embryo is placed in the uterus of a woman who is not the biological mother in order to establish a healthy pregnancy that results in the birth of a baby

Sometimes ART doesn't work, and the couple opts for **surrogacy**. With surrogacy, an individual or a couple arranges for a woman to give birth to a baby. For example, artificial insemination with sperm from the male partner could initiate the pregnancy. The egg then comes from the surrogate. Alternatively, a couple's egg and sperm could make the embryo, and the surrogate then has the pregnancy. The baby then has no genetic relationship with the surrogate.

Finally, infertility is sometimes treatable by methods that do not involve manipulating eggs or sperm outside the body. Fertility drugs such as gonadotropic hormones can stimulate ovaries and help women who have problems with ovulation. Also, if infertility is due to a physical blockage of an oviduct, for example, surgery may help.

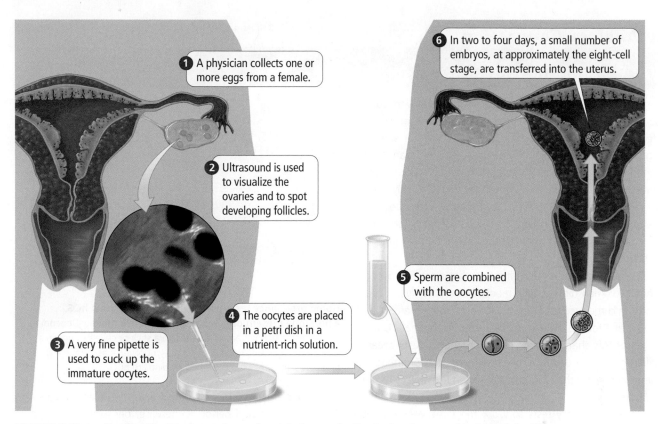

FIGURE 6.18 *In vitro* fertilization is one type of assisted reproductive technology used to treat infertility.

More than 3 million children worldwide have been born thanks to these medical breakthroughs. Starting with one IVF baby in 1978, the number of ART babies now exceeds 200,000 per year. The road to a biological child through ART can be arduous. The process is long and involved, expensive, potentially painful, and not guaranteed to succeed.

Happy families can come about in many ways. The love any of us has to give to our biological children can instead be showered on adopted children, extended family, and friends.

6.7 How Can We Tell If a Fetus or Baby Is Healthy?

In every welcome pregnancy, parents look to the future with great excitement. (See *Technology Connection: Home Pregnancy Tests* to learn how pregnancies are confirmed.) Whatever else their dreams for their child, the top priority for these parents is the baby's health. Fortunately, an array of tests can now detect possible medical problems, so that medical professionals can help and future parents can make appropriate decisions.

Blood and Urine Tests

Information about fetal health can be obtained from the mother's blood or urine. It is also gathered from measurement and observation of the fetus. In some cases, fetal cells are collected for analysis.

Technology Connection

Home Pregnancy Tests

Although it might not seem to rank up there with curing disease, home pregnancy tests were an important innovation. Before 1978, a woman wishing to know whether she was pregnant as early as possible had to see a physician. Some women would not see a doctor to find out. No doubt some did not want their doctor to know they were sexually active. But many simply wanted privacy: they wanted to learn whether they were pregnant, but not in front of someone else. And some women, worried they might be pregnant, did not have the courage to find out.

The need for a physician visit meant that more than 50% of pregnant women did not seek medical care until two or more months into the pregnancy. In other words, more than 50% of embryos and fetuses were at risk of inadequate prenatal care.

All of this changed once home pregnancy tests became available. Women can now purchase a kit at a pharmacy or supermarket. And they can take the test in the privacy of their own homes.

Home pregnancy tests all work the same way. An antibody detects hCG, a hormone present only when there is a pregnancy. Most tests rely on a stick, with a results window. Depending on the test, a woman either urinates on the stick or collects urine in a cup and then dips a stick in it. Then she waits for the results—often a symbol such as + or −. An indicator molecule causes the "pregnant" symbol to appear if the antibody binds to hCG.

The test is 99% accurate—especially if the first urine of the day is checked, when hCG levels are at their highest.

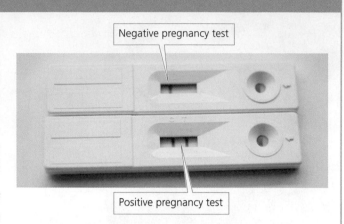

Negative pregnancy test

Positive pregnancy test

Home pregnancy tests use an antibody to detect hCG, a hormone present only when there is a pregnancy. Depending on the test, a woman either urinates on a stick or collects urine in a cup and then dips a stick in it; often a symbol, such as a double line or a plus sign, appears if the antibody binds to hCG.

The results are also more accurate if a woman waits at least a week after a missed period. Otherwise, because conception occurs days before implantation, a woman could get a negative result even if she is pregnant. Or she could get a positive result for a pregnancy that does not progress any further. Approximately 50% of all conceptions don't lead to successful pregnancy.

Regardless of the results, a woman should retest in a few days. And if the results remain positive, or if she would like to have negative results verified, the next thing to do is to see a physician. ::

Blood samples from the mother can show whether she is in good health. Physicians can determine her blood type, whether she is anemic, and whether she has been exposed to infections that she could pass on to her newborn. Of particular concern are sexually transmitted diseases, especially HIV/AIDS, syphilis, and hepatitis. Most women will also be tested to see whether they have ever had rubella (German measles) or chicken pox. Screening is also done to determine if the mother carries defective genes that could be passed on. Depending on family history and ethnicity, other tests may be offered too.

Urine tests are routine, occurring with almost every visit to the obstetrician, to monitor the mother's health. Glucose levels, for example, can show whether diabetes is present. High protein levels sometimes mean the mother has preeclampsia, or high blood pressure that endangers some pregnancies.

Screens and Diagnostic Tests

Most women will be offered **screening tests** to see whether an embryo or fetus is at risk for a particular problem. If screens are positive, they may prompt **diagnostic tests**, which are more precise. As we have already seen, diagnostic tests are also important when family history or ethnic background

screening test :: evaluation to see whether an embryo, fetus, baby, child, or adult is at risk for a particular problem

diagnostic test :: precise evaluation of risks to an embryo or fetus

amniocentesis :: a method for examining fetal cells that can be done between 14 and 17 weeks of pregnancy

chorionic villus sampling (CVS) :: procedure used to isolate fetal cells that can be done between 10 and 12 weeks of pregnancy

indicates an increase in the risk of genetic disease (see Chapter 4). Other risk factors include problems in a previous pregnancy or the mother's ill health.

Several screens during the first and second trimesters of pregnancy can assess fetal health. Because the circulatory systems of the mother and fetus are connected, sampling the mother's blood provides a look at what is in fetal blood too. Some of the blood tests measure the levels of certain molecules that are known to vary when Down syndrome, spina bifida, or other disorders are present. Ultrasound is used to measure the fetus and to detect physical problems.

If the results of screening raise a concern, the next step is diagnostic testing. **Amniocentesis**, a method for examining fetal cells, can be done between 14 and 17 weeks of pregnancy (**FIGURE 6.19**). A long needle is inserted through the abdomen into the uterus, and fluid is withdrawn. This fluid contains fetal cells that can be isolated, cultured, and examined to see whether the chromosomes are normal. Also, chromosomes can be tested to see whether genes associated with hemophilia, muscular dystrophy, or cystic fibrosis are present.

Chorionic villus sampling (**CVS**) can be used to isolate fetal cells, too (**FIGURE 6.20**). This procedure, somewhat riskier than amniocentesis, can be done between 10 and 12 weeks of pregnancy. With CVS, the fetal cells are collected from the chorion, the fetal part of the placenta.

This toolbox of screens and diagnostic tests raises important questions. How should one decide which tests to have? If tests are done, what should one do with this information? Tests like these increase the chance of a healthy birth. However, in some cases, serious defects are detected for which there is no cure. Is there any point in having this information, or is it just a burden on already anxious parents? Even in the most dire of circumstances, treatments may still be available to lessen symptoms. In addition, parents may have time before birth to prepare for a child's special needs, or in other circumstances, parents may decide to terminate the pregnancy.

These tools may also force us to evaluate what we even mean by a "defect" or "normal." What characteristics in a child are acceptable to us? Once again, we must face our ideas of what constitutes a good life. With rigorous prenatal testing, many individuals with disorders of sexual development can be identified before birth. What might Caster Semenya's fate have been with prenatal screening? Could medical treatments have helped? Would different parents have chosen to terminate? Understanding the biology is not the hard part. Figuring out the right thing to do with the knowledge can be intensely difficult.

Newborn Tests

Finally, the nine months of waiting are over, labor and delivery are complete, and the obstetrician or midwife places the baby on the mother's chest. The baby, already alert, looks into the mother's eyes. It is a moment hard to describe (**FIGURE 6.21**). And then, after the hard work of being born, the baby is whisked away briefly so that tests can begin to see whether the baby is healthy.

At one and five minutes after birth, a healthcare provider assesses the baby using the **APGAR scale**. Developed by Dr. Virginia Apgar, this test records the baby's **a**ctivity, **p**ulse, **g**rimace response, **a**ppearance, and **r**espiration (see *Scientist Spotlight: Virginia Apgar*). Each category is scored 0, 1, or 2 (**FIGURE 6.22**). The totals alert doctors and nurses if some help is needed. The baby is also measured and weighed at this early stage.

All U.S. states require newborn screening tests and procedures, although specific requirements vary from state to state. Because newborns do not yet have all the molecules necessary for blood clotting, they receive an injection of

FIGURE 6.19 Amniocentesis is a method used to examine fetal cells between 14 and 17 weeks of pregnancy. A long needle is inserted through the abdomen into the uterus, and fluid is withdrawn.

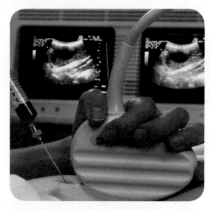

FIGURE 6.20 Chorionic villus sampling (CVS) is used to isolate fetal cells between 10 and 12 weeks of pregnancy. The procedure is somewhat riskier than amniocentesis because the fetal cells are collected from the chorion—the fetal part of the placenta.

FIGURE 6.21 When labor and delivery are complete and the obstetrician or midwife places the baby on the mother's chest; it is a moment hard to describe.

APGAR scale :: an assessment of a newborn baby conducted at one and five minutes after birth to assess a baby's activity, pulse, grimace response, appearance, and respiration

Activity (muscle tone)	Absent	Arms and legs flexed	Active movement		
Pulse	Absent	Below 100 beats per minute	Over 100 beats per minute		
Grimace (reflex irritability)	Flaccid	Some flexion of extremities	Active motion (sneeze, cough, pull away)		
Appearance (skin color)	Blue, pale	Body pink, extremities blue	Completely pink		
Respiration	Absent	Slow, irregular	Vigorous cry		
Score	0	1	2		

Total score	
0–3	Severely depressed
4–6	Moderately depressed
7–10	Excellent condition

FIGURE 6.22 At one and five minutes after birth, a healthcare provider assesses a baby using the APGAR scale, which records the baby's activity, pulse, grimace response, appearance, and respiration. The totals alert doctors and nurses if some help is needed.

vitamin K. Babies also get eye drops to prevent infection from any microbes to which they may have been exposed on the trip from the uterus into the world. In many hospitals, newborns are also given hearing tests.

Since the 1960s, blood samples have also been taken. Early detection can give physicians the opportunity to suppress at least 30 diseases or lessen the symptoms. For example, all babies are screened for PKU, a disease that interferes with metabolism. Individuals with PKU are missing an enzyme that breaks down the amino acid phenylalanine. As phenylalanine builds up, it causes brain damage. Fortunately, if PKU is detected in a baby, a phenylalanine-free diet can prevent the disease entirely.

Scientist Spotlight

Virginia Apgar (1909–1974)

When women gave birth in a hospital in the early 20th century, many of them did so under anesthesia. And then the newborn baby went straight off to the hospital's nursery, without so much as an examination. For 90% of infants, this practice was just fine. However, 9% of babies need medical attention; 1% may not survive without it. As an anesthesiologist, Virginia Apgar had assisted in thousands of deliveries. She realized that there was a real need to identify which newborns need help. Although other physicians resisted at first, her test is now used all over the world.

Born in 1909, Apgar decided in high school to become a physician. She studied zoology at Mt. Holyoke and began medical school in 1929—a month before the stock market crash and the start of the Great Depression. Despite serious financial pressure, she graduated fourth in her class at Columbia University College of Physicians and Surgeons in 1933, where she won a highly competitive surgical internship.

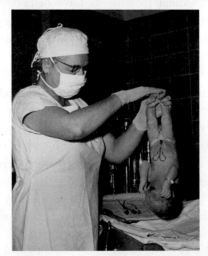

In 1952, Dr. Virginia Apgar designed a test to determine whether a newborn needed to be resuscitated. Now known as the APGAR test, it has held up over decades of use and study.

For all Apgar's achievement, Dr. Allen Whipple, her mentor, discouraged her from a career in surgery. People simply would not go to a "woman surgeon." A talented surgeon himself, Whipple understood that surgery could not improve unless anesthesiology did. In 1937, however, anesthesia was generally handled by nurses. Whipple thought that Apgar was just the person to help transform the field. She took his advice.

After a year of training, Apgar returned to Columbia P&S as director of the anesthesia division. She was also the division's *only* member until the mid-1940s. Other physicians wanted no part of a low-prestige field. Finally, in 1946, anesthesiology became a medical specialty that required residency training. In 1949, Apgar was appointed full professor, the first woman ever at Columbia P&S to reach this rank.

Apgar studied the effects on newborns of anesthesia given to their mothers during labor. She originally designed her test in 1952 as a guide for whether a newborn needed to be resuscitated. Now known as the APGAR test, it has held up over decades of use and study. Data from more than 17,000 babies have demonstrated that Apgar scores, especially taken five minutes after birth, can predict with great accuracy both newborn survival and neurological development.

In 1959, Dr. Apgar decided to leave academic medicine so she could devote herself to the prevention of birth defects. She worked diligently to educate the public and to raise money for research. Perhaps we are fortunate that she did not become a surgeon after all. ::

6.8 What Tests Are on the Horizon?

Scientists continue to develop screening methods that are less invasive and more accurate. One new idea takes advantage of the discovery that fetal DNA, not enclosed in cells, circulates in the mother's bloodstream. Scientists can collect this DNA and make many copies of it (Chapter 7). The fetal DNA can then be examined for genetic disorders.

Scientists are also exploring whether **gene therapy**, the correction of a faulty gene, could be possible for embryos (Chapter 8). We can already screen for genetic disorders in embryos produced by IVF and related methods. Could gene therapy fix these defects so that the embryos will develop into healthy babies?

gene therapy :: the correction of a faulty gene

We can all agree on the goal of healthy babies. OK, but what does *healthy* mean? Biological and technological breakthroughs appear inevitable. But will everyone have access to these technologies? Would Caster Semenya be alive today if they did, and what sex would she be? And who pays the bill? We return to these and other issues in the next two chapters.

Biology in Perspective

Reproduction is the most important thing we do; it ensures the survival of our species. And one of the first questions asked upon hearing of the birth of a baby is whether it is a boy or girl. Perhaps we are so fascinated with sex and gender because reproduction matters so much.

Like any system in the body, the reproductive system usually develops in a fixed way. So do sexual characteristics. But not always. Individuals with disorders of sexual development (DSD) are often seen as profoundly different from others. A defect in the circulatory system or a heart problem does not often come with the same stigma. Why? What is it about sex and gender that we find so defining? Reproduction seems to be about more than just biology.

Biologically, the reproductive system's job is to produce offspring. Men produce sperm, women eggs, and if hormones and everything else line up, fertilization and pregnancy result. People spend a lot of effort and money to control pregnancy. Some try to prevent it. Infertile couples try to achieve it.

Once a welcome pregnancy has been established, we have a responsibility to do what we can to protect the health of the embryo and fetus. And when the baby is born, our responsibility expands to include this new member of the human family.

Chapter Summary

6.1 How Do Males and Females Form?

Describe how chromosome and hormone instructions interact to determine sex.

- Different species employ different methods of determining sex, including chromosomes, temperature, and chemical cues.
- Human primary sex characteristics are determined by sex chromosomes. Genes in female chromosomes (XX) result in ovaries; genes in male chromosomes (XY) result in testes.
- Other gender-specific characteristics (secondary sex determination) are the result of hormones produced by the ovaries (estrogen) and testes (testosterone).

6.2 What Happens If the Hormonal Signals Are Missing or Misread?

Predict the outcome of a change in hormone production or response.

- Defective hormonal signals and production, or signals that are misread, can result in disorders of sexual development (DSD).
- In AIS, receptors for testosterone are missing or too few and individuals develop testicles instead of ovaries but have female secondary sexual characteristics.
- In one form of pseudohermaphroditism, an enzyme that converts testosterone to DHT is defective. Pseudohermaphrodites with this condition initially develop female secondary sex characteristics but at puberty develop male secondary sexual characteristics because of increased testosterone production.

6.3 How Do Men Produce Sperm?

Outline the steps of sperm production in males.

- Stem cells in the seminiferous tubules of the testes produce sperm through meiosis.
- Immature sperm leave the testes and enter the epididymis, where they complete differentiation and are stored until sexual arousal.

- When sexually aroused, sperm are pushed into the vas deferens, where they encounter substances from different glands that stimulate swimming, neutralize acidity of the vagina, and stimulate contraction in the female reproductive tract.
- The final mix of fluid and sperm is called semen.
- In response to testosterone levels, the hypothalamus releases the hormone GnRH that stimulates the pituitary gland to produce FSH and LH.
- In males, FSH stimulates sperm production and LH stimulates testosterone production.

6.4 How Do Women Produce Eggs?

Outline the steps of egg production in females.

- Women are born with all the primary oocytes they will have.
- The two ovaries present in women alternate, producing an egg each month through a process called ovulation.
- In ovulation, an egg completes meiosis I, stops at meiosis II, and bursts from the ovary. The egg is swept into the oviduct (fallopian tubes) to the uterus.
- If an egg fuses with a sperm, it completes meiosis II, producing a mature egg.
- In the ovarian cycle, the hypothalamus secretes GnRH, stimulating the pituitary to release FSH and LH.
- In women, FSH causes estrogen secretion and a layer of cells surrounding an oocyte to grow and burst out of the ovary.
- The ruptured follicle develops into the corpus luteum, which secretes estrogen and progesterone, changing the lining of the uterus in preparation for pregnancy and preventing ovulation of other oocytes.
- Without pregnancy, the corpus luteum breaks down and restarts the ovarian cycle.
- The uterine cycle has three phases: menstruation, ovulation, and luteal.

6.5 How Can Pregnancy Be Prevented?

Compare methods used to prevent pregnancy.

- Keeping the egg and sperm apart prevents pregnancy.
- Contraceptive methods include surgery (blocks passage of the egg), hormones (inhibit follicle development or prevent endometrial formation), a physical barrier (blocks passage of sperm), or other methods (kill sperm or slow the development of the endometrium).

6.6 What Causes Infertility, and How Can It Be Treated?

Predict causes of infertility and propose potential treatments.

- Infertility is the failure to conceive after one year. Its cause can be traced to males and females equally.
- Male infertility results from low sperm count or few healthy sperm. It has been linked to smoking, alcohol, drug use, and damage to testes.
- Female infertility occurs when ovulation is disrupted, oviducts are blocked, or the endometrium does not develop properly.
- Assisted reproductive technology, including artificial insemination, intrauterine insemination, and *in vitro* fertilization can be used to help infertile couples conceive a child.
- Other solutions for infertility include surrogacy and fertility drugs.

6.7 How Can We Tell If a Fetus or Baby Is Healthy?

Explain the purpose of and methods for testing fetal and newborn health.

- Blood and urine tests can find conditions that may harm a fetus.
- Screening tests help determine whether a fetus is at risk, whereas diagnostic ones are more precise tests for specific conditions. Tests include ultrasound, amniocentesis, and chorionic villus sampling.
- Our ability to detect defects is greater than our ability to treat many defects.
- The APGAR scale represents a newborn's general health and awareness immediately following birth.

6.8 What Tests Are on the Horizon?

Evaluate the ramifications of further refinements in genetic tests and treatments.

- Fetal DNA may be sampled from fetal blood. In the future, this DNA could be tested to find genetic disorders.
- Gene therapy could potentially be used to correct genetic disorders early so that they never develop.

Key Terms

amniocentesis 176
androgen-insensitivity syndrome (AIS) 161
APGAR scale 177
artificial insemination 174
assisted reproductive technology
 (ART) 174
cervical cap 172
cervix 166
chorionic villus sampling (CVS) 176
condom 172
contraceptive 171
diagnostic test 176
diaphragm 172
disorder of sexual development
 (DSD) 158
endometriosis 173
endometrium 168
epididymis 164

estrogen 161
follicle 167
gene therapy 179
genital 159
hormone 159
hypothalamus 165
in vitro fertilization (IVF) 174
infertility 172
intrauterine device (IUD) 172
intrauterine insemination 174
menstruation 168
oogenesis 166
ovary 159
oviduct 166
ovulation 167
pelvic inflammatory disease (PID) 173
pituitary gland 169
primary oocyte 166

primary sex determination 159
progesterone 167
prostaglandins 165
prostate gland 165
pseudohermaphroditism 163
screening test 176
scrotum 161
secondary sex determination 159
semen 165
spermicide 172
stem cell 164
surrogacy 174
testis 159
testosterone 160
tubal ligation 172
uterus 159
vasectomy 171

Review Questions

1. List three different methods for determining sex in species. Give an example for each.

2. Diagram the process of primary and secondary sex determination in human females. Be sure to include the hormones and organs involved.

3. Diagram the process of primary and secondary sex determination in human males. Be sure to include the hormones and organs involved.

4. What is primary sex determination?
   a. A step in sexual development in which genitalia are different with respect to the number of chromosomes associated with the male or female sex
   b. A step in sexual development that governs sex-related body characteristics
   c. A step in sexual development that determines whether ovaries or testes form
   d. A step in sexual development that determines whether external genitals form, the structure of the pelvic bone, voice tone, and locations of body hair
   e. A step in sexual development that initiates the production of hormones

5. What is androgen-insensitivity syndrome?
   a. A disorder of sexual development in which an individual who is XY does not develop as a male because their cells lack receptors to bind to testosterone
   b. A disorder of sexual development in which an individual starts life as a girl and then develops into a man at puberty
   c. A disorder of sexual development in which an individual develops both male and female genitalia
   d. A disorder of sexual development in which an individual who is XX develops as a male
   e. A disorder of sexual development in which an individual who is XX develops both male and female genitalia

6. How do the formation of eggs and sperm differ? How are the processes alike?

7. Complete the following table for these six sex hormones.

Hormone	Produced by which organ or tissue	Effect on males	Effect on females
GnRH			
FSH			
LH			
Testosterone			
Estrogen			
Progesterone			

8. How do hormones control how sperm form?

9. How do hormones control oogenesis and the uterine cycle?

10. What happens to the corpus luteum when there is a pregnancy? What happens to it when there is no pregnancy?

11. How does a bipotential embryo become male or female? (Consider both its internal and external structures.)

12. How can pregnancy be prevented? How effective is each method?

13. Which of the following is NOT a barrier method of birth control?
    a. Cervical cap
    b. Diaphragm
    c. Male condom
    d. Intrauterine device
    e. Female condom

14. What are possible causes of infertility in men? In women?

15. In which type of infertility treatment are sperm placed directly into the uterus?
    a. *In vitro* fertilization
    b. Intrauterine insemination
    c. Artificial insemination
    d. Gamete intrafallopian transfer
    e. Intracytoplasmic sperm injection

16. What prenatal tests are performed during pregnancy?

17. How is the health of a newborn evaluated?

The Thinking Citizen

1. If you were trying to develop a male birth control pill, what aspects of a man's reproductive biology might you target? Explain why.

2. What are the potential advantages for a fish species that changes sex over the course of its life cycle?

3. Some individuals are born with a defective enzyme that results in production of testosterone. If an individual is born XX with this defective enzyme, would this person have testicles or ovaries? What secondary sex characteristics would the individual have? Would the person be fertile or infertile? Be sure to explain your answers.

4. In the 1960s it came to light that some male children had XX chromosomes. An attempt was made to reassign their gender to female. These efforts were largely unsuccessful. Would you consider someone who looked male but had XX chromosomes male or female? What about individuals who are pseudohermaphrodites? Explain your reasoning for each.

5. BPA is a chemical used in plastics that mimics estrogen. How might BPA in baby bottles disrupt sperm formation? What effect might BPA have on preadolescent females?

6. A couple comes in to a clinic because they have been unable to conceive a child. What would you check on the man and on the woman? What options would you suggest if the problem was that sperm count was low and some sperm were deformed?

7. Why doesn't assisted reproductive technology (ART) always work? What are possible reasons for the failure of these methods to produce a healthy baby?

The Informed Citizen

1. Assisted reproductive technology (ART) such as *in vitro* fertilization generates embryos that will never be placed in a woman's body. What should be done with these embryos? Who should decide? Who "owns" them?

2. Is there a *right* to reproduce biological offspring? Why or why not?

3. Should health insurance companies be required to cover the costs of infertility treatments and ART? Why or why not?

4. Should parents be permitted to refuse all prenatal and newborn screening and testing? Why or why not?

5. What rights and responsibilities should sperm or egg donors or surrogates have for a child born as a result of their contribution?

6. Gender is an important category in our society. Not everyone fits neatly into "male" or "female." Who decides the gender of an individual? The person? Society? A medical professional? Is it always essential to assign gender? Why or why not?

7. Huntington's disease is a neurodegenerative genetic disorder that causes mental decline generally in middle age, along with other physical symptoms. Knowing that there is currently no treatment for Huntington's, would you test your fetus? Why or why not?

CHAPTER LEARNING OBJECTIVES

After reading this chapter, you should be able to answer the following questions:

7.1 Why Are Plants Such Good Sources of Food?
Explain how the bodies of plants are structured and specialized to store food molecules.

7.2 How Do Plants Make Food?
Summarize the light reactions and Calvin cycle phases of photosynthesis.

7.3 What Are the Goals of Genetic Engineering in Plants?
Describe the goals for genetic engineering of crop plants.

7.4 How Was Golden Rice Engineered?
Outline the steps used to produce a genetically modified organism.

7.5 How Else Is Genetic Engineering Being Used?
Describe how areas other than agriculture use genetic engineering.

7.6 What Are the Risks of Genetic Engineering?
Assess the risks of genetic engineering in terms of safety, the economy, and effectiveness.

7.7 How Ethical Is Genetic Engineering?
Evaluate the ethical considerations related to genetic engineering that confront scientists.

With modern genetic engineering, we can take genes from almost any species and insert them into almost any other species, but questions about the wisdom and safety of doing so persist. Bell peppers, *Capsicum annuum*, are just one of many plants species that have been genetically engineered.

CHAPTER 7

Plants, Agriculture, and Genetic Engineering

Can We Create Better Plants and Animals?

Becoming an Informed Citizen ...

Domesticating plants and animals is a form of genetic engineering we have been using for over 10,000 years. Today, we can take genes from almost any species and insert them into almost any other species to create new features. But is genetic engineering safe, and is it wise?

enetic modification has been going on for billions of years—for as long as living organisms have been on Earth. Every day, each organism faces challenges to its survival and reproduction. Natural selection occurs because some individuals are more successful than others at meeting their daily challenges and passing on genes to the next generation. As a result, each generation differs genetically from the previous one, and members of each species adapt by becoming better at survival and reproduction. All species evolve, and evolution is a naturally occurring form of genetic engineering.

Domesticating plants and animals is a deliberate form of genetic modification that we have been using for over 10,000 years. Human intervention leads to new combinations of genes in organisms. With modern **genetic engineering**, however, we can accomplish far more in much less time. We can take genes from almost any species and insert them into almost any other species—at least in theory. With thousands of genes and millions of species to work with, we can create an almost unlimited number of new living features. This natural treasure trove of genetic information is one of the main reasons why many scientists want to protect biodiversity on our planet: if we lose species, we also lose the genetic information they carry. But can genetic engineering be done safely, and is it wise? Consider how two scientists used genetic engineering in the hope of feeding the planet.

genetic engineering :: the process of taking genes from one species and inserting them into another species

---

case study

Golden Rice

It all began in 1984 over beers at an international agricultural meeting in the Philippines. Gary Toenniessen of the Rockefeller Foundation asked a group of plant scientists this question: *how can we use the new technology of genetic engineering to improve rice?* Since rice is the principal food staple for billions of people, making it more nutritious would benefit a significant percentage of the world's population, including many in underdeveloped countries.

The scientists put their heads together and agreed on the answer: *engineer rice so that it can be a source of vitamin A.* This idea immediately generated great enthusiasm. Why? At least 750 million children in the world have diets that are deficient in vitamin A (**FIGURE 7.1**). Vitamin A deficiency (VAD) can lead to terrible consequences. According to the World Health Organization (WHO), 250,000 to 500,000 children go blind every year because VAD damages eyes. Half these children may die each year because VAD also increases susceptibility to many infectious diseases.

Vitamin A is found in many foods: carrots, tomatoes, meat, and milk, for example. But many economically disadvantaged people do not have access to balanced nutrition, and among these people VAD is a persistent problem. Most cases of VAD occur in parts of Asia, where rice is the major source of calories. Rice fortified with vitamin A could reduce malnourishment and help save lives.

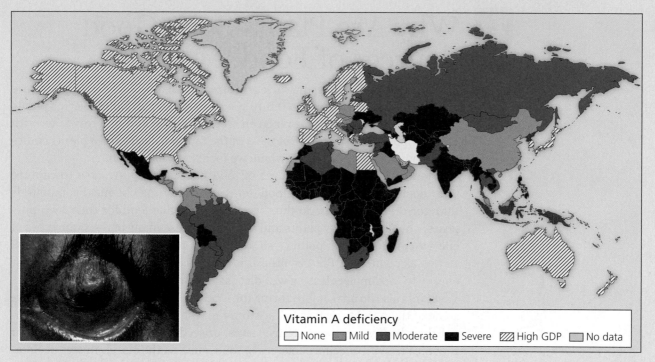

FIGURE 7.1 At least 750 million children in the world have diets that are deficient in vitamin A, but in countries with high GDP, this deficiency is not of public health significance. Insufficient vitamin A in the diet can lead to serious health problems, including xerophthalmia; left untreated, this disorder may lead to blindness.

It's a long journey from an idea hatched over beers to a scientific payoff. At the time, the only way to get rice to make a biochemical precursor of vitamin A was to take genes from a species that can make it and insert them into rice—a formidable task. When eaten, this precursor can be converted into Vitamin A by the body. Toenniessen's question lay on the table for almost a decade. A chance meeting provided the spark needed for the answer.

Ingo Potrykus, a German scientist living in Switzerland, and Peter Beyer, a German scientist living in Germany, met in 1992 on a plane as they traveled to New York City to attend a conference organized by the Rockefeller Foundation. Potrykus was an expert at manipulating genes in rice, and Beyer was a specialist in the biochemistry of vitamin A production. They realized that their complementary specialties put them in a unique position to solve the problem of making rice a source of vitamin A.

Together they wrote an ambitious proposal to insert the genes for making vitamin A into rice seeds. The Rockefeller Foundation took a chance and funded them. And in 1999, *golden rice* was born.

Potrykus and Beyer pioneered many techniques and made discoveries that advanced the emerging field of genetic engineering. However, the genetic engineering used to develop golden rice and other crops has raised complex questions about safety, ethics, and other considerations. In this chapter, we will examine how genetic engineering works and how these techniques are being applied. We will also address an important social issue: what *should* we do with our new creative powers? But first we will take a look at how plants work to better understand why they make such good targets for genetic engineering.

7.1   Why Are Plants Such Good Sources of Food?

Plants come in a bewildering variety of sizes and shapes. Flowering plants are the largest group, with over 250,000 species (**FIGURE 7.2**). Nonflowering plants such as ferns, mosses, and evergreen trees like pine and fir provide us with many important resources. But most of the food we derive from plants comes from flowering ones, which is where we focus our discussion.

Plants are an excellent source of food because they are able to capture solar energy and turn it into food—energy-containing chemical compounds that our cells can use to do their work. Plants make food for their own purposes, not ours. Luckily plants and animals can use many of the same types of energy-containing compounds, so we are able to use these molecules almost as well as plants can. Photosynthesis is the process plants use to convert solar energy to food molecules; we will discuss this later in the chapter.

Plants also do a great job storing the food molecules they produce. They use some of them right away, but they store the rest in specific locations throughout their bodies—in roots, stems, leaves, or seeds, for example. This behavior provides us with specific packages of food, such as the starch in rice or potatoes, which we can harvest and eat without having to consume the entire plant. In this section, we will discuss how the bodies of plants are structured, and how different parts of their bodies are specialized to store food molecules.

Despite their astonishing diversity, the basic structure is similar for all flowering plants. They have four types of organs: roots, stems (including branches), leaves, and flowers that produce seeds and fruit (**FIGURE 7.3**). These organs are interconnected by two vascular, or internal transport, systems. The first system of tubes, xylem, transports water and minerals throughout the plant. The second, phloem, transports food and other chemicals like hormones. Each of the four organs can be used as a storage organ for vitamins and edible compounds such as starch and sugar.

Roots

root ::: an organ that anchors a plant to its surface and enables it to absorb water and nutrients from the soil

root hair ::: tiny lateral extension of a root's outer cells that absorbs water and minerals from the soil

The main functions of **roots** are to anchor the plant to its surface and to absorb water and nutrients from the soil. Typically (but not always) roots are below ground, and they grow downward and away from the light—the opposite of what the aboveground parts do.

The larger, older parts of roots do most of the anchoring. As roots grow into the soil, the younger growing tips produce a dense array of **root hairs**: tiny lateral extensions of the root's outer cells. Root hairs absorb water and

A   B   C   D   E   F

FIGURE 7.2 There are more than 250,000 species of flowering plants. Some examples include: carrot (A), daisy (B), orchid (C), apple (D), tulip (E), and rose (F).

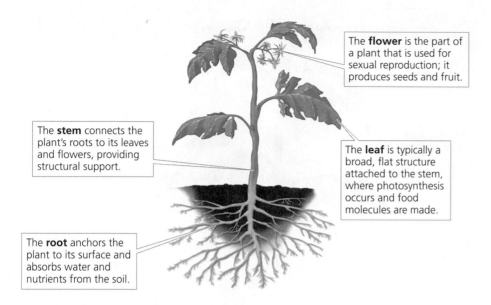

The **flower** is the part of a plant that is used for sexual reproduction; it produces seeds and fruit.

The **stem** connects the plant's roots to its leaves and flowers, providing structural support.

The **leaf** is typically a broad, flat structure attached to the stem, where photosynthesis occurs and food molecules are made.

The **root** anchors the plant to its surface and absorbs water and nutrients from the soil.

FIGURE 7.3 Flowering plants have four types of organs: roots, stems (including branches), leaves, and flowers.

minerals from the soil. Because there are so many of them, they have a very large surface area, with lots of soil contact. This allows root hairs to absorb as much water and as many minerals as possible (**FIGURE 7.4**). Rice plants, for example, have a fibrous root network that spreads out for maximum water and mineral absorption; more than 90% of the roots go no deeper than 20 cm (8 in.).

Many plants store food molecules, especially starches, in their roots. Since many plants lose their aboveground parts each winter, root storage is the only reliable location. From the plant's perspective, this food reserve supplies the energy to grow new aboveground parts in the spring, or perhaps to make flowers and seeds. From our perspective, root storage organs provide an important food source. We eat a lot of roots, including beets, carrots, parsnips, radishes, rutabagas, sweet potatoes, and turnips.

Roots are not always below ground; nor are they always recognizable as roots. Tropical strangler figs, for example, begin their lives as seeds germinating in the upper branches of other trees. They send aerial roots down to the soil, and as these roots grow, they surround the original trunk. Eventually the original tree may die and rot away, leaving a hollow cylinder of strangler fig roots as the "trunk." Also in rainforests, dirt and organic matter can accumulate on the upper branches of trees, forming what is essentially a mat of soil many meters above the forest floor. Many trees grow roots from their branches to penetrate these mats and extract rainwater and nutrients from them, just as they do from the forest floor.

Stems

Stems connect the plant's roots to its leaves and flowers. Stems provide structural support for the plant, holding all the parts together and defining the plant's overall shape. Stems allow plants to grow taller and compete for sunlight. The vascular system in stems ties the plant parts together physiologically. Their **xylem** transports water and minerals from roots to leaves and flowers, and the **phloem** carries food made in the leaves through the stems to the roots and other parts of the plant.

Although all plant stems share the same basic components, stem structure varies in different species. For example, the main stem of rice plants is round

FIGURE 7.4 Root hairs absorb water and minerals from the soil. Numerous fine root hairs on this radish seedling increase the absorptive surface area of the root.

stem :: connects a plant's roots to its leaves and flowers; it provides the structural support for the plant

xylem :: part of a stem that transports water and minerals from the roots to the leaves and flowers

phloem :: part of the stem that carries food made in the leaves through the stems to the roots and other parts of the plant

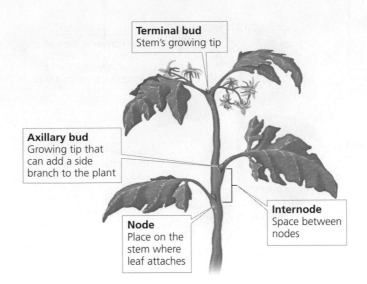

Terminal bud
Stem's growing tip

Axillary bud
Growing tip that
can add a side
branch to the plant

Internode
Space between
nodes

Node
Place on the
stem where
leaf attaches

FIGURE 7.5 Plant stems connect plant roots to leaves and flowers and provide structural support. At various points along their length, stems have nodes.

Terminal bud

FIGURE 7.6 Buds are the plant's insurance policy, a way to start up new growth if anything happens to the existing parts. If the terminal bud is damaged or does not exist, the axillary buds are released, and the plant may become bushier.

leaf :: a broad or long, flat structure attached to the stems by a stalk; it is the place in the plant where photosynthesis occurs and food molecules are made

and hollow. At various points along their length, stems have nodes (**FIGURE 7.5**). Each node has a leaf and a structure called a bud. Some buds grow into leafy stems that become the plant's side branches, while other buds grow into flowers. There is also a bud at the tip of each stem, and as this bud grows, the stem itself gets longer.

The plant's shape is determined by which buds grow and by how long each stem becomes. In many plants, the bud at the stem's tip (called the terminal bud) "dominates" other buds on the stem, preventing them from growing. This allows the plant to elongate and grow taller to reach the sunlight. If the terminal bud is damaged—eaten by an herbivore or pinched off by a gardener, for example—the side buds are released from domination and the plant may become bushier (**FIGURE 7.6**). In a sense, buds are the plant's insurance policy, a way to start up new growth if anything happens to the existing parts.

Some stems are modified as storage organs, similar to what happens to some roots. Potatoes, which store starch, are the most familiar case. Potatoes are underground stems called tubers, and each "eye" (indentation in the potato) is actually a node from which a leaf and new stem can grow. Other stems are sources of nutrients, vitamins, fiber, and other useful products. Asparagus, bamboo shoots, cactus pads, and water chestnuts fall into this category. Sugar comes from the stems of sugarcane and maple syrup from the "stem" (trunk) of sugar maple trees. Aspirin, cinnamon, curare, and quinine come from the bark of tree stems. And although we can't eat it, wood comes from the stems of trees, and fibers from stems of flax and hemp are used to make textiles.

Like roots, stems can be modified in many ways by different species of plants. In some species, underground stems called rhizomes grow horizontally just below the surface to help plants spread out. Occasionally a rhizome sends up a vertical stem, which we see as the "plant." Irises and ginger have rhizomes. Other horizontal stems, called stolons, lie just above the ground surface, like the "runners" of strawberry plants. Stems can also be broad and flat, taking on many of the functions of leaves. Cactus pads are a good example.

Leaves

Leaves typically are broad, flat structures attached to stems by a stalk; in some species, such as rice, the leaves wrap around as a sheath. Each species has a characteristic leaf shape, and often leaf shape helps to identify plants.

Leaves are the plant's biochemical factories, where photosynthesis occurs and food molecules are made. This is one of the reasons leaves are so nutritious.

Like roots and stems, leaves sometimes are modified to act as storage organs. When leaves serve as storage organs, they typically are located underground. **Bulbs** are short, underground stems with fleshy leaves that store nutrients and food. Onions and garlic are two familiar examples. The spicy chemicals they store in their bulbs prevent herbivores from eating them. Ironically, we eat them precisely because they have these spicy chemicals. In succulent plants, such as aloe, ice plant, and stone plants, found in arid places, the leaves are modified to store water.

In most cases, we eat regular aboveground leaves. We eat leaves mainly for their nutrients, vitamins, and fiber rather than for their food storage. The list of "leafy vegetables" is a long one and includes arugula, bok choy, cabbage, chard, cress, endive, lettuce, and spinach. Celery and rhubarb are actually the stalks of these plants' leaves (rhubarb leaves themselves are toxic). Many of the herbs we use to flavor our food are leaves as well.

Leaves are modified to perform many different functions (**FIGURE 7.7**). In cactus, the stems take on the photosynthetic function of leaves, and the leaves are modified as spines. In some climbing plants, all or part of the leaf is modified to produce a tendril that anchors the plant onto whatever it climbs. Some of the most dramatically modified leaves are found in butterworts, pitcher plants, sundews, and Venus flytraps. These plants are all carnivorous, and their leaves are adapted in various ways to capture and digest small insects and other prey.

bulb :: short, underground stem with fleshy leaves that store nutrients and food

Flowers

Flowers come in a delightful array of colors, sizes, shapes, and smells, and each species has its own unique blossom. Technically, flowers are not organs in their own right. Unlike other plant organs, flowers are used for only one purpose: sexual reproduction.

From an evolutionary perspective, reproduction is the goal of every living creature, and plants are no exception. In the case of plant flowers, reproductive energy is directed toward two purposes: providing offspring with a food supply to use when the offspring are very small (in seeds), and providing

flower :: part of a plant that is used for sexual reproduction; it produces seeds and fruit

In the cactus plant, the leaves are modified as spines, which protect the plant from being eaten. It is the stem that does photosynthesis.

In the Venus flytrap, leaves are adapted in various ways to capture and digest small insects and other prey.

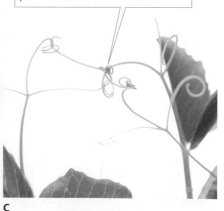

In the pea plant, leaves are modified to produce tendrils that anchor the plant to whatever it climbs on.

A   B   C

FIGURE 7.7 Leaves are the plant's biochemical factories, where photosynthesis occurs and food molecules are made. Leaves come in many different shapes and sizes and are modified to perform different functions.

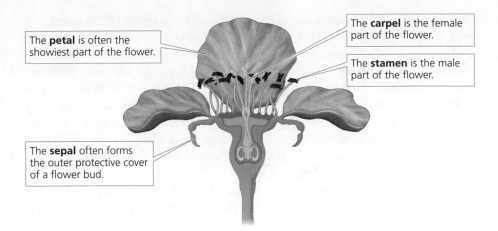

The **petal** is often the showiest part of the flower.

The **carpel** is the female part of the flower.

The **stamen** is the male part of the flower.

The **sepal** often forms the outer protective cover of a flower bud.

FIGURE 7.8 Flowers are the parts of the plant used for sexual reproduction. Flowers are derived evolutionarily from modified leaves and consist of four whorls: sepals, petals, stamens, and carpels.

sepal :: the outer protective cover of a flower bud

petal :: showiest part of the flower

stamen :: the male part of a flower

carpel :: the female part of a flower

seed :: a plant embryo surrounded by a protective coating

endosperm :: cells that surround a plant embryo and form the bulk of the seed

seed coat :: a protective cover over the seed that keeps it from drying out

animals with a food supply to entice them to disperse young plants to new homes (in fruits).

Flowers are derived evolutionarily from highly modified leaves, on a short stem. In fact, flowers consist of four sets, or whorls, of leaves, each modified for a different purpose (**FIGURE 7.8**). The outermost whorl makes the **sepals**, which often form the outer protective cover of a flower bud. The next whorl forms the **petals**, often the showiest part of the flower. The innermost whorls are, in order, the **stamens** and the **carpels**, which form the male and female parts of the flower.

While many flowers are edible (broccoli and cauliflower are examples), their main nutritional value comes from seeds, such as rice, and fruit. Their importance cannot be overemphasized: seeds and fruit form a large portion of the diet for virtually the entire human population.

Seeds A plant **seed** (**FIGURE 7.9**) contains a tiny embryonic plant, complete with a root, stem, and one or two leaves. Surrounding the embryo are cells called **endosperm**, which form the bulk of the seed. The **seed coat** forms a protective cover over the seed to prevent it from drying out.

When a plant makes a seed, it pumps the endosperm full of food molecules—sugar, starch, proteins and amino acids, oils, and other nutrients. The embryonic plant digests food in the endosperm when the seed germinates (begins to grow), and the endosperm provides energy until the young seedling is big enough to make its own food.

The endosperm contains everything a plant needs to begin its life. Luckily for us, we can use the endosperm for food as well. Corn kernels and beans are

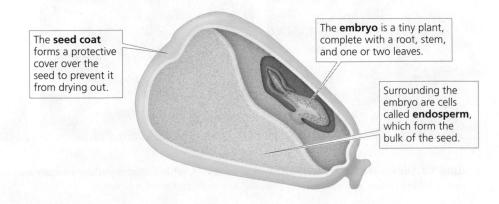

The **seed coat** forms a protective cover over the seed to prevent it from drying out.

The **embryo** is a tiny plant, complete with a root, stem, and one or two leaves.

Surrounding the embryo are cells called **endosperm**, which form the bulk of the seed.

FIGURE 7.9 A plant seed contains a tiny embryonic plant. Corn seeds house a small embryo and endosperm that is covered by a seed coat.

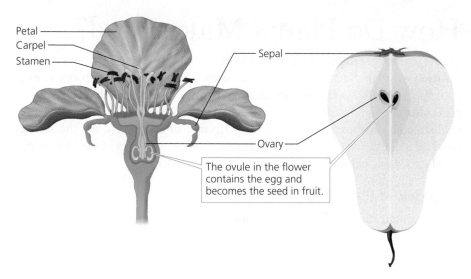

Petal
Carpel
Stamen
Sepal
Ovary

The ovule in the flower contains the egg and becomes the seed in fruit.

FIGURE 7.10 **Seeds form inside the ovary of a flower, and as a seed ripens, it develops into a fruit.** Botanically, any structure that develops from a flower's ovary is a fruit.

fruit :: any structure that develops from a flower's ovary

Dandelion fruits help seeds disperse by wind.

Coconuts are fruits that float around the ocean to disperse.

Touch-me-nots have fruits that explode and spray out seeds like tiny missiles.

Burdock fruits have tiny hooks that cling to passing mammals.

FIGURE 7.11 **Fruits help seeds disperse.**

actually seeds, and the endosperm of these two seeds alone comes close to providing a completely balanced diet. The nutrition we get from barley, beans, corn, peanuts, peas, rice, wheat, and many other "seed crops" also comes from the endosperm. We obtain many useful oils from the endosperm of corn, cotton, peanuts, safflower, and other plants. It is safe to say that a large proportion of the diet for the entire human race—and many other animal species—comes from plant endosperm.

Fruits Seeds form inside the female part of the flower, in a structure called the ovary (**FIGURE 7.10**). As the seeds ripen, the ovary develops into a **fruit**. Botanically, any structure that develops from a flower's ovary is a fruit, even things we consider to be vegetables and other structures we don't eat. Tomatoes are fruits, not vegetables, and the little tufts that help dandelion seeds blow in the wind are fruits as well.

While fruits come in many forms, they all have the same function: they help the seeds disperse (**FIGURE 7.11**). Some, like dandelion fruits, maple fruits, and cotton, help seeds disperse by wind. Coconuts are fruits that float around the ocean to disperse. Touch-me-nots have fruits that explode and spray out seeds like tiny missiles. Burdock fruits have tiny hooks that cling to passing mammals (on either fur or clothes), hitching a ride to disperse. Burdock fruits were the inspiration for Velcro.

But the structures we think of as fruits are the ones that are tasty and edible: apples, cherries, oranges, and the like—or maybe even pea pods. From the plant's perspective, the idea behind these types of fruits is simple: they are eaten by animals. The seeds pass through the digestive system unharmed, protected by their seed coat. Since plants are not mobile, they employ the mobility of animals to disperse their seeds. They reward these helpful animals with tasty and nutritious fruit to secure their services. By the time the seeds pass through, the animal has moved to a new location, effectively dispersing them. As an added bonus, the animal "plants" the seeds in a bit of fertilizer to help them get a good start in life. We have capitalized on naturally occurring fruits, selectively breeding plants to grow larger and more nutritious fruits—the ones we buy and eat today.

7.2 How Do Plants Make Food?

Plants capture solar energy and use it to make sugar, a process called **photosynthesis**. Sugar has a lot of calories, a measure of the stored energy it contains. Cells use the energy stored in sugar to perform the biochemical reactions needed for life. Plants store the extra sugar they make in food molecules, such as starch in rice grains. Directly or indirectly, all the energy used by most species on the planet comes from sunlight captured by photosynthesis. Let's examine how photosynthesis works.

Overview of Photosynthesis

We can illustrate photosynthesis with this chemical "equation":

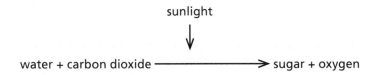

Photosynthesis is a biochemical reaction that converts water and carbon dioxide into sugar. Sunlight provides the energy for the conversion. The result of photosynthesis is that the plant stores some of the solar energy as sugar, and molecules of oxygen are released into the atmosphere.

Sugar is a carbohydrate, which literally means carbon bonded to water. You can see this in the basic formula for a carbohydrate: CH_2O. In carbohydrates, the ratio of carbon to hydrogen to oxygen is 1:2:1. Different carbohydrates have different numbers of carbon atoms, so the formula scales up as needed. The sugar made in photosynthesis is glucose, which has six carbon atoms. Its formula is $C_6H_{12}O_6$ (note the 1:2:1 ratio among the atoms).

Using chemical formulas, we can make our equation more concise and more specific:

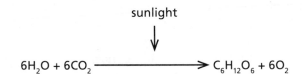

Looking at the chemical bookkeeping, we see that it takes six molecules of water and six molecules of carbon dioxide to make one molecule of glucose, and six molecules of oxygen are released. Sunlight not only provides the energy for the reactions, but some of this solar energy is stored in the chemical bonds of glucose.

How efficient is photosynthesis? The answer may surprise you. As we'll see, plants are not able to use all sunlight, and only some of the energy they capture is successfully stored in sugar. Photosynthetic efficiency varies among plants and under different conditions, but a reasonable estimate is that the sugar that plants make stores 1–5% of the energy available in sunlight. Plants provide food for many species, but they do so using only a small fraction of the available solar energy.

Light Reactions and the Calvin Cycle

Photosynthesis occurs in two stages: light reactions and the Calvin cycle. Light reactions capture solar energy, and the Calvin cycle uses this energy to make sugar (**FIGURE 7.12**).

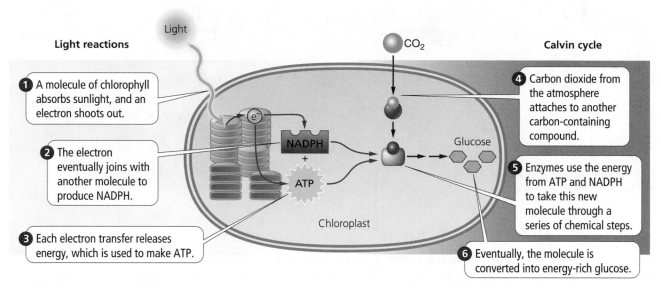

Light reactions

Light

① A molecule of chlorophyll absorbs sunlight, and an electron shoots out.

② The electron eventually joins with another molecule to produce NADPH.

③ Each electron transfer releases energy, which is used to make ATP.

Chloroplast

NADPH + ATP

Glucose

CO₂

Calvin cycle

④ Carbon dioxide from the atmosphere attaches to another carbon-containing compound.

⑤ Enzymes use the energy from ATP and NADPH to take this new molecule through a series of chemical steps.

⑥ Eventually, the molecule is converted into energy-rich glucose.

FIGURE 7.12 **Photosynthesis occurs in two stages: light reactions and the Calvin cycle.** Light reactions capture solar energy, and the Calvin cycle uses this energy to make sugar.

Light reactions use a green pigment called **chlorophyll**, which is found in all the green parts of a plant. Incoming sunlight contains all the colors of the rainbow, as well as other colors that we can't see. Chlorophyll reflects (and therefore does not use) the green wavelength but can absorb both red and blue wavelengths. When chlorophyll absorbs this sunlight, it also absorbs some of the energy it contains.

When a molecule of chlorophyll absorbs sunlight, the solar energy is transferred to an electron, which shoots out of the chlorophyll. This energized electron is the key to capturing solar energy. The electron is transferred to a series of other compounds and eventually ends up in a molecule called NADPH. During each electron transfer, a little energy is released, which is used to make another molecule called ATP. Both NADPH and ATP are storehouses of chemical energy, and both are used in the Calvin cycle to make sugar.

But there is a problem: the chlorophyll molecules now have "holes" where their electrons used to be. Chemical reactions must be balanced, and photosynthesis would eventually stop if these holes were not filled. In one of the light reactions, an enzyme splits apart water molecules, separating the hydrogen from the oxygen. The oxygen is released, eventually becoming part of the atmosphere we breathe. When the water is split, free electrons are produced, which fill the holes in the chlorophyll molecules.

The **Calvin cycle** uses chemical energy produced by the light reactions: ATP and NADPH. As a result, the Calvin cycle does not require sunlight. In the Calvin cycle, carbon dioxide from the atmosphere attaches to another carbon-containing compound. This attachment is called **carbon fixation**. Now that the carbon dioxide has been trapped chemically, enzymes take this new molecule through a series of chemical steps. During these steps, energy is added to the developing molecule. This energy comes from the ATP and NADPH generated by the light reactions. Eventually, the molecule is converted into energy-rich glucose. Glucose lies at the center of many biochemical pathways in the cell. It can be used directly for energy, or it can be converted into many other compounds that cells need.

light reaction :: the first stage of photosynthesis in which solar energy is captured

chlorophyll :: a green pigment used by plants in photosynthesis

Calvin cycle :: the second stage of photosynthesis in which the solar energy captured from light reactions is used to make sugar

carbon fixation :: the attachment of carbon dioxide from the atmosphere to another carbon-containing compound

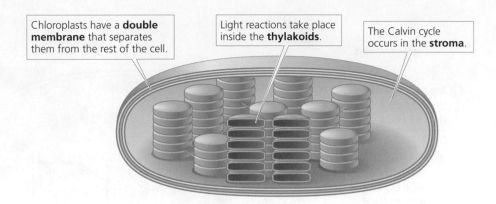

Chloroplasts have a **double membrane** that separates them from the rest of the cell.

Light reactions take place inside the **thylakoids**.

The Calvin cycle occurs in the **stroma**.

FIGURE 7.13 Photosynthesis occurs inside the chloroplasts.

chloroplast :: an organelle in a photosynthetic cell in which photosynthesis takes place

The Role of Cells in Photosynthesis

Each photosynthetic cell contains numerous organelles called **chloroplasts** (**FIGURE 7.13**). Chloroplasts have a double membrane separating them from the rest of the cell. Within the membrane are stacks of wafers, rather like stacks of poker chips. The individual wafers are called thylakoids, and the stacks are called grana (singular granum). The space inside the membrane surrounding the grana is filled with a fluid called stroma.

Photosynthesis occurs inside the chloroplasts. The thylakoids are green, and they are packed with chlorophyll molecules. The light reactions occur on the thylakoids. ATP and NADPH made by the light reactions move into the stroma. The reactions of the Calvin cycle take place in the stroma.

The Role of Leaves in Photosynthesis

Photosynthesis requires water, carbon dioxide, and sunlight. The main function of a leaf is to provide these commodities so leaf cells can photosynthesize.

The shape of a leaf promotes light capture. Most (though certainly not all) leaves are broad and thin or like rice, long and thin. Their width or length provides a large surface for capturing light. Most cells are close to the surface or just a few cells down, so all have access to sunlight and don't shade each other.

The internal structure of a leaf (**FIGURE 7.14**) helps water and carbon dioxide to circulate around the photosynthesizing cells. The inner cells are loosely packed and spongy, with space around them for gases to circulate. Water

The **phloem** carries food made in the leaves through the stems to the roots.

Xylem vessels transport water from the roots to the leaves.

The **waxy cuticle** makes the leaf waterproof and prevents water vapor from escaping inside the leaf.

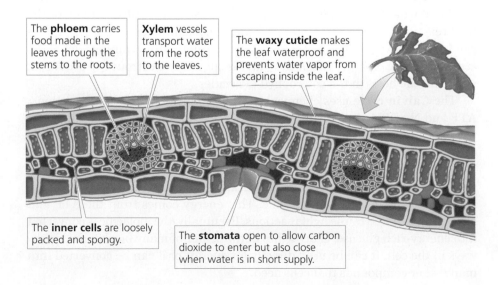

FIGURE 7.14 The internal structure of a leaf helps water and carbon dioxide circulate around the photosynthesizing cells.

The **inner cells** are loosely packed and spongy.

The **stomata** open to allow carbon dioxide to enter but also close when water is in short supply.

enters the leaf through xylem vessels in its stalk. Water evaporates into the inner part of the leaf and diffuses to photosynthesizing cells. The humidity inside a leaf is close to 100% as long as water is available.

The outer layer of a leaf helps the plant conserve water and also controls the uptake of carbon dioxide. The surface cells are packed tightly together and secrete a layer of wax. This makes the leaf waterproof, and water vapor can't escape from inside the leaf. At various places on the leaf are small pores called stomata. These pores open to allow carbon dioxide to enter and diffuse throughout the inner spongy layer. When stomata are open, water vapor can be lost by evaporation, a necessary side effect of allowing carbon dioxide in. However, stomata can close when water is in short supply, and they often close at night when darkness prevents the light reactions of photosynthesis from occurring. Closed stomata shut down photosynthesis, but they do help conserve water.

7.3　What Are the Goals of Genetic Engineering in Plants?

Plants have the capacity to do something humans cannot—they can capture energy from the sun and transform it into food molecules. As a result, we depend upon plants for our survival. In early human history, people gathered and ate edible plants. With the development of agriculture, people learned to selectively breed plants to develop desirable traits. Genetic engineering promises the possibility of generating plants with traits that cannot be acquired through breeding.

Potrykus and Beyer had very clear social goals with their genetic engineering: they wanted to distribute golden rice to low-income farmers as a humanitarian gesture. They sought to design a **genetically modified organism (GMO)** that would be widely available for use. Any farmer anywhere in the world with an annual income less than $10,000 would have access to golden rice for just the cost of the seeds—no licensing fees. Working with a corporation named Syngenta, Potrykus and Beyer set up the Golden Rice Humanitarian Board to oversee the improvement, licensing, and distribution of golden rice.

genetically modified organism (GMO) :: an organism whose genetic material has been altered

On the other hand, many corporations see huge profits in genetically engineered crops. These crops are big business in the United States, Argentina, Brazil, and other countries, although they are banned in Europe and Japan because of popular opinion and fears about safety. In the United States, well over half of the corn, cotton, soybeans, and alfalfa grown are genetically engineered strains. Given the widespread use of corn syrup, cottonseed oil, and soybeans in processed foods, it is likely that many of the foods we eat come in part from genetically engineered plants.

What exactly is it that genetically engineered crops have to sell? The answer lies mainly in the biological goals that researchers have for crop improvement: pesticide production, herbicide resistance, and increased nutritional value.

Pesticide Production

Every year we lose billions of dollars' worth of crops to insect pests and spend billions to spray crops with pesticides, chemicals that kill insects and other pests. But what if our crops could make their own pesticides?

FIGURE 7.15 The bacterium *Bacillus thuringiensis* (*Bt*) produces toxins that kill several types of pests, including the European corn borer (shown), which causes billions of dollars of corn damage each year. The *Bt* toxin is a protein that binds to the intestinal cells of vulnerable insects and prevents them from feeding, leading to quick death from starvation.

glyphosate :: a broad-spectrum herbicide; it is the active ingredient in Roundup

FIGURE 7.16 Glyphosate is the most widely used herbicide in the United States—capable of killing virtually any plant. Shown are weeds that have been treated with the herbicide.

A bacterium named *Bacillus thuringiensis* (*Bt*) produces toxins that kill several types of insect pests. The toxin is very specific and kills only certain caterpillars, beetles, and fly larvae, such as mosquitoes and black flies. This specificity is good because *Bt* does not appear to harm many beneficial insects like pollinators and predators, and it does not appear to harm us. However, the species it does kill are leaf eaters that can devastate entire fields of crops or other vegetation in short order. Insects that are vulnerable to *Bt* toxin include the cabbage worm, tomato and tobacco hornworms, European corn borer (which causes billions of dollars of corn damage annually), spruce and pine budworms, tent caterpillars, and the Colorado potato beetle (**FIGURE 7.15**).

The *Bt* toxin is a protein that binds to the intestinal cells of vulnerable insects and prevents them from feeding, leading to quick death from starvation. Because the toxin is a protein, there is a gene that codes for it. Actually, more than 50 genes for toxins have been discovered in *Bt*, carried by various forms of the bacterium.

A toxin gene provides the starting point for genetically engineering other plants. It is possible to extract the toxin gene from the bacterium and insert it into the cells of a young crop plant. The adult plant then has a working copy of the toxin gene and makes its own pesticide, with no additional work or expense on the farmer's part. The majority of corn and cotton grown in the United States has been genetically engineered to carry the gene for *Bt* toxin.

Herbicide Resistance

Weeds are an even bigger problem for crops than pests. One way we attack weeds is to spray crops with herbicides, chemicals that kill plants. But herbicides can inadvertently harm the crops—weeds and crops are both plants, after all. What if our crops were resistant to an herbicide that killed almost any other type of plant?

Glyphosate is the most widely used herbicide in the United States. Sold under the name Roundup, it is a wide-spectrum herbicide and kills virtually any plant (**FIGURE 7.16**). Once inside a plant, glyphosate interferes with a vital plant enzyme, one that makes certain amino acids. Without these amino acids, the plant can't make proteins and quickly dies.

Glyphosate is found in many different species, with slight variations. As it turns out, the enzyme in a bacterium named *Agrobacterium* is not affected by glyphosate. Genetic engineers have taken the gene for this resistant enzyme from *Agrobacterium* and inserted it into a variety of crop plants. Now these crops also make the enzyme that is not affected by glyphosate. These genetically engineered plants are marketed as Roundup Ready. If a farmer plants Roundup Ready crops, the fields can be sprayed with Roundup—which kills all the weeds and other plants but leaves the crops unharmed.

Increased Nutritional Value

Almost all the genetically engineered crops grown in the United States have been engineered for pesticide production, herbicide resistance, or both. As we will discuss in Chapter 16, we lose 25–50% of our global food crop to pests and weeds each year—a tremendous expense. If genetic engineering can reduce these losses, the economic and nutritional impact will be significant.

But there are other avenues of research for genetic engineering in plants, each with its own specific goals. One goal is to increase the nutritional value of the food we eat. Golden rice is a good example.

In developing golden rice, Potrykus and Beyer managed to insert two genes that caused the seeds to produce a molecule called beta-carotene, which is the molecular precursor of vitamin A. Our bodies have an enzyme that converts beta-carotene into vitamin A. Beta-carotene is an orange pigment, which gives color to carrots and other fruits and vegetables. Beta-carotene gave the rice a yellowish color, which is how golden rice got its name (**FIGURE 7.17**).

Other plants have been engineered to keep humans healthier and better fed. Some of these crops are commercially available, while others are still in development:

- Soybeans and canola have been engineered to provide a source of omega-3 fatty acids, compounds that may help reduce the risk of heart attack and stroke.
- Genetically engineered virus-resistant squashes are commercially available, as is an engineered form of sugar cane that produces more sugar.
- Scientists have engineered a strain of potato that makes a more edible form of starch.
- Scientists are investigating techniques for engineering drought-resistant and freeze-resistant crops.

Some genetically engineered plants have achieved commercial success. But others have flopped—like the tomato engineered to ripen slowly, allowing it to be transported greater distances after harvesting. It simply didn't taste right. The list of genetically engineered crops gets longer every year.

FIGURE 7.17 In developing golden rice, Potrykus and Beyer inserted two genes that caused the seeds to produce beta-carotene, a molecule that contains an orange pigment and that gives the rice a yellowish color (top) compared with white rice (bottom). Our bodies have an enzyme that converts beta-carotene into vitamin A.

7.4 How Was Golden Rice Engineered?

Potrykus and Beyer started the golden rice project in 1992. They had considerable expertise and 20 years of scientific progress with gene manipulation to build on, but it still took seven years to make the first successful strains. What did Potrykus and Beyer have to do to make golden rice?

Each genetic engineering project is unique in some ways. Different species and genes are used, and a different set of biochemical techniques may be needed as a result. But each project seeks to insert working copies of genes from one organism into another, and most genetic engineering projects follow the same basic steps: define the problem, clone the genes, package the genes, transform the cells, and confirm the strain (**FIGURE 7.18**).

Define the Problem

Just like us (and most other organisms), plants need vitamin A, but most plants can make their own. Rice plants make vitamin A in their leaves and other organs, but *not* in the endosperm of their seeds—the part we eat.

Potrykus and Beyer had to figure out why vitamin A was not made in seeds. They discovered that the cells of the endosperm were missing two enzymes, called *psy* and *crt1*, which are needed to make vitamin A. Although rice has the genes for these enzymes, for some reason they are inactivated in

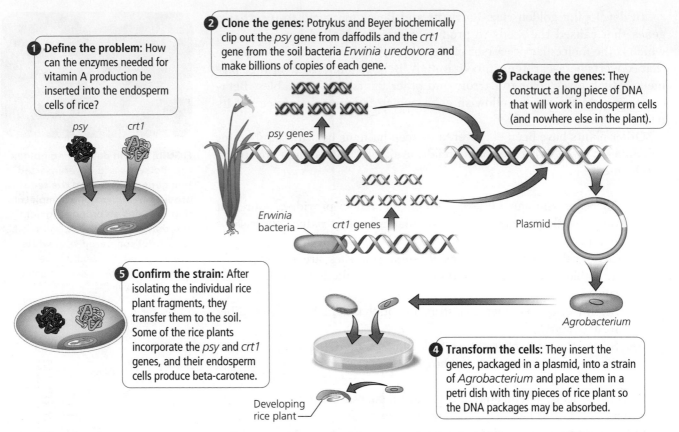

① Define the problem: How can the enzymes needed for vitamin A production be inserted into the endosperm cells of rice?

psy *crt1*

② Clone the genes: Potrykus and Beyer biochemically clip out the *psy* gene from daffodils and the *crt1* gene from the soil bacteria *Erwinia uredovora* and make billions of copies of each gene.

psy genes

Erwinia bacteria

crt1 genes

③ Package the genes: They construct a long piece of DNA that will work in endosperm cells (and nowhere else in the plant).

Plasmid

Agrobacterium

⑤ Confirm the strain: After isolating the individual rice plant fragments, they transfer them to the soil. Some of the rice plants incorporate the *psy* and *crt1* genes, and their endosperm cells produce beta-carotene.

Developing rice plant

④ Transform the cells: They insert the genes, packaged in a plasmid, into a strain of *Agrobacterium* and place them in a petri dish with tiny pieces of rice plant so the DNA packages may be absorbed.

FIGURE 7.18 Most genetic engineering projects follow the same basic steps that Potrykus and Beyer followed in the golden rice project: define the problem, clone the genes, package the genes, transform the cells, and confirm the strain.

endosperm cells. Armed with this information, Potrykus and Beyer were able to define their problem: they needed to insert working copies of the *psy* and *crt1* genes into endosperm cells of rice.

Clone the Genes

Since many organisms make vitamin A, there are a number of *psy* and *crt1* genes to choose from. The most obvious choice is to use the genes naturally found in rice. However, the mechanism that turns off these rice genes is effective, and Potrykus and Beyer were unable to turn them on. Instead, Potrykus and Beyer chose to get their *psy* gene from daffodils and their *crt1* gene from a soil bacterium named *Erwinia uredovora*.

Once the source species were selected, Potrykus and Beyer began a long and painstaking process. They had to biochemically clip out the relevant genes from the rest of the DNA in the daffodil and bacterial chromosomes, and they had to make billions of copies of each gene to have enough material to work with. This process of isolating and copying a gene is called **cloning**, and it is a central technique for a genetic engineering project on any type of organism (see *Scientist Spotlight: Kary Mullis*).

cloning :: the process of isolating and copying a gene

Package the Genes

If a gene's DNA is simply inserted into an organism, it is very unlikely to work. Enzymes inside the cell quickly break down the foreign DNA. Also, genes are

Scientist Spotlight

Kary Mullis (1944–)

Born in 1944 in North Carolina, Kary Mullis completed his bachelor's degree in chemistry at the Georgia Institute for Technology in 1966. He went on to graduate school at the University of California at Berkeley, earning his Ph.D. in biochemistry in 1973. Following two postdoctoral research positions, one at the University of Kansas Medical School and the other at the University of California San Francisco, Mullis joined Cetus Corporation in 1979 as a DNA chemist. It was there that he developed a method to produce copies of DNA quickly and efficiently.

Kary Mullis won a Nobel Prize in Chemistry for developing a method to produce copies of DNA quickly and efficiently.

Mullis was driving home after a party one night. His mind was active, and the winding roads with their double white lines got him to thinking about DNA. When a cell divides, each chromosome separates its double strands, and each of these single strands is copied. At this point, the cell has two complete sets of chromosomes—one for each daughter cell. Mullis's revelation was that you could use this process to make a copy of any *part* of a DNA strand: you don't need to copy the entire chromosome. And you could do it in a test tube. And you could make *lots* of copies.

First you define the part of the double-stranded DNA you want copied—perhaps a gene. You heat the DNA up almost to boiling to separate the two strands. Then on each strand, you attach a short piece of DNA just before the beginning and just after the end of the region you want to copy. These short DNA pieces are called primers, and they direct enzymes to where DNA copying should start and stop. The enzymes copy each DNA strand in between the start and stop markers, and the solution is cooled to allow the DNA strands to come together. Now there are two copies of the particular section of DNA you marked.

Mullis's main insight while driving that night was that you don't have to stop with just one round of copying. The enzymes don't get used up, so as long as you have enough raw materials to make DNA (you need the individual nucleotides A, T, G, and C), you can heat up the tube, copy, and cool down again and again. Each time, every DNA copy generates two copies—first two copies, then 4, 8, 16, and so on. Each copying session only takes 5–10 minutes. In half a day, you easily could go through 30 cycles—over a billion copies of the original DNA strand!

Mullis called his process the polymerase chain reaction (PCR), after DNA polymerase, the copying enzyme. PCR is the first step in every genetic engineering project. It allows you to clone the gene—make enough copies to work with. PCR is also the basis for forensic DNA analysis. Running PCR on two DNA samples makes enough DNA so the samples can be compared. This is used for paternity testing (comparing the possible father's DNA to the child's) and for determining whether DNA samples left at a crime scene match a suspect's DNA. PCR is a cornerstone of modern molecular biology. In 1993, Mullis and his colleague Michael Smith won the Nobel Prize in Chemistry for their work on PCR. ::

only one part of the biochemical network that makes an enzyme or other type of protein. For successful insertion, the gene copy must be protected and processed, and become part of a DNA package that the cell can use.

To make a gene package, extra DNA is added at the beginning and end of the gene (**FIGURE 7.19**). A section of DNA is added at the front of the gene to show where the gene starts and to provide a target where enzymes like RNA polymerase can bind to transcribe the DNA. Another section of DNA is added to the end of the gene to mark where the RNA polymerase should stop

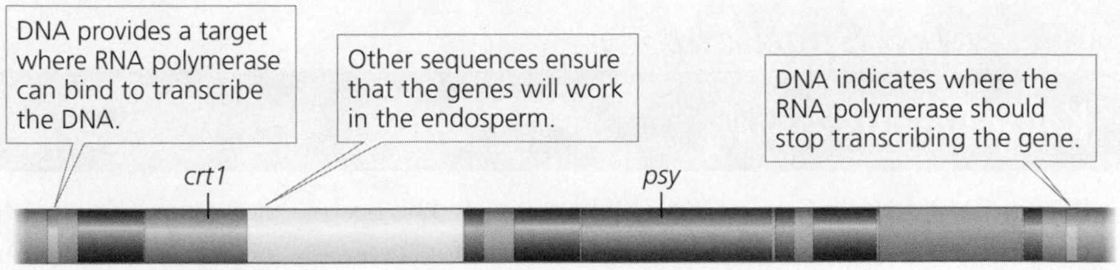

FIGURE 7.19 To make a gene package, extra DNA is added to the beginning and end of the gene. The DNA "package" used to make golden rice includes the *psy* and *crt1* genes, which are essential for vitamin A synthesis.

DNA provides a target where RNA polymerase can bind to transcribe the DNA.

Other sequences ensure that the genes will work in the endosperm.

DNA indicates where the RNA polymerase should stop transcribing the gene.

crt1

psy

transcribing the gene. Other sections of DNA may be needed as well, depending on the details of how the gene is expressed and controlled.

Once they had the cloned genes, Potrykus and Beyer had to combine them with these other DNA elements. They had to construct a long piece of DNA that contained the *psy* and *crt1* genes, their start and stop sections, markers that would indicate successful insertion into a cell, and sections of DNA that would ensure that the genes would work in endosperm cells (and nowhere else in the plant).

Potrykus and Beyer constructed biochemical assembly lines that took DNA through a series of reactions, joining together the necessary sections of DNA. When the reactions were done, they had a complete DNA package that included everything needed for the genes to work.

Transform the Cells

After much work, Potrykus and Beyer had a large supply of packaged *psy* genes from daffodils and packaged *crt1* genes from *E. uredovora*. Now they had to get these genes inside rice cells and verify that they worked in the endosperm. When a cell takes in and uses DNA from a foreign source, the process is called **transformation** (see *Technology Connection: How to Transform Cells*). Potrykus and Beyer's next step was to transform rice cells using the foreign *psy* and *crt1* gene packages.

There are several ways to transform plant cells. Potrykus and Beyer settled on a method that uses *Agrobacterium* (the same bacterium that contributed to the Roundup Ready gene). *Agrobacterium* infects plants by inserting DNA into their cells. That is, *Agrobacterium* does the job that Potrykus and Beyer needed: it transforms plant cells.

Potrykus and Beyer first transformed a strain of *Agrobacterium* by inserting their *psy* and *crt1* gene packages. Then they cut up rice plants into tiny pieces and bathed each piece in a solution of their transformed *Agrobacterium*. Many (though not all) of these rice fragments became infected by *Agrobacterium*, and some of the cells successfully absorbed the DNA packages, completing the transformation.

Confirm the Strain

Potrykus and Beyer isolated the individual rice plant fragments and placed them on a growth medium. The individual fragments grew into whole rice plants, much like taking a cutting from a plant and growing a new one. Eventually these rice plants were transferred to soil, where they flowered and produced seed.

In many cases, the seed was regular rice seed—no beta-carotene, the precursor for vitamin A. In a few cases, however, the process was successful.

transformation :: the process by which a cell takes in and uses DNA from a foreign source

Technology Connection

How to Transform Cells

Transformation starts when a cell takes in foreign DNA from the environment. If transformation is successful, the new DNA becomes a working part of the cell's genetic material. The cell now has new genes that make functional proteins. When the cell divides, the new DNA is copied and passed to daughter cells just like the original DNA.

For DNA to get into a cell, it must cross the cell membrane, and possibly a cell wall. There are several ways to do this. Many species of bacteria transform naturally. When cells die in a bacterial habitat, they often split open, releasing their contents and their DNA. Other bacterial cells can take up this DNA, getting new genes for free. Many bacteria can also be transformed easily in the lab, just by exposing them to DNA and adjusting a few conditions like temperature. This makes bacteria good subjects for genetic engineering.

Transforming plant and animal cells is a little more involved because they don't take in foreign DNA naturally. Several techniques are used:

- *Agrobacterium.* This was the technique used to transform rice cells, and it is probably the easiest way. First *Agrobacterium* cells are transformed. Once they have the desired DNA, *Agrobacterium* can inject it directly into plant cells, almost like a virus.

- *Viruses.* It is possible to genetically engineer a virus to carry extra genes used to genetically engineer a plant. When the virus infects a plant cell, it injects its DNA—and the desired genes, too—much like what happens with *Agrobacterium*.

- *Electroporation.* Exposing cells to electric currents opens holes in the cell membrane and cell wall, allowing DNA to enter. This works on bacteria also.

- *Gene "guns."* Small particles of metal (often gold or tungsten) are coated with the transforming DNA. These particles are literally shot into cells using an air gun. The DNA dissolves off the particle, which is later expelled by the cell.

Once the DNA is successfully inside the cell, it must become functional. Sometimes the transforming DNA is in circular form and is called a plasmid. Bacterial cells often

One method for inserting DNA into plant and animal cells is a biolistic particle delivery system, or "gene gun." Small particles of metal are coated with the transforming DNA and literally shot into cells.

have many plasmids floating around inside, so once a bacterial cell is transformed with a plasmid, the transformation is complete.

Plant and animal cells don't typically have plasmids. In fact, these cells interpret foreign DNA as a sign of viral infection, and enzymes usually digest it. In these cells, the introduced DNA has to become part of a chromosome, and this has to happen before the digestive enzymes find it. Cells have enzymes that cut chromosomes, insert DNA, and chemically combine the parts into a seamless stretch of DNA. These enzymes are used to repair chromosomes, but they can also insert foreign DNA into a chromosome. When this happens, the foreign DNA becomes part of the cell's original DNA, and the cell is transformed. ::

These plants had incorporated the *psy* and *crt1* genes, and their endosperm cells produced beta-carotene. Potrykus and Beyer grew and interbred these seeds in the conventional way to produce a true-breeding strain of rice in which all plants produced vitamin A. Golden rice was born.

How Else Is Genetic Engineering Being Used?

Much of the world is malnourished, and we lose a large percentage of our annual crops to weeds and pests. That's certainly one reason why a lot of genetic engineering effort is directed toward crops. Even before the advent of molecular biology techniques, humans tried to improve crops by selecting for new combinations of genes in plants (see *Life Application: From Teosinte to Maize* to learn how modern corn was developed). But agriculture is only one of several enterprises that are investing heavily in genetic engineering.

Life Application

From Teosinte to Maize

Maybe your image of genetic engineers is people in white lab coats surrounded by test tubes and equipment in a modern genetics lab. But the first genetic engineers dressed much differently and had no equipment other than their hands and the power of observation. One "research team" lived 9000 years ago in what is now southern Mexico.

These people began a genetic engineering program that converted a native plant, teosinte, into maize (or "corn"). Teosinte and maize are closely related subspecies that do not look alike at all. Teosinte is a small, highly branched grass with many "ears." Each ear has only a dozen kernels in only two rows, and each kernel is covered with a hard case, which protects it but makes it hard to digest. When the ears ripen, the kernels break apart, dispersing the seeds. The maize you are most familiar with consists of a single stalk with a single ear, although some types of sweet corn can have two to four ears per stalk. The ear contains many rows of kernels, which are not covered by a case. The kernels stay on the ears when they ripen.

If you think these are big differences, you aren't alone. For decades, botanists debated where maize came from. But genetic analysis along with archeological evidence now gives us a clear picture. Maize developed from a particular type of teosinte that grew in the Central Balsas River Valley in southern Mexico. Despite the fact that teosinte and maize look so different, most of the major differences are due to only five genes (though there are probably hundreds of minor genetic differences among these five genes).

Those ancient people were true genetic engineers: they altered the genetics of one species to produce a new subspecies that was more nutritious and easier to grow. The procedure they used was selective breeding. They selected plants that had useful traits, ate some of them, and crossed the rest with other plants with useful features. The frequency of useful features increased over time, and these engineers guided the evolution of teosinte to maize.

We can make some educated guesses about how the domestication progressed. Harvesters would naturally select for plants that kept their kernels on longer. They would have preferred plants that had the most kernels (more food) and probably the fewest ears (easier to harvest). And they certainly would have preferred kernels with the softest cases. If seed harvesters selected plants according to these preferences, over time teosinte kernels would remain fixed to the ears, fewer and larger ears would be produced per plant, and the hard kernel cases would become softer and perhaps reduced in size. Just the traits we see in modern maize.

Genetic evidence supports these possibilities. One of the five major genetic differences, for example, is a gene that makes the glume: the hard case that covers teosinte kernels. In maize, this gene produces a soft, small glume. It's the part that gets stuck in your teeth when you eat corn on the cob. These early genetic engineers did their jobs well. Today maize accounts for over 20% of human nutrition globally. ::

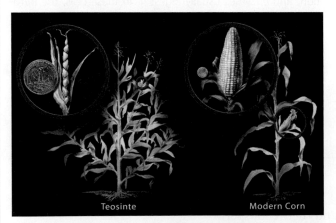

Genetic engineering has existed for thousands of years. Over 9000 years ago, ancient people used selective breeding to produce corn that was more nutritious and easier to grow than the existing native plant teosinte.

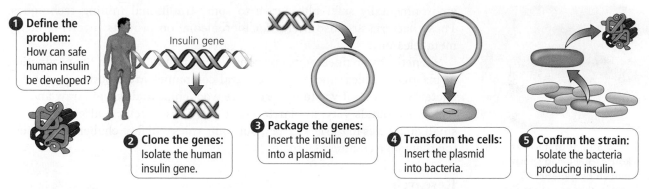

① **Define the problem:** How can safe human insulin be developed?

② **Clone the genes:** Isolate the human insulin gene.

Insulin gene

③ **Package the genes:** Insert the insulin gene into a plasmid.

④ **Transform the cells:** Insert the plasmid into bacteria.

⑤ **Confirm the strain:** Isolate the bacteria producing insulin.

FIGURE 7.20 Genetic engineering methods like those used to produce golden rice can also produce medically important products like insulin. Once bacteria-producing insulin are isolated, additional steps must be undertaken to purify it for safe use.

Medicine, industry, and research, among other areas, are looking toward genetic engineering to solve many of our biggest biological challenges.

Medicine

Diabetics used to be treated with insulin from pigs or cattle—human insulin was too expensive and was not available in large quantities. In 1978 Herbert Boyer successfully inserted the human insulin gene into a bacterium, which produced large quantities of human insulin. Today, most of the insulin used is human insulin produced by bacteria, yeast, or other organisms (**FIGURE 7.20**).

Insulin was the first drug produced by a genetically engineered organism. After several decades of research and development, genetically engineered organisms now produce a number of medically and commercially important drugs. The list includes human growth factor, a hormone used to treat dwarfism; blood clotting factors to treat hemophilia; erythropoietin to treat anemia; and vaccines like Gardasil that help prevent cervical cancer.

Bacteria and yeast are used to make most genetically engineered drugs because they are relatively easy to transform and grow quickly. The first genetically engineered drug from an animal was approved in 2009. Scientists inserted the gene for human antithrombin, which reduces blood clotting, into a goat. The protein forms in its milk, and after purification it is injected into patients at risk for life-threatening blood clots.

Industry

Although genetically engineered organisms are often designed to make things, they also can be used to break chemicals down. This occurs during **bioremediation**—the biological cleanup of pollution and contamination. Some bacteria have been engineered to break down oil spills. These organisms were used to clean up the 1989 *Exxon Valdez* spill; the Chinese government released 20 tons of genetically engineered bacteria to help clean up a massive oil spill in the Yellow Sea in 2010; and these bacteria were called upon again to help with the BP Deepwater Horizon spill in the Gulf of Mexico in 2010 (**FIGURE 7.21**).

Bacteria often live in harsh environments—hot, highly acidic, very salty, or low in oxygen. As result, they can be right at home inside many industrial processes, including deep within mines. Some bacteria have been genetically engineered to produce enzymes that dissolve metals from their ores, making the metal easier to extract. This is called **bioleaching**, and it provides an

A

B

FIGURE 7.21 Certain types of naturally occurring and genetically engineered bacteria can break down oil. Oil-eating bacteria were used in the clean up efforts of the 2010 massive oil spills in the Yellow Sea (A) and the Gulf of Mexico (B).

bioremediation :: the biological cleanup of pollution and contamination

bioleaching :: the action of bacteria that have been genetically engineered to produce enzymes that dissolve metals from their ores, making the metal easier to extract

FIGURE 7.22 Strains of mice, known as "knockouts," have had specific genes removed by genetic engineering so that scientists can better study what genes do. The mouse on the left has had the gene affecting hair growth "knocked" out.

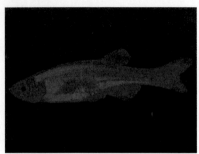

A

B

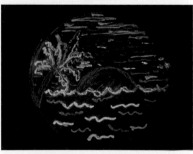

C

FIGURE 7.23 Novelties produced by genetic engineering methods include GloFish (A), blue roses (B), and fluorescent bacteria (C) that can be grown to create pictures in petri dishes.

genome :: all of an organism's DNA and every gene required to make it all work

environmentally safer alternative to some traditional mining techniques. These bacteria are also useful for bioremediation projects involving heavy metal cleanup.

Genetically engineered bacteria make compounds used in the production of plastics, textiles, and cheese. Several companies are working to develop bacteria that digest waste material like woodchips and wheat straw and excrete compounds very similar to crude oil. These bacteria already exist in small quantities in the laboratory, but companies still face challenges producing them on a commercial scale.

Research

The ability to manipulate genes gives researchers many ways to study genes and genetic diseases. For example, there are strains of mice that have been genetically engineered to be highly susceptible to cancer. These mice provide researchers with a large population of subjects on which to test treatments that may ultimately help human cancer patients.

Other strains of mice, known as "knockouts," have had specific genes *removed* by genetic engineering (**FIGURE 7.22**). Scientists try to correlate missing biological functions with the missing genes, which helps them discover what the genes do.

Novelty

Some genetically engineered organisms were developed simply for their novelty (**FIGURE 7.23**). GloFish are genetically modified zebra fish that give off a green, red, or orange glow. Originally, glowing zebra fish were developed for research, with the hope that the glow would help scientists see internal structures better. But now they have become a popular addition to home aquariums—the first genetically engineered animal you can buy in a pet store.

In 2004, after 13 years of research, two companies working together succeeded in inserting a pigment gene from a petunia into a rose. They created a blue rose, a color not found in any rose in nature.

Bioart is a new form of expression that "paints" with living organisms that have been genetically modified to glow or to produce specific colored pigments. One form of bioart uses bacteria that have had genes for fluorescent proteins inserted, similar to GloFish. These bacteria can be grown in patterns on petri dishes to produce colorful displays.

Construction of an Organism's Genome

What would it take to create an entirely new form of life? One approach would be to design and construct the organism's **genome**—all of an organism's DNA and every gene required to make it all work. Then DNA could be removed from a cell, and the entire synthetic genome could be inserted. Would this cell be a new form of life?

We are not yet able to design genes from scratch. But scientists at the J. Craig Venter Institute have taken the first steps in this process. In May 2010, the scientists announced that they had synthesized the entire genome of one species of bacteria. They made an exact copy of the genome in a test tube using genetic engineering procedures. They removed the DNA from cells of a different bacterial species and inserted the synthetic genome. These transformed cells grew and reproduced normally. And they began to take on the characteristics of the species that provided the synthetic genome. Interestingly, these cells were alive but did not have a living parent.

This is not quite the same thing as creating a new form of life. The research team did not design new genes; they copied existing ones. But they did construct the genes themselves, and they demonstrated that a synthetic genome could be inserted successfully into a recipient cell. In the foreseeable future, scientists may be able to design a variety of synthetic genomes by cobbling together genes from many different species. While none of the genes may be new, the cells containing these unique combinations of genes would be different from all existing species. In many ways, these cells could be considered new, artificially created forms of life.

7.6 What Are the Risks of Genetic Engineering?

Ultimately, genetic engineering could produce changes in society similar in scope to the Industrial Revolution. Custom-designed life forms could change the way we handle agriculture, medicine, manufacturing, and many other enterprises. Changes as sweeping as these promise large benefits, but they also involve risks. This situation produces debates, often quite impassioned, about whether the benefits are worth the risks (**FIGURE 7.24**). Golden rice provides a good example of how these debates play out.

FIGURE 7.24 Many worry that genetically engineered crops pose unacceptable risks to the food supply and the environment. A Greenpeace activist protests the location of a rice crop genetically engineered to produce human proteins for drug production.

The golden rice project ran into opposition right from the start. Many people, including a number of scientists, were concerned about the safety of genetically engineered crops. Are they safe for human consumption, and are they safe for the environment? Others were concerned that only huge businesses can produce genetically engineered plants, which gives them control over this part of the agricultural markets. Still other opponents argued that it would be better and cheaper to find ways to provide people in developing countries with a more balanced diet that automatically would provide vitamin A and other vitamins and nutrients as well.

Scientifically, golden rice is a success, but the scientific success has not paid off yet in terms of the commercial or humanitarian success Potrykus and Beyer had intended to achieve. Critics of golden rice have challenged its usefulness on three grounds: safety, economic considerations, and whether it effectively solves the social problems associated with vitamin A deficiency.

Safety

The beta-carotene produced in golden rice seeds is chemically identical to beta-carotene found naturally in other plants. Are there reasons to be concerned about safety when the beta-carotene is produced by genetic engineering? Some potential risks from consuming genetically modified food may include an unanticipated harmful reaction caused by the inserted gene, a product that converts to an unanticipated molecule, or a product or one of the reactants that triggers an allergic response.

There are biochemical differences in genetically engineered cells. A great many biochemical reactions occur in every cell, and these reactions are linked in complex ways because they have certain chemical compounds in common. The enzymes inserted into golden rice convert existing compounds into beta-carotene. But do they also participate in other biochemical reactions? If so, are they making harmful compounds that normally would not be produced? A related concern comes from the excess beta-carotene now found in the cell. Does it just sit there waiting to be eaten, or is some of it converted into other compounds that could have adverse effects on us?

Genes make proteins, and proteins can trigger allergic reactions. Genetic engineering combines genes from several species into one, which also combines potential allergens from several species into one. Every species makes tens of thousands of proteins, and genetic engineering uses genes for only a couple of them, so there is a good chance that the introduced proteins don't cause an allergic reaction. Still, some people are allergic to daffodils (or at least certain compounds in daffodils). Would these people also be allergic to golden rice? Allergies appear to be on the rise. It can be very difficult to diagnose the causes of an allergy, but many consider the possibility that the increase in genetically engineered food may be partly responsible.

Some consumers want to see the word "natural" on the label of foods they buy, and organically produced foods have gained great popularity. Critics of genetically engineered crops often refer to them as "Frankenfoods," after the cobbled-together body parts of Frankenstein's monster in Mary Shelley's novel. Some consumers are hesitant to eat genetically engineered foods and are concerned that labels currently are not required to say whether any of the ingredients come from genetically engineered sources. Another source of concern is that many of the crops we've chosen to engineer are used mainly in processed and fast foods—high-fructose syrup from corn and protein from soybeans, for example. Since processed and fast foods are so prevalent in our

How Do We Know?

Evaluating the Safety of Genetically Engineered Products

People question the safety of genetically engineered products on several levels, including the effects on humans, effects on animals, and effects on the environment. Some of the arguments for and against these products are biological, while others involve the aesthetic value of using them. Here we consider some of the guidelines used to make decisions.

The most fundamental guideline is substantial equivalence, a principle defined in 1993. The principle states that a genetically engineered product is safe provided it is "substantially equivalent" to the nonengineered product. For genetically engineered drugs, this means that the engineered drug has the same or better effectiveness and the same or more positive effects (including side effects) than the nonengineered drug.

For crops, substantial equivalence means that the nutritional content of the food and the levels of allergens and toxins fall within the range of variation for nonengineered foods. The food developer is responsible for performing tests and presenting findings to the Food and Drug Administration for approval. Approval is often granted if the engineered product is found to be substantially equivalent. If any differences are found, additional rounds of testing are required, or the developer may decide not to continue development. The same rules apply for crops intended for human or animal consumption.

Critics of genetically engineered products argue that substantial equivalence is not a strong enough criterion. Foods can be shown to be substantially equivalent on the basis of biochemical tests, for example, without doing tests to see what happens when humans eat them. There is also concern that crop developers, who intend to market the crops, are the ones responsible for testing their safety.

Proponents of genetically engineered products argue that they are tested much more thoroughly than conventionally produced products. Many foods naturally contain toxins and allergens. Consider how many people are allergic to peanuts or the gluten in wheat, but these products don't require additional testing or regulation. Also, genetic engineering involves transferring only small numbers of genes between species, and under controlled conditions. Conventional cross-breeding transfers thousands of genes, with no hope of testing the effects of all of them.

The question of safety does not yet have a clear answer. At least three major interests are active in the debate: health and safety, humanitarian efforts to increase food production and nutritional value, and the profit motives of major biotech corporations. Until consensus is reached on how to deal with genetically engineered products, our best bet as individuals is to become as knowledgeable as we can, so that we can make decisions that are right for us. ::

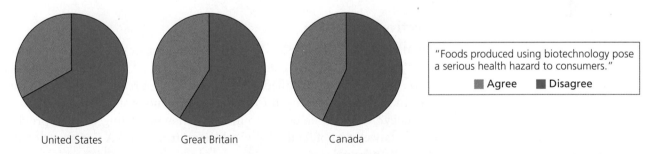

"Foods produced using biotechnology pose a serious health hazard to consumers."

■ Agree ■ Disagree

United States   Great Britain   Canada

Consumers have different perceptions about the safety of genetically modified food. In one 2005 poll, respondents from the United States were less likely than respondents from Canada and Great Britain to believe that foods produced using biotechnology posed a serious health hazard to consumers.

diets, these engineered ingredients show up in a wide variety of foods, making a "natural" diet harder to put together.

The scientists and researchers who work on golden rice believe they have tested golden rice thoroughly and that it is safe to eat. Critics argue that the tests may have measured the wrong things, or that the results have been interpreted incorrectly and that safety issues remain a concern (see *How Do We Know? Evaluating the Safety of Genetically Engineered Products*).

Economic Considerations

It takes a large corporation with good financial, biotechnical, and legal resources to produce a genetically engineered crop. When a corporation makes such an investment, it expects a profit in return. Some of the concern about golden rice comes from worry over how these economic realities will affect the ability of farmers in developing regions to afford the seed.

Corporations working on golden rice obtained over 70 patents for new techniques they developed. Syngenta, the corporation that partnered with Potrykus and Beyer, held most of these patents, but over 30 corporations patented some parts of the development process. Potrykus, Beyer, and Syngenta worked with the Golden Rice Humanitarian Board to persuade each of these corporations to forgo licensing fees on their patents for low-income farmers. Agricultural programs in developing countries and farmers making less than $10,000 annually would be able to purchase golden rice seed for the same price as other rice seed—there are no additional licensing fees. Farmers also can keep seed from one year to grow the next with no additional charge.

Perhaps Syngenta and the other corporations expected to make a profit by selling golden rice to wealthier farmers and agricultural conglomerates in developed countries. However, vitamin A deficiency is simply not a problem in these countries, so there is no market for golden rice there. Syngenta no longer has a commercial interest in golden rice, but they continue to work with the Humanitarian Board on improving the crop and developing plans for its eventual distribution.

However, the role of large corporations may not be so benign in other cases. The patents held on techniques used to make golden rice may affect the price of other genetically engineered crops that use these patented techniques. In general, it takes a large investment to produce a genetically engineered crop, and corporations protect their patents aggressively. A number of biotech corporations are suing each other over patent infringement, and some corporations have even sued individual farmers.

Effectiveness

A final concern is whether golden rice is a solution, or even a partial solution, to the problem of vitamin A deficiency in developing Asian countries and elsewhere. Is relying on a fortified rice to provide a source for the body to make vitamin A the best approach to combating malnourishment and vitamin A deficiency? Would it be better to invest money in finding ways to provide a more balanced diet of foods that would contain vitamin A and a host of other vitamins and nutrients?

Critics argue that golden rice will simply encourage low-income people to continue eating a diet consisting largely of rice. Proponents counter that golden rice is not intended as a substitute for a balanced diet, but rather as a more nutritious component of a balanced diet, like our vitamin-fortified milk and cereal.

Without a doubt, the development of golden rice is a significant scientific achievement. But after more than a decade of testing and debate, golden rice is still not commercially available, and no consensus has been reached on its potential benefits or risks. The debate over golden rice rages on, both sides are impassioned, and both do their best to make a strong case for their position.

How Ethical Is Genetic Engineering?

What happens when a genetically engineered organism is released into the environment, or if the genes it carries "escape"? Can we trust ourselves to use genetic engineering wisely? What happens when we turn genetic engineering on ourselves? These are just some of the ethical considerations confronting scientists and others as the scientific techniques of genetic engineering progress.

Genetic Engineering and Our Environment

Normally genes in nature are passed from parent to offspring, often referred to as **vertical gene transfer**. Genetic engineering uses **horizontal gene transfer**: transferring genes between individuals of the same generation, and often between individuals of different species. Horizontal gene transfer occurs naturally as well. Many species of bacteria participate in horizontal gene transfer as a regular part of their life cycle. As with genetic engineering, this provides the bacterial cell with a combination of new genes from a variety of species. If a genetically engineered bacterium were released into the environment, it is quite possible that its engineered genes could find their way into other bacteria.

Plants also engage in horizontal gene transfer, though not to the same level as bacteria. Certain viruses that infect plants can transfer genes horizontally. Pollen from a genetically engineered plant carries foreign genes. This enables engineered genes to spread beyond a farmer's field. Many plants can hybridize: different species can mate to produce hybrid offspring. Suppose that Roundup Ready corn accidentally hybridized with a weedy grass species. If the grass became Roundup Ready, then the whole strategy of using Roundup to control weeds would be in trouble. In fact, Roundup-resistant weeds have been reported in Iowa and are spreading throughout the state due to the spread of pollen and seeds into new sites. Resistance to the herbicide was essentially nonexistent before the introduction of Roundup.

Many of the concerns over genetically engineered organisms arise from the possibility of accidental or unintended consequences. A darker specter is the possibility of deliberate harm, perhaps from genetically engineered weapons. Imagine the havoc that would be wreaked if a genetically engineered virus produced a toxin as deadly as anthrax, and was as virulent as Ebola and as contagious as the flu. Genetic engineering may provide many benefits, but we must also guard against its potential misuse.

Genetic Engineering and Human Life

Genetic engineering gives us extraordinary powers. Someday we may be able to create life forms to our own specifications. This raises several ethical concerns. Do we have the wisdom to use these powers in a safe and beneficial way? Do we have the right to genetically alter naturally occurring species to suit our needs? Do we have a responsibility to do all we can to use genetic engineering to improve our food supply and agricultural systems, potentially saving millions of people from malnutrition or worse? These are difficult questions, and clear-cut answers have yet to emerge.

Perhaps the greatest ethical dilemma is the question of whether we should genetically engineer ourselves, far-fetched though this may seem. There are

vertical gene transfer :: the transference of genes from parents to offspring

horizontal gene transfer :: the transference of genes between individuals of the same generation, and often between individuals of different species

no technical reasons why genetic engineering procedures won't work on us, other than that we are a different species so the procedures need to be modified accordingly. We could use genetic engineering to extend life, improve health, and cure diseases, as we will discuss in Chapter 8. We could try to engineer for greater intelligence or athletic ability, or different aspects of beauty.

The things we could do with genetic engineering cover a spectrum from beneficial to scary. It is important that we become knowledgeable about the benefits and risks, and that we carefully consider the ethical issues. We are a long way from being able to do most of these things. But the basic techniques have already been worked out, and our technical skills increase every year.

⊞ Biology in Perspective

Can we create better plants and animals? The short answer appears to be *yes*. Because they photosynthesize and store food conveniently in their organs, plants have always been a good source of nutrition. Through thousands of years of selective breeding, we have produced many new varieties of plants and animals, more nutritious or more useful in other ways.

Genetic engineering allows us to speed up this process and make it dramatically more precise. By identifying and cloning genes and using them to transform cells, we can recombine genes (and genetic traits) from almost any species with those of any other species. The new combinations of traits we can produce are limited only by the range of traits found in nature and our technical skill. Already we have engineered vitamin-fortified crops, as well as crops that make their own pesticide or are resistant to herbicides, drought, freezing, and other hazards. Genetically engineered organisms make medicine and play important roles in industry and environmental cleanup.

While it provides potential benefits, however, genetic engineering raises concerns about economics, the environment, and safety. It also creates ethical concerns, particularly in the context of genetically engineering humans, as we will discuss in the next chapter.

Chapter Summary

7.1 Why Are Plants Such Good Sources of Food?

Explain how the bodies of plants are structured and specialized to store food molecules.

- Roots anchor plants, absorb water and nutrients, and often store resources for use later.
- Stems connect the roots to the leaves, transporting nutrients and water.

- Leaves are plants' biochemical factories but are often modified to store food, protect plants, help plants attach to surfaces, and even to catch and digest insects.
- Flowers are modified leaves that produce the egg and pollen used for reproduction.
- Seeds contain embryonic plants.
- Fruits are any structure developed from the ovary of the flower; they help disperse seeds.

7.2 How Do Plants Make Food?

Summarize the light reactions and Calvin cycle phases of photosynthesis.

- Plants capture energy and store it as chemical energy through photosynthesis.
- Photosynthesis occurs in organelles called chloroplasts and converts 1–5% of available sunlight energy into energy stored in sugar.
- Photosynthesis is divided into two phases: light reactions and the Calvin cycle.
  - In light reactions, chlorophyll captures light energy and transfers that energy to an electron, which is used to generate ATP and NADPH, which are passed to the Calvin cycle.
  - The Calvin cycle converts the energy captured by the light phase of photosynthesis into the chemical bonds of sugar.

7.3 What Are the Goals of Genetic Engineering in Plants?

Describe the goals for genetic engineering of crop plants.

- The primary goals of genetic engineering in crops include protection from insect pests, herbicide resistance, and increased nutritional value.
- Other goals include disease resistance and drought and freeze resistance.
- Many crop plants have been engineered to include a gene for a toxin produced by a bacterium (*Bt* toxin) that kills many insect plant pests.
- Golden rice was developed to produce beta-carotene in rice and relieve vitamin A deficiency in underdeveloped regions.

7.4 How Was Golden Rice Engineered?

Outline the steps used to produce a genetically modified organism.

- Define the problem: identify the genes that need to be altered.
- Clone the genes: isolate the gene to fix the problem and make lots of copies.

- Package the genes: add control regions of DNA to the cloned gene to make sure that the gene turns on properly.
- Transform the cells: put the packaged DNA into the DNA of the target organism.
- Confirm the strain: grow whole plants from the plant tissue and select transformed individuals with the new trait.

7.5 How Else Is Genetic Engineering Being Used?

Describe how areas other than agriculture use genetic engineering.

- Genetic engineering is used to produce medicine, to clean up pollution, to produce products such as plastics, to produce art and novelty organisms, and to help us understand how cells work through research.

7.6 What Are the Risks of Genetic Engineering?

Assess the risks of genetic engineering in terms of safety, the economy, and effectiveness.

- Some people have concerns about the safety of genetically modified organisms (GMOs), particularly as they relate to food allergies.
- Most genetically engineered crops have patented genes and do not allow farmers to save seed to plant the next year. This can be a significant problem for farmers in developing countries.
- Some critics claim that GMOs do not help the problems they are designed to solve.

7.7 How Ethical Is Genetic Engineering?

Evaluate the ethical considerations related to genetic engineering that confront scientists.

- Many concerns over genetically engineered organisms arise from the possibility of accidental or unintended consequences.
- Some ethical issues include the following: misuse, genetically engineered biological weapons, inadvertent horizontal transfer of herbicide resistance to weeds, and genetic engineering of humans.

Key Terms

Review Questions

1. What are the four types of plant organs?

   a. Roots, stems, leaves, and seeds

   b. Roots, stems, leaves, and fruit

   c. Roots, stems, flowers, and seeds

   d. Roots, stems, flowers, and fruit

   e. Roots, stems, leaves, and flowers

2. List the functions of each plant organ and explain how each plant organ provides food.

3. What are the four "whorls" that make up a flower?

   a. Sepals, petals, stamens, and carpels

   b. Sepals, petals, xylem, and carpels

   c. Sepals, petals, phloem, and stamens

   d. Sepals, petals, phloem, and carpels

   e. Sepals, petals, bulbs, and stamens

4. Describe the reproductive role of seeds and fruits, explaining how they are similar and how they are different.

5. What occurs during light reactions and during the Calvin cycle of photosynthesis?

6. What are several differences and similarities between selective breeding that led to the domestication of crops and "modern" genetic engineering?

7. Why did Potrykus and Beyer decide to work with rice as a way to reduce vitamin A deficiency worldwide?

8. What are the biological goals that researchers have for crop improvement?

   a. Pesticide reduction, herbicide resistance, and increased nutritional value

   b. Pesticide production, herbicide enhancement, and increased nutritional value

   c. Pesticide production, herbicide resistance, and decreased nutritional value

   d. Pesticide production, herbicide resistance, and increased nutritional value

   e. Pesticide reduction, herbicide enhancement, and decreased nutritional value

9. How do farmers benefit from crops that are resistant to herbicides like Roundup?

10. Describe two examples of genetic engineering used in medicine and suggest an organism used to produce the product.

11. Describe one example in which genetic engineering has been used "for fun": to produce an entertaining novelty.

12. What are three reasons why genetically engineered plants that produce a vitamin may not be safe to eat even if the vitamin is chemically identical to "natural" vitamins?

13. What is horizontal gene transfer? How does it compare to vertical gene transfer?

14. What are the five general steps involved in designing and carrying out a genetic engineering project?

The Thinking Citizen

1. What do the endosperm of seeds, egg yolks, and milk and other dairy products all have in common? Why are they so nutritious?

2. Some plants have a different version of photosynthesis. They open their stomates and take in CO_2 only at night. They store CO_2 internally until daylight, when they close their stomates and begin to photosynthesize. What are the benefits of this type of photosynthesis? What sort of habitat might these plants live in?

3. Many genetically engineered crops have a gene for a toxin produced by *Bacillus thuringiensis* (*Bt*). Many bacteria and other microorganisms produce toxins, pesticides, and antibiotics. Penicillin comes from a soil fungus, for example. What benefit might these organisms get from producing these compounds? Does this suggest that other useful genes might be found in nature?

4. Human growth hormone is a pituitary protein that can be used to treat dwarfism. Suppose you wanted to genetically engineer a strain of bacteria (*E. coli*) to produce human growth hormone. Outline the general steps you would take. (You don't have to know the detailed biochemical procedure.)

5. At first it may seem surprising that genes are largely interchangeable: genes that work a certain way in one species work the same way if they are inserted into a different species. This suggests that despite all their obvious differences, biochemically species are quite similar. Do you think this result provides evidence that species are related evolutionarily, all descending from a common ancestor? Why or why not?

The Informed Citizen

1. The licensing agreement for golden rice states that the crop cannot be released in a country lacking governmental or other regulations for ensuring that it does no environmental damage. However, the targets for golden rice adoption are developing countries, which can lack these types of regulatory agencies. Does this licensing requirement protect the environment, or does it present an obstacle to adopting golden rice?

2. In 1996, Monsanto tested a genetically engineered strain of soybean. They had inserted a gene from Brazil nuts that made a protein particularly rich in certain amino acids, which they hoped would make the soybean more nutritious. This strain of soybean was intended for animal consumption, not human. However, because some people are allergic to Brazil nuts and there was a chance that the soybeans could enter the human food chain, they tested whether the strain could cause an allergic reaction in humans. Some people were allergic to the soybean, and Monsanto discontinued work on the project. Is this an example of a testing and regulatory system that works well to prevent unsafe genetically engineered products from coming to the market? Or does it show that the dangers of allergies to genetically engineered crops are real and that we should not support the crops' development?

3. How do you feel about the potential to genetically engineer humans? Would you ban all research for this subject? Are there some genetic therapies or modifications you would support but others you would prohibit? What guidelines would you use to decide whether a particular proposal for human genetic engineering was ethical?

Thanks to the creativity, curiosity, and hard work of a great number of scientists, we now know things about inheritance, how genes work, and how to manipulate genes that would have been the stuff of science fiction 40–50 years ago.

CHAPTER 8

Health Care and the Human Genome

How Will We Use Our New Medical and Genetic Skills?

Becoming an Informed Citizen ...

Although we live in an exciting time and have the capacity to do wonderful things with the genetic and medical technologies that exist and are being developed, we also have the capacity to do harm. How can we ensure that we maximize the positive consequences of our actions while minimizing the negative ones?

e are at a profoundly important junction in our history. Thanks to the creativity, curiosity, and hard work of a great number of scientists, we now know things about inheritance, how genes work, and how to manipulate genes that would have been the stuff of science fiction 40–50 years ago. Even 20 years ago, some of today's routine procedures would not have seemed possible. We look now toward a horizon where potential preventions or cures for many diseases may become realities. The challenges we face are not the scientific or technical obstacles, but rather how to define the ethical boundaries for our actions.

This is not the first time society has faced the collision between what we know (or think we know), what we *can* do, and what we *should* do. Let's look at a similar scenario from the early part of the 20th century to see if it holds any helpful lessons.

case study

Carrie Buck and the American Eugenics Movement

FIGURE 8.1 When her widowed mother, Emma (right), was committed to the Colony for Epileptics and Feebleminded in 1920, Carrie Buck (left), only 14 years old, was sent to live with the Dobbses. Carrie's life with the Dobbses was clouded by her mother's reputation, and she was treated poorly.

John and Alice Dobbs seemed to be good-hearted, decent people. After all, they took in Carrie Buck (**FIGURE 8.1**) when her widowed mother, Emma, was committed to the Colony for Epileptics and Feebleminded in 1920. Carrie was only 14 years old at the time and had no other family to care for her. But although it appeared that Carrie was quite fortunate to be taken in by the Dobbses, she had actually just traded one problem for another.

Emma Buck had not been a well-respected, contributing member of society. Poor and uneducated, she went downhill after her husband died. People in her rural Virginia community thought Emma was a "worthless" person, and so she was brought before the Commission on Feeblemindedness. These local commissions were found in cities and towns across the state. Once the commission learned that Emma had been arrested for prostitution and that she suffered from syphilis, a sexually transmitted disease, they declared her "notoriously untruthful" and "suspected of being feebleminded or epileptic." While we have no way of evaluating the claims of her untruthfulness, we do know that Emma was neither feebleminded nor epileptic. Nevertheless, once sentenced, she spent the rest of her life in the colony.

Carrie's life with the Dobbses was clouded by her mother's reputation. She never felt like she was part of the family. The Dobbses expected Carrie to do chores around the house, and she did them skillfully. She also did well in school. However, once Carrie was promoted to the sixth grade, the Dobbses removed her from school and gave her more housework to do. They also "loaned" her to other families to do housework for them. Still, although life might not have been great for Carrie, it was tolerable.

Everything changed in 1923. Carrie, only 17 years old, became pregnant. She said she was raped. The Dobbses did not believe her, and we will never know what actually happened. However, the truth of how she became pregnant hardly mattered, because in small-town Virginia in this era, *any* pregnancy outside of marriage was deeply shameful. The Dobbses wanted Carrie out of the house—immediately.

Mr. Dobbs filed papers to have Carrie committed, and Justice Shackleford, who had sentenced Emma Buck, scheduled a hearing. The Dobbses claimed that Carrie was prone to outbursts and hallucinations. They said she behaved oddly. Shackleford declared Carrie feebleminded and sentenced her to the same colony to which he had committed her mother; she was to report to the colony after the baby was born. On March 28, 1924, Carrie gave birth to a daughter, Vivien, and the Dobbses took Vivien in.

In the time between Emma's and Carrie's commitments, the state of Virginia passed a eugenics law. **Eugenics** is the idea that the way to improve "human stock" (analogous to breeding better cows, for example) is to prevent biologically defective people from having children and encourage "worthy" people to have many. It is an idea based on a flawed understanding of inheritance and evolution. Virginia's eugenics law stated that a person could be ordered to be sterilized for "the good of society." Carrie was the test case. She had been declared flawed, as had her mother before her. Moreover, feeblemindedness was deemed to be passed from parent to offspring according to the laws of heredity. So was Vivien feebleminded too?

Seven-month-old Vivien (**FIGURE 8.2**) was examined by a social worker who reported, "There is a look about it that is not quite normal, but what it is I can't tell." This testimony was enough to have Vivien labeled feebleminded. The constitutionality of the Virginia law was immediately challenged in a higher court. The case, *Buck v. Bell*, finally reached the U.S. Supreme Court, where, in 1927, Justice Oliver Wendell Holmes wrote the Court's opinion upholding the Virginia law: "It is better for all the world, if instead of waiting to execute degenerate offspring for crime, or to let them starve for their imbecility, society can prevent those who are manifestly unfit from continuing their kind. The principle that sustains compulsory vaccination is broad enough to cover cutting the Fallopian tubes. Three generations of imbeciles are enough."

Carrie Buck was sterilized on October 19, 1927. She was only 21 years old. And what happened to the "imbecile" Vivien? Sadly, she died at the age of eight from an infection—but not before going to school and making the honor roll.

Why did so many people think that society's ills were biologically inherited and could be cured by controlling who was permitted to be born? We know more about genetics and inheritance now than people did during the early part of the 20th century. Do we now have the ability, dreamed of by eugenicists, to eliminate all genetic "defects"? Should we? Is eugenics simply a shameful chapter in our history, or is its goal of improving "human stock" still alive with us today in a modern form?

SOURCES: Edwin Black, *War Against the Weak: Eugenics and America's Campaign to Create a Master Race*, Dialog Press, 2012.

eugenics :: the notion that the way to improve human stock is to prevent biologically defective people from having children and encourage worthy people to have many

FIGURE 8.2 Vivien Buck (shown here with Mrs. Dobbs) was labeled feebleminded at the age of seven months after being examined by a social worker.

8.1 Do Complex Human Characteristics Have a Genetic Basis?

According to the eugenics sterilization laws that were prevalent in the United States in the first half of the 20th century, a board of experts could order someone sterilized if the person was judged to be "insane, idiotic, imbecilic, or moronic." Convicted criminals could also be ordered to be sterilized. In some states, offenses that could lead to sterilization included alcoholism, drug addiction, blindness, deafness, homelessness, or pauperism. The crux of the argument was that these defects were heritable, and therefore sterilization was the only way to stop these defects for the good of society (**FIGURE 8.3**). Were eugenicists correct in thinking that a lot of what is wrong with us is encoded in our genes? Are there "bad" genes?

Genetic Determinism

A 1998 article in *Life* magazine proclaimed: "Were you born that way? Personality, temperament, even life choices. New studies show it's mostly in our genes." This article captured the overly simplistic and largely incorrect ideas of genetic essentialism and genetic determinism.

Genetic essentialism is the notion that being human means having a human **genome**, the term for all genes in an organism. Humans do indeed have a human genome. In fact, each of the trillions of cells in the body contains a human genome. But no one would think of each individual cell as being a human person. Having a human genome is certainly not the only thing that makes us human. **Genetic determinism** is the idea that our genes determine, direct, or cause *everything* about us: our physical appearance, emotions, and behaviors. According to genetic determinism, our genes are in charge: the environment in which we live has little effect on how we grow and develop.

Common sense and a great deal of empirical evidence demonstrate that genetic determinism is simplistic and largely incorrect. Consider identical

genetic essentialism :: the notion that being human means having a human genome

genome :: the term for all genes in an organism

genetic determinism :: the idea that our genes determine, direct, or cause everything about us

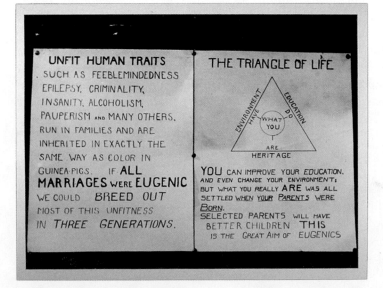

FIGURE 8.3 Eugenicists believed that many "unfit human traits" were inherited, and therefore sterilization was the only way to eliminate these defects for the good of society.

twins. They have identical genomes, but anyone who has known twins has observed that they are very different people with distinct personalities, likes and dislikes, and behaviors (**FIGURE 8.4**). As you read in Chapter 3, even conjoined twins, who share not only the same genes but nearly identical environments, are unique. Each of us inherits a genome that defines a biological potential. We are going to be humans, not giraffes or dandelions. However, the genome is responsive to the environment even while we are *in utero*, as well as after we are born. Did my mother drink excessively while she was pregnant? Did I obtain adequate nutrition, so my brain developed well during childhood? Neither of these conditions is "in our genes."

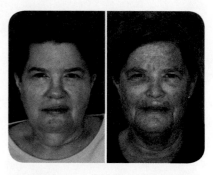

FIGURE 8.4 Identical twins can have dramatic differences. One sister (right) smoked for 16 years before quitting and spent a lot of time in the sun, while the other sister did neither.

It is fortunate for us that our genomes are not completely deterministic. If they were, we would all be dead. For example, most mammals can make vitamin C, an essential nutrient necessary for survival. Humans can't. In fact, 100% of people have a mutation that leads to a deficiency in an enzyme needed in the pathway for vitamin C synthesis (see Chapter 9). Luckily, humans have a way to survive these faulty genetic instructions: we get vitamin C from our environment by eating foods, such as citrus fruits, that are rich in this nutrient. Problem solved.

This returns us to our initial question: do complex human characteristics have a genetic basis? The answer is *not really*, or at least not in the way eugenicists or genetic determinists think. Despite reports trumpeting these latest "findings," no specific genes have been identified for aggressive behavior, altruism, bipolar psychosis, criminality, drumming, figure skating, homosexuality, mathematical ability, schizophrenia, shyness, or television viewing. We are biological beings, so of course genes play a role in many of our characteristics, but this does not mean they determine outcome.

Defining Normal

The experts who committed Emma and Carrie Buck, sterilized Carrie, and labeled Vivien feebleminded thought they had a good understanding of what it means to be normal. The actions that followed their judgment that the Bucks were defective were undeniably terrible, but were they correct that Emma, Carrie, and Vivien were not normal?

The definition of "normal" is quite difficult to articulate. It can refer to a statistical average, like the average height in a population. In this context, normal does not generally suggest a value judgment. Normal can also be used to mean how similar something is to an ideal. For instance, we might have an ideal for beauty and evaluate some people as normal or abnormal accordingly. When normal is used in this way, it is clear that normal is considered superior.

The challenge of defining normal is that in biological beings like us, there is no ideal form. Variation in both how we look and how we function *is* normal. Variation gives us the flexibility to respond to different and changing environments. Natural selection selects traits that happen to work, not traits that are perfect. *Most* individuals in a population will deviate from the average.

Even what humans view as normal varies. What we consider to be beautiful or admirable behaviors have changed over time. Women working as professionals outside the home and men taking on the role of stay-at-home parents would have been seen as seriously abnormal in 1900. Men without facial hair and thin, muscular women would have also seemed odd.

We can shift the question from what is normal or abnormal to what constitutes a diseased or disabled state. In some cases, it might be easy to answer. For example, breast cancer is a disease, not simply a variation in a trait.

But consider dwarfs who suffer no health problems but are simply short in stature. Is this a disability that needs to be prevented or remedied?

We are not any better at defining normal now than people were in Carrie Buck's day. And like them, many of us act as if much of disease and abnormality is genetic at its roots. In Carrie Buck's world, this meant that so-called experts could label someone as shiftless or apply another vague diagnosis. That individual would then be sterilized, stopping those "shiftless genes" from being passed on. Today we have genetic screening, prenatal diagnosis, and gene therapy to hunt down the "bad genes"—and perhaps try to fix them. Is this modern approach any more rational and effective than were eugenics laws? Have good outcomes resulted, or have we merely created new concerns?

8.2 What Is Gene Therapy?

Thousands of diseases and disorders can be attributed to mutations in specific genes (**TABLE 8.1**). Examples include cystic fibrosis and sickle cell disease (Chapter 4), androgen-insensitivity syndrome (Chapter 6), heart defects, cleft palate, certain types of deafness, lack of eyes, extra fingers or toes, male sex reversal, and Huntington's disease (see *Scientist Spotlight: Nancy Wexler* to learn about the search for the Huntington's gene). Serious mutations resulting in spontaneous miscarriage are not at all unusual; approximately one-half to two-thirds of all conceptions result in spontaneous miscarriage.

Prenatal screening tries to identify embryos or fetuses that have genetic or developmental abnormalities. Parents can then use this information to plan and prepare for a child with special needs or to decide to terminate the pregnancy because they do not want to pass on an unwanted trait. One of the challenges we face as a society is that prenatal screening permits us to decide which traits are acceptable to us and which should be avoided or eliminated.

Even with perfect prenatal screening, there will always be individuals born with genetic syndromes. Not all of these are predictable. For example, although the trait for achondroplasia (**FIGURE 8.5**), a type of dwarfism, is passed from parent to child, seven out of eight cases of achondroplasia are due to new mutations. No one else in the family is a dwarf.

FIGURE 8.5 Achondroplasia, a form of dwarfism, can be inherited from parents. However, most cases are due to new mutations (meaning that no one else in the family suffers from the disorder).

Table 8.1 :: Examples of Diseases or Disorders That Can Be Attributed to Specific Gene Mutations

Disease or Disorder	Chromosome Location
Huntington's disease	4p
Juvenile epilepsy	6p
Cystic fibrosis	7q
Wilms tumor-1	11p
Retinoblastoma	13q
Polycystic kidney disease	16p
Cataract	16q
Malignant hyperthermia	19q
Alzheimer's disease	21q
Retinitis pigmentosa	Xp

Scientist Spotlight

Nancy Wexler (1945–)

A small village in Venezuela called Lake Maracaibo holds the key to understanding Huntington's disease (HD). Nancy Wexler began studying the people in this village in 1979 because it had been reported that they have an extremely high incidence of this neurological disorder. She has spent her life trying to solve the problem of HD and other genetic syndromes, and her interest is personal. Wexler's mother was diagnosed with HD in 1968. Wexler's maternal grandfather and two uncles died of the disease. And she and her sister are at risk for developing it.

Nancy Wexler was born in 1945 to Milton Wexler, a psychoanalyst, and Leonore Sabin Wexler, a geneticist. Wexler went to Radcliffe College and graduated with a bachelor's degree in social relations and English. However, her mother's diagnosis shifted Wexler's career plans. She began studying HD in graduate school. Her father founded the Hereditary Disease Foundation in 1968. The Wexler family made fighting HD their number one priority.

Her work in Venezuela turned out to be pivotal. Wexler collected medical and family history information from more than 18,000 people. She also collected 4000 blood samples and sent them to her collaborator James Gusella at the Massachusetts General Hospital. He used the blood samples to map the location of the HD gene to the tip of chromosome 4. In 1993, an unprecedented collaboration among a half dozen research laboratories led to the identification of the actual HD gene. The willingness of this Huntington's Disease Collaborative Research Group to work together instead of competing hastened this discovery. This work has also led to the development of a test that can detect HD long before any symptoms appear. However, since there is no cure for HD, not all people who are at risk opt to take it. In fact, Wexler

Nancy Wexler has spent her life trying to solve the problem of Huntington's disease (HD) and other genetic syndromes. Her work in Venezuela led to the identification of the HD gene.

chose not to have the test done or to have children. She recognizes how difficult the decision can be and has wondered whether she might have been happier had she learned that she would not develop the disease. However, she understands firsthand that knowledge of a disease's certainty may be more than a person wants to live with.

Wexler continues her work on HD and other hereditary diseases. She is a professor at Columbia University and the president of the Hereditary Disease Foundation. She also returns to Lake Maracaibo every year to work with HD patients. Wexler's research has revealed that all of the people there who suffer from HD are related to one woman who died of the disease. Her work has also helped to map some other disease genes, including those for familial Alzheimer's and a certain type of kidney cancer. ::

Some people argue that the emphasis on prenatal screening to prevent certain births shows a lack of understanding of disability. They argue that the problems experienced by the disabled are due not to the physical challenge itself, but rather to the response of others to the disability. While this may be true in some situations, it would be hard to make the case that suffering from a disease like cystic fibrosis or muscular dystrophy occurs principally because of the way that others act. Most people think it would be great to have a way to cure such diseases. Gene therapy is designed to do just this. There are two types of gene therapy: somatic gene therapy and germ-line gene therapy.

Somatic Gene Therapy

The idea of **somatic gene therapy** is pretty simple: cure a genetic disorder by inserting a normally functioning gene into the patient. The actual implementation of the idea has been difficult, however.

somatic gene therapy :: the insertion of a normally functioning gene into a patient to cure a genetic disorder

FIGURE 8.6 Individuals with severe combined immunodeficiency (SCID) lack normal immune function. They must be protected from exposure to anything that could cause an infection.

lymphocyte :: a type of white blood cell

immune response :: the body's ability to fight off infection

germ-line gene therapy :: the correction of genetic problems in the eggs and sperm so that harmful mutations can never be passed on

The first successful gene therapy was done in 1990 to help children who suffered from severe combined immunodeficiency (SCID) (**FIGURE 8.6**). These children are missing an enzyme called adenosine deaminase (ADA) that is needed for normal lymphocyte function. **Lymphocytes** are a type of white blood cell essential for the body's **immune response**, or ability to fight off infection (Chapter 14). Children with SCID are vulnerable to infection from *everything* and must therefore live in a sterile environment to survive.

Scientists took immature lymphocytes and lymphocyte stem cells from SCID patients and cultured them to grow a population of cells in the lab. Lymphocyte stem cells divided to make new lymphocytes. The scientists isolated the ADA genes from healthy cells and inserted them into the genomes of harmless viruses. They then infected the cultured cells with the viruses. Once the cells were infected, the ADA genes popped out of the viruses and inserted themselves into the cells' genomes. The cells that had incorporated the ADA gene were injected back into the patient (**FIGURE 8.7**). In successful cases, the new gene produces the missing enzyme, treating the disease.

Somatic gene therapy may indeed hold the promise of curing some diseases besides SCID. But as with any new medical treatment, there are risks both technical and ethical. We will discuss these risks later.

Germ-Line Gene Therapy

Somatic gene therapy is similar to other types of medical treatments: something is not working properly in a patient, and physicians try to repair it. **Germ-line gene therapy** aims to correct genetic problems in the germ line—eggs and sperm—so that harmful mutations can never be passed on. This technique opens a new world that some think is exciting and wonderful and others consider perilous and wrong. Imagine being able to eliminate the gene for cystic fibrosis, for example, from a family so that no descendants would ever inherit this gene or develop the disease. It's as if the authorities could

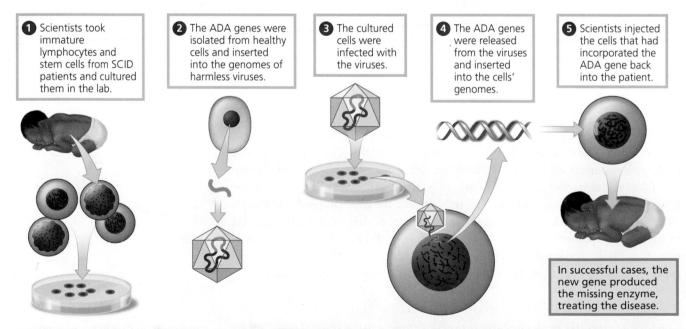

1. Scientists took immature lymphocytes and stem cells from SCID patients and cultured them in the lab.

2. The ADA genes were isolated from healthy cells and inserted into the genomes of harmless viruses.

3. The cultured cells were infected with the viruses.

4. The ADA genes were released from the viruses and inserted into the cells' genomes.

5. Scientists injected the cells that had incorporated the ADA gene back into the patient.

In successful cases, the new gene produced the missing enzyme, treating the disease.

FIGURE 8.7 Somatic gene therapy involves the insertion of a normally functioning gene into a patient and has been used to cure patients with severe combined immunodeficiency (SCID). These children are missing an enzyme called adenosine deaminase (ADA) that is needed for normal lymphocyte function.

have eliminated Emma Buck's "feeblemindedness gene" (if such a thing actually existed) before it doomed both Carrie and Vivien.

So far, no one has tried to do germ-line gene therapy on humans. It has been done in other mammals, however, and is actually a routine research technique in laboratory mice. Also, the rhesus monkey ANDi (the backward abbreviation for "*i*nserted *DNA*") has been altered so that all of its cells have the DNA and RNA encoding for the molecule GFP, green fluorescent protein. However, for some as yet unknown reason, the GFP RNA is not translated. As a result, the GFP protein itself is not present. Nevertheless, all of ANDi's descendants will have the GFP gene in their genomes. Strangely, two other monkeys that also had the GFP gene and RNA in every cell, but which were stillborn, did glow green, meaning they did make the GFP protein (**FIGURE 8.8**). Scientists are not sure why ANDi and these two monkeys were different in this regard.

The goal of this research was not to make "glow monkeys." Scientists wanted to develop methods to insert human disease genes into primates so that the disease could be studied in organisms more similar to humans than mice or rats. Some people argue that it is unethical to genetically engineer animals in this way for this purpose. Even though the scientists who produced ANDi do not support the extension of their work to humans, many are concerned that it is just a matter of time until we engineer our own germ lines.

Besides the ethical questions, significant technical and scientific ones remain. The process of genetically engineering the germ line is extremely inefficient. The experiment that produced ANDi began with more than 200 fertilized eggs. From these, only 40 embryos were produced. The 40 embryos resulted in five pregnancies. And only three live births occurred: ANDi and two monkeys that did not incorporate the GFP gene into their genomes. If this research were to be done in humans, and at this time no one is trying to do so, it would entail research with human embryos and similar inefficiency, at least at first.

For an idea of what would need to be done to alter human germ lines, let's look at how ANDi was actually produced (**FIGURE 8.9**). Scientists inserted the

FIGURE 8.8 Germ-line gene therapy aims to correct genetic problems in the eggs and sperm so that harmful mutations can never be passed on. The rhesus monkey ANDi was altered so that all of its cells encode for the molecule GFP.

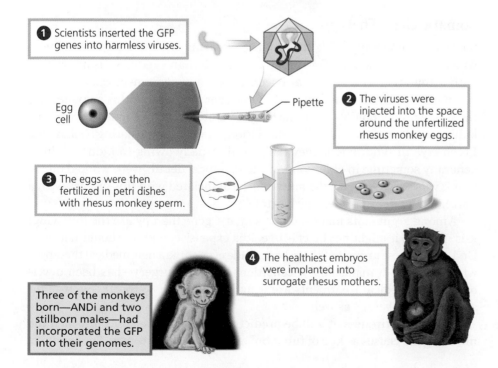

1 Scientists inserted the GFP genes into harmless viruses.

Egg cell

Pipette

2 The viruses were injected into the space around the unfertilized rhesus monkey eggs.

3 The eggs were then fertilized in petri dishes with rhesus monkey sperm.

4 The healthiest embryos were implanted into surrogate rhesus mothers.

Three of the monkeys born—ANDi and two stillborn males—had incorporated the GFP into their genomes.

FIGURE 8.9 The germ-line gene therapy procedure used to produce ANDi was not efficient. Out of more than 200 attempts, only one live monkey was born with the GFP gene incorporated into its genome.

GFP genes into harmless viruses and injected the viruses into unfertilized rhesus monkey eggs. The eggs were then fertilized with rhesus monkey sperm. The embryos that divided twice to produce four cells were next implanted into surrogate rhesus mothers. The surrogates had undergone hormone treatments so that they would accept a pregnancy. A tiny number of surrogates carried their babies to term. Three of the monkeys that were born—ANDi and two stillborn males—had incorporated the GFP into their genomes.

Even though the technical issues remain formidable, many scientists and bioethicists are concerned that we will eventually be able and willing to alter human germ lines. Are there benefits that might result, and what are the risks?

8.3 What Are the Benefits and Risks of Genetically Altering Humans?

When Carrie Buck was sterilized in 1927, the experts truly thought they were doing something that was both scientifically sound and good for society. These were mainstream ideas. The geneticists who were convinced that characteristics like being shiftless or loving the sea were heritable were not on the extreme edge of their profession. Are we more knowledgeable now? Are we sufficiently knowledgeable to alter genes in people in a way that always results in a beneficial outcome?

Scientists and the general public view somatic gene therapy and germ-line gene therapy as quite different from each other. Somatic gene therapy aims to fix or replace a defective gene in *an individual*. This alteration is not passed on to subsequent generations. In contrast, germ-line gene therapy is designed to make genetic changes that will be inherited by *all* descendants. The ethical challenges for these two forms of gene therapy are not the same. Neither are the potential benefits and risks.

Somatic Gene Therapy

Somatic gene therapy holds the promise of curing genetic disorders, some of which are quite terrible. For example, Lesch-Nyhan syndrome is associated with a mutation in a gene that encodes for an enzyme necessary to break down uric acid. Individuals with this disorder are missing this enzyme, and the waste product builds up in the body. Symptoms include muscle problems, kidney malfunction, mental retardation, and self-mutilating behaviors. Lesch-Nyhan syndrome is inevitably fatal, usually owing to kidney failure, generally sometime in childhood. If the technical details were worked out so that it was safe to replace the mutant gene associated with this disease with a normally functioning one, why hesitate?

Among arguments made against somatic gene therapy are the following: it is not safe, it might not be effective, it is expensive, and we should not "play God." All of these arguments have been made every time a new medical therapy—whether blood transfusions, vaccinations, or even surgery—has been developed. New medical treatments are usually risky at first, and this is true for somatic gene therapy as well.

As for effectiveness, it will be predictably variable. Diseases in which the mutant gene causes a "loss of function" are much easier to cure. For example,

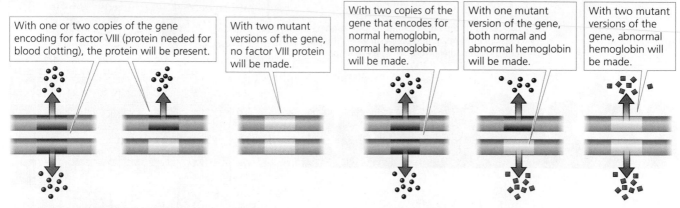

A With a **loss-of-function disease** such as hemophilia, it is necessary to ensure only that the normal protein is made because the mutant allele makes no protein.

B With a **gain-of-function disease** such as sickle cell, the production of bad proteins must be stopped in addition to ensuring that the correct protein is made.

FIGURE 8.10 Loss-of-function diseases (A) are much easier to cure with somatic gene-line therapy than gain-of-function diseases (B).

the gene associated with **hemophilia** fails to direct the production of a clotting factor. As a result, an individual with hemophilia cannot form blood clots in the event of an injury. If a patient with hemophilia received somatic gene therapy to insert the functional gene, he or she would be able to make the clotting factor and would be cured (**FIGURE 8.10**). In contrast, mutant genes that produce a "gain of function" will be very difficult—maybe impossible—to treat. Sickle cell disease, for instance, is the result of a mutant gene that produces abnormal hemoglobin. To correct this problem with somatic gene therapy, it would be necessary to insert the functional gene *and* to disable or eliminate the mutant genes in all blood cells.

Somatic gene therapy is expensive. But so are all new medical technologies, especially when they are first developed and implemented. This is true for heart surgery, organ transplantation, chemotherapy, and many other treatments that are routine in modern hospitals today. As for "playing God," one could argue that everything we do to heal, treat, or overcome illness, disability, or accident is playing God. Before the advent of open heart surgery to bypass clogged arteries, many people died at a much younger age than they do today.

The most successful somatic gene therapy so far has been the treatment of severe combined immunodeficiency (SCID). But scientists do not have complete control over where an introduced gene inserts, and the consequences are potentially lethal.

Even though a great deal of scientific work still needs to be done, it seems likely that development of somatic gene therapy will continue. It is a medical intervention designed to help. And who wouldn't like a world in which people are free from the suffering caused by cancer, cystic fibrosis, muscular dystrophy, and a host of other diseases?

Germ-Line Gene Therapy

What happened to Carrie Buck and the tens of thousands of individuals sterilized under the provisions of the eugenics laws was the ultimate attack on germ lines. Because of prejudices and beliefs about what could be inherited, the experts and legal authorities simply put an end to the germ line altogether. To ensure that no "bad" genes were passed on, they prevented the passing of *all* genes into the next generation.

hemophilia :: a disease in which an individual cannot form blood clots in the event of an injury

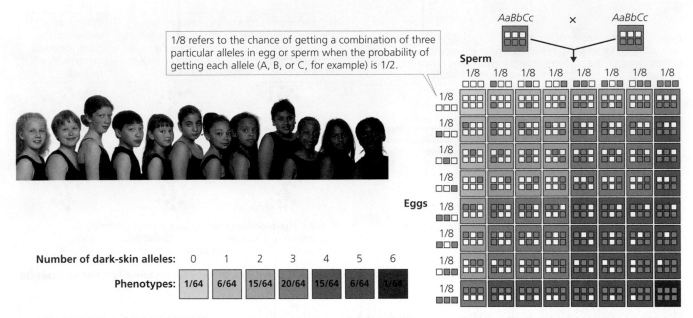

1/8 refers to the chance of getting a combination of three particular alleles in egg or sperm when the probability of getting each allele (A, B, or C, for example) is 1/2.

Number of dark-skin alleles:	0	1	2	3	4	5	6
Phenotypes:	1/64	6/64	15/64	20/64	15/64	6/64	1/64

FIGURE 8.11 Genes do not act alone as individual entities. Some traits, such as skin color, are influenced by multiple genes.

FIGURE 8.12 Many genes influence more than one trait. Albino organisms are unable to make the pigment melanin because of a single mutation, which affects the traits of coat color, eye color, and vision, among other characteristics.

genetic counseling :: an approach used to inform couples about their relative risk of passing on certain specific genes associated with a disease

Germ-line gene therapy is not nearly as crude. Nevertheless, it does share two things in common with eugenics: (1) confidence about the deterministic genetic basis of characteristics and (2) confidence about defining desirable and undesirable traits. Are these confidences well grounded?

The relationships between individual genes and specific characteristics, physical or otherwise, are not straightforward. Genes don't act alone as individual entities. Some traits—such as height, weight, skin color, eye color, and susceptibility to cancer—are influenced by multiple genes (**FIGURE 8.11**). Also, many individual genes influence more than one trait (**FIGURE 8.12**). For instance, the globin gene affects both how well red blood cells carry oxygen and the shape of the cells themselves. And the environment influences the way many traits develop. For example, genes that affect how the brain and face form during embryogenesis and fetal development are sensitive to whether a mother drinks alcohol during pregnancy. Do we have adequate knowledge to alter genes for all of the descendants of an individual? Even if all of the technical details were worked out and we could alter germ lines safely, is it ethical to do so?

Many people argue that we have safe and effective alternatives to germ-line gene therapy. **Genetic counseling** can inform couples about their relative risk of passing on certain specific genes associated with a disease. Prenatal diagnosis of embryos can identify whether any of these harmful, disease-causing mutations are present. If the embryos are the result of *in vitro* fertilization (IVF), which we studied in Chapter 6, healthy embryos can be selected for implantation. A couple could opt for sperm or egg donation from individuals who are not at risk for passing on these particular defective genes. And adoption is always an option.

Many people consider the following ethical issues surrounding germ-line gene therapy to be so serious that efforts to develop this technology for people should stop:

- First, safety is by no means certain, and a mistake is permanent, affecting more than one person. The insertion of a gene can unexpectedly disrupt the functioning of another gene, sometimes producing grave consequences.

- Second, we are not sure which genes are "good" and which are "bad." In some cases, this depends on the environment. For example, the sickle cell mutation has the potential to produce sickle cell disease in individuals who get two copies of the gene. However, a person with only one copy enjoys protection from malaria. Whether or not having a copy of the sickle cell gene is beneficial or harmful depends on whether a person lives in an area where malaria is rampant.

- Third, as with somatic gene therapy, we need to distinguish between medical treatment and enhancement. Is it appropriate for parents to make specific choices about their children's genomes? Is it acceptable to "design" children according to the parents' tastes or needs? And what if "genetic fashion" were to change and as a society we stopped favoring a trait that we had once valued? (See *Life Application: Sex Selection* to learn about choices some parents are making about their children.)

- Finally, this type of therapy has the potential to alter how we feel about the sick and disabled in society. If we were able to design people, would we judge a person's family harshly because they hadn't prevented this affront to perfection? What about poor people who could not afford access to this technology? Would we see a growing gap between the rich and poor based on their biology?

Life Application

Sex Selection

Who would have predicted that the availability of inexpensive ultrasound machines would result in increased crime, sex trafficking, and social unrest? Yet that is exactly what is happening in several places in the world, especially China and India, where some people engage in sex selection—choosing whether to allow a female embryo or fetus to come to term and be born.

Male children are preferred in many cultures for varying reasons. In China, for example, sons are important for the care and protection of the elderly. They are also considered essential for continuing the family line. In India, sons are seen as a financial asset to the family. It costs a family money, in the form of a dowry, to "marry off" a daughter, and when she does marry, she joins the husband's family.

The shift in sex ratio is dramatic in both of these countries. In some Chinese provinces, 120–130 boys are born for every 100 girls. A study in India evaluated 1.1 million households and collected data for 133,738 births. They found that for families that already had one girl, the ratio of boys to girls for the second child was 1000 to 759. If the first child was a boy, the sex ratio for the second child was 1000 boys to 1102 girls. As long as they had one boy, girls were acceptable.

This pattern of sex selection is not confined to China and India. The 2000 U.S. census showed a similar bias in U.S.-born children of Chinese, Korean, and Asian Indian parents. The bias in favor of sons was especially dramatic in families in which the first two children had been girls: the number of boys born in this third position outnumbered girls by 50%.

This shift in gender ratio worldwide is due to the selective abortion of females. A 2002 study showed that 300 out of 820 women in a central Chinese village had abortions, and 70% of the aborted fetuses were female. A 1985 survey in Bombay (now Mumbai), India, reported that of the female fetuses aborted, 96% occurred after the sex had been determined by amniocentesis.

The gender gap created by these practices is very serious. The 2000 census in China showed that there were 19 million more boys than girls in the newborn to 15-year-old age bracket, and the 2010 census showed a worsening of this skewed gender ratio. By 2014, there will be 40–60 million "missing" women. In India, 500,000 fewer girls than boys are born each year. This has resulted in a "deficit" of 10 million females over two decades. And, similar to China, the 2010 census shows that the gender ratio in India is worsening. What will the "extra" men do for wives?

China and India are trying to remedy the "missing girls" problem. They can't fix the unbalanced sex ratios of the

(Continued)

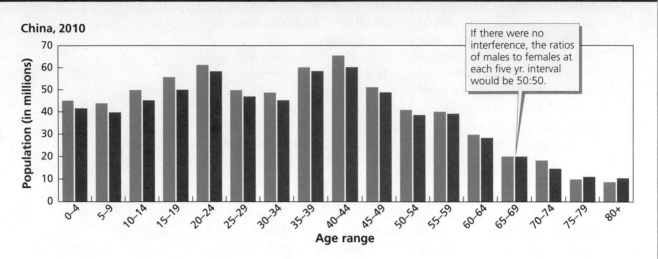

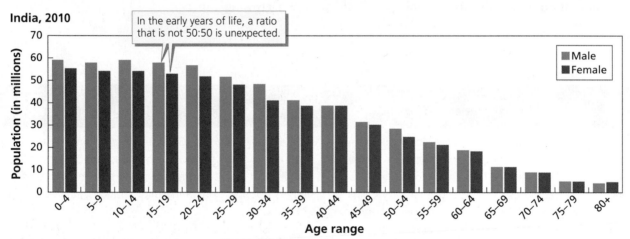

Male children are preferred in many cultures for varying reasons. Both China and India engage in sex selection, and the shift in sex ratio has been dramatic in both countries.

generations that are already born, but they are taking steps to rebalance sex ratios for future generations. In India, for example, prenatal sex determination has been illegal since 1994. Also, the practice of paying dowries is slowly being shifted to "bride money," in which the groom pays the bride's family. In China, steps have been taken to make it more desirable to have a baby girl, including elimination of school fees for girls; the awarding of an annual pension to couples who limit their offspring to two girls; and the preferential treatment of parents of girls in health care, housing, and employment. But even if these efforts are successful, it will take decades to see the effects. ::

8.4 How Can Stem Cells and Cloning Be Used to Alter People?

The case of Carrie Buck and the modern development of gene therapy technologies illustrate our long-standing confidence, perhaps misplaced, that we really do know how to define desirable and undesirable traits. The eugenics movement in the United States is a terrible lesson about what can happen when actions go beyond actual knowledge and are fueled by prejudice (particularly targeted at the poor).

Gene therapy is not the only technology that is being developed to permanently change the biology of an individual at a very fundamental level. The use of stem cells and cloning also offer the hope of curing disease, as well as the risk that we might produce outcomes with unintended consequences. Do we have adequate knowledge to develop and use these technologies safely? Do we know enough to judge whether we should?

Stem Cells

The basic common feature shared by all **stem cells** is the ability for self-renewal. At each division, a stem cell can give rise to another stem cell as well as a more specialized cell (**FIGURE 8.13**). But not all stem cells are the same. They come from varied sources and differ in their developmental flexibilities. Some stem cells can be directed to make any type of cell in the body, whereas others are capable of producing only a much more limited range of cells.

How Stem Cell Research Works The hope engendered by stem cell research is that scientists will be able to develop methods to use stem cells to replace neurons lost by Parkinson's disease or spinal injury; replace diseased pancreatic cells and cure diabetes; or repair heart damage by adding new cells to the heart, to cite some examples. The concerns about stem cells center mostly on the sources of the stem cells themselves.

Stem cells can be collected from many sources: early embryos; fetal cells; umbilical cord blood; and babies, children, or adults—these are usually referred to as **adult stem cells**. Stem cells collected from early embryos have the greatest developmental flexibility, and those from babies, children, or adults have the least.

Stem Cells in Human Development Embryo development starts with a single cell, the zygote, which was produced when an egg and sperm fused together (Chapter 3). That cell is capable of producing any of the cells in the developing embryo; it is **totipotent**. This initial cell undergoes divisions and up until

stem cell :: a cell that can renew itself and give rise to other cells

adult stem cell :: a cell from an adult that can renew itself, but that is limited in its flexibility

totipotent :: a cell that is capable of producing any of the cells in a developing embryo

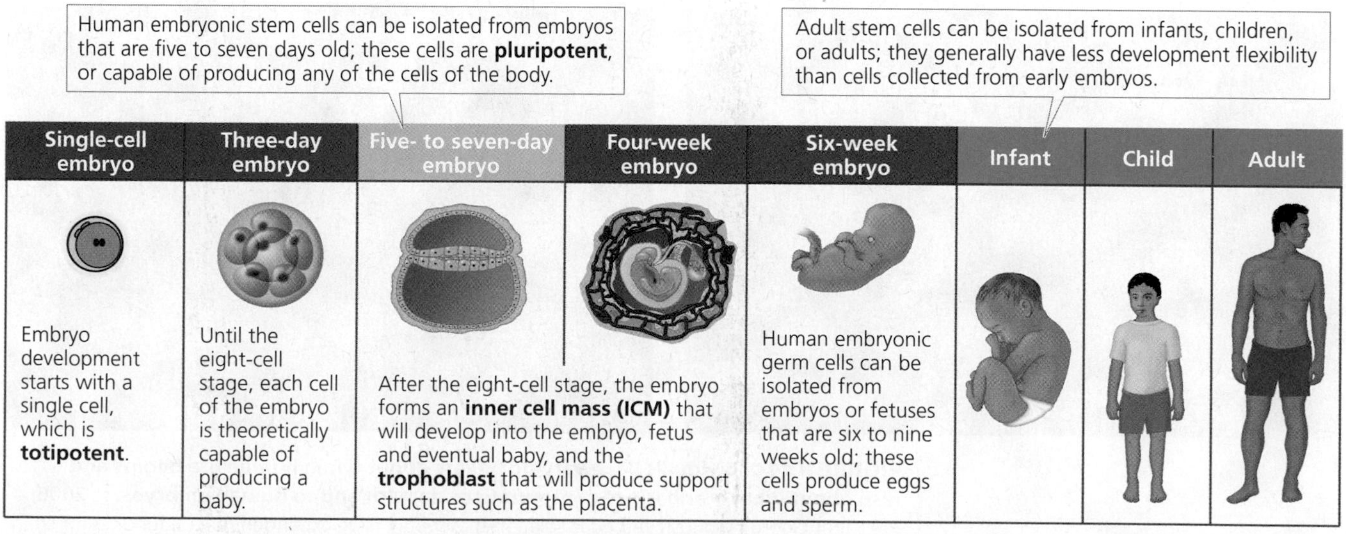

Human embryonic stem cells can be isolated from embryos that are five to seven days old; these cells are **pluripotent**, or capable of producing any of the cells of the body.

Adult stem cells can be isolated from infants, children, or adults; they generally have less development flexibility than cells collected from early embryos.

Single-cell embryo	Three-day embryo	Five- to seven-day embryo	Four-week embryo	Six-week embryo	Infant	Child	Adult
Embryo development starts with a single cell, which is **totipotent**.	Until the eight-cell stage, each cell of the embryo is theoretically capable of producing a baby.	After the eight-cell stage, the embryo forms an **inner cell mass (ICM)** that will develop into the embryo, fetus and eventual baby, and the **trophoblast** that will produce support structures such as the placenta.		Human embryonic germ cells can be isolated from embryos or fetuses that are six to nine weeks old; these cells produce eggs and sperm.			

FIGURE 8.13 A stem cell is self-renewing, and at each division it can give rise to another stem cell as well as a more specialized cell.

human embryonic stem cell :: a cell that comes from an inner cell mass and can be isolated from an embryo that is five to seven days old and grown in a lab

pluripotent :: a cell that is capable of producing any of the cells in the body but none of the support structures needed for the development of a baby

human embryonic germ cell :: a cell that comes from a six- to nine-week-old embryo and fetus; these cells ultimately form the cells that produce eggs and sperm

the eight-cell stage, each cell of the embryo is totipotent—theoretically capable of producing a baby. After this stage, the embryo forms two populations: the inner cell mass (ICM) that in normal development will form the embryo, fetus, and eventual baby, and the trophoblast that will eventually produce support structures such as the placenta.

The ICM cells are the source of **human embryonic stem cells**. They can be isolated from embryos that are five to seven days old and grown in a lab. These cells are **pluripotent**, or capable of producing any of the cells of the body but none of the support structures needed for the development of a baby. They can't help form the placenta and, therefore, are not remotely capable of producing a baby. Nevertheless, embryonic stem cells have tremendous flexibility. (See *How Do We Know? How Human Embryonic Stem Cells Can Be Directed to Form Specialized Cells.*)

Rivaling embryonic stem cells in their potential are the **human embryonic germ cells** collected from six- to nine-week-old embryos and fetuses. During normal development, embryonic germ cells ultimately form the cells that produce eggs and sperm. Embryonic germ cells, too, are pluripotent; they behave similarly to embryonic stem cells in experiments.

As development unfolds, the versatility of stem cells becomes more limited. In adults, for example, many types of stem cells are restricted, meaning that at each division they can make either another stem cell or *one* type of specialized cell. For instance, certain blood stem cells are limited in this way. Stem cells collected from umbilical blood have less developmental potential than embryonic stem cells but more than adult stem cells.

Concerns About the Sources of Stem Cells Embryonic stem cells have tremendous potential for the medical treatment of ailments like Alzheimer's disease, heart disease, brain injury, blood cancers, immune system disease, and many others. As a consequence, many individuals support human embryonic stem cell research wholeheartedly. However, a significant number of people oppose human embryonic stem cell research because it destroys human embryos. Individuals have varying beliefs about when human life begins and also about what is and is not appropriate to do with and to human embryos (**FIGURE 8.14**).

FIGURE 8.14 Individuals have varying beliefs about when human life begins and also about what is and is not appropriate to do with and to human embryos. In 2006, President George W. Bush vetoed legislation that would have expanded federal funding for stem cell research. In 2009, President Barack Obama signed an executive order restoring full federal funding of stem cell research.

How Human Embryonic Stem Cells Can Be Directed to Form Specialized Cells

Scientists have been studying and manipulating the cells of mammalian embryos, especially mice, since the 1950s. This research has led to the understanding that embryonic cells are totipotent at the very earliest stages, but cells of the inner cell mass are pluripotent. Many details of how to direct embryonic stem cells to form particular specialized cells in culture have also been determined. This progress is excellent if all we want to do is to heal the illnesses and disabilities of mice, but what about humans? Can human embryonic stem cells be directed to make specific types of cells?

The answer is *yes*. The first published report of human embryonic stem cells directed to make specialized human tissues appeared in 2001. Daniel Kaufman and his colleagues at the University of Wisconsin–Madison succeeded in getting human embryonic stem cells to make precursor cells for blood cell production.

Kaufman isolated human embryonic stem cells and cultured them in media containing fetal bovine serum and either mouse bone marrow cells or mouse yolk sac endothelial cells.

The idea was to mimic the cocktail of chemical signals and cell communications that happen during normal development. To everyone's delight, the human embryonic stem cells produced hematopoietic precursor cells, which in normal embryos will make different types of blood cells. Kaufman cultured these hematopoietic precursor cells in various hematopoietic growth factors—the same molecules that direct blood cell development in the bone marrow. Amazingly, myeloid, erythroid, and megakaryocyte colonies grew. These give rise to white blood cells, red blood cells, and platelets, respectively—all the solid components of blood.

Kaufman's findings are important for two reasons. First, they add to our basic knowledge of how to manipulate human embryonic stem cells. Second, they may allow for the development of a new way to generate blood cells for transfusion and for transplantation to treat anemias and blood cancers. Given the shortage of the blood supply and the difficulty in finding bone marrow donors, this would be a welcome breakthrough. ::

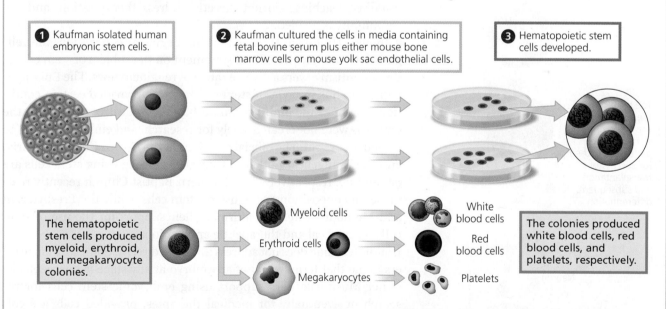

❶ Kaufman isolated human embryonic stem cells.

❷ Kaufman cultured the cells in media containing fetal bovine serum plus either mouse bone marrow cells or mouse yolk sac endothelial cells.

❸ Hematopoietic stem cells developed.

The hematopoietic stem cells produced myeloid, erythroid, and megakaryocyte colonies.

Myeloid cells → White blood cells

Erythroid cells → Red blood cells

Megakaryocytes → Platelets

The colonies produced white blood cells, red blood cells, and platelets, respectively.

The first published report of human embryonic stem cells being directed to make specialized human tissues occurred in 2001. Daniel Kaufman and his colleagues at the University of Wisconsin–Madison succeeded in getting human embryonic stem cells to make precursor cells for blood cell production.

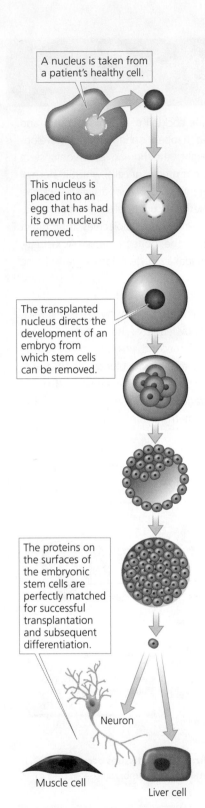

A nucleus is taken from a patient's healthy cell.

This nucleus is placed into an egg that has had its own nucleus removed.

The transplanted nucleus directs the development of an embryo from which stem cells can be removed.

The proteins on the surfaces of the embryonic stem cells are perfectly matched for successful transplantation and subsequent differentiation.

Neuron

Muscle cell

Liver cell

FIGURE 8.15 Therapeutic cloning is used to produce healthy embryonic stem cells for the medical treatment of individuals. In somatic cell nuclear transfer therapy (SCNT), a nucleus is taken from a somatic cell and placed in an oocyte to produce embryonic stem cells.

At issue are two questions: when does a human life begin, and what is appropriate to do *with* and *to* a human embryo? Neither question is easy to answer because many of the arguments are based in religious beliefs. And since we are a pluralistic society open and respectful to all faith traditions, it is difficult to envision how to resolve our differences. It may help to look at how human embryonic stem cells are collected.

Human embryonic stem cells can be collected from five- to seven-day-old embryos that are composed of 100–150 cells. Some argue it is acceptable to generate human embryonic stem cells by combining donated eggs and sperm in the lab expressly for this purpose. Others maintain that creating embryos as a source of stem cells is unacceptable, but using "extra" embryos left over in IVF clinics is fine. (There are more than 400,000 frozen embryos stored in the United States.) And some say that no source is acceptable.

People of all religions agree that human life is special and should be respected, valued, and protected. But they do not agree on whether embryonic stem cells should be used for research or for therapeutic reasons. Here are just *some* examples of the range of beliefs about this issue:

- According to Jewish teachings, a fertilized egg does not have "personhood." Therefore, it is acceptable to use it for stem cell collection for medical research or ultimately therapeutic treatment.
- Roman Catholics believe that an embryo is a human being from the moment of conception. Therefore, they maintain that human embryonic stem cell research is prohibited. The Catholic Church does, however, support adult stem cell research.
- Buddhist teachings do not directly address this question, and the community is divided.
- Although Hinduism teaches that life begins at conception, the religion does not have an official statement on the subject of stem cells.
- Protestantism covers a diverse range of religious views. The Episcopalian Church is in favor of stem cell research provided certain conditions are met: the embryos would have been destroyed anyway, the embryos were not created solely for research, and embryos are not to be bought or sold. The official positions of the United Church of Christ and the Evangelical Lutheran Church regarding using stem cells are *yes* and *no*, respectively. The Southern Baptist Church recently reaffirmed its opposition to the use of stem cells, while the Presbyterian and Methodist Churches affirmed their support for the use of stem cells for medical and therapeutic applications.
- Islamic teaching does not address this issue directly. Some Muslim leaders argue that termination of an embryo at any stage is unacceptable. Other Muslim leaders support using embryonic stem cells for research or eventually for medical therapies, provided cells are collected before "ensoulment"—the moment when an embryo or fetus attains full human rights. In either case, research on embryonic stem cells obtained from extra IVF embryos *or* from embryos created for research is acceptable.

Noting the lack of a clear consensus on the issue, the National Council of Churches officially neither endorses nor condemns the use of human embryonic stem cells. Perhaps considering several scientific facts may help to shed light on this question. First, human embryonic stem cells are isolated from embryos at such an early stage that no specialization or differentiation has occurred. In this view, the cells contain human genomes but are not persons. Second, human embryonic stem cells are not even "potential persons." The isolated cells could never give rise to an entire embryo because they can't produce any of the tissues needed to support the development of an embryo, fetus, or baby.

Is it possible to find a resolution for these dramatically varying perspectives? Should religion play a role in how we make research decisions? What do we do when religions come to different conclusions about issues that may affect everyone?

Cloning

The overarching goal of eugenics was to improve the biology of human beings. Carrie Buck's experience demonstrates "negative" eugenics—stopping the mating of so-called defectives. "Positive" eugenics is illustrated by germ-line gene therapy. Here the effort is to improve the genetic lineage for all descendants. One more potential manifestation of positive eugenics is human cloning, which is distinct from the gene cloning discussed in Chapter 7.

Just as gene therapy can be divided into somatic gene therapy (treatments meant to cure an individual) and germ-line gene therapy (efforts to fix "bad" genes for descendants), cloning can be divided into two categories. **Therapeutic cloning** refers to producing embryos as sources of healthy embryonic stem cells for the medical treatment of individuals. **Reproductive cloning** means creating an embryo with the intention of producing a baby. Any features present in the clone are intended to be passed on to future generations.

Therapeutic Cloning Using stem cells to treat or cure a disease can sometimes produce undesirable side effects. All cells have proteins on their surfaces that permit the body to differentiate between "self" and "nonself" to determine whether transplanted organs, tissues, or cells will be accepted or rejected. Human stem cells collected from an embryo will not have the same protein recognition markers as those of the patient's cells and therefore may be rejected when transplanted. The technical solution to this problem is therapeutic cloning done by **somatic cell nuclear transfer (SCNT)** (**FIGURE 8.15**).

In SCNT, a nucleus is taken from a healthy cell in the patient's body. This nucleus is placed into an egg that has had its own nucleus removed. The transplanted nucleus then directs the development of an embryo from which embryonic stem cells can be removed. Because the transplanted nucleus came from the patient, the proteins on the surfaces of the embryonic stem cells are perfectly matched for successful transplantation. Embryos created in this way are intended for the medical treatment of patients, not for reproduction.

Reproductive Cloning The express goal of reproductive cloning is to create embryos that will develop into offspring. As with SCNT, a nucleus taken from a "parent" is placed into an egg that had its nucleus removed. The embryo that results is implanted into a surrogate mother. An organism produced in this way will be genetically identical to the parent that donated the nucleus.

Several nonhuman animals—including a cat, a dog, cattle, a horse, a goat, a mouse, sheep, and a rabbit—have been cloned. But there seems to be little or no support for cloning humans. In fact, reproductive cloning of humans is illegal in more than 25 countries (but not in the United States; although some state laws ban the practice, no federal laws ban human reproductive cloning).

therapeutic cloning :: producing embryos as sources of healthy embryonic stem cells for the medical treatment of individuals

reproductive cloning :: creating an embryo with the intention of producing a baby

somatic cell nuclear transfer (SCNT) :: therapeutic cloning in which a nucleus is taken from a healthy cell and placed into an egg that has had its own nucleus removed; the transplanted nucleus then directs the development of an embryo from which embryonic stem cells can be removed

What Are the Benefits and Risks of Stem Cell Research?

When it comes to the medical role that stem cells can play in the future, there is generally more support for such research than support for cloning. What has the research revealed so far about the benefits and risks of using stem cells?

Benefits

Most of the experiments done to understand stem cells were performed on mice. Mouse embryonic stem cells have been directed to form pancreatic cells that make insulin, muscle cells that contract, and functional neural cells. Similarly, human embryonic stem cells have been directed to make specific cell types in culture.

Research in whole animals has also produced promising results. For example, when injected into the brains of rats that suffered from a disease akin to Parkinson's disease in humans, human embryonic stem cells differentiated into neurons. Parkinson's disease is characterized by the failure of certain neurons to secrete the chemical signal dopamine. The result is that brain cells deteriorate, with symptoms that include tremors, rigid movements, and slow speech. In rats, the neurons that formed from the injected human embryonic stem cells not only produced dopamine, but also reduced the Parkinson's-like symptoms. Similarly, human embryonic stem cells that differentiated into heart cells were able to repair a heart defect in pigs. There are many technical details to be worked out, but it does seem plausible that stem cells may be able to cure and treat disease.

Human embryonic stem cells have several other uses besides the treatment of disease. First, they are a basic research tool to learn more about normal human development and how birth defects occur. Second, by altering specific genes of embryonic stem cells, it will be possible to create human disease models at the cellular level instead of relying on mouse or other mammalian versions. Third, currently, pharmaceutical drugs are initially tested on nonhuman organisms. Embryonic stem cell cultures would allow scientists to test drug effects on specific types of human cells.

Risks

As with any new technology, stem cells are not without risk. First, the procedures are not completely safe as yet. Transplanted stem cells are not always perfectly matched to patients, and the cells may be rejected. For example, adult stem cells are routinely used to treat certain blood cancers. If these stem cells are rejected, the patient may develop host-versus-graft disease wherein the immune system attacks other parts of the body. Also, the dosage of embryonic stem cells needs to be carefully controlled or tumors may form. This occurred when neurons developed from human embryonic stem cells that were transplanted into rat brains.

Besides the safety issues surrounding the actual use of stem cells, the donation of eggs needed for SCNT is also potentially dangerous. A large number of eggs will be needed for research and implementation of medical treatments. To donate eggs, women must undergo hormone treatments so they will "superovulate," or release multiple eggs. About 5% of women who go through this process develop ovarian hyperstimulation syndrome; 1–2%

experience a severe form of this syndrome, leading to possible kidney failure, lung failure, shock, or ruptured ovaries. Although it might seem that a woman fully informed about the risks can make her own decision about whether to donate eggs, some people worry that poor women might be exploited or pressured into selling eggs.

Another concern about stem cell and cloning technologies is that they could be used for enhancement rather than for medical or therapeutic purposes. Even individuals supportive of therapeutic cloning worry that the knowledge gained in this research could be used to make human reproductive cloning a reality.

Finally, some people are concerned that these new technologies could extend the human life span beyond what is beneficial to society. They argue that having an overly long life, relatively free of worry about "unfixable" ailments, might lead to a less meaningful life since there is less concern about dying any time soon. Also, an able-bodied healthy but older generation would not have to get out of the way for the next, so perhaps innovation would suffer. Or maybe we need to rethink how we shape our lives so that fulfillment is possible at *any* age.

8.6   What Other Challenges Result from Advances in Medical Technology?

This leads us back to where we began—with Carrie Buck. A group of powerful people used the claim of scientific expertise, political power, and ideas about what was desirable or normal to put an end to her germ line forever. They made sure that no more "imbeciles" would be born in the Buck family.

The eugenics law, with mandatory sterilization, was not unique to Virginia. More than 30 states enacted them by the 1930s. And the laws remained on the books for several decades. For example, North Carolina's eugenic sterilization law, although not enforced, was updated in 1973 and 1981. Also, forced sterilizations continued at least into the 1970s: between 1972 and 1976, 3406 women and 142 men were sterilized in just four hospitals in a program of Native American sterilization implemented by Indian Health Service physicians.

Today we don't define "desirable" in as coarse a manner. Yet we do make judgments about which traits and physical characteristics are good and which are to be avoided. And we still face challenges over how to avoid the undesirable characteristics and how to pay for the privilege.

Privacy

The ability to identify specific disease genes and other DNA sequences in people has the potential to create a new body of information about each of us (**FIGURE 8.16**). What disease genes do I carry? What risk does my genome pose for me or my children? (See *Technology Connection: Who's the Daddy?*) Do my genes suggest that I have a predisposition for cancer, heart disease, schizophrenia, or alcoholism? Who should be allowed access to this private information about me? My doctor? Employer? Insurance company? The police or government?

Blood from the heel of a newborn baby is taken via a Guthrie test that screens for a rare genetic disease, phenylketonuria (PKU), which affects around 1 in 16,000 people and can cause severe brain damage.

According to the Health Insurance Portability and Accountability Act (HIPAA), patients have significant legal rights and control over their personal medical information and records.

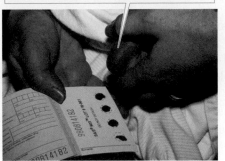

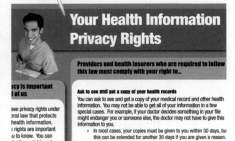

FIGURE 8.16 Genetic testing is available for many diseases and disorders. However, such testing and results raise concerns about privacy and who should be allowed access to this private information.

Technology Connection

Who's the Daddy?

Although it's generally easy to identify a baby's mother at birth, it can sometimes be a challenge to identify the father. In 1944, for example, the actress Joan Berry sued Charlie Chaplin, a famous silent film star, for paternity. Chaplin denied that he was the father of her baby, and the case went to court. Chaplin, Berry, and the baby all had their blood types tested. The results were unequivocal—Chaplin was not the father. Three physicians testified at the trial on Chaplin's behalf, but the judge refused to allow blood type as evidence. Chaplin was judged to be the father and ordered to pay child support. This case led to a change in the law regarding paternity and blood tests. But blood tests can only *rule out* paternity, not confirm it. More informative tests were needed to identify the match with more certainty.

Fast forward 60 years. Paternity testing is now almost foolproof, thanks to the development of DNA testing in the late 1980s. Here is how it works.

Each of us has 46 chromosomes—23 inherited from our mother, 23 from our father. Consequently, each of us has a unique combination of chromosomes. A comparison of the DNA base sequence (A, G, T, and C) of any two human beings is more than 99.9% identical. The 0.1% translates into approximately 3 million bases that are different from person to person. The sequence of this 0.1% of DNA is unique to an individual. DNA testing examines these sequences to determine identity. In the case of paternity testing, DNA samples are taken from the mother, baby, and potential father. The DNA is chopped up into pieces using enzymes that break the DNA

strand at specific sequences of bases. The fragments that result are particular to an individual. Next, these DNA fingerprints from the baby, mother, and potential father are compared to each other to see whether they match. Paternity can be confirmed or ruled out with greater than 99.9% accuracy.

Why go through all this trouble to learn the identity of a father? There are several good reasons: to establish legal rights such as eligibility for the father's Social Security or veterans' benefits; to obtain family medical history for better medical care; and to establish a caring and supportive relationship between father and child. ::

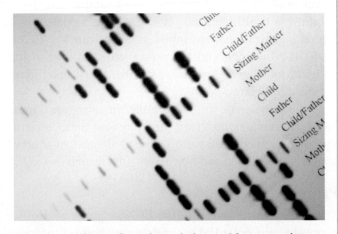

Paternity can be confirmed or ruled out with greater than 99.9% accuracy. In both visible paternity investigations on this autoradiograph, the supposed father is "included," meaning that his DNA bands match those of the child.

Accessibility

If the much-hoped-for medical cures from stem cells and therapeutic cloning become a reality, will they be available to everyone? And if so, how? These cures are likely to be expensive, so it is hard to imagine how poor people will be able to gain access. Even today, in countries with privatized health care like the United States, poor people are unable to obtain the same medical treatment enjoyed by the wealthy. In countries where there is not a well-developed medical system with hospitals, clinics, and an adequate number of physicians and nurses, there is little or no access to the medical technologies that already exist.

In fact, it is reasonable to ask whether spending money on the development of advanced technologies like stem cells and therapeutic cloning is a good use of resources. Would it be wiser to provide basic health care to all? Are there adequate resources to do both?

Danger of a New Eugenics Movement

We have established an entire system of genetic counseling and prenatal and postnatal screening. Is our scientific knowledge more accurate today compared to Carrie's time? Undoubtedly, yes, but there is still a great deal to learn. Has our judgment improved since the 1920s? Or are we creating a new eugenics movement? These questions are harder to answer.

Genetic screening makes sense to most people. The results provide parents with information about their child's potential future. If an embryo or fetus has a "problem," the parents can decide whether they wish to continue the pregnancy. In effect, they are deciding which traits are acceptable and which are not. Not everyone agrees about these judgments.

In a survey conducted by the Shriver Center across 37 countries (the results of which were published in 2002), 82% of U.S. geneticists supported the abortion of fetuses with Down syndrome (**FIGURE 8.17**) and 92% thought it advisable to abort fetuses with severe spina bifida. Abortion of fetuses with

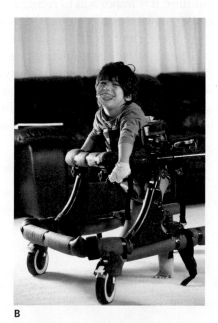
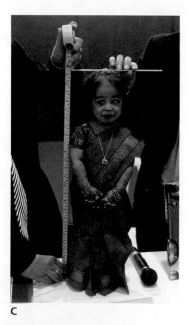

A   B   C

FIGURE 8.17 In a survey conducted by the Shriver Center across 37 countries, the majority of U.S. geneticists advised the termination of pregnancy for fetuses with Down syndrome (A), spina bifida (B), or dwarfism (C). The public generally did not agree.

achondroplastic dwarfism was advised by 56% of the geneticists. However, the general public was not so eager to support the termination of fetuses unless there was the certainty of extremely severe mental retardation.

Most people would probably agree on the merits of prenatal screening for the detection of invariably fatal genetic diseases, but even here the choice is not easy. What if the fetus has a fatal genetic disease like Huntington's, which generally does not reveal symptoms of neurological degeneration until the third or fourth decade of life? If a person dies at 50 years, was that life not worth living? Should such a person not have children?

At its extreme, genetic counseling and selective abortion could create a world in which wealthy families have less risk of having disabled children—simply because they can afford the tests. This world could be divided into genetic "haves" and "have-nots." Or we could try to use these tools more equitably. We could recognize that because of mutation, there will always be genetic disease and that meaningful, satisfying lives come in many forms.

 # Biology in Perspective

Most people are confused about what genes can and can't do. At its most basic, the DNA of genes has the information to direct the production of a protein. The protein has some particular function in a cell. Somehow this simple scenario plays a critical role in how we look, function, and act. It is also important for influencing whether we are healthy or sick.

Genes don't act alone, however. They form interacting networks. The proteins that are produced may in turn regulate gene function, and they may work together with other proteins too. The whole process is sensitive to the environment, including many of the behavioral and lifestyle choices we make. We have learned so much about gene structure and function that it is now possible to change the genetic makeup of cells, organisms, and even entire families. It is important to recognize that for every question answered by discovery, new ones, often unanticipated, are uncovered. As much as we know, there is even more to learn.

The ability to manipulate genes in people, collect stem cells from human embryos, and clone human embryos has crowded us into a space where science is forcing us to define ethical and moral boundaries to guide this work. The question remains: how do we define universal ethical standards so we can decide whether or how to move forward?

Chapter Summary

8.1 Do Complex Human Characteristics Have a Genetic Basis?

Assess the legitimacy and risks of genetic determinism to society.

- Genetic determinism is the idea that genes determine, direct, or cause all traits.
- In the early 20th century, eugenics laws designed to improve the "human stock" forcibly sterilized people perceived as defective.
- In contrast to genetic determinism, we now know that traits are the result of the interaction of the environment and genetics.
- We are beginning to be able to identify maladaptive genes; gene therapy and cloning have the potential to cure many diseases but also hold the inherent risk of defining a new "normal."

8.2 What Is Gene Therapy?

Distinguish between somatic and germ-line gene therapies.

- Somatic gene therapy cures genetic disorders by inserting normally functioning genes into patients.
- Germ-line gene therapy alters the genes in germ cells (eggs or sperm) prior to fertilization.

8.3 What Are the Benefits and Risks of Genetically Altering Humans?

Contrast the benefits and risks of somatic and germ-line gene therapies.

- Somatic cell therapy has the potential to cure many genetic disorders by replacing specific cells in an individual, but those changes are unlikely to be passed on to offspring.
- Germ-line gene therapy can eliminate maladaptive genes from embryos so that all of the cells of the individuals have the trait.
- For both therapies, it can be difficult to predict the impact of new or changed genes on other genes.
- Some potential risks specific to germ-line gene therapy include the loss of potentially valuable variation through designer babies and the development of an enhanced genetic class.

8.4 How Can Stem Cells and Cloning Be Used to Alter People?

Compare the different methods and uses of stem cells and cloning.

- Stem cells offer the promise of curing many diseases because they can replace damaged, dysfunctional, or missing cells.

- Totipotent cells are taken from embryos up to the eight-cell stage and can develop into any cell type. Pluripotent cells are taken from the inner cell mass of a five- to seven-day-old embryo and can differentiate into many but not all cell types and consequently cannot grow into a baby.
- Human embryonic germ cells are cells taken from six- to nine-week-old embryos; germ cells form eggs or sperm.
- Cloning is the creation of a genetically identical copy of an organism. In therapeutic cloning, healthy embryos are created for medical treatment. Reproductive cloning creates embryos to form a baby.

8.5 What Are the Benefits and Risks of Stem Cell Research?

Use information on risks and benefits to evaluate the ethics of stem cell research.

- Human embryonic stem cells may be the key to curing many diseases.
- We currently use stem cells to learn about human development, and in the future we may have cells that model disease and enable us to test drugs, reducing the use of laboratory animals.
- Technical risks for stem cells include possible rejection by patients, cancer, and tumors.
- Egg donation to generate embryos requires hormone treatments that can be dangerous to the donor.
- Stem cells and cloning could be used for enhancement, rather than medical or therapeutic purposes, and may extend life beyond the normal life span, putting a possible burden on society.

8.6 What Other Challenges Result from Advances in Medical Technology?

Assess the ethical challenges we face as a result of advances in genetic technology.

- Access to genetic screening data could be used by governments, insurance companies, or corporations to discriminate against individuals.
- At the other extreme, individuals with resources to genetically enhance themselves and their children may create a genetically enhanced class.
- Our ability to identify genetic traits is more advanced than our ability to fix genetic defects.

Key Terms

adult stem cell 231
eugenics 219
genetic counseling 228
genetic determinism 220
genetic essentialism 220
genome 220
germ-line gene therapy 224

hemophilia 227
human embryonic germ cell 232
human embryonic stem cell 232
immune response 224
lymphocyte 224
pluripotent 232
reproductive cloning 235

somatic cell nuclear transfer (SCNT) 235
somatic gene therapy 223
stem cell 231
therapeutic cloning 235
totipotent 231

Review Questions

1. What is genetic essentialism?
   a. The notion that our genes determine, direct, or cause everything about us
   b. The notion that being human means having a human genome
   c. The notion that there are just a few essential things about our genes that make us human
   d. The notion that there are several things about our genes that make us human
   e. The notion that all living organisms are unique

2. What is genetic determinism?
   a. The idea that our genes determine, direct, or cause everything about us
   b. The idea that being human means having a human genome
   c. The idea that there are just a few essential things about our genes that make us human
   d. The idea that there are several things about our genes that make us human
   e. The idea that all living organisms are unique

3. Describe the process of somatic gene therapy.

4. Describe the process of germ-line gene therapy.

5. What are four ethical concerns associated with germ-line gene therapy?
   a. Safety, distinguishing between medical treatment and enhancement, privacy, and changing how we think about the sick and disabled
   b. Safety, accessibility, privacy, and changing how we think about the sick and disabled
   c. Privacy, accessibility, safety, and difficulty distinguishing between good and bad genes
   d. Privacy, safety, a misunderstanding about what genes can and can't do, and accessibility

   e. Safety, difficulty distinguishing between good and bad genes, distinguishing between medical treatment and enhancement, and changing how we think about the sick and disabled

6. How do the benefits and risks of somatic gene therapy and germ-line gene therapy compare to one another?

7. What are stem cells?

8. What is the difference between totipotent and pluripotent stem cells?
   a. Totipotent cells are capable of producing any of the cells in the body but none of the support structures needed for a baby's development; pluripotent cells are capable of producing any of the cells in a developing embryo.
   b. Totipotent cells will form the embryo, fetus, and eventual baby; pluripotent cells generally produce support structures such as the placenta.
   c. Totipotent cells are capable of producing any of the cells in a developing embryo; pluripotent cells are capable of producing any of the cells in the body but none of the support structures needed for a baby's development.
   d. Totipotent cells produce the support structures of the embryo such as the placenta; pluripotent cells form the embryo, fetus, and eventual baby.
   e. There is no real difference between totipotent and pluripotent stem cells; they are synonyms.

9. Describe the process of therapeutic cloning using somatic cell nuclear transfer (SCNT).

10. How do the objectives of therapeutic and reproductive cloning differ?

11. What are the potential benefits that may derive from (a) gene therapy, (b) the use of stem cells, and (c) cloning?

12. What are the potential risks that may derive from (a) gene therapy, (b) the use of stem cells, and (c) cloning?

13. What are some new challenges that result from our new medical technology?

The Thinking Citizen

1. Considering the question from a strictly scientific standpoint, could eugenics be used effectively to improve humans? Why or why not?

2. Explain, with examples, why it is more difficult to correct gain-of-function mutations than loss-of-function mutations in patients.

3. Is there a genetic basis for poverty? Why does it often seem to "run in families"?

4. How do you define "normal"? Explain your reasoning.

5. Should we try to eliminate all genetic defects? Why or why not? Is this possible to do? Why or why not?

The Informed Citizen

1. Should the reproductive cloning or germ-line gene therapy of humans be permitted? Why or why not?

2. In what ways are modern methods to manipulate genes in people different from eugenics? In what ways are they similar?

3. Is it acceptable to use gene therapy for enhancement? Would it be appropriate to insert genes so that a person could be taller, stronger, faster, or prettier? Where do we draw the line between treatment and enhancement? Is being short a medical condition? What about being bald?

4. What solutions can you suggest to solve the problem of providing equal access to medical treatments and technologies for all? Is it possible to do so?

5. What do you think are the potential consequences of a world in which there are genetic "haves" and "have-nots"?

6. What do you think are the potential positive and negative consequences of longer life spans?

7. Given that we have finite resources, should we spend our money on basic health care for all right now or on research to develop the cures of the future? Explain your reasoning.

8. Who should have access to your genetic medical record? Why?

9. What role, if any, should religion play in medical research decisions? How do we resolve differences in judgments about issues that affect all of us when these opinions are based in religion?

CHAPTER LEARNING OBJECTIVES

After reading this chapter, you should be able to answer the following questions:

9.1 How Does Your Body Reflect an Evolutionary History?
Provide three examples demonstrating the effect of evolutionary history on human structure or function.

9.2 What Convinced Darwin of the Fact of Evolution?
Detail the discoveries that led Darwin to change his mind about evolution and describe the steps of natural selection.

9.3 How Do Humans Adapt to Their Environment?
Use specific examples to describe how humans adapt to their environments.

9.4 How Does Natural Selection Produce Adaptations?
Use natural selection to describe the process whereby a population adapts to a specific environment and outline ways that adaptation might be limited.

9.5 What Are Random Events in Evolution?
Describe the potential role of random processes in evolution and differentiate natural selection from genetic drift.

9.6 What Is the Evidence for Speciation?
Summarize the evidence for speciation.

9.7 How Do New Species Arise?
Outline the stages by which new species arise from a single species.

9.8 Why Is It So Difficult for the Public to Accept Evolution?
Assess reasons why some public resistance to evolution still exists despite overwhelming scientific evidence.

Evolution helps us understand many of the most prevalent facts of nature: why organisms fit in so well with their environments, why there are so many species, and why some species are so similar while others are not.

Evolution

How Do Species Arise and Adapt?

Becoming an Informed Citizen …

Evolution is the central unifying theme of biology; it explains how all living organisms are related and how existing species adapt to their environments and new species arise. For the majority of scientists, evolution is an accepted fact of nature; yet, in the United States, controversy and confusion over the theory remain.

*n Chapter 1, we pointed out that **evolution**—the theory that explains how all living organisms are related and how existing species adapt to their environments and new species arise—is the central unifying theme of biology. All of the biological concepts we discuss in this book and more are shaped by evolution. Our discussion so far has focused on how these biological processes work. Evolution helps us understand why they work the way they do. Evolution also helps us understand many of the most prevalent facts of nature: why organisms fit in so well with their environments, why so many species exist, and why some species are so similar while others are not.*

In this chapter, we complete the story of what evolution is and how it works. We point out interesting and puzzling aspects of the natural world and show how they can be accounted for by evolution operating over long periods of time. Understanding evolution is vital to understanding nature, biology, and medicine. For the majority of scientists and many other people, the idea of evolution is no more controversial than any other fact of nature. Still, controversy (and confusion) over evolution is common in the United States, and we will examine why.

evolution :: theory that explains how all living organisms are related and how existing species adapt to their environments and new species arise

case study

Lactose Intolerance and the Geographic Variation of Human Traits

Do you like ice cream? How about a nice grilled cheese sandwich or yogurt for lunch, or perhaps some milk and cookies before bed? If you enjoy this sort of thing, it might surprise you to learn that for *most* adults, eating food like this has disastrous consequences: intestinal cramping, bloating, diarrhea, and embarrassing flatulence.

Dairy products contain a sugar called lactose. Humans—and all mammals—drink their mother's milk as infants, and most infants are able to break down lactose in breast milk. But the majority of humans lose this ability around two years of age, around the time they are weaned. Worldwide, about 75% of us can't digest dairy products, and only 25% can eat that ice cream sundae without suffering uncomfortable side effects.

Lactose is broken down by an enzyme called lactase (**FIGURE 9.1**). Almost all of us can make lactase, but its production is controlled by another gene. In humans, this gene shuts down lactase production in adults. But there is a recessive **mutation**, or change in DNA, that allows lactase production to continue on into adulthood. This mutation is not common worldwide, which is why most adults can't digest dairy products. Interestingly, this mutation is common in two groups of people: Northern Europeans and their descendants, and several ethnic groups from equatorial Africa (**FIGURE 9.2**). Using genetic techniques, scientists can estimate when this recessive mutation first appeared. Among Europeans, it became common about 10,000 years ago. How can we explain the geographic distribution and age of this mutation?

For some genetic traits that vary geographically, the consequences are more severe than indigestion. For example, one way to treat malaria is to give

mutation :: change in DNA

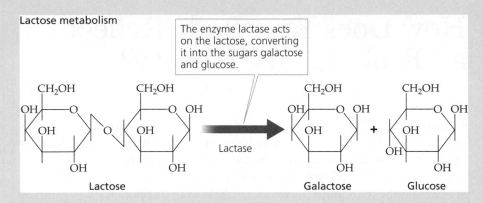

Lactose metabolism

The enzyme lactase acts on the lactose, converting it into the sugars galactose and glucose.

Lactose → Lactase → Galactose + Glucose

FIGURE 9.1 Lactose, the sugar found in dairy products, is broken down by the enzyme lactase.

the patient "oxidizing drugs." These drugs put metabolic stress on red blood cells—the cells that the malaria parasite infects. The stress is bad for the blood cells, but it's even worse for the parasite. The treatment is calibrated to kill a maximum number of parasites and a minimum of red blood cells. But when this treatment is given to people of African and Mediterranean descent, it kills about 10% of these individuals. People from other ancestries are not subject to the potential lethal effects of these drugs. Why are people from areas with a high rate of malaria so susceptible to this drug treatment, while people from other areas are not?

There are many traits that show geographic **variation**, which means the traits occur in people in some regions of the world but not in others: skin color, susceptibility to various types of cancer, cholesterol levels, and the ability to metabolize alcohol, to name just a few. Why do these differences occur, and why are they tied to geography? All of these traits are examples of relatively recent human evolution, as our ancestors adapted to their local environment in different parts of the world.

variation :: traits that differ, often by geographic location

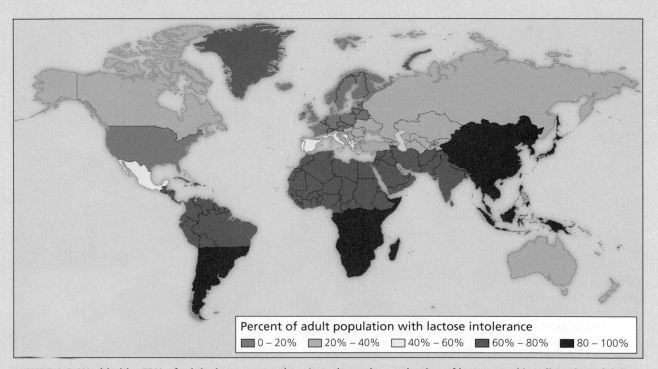

Percent of adult population with lactose intolerance
■ 0 – 20% ■ 20% – 40% □ 40% – 60% ■ 60% – 80% ■ 80 – 100%

FIGURE 9.2 Worldwide, 75% of adults have a gene that shuts down the production of lactase, making digestion of dairy products difficult. Even though this map depicts lactose intolerance as being low overall in the United States, there are large populations of people who are lactose intolerant.

How Does Your Body Reflect an Evolutionary History?

Suppose you wanted to learn about the borders of the nations of the Middle East (**FIGURE 9.3**). Some of these borders make perfect sense, as they follow natural features like mountains or perhaps rivers. But many other borders appear arbitrary and don't seem to have an obvious basis, geographical or otherwise. How can you explain why the borders are where they are?

To fully understand, you would need to study history. You would have to learn about the founding of Israel after World War II, colonial rule in Iraq by the British following World War I, and a series of other events reaching far back into history. If you want to understand why things are the way they are, including political borders, you need to consider not just the current situation but also the historical events that caused it.

Life is like this, too. Significant events in the history of life help to explain current life. Let's look at some examples from a familiar source: your own body.

Human (and Mammalian) Testes Hang Loose

Given male testes' importance in reproduction and sensitive nature, it is no surprise that most male vertebrates keep their testes protected inside their bodies. Mammals, and therefore humans, are exceptions to this rule. As you know, our testes are contained within an external scrotum adjacent to the penis (Chapter 6). This seems to be a convenient arrangement: the testes produce sperm, and the penis delivers them, all in one nicely localized package.

Things are not exactly as they seem, however. The vas deferens, or sperm duct, does not go directly from the testes to the penis. Instead, it leaves the scrotum, travels anteriorly, curves around the pelvis, loops around the ureter (which transports urine from the bladder to the penis), and then heads back down to join with the penis (**FIGURE 9.4**).

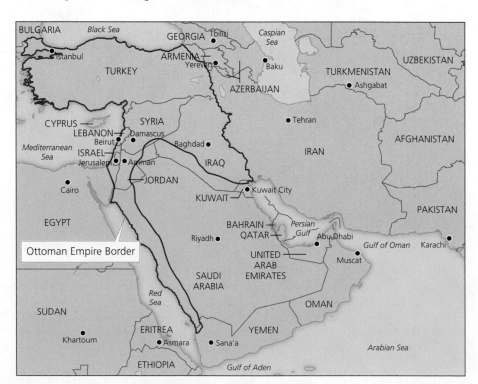

FIGURE 9.3 Significant events in the history of life help to explain current life, just as significant past world events help to explain current world affairs. After World War I, the victorious countries dissolved the Ottoman Empire and redrew the boundaries of countries in the Middle East, setting the stage for many of the current hostilities and conflicts.

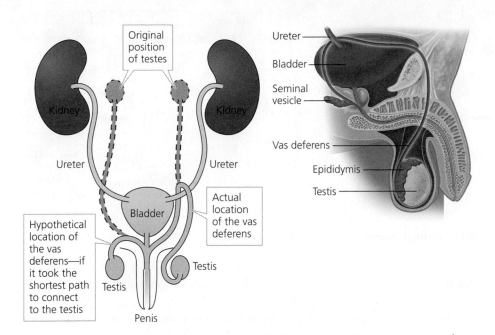

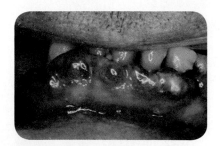

FIGURE 9.4 The vas deferens, or sperm duct, takes a circuitous route from the testes to the penis rather than a more direct one. There is no particular functional reason for where the vas deferens is; rather, it is a legacy from earlier ancestors.

Such a circuitous route! The vas deferens may be over 10 inches long, even though only a few inches would be needed. Our bodies do make use of this detour, however. Sperm cells mature during their passage through the vas deferens, and this route allows more room for the seminal vesicle and prostate gland that add seminal fluid to the sperm. Even so, it is hard to imagine that a more efficient route for sperm could not be found. And there is still the question of why the testes in mammals are external in the first place.

If we recognize that life has a history, and that all organisms are related because of this history, some pieces of the puzzle fall into place. In vertebrates (which include humans and all other mammals), the testes form inside the abdominal cavity near the liver during embryological development. In most vertebrates, that is where they stay, protected from harm.

However, mammals maintain a constant high body temperature, and sperm cells cannot develop properly at these temperatures. Male mammals have developed an external scrotum to house the testes and keep them cooler than the rest of the body. But in mammals, the testes form initially far from the scrotum, just as they did in our nonmammalian ancestors. This problem is solved by having the testes "descend" from the abdomen into the scrotum, and as they travel they drag the vas deferens along with them. In the process, the vas deferens loops over the ureter, the outlet tube from your bladder, and settles into the circuitous route just described. There is no particular functional reason for where the vas deferens is. It is a legacy from earlier ancestors, a product of life's historical events.

You Can Get Scurvy, but Your Pet Can't (Unless You Have Guinea Pigs)

Scurvy is a condition caused by a lack of vitamin C in your diet. It is not a pretty sight. Scurvy results in spongy, bleeding gums and eventual tooth loss, and in advanced cases patients develop open, pus-filled wounds (**FIGURE 9.5**). Scurvy has been recognized since ancient times, and many herbal cures are known, all of which are rich in vitamin C. British royal surgeon James Lind is credited with discovering in the 1750s that scurvy could be prevented by

FIGURE 9.5 Scurvy is a condition caused by lack of vitamin C that results in spongy, bleeding gums and eventual tooth loss. The vast majority of mammals do not need vitamin C in their diets because their cells can make their own vitamin C, but humans inherited a pseudogene that destroyed this function.

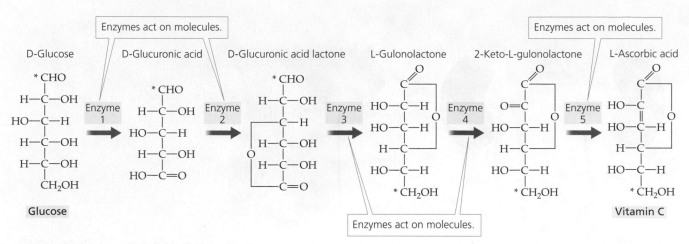

FIGURE 9.6 Vitamin C is made from a modified form of glucose. A series of enzymes alter this molecule in an assembly-line process, which results ultimately in the production of vitamin C.

eating citrus fruit, which resulted in the British Navy's policy of stocking its ships with limes and other fruits.

You know you need to add vitamin C to your diet. But you may not know that this is unnecessary for your cat, dog, or horse. In fact, the vast majority of mammals do not need vitamin C in their diets because their cells can make their own vitamin C. Vitamin C is made from a modified form of glucose, a simple sugar. A series of enzymes alter this molecule in an assembly-line process, which results ultimately in the production of vitamin C (**FIGURE 9.6**).

Most mammals have working copies of all enzymes, but humans (and all other primates) are missing the enzyme that performs the last step. However, humans are not missing the gene that makes this enzyme. The gene is in our chromosomes, but it carries a mutation that renders the enzyme it makes nonfunctional. All other primates carry exactly the same mutation in this gene and therefore lack the ability to make vitamin C.

pseudogene :: a gene carrying a mutation that makes its products nonfunctional

Genes carrying mutations that make their products nonfunctional are known as **pseudogenes**. Modern molecular techniques allow us to examine the DNA sequence for many or all of an organism's genes. It turns out pseudogenes are surprisingly common. Some estimates for mammals run as high as 20,000 pseudogenes in each species—almost as many pseudogenes as regular, functional genes. What does the presence of a specific pseudogene shared only by primates, including humans, say about the genetic relatedness among the members of this group?

A mutation that prevented a species from making its own vitamin C would be lethal in most species, unless that species happened to have a diet already rich in vitamin C. Primates eat a lot of fruit and therefore get plenty of vitamin C. The fact that all primates have the same pseudogene indicates that the mutation that destroyed its function occurred in the original ancestor species of all the primates. We have all inherited this pseudogene from this ancestor.

Your Eye Is Organized Backward

The human eye is essentially identical in structure to the eyes of all vertebrates. It is a remarkable organ, one that works exceedingly well and plays an important role in the life of all vertebrates that are able to see. But the vertebrate eye does have one peculiar feature: it apparently is organized backward.

The light you see enters your eye and hits the photoreceptors—rods and cones—in the retina. These receptors send impulses to nerves that carry the

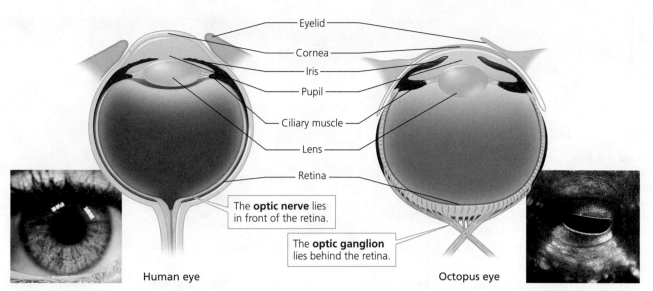

FIGURE 9.7 **The human eye is nearly identical in structure to the eye of certain vertebrates, except that it is organized backward.** In humans, the optic nerve lies in front of the retina, resulting in a blind spot, whereas in the eyes of a few types of invertebrates, such as octopi, the optic nerves are behind the retina.

signal to the brain for interpretation. However, these transmission nerves lie *in front of* the retina itself. Light must filter through these nerves in order to hit the retina. Furthermore, these nerves must at some point exit the eye to reach the brain, resulting in a hole, or "blind spot" in the retina.

We cannot fault the eye for its functionality, but the design seems odd. Squid, octopi, and cuttlefish have eyes that work as well as ours, but in these species the transmitting nerves lie behind the retina (**FIGURE 9.7**). No light filtering occurs, and there are no shadows or need for a blind spot in the retina, which means that in some ways they actually see better than we do.

The human eye is an example of excellent functionality supported by a less-than-excellent design. It is likely that the eye developed over a long period of time through a number of small steps. Each step would improve the function of the eye a bit. Each new step would have to be based on earlier ones—which means that each step *constrains* what the next step can be. Over time, the eye could become highly functional, but its structure would reflect the historical constraints that led to it. It wouldn't be a "perfect" organ. The end result of such a process is exactly what we see in the human eye: excellent functionality stemming from a historically constrained structure.

The point of this section is that life does indeed have a history. If we recognize this fact, we can understand many features of life (for example, the structures of our own bodies) that otherwise make little sense. Recognizing that life has a history, and that historical changes and relationships account for the features of life we observe today, is the same thing as recognizing that life has evolved. This is one line of evidence (and there are many others) that has convinced biologists that evolution is simply a fact of nature.

9.2 What Convinced Darwin of the Fact of Evolution?

In the early 1800s, the prevailing scientific view of nature was that God created all forms of life in much the same form as they appear today. This view

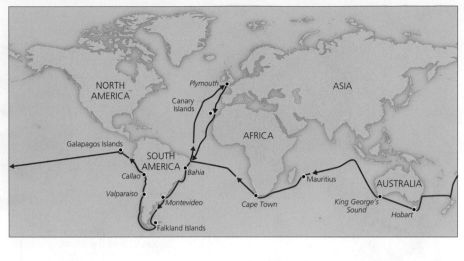

FIGURE 9.8 During his five-year voyage aboard the H.M.S. *Beagle* (1831–1836), Darwin made extensive studies of the plants and animals living in South America. He began to see evidence that eventually would lead him to the idea of evolution.

was not universally held, however. Ideas about evolution had been circulating for quite some time, and advances in geology pointed to a much more dynamic, and older, Earth. But as a young man, Charles Darwin (1809–1882) held to the prevailing view.

The Voyage of the *Beagle*

Darwin's perspective began to change during his five-year voyage aboard the H.M.S. *Beagle* (1831–1836) (**FIGURE 9.8**). In his role as ship's naturalist, Darwin made extensive studies of the plants and animals living in South America. As Darwin learned more about this new part of the world, he began to see evidence that eventually would lead him to the idea of evolution: that species could change and new species could arise from existing ones.

One of these lines of evidence involved **biogeography**, the study of where different species live throughout the world. South America has rain forests, as do Africa and Southeast Asia. South America also has extensive prairies, foothills, and mountain habitats like those found on other continents. If species were created specifically to live in a particular habitat like a rain forest, you might expect that species in different rain forests would be related, or perhaps that the same species would be found in all of them.

Instead, what Darwin found was that the rain forest species in South America were more similar to the non–rain forest species of South America than they were to species from other rain forests (**FIGURE 9.9**). The same held true for other habitat types. Apparently, living on the same continent was a better predictor of relatedness than the habitat in which a species lived. Darwin also discovered that the fossil species found in South America bore a close relationship to the current living species, though some modifications had occurred (**FIGURE 9.10**).

Darwin reasoned that one way to explain these facts is to infer that species have a history. They have changed over time, which accounts for the fossil evidence of both relatedness and differences of past and current species. As successful species spread out across a continent, they change in ways that help them to survive in the new habitats they encounter. Species on the same continent may, therefore, be related even though they live in different habitats. Darwin's key inferences in both cases were that species change over time, and that these changes take a long time to occur.

Another line of evidence to support Darwin's growing consideration of evolution came from his celebrated visit to the Galapagos Islands, about

biogeography :: the study of where different species live throughout the world

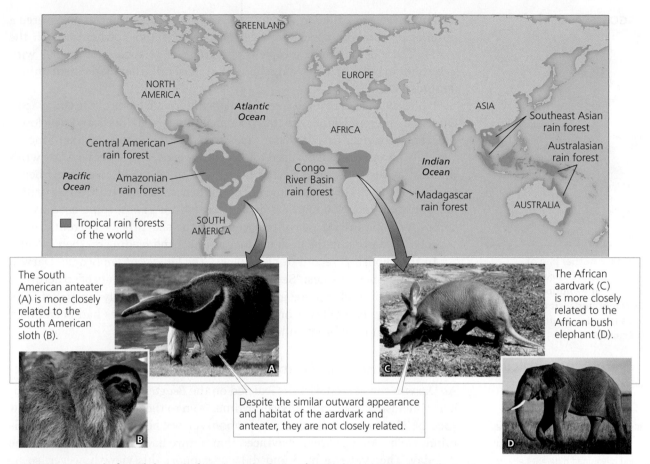

The South American anteater (A) is more closely related to the South American sloth (B).

The African aardvark (C) is more closely related to the African bush elephant (D).

Despite the similar outward appearance and habitat of the aardvark and anteater, they are not closely related.

FIGURE 9.9 **Darwin found that rain forest species in South America were more similar to the non–rain forest species of South America than they were to species from other rain forests.** Even though they might look similar, the aardvark found in an African rain forest and the anteater found in a South American rain forest are not closely related.

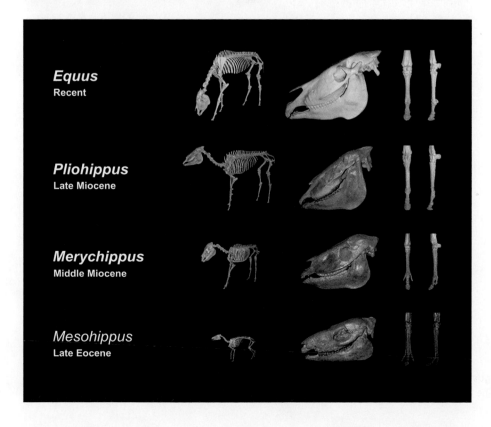

Equus
Recent

Pliohippus
Late Miocene

Merychippus
Middle Miocene

Mesohippus
Late Eocene

FIGURE 9.10 **Darwin discovered that the fossil species found in South America bore a close relationship to the current living species, though some modifications had occurred.** The fossil record of horse evolution (although not studied by Darwin) also shows modification of ancestral species.

GALAPAGOS ISLANDS

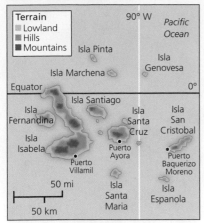

FIGURE 9.11 Another line of evidence to support Darwin's consideration of evolution came from his celebrated visit to the Galapagos Islands, about 600 miles west of Ecuador.

endemic :: a species that is specific to a location and found nowhere else in the world

Darwin's finches :: the finches on the Galapagos Islands that occupied different habitats and developed different behaviors

600 miles west of Ecuador (**FIGURE 9.11**). On the islands, Darwin discovered a number of **endemic** species, species that were found nowhere else in the world. The most famous example is the group of 14 species of finches with variations in bill structure, body shape, and short tails that made them unique (**FIGURE 9.12**).

Different species of finch were found on different islands of the Galapagos. On these different islands, finches occupied different habitats and developed different behaviors that in some cases were decidedly unfinchlike. Finches typically have strong beaks used for crushing seeds, and several of **Darwin's finches** (as the Galapagos finches have come to be called) used their beaks for this purpose. However, some of these finches also ate fruit, some drilled holes in trees like woodpeckers in their search for grubs, some used cactus spines to probe insects out of crevices, and the sharp-beaked ground finch was observed pecking the flesh of other animals and feeding on their blood.

In the published record of his journey, *The Voyage of the* Beagle, Darwin writes about these finches: "Seeing this gradation and diversity of structure in one small, intimately related group of birds, one might really fancy that from an original paucity of birds on this archipelago, one species had been taken and modified for different ends."

Darwin's Theory of Natural Selection

As Darwin considered what he learned on the *Beagle*, he came to an inescapable conclusion: species change over time, even to the point of becoming new species. There is an important lesson about science here: Darwin changed his mind. He began his career convinced that nature had been created as we see it today. The evidence he found did not support this view, however, so he

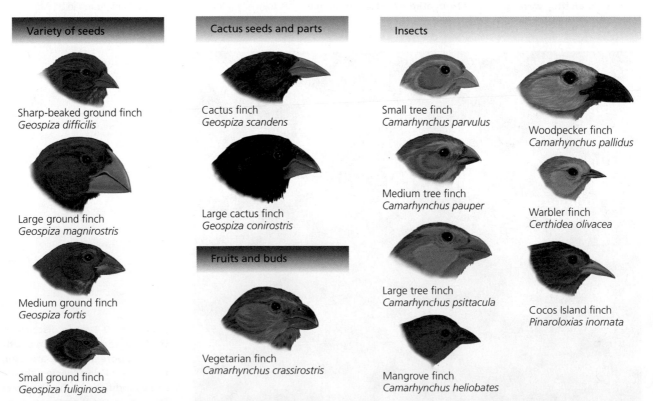

FIGURE 9.12 The 14 species of Galapagos finches (or Darwin's finches, as they are called) are endemic species, which means they are found nowhere else in the world. The finches evolved in terms of body shape, tail size, and bill structure (shown categorized by diet) to adapt to their environment.

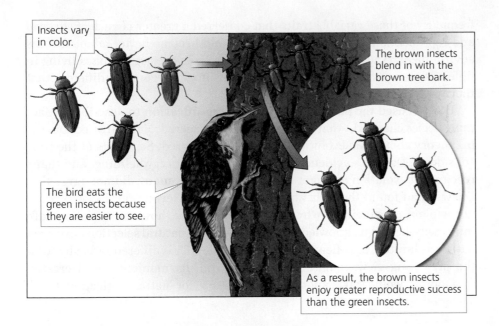

Insects vary in color.

The brown insects blend in with the brown tree bark.

The bird eats the green insects because they are easier to see.

As a result, the brown insects enjoy greater reproductive success than the green insects.

FIGURE 9.13 Natural selection is based on three observable facts of nature: individuals in a population vary, at least some of these variations are inherited, and many more offspring are born than survive to adulthood.

changed his views to accommodate the evidence. This is the way science works: evidence trumps belief and prior views.

Darwin realized his new ideas were radical. He understood the need to compile evidence that species change and to discover the mechanism by which the change occurred. He spent the next 20 years on this task, and in 1859 he published his most famous and influential book: *The Origin of Species by Means of Natural Selection or the Preservation of Favoured Races in the Struggle for Life*.

Natural selection, the mechanism whereby individuals with certain heritable traits have an increased chance of surviving and producing offspring, is one of those rare ideas that is both simple and earth-shattering. The pieces of this idea were available for all to see. It is a tribute to Darwin's intellect that he was able to put all the pieces together and understand their significance.

Natural selection is based on three observable facts of nature (**FIGURE 9.13**):

1. Individuals in a population vary; they differ from one another (variation).

2. Variations are inherited; parents pass some of their variation on to their offspring (inheritance).

3. In any population, many more offspring are born than survive to adulthood (differential survival and reproduction).

Point 3 raises an important question: if many (in some species, most) of the offspring die, which ones survive? Perhaps mortality is simply random, and every individual has the same chances. But if individuals vary (point 1), maybe some are better at survival than others, at least in their particular environment. Certain individuals with particular variations would have a greater chance of survival and therefore a greater chance of reproducing. And when they reproduced, these individuals would pass on to their offspring the variations that helped them survive (point 2).

The logic was simple, but the conclusion was startling: species would change over time. Each generation would differ from the one before it, and it would differ in a specific way. The next generation would have a higher

natural selection :: the mechanism whereby individuals with certain heritable traits have an increased chance of surviving and producing offspring

frequency of those variable traits that conferred a greater chance of survival and reproduction. Over time, a species would become better adapted to its environment. And if two populations of the same species, perhaps living in different areas, adapted independently, they might become so different that they would split into two different species.

descent with modification :: a phrase used by Darwin to describe the fact that species changed and that evolution occurred

Darwin used the phrase **descent with modification** to describe the fact that species changed—that is, that evolution occurred. Natural selection was his theory about *how* evolutionary changes occurred. Scientists at the time considered Darwin's evidence for evolution to be overwhelming, and many recognized evolution as a fact of nature. However, many did not agree that Darwin had found the right theory to account for evolution. Darwin could not explain convincingly why organisms vary and how the variations are inherited—he had no knowledge of genetics. Also, natural selection portrayed nature as limited, cold-hearted, and dog eat dog. This differed radically from the prevailing view of nature as a bountiful, harmonious world created by God. It would take another half century for natural selection to be accepted.

The Modern Synthesis

The early part of the 20th century witnessed many advances in biology. Mendel's work with pea plants (Chapter 4) was rediscovered, and a science of genetics emerged. It soon became clear that genetics solved several of the problems that plagued Darwin. Traits were controlled in part by genes. Genes existed in several different forms, or alleles, which helped to explain why organisms vary. And genes passed from parent to offspring according to genetic rules of inheritance. Natural selection explained how **allele frequency**, the number of alleles present in a population, changes over time. The integration of genetics and evolutionary biology was known as the **modern synthesis**, and one of its main architects was Sir Ronald Aylmer Fisher (see *Scientist Spotlight: Sir Ronald Aylmer Fisher*).

allele frequency :: the number of alleles present in a population

modern synthesis :: the integration of genetics and evolutionary biology; one of its main architects was Sir Ronald Aylmer Fisher

One of the discoveries from the modern synthesis was that mathematical equations could be developed that predicted how a population would evolve under a particular set of circumstances. These equations could predict, for example, how rapidly a beneficial allele would increase in frequency throughout a population, or how long it would take for a disadvantageous allele to be eliminated (**FIGURE 9.14**). Scientists tested these mathematical predictions experimentally in the lab and in the field. They found that natural selection

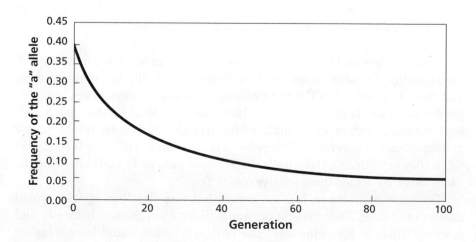

FIGURE 9.14 One of the discoveries from the modern synthesis was that mathematical equations could be used to predict how a population would evolve under a particular set of circumstances. For instance, equations could be used to predict how long it would take for a disadvantageous allele to be eliminated from a population.

Sir Ronald Aylmer Fisher (1890–1962)

R. A. Fisher was one of the main architects of the modern synthesis. Fisher was born February 17, 1890, in London. He showed talent in mathematics as a boy, and in 1909 he entered Cambridge University to study mathematics and astronomy. He graduated with distinction in 1912. He tried to enlist in the army during World War I but was rejected due to poor eyesight. Instead, he taught mathematics and physics at local schools. Fisher married Ruth Eileen in April 1917. Eileen was barely 17 at the time, and because her mother objected to her early marriage, they married in secret. Fisher and Eileen had two sons and seven daughters, though one died in infancy.

In 1919, Fisher accepted a job as statistician at Rothamsted Agricultural Experiment Station, where he developed statistical techniques to help agriculturists design and interpret experiments involving crop breeding. In 1930, Fisher was appointed Galton Professor of Eugenics at University College. He published several influential books, including *The Genetical Theory of Natural Selection* in 1930, and has been described as "Darwin's greatest twentieth-century successor." He also developed a number of important statistical tests that scientists use to analyze data.

It is a stunning achievement for any one person to reach the pinnacle of two different fields, but that is exactly what Fisher did in biology and mathematics. He was elected to the Royal Society in 1929, and he received the society's Royal Medal in 1938 and its Darwin Medal in 1948. Queen Elizabeth dubbed him a Knight Bachelor in 1952. Fisher retired in 1957 and moved to Adelaide, Australia, where he died July 29, 1962.

Fisher was an intense and brilliant man whose temper frequently got the better of him. He was intolerant of

R.A. Fisher was one of the main architects of the modern synthesis—the integration of genetics and evolutionary biology.

mistakes in other people, and his teaching and writing were difficult for most people to understand. His contributions to statistics and biology were significant, far reaching, and long lasting.

Fisher's interest in eugenics (see Chapter 8) began at an early age. At Cambridge, he formed the Cambridge Eugenics Society and became its first president. His wife shared his views on eugenics, and because both felt they had desirable traits worth promoting, they decided to have a large family. Fisher was a fierce opponent of contraception, especially among the educated and wealthy. He was concerned that members of the educated and upper classes weren't having enough children and would soon be overrun by more rapidly breeding members of lower classes. Fisher is remembered as a brilliant scientist, but on the personal level most people today would regard his views as extreme. ::

worked just as Darwin thought it did, and just like the equations said it should. It took almost 70 years, but by the 1940s scientists accepted natural selection as an important and powerful scientific theory, supported by mountains of theoretical and empirical evidence.

9.3 How Do Humans Adapt to Their Environment?

Let's return to the examples in our case study. How can we explain the geographic distribution and the age of the recessive mutation that allows lactose production to persist into adulthood? And how can we explain why some people of African and Mediterranean descent are killed by certain treatments for malaria when they come from regions where malaria is prevalent?

Lactase Persistence

We know that mutations occur randomly. The particular mutation that enables lactase production to continue into adulthood has occurred many times throughout the 200,000-year history of our species, and it has occurred across the globe, everywhere humans live. But the mutation didn't "take," that is, it didn't increase in frequency, until it happened to occur in Northern Europe about 10,000 years ago. Why?

Consider what might happen to the mutation if it occurred in a typical settlement of our earliest ancestors. For an infant feeding on breast milk, lactase production is essential. But after the child is weaned, there is no more milk to drink. There are no domesticated dairy cows, no cheese, no butter, and no yogurt—nothing but game, nuts, maybe berries, and other available plant material. Under these conditions, a mutation that allowed an adult to produce lactase would not provide any advantage. It could even be the case that lactase interfered with the digestion of nondairy food. Regardless, there would be no particular reason for the mutation to increase its frequency in the population.

The picture changes dramatically when domesticated cows and goats are around. Their milk provides an untapped reservoir of food for humans who can digest it. In times of famine (and there were plenty of those), an adult who could produce lactase might survive and pass the mutation on to his or her offspring. An adult lacking the mutation might starve. Under these conditions, the mutation for lactase persistence would confer an advantage and likely would increase in frequency throughout a population that kept dairy animals.

When and where did the domestication of dairy animals first occur? It's no surprise that one center of dairy farming originated in Northern Europe about 10,000 years ago (**FIGURE 9.15**). Eventually the mutation for lactase persistence occurred in this population and increased in frequency. Among Northern Europeans and their descendants in North America and Australia,

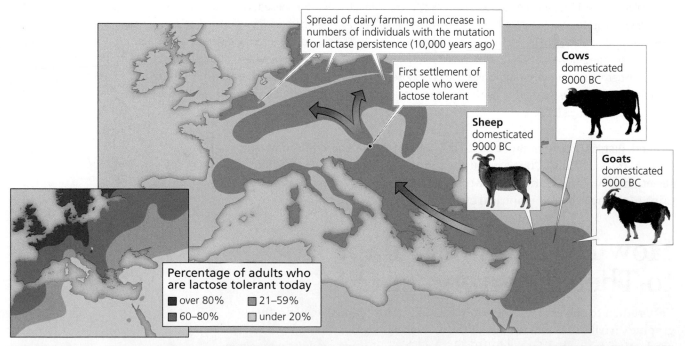

FIGURE 9.15 The mutation for lactase persistence originated in many places in the world but became established and increased its frequency in places where dairy farming thrived. One center of dairy farming originated in Northern Europe about 10,000 years ago, and among Northern Europeans and their descendants in North America and Australia, almost all individuals can digest lactose as adults.

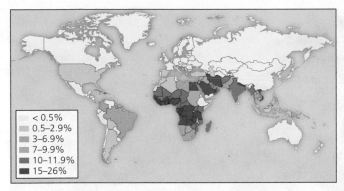

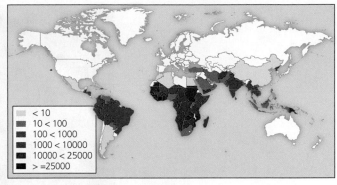

Prevalence of G6PD deficiency   Prevalence of malaria infection

FIGURE 9.16 In areas where malaria is prevalent, an adaptation—a deficiency in the enzyme glucose-6 phosphate dehydrogenase (G6PD)—reduces a red blood cell's ability to use sugar as an energy source and hinders reproduction of any infecting malarial parasites. This is the same effect and result achieved using the oxidizing drugs we have developed to treat malaria.

almost all individuals can digest lactose as adults. The Tutsis and Fulanis of equatorial Africa have a high frequency of persistent lactase, and they are also among the few African people who have a long history of herding dairy cattle.

Adult-persistent lactase is a human **adaptation**—a change by a population to better align with specific environmental conditions. It is an example of how natural selection acts on human populations and causes evolutionary change.

adaptation :: a change by a population to better align with specific environmental conditions

Malaria and Oxidizing Drugs

In the opening case study, we also described a situation in which some people of African and Mediterranean descent are killed by certain treatments for malaria. Given that these people come from regions where malaria is a problem, why would they respond this way to treatment?

The answer is that malaria, like all major diseases, sets up conditions for natural selection in our species. In Chapter 4, we discussed one human adaptation to help fight malaria: the sickle cell trait. But in high-malaria areas, a whole host of adaptations to fight malaria have evolved.

One of these adaptations is a mutation—a deficiency in the enzyme glucose-6-phosphate dehydrogenase (G6PD)—that reduces a red blood cell's ability to use sugar as a metabolic energy source (**FIGURE 9.16**). This mutation puts the red blood cell under metabolic stress and hinders reproduction of any infecting malarial parasites—the same effect and result as the oxidizing drugs we have developed to treat malaria.

But when a person with this genetic adaptation takes an oxidizing drug to treat malaria, the red blood cells are put under a double dose of metabolic stress. Many red blood cells may burst, and the patient may die. It's interesting to see that humans have adapted just like other animals. But it's even more important to understand that these adaptations, and our evolutionary history in general, have medical consequences that we need to consider to provide good care.

9.4 How Does Natural Selection Produce Adaptations?

We have presented several examples of human adaptations. Lactase persistence is advantageous in populations where dairy animals are kept, and the

FIGURE 9.17 Peter and Rosemary Grant have been studying the Galapagos finches continuously since 1973.

metabolic stress adaptation is advantageous in regions with malaria, although it can be lethal with the wrong medication. Examples like these provide evidence that humans evolve and adapt just like other organisms. However, they do not really show the details of how natural selection acts to produce adaptation. Detailed studies of natural selection take a great deal of time and effort—over 40 years, in the case of Peter and Rosemary Grant's work.

The Grants' 40-Year Study of Natural Selection

Peter and Rosemary Grant have been studying the Galapagos finches continuously since 1973 (**FIGURE 9.17**). The work we describe here took place between 1975 and 1978 on Daphne Major, which, despite its name, is one of the smallest islands in the archipelago. The medium ground finch (**FIGURE 9.18**) lives on Daphne Major, with an average population size of 1200 birds. With great patience, the Grants captured every finch and gave them each distinctive leg bands. Scientists on the island know each bird individually. They can tell which birds mate and with whom. They know who the parents are and who the young are, so they are able to see how genetic traits are passed on. They can tell when each bird is born and when it dies. The Grants' 40-year effort to study the medium ground finch and other inhabitants of the Galapagos is a remarkable achievement, and one that has provided numerous tests of natural selection.

Medium ground finches are seed eaters, and they use their beaks to pick up and crush the seeds they eat. Beak size is variable in this finch. Birds with larger beaks are able to crush and eat larger, tougher seeds. Birds with smaller beaks are better able to handle smaller seeds. Daphne Major is a dry volcanic island with a typical rainfall of only 5 inches per year. The vegetation is mainly dry forest, scrub, and cactus. Still, Daphne Major supports a diversity of plants, and a variety of seeds are typically available for the finches to eat.

Between 1976 and 1977 a drought hit Daphne Major, and less than an inch of rain fell during the "rainy" season. This drought severely reduced the number of seeds available, and over the course of the drought more than 80% of the medium ground finches died (**FIGURE 9.19**). The few soft seeds produced were quickly eaten, and for much of the drought, mostly large, hard seeds were available (**FIGURE 9.20**).

FIGURE 9.18 The medium ground finch lives on Daphne Major. Medium ground finches are seed eaters, and they use their beaks (which are variable in size) to pick up and crush the seeds they eat.

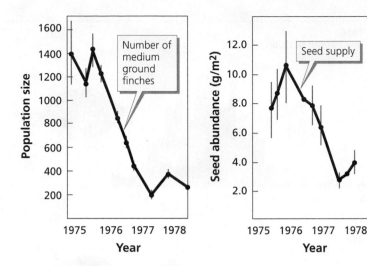

FIGURE 9.19 **Between 1976 and 1977, a drought hit Daphne Major.** This drought severely reduced the number of seeds available, and over the course of the drought more than 80% of the medium ground finches died.

The drought on Daphne Major set the stage for a decisive test of Darwin's theory of natural selection. We can see that there is variation in beak depth in the finch population. The Grants analyzed data (not reported here) to verify that the variation is inherited: parents with larger beaks tend to have offspring with larger beaks. The drought reduced the food supply and set up a struggle for survival, and many finches died.

The Grants routinely capture their finches and take measurements of their weight, size, and beak depth. **FIGURE 9.21** shows the distribution of beak depths before and after the drought. As you can see, the average beak depth grew from 9.4 mm to 10.1 mm, an increase of 7% in just a few seasons. In an environment with mostly large, hard seeds, birds with bigger beaks have the advantage. This change in average beak depth was due to a higher rate of mortality among smaller-beaked birds.

To demonstrate evolutionary change and adaptation, however, we need to look at changes that occur in the next generation. **FIGURE 9.22** shows the distribution of beak depth for birds that hatched in 1976, before the drought, and for birds that hatched after the drought in 1978. Again, beak depth grew, from about 8.9 mm to 9.7 mm, a 9% increase. This demonstrates that not only did larger-beaked birds survive the drought better, but they also passed on their alleles for larger beaks to their offspring. Evolution occurred in the population of medium ground finches. Over several seasons, the average beak depth increased as the population adapted to drought conditions.

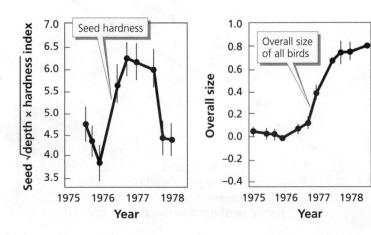

FIGURE 9.20 **For much of the drought, mostly large, hard seeds were available, and the average size of the birds increased during that time as well.**

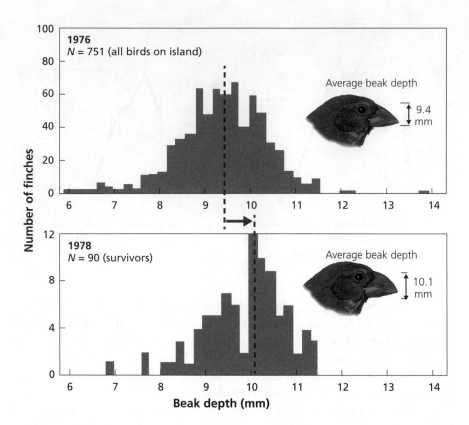

FIGURE 9.21 Between 1976 and 1978, the average beak depth of the finches grew from 9.4 mm to 10.1 mm, an increase of 7%. This change in average beak depth was due to a higher rate of mortality among smaller-beaked birds.

It is important to note that no individual finch changed during the drought: once a finch reaches maturity, its beak depth stays the same. Adaptation is a property of populations, not individuals. It is also important to note that adaptation is not necessarily a unidirectional process. Droughts are common on Daphne Major, and they are interspersed with wetter seasons. During drought, food is scarce, larger seeds predominate, and beak depth increases. During wetter years, the reverse is true: small seeds predominate and beak depth decreases. Over time, beak depth fluctuates both up and

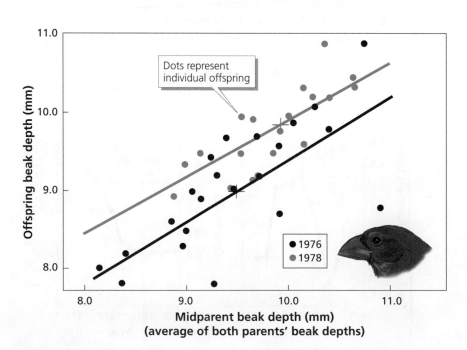

FIGURE 9.22 Larger-beaked ground finches survived the drought better, and they also passed on their alleles for larger beaks to their offspring. The beak depth for ground finches that hatched in 1976 before the drought was about 8.9 mm, and the beak depth for ground finches that hatched after the drought in 1978 was about 9.7 mm, a 9% increase.

down. As is true of evolution in general, there is no overall direction, simply a response to local conditions.

Fitness and Natural Selection

Fitness is a simple-sounding term that can have several different meanings. We say that an organism "fits" its environment well if it has adaptations that promote its survival and reproduction. Evolutionary biologists often use the term to refer to an organism's reproductive potential or chances of survival. It is common to say that natural selection selects the fittest individuals, although this is sloppy logic because it doesn't really say what fitness is—other than that it is what natural selection selects.

The modern synthesis brought together evolutionary biology and genetics, and this suggested a specific definition of fitness. Fitness is the number of alleles, or forms of a gene, an organism passes on to the next generation. This is a clear definition, and it has a nice scientific feature: under many conditions, fitness can be measured in the laboratory or in the field. The Grants, for example, were able to measure finch fitness precisely.

Natural selection acts to increase fitness in a population. Adaptations increase survivorship and/or reproduction, which in turn increases the number of genes an organism leaves in the next generation. This helps to explain the tendency for parents to care for their offspring. By so doing, they are helping to pass on their genes.

According to the rules of sexual reproduction, your children share half their genes with you. But your children are not the only individuals with whom you share genes. You share on average half your genes with your siblings, about a quarter with your grandchildren, and an eighth with your cousins (**FIGURE 9.23**). Our definition of fitness rewards passing on genes. But it doesn't matter who happens to have those genes.

fitness :: the number of alleles, or forms of a gene, an organism passes on to the next generation

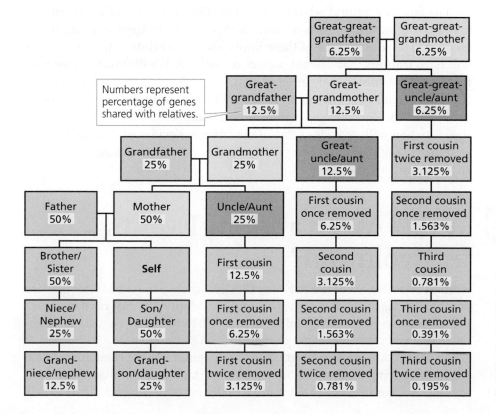

Numbers represent percentage of genes shared with relatives.

FIGURE 9.23 An organism maximizes its fitness by passing genes on to the next generation, but it does not matter who happens to have those genes. You share on average half your genes with your children and your siblings, about a quarter with your grandchildren, and about an eighth with your cousins.

kin selection :: selection for behaviors that help your relatives survive and reproduce

Although we pass on our genes directly through our children, we also share genes with all of our relatives. Therefore, there is an evolutionary advantage to helping your siblings, grandchildren, and cousins pass on their genes as well. Such activity is common in humans, and also in many other animal species. If fitness were limited to genes passed on directly through your own descendants, it would be hard to explain why you would help anyone else. But if fitness is more inclusive, and can refer to genes passed on by your kin, this type of social behavior becomes easier to explain. Selection for behaviors that help your relatives survive and reproduce is known as **kin selection**.

Limits of Natural Selection

The more you learn about adaptation, the stronger the temptation becomes to infer that given enough time, natural selection can do almost anything. To see why this is not the case, it is important to understand the factors that limit and constrain what natural selection can do.

Natural selection can't anticipate or plan for future events. It therefore can't have goals or make progress toward goals. Natural selection responds only to the current situation, and whenever that changes, the traits that are selected change as well.

This is a photo-edited image; pigs do not have wings!

FIGURE 9.24 Natural selection can only act on the variation present in a population. While it might be useful for pigs to have wings, if the available range of genetic variation doesn't give natural selection something to work with, they never will.

Natural selection can only select from the variation that is present in a population. This variation is produced by random processes like mutation; natural selection can't produce specific variation, even if it would be very useful. It might be useful for pigs to have wings, but if the available range of genetic variation doesn't give natural selection something along these lines to work with, they never will (**FIGURE 9.24**). Natural selection works under historical constraints as well. We inherit a lot from our ancestors and from our ancestral species. This inheritance constrains the types of organisms we can be.

The fact that natural selection has limitations and constraints is not surprising. Any natural process must work this way, since there is no agency to guide it. The consequence of these limits and constraints is that organisms will have features that are not perfect or optimal. We discussed three such features of the human body in Section 9.1: the male testes that exist outside of the body, the missing enzyme to produce vitamin C, and the backward organization of the eye. Design imperfection in features like these, even though they function well, provides some of the strongest evidence that the features were shaped by a natural process like selection rather than being created by a supernatural designer.

9.5 What Are Random Events in Evolution?

Natural selection is not the only way that evolution occurs. Natural selection accounts for one particular type of evolution: adaptation. But random events can also cause evolution. Random events don't change a population in any particular way; they just cause change.

Mutations are random events. Recall that mutations are random mistakes made in copying DNA when a cell divides. Although mutations are rare, organisms have so much DNA that virtually every newborn individual will have a couple of mutations. Random mutation is the first major cause of genetic change in populations and is one of the reasons populations have the variation required for natural selection to work.

Natural selection will increase the frequency of beneficial mutations and reduce the frequency of harmful mutations in a population. But some mutations are neutral: neither beneficial nor harmful. Natural selection does not act on these mutations. Still, the frequency of these mutations does not simply stay the same year after year. Its frequency changes randomly over generations, like a series of coin flips: heads it increases, tails it decreases. This process of random changes in the frequency of an allele is called **genetic drift**, and it is a second major cause of genetic change in populations (**FIGURE 9.25**).

One special case of genetic drift is **founder's effect**. Often a few individuals of a population leave (or are forced to leave) and begin a new population. For example, a flood or forest fire might make a small population of squirrels seek a new place to live. By chance, it may happen that these founders differ genetically from the main population. If so, the new population starts off being genetically different from the old. With a different genetic starting point and different conditions in its new habitat, the new population may adapt in new and different ways (**FIGURE 9.26**).

genetic drift :: random changes in the frequency of an allele

founder's effect :: a special case of genetic drift in which a few individuals of a population leave or are forced to leave and begin a new population

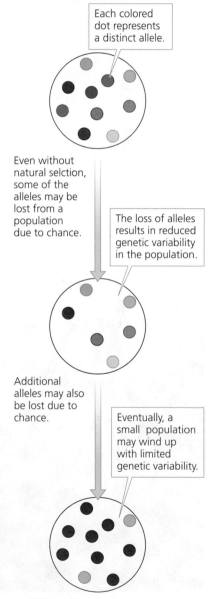

Each colored dot represents a distinct allele.

Even without natural selction, some of the alleles may be lost from a population due to chance.

The loss of alleles results in reduced genetic variability in the population.

Additional alleles may also be lost due to chance.

Eventually, a small population may wind up with limited genetic variability.

FIGURE 9.25 Genetic drift leads to random changes in the frequency of an allele as a consequence of random or chance events. It can be a major cause of genetic change in populations.

Ancestral population

Some salamanders migrated east.

Some salamanders migrated west.

The salamanders found themselves separated by mountains and could not easily mingle.

Genetic drift

Eventually the salamanders moved far enough south and past the mountains that reconnection was possible.

The salamanders had changed so significantly that they could no longer mate.

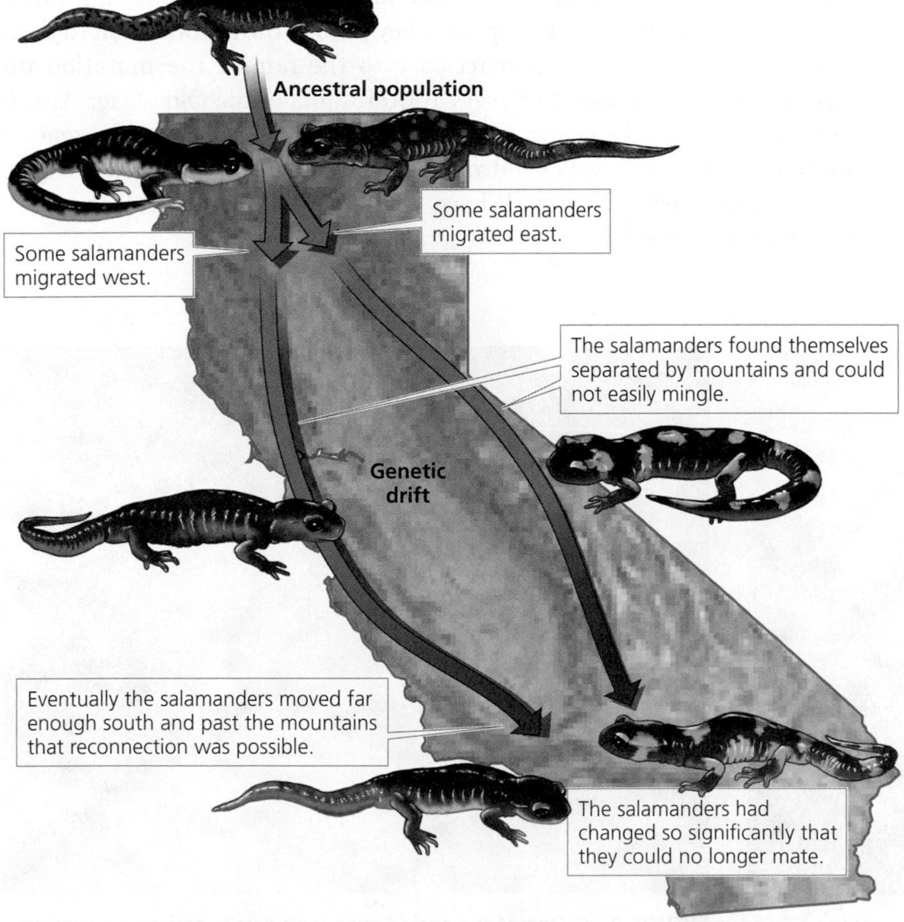

FIGURE 9.26 In a special case of genetic drift known as founder's effect, a few individuals of a population may leave (or be forced to leave) and begin a new population. If by chance these founders differ genetically from the main population, the new population may adapt in new and different ways.

Founder's effect has been important in human evolution. As our earliest ancestors dispersed from Africa, they traveled as small groups and founded small settlements. When a settlement got too large, a small number of people would leave to start a new settlement. The native residents of Asia, Australia, Europe, and the Americas all descended from small groups of people that started off genetically different from one another. Combined with adaptation to the different environmental conditions in these regions, founder's effect helps to explain why people from these regions differ in skin color, height, and even in the prevalence of certain diseases such as sickle cell disease (see Chapter 4).

There are a number of more recent examples of founder's effect, too. One involves the Amish of eastern Pennsylvania (**FIGURE 9.27**). The Amish descend from a small group of perhaps 200 German immigrants, dating back to the early 1700s. The Amish practice a traditional lifestyle and tend to marry among themselves. As a result, the genetic makeup of the Amish comes largely from their small group of ancestors.

Two of the early Amish immigrants were Samuel King and his wife, who arrived in Pennsylvania in 1744. One or both of the Kings carried a mutation for Ellis-van Creveld syndrome (**FIGURE 9.28**). Ellis-van Creveld syndrome has several symptoms, usually seen in newborns: polydactyly (more than five fingers or toes), a narrow chest, signs of dwarfism, and several types of heart defects. The Kings were only one couple, but in a small population a single couple can have a large genetic impact. They passed the mutation on to their offspring, and as other Amish married into the family, the mutation increased in frequency. Today 5 of every 1000 infants of the Old Order Amish in Pennsylvania have the syndrome, compared to 1 in 60,000 in the general population. Ellis-van Creveld syndrome is not an adaptation, and there is no particular reason why the Amish have it—it was just the luck of the draw when the Kings moved in.

FIGURE 9.27 **The Amish of eastern Pennsylvania descend from a small group of perhaps 200 German immigrants, dating back to the early 1700s.** They practice a traditional lifestyle and tend to intermarry, so their genetic makeup comes largely from their small group of ancestors.

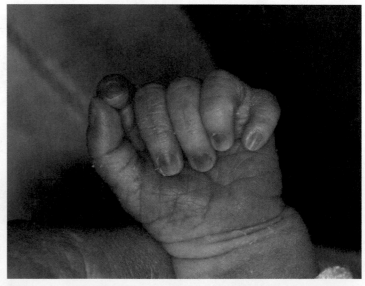

FIGURE 9.28 **One of the early Amish immigrants carried a mutation for Ellis-van Creveld syndrome, which may result in infants who are polydactyly (having more than five fingers or toes).** Today, 5 of every 1000 infants of the Amish in Pennsylvania have the syndrome, compared to 1 in 60,000 in the general population; this is an example of founder's effect.

9.6 What Is the Evidence for Speciation?

There is no doubt that species change genetically—that is, that they evolve. Scientists like the Grants and many others have painstakingly documented the actions of natural selection. They have shown that natural selection causes genetic changes in a population that allow it to better adapt to the environment. Similarly, lactase persistence in adults was an advantage in populations where dairy foods were staples. Random genetic changes occur as well. But a changed species can still be the same species. Do we have evidence that brand new species can appear through evolution?

The answer is yes. Scientists have a good deal of evidence to show that new species branch off from existing species, a process called **speciation**. We also have evidence that speciation occurs regularly and continuously under the right conditions. It's true that speciation can take thousands of years, which makes direct observation of the process from start to finish a difficult or impossible task. But when speciation occurs, it leaves behind many clues. Scientists put together and analyze these clues, much the same way a detective uses clues to solve a crime, even if no one witnessed it.

speciation :: a process by which species branch off from existing species to form new species

Biological Species Concept

Two individuals are considered to be in the same species if they can interbreed. This has several important biological consequences. Individuals of the same species are genetically linked: they share the same set of genes. If a mutation (or an adaptation) occurs in one individual, it could increase its frequency in different families of the same species by interbreeding. On the flip side, individuals of different species cannot interbreed. The species are genetically isolated. Because they can't interbreed, mutations in one species cannot spread to the other. Each species follows its own independent evolutionary path. This is called the **biological species concept**, and it defines species as genetically distinct, independently evolving units.

biological species concept :: it defines species as genetically distinct, independently evolving units

The biological species concept works well in many cases. Dogs and cats look very different. They are different species, and they can't interbreed. **FIGURE 9.29** shows four corn snakes. They look very different as well, but they can interbreed. Because they can, biologists consider them to be variant forms of the same species.

But the biological species concept doesn't work in all cases. Dogs and wolves can interbreed, as can lions and tigers, whales and dolphins, and

Cats and dogs are distinct species. Although they may look cute together in this photo, they cannot interbreed.

Although each of these corn snakes looks very different, they are all the same species and can interbreed.

FIGURE 9.29 According to the biological species concept, each species follows its own independent evolutionary path, and individuals of different species cannot interbreed.

A Zorse—male zebra mates with female horse; offspring are generally sterile.

B Wholphin—male whale mates with female dolphin; offspring are generally sterile.

C Liger—male lion mates with female tiger; offspring are generally sterile.

FIGURE 9.30 **The biological species concept does not apply in all cases, however.** Zebras and horses can breed to form a zorse (A), whales and dolphins can breed to form a wholphin (B), and tigers and lions can breed to form a liger (C), among many other examples.

horses and zebras, if they get the opportunity (**FIGURE 9.30**). Different plant species interbreed regularly, producing hybrids. Some one-celled organisms reproduce asexually and generally don't "breed" at all in the regular sense. Breeding is also not an option for a fossil. Scientists can't use the biological species concept for a species identified only from fossils.

We'll use the biological species concept for our discussion, recognizing that it doesn't apply universally to all organisms. In a way, the fact that species are difficult to define is good evidence that new species arise when an existing species splits in two. If speciation is ongoing, you'd expect to see some cases of species just starting to split, others that are far along in the process, and everything in between. This explains why similar species like dogs and wolves can still interbreed (though at some point in the future they may lose this ability), while others like horses and donkeys technically can interbreed, but their offspring are sterile. If every species was created separately, you'd expect that it would be much easier to define a species and tell one apart from another.

Evidence for Speciation

One intriguing "historical" record lies in the sequence of nucleotides in an organism's DNA. DNA is passed down through the generations relatively intact, but over time mutations accumulate that change, add, or remove nucleotides. If we compare the DNA sequence for a specific gene in two different species and find that there are few differences, we conclude that the species are closely related. More differences indicate more distant relatedness. By taking into account the various degrees of similarity among a group of species, it is possible to construct an evolutionary tree that shows one way in which the group may have evolved. (See *How Do We Know? Constructing Evolutionary Trees.*)

A group of scientists performed a study of this type on 17 species of snapping shrimp, marine crustaceans that resemble crayfish or small lobsters. All of these species of shrimp came from the waters off the coast of Panama. The scientists extracted DNA from each species, determined the sequences of specific regions, and constructed an **evolutionary tree**, or a diagram that took into account the degrees of similarity among the species based on their

evolutionary tree :: diagram that takes into account the degrees of similarity among a species based on their physical characteristics and genes

How Do We Know?

Constructing Evolutionary Trees

One of the most important consequences of speciation is that all species share a common ancestor. For closely related species, the common ancestor was relatively recent, while more distantly related species have a common ancestor further back in time. An evolutionary tree is therefore very similar to a family tree, or pedigree. You and your siblings are closely related because your common ancestor is your parent. You and your cousin share a grandparent as a common ancestor, but you are less closely related because the grandparent lived before your respective parents.

Evolutionary biologists construct evolutionary trees to study the lineages of species just as genealogists construct trees to study families. There are many ways to construct an evolutionary tree. One of the best ways is to group together species whose DNA is most similar. This makes sense because relatedness, whether between species or individuals in a family, is determined by how similar the DNA is.

DNA sequences can be represented by strings of four characters: A, T, C, and G. Each character represents one of the four nucleotides present in DNA. We start with the assumption that the species under study all had a common ancestor, with a single ancestral DNA sequence. The sequences we observe in the species differ from that of the ancestor because they have undergone mutations. Mutations include

exchanging one nucleotide for another. They also include the loss or insertion of a nucleotide—the DNA sequences in the descendants can be different lengths. Finally, we assume that species that share mutations are more closely related than those with different mutations.

We can construct an evolutionary tree by grouping together the species with the most similar DNA, adding in species with slightly less similar DNA, and continuing until all species are part of a group. The process is tedious and mathematical and requires considerable trial and error to get it right. That's why we have developed computer programs to do the work for us. The simple example in the figure illustrates some of how the process works.

Step 1 of the figure shows a DNA sequence from five species. You can see that there are similarities and differences between the DNA sequences. Step 2 shows an alignment: "blanks" that represent the loss of a nucleotide have been added to make the sequences match up better. Once the program computes the alignment, it forms a tree by going through the grouping process just described, placing the most similar DNA sequences on the closest branches in the tree. Step 3 shows the tree that results from this particular set of DNA sequences. ::

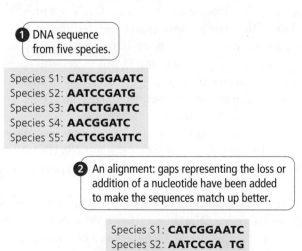

① DNA sequence from five species.

Species S1: **CATCGGAATC**
Species S2: **AATCCGATG**
Species S3: **ACTCTGATTC**
Species S4: **AACGGATC**
Species S5: **ACTCGGATTC**

② An alignment: gaps representing the loss or addition of a nucleotide have been added to make the sequences match up better.

Species S1: **CATCGGAATC**
Species S2: **AATCCGA TG**
Species S3: **ACTCTGATTC**
Species S4: **AA CGGA TC**
Species S5: **ACTCGGATTC**

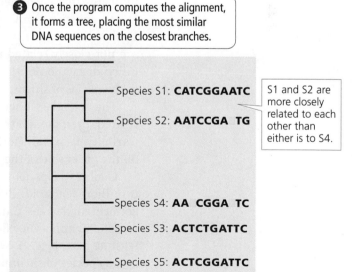

③ Once the program computes the alignment, it forms a tree, placing the most similar DNA sequences on the closest branches.

Species S1: **CATCGGAATC**
Species S2: **AATCCGA TG**

S1 and S2 are more closely related to each other than either is to S4.

Species S4: **AA CGGA TC**
Species S3: **ACTCTGATTC**
Species S5: **ACTCGGATTC**

Evolutionary biologists construct evolutionary trees to study the lineages of species just as genealogists construct trees to study families. One of the best ways to construct a tree is to group together species whose DNA is most similar.

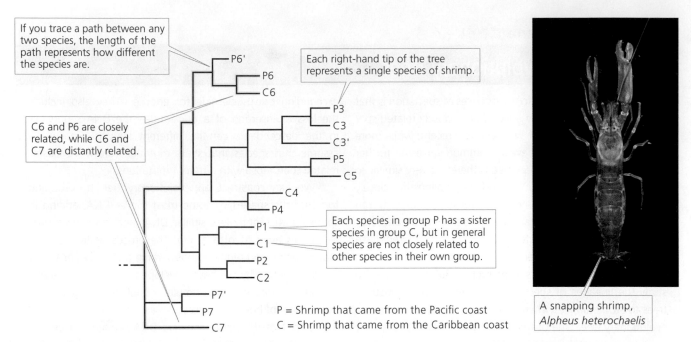

If you trace a path between any two species, the length of the path represents how different the species are.

Each right-hand tip of the tree represents a single species of shrimp.

C6 and P6 are closely related, while C6 and C7 are distantly related.

Each species in group P has a sister species in group C, but in general species are not closely related to other species in their own group.

P = Shrimp that came from the Pacific coast
C = Shrimp that came from the Caribbean coast

A snapping shrimp, *Alpheus heterochaelis*

FIGURE 9.31 An evolutionary tree is a diagram showing the degrees of similarity among species based on DNA sequences in genes. A group of scientists studying 17 species of snapping shrimp from waters off the coast of Panama created an evolutionary tree and discovered that not all shrimp were closely related, despite living in close proximity.

physical characteristics and genes (**FIGURE 9.31**). Each right-hand tip of the tree represents a single species of shrimp. If you trace a path between any two species in the tree, the length of the path represents how different the species are with respect to DNA sequence. The 17 species of shrimp are divided into two subgroups, labeled P for the shrimp that came from the Pacific coast and C for the shrimp found off the Caribbean coast. The species sort out mainly into pairs, with one member from P and one from C in each pair. This indicates that each member of the pair is more closely related (according to DNA sequence similarity) to the other than either is to any other species. Each species in group P has a sister species in group C, but in general species are not closely related to other species in their own group. (For a more detailed discussion of the process the scientists used to extract the DNA from each species of shrimp, see *Technology Connection: Genbank*.)

The scientists uncovered a puzzle. Pacific coast shrimp are not closely related, even though they live in close proximity. The same holds true for the Caribbean shrimp. However, each species is closely related to a species found on the other side of the Isthmus of Panama.

Geology helps us to solve this puzzle. The Isthmus of Panama is about 3 million years old. Prior to its rising, North and South America were not connected, and the Caribbean and the Pacific Oceans were confluent in this region, meaning they flowed together and were not separated by a landmass (**FIGURE 9.32**, p. 272). We can assume at this time that there were only six or seven species of shrimp, which we could label 1–7 without any Ps or Cs. After Panama rose from the sea, each of these species ended up separated, with about half their population in the Caribbean and the other half in the Pacific and with no chance of contact between them.

At this point, the process Darwin described as descent with modification began to occur. Each half of the original species began to diverge genetically from the other, adapting to different conditions on either side of the isthmus.

Technology Connection

Genbank

Using current technology, scientists can extract DNA from the cells of the organisms they study and rapidly determine the sequence of nucleotides in each segment of DNA. This leads to an interesting question: what should be done with all this sequence information?

The National Center for Biotechnology Information (NCBI) has the answer. They have assembled a cluster of supercomputers to host a huge database called GenBank, which stores the DNA sequences and other information as well. Any scientist can upload his or her sequence data to GenBank. There is an online form to record the necessary information, and the process is not much more involved than making an online purchase. In addition to DNA sequence data, GenBank also stores protein sequence data. It has over 15 million sequence records, and the size increases virtually every day.

Impressive as it is, GenBank is just the tip of the biotechnology iceberg. NCBI provides a number of online tools to help scientists access and analyze this storehouse of genetic information.

Entrez is a cross-database search program that allows you to look up sequence information by keyword search. If you want to know the DNA sequence for the human hemoglobin gene, simple go to Entrez and type in "human hemoglobin."

Another program, BLAST (Basic Local Alignment Search Tool), is also a search engine, but of a very different kind. If you give BLAST a DNA (or protein) sequence, it will search the entire GenBank database and return all the sequences that "match" yours. How does this work? A DNA sequence is represented by a long string of the characters ATG and C. The records in GenBank might be hundreds or thousands of characters long, and your search sequence can be long as well. Identical matches between the search sequence and a GenBank record are easy to detect—but hardly ever happen. Usually BLAST finds sequences in GenBank that are "close" to the one you submitted.

The concept of what it means for two strings of characters to be "close" is sophisticated and complex, an advanced version of what your spell checker does when it looks for alternatives to the words you misspell. The end result is that if you find the DNA sequence for human hemoglobin in Entrez, you can use this sequence to search GenBank using BLAST, and in a few seconds you will receive a record of all the hemoglobin genes from all the organisms whose hemoglobin gene is like ours. This can be the starting point for an in-depth study of the evolution of the hemoglobin gene.

NCBI also provides a number of programs to view DNA and protein sequences, a viewer to visualize protein molecules in 3-D with rotations, and tutorials on how to use and interpret all this information. And all of this is free to any scientist—or any person—anywhere in the world if you have an Internet connection. If you would like to learn more, visit the NCBI website: http://www.ncbi.nlm.nih.gov. ::

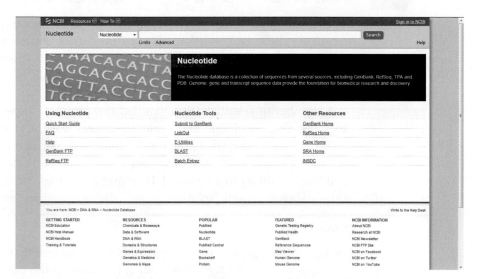

The National Center for Biotechnology Information (NCBI) has assembled a huge database called GenBank, which stores DNA sequences. Any scientist can upload his or her sequence data to GenBank.

10 million years ago

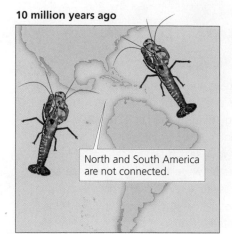

North and South America are not connected.

Ten million years ago, North and South America were not connected, and the Caribbean and the Pacific Oceans flowed together; there were likely only six or seven species of shrimp at that time.

3 to 5 million years ago

The Isthmus of Panama rose from the sea to connect North and South America.

Three to five million years ago, the Isthmus of Panama rose from the sea connecting North and South America and likely separating the shrimp. Descent with modification occurred, and the species diverged genetically as they adapted to different conditions on either side of the isthmus.

Present day

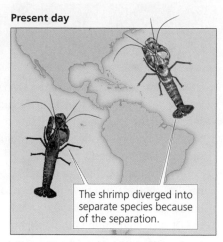

The shrimp diverged into separate species because of the separation.

Three million years later, these populations have diverged into separate species but still share the greatest degree of similarity with their sister species on the other side.

FIGURE 9.32 The geological history of the Isthmus of Panama explains why some species of Pacific coast snapping shrimp were more closely related to species of Caribbean coast snapping shrimp than they were to other species of Pacific coast snapping shrimp.

Three million years later, these populations have diverged into separate species. But they still share the greatest degree of similarity with their sister species on the other side.

As we will see, it takes a long time for species to form, and the process is not directly observable. But examples like this one provide strong evidence that a species can split in two, thereby generating new species.

9.7 How Do New Species Arise?

Darwin recognized that different populations of the same species often live under very different conditions. He reasoned that these populations would evolve independently, and that each would adapt to the specific environment in which they lived. Ultimately, the populations could become so different that they would be different species.

A number of scientists have studied the process of speciation, and it is clear now that speciation can occur in several ways. But Darwin's view of speciation is a good start. Based on the biological species concept, we have to account for one more fact: if one species splits into two, the two new species must lose the ability to interbreed. How might this happen? Speciation occurs in three stages: genetic isolation, genetic divergence, and secondary contact.

Genetic Isolation

The first stage is genetic isolation (**FIGURE 9.33**). Something must divide populations in such a way that interbreeding is cut off. In the case of the snapping shrimp, the dividing feature was the Isthmus of Panama. The Grand Canyon has divided a number of populations between its north and south rims. Smaller features like a river may divide populations if the animals are not very mobile or cannot swim. In all these cases, some sort of geographical feature arises to physically separate populations.

FIGURE 9.33 The first stage of speciation is genetic isolation: something divides a population in a way that cuts off interbreeding. The division may be caused by a geographical feature such as an isthmus or a river that arises to physically separate the population, or it may be caused by dispersal, whereby a small number of individuals migrate to a new habitat.

Isolation can also occur through dispersal, in which a small number of individuals migrate to a new habitat and form a new population. If the new population is sufficiently distant and the journey between the populations is difficult, then interbreeding cannot occur and the new population is genetically isolated from the old. The dispersal of finches from the mainland to the Galapagos Islands is a good example of this type of isolation.

Genetic Divergence

The second stage is genetic divergence (**FIGURE 9.34**). Founder's effect starts the divergence. The founding members of each population are likely to differ, simply by chance, so the populations start off genetically different from each other.

In each of the separated populations, genetic changes begin to accumulate. Some of these changes are due to the fact that different mutations, and therefore different forms of variation, occur. Natural selection will increase the frequency of some of these mutations and decrease others. This will

S. aberti aberti

S. aberti kaibabensis

A. harrisi

A. leucurus

South Rim

North Rim

FIGURE 9.34 The second stage of speciation is genetic divergence: the founding members of each population are likely to differ by chance, and genetic changes begin to accumulate through natural selection. Over time, the populations diverge genetically—squirrels who reside in the different rims of the Grand Canyon are one example of this—and possibly morphologically (structurally) and behaviorally too.

happen differently across populations since the environmental conditions are likely to be different. Over time, the two populations diverge genetically, and maybe morphologically (structurally) and behaviorally, too.

Secondary Contact

In the third stage, the separated populations may come back into contact. Recall that the biological species concept is based on whether individuals can interbreed. Secondary contact provides the opportunity for interbreeding and determining whether speciation has occurred during the time the populations were separated. Whether speciation occurs depends on what happens after contact is reestablished (**FIGURE 9.35**).

It might be the case that in spite of their genetic divergence, the two populations are still able to interbreed successfully. In this case, they probably will do so, and eventually the populations will merge—no speciation will have occurred.

It is also possible that genetic divergence will be so great that members of the populations will not even try to mate. This appears to have occurred with the snapping shrimp. In this case, the two populations satisfy the criterion of the biological species concept, and speciation is complete.

There is a third possibility: "incomplete speciation." Members of the two populations may interbreed and produce hybrid offspring, but the hybrids may not be as successful as nonhybrids. After all, the parents have genes that are adapted to different conditions. The hybrid combines these different genes and may find itself less well adapted to either of its parents' habitats. The hybrids may not be as strong or as fast, and they may die young. Or they might be perfectly viable but less fertile or even sterile, like a mule. In such cases, the parents will end up spending time and energy making and raising their offspring, but they will have either fewer offspring (if hybrid offspring are less viable) or fewer grandchildren (if hybrids are less fertile or sterile).

Natural selection will favor parents that mate with their own kind, and it will work against parents that produce hybrids. Over time, the tendency to produce hybrids becomes rarer and the frequency of hybrids declines.

In spite of the genetic divergence, the populations may still be able to breed successfully; for instance, a coyote and a dog will produce a coydog, which is fertile.

The genetic divergence may be so great that members of the populations will not even try to mate; this is the case with two species of meadowlark whose songs are so different.

Members of the population may interbreed and produce hybrid offspring; but they may not be as successful; for instance, the liger and the zorse are infertile hybrids.

FIGURE 9.35 The third stage of speciation is secondary contact: the separated populations come back into contact. Secondary contact provides an opportunity for interbreeding, and whether speciation occurs depends on what happens after contact is reestablished.

Natural selection that acts to reduce hybrid formation has a special name: **reinforcement**. Reinforcement might select each population to develop different mating rituals, or to mate at different times or in different places, or to develop some type of distinctive difference so that prospective parents don't mistake a member of the other population for a suitable mate. The result of reinforcement is that individuals mate within their own group, interbreeding no longer occurs, and, by the biological species concept, true species have formed.

reinforcement :: natural selection that acts to reduce hybrid formation

9.8 # Why Is It So Difficult for the Public to Accept Evolution?

The short answer is this: it's not. Our *Life Application: Public Acceptance of Evolution* feature details public views of evolution from 34 countries. If it came down to a vote, evolution would "win" in 30 of them (though in a few countries, it would be close). But in the United States, as well as Turkey, Cyprus, and Latvia, public acceptance of evolution is low. About 40% of U.S. residents report they have doubts about evolution, 40% accept it, and 20% have not been able to form an opinion. There has been an equal split between those who doubt and those who accept for many years, although the undecideds have increased recently.

While all generalizations have exceptions, the scientific community views evolution as a fact of nature, as well supported as the fact that the Earth revolves around the sun. Many scientists are deeply religious and do not feel their views on evolution and religion conflict. A number of religions have formally adopted the position that evolution is compatible with their spiritual teaching. But not all religions have done so, and the arguments against evolution are largely religious.

The co-option of evolution to support social and political ideology can have terrible consequences. The theory of natural selection suggests that the "natural" struggle for existence leads to the "survival of the fittest." The strong naturally replace the weak. Adopting this view seems to justify aggression, exploitive forms of capitalism, and programs to improve the strength and purity of a human race through selective breeding and pogroms (large-scale massacres of ethnic groups) (**FIGURE 9.36**, p. 277). Some of the concerns about evolution reflect the fear that we would apply its principles to our own society and to the ways that we interact with one another. Most scientists would caution against applying the principles of evolution to our society because human behavior, relationships, and society are too complex to be explained by evolutionary or biological knowledge alone.

Regardless of how we use or misuse it or what its consequences are, evolution is still with us as the unifying theme of life. The significance of evolution lies in its ability to explain what we see in the natural world around us: why some species are more closely related than others, and why each species is adapted to its habitat, for example. There is also a deeper question about how we make decisions about what is true and what is not. Science has presented us with over a century's worth of evidence from every scientific discipline, all of which points to the fact of evolution. Reconciling personal belief and scientific

Life Application

Public Acceptance of Evolution

A survey of 34 countries revealed that the United States has a significantly lower public acceptance of evolution than any of its European allies or Japan, and the second-lowest level of all the countries surveyed. Surveys like these have been conducted in the United States for 30 years. A June 2012 Gallup poll revealed that only 15% of Americans believe in atheistic evolution, whereas 32% believe that God guides evolution and 46% of Americans believe in creationism.

As they are on so many issues, the people of the United States are split over evolution, in spite of the many changes to science education and in spite of the dramatic advances in the field of evolutionary biology.

One reason for this difference is that the United States has a stronger presence of Christian sects that believe in biblical literalism. Some mainstream Christian religions have reconciled their teachings with evolution, but this is difficult, or perhaps impossible, if one believes the Bible should be taken literally.

Second, evolution has become politicized in the United States as political parties adopt religious positions in the hope of securing votes from like-minded citizens. For example, in a number of recent state-level elections, the Republican platform has called for teaching alternatives to evolution in public schools. When seven Republican presidential candidates were questioned about it during the 2012 election season, five rejected evolution and two accepted it. This shows an even lower level of acceptance among the candidates than in the general public.

The conflict over teaching evolution in public schools began with the Scopes trial in 1925 and continues to this day. Over the past several years, as Kansas State Board of Education members have been voted in and out of office, the board has oscillated between teaching evolution and teaching intelligent design. In 2001, Alabama began inserting stickers into biology textbooks warning that evolution was "only a theory," and Georgia followed suit in 2002. In 2004, the school board of Dover, Pennsylvania, added intelligent design to its K–12 biology curriculum, a move that was struck down in federal court a year later. A large number of homeschool curricula (though by no means all) are structured to teach creationism and intelligent design rather than evolution.

We think that education, particularly of science teachers, can ease these tensions. As more people learn more about science in general and evolution in particular, fewer disagreements will be based on misconceptions. ::

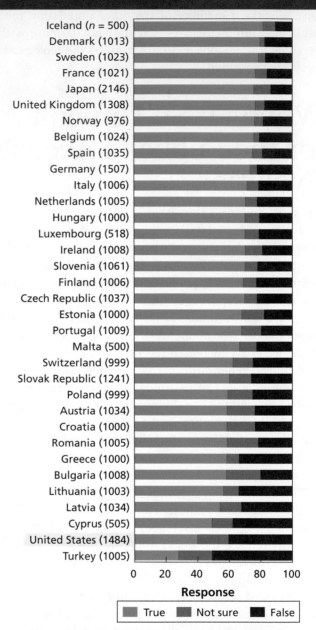

n = number of people surveyed

A survey of 34 countries revealed that the United States has a significantly lower public acceptance of evolution than any of its European allies or Japan, and the second-lowest level of all the countries surveyed. Adults were asked to respond to the following statement: "Human beings, as we know them, developed from earlier species of animals."

FIGURE 9.36 The co-option of evolution to support social and political ideology can have terrible consequences, as it did when the Nazis used the theory of natural selection to justify large-scale massacres of ethnic groups in an effort to "purify" the human race. However, regardless of how we use or misuse it, or what its consequences are, evolution remains with us as the unifying theme of life.

fact has always been a challenging endeavor. But to avoid evolution, one must argue that science is wrong, that evidence doesn't matter, and that personal belief carries more weight than evidence.

Biology in Perspective

How do new species arise and adapt to their environment? Darwin's work provided the first part of the answer, and work by scientists who came later has confirmed and extended his ideas. The evolutionary framework that Darwin and others have developed helps to explain two of the most prominent, observable facts of nature: the diversity of species and the fit of species to their environment.

As a result of evolution, living things have a history. The pattern of geographic variation in humans reflects the way we adapted to local conditions as we dispersed throughout the planet. Some of the puzzling aspects of our own body make sense when we think of them as legacies from earlier ancestors. Understanding why our body is the way it is has medical and social implications, topics we take up later.

For some, the concept of evolution continues the long-standing conflict between science and religion. The consequences of this conflict can be far reaching and can affect education and politics. The consequences also can be polarizing. Education about both religion and evolution is the best way to reconcile belief with science and to defuse the conflict.

Chapter Summary

9.1 How Does Your Body Reflect an Evolutionary History?

Provide three examples demonstrating the effect of evolutionary history on human structure or function.

- The vas deferens does not go directly from the testes to the penis; instead, it takes a circuitous route that is a legacy from earlier ancestors.
- The vast majority of mammals have enzymes that enable them to make their own vitamin C, but humans and other primates inherited a pseudogene that renders the enzyme nonfunctional.
- The vertebrate eye reflects historical constraints; the optic nerve lies in front of the retina, which leaves a "blind spot." .

9.2 What Convinced Darwin of the Fact of Evolution?

Detail the discoveries that led Darwin to change his mind about evolution and describe the steps of natural selection.

- Through biogeography, Darwin found that the rain forest species in South America were more similar to the non–rain forest species of South America than they were to species from other rain forests.
- On the Galapagos Islands, Darwin discovered a number of endemic species, including 14 finches with variations that made them unique; Darwin proposed that "one species had been taken and modified for different ends."
- Darwin's theory of natural selection holds that individuals with certain heritable traits have an increased chance of surviving and producing offspring because of variation, inheritance, and differential survival and reproduction.

9.3 How Do Humans Adapt to Their Environment?

Use specific examples to describe how humans adapt to their environments.

- Humans have adapted to their environment in many ways, including the development of lactase persistence and G6PD deficiency.
  - Although most adult humans cannot digest lactose, some Africans and persons of Northern European descent can.
  - The ability to digest lactose in milk is correlated with the origin of domestication of cattle about 10,000 years ago in Northern Europe and Africa.
- In high-malaria areas, a host of adaptations to fight malaria have evolved: one is the sickle cell trait; another is a mutation that puts red blood cells under metabolic stress.
  - A deficiency in the enzyme glucose-6-phosphate dehydrogenase puts red blood cells under metabolic stress and also prevents malarial parasites from reproducing.

9.4 How Does Natural Selection Produce Adaptations?

Use natural selection to describe the process whereby a population adapts to a specific environment and outline ways that adaptation might be limited.

- Natural selection acts to increase fitness in a population.
- Adaptations increase survivorship and reproduction, which increases the number of genes an organism leaves to the next generation.
- Natural selection has limitations, however: it cannot anticipate or plan for future events, and it can only select from the variation that is present in a population.

9.5 What Are Random Events in Evolution?

Describe the potential role of random processes in evolution and differentiate natural selection from genetic drift.

- Random mutation is the ultimate source of variation required for natural selection to work.
- Virtually every newborn individual will have at least a couple of mutations.
- Genetic drift refers to random changes in the frequency of an allele; in a special case known as founder's effect, a few individuals of a population leave and found a new population. The variation in the starting population limits the potential variations of future generations.

9.6 What Is the Evidence for Speciation?

Summarize the evidence for speciation.

- The biological species concept defines a species as a group of individuals that can breed and have viable young.
- Scientists can extract DNA from different species and compare the sequences for a specific gene on an evolutionary tree.
- In species with few genetic differences, the species are closely related, whereas more genetic differences indicate more distant relatedness.

9.7 How Do New Species Arise?

Outline the stages by which new species arise from a single species.

- Genetic isolation: something divides populations to stop interbreeding.
- Genetic divergence: over time, the populations diverge genetically, structurally, and behaviorally.

- Secondary contact: the separated populations may come back into contact and interbreed successfully (no speciation has occurred), interbreed and produce less-successful hybrids ("incomplete speciation"), or not breed at all (speciation has occurred).

9.8 Why Is It So Difficult for the Public to Accept Evolution?

Assess reasons why some public resistance to evolution still exists despite overwhelming scientific evidence.

- Not all religions have adopted the position that evolution is compatible with their spiritual teaching.
- Evolution has been co-opted in the past to support social and political ideology that justifies selective breeding and large-scale massacres of ethnic groups.
- Some people fear that evolutionary principles will be applied to the ways we interact with one another on individual and societal levels.

Key Terms

adaptation 259
allele frequency 256
biogeography 252
biological species concept 267
Darwin's finches 254
descent with modification 256
endemic 254

evolution 246
evolutionary tree 268
fitness 263
founder's effect 265
genetic drift 265
kin selection 264
modern synthesis 256

mutation 246
natural selection 255
pseudogene 250
reinforcement 275
speciation 267
variation 247

Review Questions

1. Why are most adults unable to eat dairy products, at least not comfortably?

    a. They are missing the lactose gene.

    b. They have lost the ability to break down the sugar lactose.

    c. They have gained the ability to break down the sugar lactose.

    d. They have a gene that increases lactase production.

    e. They have a gene that increases lactose production.

2. List one puzzling feature about the following: human testes, our inability to make vitamin C, our eyes.

3. Briefly give an evolutionary argument for each of the items in question 2.

4. Briefly explain one line of evidence that helped convince Darwin that evolution occurred.

5. On what three observable facts of nature is Darwin's theory of natural selection based?

    a. Individuals in a population vary, variations are inherited, and only the fittest individuals survive.

    b. Individuals in a population vary, individuals in the same species may interbreed, and variations are inherited.

    c. Individuals in a population are all the same species, variations are inherited, and many more offspring are born than survive to adulthood.

    d. Individuals in a population vary, variations are inherited, and in any population, many more offspring are born than survive to adulthood.

    e. Individuals in a population vary, variations are caused by genetic mutations, and more offspring survive into adulthood than die at birth.

6. What was the modern synthesis?

7. Briefly explain how natural selection operated on the medium ground finch during the drought on Daphne Major. How did the birds differ before and after the drought?

8. What is fitness? Explain why fitness involves all your relatives, not just your offspring.

9. Can natural selection fashion perfect structures or adaptations? Explain briefly, with examples.

10. What is genetic drift? What is founder's effect?

    a. Genetic drift is the process by which the frequency of mutations changes randomly over time; founder's effect is a case of genetic drift when a few individuals of a population leave and found a new population.

    b. Genetic drift occurs when a few individuals of a population leave and found a new population; founder's effect is the frequency of random mutation changes over time.

c. Genetic drift is the process by which social behavior becomes easier to explain; founder's effect is a case of genetic drift when a few individuals of a population leave and found a new population.

d. Genetic drift is the process by which inherited mutations cause individuals to be more dissimilar from their descendants over time; founder's effect is a direct consequence of genetic drift.

e. Genetic drift occurs when species become isolated from one another because of geographic changes; founder's effect is when a few individuals of a population leave and found a new population.

11. Define the biological species concept. Provide one example of a case in which the definition does not seem to apply.

12. List the three steps required for speciation to occur.

a. Genetic divergence, secondary contact, and reinforcement

b. Genetic isolation, genetic divergence, and genetic resurgence

c. Genetic mutation, genetic divergence, and genetic resurgence

d. Genetic isolation, genetic convergence, and secondary contact

e. Genetic isolation, genetic divergence, and secondary contact

13. What is reinforcement? Why is reinforcement considered a type of natural selection?

14. In the United States, approximately what percentage of people accept that evolution occurs? Don't accept? Aren't sure? How does this compare to other countries?

15. What is an evolutionary tree? How does it show the degree of relatedness between two species? Briefly explain one way to construct an evolutionary tree.

The Thinking Citizen

1. Select a human trait that varies across different geographic locations. Two good candidates are skin color and the ability to metabolize alcohol. Do some research to make a map of how the trait varies throughout the world. Propose a hypothesis about why the trait varies, and why it is specific to certain geographic regions.

2. Consider a plant species that grows in an area that normally has a lot of rain. However, for several years, there has been a drought and water has been scarce. Describe how the plants could adapt to the drought through the action of natural selection. Be sure to point out the importance of (a) the fact that the plants vary initially in their drought tolerance (or lack thereof), (b) the fact that drought tolerance (or intolerance) can be passed on to offspring, and (c) the fact that many of the young plants produced in a new generation will not survive.

3. A number of bird species have a behavior called "helpers at the nest." Young birds stay with their parents for a number of years, helping to raise their siblings. But the helpers don't have young of their own, at least not in their early years. Are helpers sacrificing their evolutionary fitness by helping? Do helpers get any evolutionary benefit themselves from helping?

4. Lions and tigers are considered separate species. However, they can interbreed. A tigon is the offspring of a male tiger and a female lion. A liger is the result of the opposite cross: a male lion and a female tiger. Male tigons and ligers are probably sterile, while females of both hybrids can have offspring. Do lions and tigers conform to the biological species concept? Which of the three stages of speciation has not occurred in these species? You might want to consult a range map showing where lions and tigers live. If this stage were to occur, what do you predict about the final outcome?

5. Suppose a DNA analysis of four species reveals the following sequences:

Species 1	CCATGAT
Species 2	CCGCGAC
Species 3	CCTTGAT
Species 4	CCCCGAC

Construct an evolutionary tree that shows how these species are related. Recall that to construct an evolutionary tree, you group the sequences into pairs according to how similar their DNA is. Then you connect the pairs, working backward to construct the tree. Which pairs of species are most closely related?

The Informed Citizen

1. Before we became such accomplished global travelers, people tended to be born, grow up, and die in the same locale. Doctors could focus their treatments on diseases and conditions that were specific to their locale. Now that we travel and mix much more, this approach does not work as well. From an evolutionary point of view, why does a person's geographic origin affect the way he or she responds to medical treatment? Should doctors know about these evolutionary differences in order to treat patients better?

2. In 2011, the United States Department of Agriculture (USDA) unveiled MyPlate, which recommends that dairy products, including soy alternatives to cow's milk, be part of a healthy diet (**FIGURE 9.37**). Prior to that, the USDA had recommended a food pyramid expressly promoting the consumption of two to three servings of milk or milk-based products per day. However, approximately 75% of humans are at least somewhat lactose intolerant, and the incidence is even higher in some ethnic groups such as Native Americans, Asian Americans, African Americans, and Latinos. In fact, the group with the lowest incidence of lactose intolerance is that with Northern European ancestry. What role do you think these factors may have played in the government's decision to change its recommendations? What other factors may have influenced the change in recommendations? Given the known high incidence of lactose intolerance, why do you suppose the government waited so long to make the change? What reasons can you think of to explain why milk-based dairy products have been so strongly promoted for so long?

3. A number of grass-roots movements seek to remove the teaching of evolution from science classes in public schools, downplay the importance of evolution, or require teaching alternatives like creationism and intelligent design. How should we decide what to teach? Who

FIGURE 9.37

should decide what we teach? Is it best to allow everyone to believe what they want and teach accordingly? Is it wrong to teach things in science class that scientists agree are not true?

4. We now have some ability to screen for the genes that are implicated in certain diseases such as cystic fibrosis, Tay Sachs disease, and sickle cell anemia. Should we encourage individuals who have these traits to avoid having children? Why or why not? If we were able to identify genes associated with desirable traits and could screen for those, would it be a good idea to encourage individuals who have those genes to have children? Why or why not? Who decides what constitutes a desirable or undesirable trait? What consequences might result if we encouraged or discouraged reproduction in particular individuals? Is it possible to influence the direction of our own evolution? Should we even try?

CHAPTER LEARNING OBJECTIVES

After reading this chapter, you should be able to answer the following questions:

10.1 In What Ways Is Your Body an Ecosystem?
Recognize and provide examples of the different types of species interactions.

10.2 Why Do Diseases Evolve Resistance to Antibiotics?
Describe the process by which some disease organisms increase their resistance to antibiotics.

10.3 Why Are Some Diseases More Deadly Than Others?
Use the trade-off hypothesis to explain and predict the spread and impact of a disease.

10.4 Where Do New Diseases Come From?
Consider the processes of mutation and natural selection and predict likely sources of new disease.

10.5 How Can Evolution Help Us Control Disease?
Propose a strategy to reduce the virulence of a disease using evolutionary medicine.

One reason we get sick is because pathogens, such as bacteria, invade our bodies and do damage. But thousands of species of organisms live in and on our bodies and may even keep us healthy.

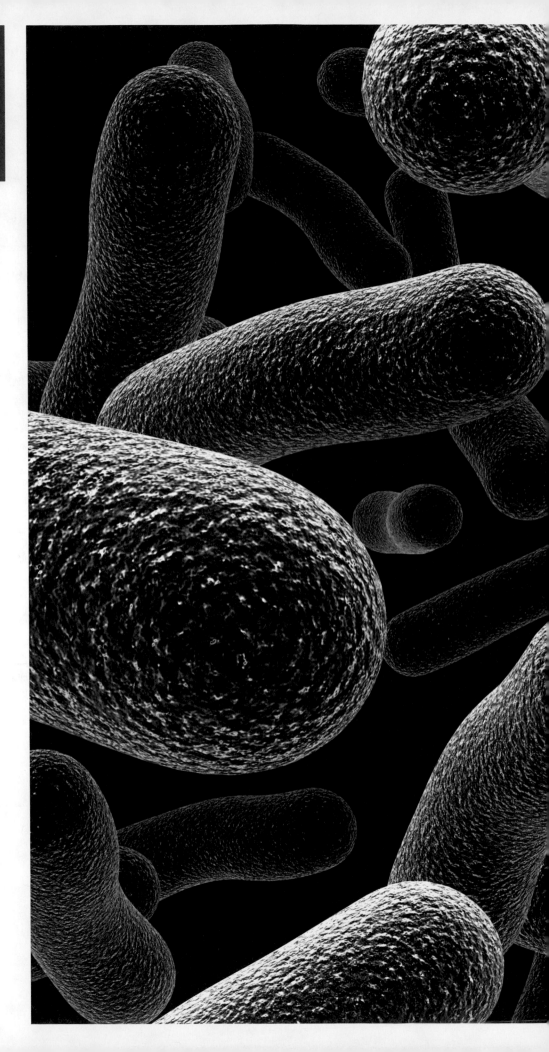

CHAPTER 10

The Evolution of Disease

Why Do We Get Sick?

Becoming an Informed Citizen ...

Pathogens evolve to overcome our attempts at disease control, and new diseases emerge continuously for which we have no natural immunity, antibiotics, or vaccines. A better understanding of the evolutionary dynamics of disease may provide new approaches to public health.

pathogen :: disease-causing organism, such as bacterium, fungus, or worm, that invades the body and does damage

hy do we get sick? It's an interesting and surprisingly complicated question. One reason we get sick is because **pathogens**, *disease-causing organisms such as bacteria, fungi, or worms, invade our bodies and do damage. Although not considered living organisms, some viruses are also pathogenic. However, as we'll see, thousands of species of organisms, especially bacteria, live in and on our bodies and do not make us sick—they may even help us stay healthy. In this chapter, we discuss how diseases evolve, in some cases to become even more harmful to us, and also what evolution can tell us about how diseases can be controlled.*

First, we will look at the disease that has probably killed more humans than any other: malaria. We began our study of malaria in Chapter 4, but here we address a different question: how has malaria evolved, and how have we evolved in response?

case study

Deadly Malaria

Malaria is a disease suffered by about 300 million people annually, and about 1.5 million die from it each year—almost three people each minute. Malaria has been with us throughout the entire history of our species, perhaps 200,000 years. If you consider all human deaths from all causes, malaria may have killed more than half of us.

There are four species of *Plasmodium* that infect humans (and another 170 species that infect other animals). About 7000 to 10,000 years ago, one of these species, *Plasmodium falciparum*, underwent a massive population explosion. *P. falciparum* is by far the most deadly species, accounting for over 50% of all cases of malaria (more than 75% of those in Africa) and 99% of all malarial deaths worldwide. Why is malaria so deadly?

One reason is that *Plasmodium*, the pathogen that causes malaria, is not as easy to kill as it was in the past. Quinine was first used to treat malaria in 1820. Since then, we have developed a dozen different drugs to treat the disease, but none is effective anymore against all types of malaria. The reason for that is that different types of malaria are caused by different strains of *Plasmodium* (a strain is a type of genetic variation within a species), and some strains of *Plasmodium* have evolved **resistance** to multiple drugs. Resistance refers to the ability of an organism to survive exposure to a drug that had previously been able to kill or disable it. The World Health Organization reports that resistance to antimalarial drugs is now one of the major challenges to our attempts to control the disease.

We humans have not been idle in the evolutionary fight against malaria. Before fighting with drugs and medical practices, we "fought back" the way all organisms do: we evolved resistance. Recall that the malaria parasite causes damage by infecting red blood cells, reproducing inside them, and bursting the cells to release new parasites that continue the damage (**FIGURE 10.1**). All our natural defenses to malaria are mutations that make red blood cells less

resistance :: the ability of an organism to survive exposure to a drug that previously had been able to kill or disable it

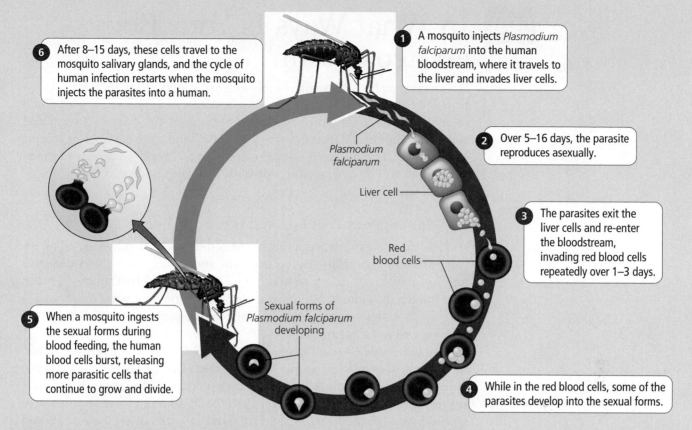

6 After 8–15 days, these cells travel to the mosquito salivary glands, and the cycle of human infection restarts when the mosquito injects the parasites into a human.

1 A mosquito injects *Plasmodium falciparum* into the human bloodstream, where it travels to the liver and invades liver cells.

Plasmodium falciparum

2 Over 5–16 days, the parasite reproduces asexually.

Liver cell

3 The parasites exit the liver cells and re-enter the bloodstream, invading red blood cells repeatedly over 1–3 days.

Red blood cells

Sexual forms of *Plasmodium falciparum* developing

5 When a mosquito ingests the sexual forms during blood feeding, the human blood cells burst, releasing more parasitic cells that continue to grow and divide.

4 While in the red blood cells, some of the parasites develop into the sexual forms.

FIGURE 10.1 The malaria parasite, *Plasmodium falciparum*, causes damage by infecting red blood cells, reproducing inside them, and bursting the cells to release new parasites that continue the damage.

vulnerable to *Plasmodium.* In Chapter 4 you learned how the sickle cell mutation does this. There are actually four different mutations that cause red blood cells to sickle. There are about a dozen other mutations that affect red blood cells in a way that makes them partially resistant to malaria. Like sickle cell, these mutations cause physiological problems in people who inherit two copies of the mutated gene, and they are found only in regions where malaria is common.

Scientists can study the evolutionary history of our resistance mutations by looking at the DNA sequence of the genes. It turns out that these mutations date back to around 7000 to 10,000 years ago—the same time the deadly form of malaria began to spread.

So why is malaria deadly? One of the reasons has to do with the evolution of resistance. Another has to do with 7000-year-old changes in human behavior that allowed a deadly disease to spread. Before we explore these reasons in more depth, however, we'll take a closer look at the many species that inhabit our bodies. Some cause disease, some don't, and some can go either way depending on conditions.

In What Ways Is Your Body an Ecosystem?

The space your body takes up contains an estimated 100 trillion cells. However, only 10 trillion of them genetically are yours. The other 90 trillion belong to bacteria and other one-celled organisms, fungi such as yeast and mold, and multicellular organisms like mites and a whole host of worms (**FIGURE 10.2**). If you were to somehow remove all "your" cells, these foreign cells would still form a recognizable outline of your body and many of your internal organs.

The Many Species That Live and Evolve in Your Body

Medical researchers are beginning to study the species that live in and on us. Most of these organisms live on our skin and in our mouth and digestive tract, urinary tract, and nose and lungs. The National Institutes of Health have started the Human Microbiome Project to figure out what these species are—the first step toward understanding how they affect our health. What we have learned so far is surprising.

One 2009 study found 182 species of bacteria living on the forearm alone, 30 of which were new to science. When scientists examine other regions of the skin, they discover many hundreds of other species. Furthermore, the species are very different on different parts of your body. Your dry forearm is home to species adapted to a "desert," species adapted to warm, moist underarms and the groin form "tropical" communities, and still other communities live in oily areas like hair follicles and the crease of your nose. Scientists have discovered over 700 species that live in the human mouth and over 800 species that live in the colon. Other parts of your body support similar numbers (see *How Do We Know? The Many Species That Live On You*).

ecology :: the study of the interactions among species and their physical environments

Ecology is the study of the interactions among species and their physical environments. We all have many hundreds of species living with us, but none of us has all the species just mentioned. A few species tend to be found on all of us all the time. But each of us has our own particular set of species. These species form ecological communities that differ throughout our bodies, and the

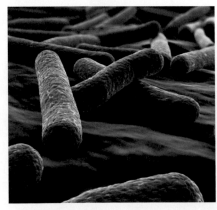

A *E. coli* bacteria

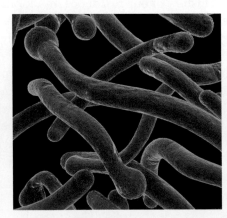

B *Candida albicans* fungi (mouth, vagina, and skin)

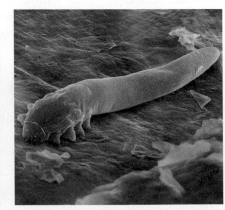

C Follicle mite

FIGURE 10.2 The body is a complex ecosystem. In addition to its own cells, bacteria (A), fungi such as yeast and mold (B), and multicellular organisms like mites (C) live on and within the human body.

The Many Species That Live on You

Many biology courses have a lab where students study the species that grow on them. Students run a cotton swab over a certain region of skin, perhaps their armpit or the crease of their nose, and then rub the swab over a petri dish containing culture medium. A few species of bacteria or mold will grow, but it turns out that most of the species that live on your body are highly adapted to the conditions they find there and simply don't grow well or at all on a petri dish. Scientists missed the vast majority of species simply because they couldn't grow them in a lab.

Now scientists are finding other ways to learn about these organisms. A swab from your skin, a small amount of feces, or a cough contains a whole community of microorganisms. Scientists collect the sample, break open all the cells, isolate the ribosomes, and extract the ribosomal RNA (see Chapter 3). Each species has its own specific type, or sequence, of ribosomal RNA. When scientists count how many types of ribosomal RNA they find, they know how many different species there were in the original sample. They can also try to match the RNA to a database of known RNA types, and if they find a match, they can even tell which species were present.

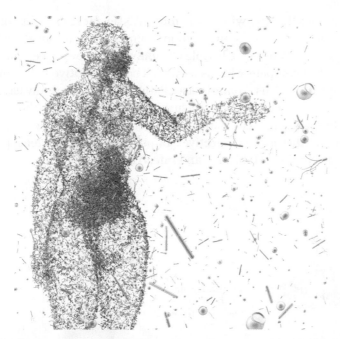

Many biology courses have a lab where students study the species that grow on them. Today, scientists are finding other ways to learn about these organisms without growing them in the lab.

Techniques like these don't tell us much about these microorganisms, how they live, or what they do. But we have learned that there are thousands of species that live on and in us—many more than we expected. This changes our perspective and shows us that our bodies truly are ecosystems. It also changes our notion of what it means to be "clean." ::

species compositions of these communities change over time. Within these communities, individuals survive, reproduce, and compete for resources, just as they do in any biological community. As a result, the species evolve over time, adapting to one another and to the conditions in our bodies. Your body is an ecosystem that provides numerous habitats and supports the evolution of many species.

The Ecology of Our Resident Species

We are in the early stages of understanding the life on us and how it affects our health. Many of our resident species appear to have no effect on us at all. Eyebrow mites, for example, eat dead skin cells and excess oil from hair follicles, but most of the time we don't even know they exist. This ecological interaction is called **commensalism**: one species benefits from the interaction, and the other is neither helped nor harmed.

We do know that our resident species compete with one another for food and space. This competition can be fierce. In the digestive tract, for example, many species secrete toxic compounds that kill other species. Others use up critical nutrients or make the habitat more acidic, interfering with the growth of competing species. Some even form thin films on our intestinal surfaces that prevent other species from sticking there. These free-floating species can be washed away as food moves through us, again removing competitors.

Many species that live with us are **mutualistic**, meaning that both we and the species benefit from the interaction. Some species in our digestive tract help us digest our food, for example (**FIGURE 10.3**). They also eat our food, so technically they compete with us as well. The competitive behavior of some species helps protect us from disease. Pathogens often are invasive species. Before they can infect us, they must survive competition with the species that normally live with us. The competitive behavior our resident species use against one another is often effective against invading pathogens, preventing them from becoming established and infecting us.

commensalism :: an ecological interaction in which one species benefits and the other is neither helped nor harmed

mutualistic :: an ecological interaction in which both species benefit

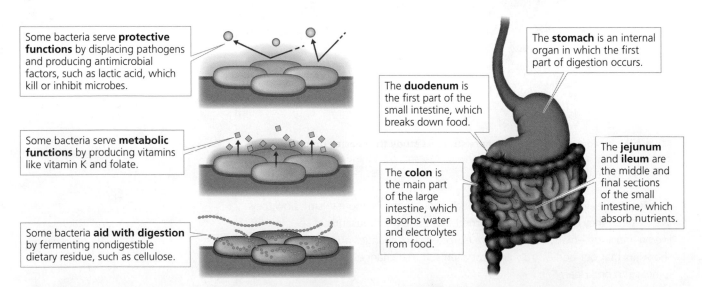

Some bacteria serve **protective functions** by displacing pathogens and producing antimicrobial factors, such as lactic acid, which kill or inhibit microbes.

Some bacteria serve **metabolic functions** by producing vitamins like vitamin K and folate.

Some bacteria **aid with digestion** by fermenting nondigestible dietary residue, such as cellulose.

The **duodenum** is the first part of the small intestine, which breaks down food.

The **colon** is the main part of the large intestine, which absorbs water and electrolytes from food.

The **stomach** is an internal organ in which the first part of digestion occurs.

The **jejunum** and **ileum** are the middle and final sections of the small intestine, which absorb nutrients.

FIGURE 10.3 Many species that live with us are mutualistic, meaning that both we and the species benefit from the interaction. In the gut, for example, bacteria prevent colonization by harmful bacteria, produce vitamins, and help with digestion.

Species That Cause Disease

A species living with us may be commensal or mutualistic, but either way it will still be looking out only for itself. If conditions are right—or from our perspective, wrong—many of these species will change their behavior and attack us. If eyebrow mites experience a population explosion, for example, they can cause inflammation and itching of the skin around our hair follicles. Bacteria living in our digestive tract initially may be helpful. But if their population becomes too large, they may begin to secrete toxins or invade our cells, causing disease.

One example of this behavior is seen in the bacterium *Pseudomonas aeruginosa*. This species lives almost anywhere: in soil and water, on the surface of nonsterile medical equipment, and on your skin. It can live on you and cause no damage. However, *Pseudomonas* can be a serious problem for individuals who are already sick or who have a weakened immune system. In these cases, *Pseudomonas* can cause pneumonia, urinary tract infections, septic shock, and even tissue **necrosis** or death. *Pseudomonas* cells secrete a chemical compound that acts as an alert signal. When enough bacteria are present and the concentration of the chemical becomes large enough, *Pseudomonas* begins to produce a toxin that kills our cells. *Pseudomonas* bides its time until it has a **quorum**: enough cells to overwhelm our body's defenses. Many species of bacteria engage in quorum sensing, waiting until their numbers are large before switching to behaviors that cause disease (**FIGURE 10.4**).

Some species don't have these switches: they always cause disease. Viruses fall into this category. A virus generally can't reproduce without killing the cell it infects. And many species that cause disease are true invasive species. They don't normally live on us, and we "catch" them from other sources: an infected person or contaminated soil or water, for example. These species simply use our bodies as a resource for their own reproduction. We may inhale or swallow the invading species, or they may burrow through our skin or the lining of our digestive tract. They can also get in through wounds in our body's surfaces. Some enter the bloodstream, where they can travel throughout our body to particular locations or organs where they can reproduce. Their reproduction uses our body's energy and can damage our organs and tissues, which causes the disease.

In the rest of this chapter, we will look at how disease organisms evolve, and how their adaptations affect the severity of the disease and the way organisms transmit from host to host.

necrosis :: tissue death

quorum :: a population of bacteria large enough to overcome the body's defenses

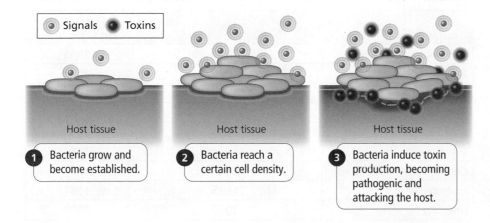

| Signals | Toxins |

① Bacteria grow and become established.

② Bacteria reach a certain cell density.

③ Bacteria induce toxin production, becoming pathogenic and attacking the host.

Host tissue

FIGURE 10.4 Whether species that live with us are commensal or mutualistic, many will change their behavior and attack us under certain conditions. Some bacteria engage in quorum sensing, becoming pathogenic only after they have reached a high enough cell density.

Why Do Diseases Evolve Resistance to Antibiotics?

antibiotic :: drug used to kill a pathogen

We mentioned earlier that some strains of *Plasmodium falciparum* have evolved resistance to at least one of a dozen **antibiotics**, drugs used to kill pathogens. Resistance evolves in every species that confronts antibiotics, including prokaryotic pathogens like bacteria, single-celled eukaryotic pathogens such as *Plasmodium*, pathogenic worms, pathogenic insects, and even pathogenic viruses. And they all evolve resistance for the same reasons. We'll discuss resistance in bacteria, but the principles apply to all species.

Alexander Fleming, a Scottish scientist, ushered in the age of antibiotics. In 1928, Fleming discovered that one of his bacterial cultures had been contaminated by the mold *Penicillium notatum*. Many scientists simply would have discarded the contaminated plate, but Fleming noticed something peculiar: there was a clear ring around the mold where the bacteria did not grow. Fleming performed several experiments and discovered that the mold produced a compound that could kill bacteria. He named this compound penicillin, after the mold that produced it.

Widespread use of penicillin began in 1943, and during World War II it saved the lives of many thousands of soldiers. Since then, we have developed many other antibiotics, and they have continued to cure infections and save lives. One disturbing pattern, however, is that only a few years after a new antibiotic is introduced, new strains of disease evolve that are resistant to it (**TABLE 10.1**).

How Resistance Evolves

Resistance evolves in a very simple manner. When you apply antibiotics to a population of bacteria, most of the cells are vulnerable and they die. However, a few cells will not die. Instead, they will reproduce and form a population of resistant bacterial cells (**FIGURE 10.5**). This is a classic and simple example of natural selection. Resistance evolves like any other trait. Environmental conditions favor some traits over others, and individuals with these traits become more common. And the evolution of resistance is not limited to bacteria. For example, insects develop resistance to insecticides, which presents problems in pest control. For a discussion of how the mosquitoes that carry malaria developed a resistance to DDT, see *Life Application: Malaria and DDT*.

① Antibiotics are applied to a population of bacteria.

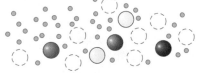

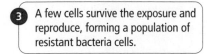
② Most of the cells are vulnerable and they die.

③ A few cells survive the exposure and reproduce, forming a population of resistant bacteria cells.

FIGURE 10.5 Antibiotic resistance evolves like any other natural trait; it is an example of natural selection.

Table 10.1 :: Timeline of Antibiotic Resistance in Bacteria

Antibiotic	Year Introduced	Year Resistance First Observed
Penicillin	1943	1945
Chloramphenicol	1949	1950
Erythromycin	1952	1956
Methicillin	1960	1961
Cephalothin	1964	1966
Vancomycin	1958 (not used much until 1980s)	1986
Cephalosporin (two types)	1979, 1981	1987
Carbapenems	1985	1987
Linezolid	2000	2002

Life Application

Malaria and DDT

If we could kill all the mosquitoes that carry malaria, we would exterminate the disease. DDT (dichlorodiphenyltrichloroethane) is a powerful insecticide that could be used to do the job. During World War II, DDT was used widely to control mosquitoes, lice, and other insects that spread disease among soldiers and civilians in war-torn areas. DDT helped eliminate malaria in many developed countries, although using screened doors and windows, draining swamps, and providing access to better medical care did most of the work.

After the war, DDT was put to a new use: it was sprayed on crops to reduce insect damage. It soon became clear, however, that DDT was linked to many environmental issues. It affected not only insects, but also mammals and birds that ate these insects. Some studies found links between DDT and human birth defects and cancer. In 1962, Rachel Carson published *Silent Spring*, a popular book that chronicled the effects of DDT and touched off the environmental movement. Ten years later, the United States banned DDT, and many countries have followed suit.

Sometimes the science gets lost when issues of public health become entangled with politics. Environmentalists point to the adverse effects of DDT and the damage it can do when it builds up in the environment. Others argue that banning the insecticide is an example of government intrusion, and that millions of children in regions with malaria might be saved if DDT were used to control mosquitoes.

The science behind DDT is clear, however. Spraying DDT will reduce the mosquito population, and this will reduce malaria—initially, at least. But mosquitoes rapidly develop resistance to DDT, as do many insect pest species. DDT persists in the environment long after the insects break free of its effects. It continues to harm mammals and birds and exposes humans to the health risks. Worse, after the mosquitoes become resistant, malaria returns.

Science tells us about the positive and negative consequences of using DDT. But science alone can't help us decide whether the positives outweigh the negatives. In the midst of a particularly severe outbreak of malaria, spraying DDT could save many people. Is its use in such cases justified, in spite of the fact that its benefits are temporary and its disadvantages will persist long term? These questions are hard to answer, but they become even more difficult when political ideology influences the decision-making process. ::

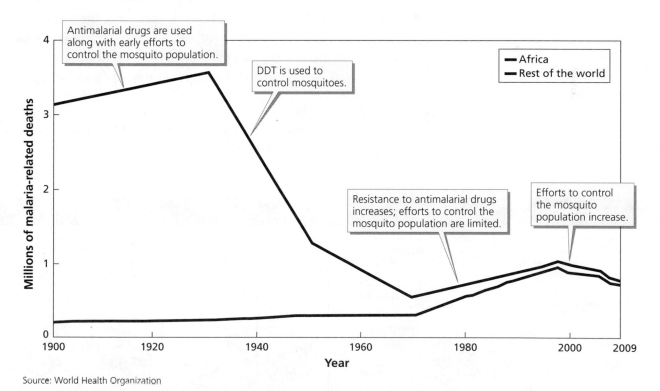

Source: World Health Organization

Although significant DDT spraying led to an initial decrease in malaria-related deaths beginning in the 1930s, mosquito resistance to DDT became more apparent starting in the 1970s.

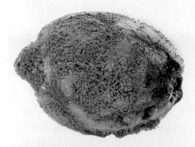

A *Penicillium* grows on a lemon.

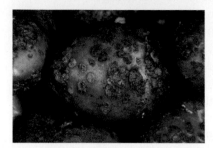

B *Streptomyces* scabs grow on potatoes.

FIGURE 10.6 Most antibiotics are naturally occurring chemicals made by bacteria, molds, and other species. The fungus *Penicillium* excretes penicillin, which kills bacteria in its vicinity (A); the bacterium *Streptomyces* produces streptomycin, which inhibits bacterial growth (B).

resistance gene :: gene that has mutated in such a way that the proteins it makes take on a different function or perform old functions in a new way

But that still leaves several questions about the evolution of resistance. Why are some bacterial cells resistant even before the antibiotics are applied? What characteristics do these cells have that makes them resistant? If resistance is beneficial, why aren't all bacteria resistant? If using antibiotics results in resistant bacteria, why do antibiotics continue to be effective, and how long will this last?

Where Resistant Bacteria Come From

Most antibiotics are naturally occurring chemicals made by bacteria, molds, and other species (**FIGURE 10.6**). Although we can't see them, these microorganisms live in densely populated conditions in the soil, in aquatic habitats, and in you. Competition for food and other resources is very strong. *Penicillium* mold has an advantage: by secreting penicillin, it kills its competitors and keeps more food for itself. Many other species produce antibiotics of their own and get the same advantage.

A species that produces an antibiotic must be resistant to it; otherwise, it would kill itself. This logic is a sort of "proof" that if antibiotics can evolve, resistance must be able to evolve as well. If antibiotic-producing species evolve resistance to their own antibiotics, there is no reason that species that don't produce antibiotics can't do the same.

Resistant bacteria have one or more genes that make them that way. These are known as **resistance genes**. These genes have mutated in such a way that the proteins they make take on a different function or perform old functions in new ways. There are four ways that resistance genes, and the proteins they make, can produce a resistant cell (**FIGURE 10.7**):

1. They can change the cell membrane so the antibiotic cannot get in.
2. They can modify an enzyme so that it breaks down the antibiotic.
3. They can alter the structure of the antibiotic's "target" (what it attacks) so that it is no longer vulnerable to the antibiotic.
4. They can form a "pump" that pumps out the antibiotic before it can cause damage.

Mutation is the reason that every bacterial population has a few resistant cells in it, even when no antibiotics are present. Antibiotics don't cause

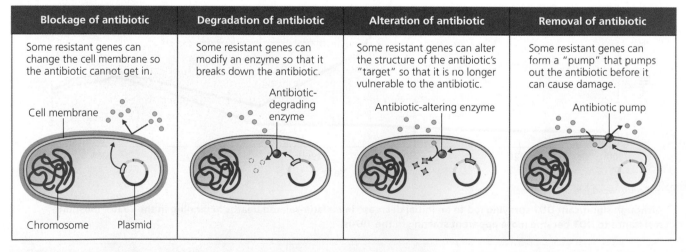

Blockage of antibiotic	Degradation of antibiotic	Alteration of antibiotic	Removal of antibiotic
Some resistant genes can change the cell membrane so the antibiotic cannot get in.	Some resistant genes can modify an enzyme so that it breaks down the antibiotic.	Some resistant genes can alter the structure of the antibiotic's "target" so that it is no longer vulnerable to the antibiotic.	Some resistant genes can form a "pump" that pumps out the antibiotic before it can cause damage.

FIGURE 10.7 There are four mechanisms that make bacteria resistant to antibiotics.

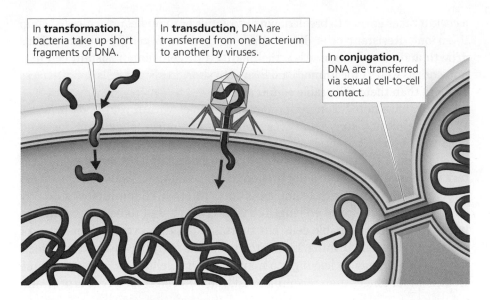

In **transformation**, bacteria take up short fragments of DNA.

In **transduction**, DNA are transferred from one bacterium to another by viruses.

In **conjugation**, DNA are transferred via sexual cell-to-cell contact.

FIGURE 10.8 Horizontal gene transfer spreads antibiotic resistance genes through a bacterial population. Genes enter bacteria through transformation, transduction, or conjugation.

mutations for resistance; they only select for resistance and increase the number of resistant cells. Like all mutations, mutations that make resistance genes are extremely rare and occur only due to chance. However, microbial competitors have been producing antibiotics for millions, maybe billions, of years, so there has been plenty of time for these mutations to occur. Also, bacterial populations are unimaginably large: billions or trillions of cells in each individual host or patch of soil. Every time a cell divides, there is the chance of a mutation. With so many cell divisions over such a long period of time, some mutations for resistance are bound to occur.

Resistance to Multiple Antibiotics

Mutation alone is too slow to account for the rapid appearance of resistant strains. But bacteria have other ways of acquiring resistance genes. Some bacteria can get DNA directly from their neighbors, a process known as **horizontal gene transfer** (**FIGURE 10.8**). DNA can spread through a bacterial population the way news and gossip move through a human community. There are three ways that DNA can be transferred horizontally: transformation, transduction, and conjugation. With **transformation**, DNA released into the environment by bacteria is physically absorbed by other bacteria. **Transduction** involves the transfer of bacterial DNA from one bacterium to another by a virus. In **conjugation**, two bacterial cells physically join together and transfer DNA from one to the other. If one of the bacterial cells has a resistance gene, it can pass it along to other cells, creating many resistant bacteria—and all their descendants will be resistant as well. Worse still, horizontal gene transfer allows many bacteria to become resistant to several antibiotics. With horizontal gene transfer, the spread of resistance occurs much more quickly than by mutation alone.

Not every strain of bacteria is resistant, and not all resistant bacteria are resistant to all antibiotics. Antibiotics continue to be effective against some strains of bacteria, and if one antibiotic doesn't work, you can try another one. But we have already reached the point where it is common for a tried-and-true antibiotic to fail because of resistant strains.

Why Not All Bacteria Are Resistant to Antibiotics

In any environment where antibiotics exist, resistance is a strong evolutionary advantage. But everything in nature, including antibiotic resistance, comes at

horizontal gene transfer :: the process by which bacteria can get DNA directly from a neighbor

transformation :: DNA released into the environment by bacteria is physically absorbed by other bacteria

transduction :: the transfer of bacterial DNA from one bacterium to another by a virus

conjugation :: two bacterial cells physically join together and transfer DNA from one cell to the other

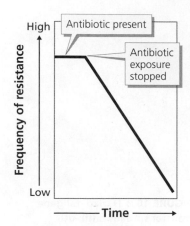

FIGURE 10.9 Bacteria that are resistant to antibiotics become less prevalent in the population when antibiotic exposure ceases.

a cost. It takes energy to break down antibiotics or to make molecules to pump them out. Resistant cells divert this energy from other functions such as growth and reproduction. Also, the modified proteins in a resistant cell may not work as well as the unmodified forms. In short, resistant cells are often less efficient than their vulnerable counterparts.

As we have noted, competition among species within your body is fierce. When antibiotics are absent, resistance is no longer an advantage. And because they are energetically less competitive, resistant strains are outcompeted by more efficient nonresistant strains. Eventually a population of resistant bacteria gives way to a population of nonresistant cells (**FIGURE 10.9**).

This is why it is critically important that you take your antibiotics exactly as prescribed. The specified length of time is calculated to be long enough to kill most of the vulnerable cells but short enough to prevent a population of resistant cells from forming. If you stop your pills early because you start feeling better, you may be leaving too many bacterial cells alive. If you continue treatment for longer than recommended, you may select for a resistant strain.

Antibiotics in the Environment

If resistant bacteria are eliminated in the absence of antibiotics, why is resistance becoming a major global health threat? It is because antibiotics are now available almost everywhere (**FIGURE 10.10**). Besides administering antibiotics in individual patients, hospitals themselves are a huge reservoir of antibiotics.

A high concentration of many kinds of antibiotics is found in hospitals, and as a result hospitals harbor many resistant strains. Over 200,000 patients annually are infected with resistant strains of bacteria, and the Centers for Disease Control and Prevention (CDC) estimates that over 90,000 deaths occur as a result. Since antibiotics are always present, the vulnerable strains of bacteria cannot displace the resistant ones. Hospitals also provide many hosts—some already weakened and sick, and therefore highly susceptible to infection.

Antibiotics are also used in agriculture, which means they can be found in our food, soil, and water. Over half the antibiotics used in the United States are given to livestock, poultry, and crops. Some of these antibiotics are used to treat infection or, in the case of crops, to prevent bacteria from spoiling the food. But farmers often routinely give low doses of antibiotics to healthy livestock and poultry in their feed. These low doses kill some of the bacteria in the digestive tract. This means more food goes to the animal (not to the bacteria that live in it), and it grows larger in a shorter period of time.

FIGURE 10.10 The amount of antibiotics used in agriculture and for human health is staggering. In 2009, almost 29 million pounds of antibiotics were administered to livestock and more than 7 million pounds to humans.

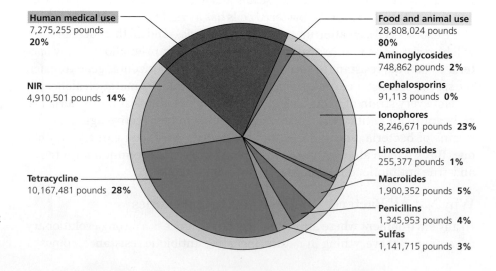

Human medical use
7,275,255 pounds
20%

Food and animal use
28,808,024 pounds
80%

Aminoglycosides
748,862 pounds **2%**

Cephalosporins
91,113 pounds **0%**

NIR
4,910,501 pounds **14%**

Ionophores
8,246,671 pounds **23%**

Lincosamides
255,377 pounds **1%**

Macrolides
1,900,352 pounds **5%**

Tetracycline
10,167,481 pounds **28%**

Penicillins
1,345,953 pounds **4%**

Sulfas
1,141,715 pounds **3%**

As antibiotics become more common in bacterial habitats, resistant strains of bacteria become more common as well. Many medical researchers predict that eventually antibiotics will become ineffective for treating bacterial infection, unless we can find ways around the problem of resistant strains.

10.3 Why Are Some Diseases More Deadly Than Others?

Some infectious diseases can be fatal. Malaria, tuberculosis, and AIDS are responsible for half of the deaths caused by infectious disease each year. Other diseases, like the common cold or chicken pox, cause discomfort but are rarely fatal. Pathogens vary quite a bit in their **virulence**—their ability to cause damage or death. But the one characteristic they have in common is that each pathogen is adapted to its way of life. These adaptations are the result of natural selection acting to maximize the pathogen's survival and reproduction. (See *Scientist Spotlight: Paul W. Ewald* for one scientist's

virulence :: a measure of a pathogen's ability to cause damage or death

Scientist Spotlight

Paul W. Ewald (1953–)

Not many people have redirected their careers because of a case of diarrhea, but that is exactly how Paul Ewald, a professor of biology and director of the Program on Disease Evolution at the University of Louisville, became interested in evolutionary medicine.

Ewald was born in Wilmette, Illinois. His mother was a psychologist and his father was a physicist. Ewald earned his bachelor's degree from the University of California, Irvine, in 1975 and his Ph.D. in 1980 from the University of Washington.

It was in 1977, during his graduate fieldwork studying bird behavior, that Ewald suffered a three-day bout of diarrhea. While he was sick, he wondered about his role as a host for the pathogen that was causing him such uncomfortable physical distress. As Ewald remembers thinking, "There's some organism in there and this diarrhea might be my way of getting rid of the organism—or it might be the organism's way of manipulating my body."

Ewald considered that the organism might be ensuring its success of finding another host to infect by passing efficiently out of his body. He thought about how he should treat his symptoms. "If it's a manipulation and you treat it, you are avoiding damage, but if it's a defense and you treat it, you sabotage the host."

As a result of this experience, Ewald changed his research focus to host-pathogen interactions. He realized that medicine could be enriched by evolutionary theory. In the case of infectious disease, for example, it is essential to look at the disease not simply from the standpoint of the host but also from that of the pathogen.

Paul W. Ewald is one of the pioneers in the field of evolutionary medicine.

Ewald is one of the pioneers in the field of evolutionary medicine, which rests on the idea that illness can be understood fully only through inclusion of a Darwinian evolutionary perspective. His ongoing research focuses on the questions of whether some chronic diseases—such as mental illness, cancer, Alzheimer's, and heart disease—are caused by low level, long-term infections and whether pathogens can be made less virulent by altering their behaviors or transmission patterns. ::

Source: http://www.scientificamerican.com/article.cfm?id=a-host-with-infectious-id.

FIGURE 10.11 Evolution selects for pathogens that do minimal harm to their hosts, thereby, prolonging their own life span. The myxoma virus is a deadly pathogen that was introduced to control the rabbit population in Australia; today it kills only about 50% of the rabbits that catch it, and infected rabbits live almost a month.

experience with the tension between the needs of pathogens and hosts.) In this section, we will look at some of the reasons why diseases evolve and adapt in different ways.

Why Some Diseases Become Milder over Time

As the old saying goes, "Never bite the hand that feeds you." But this is exactly what pathogens do. They take energy and nutrients away from their hosts, and when their population size grows, they damage cells and internal organs. This seems shortsighted because pathogens are completely dependent on their host for their very lives. If the host dies, the pathogen dies with it.

Evolutionary biologists recognized this paradox early. They hypothesized that evolution would select for pathogens that did minimal harm to their hosts, thereby prolonging their own life span (or the life span of their descendants). Pathogens that were highly virulent were thought to be out of balance evolutionarily. Most likely these pathogens caused diseases that were new to humans, and evolution had not yet had time to select for milder forms.

There was some experimental evidence to support this view. The best example, which comes from Australia, involves rabbits and a virus called myxoma (**FIGURE 10.11**). In 1859, Australian ranchers imported rabbits from England to hunt for sport. But rabbits, being rabbits, soon reproduced uncontrollably. Within 20 years, rabbits became a major pest, destroying crops and displacing native species.

Myxoma virus, which causes a lethal disease in rabbits, was discovered in Uruguay around 1900. During the 1950s, the Australian government released a plague of myxoma virus that spread throughout the rabbit population. Myxoma virus was virulent and very effective at controlling the rabbit population. It killed 99% of the rabbits that caught it. Their lungs filled with fluids, their eyelids filled with pus and swelled shut, and death occurred within a week. The rabbit population dwindled from around 600 million to 100 million in only a couple of years.

Since then, both the rabbit population and the virus have evolved, adapting to each other. The most virulent strains of the virus were selected against because they killed their hosts before the rabbits could spread the disease to others. The most vulnerable rabbits died off quickly, leaving rabbits that had some resistance to the virus. The net result is that myxoma has evolved to become less virulent, as the theory predicted it would. Today, myxoma kills only about 50% of the rabbits that catch it, and infected rabbits live almost a month—plenty of time for the virus to spread. The rabbit population is estimated to be about a quarter of what it was at its peak. This is still too many rabbits, but exterminating rabbits has proven to be a public relations nightmare for the Australian government, especially around Easter.

The Trade-Off Hypothesis of Reproduction Versus Transmission

In some cases, it is true that pathogens are selected for traits that cause their hosts less harm. But we know now that this is not a general rule. Human malaria, after all, is at least 200,000 years old, and it remains deadly. Interpreting the evolution of disease involves two ideas. First, natural selection selects for increased reproductive success however it occurs. There is no selection for species to be nice to one another unless this happens to increase reproductive success as well. Second, pathogens have a two-pronged challenge for their success. They evolve to compete and reproduce successfully

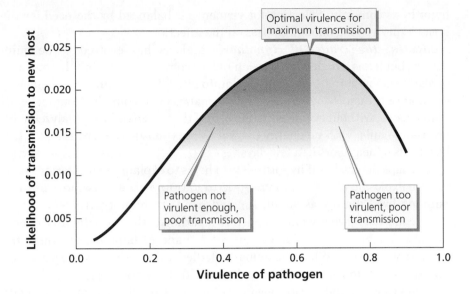

FIGURE 10.12 **Pathogens evolve to compete and reproduce successfully within a host, but they must balance this against the need for transmission to a new host.** If a pathogen becomes too virulent, the host may die before transmission succeeds.

within a host, but they must balance this against the need for **transmission** to a new host (**FIGURE 10.12**).

There is a trade-off between reproduction within a host and transmission to a new one. If this trade-off favors a milder disease to increase transmission, the selected pathogen will be milder. This occurred with rabbits and myxoma virus. However, if the trade-off favors causing serious damage to the host, then the pathogen evolves to become more severe and deadly. The evolution of disease can take either path. The key is to understand which path is selected for and why. We'll look first at how pathogens evolve inside their host. Then we'll examine the trade-off hypothesis with respect to deadly diseases such as malaria, more mild ailments like the common cold, and sexually transmitted diseases.

How Disease Evolves in Your Body Original infections may involve only a few bacterial cells or viral particles. These invaders reproduce vigorously, producing billions of new bacteria or viruses. Almost all the damage done by disease organisms is related to this population growth. A larger population steals more energy from the host and does more damage to cells and organs. More invaders also mean a stronger response from your body's natural defenses: a higher fever, more mucus, or more inflammation, for example. Sometimes your body's defenses overreact and increase the severity of the disease. The overreaction can even kill you.

As the pathogen population grows, it undergoes evolutionary change. Mutations occur, and genetically different strains of pathogen emerge. Some of these strains may be better at infecting cells or utilizing energy. These strains will reproduce more quickly, causing more damage and outcompeting strains that grow more slowly.

Strains with the largest population size have another advantage: they are most likely to be transmitted to a new host (**FIGURE 10.13**). The most numerous strains will have the largest number of representatives in each cough, tick or mosquito bite, or bowel movement. Therefore, the most numerous strains have the best chance of infecting another host.

So far, it looks like natural selection always will select for the fastest-growing, most virulent strain of pathogen. If this were universally true, diseases would evolve to become very bad indeed. But this is where the trade-off

transmission :: the passage of a pathogen from one host to another

FIGURE 10.13 **Strains with the largest population size are most likely to be transmitted to a new host.** A person infected with a large population of influenza virus, for example, can easily transmit the pathogen to another person.

hypothesis comes in. Selection for virulence is balanced by the need for the host to survive long enough to pass on the infection.

Testing the Trade-Off Hypothesis Malaria has evolved to be quite deadly, but it has a high transmission rate to compensate. Recall that malaria spreads among humans through mosquito bites. This is a direct and efficient method of transmission. A person with malaria can come in "contact" with many people without ever leaving the bed. If the patient is half paralyzed with fever and chills, with no energy to swat hungry mosquitoes, he might infect hundreds of neighbors in a few days.

Perhaps the trade-off hypothesis also helps to explain the mysterious population explosion of *P. falciparum* 7000 to 10,000 years ago. Before this time, humans lived mainly as small, isolated bands of hunter-gatherers. Under these conditions, what would happen to a pathogen that mutated to a deadly form? It might quickly wipe out an entire band of humans, but there the damage would stop. When the humans all died, this strain of pathogen would die out as well, much like a hot fire that quickly burns out its fuel.

Sometime around 10,000 years ago, humans developed agriculture and our society changed forever. With a more reliable source of food, people began living together in greater numbers. Agriculture also requires irrigation, which leaves open puddles of standing water very close to where people live—a haven for malaria-carrying mosquitoes. Under these conditions, a deadly pathogen has many hosts to infect and may be able to maintain itself in the human population. Certainly we have no proof that agriculture permitted a deadly form of malaria to evolve, but it's a plausible idea.

The trade-off hypothesis predicts that the most severe diseases will be those that spread to people most easily and efficiently. Diseases that spread by mosquitoes, fleas, and contaminated water all fall into this category. The bubonic plague that killed approximately one-third of the people in Europe during the Middle Ages was spread by fleas, and cholera is a classic example of a deadly disease spread by drinking contaminated water (see Chapter 14).

Diseases like the common cold or "childhood" diseases like mumps, measles, and chicken pox spread by airborne viruses or by contact with droplets of fluids from infected people. These diseases spread when infected people are up and walking around, coming in contact with uninfected people. Strains of pathogens so virulent that infected people quarantine themselves in bed would be at a disadvantage (**FIGURE 10.14**). Many diseases that spread this way are indeed milder than malaria, bubonic plague, and cholera.

While a number of examples and cases support the trade-off hypothesis, it is by no means universally true. Tuberculosis is spread much the same way that colds are spread, yet tuberculosis is a much deadlier disease. Influenza is another example of a sometimes fatal disease spread by direct contact. Still, the general pattern is that the deadliest diseases have the highest rates of transmission, and that in many cases a lower rate of transmission selects for a milder disease.

Sexually Transmitted Disease There is one last class of disease that depends on a "healthy" host for its spread. Imagine a sexually transmitted disease (STD) that was so severe it immobilized you within 24 hours of infection, or caused obvious symptoms like open sores and pustules, and killed you in a couple of days. What do you think about this disease's chances of being spread? For an STD, the ability of the pathogen to spread is tied directly to the host's ability to mate. Any STD that interferes with its host's reproduction will be strongly selected against.

Highly virulent diseases have less opportunity to spread to others.

Less virulent diseases more easily spread to others.

FIGURE 10.14 The trade-off hypothesis predicts that the most severe diseases will be those that spread to people most easily and efficiently. Highly virulent diseases do not spread easily unless helped by vectors such as mosquitoes or fleas.

Does this allow us to make predictions about the severity and contagiousness of STDs? The trade-off hypothesis states that like any pathogen, STDs will be selected for competitive ability and fast growth rate within the host. The strain with the largest population has the best chance of transmission to a new host. However, STDs can't cause too much damage too quickly, or their host will be unable to mate often enough. STDs must travel a fine evolutionary line and achieve a balance between these two constraints to be successful. The trade-off hypothesis predicts that STDs will have a slow onset, stay in the host's system for a long period of time, and be fairly mild, at least in the initial stages.

Chlamydia is the most common STD in the United States, which by evolutionary standards means it is the most successful STD (**FIGURE 10.15**). At least 2 million people are infected with chlamydia. But it's hard to get an accurate count because about half the men and three-quarters of the women infected with chlamydia have no symptoms at all. For this reason, chlamydia is sometimes called a "silent" infection. When symptoms do occur, it takes up to three weeks for them to appear—ample time for the disease to be spread among a population of sexually active people. Chlamydia infections can lead to pelvic inflammatory disease, scarring of fallopian tubes, inability to become

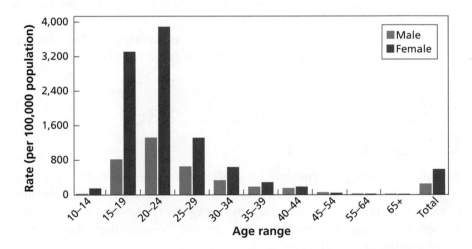

A Chlamydia rates by age and sex, United States, 2011

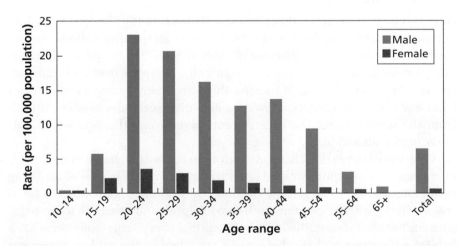

B Syphilis rates by age and sex, United States, 2011

FIGURE 10.15 Chlamydia and syphilis are common STDs, which by evolutionary standards means they are successful. Because few or no symptoms appear early on in the disease, they spread easily throughout the population, with chlamydia being comparatively high in females (A) and syphilis being more prevalent in males (B).

ectopic pregnancy :: a pregnancy in which an embryo develops in the fallopian tube rather than in the uterus; ectopic pregnancies cannot develop successfully to term and if left untreated can lead to rupture of the fallopian tubes

pregnant, or **ectopic pregnancy**, in which an embryo develops in the fallopian tube rather than in the uterus. Ectopic pregnancies cannot develop successfully to term and if left untreated can rupture fallopian tubes.

Syphilis is another STD that may not show symptoms in its early stages. During the first two months after infection, some people develop a series of sores and rashes that begin at the site of sexual contact and later spread to other parts of the body. But many people with syphilis don't develop these sores and rashes, or the breakouts may occur at locations that aren't easily visible. The lack of symptoms results in two major problems. First, infected people won't seek medical treatment, which means that the disease may progress to its final, and lethal, stage. Second, infected people unknowingly spread the disease to their sexual partners.

The final stages of syphilis can take decades to develop. If the disease is not treated, syphilis bacteria eventually invade the tissues of your circulatory system, nervous system, liver, skin, or bones. Depending on where the bacteria invade, syphilis can present a wide array of symptoms in its later stages, all leading to disability and death. But it is possible to be contagious and spread the disease for 20, 30, or even 50 years before symptoms appear. This is bad for infected people and their sexual partners, but good for the pathogen. The absence of early symptoms in some people and the delayed onset of the really bad symptoms is an adaptation of the syphilis bacteria that helps the disease to spread.

Many STDs follow the pattern of symptoms shown by chlamydia and syphilis. One of the most devastating, and most recent, is AIDS.

 # Where Do New Diseases Come From?

New diseases emerge all the time. Most of these diseases are known only from small outbreaks in localized areas. But there is always concern that some new disease, like HIV/AIDS, will become the next global epidemic. In this section, we will look at the sources of new disease and the conditions they need to become established and spread.

Sources of New Diseases

New diseases come from three sources (**FIGURE 10.16**). First, species that normally live on your body can evolve a new strain that causes disease. Second, you can catch the diseases of other species. Third, species that normally live in soil or water can invade your body. There are over 1400 different species that cause disease in us, and there are many examples of diseases from each of these sources. However, new diseases come overwhelmingly from the second source: pathogens jump to us from other species, mainly other mammals and birds.

We pointed out earlier that your body is an ecosystem, home to thousands of species that are quite well adapted to humans. This is true of all living organisms, especially larger, multicellular ones. Every organism has a set of species living with it, some neutral, some helpful, and others that are parasites and cause disease. Biologically speaking, every large body presents a habitat and resources that other species can adapt to and exploit. Pathogens

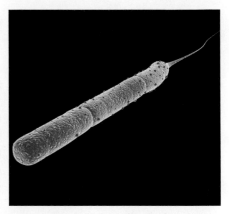

A The avian flu virus (normally harmless to humans) can become highly virulent, if combined with a particular human virus.

B Ringworm (shown on this dog's face and chest) can be spread to humans from animals.

C Cholera bacteria can often be found in water in places with poor water treatment and sanitation.

FIGURE 10.16 New diseases can emerge from three sources: mutation of previously harmless microbes (A), transmission from other species (B), or transmission from the environment (C).

are a universal and natural result of evolution. And pathogens that jump from one species to another are not terribly different from a species that migrates to a new region—a fairly common occurrence. If a species, disease-causing or not, can adapt to the new conditions it encounters, it can thrive in its new home.

New diseases from other species often need some help to make the jump to humans. This help could come in the form of a mutation in the disease organism that allows it to infect humans. Or it could come from a change in human behavior and social practices that allows the disease to become established in human populations. Let's look at how the jump might occur.

Stages of a New Disease

Scientists who study evolutionary medicine have identified four stages that a new infectious disease goes through when it makes the jump to humans (**FIGURE 10.17**). These four stages form a sort of pyramid. At the base—stage 1—we are exposed to many diseases all the time. Only some of these are able to

Exposure:
Something occurs that brings the pathogen into contact with us.

Infection:
A pathogen invades us and reproduces inside.

Transmission:
A small number of microbes that infect us can be transmitted to another person.

Epidemic:
Fewer still microbes can establish an infection that spreads through a large population.

FIGURE 10.17 Scientists who study evolutionary medicine have identified four stages that a new infectious disease goes through when it makes the jump to humans: exposure, infection, transmission, and epidemic.

progress to stage 2 and infect us. Fewer still are able to spread from person to person in stage 3. And, luckily for us, only a small number reach stage 4 and have the potential to cause epidemics.

Exposure The first stage is exposure. Something must occur that brings the pathogen in contact with humans. Contact might come from working with domestic animals or hunting wild ones, being bitten by insects, drinking contaminated water, or coming into contact with contaminated soil. Our behavior can increase our chances of exposure. As we domesticated more and more animals, for example, we dramatically increased our exposure to their diseases. We are exposed to new pathogens all the time, but our immune system does an excellent job of protecting us. Only a few of these pathogens are able to progress to the next stage.

Infection The second stage is infection. Infection occurs when a pathogen is able to invade our body and reproduce inside us. This can be a difficult step. The pathogen is well adapted to the particular conditions inside its original host, and the conditions inside a human are different. The pathogen may not be able to get inside a human. Or, if it does, it may not be able to reproduce, or it may reproduce slowly, giving our natural defenses time to destroy it before it can do damage.

It is unlikely that a nonhuman pathogen will be able to infect a human successfully. However, every time a human is exposed, the pathogen gets another chance to infect. With many opportunities, the chances of successful infection get better. Also, genetic variation among pathogens means that some of them may be a little better than others at infecting humans. Likewise, humans vary genetically, and some humans may be slightly less resistant than average. Eventually, and probably regularly, the right combination occurs: infection by a strain of pathogen that is slightly better at infecting humans, in an individual human who is slightly more susceptible to it. When this occurs, the infection may take hold and the pathogen will be able to reproduce in a human.

It is also worth noting that many animal diseases infect more than one host species. These pathogens already have the genetic flexibility to infect multiple hosts, and for them the jump to humans may not be as difficult.

Transmission to Others Even if a new disease is able to infect a human, transmission to a new human host may prove difficult. First, the pathogen must be able to reproduce many new offspring, perhaps billions in the case of viruses or bacteria. Second, these pathogens must be able to make their way to the appropriate locations and tissues to get out of the body, maybe to the lungs to be coughed out or to the digestive system to be eliminated with feces. Finally, the pathogens must be able to evade our natural defenses long enough for these events to occur.

Epidemics The final stage in the emergence of a new disease is the ability to spread widely and cause an **epidemic**, in which a greater than expected number of people over a larger area are infected. In a sense, epidemics are the mark of an evolutionarily successful pathogen. An epidemic indicates that the pathogen is well adapted to reproducing in its human host, and also well adapted to spreading among hosts. As the trade-off hypothesis predicts, the disease has "found" the proper balance between these two activities.

One way a disease can reach this stage is through natural selection. Within the host, natural selection selects for the most successful strains of the disease, those that are best at reproduction and spreading to another host.

epidemic :: spread of a disease over a larger area and number of people than expected

This process takes time, and there may well be a number of false starts, in which the disease makes it through only one of the first three stages. But with enough tries, enough mutation, and enough natural selection, eventually a strain may succeed in reaching this stage.

A second way to reach this stage is through changes in human behavior and ecology. We've already suggested (though certainly not proven) that the deadly form of malaria may have become possible only after the development of agriculture. Packing more and more people together in cities, sexual practices, and many other changes in human behavior and society—all these can lead to situations that favor the spread of disease epidemics.

HIV/AIDS

Acquired immune deficiency syndrome (AIDS) is the disease caused by the human immunodeficiency virus (HIV) (**FIGURE 10.18**). AIDS was first reported in the United States in 1983, but it has been in the human population for about a century. HIV attacks and kills the blood cells that help to make **antibodies**, compounds that kill pathogens. While your body can make more of these cells, the virus eventually kills them off. Without these cells, you have an "immune deficiency," which makes you more vulnerable to disease. When a person dies from AIDS, it is usually due to other diseases that the patient can't fight, even though these diseases may not be life-threatening to a person with a healthy immune response.

There are two main forms of HIV. HIV-1 is the more virulent form and has the higher mortality rate. HIV-1 jumped to humans from chimpanzees. HIV-2 is less virulent. People infected with HIV-2 live longer and show

antibody :: a molecule that binds specifically to an antigen and kills pathogens

FIGURE 10.18 AIDS was first reported in the United States in 1983, but it has been in the human population for about a century. HIV-1 jumped to humans from chimpanzees (A), while HIV-2 originated in a monkey called the sooty mangabey (B).

FIGURE 10.19 **HIV entered the human population when humans ate infected chimpanzees.** European colonialism and the ivory trade in Africa played a role in the spread of HIV to a wider area and a greater number of people.

symptoms later, but the disease is still fatal. HIV-2 originated in a monkey called the sooty mangabey.

HIV is a relatively new disease that has progressed to stage 4. It has adapted to humans and has become a worldwide epidemic. So far, it has caused about 25 million deaths, and approximately 40 million people are thought to be infected.

We are not exactly sure how HIV made the jump to humans. We know HIV originated in Africa. Historically, people were probably first exposed when they hunted chimpanzees for food. Butchering and eating chimps in the wild exposed hunters to the blood of chimps infected with their version of the virus. Exposure and "false start" infections may have been going on for thousands of years, but HIV never progressed to stages 3 and 4.

Sometime in the late 1800s or early 1900s, changes occurred that allowed these infections to become established and gave the virus time to adapt to its human host. The list of possibly relevant changes is long. It might have been larger human settlements that allowed the virus to spread more easily. It might have been a result of large work camps set up by European colonial powers to build railroads into the interior of Africa (**FIGURE 10.19**). Over-crowding, physical stress, and malnourishment might have provided the virus with a large pool of vulnerable hosts. Medical practices of the time may also have played a role. In efforts to control tropical diseases, many people were vaccinated. But because syringes were expensive (and at that time not disposable), it was common practice to inject large numbers of people using the same needle, thus helping to spread the virus from person to person. Whatever the actual cause or causes, HIV adapted to humans and has become a global plague.

10.5 How Can Evolution Help Us Control Disease?

Antibiotic resistance shows that we are able to influence the evolution of our diseases. In this particular case, our influence has had a negative impact on our health. Is it possible that we could influence the evolution of diseases in a positive way, one that is beneficial to our health? A number of scientists believe that we can. They believe that integrating evolutionary and medical research can bring about benefits similar to those achieved in biology and chemistry.

This idea is new and somewhat controversial. The medical professions have not adopted all the ideas we'll discuss in this section, and a great deal of testing remains to determine whether these ideas are correct and feasible. But this new field of science shows great promise, and it is an excellent way to illustrate the practical advantages of evolutionary biology.

Antibiotic Resistance

One of the most obvious recommendations is to reduce our use of antibiotics, target their use more narrowly on the most severe diseases, and limit drastically the use of antibiotics for nonhealth-related activities like agriculture. This will not be an easy process. The pharmaceutical industry makes billions of dollars from antibiotics, and ironically (though certainly not by design), antibiotic resistance drives the unending search for newer antibiotics and continued profits. Doctors are under pressure from patients to prescribe antibiotics for relatively minor infections. And the use of antibiotics to protect crops from rotting and to produce larger livestock more quickly has become an economic necessity for many farmers.

A new antibiotic provides an important, but temporary, advantage in biological competition. Antibiotics will likely continue to be important in the way we treat disease. But evolution tells us that antibiotics alone cannot be a long-term strategy for controlling disease.

Vaccinations

The administration of material to provoke the development of immunity to a specific pathogen, or **vaccination**, has been successful in controlling certain diseases, including many childhood diseases like mumps, measles, whooping cough, polio, and most recently chicken pox. The only disease we have driven extinct—smallpox—was eliminated through a massive global vaccination program.

Still, many diseases are difficult or impossible to control by vaccine. Vaccines stimulate your immune system to recognize and attack invading pathogens (**FIGURE 10.20**). But some pathogens, like the influenza virus and HIV, mutate so rapidly that your immune system can't recognize the new strains. This is why you need a flu shot every year, and why scientists so far have been unable to develop a vaccine for HIV.

Evolutionary principles may help us design more effective vaccines (see *Technology Connection: How Vaccines Are Made*). One interesting idea is to use vaccines to target the disease-causing *behavior*, not the *organism* itself.

vaccination :: the administration of material to provoke the development of immunity to a specific pathogen

How Vaccines Are Made

You probably know that there are a number of diseases that you catch only once, like measles or chicken pox. After the disease clears, you are immune for a long time—perhaps your entire life. Vaccines "trick" your immune system into thinking you have a disease. Your body still develops immunity to the disease—but you don't have to go through the trouble of actually being sick.

Your immune system recognizes foreign cells and viruses that invade your body (Chapter 14). It does this by recognizing foreign proteins on the cell or viral surface. This is a good system. Foreign protein indicates the presence of foreign DNA, and foreign DNA means that an external invader is present in your body. Once a foreign protein is detected, some of your immune cells make antibodies: proteins that stick, or bind, to the foreign protein and therefore to the foreign cell or virus. Bound antibodies form a "tag" that marks the invading pathogens. Other immune cells seek out tagged pathogens and ingest and digest them. The remarkable thing is that each antibody is specific to a particular protein (i.e., a particular pathogen), and that your immune system "remembers" this protein. Having made the antibody once, the immune system can quickly make more if you are ever infected with the same disease again. This is why immunity lasts so long.

Vaccines are composed of surface proteins from a specific pathogen. In some cases, the vaccine contains the protein alone. In other cases, the vaccine may contain fragments of the pathogen's cells or whole dead cells. When these are injected, your immune system responds to the protein and makes antibodies—and immunity. But the injection does not contain anything that can reproduce and make you sick.

One drawback to these types of vaccines is that the dead or fragmented microbes don't stimulate your immune system quite as well as the real, working pathogen. Of course, injecting you with a working pathogen is not an option. But we can use evolutionary principles to design safe "live" vaccines. The pathogen is grown for many generations in a nonhuman host: maybe another animal, maybe a chicken egg, or perhaps human cells grown in a lab. Through natural selection, the pathogen adapts to these conditions, but in doing so, it becomes less adapted to humans. These pathogens form the vaccine. When injected, they can reproduce slowly in us, and they have the same proteins as the normal disease. They do an excellent job of stimulating the immune system to produce antibodies, but they are not strong enough to make us sick.

Live vaccines can be cheaper to make. They provide better immunity, so booster shots may not be needed or are needed less frequently. And sometimes these vaccines can be given orally or through a nasal spray rather than by an injection. Live vaccines are available for mumps, chicken pox, measles, polio, and other diseases. The nasal form of flu vaccine is a live vaccine. ::

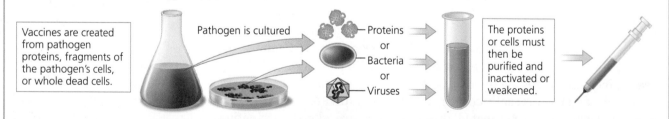

Vaccines are created from pathogen proteins, fragments of the pathogen's cells, or whole dead cells.

Pathogen is cultured

Proteins
or
Bacteria
or
Viruses

The proteins or cells must then be purified and inactivated or weakened.

Vaccines are composed of surface proteins from a specific pathogen. The immune system responds to the protein in the vaccine by making antibodies against the protein that results in immunity, but the injection does not contain anything that can reproduce and make you sick.

According to some scientists, the diphtheria vaccine is a good example of this approach. The bacteria that cause diphtheria produce a toxin that kills cells in our upper respiratory tract. These dead cells may release nutrients that the bacteria can use—an advantage to making the toxin. However, the toxin is a protein that takes energy to make—perhaps as much as 5% of the cell's total energy budget. The diphtheria vaccine stimulates the immune system to make antibodies to the *toxin*, not to the bacterial cell. The cells continue to live, but they are harmless.

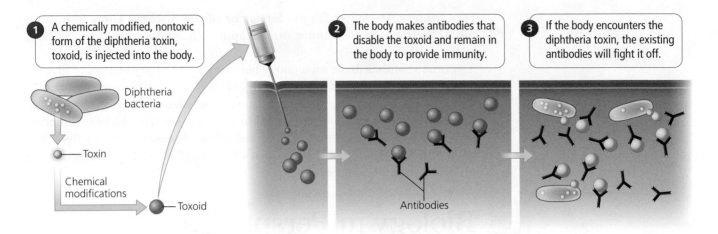

FIGURE 10.20 Vaccination stimulates the immune system to make antibodies that disable a pathogen and remain in the body to fight the pathogen, should it appear again. Vaccination for diphtheria targets the diphtheria toxin.

But not all diphtheria cells make toxin. Those that don't have a competitive advantage: they don't spend 5% of their energy making the toxin. In vaccinated people, the cells that make the toxin lose their advantage since the toxin is destroyed. Cells that don't make the toxin save energy, grow faster and reproduce more quickly, and eventually outcompete the cells that do. Instead of trying to kill all cells, only to have them evolve resistance, this approach uses natural selection to select for cells that cause no damage. Just a few scientific studies have shown this effect, but if it turns out that the process works, it opens up new possibilities for vaccine design.

Controlling the Spread of Disease to Select for Milder Forms

The final form of disease control is a direct application of the trade-off hypothesis. Diseases that spread rapidly and easily, and don't require the infected person to be up and about, evolve to be severe and lethal. Diseases that require personal contact of some form to spread are often less life-threatening. If the trade-off hypothesis is valid, then reducing a disease's transmission rate has two important consequences. In the short term, it protects people from becoming infected. But in the long term, reducing the spread selects for milder strains of the disease.

Let's consider how this might play out with AIDS. In a population of intravenous drug users who regularly share needles or a population in which sexual encounters are frequent and partners change frequently, HIV spreads quickly. There is less of an advantage to HIV to keep the host alive for long periods of time. And there is more of an advantage for fast-reproducing strains of HIV that exploit the host severely, increasing their population sizes and therefore their chance of being transmitted to the next host. Under these conditions, the risk is not simply that you'll catch HIV, but that you will catch a particularly virulent strain.

The logic works the other way as well. Behavioral and social practices that make it more difficult for HIV to spread also make it more difficult for virulent strains of HIV to persist. As a personal health choice, it clearly makes sense to practice safe sex and to limit the number of partners you have. An added benefit is that if these behaviors become widespread, we are also changing the evolutionary conditions under which

HIV evolves. We select for milder forms of the disease. There is no way for HIV to develop "resistance" to this approach. And it works for all STDs, not just HIV.

Some scientists make an analogy between evolutionary control of disease and the domestication of animals and plants. Domestication involves taking wild, and possibly dangerous or poisonous, species and selecting them so they evolve into forms that are useful to us. Since we seem to have difficulty driving diseases extinct (out of 1400 infectious diseases, only smallpox has been exterminated), perhaps the idea of "taming" diseases by selecting for milder forms makes more sense.

 # Biology in Perspective

We are never alone; our bodies are complex ecosystems. Living on or in us are bacteria, fungi, mites, and worms. Most of the time, our coexistence is peaceful—a balance is maintained so that neither we nor our co-inhabitants are harmed. But sometimes a pathogenic organism can overtake the benign microbes or our bodies can malfunction, such that a previously harmless microbe becomes a pathogen. Such events are not rare.

In fact, one-quarter of all human deaths are caused by infectious disease. This figure was much higher in the past. Through better sanitation and medical care, we have made great strides in reducing deaths due to infection. But our relationship with pathogens is a dynamic one because pathogens evolve through natural selection to overcome our attempts at disease control.

Many pathogens have evolved resistance to antibiotics, and some are starting to get around our vaccines as well. New diseases emerge continuously for which we have no natural immunity, antibiotics, or vaccines. As earlier treatments and approaches to disease control become less effective, a better understanding of the evolutionary dynamics of disease may provide new approaches to public health. The field of evolutionary medicine, while still in its early stages, may provide the next big advance in health care.

Chapter Summary

10.1 In What Ways Is Your Body an Ecosystem?

Recognize and provide examples of the different types of species interactions.

- Species that share the same space can compete for resources or interact to the benefit of one or both species through either commensalism or mutualism.
- Competition from nonpathogens can prevent pathogens from infecting us by taking up space and resources so that the pathogens do not have access.

10.2 Why Do Diseases Evolve Resistance to Antibiotics?

Describe the process by which some disease organisms increase their resistance to antibiotics.

- Resistance to antibiotics evolves because of variation.
- Resistant bacteria have mutations that produce proteins that take on different functions or perform old functions in new ways.

- In addition to mutations, some bacteria can acquire resistance genes from their neighbors through horizontal gene transfer.
- Resistance comes at a cost because resistant strains are less efficient.
- Overuse of antibiotics may be the cause of increased resistance to antibiotics.

10.3 Why Are Some Diseases More Deadly Than Others?

Use the trade-off hypothesis to explain and predict the spread and impact of a disease.

- The trade-off hypothesis states that there is a balance between virulence of an organism and its transmission.
- Severe diseases will be those that spread to people most easily.
- Pathogens with lower transmission rates usually have lower virulence.
- Sexually transmitted diseases will not spread if they interfere with the host's reproduction.

10.4 Where Do New Diseases Come From?

Consider the processes of mutation and natural selection and predict likely sources of new disease.

- New diseases come from three sources:
  - o Species that normally live on the body can evolve to become virulent.
  - o Disease can be caught from another species.
  - o Species that normally live in the soil or water can invade the body.
- Most diseases come from other species.
- The stages of a new disease are exposure, infection, transmission, and epidemic.
- HIV/AIDS, an example of a new disease, may have initially occurred from exposure to the blood of chimps in the wild. HIV originated in chimpanzees (HIV-1) and a monkey called the sooty mangabey (HIV-2). Crowding, malnourishment, and physical stress may have enabled initial HIV/AIDS infection with additional transference through reuse of contaminated needles.

10.5 How Can Evolution Help Us Control Disease?

Propose a strategy to reduce the virulence of a disease using evolutionary medicine.

- We should strive to reduce our use of antibiotics when we can.
- We should use vaccines to attack the disease-causing behavior rather than the disease to eradicate quickly mutating pathogens.
- We should change human behavior to reduce the rate of transmission.

Key Terms

antibiotic 290
antibody 303
commensalism 288
conjugation 293
ecology 286
ectopic pregnancy 300
epidemic 302

horizontal gene transfer 293
mutualistic 288
necrosis 289
pathogen 284
quorum 289
resistance 284

resistance gene 292
transduction 293
transformation 293
transmission 297
vaccination 305
virulence 295

Review Questions

1. What is the Human Microbiome Project?

   a. An entity started by the National Institutes of Health to figure out the sources of disease

   b. A study that tracked the process by which a disease becomes an epidemic

   c. An entity started by the National Institutes of Health to figure out all of the species that live in and on us

   d. A study that confirmed there are over 700 species that live in the human mouth

   e. A study that confirmed there are over 800 species that live in the colon

2. What is commensalism?

   a. An ecological interaction in which one species benefits and the other is neither helped nor harmed

   b. An ecological interaction in which neither species benefits

   c. An ecological interaction in which both species benefit

   d. An ecological interaction in which one species benefits and the other is harmed

   e. An ecological interaction in which a parasite harms a host

3. If most of the species that live on the human body can't be grown in a laboratory, how do we know how many species live there?

4. Briefly describe why natural selection results in bacteria that are resistant to antibiotics.

5. Why does the number of resistant strains of bacteria often decrease when antibiotics are not present?

6. Briefly explain the reasoning that suggests diseases should evolve to become milder and less of a threat to their hosts.

7. Briefly outline the trade-off hypothesis. According to this hypothesis, what are the conditions that select for severe or life-threatening diseases? What conditions select for milder diseases?

8. How many new diseases have been reported since 1975?

9. What are the four stages a newly emerging disease goes through?
   a. Exposure, passage, transduction, epidemic
   b. Exposure, commensalism, transmission, epidemic
   c. Exposure, infection, commensalism, epidemic
   d. Exposure, infection, transmission, commensalism
   e. Exposure, infection, transmission, epidemic

10. How does a new disease evolve as it progresses through these four stages?

11. What are the three sources of new disease?
   a. Mutated forms of existing disease, infection, other species
   b. Mutated forms of existing disease, other species, the environment
   c. Infection, other species, the environment
   d. Mutated forms of existing disease, transmission, the environment
   e. Infection, transmission, the environment

12. Which of these sources of new disease supplies the most new diseases?

13. How does a vaccine protect you from a disease?

14. The vaccine for diphtheria has a different "target" than most vaccines. What is the target of the diphtheria vaccine? Why does this difference help to prevent the diphtheria bacteria from evolving resistance?

15. What are the two beneficial, health-related consequences of reducing the rate at which a disease spreads?

The Thinking Citizen

1. In this chapter, you learned that viruses, bacteria, and insects are capable of evolving resistance rapidly to antibiotics or insecticides. Do humans evolve resistance to disease? Discuss some of the evidence that suggests we do. If we do evolve resistance, we evolve it much more slowly than do insects and microorganisms. Why might this be?

2. Go online and look up some facts about several diseases. In particular, find out how they are spread and how life-threatening they are. What sort of support did you find for the trade-off hypothesis described in this chapter? Do the more severe diseases tend to spread by insects or contaminated water, for example?

3. We humans are very good at driving species extinct, even when that isn't what we are trying to do. However, we seem to be very poor at driving diseases extinct, even though we devote considerable effort to it. Of 1400 known infectious diseases, we have exterminated only one: smallpox. List and briefly discuss several reasons why it is difficult to drive a disease extinct.

4. There are a number of diseases, such as rabies and Lyme disease, that infect humans and cause illness but which apparently cannot be spread from one human to another. Is it possible that humans are the main host for such diseases, or are we infected more "by accident"? In terms of emerging diseases, what stage are such diseases at? What are the chances that diseases like these will reach a higher stage?

5. The European conquest of North and South America was aided considerably by disease. Europeans brought with them diseases—such as smallpox, typhus, influenza, diphtheria, and measles—that killed many of the native people of the New World. Why were these diseases common in Europe but absent in the New World before the Europeans came? HINT: consider the differences in social structure of people in both regions.

The Informed Citizen

1. Suppose you are in charge of a poor and underdeveloped state in an African country. A severe outbreak of malaria occurs, threatening many people, especially children. Would you spray DDT to reduce the mosquito population? If so, would you spray as an emergency procedure or as part of a continuing program of mosquito control? What sorts of medical, economic, and ethical considerations would you take into account in making your decision?

2. What should we do about antibiotic use that is not health-related? A lot of antibiotics are used to keep crops from spoiling and to increase the growth of livestock. These practices have important economic benefits to farmers and to consumers who buy their products. But they increase the levels of antibiotics in our environment and lead to resistance. What do you think should be done about using antibiotics this way?

3. Methicillin-resistant *Staphylococcus aureus* (MRSA) is a strain of staph bacteria that is resistant to a number of antibiotics. MRSA is found primarily in hospitals and is becoming a particularly difficult infection to treat. Norway has one of the lowest rates of all infections of any country, and in particular has the fewest people infected with MRSA. Doctors in Norway prescribe fewer antibiotics than in any other developed country. How are these two facts related? Do some research on antibiotic use in Norway and find out the positive and negative effects of reduced usage. Do you think a similar policy of antibiotic usage would work in the United States? Why or why not?

CHAPTER LEARNING OBJECTIVES

After reading this chapter, you should be able to answer the following questions:

11.1 What Is Homeostasis?
Explain why organisms maintain homeostasis.

11.2 How Do Homeostatic Systems Work?
Identify the three components of a homeostatic system and explain the role of negative feedback.

11.3 How Does Your Body Sense Temperature?
Contrast the role of core and skin temperature sensors.

11.4 How Does Your Body Adjust Temperature?
Explain how the body uses effectors to change body temperature.

11.5 How Does Your Body Maintain a Constant Temperature?
Assess how sensors in the hypothalamus, fever, and behavior help your body maintain a constant temperature.

11.6 How Does Your Body Regulate Its Fluids?
Describe how the body uses osmosis and diffusion to regulate its fluids.

11.7 What Does the Kidney Do?
Outline the steps involved in making urine.

11.8 How Does the Kidney Maintain Water Balance?
Predict the impact of drugs and the environment on water homeostasis.

Our bodies work best when internal conditions, including body temperature, the amount of water in our tissues and organs, and dissolved chemicals in the blood, are kept in balance.

CHAPTER 11

Homeostasis

Why Is It Important That the Body Maintain Its Internal Balance?

Becoming an Informed Citizen …

All of the body's systems depend on an internal balance, and the chemical reactions in our bodies are fine-tuned to work most efficiently under specific conditions. The brain keeps internal conditions at normal levels most of the time, but when the balance shifts, things can go very wrong.

ur bodies work best when internal conditions are kept at their normal levels. Body temperature, the amount of water in our tissues and organs, the amount of dissolved chemicals in the blood—the brain (principally) keeps all of these settings and many others within a very narrow range. And the chemical reactions in our bodies are fine-tuned to work most efficiently under these conditions.

In this and the next few chapters, we will look at the body's different systems and how they function. All of the body's systems—including the circulatory and respiratory systems, the nervous system, and the immune system—depend on an internal balance. When the balance shifts, things can go very wrong.

This chapter examines the body's control systems. How exactly does the body control its internal conditions? And what happens if conditions get out of control? For Max Gilpin, an athlete on a hot day who was already pushing his body as hard as he could, losing control proved fatal.

case study

Max Gilpin

For most high school students, the last weeks of summer are a lighthearted, bittersweet time. But for students who play varsity football or other fall sports, it's time to put on the pads and get back into shape. No matter how much you worked out over the summer, this is a grueling process.

Max Gilpin was a sophomore on the Pleasant Ridge Park football team in Louisville, Kentucky. On August 20, 2008, he and his teammates were on the practice field. It was hot and humid, with a heat index of 94°F. Coach David Jason Stinson ran a tough practice, which included "gassers": sprints designed to develop stamina. According to a *USA Today* report, Stinson told his team that they would run until someone quit the team, and eventually someone did. Before that, though, Gilpin collapsed on the field.

As an offensive lineman, Gilpin was big: about 200 pounds, though only 15 years old. Making a 200-pound body run sprints takes a lot of effort, and as muscles do the work, they generate a great deal of heat. The body has a number of ways to get rid of excess heat, including its main method: sweating. But sweat involves water loss, which creates its own problems, and if you lose too much water, you lose the ability to sweat—and to cool off. Most of the time, heat flows naturally out of the body, but when it's hot and humid, it is much harder for the body to lose excess heat. Under conditions of heat, humidity, and exertion, the body can overheat and dehydrate—often with grave consequences.

After his collapse, Gilpin was taken to the hospital. His core body temperature registered 107°F. After three days in the intensive care unit, Gilpin died from heat-related organ failure, or heat stroke.

Following Gilpin's death, Coach Stinson was charged with reckless homicide. Although other high school football players had died as a result of heat stroke, Stinson was the first coach to be charged in a player's death (**FIGURE 11.1**).

The trial lasted three weeks. Players testified that the sprints were unusually severe. Several players who vomited were denied water, although there were regular water breaks during practice. No autopsy was performed on

FIGURE 11.1 A prosecutor at David Jason Stinson's trial shows quotes from other players about the rough conditions on the day that Max Gilpin collapsed at football practice. Stinson was the first, although not the last, coach to be charged in a player's death.

Gilpin, so there was no evidence that he was dehydrated—a fairly sure sign of heat stroke. The defense argued that Gilpin's death was not caused by the practice itself, but by Adderall, a drug that Gilpin was taking for his attention deficit hyperactivity disorder (ADHD). In the end, the jury found Stinson not guilty, and he was acquitted of all charges.

Our bodies are designed to work within a specific range of temperatures, with just the right amount of water. How does our body maintain conditions within this range? Why is this regulation beneficial? We will address these and other related questions throughout this chapter.

●●●

11.1 What Is Homeostasis?

Humans have a remarkably constant body temperature. People vary somewhat, but we maintain an average temperature of 98.6°F. This is a familiar number, but in science it is more common to use the metric Celsius scale. On this scale, body temperature is 37°C (we'll use the "C" whenever we refer to temperature using the Celsius scale). As long as we are in good health, our body temperature varies by less than 1° from the average—even though daily outdoor temperatures may vary by 50°F (10°C) and yearly temperatures by 100°F (38°C) or more.

Our constant body temperature is an example of **homeostasis**, the ability of an organism to maintain constant internal conditions in the face of fluctuating external conditions. Besides temperature, our bodies show many examples of homeostasis (**FIGURE 11.2**):

homeostasis :: the ability of an organism to maintain constant internal conditions in the face of fluctuating external conditions

- The amount of water in the blood
- The concentration of salts and waste products in the blood
- The concentration of oxygen in the blood
- The concentration of glucose in the blood
- The (resting) heart and breathing rates
- Blood pressure

All of these values are kept within very specific ranges in the body. If something happens that changes one of these values, the body has mechanisms

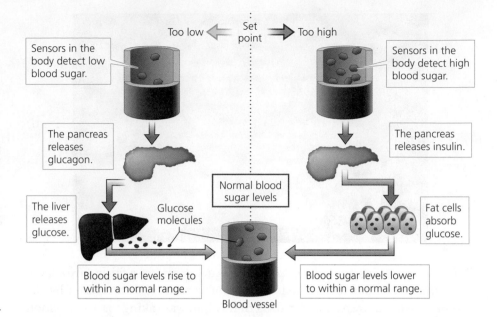

FIGURE 11.2 Homeostasis is the ability of an organism to maintain constant internal conditions in the face of fluctuating external conditions. For example, in a healthy person, the body keeps the amount of glucose in the blood within specific ranges.

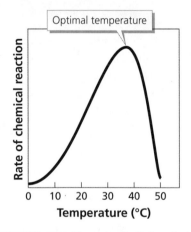

FIGURE 11.3 The rate of all chemical reactions in the human body depends on temperature. The human body's metabolic processes require enzymes, and most enzymes function most efficiently at 37°C.

that bring the value back within the range. For example, if you become dehydrated, the amount of water in your blood falls below the normal level. Your body responds by making you thirsty and encouraging you to drink. If you drink a lot of fluid, your body responds with more frequent urination, again returning the fluid in your blood to normal levels. Homeostasis involves not only having constant internal conditions, but also having control mechanisms to keep conditions constant.

Every metabolic process in your body works fastest or most efficiently under a specific set of conditions. If your body can maintain these conditions with little variation, then your metabolic processes are always working at their peak levels.

Temperature is a good example. The rate of all chemical reactions depends on temperature (**FIGURE 11.3**). Most enzymes in the human body work most efficiently at 37°C, and since our body temperature doesn't vary, our enzymes are always working at their best.

Mammals and birds share the ability to maintain a constant body temperature, but for most other organisms, body temperature varies along with changes in the outside temperature. Snakes, lizards, frogs, and insects, for example, cool down at night. Their metabolic reactions don't function best at these cooler temperatures, and as a result they are sluggish in the mornings until the sun warms them up.

Illness or extreme conditions can overwhelm the body's homeostasis. People with diabetes lose their ability to regulate blood glucose levels, for instance. And, as we saw in the case of Max Gilpin, intense exertion can cause heat to build up faster than the body can get rid of it, which can prove fatal. Our homeostatic mechanisms are good, but they do have limits.

11.2 How Do Homeostatic Systems Work?

The control principles that make a homeostatic system work are the same, whether the system is part of a living organism or designed by a human

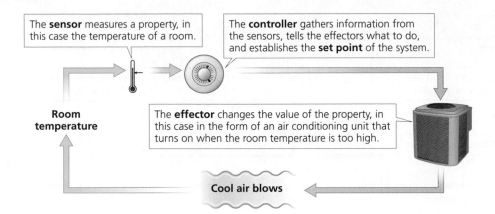

The **sensor** measures a property, in this case the temperature of a room.

The **controller** gathers information from the sensors, tells the effectors what to do, and establishes the **set point** of the system.

Room temperature

The **effector** changes the value of the property, in this case in the form of an air conditioning unit that turns on when the room temperature is too high.

Cool air blows

FIGURE 11.4 The control principles that make a homeostatic system work are the same, whether the system is part of a living organism or designed by a human engineer. Homeostatic systems include a sensor, effector, and controller.

engineer. This means that we can learn about homeostatic principles in simple, general terms. Then we can apply these principles to biological systems, even though the details of homeostasis in living systems are often more complex.

Homeostatic Systems Have Three Parts

The first part of the homeostatic system is called the **sensor** (**FIGURE 11.4**). It measures the property being regulated, such as temperature or water level in the blood. The sensor is necessary because to maintain, for example, temperature at a particular level, you need to be able to measure the current temperature.

The second part of the homeostatic system is called the **effector**, which has the ability to change the value of the system. Typically there are two types of effectors: the *positive effector* increases the value of the system, while the *negative effector* decreases it. In the homeostatic control of temperature in a building, for example, one effector raises temperature (the furnace), while another lowers it (the air conditioning). For water balance, as we noted earlier, one effector controls input of water (drinking) and the other controls the output (urination).

The third part of the homeostatic system is called the **controller**. The controller registers information from the sensors and sends signals to the appropriate effectors to tell them what to do. The controller also establishes what the desired, or normal, value of the system should be. This value is called the **set point**. For a human, the temperature set point is 37°C (98.6°F).

Homeostasis Is Achieved Through Negative Feedback

The simplest way to keep a system at a set point is this: if the system is above the set point, make it lower; if it is below the set point, make it greater. This process is known as **negative feedback** because whichever way the system deviates (higher or lower) from the set point, you apply force in the negative (opposite) direction to bring it back in line.

We can describe negative feedback by showing what each of the three parts of the homeostatic system do (**FIGURE 11.5**):

1. The controller reads the signal from the sensor.
2. Based upon the reading, the controller does one of three things:
    a. If the signal is above the set point, it activates the negative effector.
    b. If the signal is at the set point, it does nothing.
    c. If the signal is below the set point, it activates the positive effector.

sensor :: the first part of the homeostatic system; it measures the property being regulated, such as temperature or water level in the blood

effector :: the second part of the homeostatic system; it has the ability to change the value of the system, typically by either increasing or decreasing it

controller :: the third part of the homeostatic system; it registers information from the sensors and sends signals to the appropriate effectors

set point :: the desired, or normal, value of a homeostatic system; for a human, the temperature set point is 37°C

negative feedback :: the process of keeping a system at the set point; whichever way the system deviates, a force is applied in the opposite direction to bring it back in line

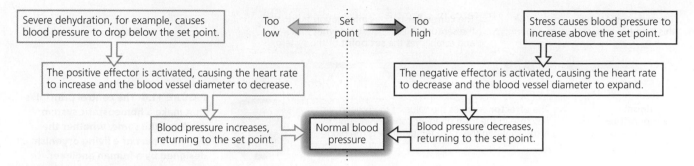

FIGURE 11.5 Negative feedback is used to keep a system at a set point. For example, if blood pressure becomes too high or too low, effects cause it to decrease or increase, respectively.

This process is repeated continuously. By selectively activating the positive or negative effector, the controller keeps the system at the set point and brings it back if it deviates.

11.3 How Does Your Body Sense Temperature?

In Sections 11.3, 11.4, and 11.5, we discuss how the body senses, adjusts, and maintains temperature. We begin by examining the sensors—the parts of the body that measure temperature.

Core Temperature Sensors

hypothalamus :: a cone-shaped region at the base of the brain that is in charge of homeostasis for many of the body's systems

The **hypothalamus** is a cone-shaped region at the base of the brain, just above the brain stem (**FIGURE 11.6**). The narrow end of the hypothalamus connects with the pituitary gland, which secretes many types of hormones, the chemicals that regulate the organs of the body (Chapters 3 and 6). The hypothalamus is in charge of homeostasis for many of the body's systems.

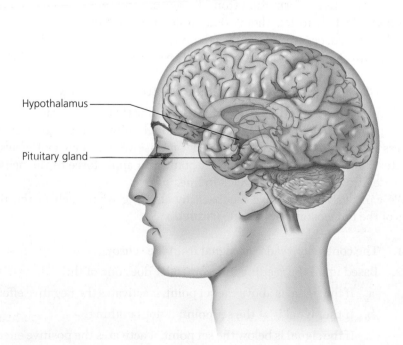

FIGURE 11.6 The hypothalamus is in charge of homeostasis for many of the body's systems. The hypothalamus connects with the pituitary gland, which secretes hormones that regulate organ functions.

We mentioned earlier that the rates of chemical reactions depend on temperature. Generally, chemical reactions speed up at higher temperatures and slow down at lower temperatures. This change in reaction rate is the key to measuring body temperature.

The temperature of the blood deep inside the brain is a good indicator of body temperature. If core body temperature increases or decreases, the temperature of the blood follows along. Inside the hypothalamus are specialized nerve cells that are bathed in blood. If blood temperature lowers, it lowers the temperature of these nerve cells. The chemical reactions in these nerve cells slow down, and they fire at a slower rate. If the temperature of the blood increases, the nerves warm up and fire more rapidly. These nerves are the core temperature sensors, and they signal body temperature by how fast they fire.

Skin Temperature Sensors

Your skin is what puts you in contact with the outside environment. Not surprisingly, there are specialized nerve cells throughout your skin that detect many environmental conditions, including temperature (**FIGURE 11.7**). We do not yet understand how the temperature sensors in your skin work. We do know they work differently from the temperature sensors in your hypothalamus. We also know that you have different sets of nerves to detect cold and warmth. When these nerves fire, they send information about the external temperature to the temperature controller: your brain.

The temperature sensors in your skin provide advance warning about possible changes in body temperature. When you walk outside on a cold (or hot) day, your body doesn't cool down or heat up right away. But this advance information allows your body to make adjustments *before* changes in body temperature occur. These adjustments help prevent changes in body temperature, and prevention is generally easier than allowing temperature change and then correcting it.

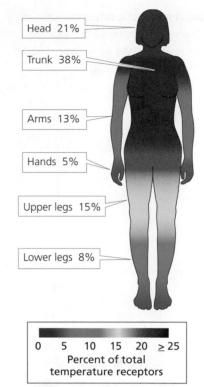

FIGURE 11.7 The skin contains specialized temperature sensors all over the body, although not evenly distributed. They provide advance warning about possible changes in body temperature.

How Does Your Body Adjust Temperature?

Your body acts as a furnace, constantly generating heat. It is fairly easy to crank up the furnace, through physical activity, for example. Turning the furnace down to get rid of excess heat is usually a much bigger problem. This is one of the reasons that even a small increase in body temperature—less than 10°F or 4°C (in Max Gilpin's case)—can be dangerous or fatal.

The body uses several effectors to change temperature. Some work directly to affect how much heat is produced. Other effectors work indirectly, either by conserving heat within the body or by allowing it to escape—rather like opening or closing windows.

The Metabolic Furnace

Your body generates heat through chemical reactions. All of your 10 trillion cells continuously engage in chemistry. Many reactions give off only a tiny bit of heat. But when you add up all the reactions in all the cells, the body is capable of generating a lot of heat. These chemical reactions collectively are known as **metabolism**, and metabolism is measured by how much heat, in Calories, you use.

metabolism :: all of the chemical reactions in cells that sustain life

basal metabolic rate :: the number of Calories you need to maintain your body's processes when you are resting and fasting

Your **basal metabolic rate** is the number of Calories you need to maintain your body's processes when you are resting and fasting. As your body burns these Calories, it generates enough heat to maintain a normal temperature as well. Your metabolic rate increases above the basal rate whenever your body does work—for example, digesting food, physical activity, even thinking. An increased metabolic rate means that more heat is generated, and your body must get rid of this heat to maintain its constant temperature.

Effectors That Cool the Body

Just as water runs downhill, heat always flows from a warmer object to a cooler one. This is a law of physics. Since our bodies are usually warmer than our environments, we tend to lose heat through all our external surfaces, mainly skin and lungs. The effectors that cool the body take advantage of this natural tendency to help us lose heat more quickly.

The blood circulating in the body performs many functions, but one of them is to transfer heat. Blood warms up in areas of high metabolic activity, like muscles and active organs, and transfers this heat to cooler, less active regions (**FIGURE 11.8**). One of the body's cooling effectors directs warm blood to the skin, where it comes in close contact with the cooler external environment. As cooler air flows around the body, it picks up heat from the warmer blood and carries it away. This type of heat transfer is called **convection**: heat is transferred from a warmer object to a cooler, moving fluid, such as liquid or gas, which surrounds it.

convection :: heat transfer from a warmer object to a cooler, moving fluid, such as liquid or gas, which surrounds it

The amount of heat transferred depends on two things. More heat is transferred if the difference in temperature between the object and fluid is greater or if the fluid is moving faster. This is why cool air cools better than warm air, and why moving air from a fan cools better than slowly moving air. On the downside, it also explains wind chill, why you are colder on a windy day than on a calm day even if the temperature is the same. Finally, some fluids are better than others at convection. Water is very good at convective cooling, as you know if you have ever jumped into an ocean, pond, or pool on a hot day.

evaporation :: heat transfer by vaporization of a liquid, such as sweat, to a gas

A second cooling effector is sweating. Sweating takes advantage of a second type of heat transfer, **evaporation** (Figure 11.8). Sweat contains a variety of dissolved salts and other compounds, but it is mostly water. Heat is needed to evaporate water, and when your sweat evaporates, it takes this heat

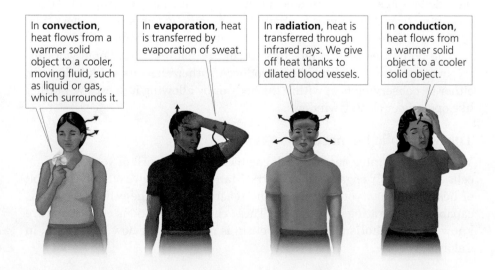

In **convection**, heat flows from a warmer solid object to a cooler, moving fluid, such as liquid or gas, which surrounds it.

In **evaporation**, heat is transferred by evaporation of sweat.

In **radiation**, heat is transferred through infrared rays. We give off heat thanks to dilated blood vessels.

In **conduction**, heat flows from a warmer solid object to a cooler solid object.

FIGURE 11.8 The human body regulates heat transfer and body temperature through convection, evaporation, radiation, and conduction.

from your body. Sweat is produced by sweat glands in your skin. Adults on average have 2–3 million sweat glands distributed over the body. However, sweat glands are concentrated in areas of the skin that have a good blood supply and tend to stay warm. These two cooling effectors can work together: blood is directed to areas of the skin where the sweat glands are located to increase temperature and help sweat to evaporate more quickly.

The lungs are also a source of evaporative heat loss. As water evaporates from the moist surfaces deep inside the lungs, it takes heat away from the body. When you exhale, this heat is lost to the external environment.

Humidity determines how well heat loss through sweating works. When humidity is low, sweat evaporates quickly and cooling is rapid. When humidity is high, however, the air is almost saturated with water vapor and sweat evaporates slowly, if at all. When sweat pours off the body in liquid form without evaporating, it is not nearly as effective at cooling.

Heat stroke occurs when the body heats up faster than the cooling effectors can cool it down. Recall that the rate of cooling from convection depends on the difference in temperature between the body and the air. On a hot day when the air approaches body temperature, the rate of heat loss will be low. If the outside temperature is above body temperature, it actually works in reverse: the body picks up heat from the air. If the day is also humid, the effectiveness of sweating for evaporative heat transfer drops as well. You lose water and salt in sweat, but it doesn't evaporate, and you end up dehydrated but not cooled. Hot, humid days—exactly the type that Max Gilpin trained on—are the most dangerous. He cranked up his metabolic furnace by running "gassers" on a hot, humid day. Gilpin's body generated excess heat but was less effective at getting rid of it. Heat stroke is the result. (See *Life Application: How to Avoid Dehydration* for a discussion of ways to be safe in the heat.)

Effectors That Warm the Body

One warming effector is shivering. Shivering works by increasing your metabolic rate—by turning up the metabolic furnace. Your muscles perform rapid, short contractions, which increases the amount of heat they produce. This heat is used to warm your body.

Life Application

How to Avoid Dehydration

When you exert yourself under hot, humid conditions you sweat and can lose a great deal of water. But in your sweat, you also lose a host of dissolved chemicals: salt, potassium, calcium, magnesium, and even small amounts of metals that are used sparingly throughout your body. You are also burning up Calories, which is why your body overheats. Most of these Calories come from carbohydrates, like sugars, that your body temporarily stores from the food you eat.

We've seen that working out in hot, humid conditions can lead to heat stroke. Luckily, in most cases the consequences are not so severe. But exertion does almost always lead to mild dehydration and a feeling of being run down and exhausted.

Partly this is due to water loss. But the loss of chemicals upsets the osmotic balance of your body fluids, which can stress your cells. Also, burning off many of your stored Calories can leave you feeling tired and interfere with your concentration. Competing athletes can't afford any of these symptoms.

In the fall of 1965, the University of Florida football team was struggling with these problems. The coaches made sure the players had plenty of water, but this didn't revive them— they still wore out quickly from practicing and playing in the hot Florida sun. The coaches contacted four scientists at the university—Dr. Robert Cade, Dr. Dana Shires, Dr. Alejandro De Quesada, and Dr. H. James Free—to help solve their problem.

(Continued)

Life Application

How to Avoid Dehydration (*Continued*)

Working on their own time, these scientists quickly realized that water alone didn't solve the problem because not only water was lost when the players sweated. The scientists developed a liquid that was essentially artificial sweat, with some carbohydrates to replace what the athletes burned off. Nutritionally, it made a lot of sense. Sadly, it tasted lousy. Their solution was to add some lemon juice to mask the bad taste.

Apparently, however, the drink worked wonders. The University of Florida team went on to a 7–4 season in 1965, and in 1966 it went 9–2 and won the Orange Bowl for the first time. Also that year, quarterback Steve Spurrier won the Heisman Trophy. If you know the mascot for the University of Florida, then you can probably guess the name of the drink. The team is the University of Florida Gators, and the drink is Gatorade.

We are not suggesting that a sports drink can win the Orange Bowl. A Heisman quarterback helps as well. But Gatorade became an instant hit. In 1967, Gatorade became the official sports drink of the NFL, and when the Kansas City Chiefs won the Superbowl in 1969, they gave credit to Gatorade. Many an NFL coach has been drenched in Gatorade after winning a championship game. ::

In 1967, Gatorade became the official sports drink of the NFL. Many an NFL coach has been drenched in Gatorade after winning a championship game.

Of course, extra heat won't help if the body doesn't retain it—that would be like turning up the heat and opening the windows. Other warming effectors serve to reduce the amount of heat the body loses. Recall that as long as body temperature is above the outside temperature, some heat will be lost. But warming effectors slow down this heat loss and allow normal body temperature to be maintained except under extremely cold conditions.

Controlling the blood supply can also be a warming effector. If the blood vessels near the skin and your extremities are partially closed down, less blood flows through your skin. This lowers the temperature of the skin, making it closer to that of the outside environment. Reducing the temperature difference reduces convective heat loss. The body's core stays warm, even if the temperature of the outer body decreases.

Goose bumps are also a warming effector, although they no longer work for humans. Goose bumps are caused by the contraction of tiny muscles near your hair follicles (**FIGURE 11.9**). In other mammals, goose bumps cause the hairs to stand erect over the skin. This makes the fur deeper, and it traps air in still pockets by the skin. The still air acts as insulation to reduce the loss of heat through convection. Humans no longer have fur, so goose bumps don't warm the body. They are simply a relic of our evolutionary past.

In addition to convection and evaporation, another form of heat transfer is **radiation** (Figure 11.8). All warm objects give off a sort of "glow," in infrared

radiation :: heat transfer through infrared rays

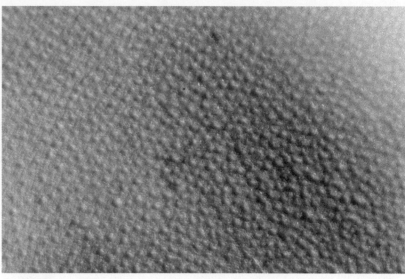

A   B

FIGURE 11.9 Goose bumps are warming effectors caused by the contraction of tiny muscles near hair follicles. In furred animals, goose bumps reduce heat loss by trapping still air in their raised hairs (A), but because humans no longer have fur, goose bumps are simply a relic of our evolutionary past (B).

rays. Infrared rays are like visible light except that their wavelength is too long for our eyes to see. (We can see the glow with special tools like night-vision goggles.) As these infrared rays leave, they carry heat energy away with them, cooling the body.

No matter how hot or humid it is outside, every warm body loses heat through radiation all the time. It is one of the ways bodies cool, although we don't need an effector to make this happen. Insulation is a way to slow radiation heat loss, whether in a house or in a living body. Air trapped in fur is one form of insulation. But the best form of insulation in humans (and mammals in general) is fat. Just below the skin, mammals have a layer of fat that acts as insulation to reduce radiation heat loss. This insulation is one of the key reasons that mammals are able to maintain a constant body temperature. Snakes, lizards, frogs, and fish all produce metabolic heat, but they lack the insulation to maintain it, so their body temperatures vary as the outside temperature changes.

Insulation is not considered a warming effector because it can't be turned on and off in the short term. Over the course of a year, however, the amount of insulation in humans, and many other mammals, changes. Fat tends to build up in preparation for winter. Some of this fat is energy storage, and it is burned during the winter to raise metabolic rate and warm the body. Some of it, however, is also used as insulation to reduce radiation heat loss.

11.5 How Does Your Body Maintain a Constant Temperature?

The controller for temperature regulation is found in the hypothalamus. In fact, the hypothalamus is the controller for almost all the homeostatic systems in the body. *How Do We Know? Hypothalamic Control of Temperature* describes the research done to understand hypothalamus function.

Hypothalamic Control of Temperature

Today we know that the hypothalamus controls body temperature, along with many other homeostatic systems. But how was this discovered? After all, you can't feel your hypothalamus, and you don't sense its actions when your body reacts to hot or cold conditions. Some early hypotheses suggested that the skin was responsible for regulating temperature. But other scientific studies suggested the control center was in the brain. How could you tell which hypothesis was correct?

During the 1960s, Dr. H. T. Hammel and his lab group performed a series of experiments to see how dogs controlled their body temperature. To set up his experiments, Hammel first operated on his dogs to insert thin tubes into their brains that ended at various points around the hypothalamus. The end of each tube in the brain was closed, so nothing could get in or out. The other end of each tube was open and slightly above the surface of the skull. This arrangement gave the researchers a channel to the hypothalamus. The dogs woke after surgery and were allowed to recover, and they showed no ill aftereffects.

To perform his experiments, Hammel placed dogs in a room with a "neutral" thermal environment—neither hot nor cold. Neutral for dogs is a little warmer than for us—a dog's body temperature is around 101°F (38.3°C). The dogs were awake and resting normally. Then Hammel introduced fluid into the tubes leading to the hypothalamus. Hammel found that if the fluid was cool (4°C lower than body temperature), the dog began to shiver and the blood vessels in its skin constricted to reduce blood flow. That is, when the dog's hypothalamus became chilled, the dog's body took physiological steps to warm it up. In fact, the shivering and restricted blood flow worked. These effectors actually caused the dog's body temperature to *increase*—but still it shivered.

Similarly, when warm fluid (4°C above body temperature) was introduced around the hypothalamus, the dog began to pant and the blood vessels in its skin dilated. As in the other experiments, these effectors caused the dog's body temperature to decrease, but the panting continued as if it were hot.

These results showed conclusively that the hypothalamus could sense changes in temperature, and it responded to temperature changes in an appropriate way to cool or heat the body as needed. Also, since the dog was in a room that was neither hot nor cold, there were no temperature signals coming in from the skin. This showed that no information from the skin was needed to make the dog pant, shiver, or adjust its blood flow to the surface.

These results pointed future scientists in the right direction. Armed with this information, they were able to discover many details about how body temperature is controlled. We still don't have a complete story, but each new study brings us closer. ::

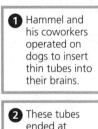 **1** Hammel and his coworkers operated on dogs to insert thin tubes into their brains.

2 These tubes ended at various points around the hypothalamus, with open ends.

3 Hammel placed the dogs into a room with a "neutral" thermal environment.

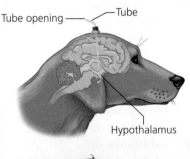 Tube opening — Tube

Hypothalamus

4 When Hammel introduced cool fluid into the tube, the dog began to shiver and the blood vessels in its skin constricted.

5 When Hammel introduced warm fluid, the dog began to pant and the blood vessels in its skin dilated.

The hypothalamus could sense changes in temperature with no information from the skin. Therefore, the control center for temperature must be in the hypothalamus.

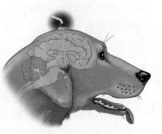

During the 1960s, Dr. H. T. Hammel and his lab group performed a series of experiments to see how mammals control body temperature. Their hypotheses were that skin regulates temperature and that the temperature control center is in the brain.

Sensors and the Hypothalamus

The temperature-control region of the hypothalamus receives constant input from two sets of nerves: the temperature sensors in the hypothalamus itself and the temperature sensors in the skin (**FIGURE 11.10**). These inputs tell the hypothalamus how warm the body is and what the outside temperature is. The hypothalamus also controls the temperature set point: 98.6°F or 37°C in humans.

The hypothalamus then performs the procedure we outlined in Section 11.2. The hypothalamus compares the body temperature to the set point. If body temperature is above the set point, the hypothalamus sends nerve signals to the cooling effectors, which divert blood flow to the skin, and sweating begins. If the body temperature is below the set point, the hypothalamus sends nerve signals to the warming effectors, which reduce blood flow to the skin and induce shivering. If the body temperature is close enough to the set point, the hypothalamus does nothing. This cycle repeats constantly to keep the body at the normal, set point temperature.

Temperature sensors in the skin allow the hypothalamus to be more pro-active in its temperature control. Skin sensors can report temperatures that may cause the body to become too hot or too cold. Based on this information, the hypothalamus can signal the proper effectors to act to prevent a change in body temperature before it occurs.

Fever

When you have an infection, you may also get a fever, a temporary increase in body temperature. One benefit of maintaining a constant body temperature is that all your systems can be fine-tuned to work specifically at this tempera-ture. One disadvantage is that parasites that infect us are also fine-tuned to live at this temperature. Raising body temperature through fever may prove harmful to the parasites, though usually not to us.

When your immune system detects that you have an infection, certain blood cells produce a special chemical signal (**FIGURE 11.11**). This chemical binds to cells in the hypothalamus, causing it to raise the set point. Although the target body temperature is higher, control of body temperature stays the

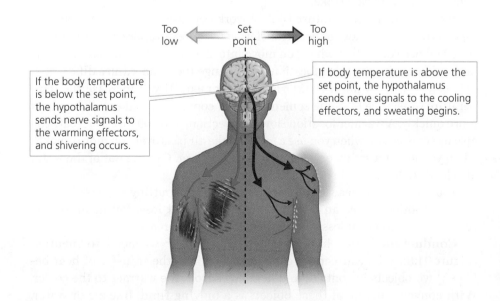

If the body temperature is below the set point, the hypothalamus sends nerve signals to the warming effectors, and shivering occurs.

If body temperature is above the set point, the hypothalamus sends nerve signals to the cooling effectors, and sweating begins.

FIGURE 11.10 The hypothalamus receives constant input from temperature sensors present in both the skin and the hypothalamus itself. It also controls the temperature set point and blood vessel dilation and constriction.

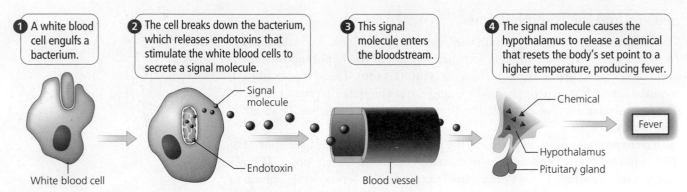

FIGURE 11.11 When your immune system detects that you have an infection, certain blood cells produce a special chemical signal that binds to cells in the hypothalamus, causing it to raise the set point.

same, and the same effectors are used. That's why fevers may be accompanied by shivering, chills, and sweating.

In most cases, fever is beneficial and helps your body to clear an infection. Taking drugs to reduce fever in these cases serves mainly to make you more comfortable. Sometimes a fever can get out of control, and, like heat stroke (and for the same reasons), it can be fatal. Despite the similarities, however, fever and heat stroke are very different processes. Fever is a controlled increase in temperature as a result of an increasing temperature set point. In heat stroke, the set point does not change, but heat builds up in the body in an uncontrolled way faster than the body can get rid of it.

Conscious Behavior

These physiological, homeostatic procedures work on their own, with no conscious effort on your part. But clearly you are aware when you are hot or cold. In addition to physiology, your conscious behavior helps to control body temperature. When you are cold, you can exert yourself to warm up, and if you are overheated, you can lie down and relax. Your behavior can work against you, too, if you don't listen to these cues. Despite all the physical symptoms that accompanied their overexertion, Max Gilpin and his teammates continued to work out in the heat and humidity, behavior that led to dehydration and heat stroke.

Your efforts at temperature control work because you again take advantage of the physical laws of heat transfer: convection, evaporation, and radiation. You use convection when you move into the shade on a hot day, or inside on a cold one. By changing location, you change the temperature differential between your body and the external environment. If you increase the temperature difference by going someplace cool, convection makes you lose heat more quickly. A warmer location slows convection, and you lose less heat. You also use convection when you use a fan to make air pass over you more quickly, when you go for a swim, or when you take a shower. Getting out of the water also leads to heat loss through evaporation.

You can control heat loss through radiation by putting on or taking off clothes. Clothes act as an insulation barrier to heat loss. Better insulation reduces radiation heat loss, and poorer insulation increases it.

conduction :: heat transfer from a warmer solid object to a cooler solid object

Conduction is one last form of heat transfer we need to mention (Figure 11.8). Like convection, conduction involves the transfer of heat between two objects in contact, and heat flows from the warmer to the cooler. With convection, one of these objects is a moving fluid, like air or water.

A

B

FIGURE 11.12 Birds and mammals use both physiology and behavior to control and keep their body temperature constant. Most other animals rely exclusively on behaviors to control their body temperatures, such as the turtle basking in the sun to increase body temperature (A) and the frog resting in the shade to keep cool (B).

With conduction, both objects are solid, and neither moves relative to the other. Ice packs and heating pads are good examples of conduction, as is lying on warm sand after a dip in the ocean.

Birds and mammals use both physiology and behavior to control their body temperature and to keep it constant. Most other animals lack the physiological controls and rely exclusively, but still effectively, on behaviors to control their body temperatures (**FIGURE 11.12**). These animals have a much lower metabolic rate, so they do not generate as much heat. They also lack the insulation to retain the heat they produce.

11.6 How Does Your Body Regulate Its Fluids?

Along with temperature control, our bodies have a second critical homeostatic system that maintains sufficient fluid in the body. Dehydration certainly contributed to Max Gilpin's heat stroke. He had trained so hard and sweated so much that his body fluids were depleted. He also did not replenish the fluid loss by drinking an adequate amount of water. Why is an imbalance of fluid intake and loss such a potential problem? Why are we so dependent on having adequate water in our bodies?

Life evolved first in the sea. Cells, and later animals, evolved a physiology based on living in an aquatic habitat. Water was everywhere, and although marine organisms do face some issues with water balance, they don't usually need to worry about dehydration.

This situation changed when organisms moved from the sea onto land. Their interior cellular environment stayed moist, preserving the conditions to which organisms were adapted. Maintaining this moist interior is quite a task on dry land, however. Evaporation is a challenge to terrestrial organisms, and if water loss through evaporation is not controlled, an organism on land will dehydrate. Successful terrestrial organisms evolved adaptations that helped maintain water balance on dry land and in the dry air.

Before we discuss adaptations for maintaining water balance, we must introduce two chemical processes—diffusion and osmosis.

Diffusion

At the cellular level, our internal environment consists mainly of molecules dissolved in water. Dissolved molecules are always in motion. The motion of any particular molecule is random. Taken as a group, molecules will spread out from an area where there are lots to areas where there are fewer, like a drop of ink that spreads out in water. This movement, called **diffusion**, is the tendency of molecules to move "down" a concentration gradient, from high concentration to low (**FIGURE 11.13**).

Cells are surrounded by thin membranes, separating them from each other and from the fluid-filled spaces between cells. Some molecules can pass through a cell's membrane, while others can't. If a molecule can cross a membrane, we say the membrane is **permeable** to that molecule. Most membranes are *selectively* permeable, letting in some molecules while keeping others out.

Permeability depends on properties like the molecule's size and whether it carries an electrical charge. Some molecules, such as gases like oxygen and carbon dioxide, can dissolve in the membrane and cross it easily. Larger molecules and molecules that carry an electrical charge have more difficulty crossing a membrane. Cells with particular needs for such molecules have special proteins on their membranes. Some proteins form pores that allow the molecules to diffuse through. Other proteins bind to the molecules, pulling them into or out of the cell.

Diffusion and permeability interact. Suppose that a cell has a high concentration of molecules on one side of a membrane and a low concentration of molecules on the other. If the membrane is permeable to the molecule, the molecules will diffuse across the membrane. As always, molecules will diffuse down the concentration gradient, from the region of higher concentration to lower. Eventually, the concentration on both sides of the membrane will become equal. If the membrane is not permeable to the molecule, it can't cross the membrane and the concentration difference stays the way it is.

Diffusion in some ways is like water running downhill or heat flowing from a warmer to a cooler object. Diffusion happens spontaneously and does not require any input of energy to make it work.

Sometimes a cell faces the problem of moving molecules inside (or outside) a membrane *against* a concentration gradient, from low concentration to

diffusion :: movement of molecules from a higher concentration to a lower concentration

permeable :: characteristic of a cell membrane through which molecules may pass

Initially present at a high concentration, the milk molecules poured into this cup will spontaneously diffuse through the coffee as they move to areas of lower concentration.

If present on either side of a semi-permeable membrane, molecules will diffuse until the concentrations are equal on both sides of the membrane.

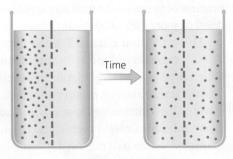

Time

FIGURE 11.13 Molecules diffuse from areas of high concentration to areas of lower concentration.

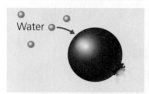

If a red blood cell is placed in distilled water, the water will move into the cell.

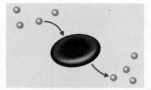

If a red blood cell is placed in a salt solution that mirrors the solution inside the cell, water will move in and out of the cell at equal rates.

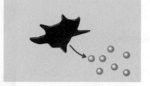

If a red blood cell is placed in a concentrated salt solution, the water will move out of the cell.

Water

The red blood cell will swell and burst.

The red blood cell maintains its normal shape.

The red blood cell will shrink, shrivel, and die.

FIGURE 11.14 Osmosis is the movement of water across a semipermeable membrane from an area of low solute concentration to one of higher concentration.

high. This may happen if the cell has specialized proteins that bind to the molecule to move it across the membrane. Because this process goes against the natural direction of movement, it is known as **active transport**. Active transport requires energy.

Osmosis

Water is by far the most abundant molecule inside and in between cells. It is a small molecule, and it travels easily into and out of cells through specialized pores present in cell membranes.

Osmosis is the movement of water through a semipermeable membrane from an area where dissolved molecules, called solutes, are in low concentration to an area where they are in higher concentration (**FIGURE 11.14**). For example, if the concentration of salt molecules inside of cells is greater than that of the surrounding fluid, water will move into cells. On the other hand, if the inside of the cell is less concentrated than the surrounding fluid, water will flow out of the cell.

At the cellular level, it is critical that cells and the surrounding fluid maintain the proper concentration of dissolved molecules. If the surrounding fluid is too concentrated, water will leave cells, causing them to shrivel up and die. If the fluid inside cells is too concentrated, water will flow in, swelling the cells and eventually causing them to burst. This is one of the reasons your body has a homeostatic system that carefully regulates the amount of water and dissolved molecules in the blood and in the fluid between cells. The organ that does this work is the kidney.

active transport :: the movement of a molecule or ion against the concentration gradient from low concentration to high; requires energy

osmosis :: the movement of water through a semipermeable membrane from an area where solutes are in low concentration to an area where they are in higher concentration

kidney :: a fist-sized organ located near the spine and just above the waist; its main job is to make urine

11.7 What Does the Kidney Do?

The **kidneys** are paired, fist-sized organs located near the spine and just above the waist (**FIGURE 11.15**). Their main job is to make urine. In the process, the kidneys also control water loss and maintain the proper concentration of many dissolved molecules that circulate in the blood. As described in the *Scientist Spotlight: Homer William Smith*, a lot of the basic information about kidneys comes from the research of Homer William Smith.

Urine consists mainly of metabolic wastes dissolved in water. These are not the wastes you make from undigested food that are removed from the body in feces (see Chapter 15). Metabolic wastes are made by the chemical reactions that occur inside your cells. Sometimes the byproducts of these

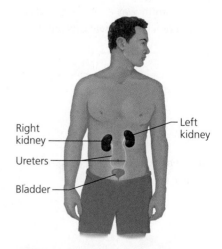

Right kidney

Left kidney

Ureters

Bladder

FIGURE 11.15 Kidneys are paired, fist-sized organs located near the spine and just above the waist. Kidneys make urine, which collects in the renal pelvis and is sent to the bladder.

Scientist Spotlight

Homer William Smith (1895–1962)

Perhaps you've heard the expression "You are what you eat." Homer W. Smith, often referred to as the father of kidney research, had a different take: "In the last analysis, composition of the plasma [blood] is determined not by what the body ingests, but by what the kidneys retain and what they excrete." In Smith's view, you are what you don't get rid of in your urine.

Homer Smith was born in Denver, the youngest of six children. Smith developed an early fondness for science, natural history, experimentation, and tinkering with gadgets. He indulged his interests in a shed his father built for him in the backyard. His mother died when he was seven. As a youngster, Smith stuttered badly and was quiet and withdrawn. He eventually overcame his stutter and became a much-sought-after speaker as a scientist. But his early years shaped him into a fiercely independent, introspective man.

Smith graduated from the University of Denver with a degree in chemistry. During World War I, he enlisted and was assigned to a research unit to study the effects of toxic gases used in the war. This spurred his interest in biology, particularly in physiology. After the war, Smith completed his graduate degrees and worked for several research organizations. He eventually made his professional home at New York University, where he was appointed professor of physiology and director of the physiological laboratories.

Smith turned his attention to the kidney, and for 30 years his work and methods completely defined this field of study. Smith developed noninvasive ways to study the kidney in living organisms. He figured out many of the details of how the kidney filters, reabsorbs, and secretes compounds.

Smith was a staunch advocate of the scientific method and precise observations and measurements. But he realized

Homer W. Smith, often referred to as the father of kidney research, believed that blood composition was determined not by what the body ingests but by what the kidneys retain and excrete.

that a disconnect existed between laboratory research and medical research. To bridge this gap, Smith worked with medical professionals, training them in the scientific method. This collaboration brought both scientific research and medical research to bear on the problem of kidney disease, which yielded many improved treatments and much understanding of kidney health. ::

Source: http://www.jci.org/articles/view/23150.

reactions are toxic and must be removed from the body before they damage cells and organs.

The most abundant and toxic metabolic wastes are those that contain nitrogen. Chemical reactions involving amino acids and proteins generate a lot of nitrogen-containing waste—including ammonia. Your liver converts ammonia into a chemical called **urea,** the main nitrogen-containing waste molecule (much less toxic), and the kidneys remove urea from the blood and put it in urine to eliminate it from the body.

urea :: nitrogen-containing compound found in urine

How does the kidney work? The answer illustrates an important biological concept: the *anatomy*, or structure, of an organ allows it to carry out its particular job, or *function*.

The Anatomy of the Kidney

If you look at a kidney sliced in half, you'll see three sections (**FIGURE 11.16**). The outer edge is the cortex. The inner region is called the medulla; the tissue

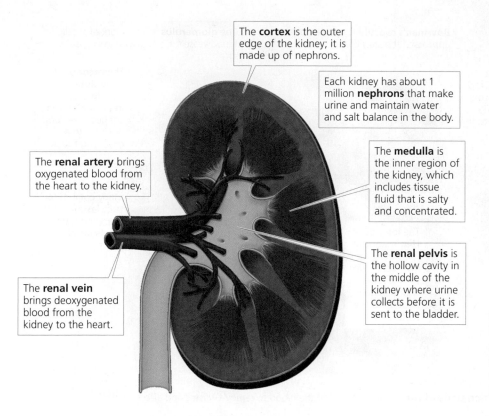

The **cortex** is the outer edge of the kidney; it is made up of nephrons.

Each kidney has about 1 million **nephrons** that make urine and maintain water and salt balance in the body.

The **renal artery** brings oxygenated blood from the heart to the kidney.

The **medulla** is the inner region of the kidney, which includes tissue fluid that is salty and concentrated.

The **renal pelvis** is the hollow cavity in the middle of the kidney where urine collects before it is sent to the bladder.

The **renal vein** brings deoxygenated blood from the kidney to the heart.

FIGURE 11.16 A kidney has three sections: the cortex, the medulla, and the renal pelvis. The cortex and medulla work together to make urine, which collects inside the renal pelvis before being sent to the urinary bladder.

fluid there is very salty and concentrated. In the middle is a hollow cavity called the renal pelvis. The cortex and medulla work together to make urine, which collects inside the renal pelvis before being sent to the urinary bladder.

If you were to look at a thin slice of kidney under a microscope, you'd see that the cortex is composed of long, thin tubes (which are referred to as tubules since they are so small). These tubules are called **nephrons**, and each kidney has about 1 million of them—all together, almost 45 miles of tubes are packed into this fist-sized organ. The nephrons do all the work to make urine and to maintain water and salt balance in the body.

Each nephron begins with an open bulb called Bowman's capsule (**FIGURE 11.17**). Inside Bowman's capsule is a network of small blood vessels, the glomerulus. Bowman's capsule leads to the first convoluted tubule, which travels a short distance through the kidney's cortex. The convoluted tubule connects to a long, U-shaped tube, called the loop of Henle, which travels from the cortex deep into the medulla and back up to the cortex. The loop connects with a second convoluted tubule in the cortex. Finally, the second convoluted tubule connects with the collecting tubule, which again dives down through the medulla. The collecting tubules of many nephrons merge to form collecting ducts, which empty into the renal pelvis. The entire nephron is enmeshed with a network of small blood vessels, a capillary bed.

The Function of the Kidney

The long tubules of the nephron operate like an assembly line. Each region of the nephron has a different job to do and specialized cells to do it. As the nephron winds through the cortex and medulla, conditions outside the nephron change dramatically, enabling each part of the nephron to do its job, with the help of osmosis, diffusion, and active transport. Kidney function can be broken down into three parts: filtration, reabsorption, and secretion.

nephron :: a tubule in the kidney that works to make urine and to maintain water and salt balance in the body; each kidney has about a million such tubes

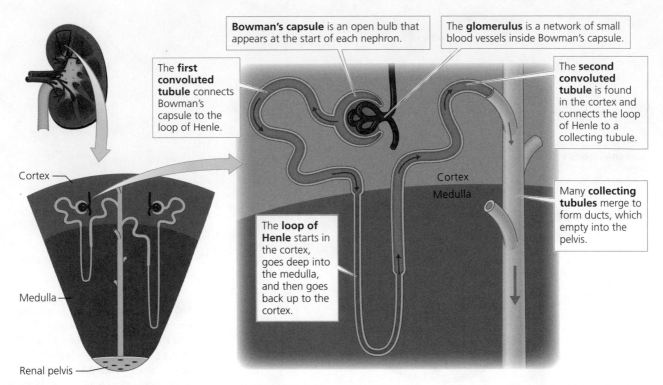

The **first convoluted tubule** connects Bowman's capsule to the loop of Henle.

Bowman's capsule is an open bulb that appears at the start of each nephron.

The **glomerulus** is a network of small blood vessels inside Bowman's capsule.

The **second convoluted tubule** is found in the cortex and connects the loop of Henle to a collecting tubule.

The **loop of Henle** starts in the cortex, goes deep into the medulla, and then goes back up to the cortex.

Many **collecting tubules** merge to form ducts, which empty into the pelvis.

Cortex

Medulla

Renal pelvis

Cortex

Medulla

FIGURE 11.17 Nephrons are the filtration units of the kidney. They filter the blood to remove wastes and also selectively reabsorb beneficial materials and water.

filtration :: process that occurs in Bowman's capsule; solid materials are removed from a liquid

reabsorption :: process in which compounds are removed from the filtrate and returned to the body

secretion :: compounds diffuse out of the blood and are moved by transport proteins into the filtrate

Filtration occurs in the first part of the nephron, Bowman's capsule. Blood is pumped into the small vessels of the glomerulus under moderately high pressure. These vessels have small pores, which act like a filter. High pressure forces blood out of the pores. The pores are too small for blood cells and other large molecules to get through, but about 20% of the fluid is forced out into Bowman's capsule, along with smaller molecules like salts, sugars, urea, and amino acids. There it forms a fluid called the filtrate.

Some of the compounds in the filtrate are needed by the body. **Reabsorption** occurs when these compounds are removed from the filtrate and returned to the body. Cells along the nephron tubules use a combination of diffusion and active transport to move the compounds out of the filtrate into the tissues of the kidney. Some compounds stay in the kidney, but others diffuse into the nearby blood vessels and are retained inside the body.

Secretion is the opposite of reabsorption. Some compounds that the body doesn't need may get past the filter and remain in the blood. These compounds diffuse out of the blood vessels as they pass by the nephron. Transport proteins grab these compounds and move them into the filtrate.

How the Kidney Makes Urine

The kidney makes urine by first making the filtrate (**FIGURE 11.18**). The filtrate contains a lot of water, metabolic waste in the form of urea, and many other small molecules like salts, sugars, and amino acids. Some of these compounds are useful and will be reabsorbed, while others are not and will eventually become part of the urine.

As the filtrate leaves Bowman's capsule, it enters the first convoluted tubule, where a great deal of reabsorption occurs. In fact, most of the filtrate is retained in the body. Almost all the useful compounds—water, salt, potassium, glucose, amino acids, and others—are returned to the bloodstream in the first convoluted tubule.

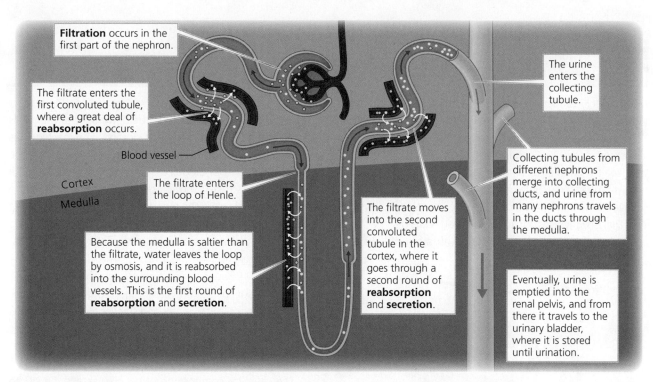

Filtration occurs in the first part of the nephron.

The filtrate enters the first convoluted tubule, where a great deal of **reabsorption** occurs.

Blood vessel

Cortex

Medulla

The filtrate enters the loop of Henle.

Because the medulla is saltier than the filtrate, water leaves the loop by osmosis, and it is reabsorbed into the surrounding blood vessels. This is the first round of **reabsorption** and **secretion**.

The filtrate moves into the second convoluted tubule in the cortex, where it goes through a second round of **reabsorption** and **secretion**.

The urine enters the collecting tubule.

Collecting tubules from different nephrons merge into collecting ducts, and urine from many nephrons travels in the ducts through the medulla.

Eventually, urine is emptied into the renal pelvis, and from there it travels to the urinary bladder, where it is stored until urination.

FIGURE 11.18 In making urine, kidneys perform the functions of filtration, reabsorption, and secretion. In filtration, small molecules are removed from the blood and passed into the nephron, leaving a liquid; in reabsorption, needed compounds are removed from the filtrate and returned to the body; in secretion, some compounds diffuse out of the blood and are moved by transport proteins into the filtrate.

The filtrate leaving the first convoluted tubule is well on its way to becoming urine. Needed compounds and water have been removed, and unneeded compounds like urea either remain or have been added back because they temporarily escaped filtration. The remaining parts of the nephron fine-tune the composition of the urine (mainly the second convoluted tubule) and remove even more water to conserve fluid and to produce a concentrated urine.

As the filtrate leaves the first convoluted tubule, it enters the loop of Henle. The descending part of this loop carries the filtrate deep within the medulla of the kidney. The ascending part of the loop takes the filtrate back up into the cortex. Recall that the medulla is salty. The cells of the descending loop actively transport salt out of the filtrate into the medulla's tissues to keep it this way. Because the medulla is saltier (more concentrated) than the filtrate, water leaves the loop by osmosis, and it is reabsorbed into the surrounding blood vessels. At the bottom of the loop, the filtrate is concentrated. It contains lots of urea and other waste products and relatively little water.

The filtrate moves up the ascending loop of Henle into the second convoluted tubule in the cortex. Here the filtrate goes through a second round of reabsorption and secretion. More water and salt are reabsorbed. Other unneeded compounds that got past the filter, such as any residual urea in the blood, are secreted into the nephron. At this point, the kidney has reclaimed all the compounds needed by the body, and the remaining waste products are dissolved in water to form the urine. As described in *Technology Connection: Dialysis*, a person with poor kidney function may need to undergo **dialysis**, a process in which blood is circulated outside the body to filter out impurities.

As the urine leaves the second convoluted tubule, it enters the collecting tubule. Collecting tubules from different nephrons merge into collecting ducts, and urine from many nephrons travels in the ducts through the medulla. Eventually urine is emptied into the renal pelvis, and from there it travels to the urinary bladder, where it is stored until urination.

dialysis :: process in which blood is circulated outside the body through an artificial kidney to remove impurities

Technology Connection

Dialysis

Many things can damage a kidney, including smoking, drinking, disease, and chemotherapy drugs used to treat cancer. Kidneys have a lot of reserve capacity, so if one fails, the other can compensate. You can live with only one kidney, which is why healthy people can sometimes donate a kidney to a relative who needs one. But you must have at least one kidney to survive. If the kidneys fail, one treatment option is dialysis.

Dialysis involves circulating blood outside the body, about half a liter at a time, through an artificial kidney, or

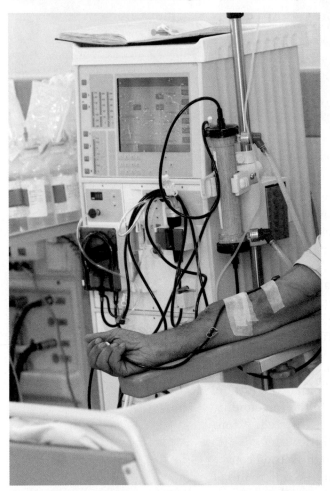

Dialysis involves circulating blood outside the body, about half a liter at a time, through an artificial kidney, or dialysis machine.

dialysis machine. Many of its features are designed to protect the blood as it circulates outside the body. Special chemicals are added to the blood to prevent clotting. The machine has special filters to ensure no air bubbles are introduced into the blood: air bubbles in blood, called embolisms, can cause a heart attack or stroke if they block a critical vessel. There also are valves to make sure the blood is pumped into the body at the correct pressure.

Most of the features of a dialysis machine are designed to mimic the action of the kidneys. The dialysis machine pumps blood through many fine tubes. The walls of the tubes are made of a thin membrane, covered with small pores. These pores allow water and smaller molecules to pass but are too small to allow blood cells and larger molecules out. These tubes are located in a chamber filled with a fluid called the dialysate.

The chemical composition of the dialysate is the key to effective dialysis. A doctor first analyzes the patient's blood composition. Then the dialysate is prepared with exactly the right concentration of all chemicals to cleanse the blood. Essentially, the entire process works by diffusion and osmosis. Waste materials are in low concentration in the dialysate, so they diffuse out of the blood. Needed chemicals are in high concentration, and they diffuse into the blood. The overall concentration of the dialysate is adjusted so water flows into or out of the blood by osmosis, as needed. After the blood flows through the machine, it has been cleansed of waste, any needed chemicals have been added, and the water level has been adjusted, just as in a living kidney.

Dr. Willem Kolff was the first to attempt dialysis, in 1943 during the Nazi occupation of the Netherlands. Materials were hard to come by, and Kolff constructed his machine with the parts he could find: sausage casings, beverage containers, a washing machine, and other creative materials. Kolff worked with 16 patients. All except one died, but the survivor lived another seven years.

Dialysis is much more reliable today, and it has extended many lives. Dialysis is inconvenient, however, requiring three 3- to 5-hour long sessions a week at a dialysis center. ::

You can see that the anatomy of the nephron is what allows it to perform its urine-producing function. The long tubule forms an assembly line. Along this line, specialized cells in different regions act upon the filtrate, taking compounds out and putting compounds in, converting the filtrate to urine. In the loop of Henle, the tubule dives down from the cortex to the salty medulla to take advantage of osmosis, conserving water. The nephron takes a second dive into the medulla with the collecting ducts. This is where the kidney uses homeostatic control to maintain water balance.

How Does the Kidney Maintain Water Balance?

Water balance means balancing water input with water output to keep the body at the correct level of hydration. The kidney is part of a homeostatic system that maintains balance of both water and salt in the body.

How Water Enters and Leaves the Body

Humans obtain water in only two ways. We get some water from the food we eat. Many foods, especially fruits and vegetables, contain water. Some water is also produced from chemical reactions in our cells as we digest food and do other chemical tasks. But most of our water comes from drinking (**FIGURE 11.19**).

Since we live in a dry, terrestrial habitat, we have several ways of losing water. We lose water every time we exhale—about half a liter per day. The water vapor inside our moist lungs is lost to the atmosphere, as you can see when you breathe out on a cold day or blow on glass to fog it. Some water evaporates through our skin.

We lose water in our feces, though usually not too much, perhaps 200 milliliters a day. However, diarrhea can cause significant water loss. Many diseases, such as cholera (Chapter 14), cause diarrhea. Modern medicine and clean water systems have helped to control diarrhea-causing illness in developed countries. But water loss through diarrhea remains a major cause of death in developing countries and is the second-most common cause of infant death worldwide.

We lose water when we sweat. Under extreme conditions, a person may lose up to a liter of water an hour by sweating, though usually not nearly so much. Water loss by sweating is a big risk factor for heat stroke, which is one of the reasons why practices like what Max Gilpin and his teammates experienced can be so dangerous. Sweating is one of the main lines of defense against a rise in body temperature. But if you sweat too much, your body dehydrates. When you run out of water, you can't sweat and your body loses its ability to control temperature. Heat stroke is a likely result unless the conditions that are causing body temperature to rise are stopped. Moreover, exertion-induced heat stroke is a significant cause of kidney failure itself, thus impairing the body's ability to balance fluid even more seriously. Water balance and temperature regulation are actually interdependent. As Max Gilpin lost water, his ability to regulate body temperature was compromised. The loss of body fluid caused his blood pressure to drop, which in turn made his heart work harder, thus generating even more heat. If you must be active during hot, humid

We obtain some water from the foods we eat, such as fruits and vegetables.

Most of our water comes from drinking.

We lose water through sweat, feces, and urine.

FIGURE 11.19 Water balance means regulating input and output to keep the body at the correct level of hydration.

weather, keeping your body hydrated is the most important step you can take to prevent heat stroke.

Finally, the body loses water through urine, about 1.5 liters each day. Other forms of water loss are not really optional. You can't stop breathing, and when you are hot, sweating is necessary to cool your body down. But it turns out the amount of water you lose through urine (as well as the amount of salts and other compounds) can be adjusted up or down to maintain proper levels in the body. The kidney is an important component of this homeostatic system.

Osmotic Sensors

Recall that homeostatic systems have three parts: sensors that detect the state of the system, effectors that change it, and a controller that coordinates the two. As with temperature regulation, blood is used as an indicator of water level in the body. If the body has too much water, the blood becomes diluted. If the body has too little water, the blood becomes too concentrated.

The sensors for this homeostatic system are specialized nerve cells in the hypothalamus. These nerve cells measure the concentration of the blood using osmosis. If the blood is too concentrated, water flows out from the nerve cells, and they shrivel a little bit. This change in size causes the nerve cells to fire, indicating that the blood is too concentrated and more water needs to be retained. If the blood is not too concentrated, or even if it is a little too dilute, the nerve cells don't shrivel and they don't fire. The absence of a signal also provides information.

The Water Balance Effector

antidiuretic hormone (ADH) :: a water balance effector that reduces the volume of urine produced and thereby the amount of water lost in urine

pituitary gland :: a pea-sized structure located in the brain connected to and just below the hypothalamus that produces many of the body's hormones, including antidiuretic hormone

The water balance effector is a small molecule called **antidiuretic hormone (ADH)**. A diuretic is a drug that increases the volume of urine produced. As the name implies, antidiuretic hormone reduces the volume of urine produced, thereby reducing the amount of water lost in urine (**FIGURE 11.20**).

Ant-diuretic hormone (ADH) is produced by the **pituitary gland**, a pea-sized structure located in the brain connected to and just below the hypothalamus. The pituitary gland produces many of the body's hormones, including growth hormone and several hormones related to the production of eggs and sperm.

When the pituitary gland makes ADH, it is released into the bloodstream. Like all hormones, ADH has a specific target. ADH binds to the cells in the collecting tubules of the kidney. Normally the collecting tubules do not allow water to move in or out. However, when ADH binds to these cells, it causes

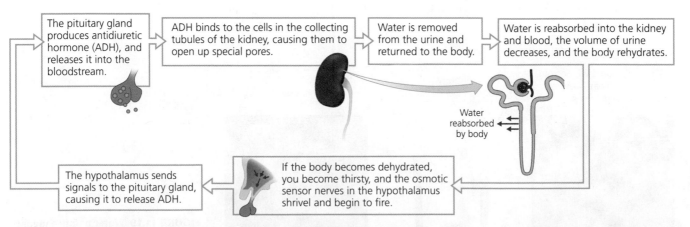

FIGURE 11.20 The hypothalamus controls homeostasis for water balance, as it does for temperature and most of the body's other homeostatic systems.

them to open up special pores in their cell membranes. These pores allow water to move across the wall of the collecting tubules.

Which way will water move? Recall that the collecting ducts travel through the kidney's medulla on their way to the renal pelvis. The tissues of the medulla are concentrated, even more so than the urine. As a result, when ADH opens the pores in the collecting tubules, water is reabsorbed into the kidney and blood, and the volume of urine decreases. Less water is lost.

The Water Balance Controller

The hypothalamus controls homeostasis for water balance, as it does for temperature and most of the body's other homeostatic systems. If the body becomes dehydrated, the osmotic sensor nerves in the hypothalamus shrivel and begin to fire. This signals to the hypothalamus that more water needs to be conserved. The hypothalamus sends signals to the pituitary gland, causing it to release ADH. ADH travels to the kidney's collecting tubules, where it causes water to be removed from urine and returned to the body, as described earlier.

Eventually the water retained by the kidney rehydrates the body. The blood becomes less concentrated and salty, the osmotic sensor nerves swell and stop firing, and the hypothalamus stops signaling the pituitary to release ADH. Water balance is achieved, and the system returns to normal.

This homeostatic system is very effective, and normally the water levels in the body vary by only a little. Under conditions in which you lose more water than usual, as was the case for Max Gilpin and his teammates, the release of ADH may not conserve enough water to maintain the normal level. If the hypothalamus detects that the dehydration is continuing even though ADH is released, it initiates one last response: it makes you thirsty. It's not exactly understood how this happens, but it is clear that thirst comes on only after ADH has failed to provide the correct level of hydration.

Disrupting the Water Balance

Drugs can disrupt the homeostasis of water balance. Alcohol, for example, binds to sites on the kidney's collecting tubules, the same sites where ADH binds (**FIGURE 11.21**). When alcohol binds, it interferes with ADH binding. If ADH is prevented from binding to the tubules, more water remains in the urine. Alcohol defeats the homeostatic control of water balance, and too much water is lost in urine. Alcohol is therefore a diuretic, and the dehydration it causes is one of the factors that contribute to a hangover the next day.

Alcohol is not the only chemical that disrupts water balance. Too much salt in your diet results in the hypothalamus signaling for greater water

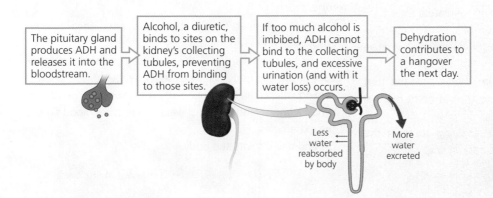

The pituitary gland produces ADH and releases it into the bloodstream.

Alcohol, a diuretic, binds to sites on the kidney's collecting tubules, preventing ADH from binding to those sites.

If too much alcohol is imbibed, ADH cannot bind to the collecting tubules, and excessive urination (and with it water loss) occurs.

Dehydration contributes to a hangover the next day.

Less water reabsorbed by body

More water excreted

FIGURE 11.21 Drugs can disrupt the homeostasis of water balance, as in the case of alcohol.

retention. This can raise blood pressure and lead to circulatory problems (Chapter 12). Many of the choices we make about diet, drugs, and our activities have the power to affect the body's homeostasis (Chapter 15).

 # Biology in Perspective

How does the body maintain its internal balance? The process, called homeostasis, uses the logic of negative feedback, a control center, and a variety of sensors and effectors to maintain body temperature, water balance, and a host of other factors at a constant (or near-constant) value. Homeostatic control uses muscles, the pituitary gland, the circulatory system, kidneys, and many other parts of the body. The hypothalamus is the master control center that oversees the regulation of many homeostatic systems.

If internal conditions are kept within a narrow range, the body's organs, cells, and biochemistry can be fine-tuned to work their best within this range regardless of external conditions. Homeostatic control has its limits, however. Under extreme conditions, it can fail, and body temperature could rise or fall uncontrollably or the body could dehydrate. Loss of homeostatic control can have dire consequences. But under most conditions, homeostasis works remarkably well and keeps us active and alive.

Chapter Summary

11.1 What Is Homeostasis?

Explain why organisms maintain homeostasis.

- Homeostasis is the ability of an organism to maintain constant internal conditions in the face of varying external conditions.
- Mammals and birds maintain constant body temperature, but most other organisms do not.

11.2 How Do Homeostatic Systems Work?

Identify the three components of a homeostatic system and explain the role of negative feedback.

- Homeostatic systems have three parts:
  o Sensors measure the property being regulated.
  o Effectors either increase the value of a system (positive effector) or decrease the value of a system (negative effector).
  o Controllers read information from the sensor and turn on effectors to return a system to the set point (normal value).
- Controllers use negative feedback, applying force in the opposite direction to bring the value back in line.

11.3 How Does Your Body Sense Temperature?

Contrast the role of core and skin temperature sensors.

- Sensors in the hypothalamus and skin control body temperature.
- Nerves in the hypothalamus sense core body temperature and fire faster at higher temperatures.
- Temperature sensors in the skin alert the body to the need to alter behavior and avoid excessive heat or cold before the core body temperature changes.

11.4 How Does Your Body Adjust Temperature?

Explain how the body uses effectors to change body temperature.

- Humans lose heat though their skin and lungs through convection and evaporation.
- If more heat is generated by the body than can be dissipated by cooling off, the body will increase its temperature and fail to cool down; this can happen during exercise on hot, humid days.
- When a body is too cold, it shivers and reduces blood flow to the extremities.
- Fat and hair insulate the body.

11.5 How Does Your Body Maintain a Constant Temperature?

Assess how sensors in the hypothalamus, fever, and behavior help your body maintain a constant temperature.

- The temperature control region of the hypothalamus receives input from nerves in the hypothalamus and the skin.
- The hypothalamus compares the temperature to the set point and reduces blood flow to the skin and increases shivering if the temperature is below the set point, or it induces sweating and increased blood flow to the skin if the temperature is above the set point.
- During a fever, your immune system detects an infection, and certain blood cells produce a special chemical signal that binds to cells in the hypothalamus, raising the body temperature set point.
- We also control body temperature through conscious behavior, such as moving into the shade, adding or taking off clothing, and slowing down.

11.6 How Does Your Body Regulate Its Fluids?

Describe how the body uses osmosis and diffusion to regulate its fluids.

- Diffusion is the tendency of a solute to move along a diffusion gradient, from high concentration to low concentration.
- Cells are surrounded by selectively permeable membranes that allow some molecules to pass through but prevent the movement of other molecules.
- In active transport, a cell uses energy to move a molecule across the cell membrane against a diffusion gradient.
- In osmosis, water moves across a semipermeable membrane from an area of low solute concentration to an area of high solute concentration.

11.7 What Does the Kidney Do?

Outline the steps involved in making urine.

- Kidneys control water loss and maintain the proper concentration of many dissolved molecules in the blood by making urine.

- Cells produce metabolic waste, and nephrons are specialized tubes in the kidneys that help extract the waste products from blood using a combination of active transport and osmosis.
  - o Nephrons begin at Bowman's capsule, where high pressure filters much of the fluid out of the blood.
  - o After Bowman's capsule, useful compounds are reabsorbed into the blood through a combination of osmosis and active transport.
  - o The nephron then travels through the loop of Henle, which goes through a salty portion of the kidney, the medulla. Water leaves the nephron through osmosis, while active transport removes salt. On the return portion of the loop, waste products are secreted through active transport into the nephron.
  - o The concentration of waste products in the nephron travels to the collecting tubules that merge into collection ducts; the collection ducts are used to control homeostasis.

11.8 How Does the Kidney Maintain Water Balance?

Predict the impact of drugs and the environment on water homeostasis.

- The body maintains water homeostasis primarily through urine production.
- The hypothalamus produces the water balance effector antidiuretic hormone (ADH), which reduces the volume of urine produced.
- ADH binds to the collection tubules in the kidneys, reducing the amount of urine produced.
- Usually the release of ADH is enough to maintain homeostasis, but the hypothalamus can also stimulate thirst.
- Some compounds, such as alcohol, attach to the ADH receptors and prevent ADH from binding to the collection tubules, which can lead to dehydration.

Key Terms

active transport 329
antidiuretic hormone (ADH) 336
basal metabolic rate 320
conduction 326
controller 317
convection 320
dialysis 333
diffusion 328
effector 317

evaporation 320
filtration 332
homeostasis 315
hypothalamus 318
kidney 329
metabolism 319
negative feedback 317
nephron 331
osmosis 329

permeable 328
pituitary gland 336
radiation 322
reabsorption 332
secretion 332
sensor 317
set point 317
urea 330

Review Questions

1. What is homeostasis?

2. List the three components of a homeostatic system.

   a. Sensor, effector, hypothalamus

   b. Sensor, controller, hypothalamus

   c. Sensor, effector, controller

   d. Effector, controller, hypothalamus

   e. Effector, controller, convection

3. Briefly define the function of each component of the homeostatic system.

4. What is negative feedback, and how does this process help to keep values (like body temperature) constant or within a small range of values?

5. List two ways your body can tell (sense) temperature.

6. What is basal metabolic rate?

   a. The number of Calories you need to maintain your body's processes when resting and fasting

   b. The number of Calories you need to maintain your body's processes when working out and performing other strenuous activities

   c. The chemical reactions that are collectively capable of generating a lot of heat

   d. The process of transferring heat from a warmer object to a cooler object.

   e. The rate at which your heart pumps to deliver blood throughout the body

7. List one way your body can cool itself, and one way it can warm itself.

8. How is a fever different from an elevated body temperature due to heat stroke?

9. Which of the following is NOT a method of heat transfer?

   a. Convection

   b. Radiation

   c. Evaporation

   d. Conduction

   e. Fever

10. Briefly define the four methods of heat transfer.

11. What part of the brain is most involved with homeostatic control?

    a. Temperature sensors

    b. Nerves

    c. The brain stem

    d. The hypothalamus

    e. The pituitary gland

12. Define diffusion and osmosis. Explain clearly how they differ.

13. List the three main regions of the kidney. List the parts of a nephron.

14. What is urine? Why do our bodies make it?

15. Briefly explain the three functions of the kidney. Explain how each function contributes to making urine.

16. What conditions cause the hypothalamus to release antidiuretic hormone?

17. What effect does antidiuretic hormone have on the kidney? How does this affect the body's overall water balance?

The Thinking Citizen

1. Which group has more nocturnal species: mammals or reptiles? Do some research to get a general idea (not an exact count). What physiological factors might account for the difference?

2. Describe in general terms how a homeostatic system keeps levels close to their set points. Frame your answer in terms of the system components: sensors, effectors, and controllers. Explain the role of negative feedback in maintaining control.

3. You have probably heard that several layers of clothing provide better heat retention than a singly bulky garment. Why is this the case? In your answer, explain whether the reduction in heat loss is due to a reduction in convection, evaporative heat loss, radiation, or conduction (or some combination of these).

4. Imagine a group of soldiers on a march across the desert. Describe how their homeostatic systems for body temperature control and water balance would react. Which sensors would be "firing"? Which effectors would be turned on?

5. Outline why it is usually easier to retain body heat than to lose body heat, and why overheating is generally a more serious problem than underheating.

6. Consider the problems of water balance in two different people. The first person has severe diarrhea. The second has just eaten a bag of salty potato chips, with some pretzels on the side. Although these situations are different, will the stress each person places on their water balance system be similar or different? Describe for each case how the homeostatic system will respond to maintain water balance.

The Informed Citizen

1. A number of sports organizations, including the National Athletic Trainers Association, have published guidelines about preventing heat-related injury in athletes. These guidelines recommend that late-summer practices begin slowly (only one practice a day, and without shoulder pads), with plenty of opportunities for water breaks. However, these guidelines are not mandatory, and coaches are not required to follow them. Many coaches and players believe that rigorous practice, regardless of outside conditions, is necessary to build toughness and team bonding. Discuss the way early practices should be run, from the perspective of a player, a coach, a parent, and a school administrator. In the event of heat-related injury, should the coaches be held responsible? Why or why not?

2. In many sports, being large is an advantage. However, for some sports, like wrestling, there is often an incentive to be in a smaller weight class. One way athletes "make weight" is to take diuretics. Why does this practice reduce weight? What physiological problems are risked by this practice?

CHAPTER LEARNING OBJECTIVES

After reading this chapter, you should be able to answer the following questions:

12.1 **How Much Oxygen Do You Need?**
Explain how oxygen needs vary depending on circumstances.

12.2 **How Is the Circulatory System Structured?**
Describe the role of the different components of the circulatory system.

12.3 **How Is the Respiratory System Structured?**
Describe the role of the different components of the respiratory system.

12.4 **How Do You Breathe?**
Explain how and where oxygen and carbon dioxide diffuse into and out of blood and the role of hemoglobin to facilitate oxygen transport.

12.5 **How Does Blood Pressure Vary Throughout Your Body?**
Describe the differences in blood pressure in different positions in the circulatory system as well as the mechanisms that ensure blood flow.

12.6 **How Are the Respiratory and Circulatory Systems Controlled?**
Assess the effect that changes in the blood and nervous system have on respiration, heart rate, and blood pressure.

12.7 **What Happens When the Circulatory and Respiratory Systems Malfunction?**
Evaluate medical advice concerning the treatment and prevention of atherosclerosis, heart attacks, and asthma.

While most of us can't hold our breath for much more than a couple of minutes, some athletes from the sport of free diving can do so for several minutes. Ultimately, every chemical reaction in our bodies depends on oxygen, and if our hearts or brains lose oxygen, even for a few minutes, the consequences may be fatal.

CHAPTER 12

Circulation and Respiration

What If Your Body Doesn't Get the Oxygen It Needs?

Becoming an Informed Citizen . . .

Under most conditions, your body does a good job of providing the oxygen you need. The respiratory and circulatory systems have evolved anatomical adaptations to ensure sufficient oxygen delivery. As good as they are, however, these systems do have limits, especially when they are stressed.

Ow long can you hold your breath? In November 2012, Stig Severinsen, a native of Denmark, held his for 22 minutes—a world record at the time. Severinsen employed a number of special techniques from the sport of free diving—underwater diving that does not involve the use of scuba gear. Most of us, however, can't hold our breath for much more than a couple of minutes. Why is oxygen so critical?

Oxygen is not stored in any great quantity in your body, but it is continuously used by every cell. Oxygen is the chemical that allows your cells and organs to make use of the energy in the food you eat. Ultimately, every chemical reaction in your body depends on oxygen. Without oxygen, metabolically active cells shut down and begin to die within minutes. If either your heart or your brain loses oxygen even for a few minutes, the consequences are severe and often fatal.

circulatory system :: carries the blood throughout the body

respiratory system :: enables you to breathe (take in oxygen and expel carbon dioxide)

Your **circulatory system** carries blood throughout the body, while your **respiratory system** enables you to breathe. These systems join forces to keep your body supplied with oxygen. Various sensors throughout your body report back to your brain about oxygen levels and requirements, and your brain responds by increasing or decreasing your heart and breathing rates. Normally, the circulatory and respiratory systems provide ample oxygen to support bodily functions, and your brain regulates the delivery. But what happens when they are stressed—for example, by the oxygen-demanding exploits of elite endurance athletes?

case study

Blood Doping at the Tour de France

It was quite a scandal. One day before the 2006 Tour de France was due to start, Jan Ullrich of Germany, Ivan Basso of Italy, and 11 other cyclists were suspended by their teams and prohibited from competing. Prior to the

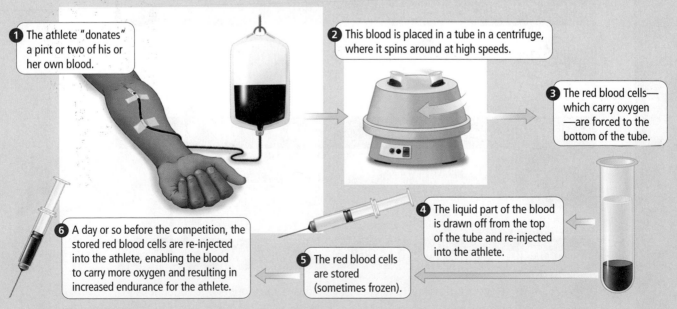

FIGURE 12.1 Blood doping is a potentially risky technique that increases the number of red blood cells circulating in a person's body. Because this banned practice is so prevalent in professional sports, officials routinely test athletes before competition.

suspension, these heavy favorites seemed poised to contend for the trophy. What happened?

A Spanish investigation team linked the suspended cyclists to a blood-doping ring in Madrid. Blood doping (**FIGURE 12.1**) is a prohibited and potentially risky technique that enables your blood to carry more oxygen. It begins like a blood donation. The athlete "donates" a pint or two of his or her own blood. This blood is placed in a tube in a centrifuge, where it spins around at high speeds, like the spin cycle of a washing machine. The red blood cells—which carry oxygen—are forced to the bottom of the tube. The liquid part of the blood is drawn off from the top of the tube and re-injected into the athlete. The blood cells are stored, perhaps frozen. As with any case of blood donation, the athlete makes more red blood cells to replace the ones taken out, and the red blood cell level soon returns to normal. A day or so before the competition, the "doping" occurs. The stored red blood cells are re-injected into the athlete. These additional red blood cells allow the blood to carry more oxygen.

Blood doping is prohibited by all major sports federations, and it is illegal in many countries. But Ullrich and Basso were not the last cyclists or athletes to be caught up in a doping scandal. In fact, in October 2012, Lance Armstrong was stripped of all seven of his Tour de France wins after evidence came to light in a U.S. Anti-Doping Agency report that he had used performance-enhancing drugs and engaged in blood doping (**FIGURE 12.2**). In a 2013 interview with Oprah Winfrey, Lance admitted to blood doping and use of other performance-enhancing substances in all seven of his Tour de France wins. He also said that he did not believe it would have been possible for him to win all seven times if he had not engaged in doping.

These scandals raise several questions. First, why would front-running athletes risk their careers by doping? Second, and more relevant biologically, why would they consider blood doping to be necessary? Are there limits to the amount of oxygen our circulatory and respiratory systems can provide, limits that are tested by the grueling mountain climbs of the Tour de France? Some facts will help us answer these questions.

FIGURE 12.2 After years of denial, Lance Armstrong admitted that he had engaged in blood doping for many years, including during his seven Tour de France victories.

12.1 How Much Oxygen Do You Need?

The answer to this question changes dramatically depending on what you are doing. At rest, a typical adult might require between 0.2 and 0.3 liters of oxygen each minute (1 liter is a little more than a quart). During active exercise, your need for oxygen may increase 10 times, to around 2 to 3 liters of oxygen each minute. World-class athletes like those on the Tour de France may increase their oxygen demand by 30 times their resting rate, requiring between 6 and 9 liters of oxygen a minute.

How much oxygen per minute can your respiratory and circulatory systems deliver? This depends on how much oxygen there is in the air, how much oxygen you breathe in, and how quickly your blood circulates.

Oxygen comprises 21% of the air you breathe. A typical breath takes in about 0.5 liters of air, or approximately 0.1 liters of oxygen. At rest (not physically active), a typical adult breathes about 12 times a minute, so about

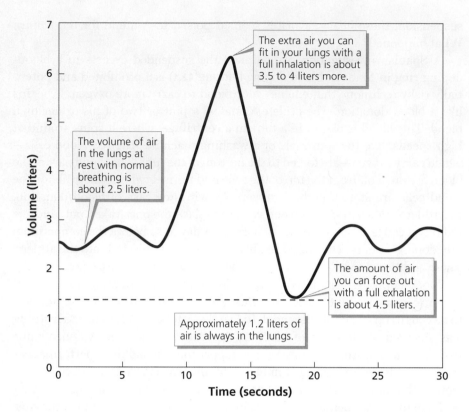

FIGURE 12.3 On average, a breath takes in about 0.5 liters of air, or approximately 0.1 liters of oxygen. At rest, during normal breathing, about 1.2 liters of oxygen enter the lungs of a typical adult each minute; this is six times as much as can be used.

hemoglobin :: an iron-containing compound that binds to oxygen and transports it throughout the body

1.2 liters of oxygen enter your lungs each minute at rest—about six times as much as you use (**FIGURE 12.3**). During strenuous exercise, an adult might breathe 30 times a minute, and the breaths are deeper, perhaps 2 liters each. This provides about 12 liters of oxygen per minute. World class athletes might breathe 60 times a minute and are probably getting at least 24 liters of oxygen into their lungs each minute.

Now, let's consider how quickly your blood circulates. At rest, your heart typically beats about 60 times a minute. During this minute it pumps about 5 liters of blood. But 5 liters is about all the blood you have. This means that at rest, all your blood—and all the oxygen it contains—circulates through your body once every minute on average. During activity, when your heart rate speeds up, your blood and oxygen circulate even more quickly. If all the oxygen in our lungs actually made it into our blood and to our cells, we would get over a liter a minute, and possibly as much as 24 liters a minute. This is much more than we use, and it even gives us a reserve capacity for emergencies.

Unfortunately, not all of this oxygen makes it from the lungs to the bloodstream. One chemical property of oxygen is that it does not dissolve very well in water—or in blood (**FIGURE 12.4**). In fact, a liter of blood (the liquid part) can dissolve only 5 *milliliters* of oxygen. At this rate, all the blood in your body could carry only 25 milliliters, about 5 teaspoons, of oxygen. This is only one-tenth of the oxygen you need, even at rest. Our bodies get around this restraint by using red blood cells to carry oxygen. Red blood cells are packed with **hemoglobin**, an iron-containing compound that latches on to oxygen and transports it throughout the body. With a normal complement of red blood cells, a liter of blood can hold 200 milliliters of oxygen. Your 5 liters of blood circulate quickly enough to deliver a liter of oxygen per minute at rest, much more than your body actually uses.

As your heart and breathing rates speed up during exercise, your blood circulates faster, and more oxygen is delivered to your muscles. But eventually

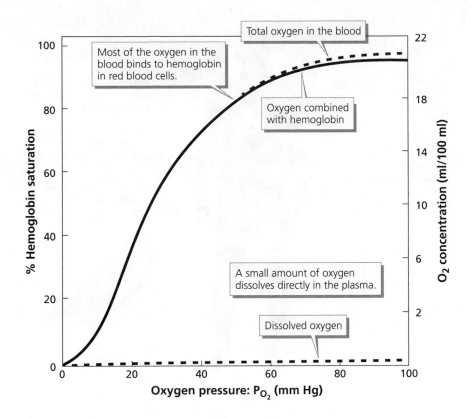

FIGURE 12.4 If all the oxygen in our lungs actually made it into our blood and to our cells, we would take in and use over a liter a minute. Unfortunately, not all of this oxygen makes it from the lungs to the bloodstream because oxygen does not dissolve very well in water or blood.

you reach a limit where you start running out of breath. You don't have enough red blood cells to carry the extra oxygen you need. One of the results of honest physical training is that you increase the number of red blood cells and therefore the amount of hemoglobin in your body. Your body can deliver more oxygen, and you have more "wind." Blood doping simulates this effect. It provides you with more red blood cells without you having to work for them. The advantage to a Tour de France rider is considerable, but so are the risks, as we shall see.

The number of red blood cells is only one of several constraints on our circulatory and respiratory systems. The anatomy and physiology of these systems shows how they have adapted to overcome these constraints and provide us with the oxygen we need.

12.2 How Is the Circulatory System Structured?

The circulatory system is the body's transportation system (**FIGURE 12.5**). It consists of a series of vessels, larger ones that branch and branch again into ever smaller and more numerous vessels, until the vessels become small enough and numerous enough to reach every cell. In some ways, the circulatory system is like our system of interstates, state highways, and smaller roads. No one lives on the interstate (and no cells get oxygen directly from the major vessels), but ultimately every home connects to a road. One major difference is that in the circulatory system, all "roads" are one-way.

Branching structures are a common feature in biological systems. Branching allows structures to get progressively smaller and to reach specific destinations. It also dramatically increases the total surface area of the

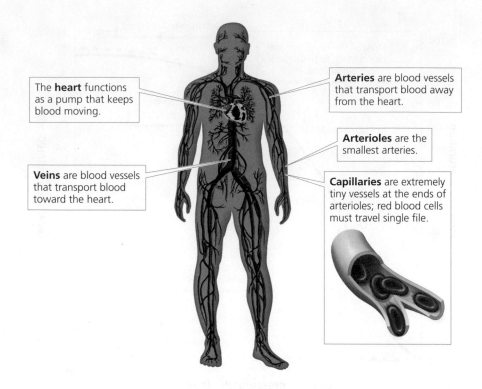

The **heart** functions as a pump that keeps blood moving.

Arteries are blood vessels that transport blood away from the heart.

Arterioles are the smallest arteries.

Veins are blood vessels that transport blood toward the heart.

Capillaries are extremely tiny vessels at the ends of arterioles; red blood cells must travel single file.

FIGURE 12.5 The circulatory system is the body's transportation system. It consists of a series of vessels that branch into smaller vessels, bringing oxygen to cells and removing carbon dioxide.

structures. Surface area refers to how much area is exposed on the surface of an object. Surface area is important in biology because any time a chemical moves from one part of the body to another, it has to move across a surface. In our case, when oxygen moves from the bloodstream to the cells, it must cross the surface of a blood vessel. A larger surface area means more room for the transfer—that is, more oxygen delivered. The branching structure of the circulatory system provides for one of its main functions: the transport and distribution of oxygen and other chemicals.

The "roadways" in the circulatory system consist of arteries, capillary beds, and veins. In addition, the system requires a pump (the heart) and a fluid to circulate (blood).

The Heart

Every minute you are alive, your heart beats, probably somewhere around 60 times (and more than this when you were younger or when you are active). If you live to be 85, your heart is likely to beat more than 2 billion times. Try clenching and unclenching your fist 100 times—then imagine repeating that 20 *million* times, and you'll have some idea of what your heart accomplishes.

Your heart's main function is to act as a pump. It is comprised of two biological pumps lying side by side that enable it to perform this function (**FIGURE 12.6**). Each pump consists of two hollow chambers with muscular walls. The first chamber receives the circulating fluid and acts as a reservoir that fills the second chamber. The second chamber has strong muscular walls that pump the fluid along when they contract.

The right-hand pump of the heart receives "used" blood from your body and pumps it to the lungs by way of the **pulmonary circuit**. The left-hand pump receives fresh blood from the lungs and pumps it to the rest of the body by way of the **systemic circuit**.

pulmonary circuit :: the mechanism by which the right-hand pump of the heart receives used blood from the body and pumps it to the lungs

systemic circuit :: the mechanism by which the left-hand pump of the heart receives fresh blood from the lungs and pumps it to the rest of the body

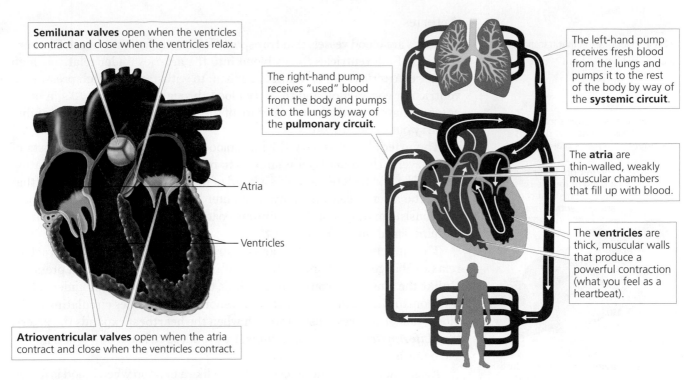

Semilunar valves open when the ventricles contract and close when the ventricles relax.

The right-hand pump receives "used" blood from the body and pumps it to the lungs by way of the **pulmonary circuit**.

The left-hand pump receives fresh blood from the lungs and pumps it to the rest of the body by way of the **systemic circuit**.

Atria

The **atria** are thin-walled, weakly muscular chambers that fill up with blood.

Ventricles

The **ventricles** are thick, muscular walls that produce a powerful contraction (what you feel as a heartbeat).

Atrioventricular valves open when the atria contract and close when the ventricles contract.

FIGURE 12.6 The heart is comprised of two side-by-side biological pumps that supply oxygenated blood to the body and send deoxygenated blood to the lungs. It has valves that ensure the correct direction of blood movement.

Since each pump has two chambers, your heart has a total of four chambers. The input, or reservoir, chambers are called atria (singular **atrium**), and the muscular pumps are called **ventricles**. The atria are thin-walled, weakly muscular chambers that fill up with blood from the body. When the heart beats, the atria contract, and the blood flows into the corresponding ventricle. Shortly after, the ventricles contract, forcing blood out into the body. The ventricle walls are thick and muscular, and they produce a powerful contraction. Ventricular contraction is what you feel as your heart beats, and it is responsible for the pulse you feel throughout your body.

A series of valves in the heart keep blood moving in the right direction. The **atrioventricular valve** lies between the atrium and its corresponding ventricle. When the atrium contracts, this valve opens to allow blood to flow into the ventricle. When the ventricle contracts, this valve closes, preventing blood from backing up into the atrium.

A **semilunar valve** lies between each ventricle and the vessel it uses to send blood from the heart. These semilunar valves are shaped rather like crescent moons. They open when the ventricle contracts to allow blood to leave the heart, and they close when the ventricle relaxes to prevent blood from being sucked back in.

If you've ever tried to start a car on a cold morning, you know that it's hard for the engine to turn over when the oil is cold and viscous (thick). Blood doping has a similar effect on the heart. Because blood doping increases the number of red blood cells in the blood, it makes the blood considerably more viscous and harder to pump. This extra strain on the heart can have a number of dire consequences, including an increased risk of heart attack. It is also possible that as the heart labors to pump the thickened blood, the circulation rate could decrease, actually *reducing* the amount of oxygen circulated.

atrium :: an input, or reservoir, chamber of the heart

ventricle :: a muscular pump of the heart

atrioventricular valve :: keeps blood moving in the right direction; lies between the atrium and its corresponding ventricle

semilunar valve :: opens when the heart contracts to allow blood to leave the heart and closes when the ventricle relaxes to prevent blood from being sucked back in; lies between each ventricle and the vessel it uses to send blood from the heart

Arteries

artery :: blood vessel that transports blood away from the heart

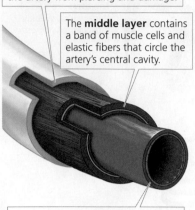

The **outer layer** consists mainly of a thick, fibrous coating that protects the artery from piercing and damage.

The **middle layer** contains a band of muscle cells and elastic fibers that circle the artery's central cavity.

The **inner layer** is smooth to reduce friction with the circulating blood.

FIGURE 12.7 Arteries are blood vessels that transport blood away from the heart. They are thick so that they can withstand the high pressure of ventricular contraction.

Arteries are blood vessels that transport blood away from the heart. The contraction of the ventricles forces blood into the arteries under relatively high pressure. Arteries have two functions then: to withstand the high pressure of ventricular contraction and to carry blood throughout the body. As in other cases we have discussed, the structure of the arteries is the key to their ability to carry out these functions.

The arterial wall is very thick by blood vessel standards. It consists of three layers. The inner layer is smooth to reduce friction with the circulating blood. The middle layer contains a band of muscle cells and elastic fibers that circle the artery's central cavity. The outer layer also contains elastic fibers but consists mainly of a thick, fibrous coating that protects the artery from piercing and damage (**FIGURE 12.7**).

The muscle cells give the artery strength. They also can contract or relax to change the volume of the artery and help to regulate blood pressure. Unlike the muscles in your arms and legs, these muscles are not under voluntary control. Instead, they contract in response to hormones circulating in the blood, and also in response to stretch when the heartbeat expands the artery. (See *Life Application: How Erections Are Produced* to learn how blood flow to the penis is regulated.)

Elastic fibers allow the artery to stretch like a balloon when blood is forced into it. These fibers absorb some of the shock wave caused by the ventricular contraction, thus protecting smaller arteries farther away from the heart. When the ventricle relaxes, the stretched arteries snap back to their normal size, which helps to move the blood along.

The second function of arteries is transport; arteries can perform this function because of their branching structure. The systemic circuit, for example, begins with a large artery called the aorta. The aorta starts off toward the head, then arches left and downward just under the spine. Along its length, arteries branch off to supply the head, arms, internal organs, and legs. These smaller arteries continue to branch, becoming still smaller and more numerous. The smallest of these are called arterioles, and where they end, the next part of the circulatory system begins.

Capillary Beds

capillary :: an extremely tiny vessel through which red blood cells travel one at a time; located at the ends of arterioles

At their ends, arterioles break up into a highly branched network of extremely tiny vessels called **capillaries**. The network itself is called a capillary bed. Capillaries are so small that red blood cells travel through them one at a time. The walls of a capillary are only one-cell thick. At this point, the blood supply is as close as it can be to organs, muscles, and other cells in the body—only one cell away (**FIGURE 12.8**, p. 352).

While one of the functions of the arteries is transport, the function of the capillary bed is distribution. In the capillary beds, oxygen, nutrients, and chemical signals are delivered to cells, and carbon dioxide and wastes are picked up. There are capillary beds everywhere in your body. Even your heart and larger arteries have capillary beds, since the muscles and other cells in them need oxygen and food.

sphincter :: tiny band of muscles that forms a ring around the entrance to the capillary bed; it contracts and relaxes to control blood flow

At places where the arterioles transition into capillaries, you can find **sphincters**—tiny bands of muscles that form a ring around the entrance to the capillary bed. Like the muscles in the arteries, these sphincter muscles operate involuntarily. They contract to shut down blood flow to a particular capillary bed or relax to allow blood in. After a meal, for example, blood can

Life Application

How Erections Are Produced

In 1983, at the annual meeting of the American Urological Association in Las Vegas, Dr. Giles Brindley made medical history. During his presentation to a roomful of distinguished scientists and doctors, Brindley dropped his pants to reveal an erection, then walked through the audience and encouraged them to examine it. In this case, what happened in Vegas did not stay in Vegas.

At the time, the conventional medical wisdom was that erections were a psychological issue and that problems with producing one were psychological as well. Brindley felt otherwise and went to great lengths to prove it. Before his presentation, he injected his penis with a muscle relaxant. This produced his erection, and at the conference Brindley demonstrated clearly and scientifically that physiology alone can produce an erection.

Why do muscle relaxants produce an erection? And how is this related to the circulatory system?

The circulatory anatomy of the penis is what allows it to become erect. Blood enters the penis through two major arteries, and it flows through three sponge-like chambers. The muscles in these arteries normally are partially contracted, so blood enters at a moderate rate. However, if these arterial muscles become relaxed, the artery dilates and blood enters at a very high rate—much faster, in fact, than it can leave. Blood fills the spongy chambers, and the pressure causes them to expand. This expansion produces the erection.

When Brindley injected a muscle relaxant, he caused the arteries in his penis to relax artificially and produce an erection.

Normally, this job is handled by the brain. When proper stimulation is received, the brain sends a signal down a nerve to the penis. There the nerve releases nitric oxide, which triggers several biochemical transformations that eventually produce a muscle relaxant—and an erection.

Brindley's work showed that erectile dysfunction (ED) could be treated by drugs. His use of a muscle relaxant was not an effective treatment—the injection is painful, and the relaxant may affect tissues other than the penis. The new ED drugs target the arteries of the penis in a clever way. One of the main causes of ED in older men is that their bodies don't produce enough of the muscle relaxant. Worse, there is an enzyme called phosphodiesterase (PDE) that breaks down the relaxant to bring the erection to a natural end. ED drugs prevent PDE from working. Since the relaxant is not broken down, eventually enough accumulates to cause an erection. Your body actually makes several forms of PDE. But the PDE made in the penis is made nowhere else in the body, so the effects of the ED drugs are limited (mainly) to the penis.

ED drugs can have side effects, some of them serious, like blurred vision and hearing loss—check the "fine print" in the ads and commercials. One particularly severe side effect is a prolonged erection—over four hours. Men who don't need ED drugs may be tempted by such a display of stamina. But remember that during an erection, blood is not *circulating* through the penis. No oxygen is being delivered, and metabolically active cells begin to die. You really can have too much of a good thing. ::

Blood enters the penis through two major arteries and flows through three sponge-like chambers surrounded by partially contracted muscles.	If these muscles become relaxed, the arteries dilate and blood enters at a much higher rate, expanding the spongy chambers and producing an erection.	Normally, upon proper stimulation, the brain sends a signal down a nerve to the penis and releases nitric oxide, which eventually produces a muscle relaxant—and an erection.

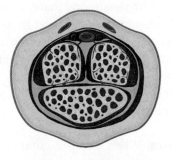
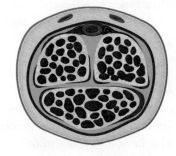
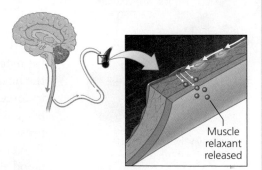

Muscle relaxant released

Dr. Giles Brindley demonstrated clearly and scientifically that physiology alone can produce an erection.

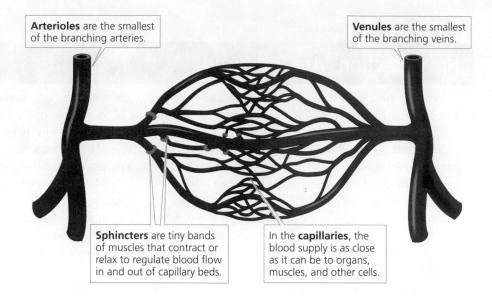

Arterioles are the smallest of the branching arteries.

Venules are the smallest of the branching veins.

Sphincters are tiny bands of muscles that contract or relax to regulate blood flow in and out of capillary beds.

In the **capillaries**, the blood supply is as close as it can be to organs, muscles, and other cells.

FIGURE 12.8 In the capillary beds, oxygen, nutrients, and chemical signals are delivered to cells, and carbon dioxide and wastes are picked up.

microcirculation :: the process by which different capillary beds open and close to divert blood where it is needed throughout the body

vein :: a vessel that transports blood toward the heart

be diverted to your digestive organs by opening these capillary beds. If you are playing a sport or running away from something, blood can be diverted to the large muscles of your legs and arms. At any given time, most of your capillary beds are shut. As your activities change throughout the day, different capillary beds open and close to divert blood where it is needed. This process is called **microcirculation**.

You might wonder why all capillary beds aren't simply open all the time. The answer is that you don't have enough blood. If all the capillary beds were opened, they would fill with blood, draining all the blood from your heart and larger vessels. With no blood for the heart, circulation would cease. Opening all the capillary sphincters can sometimes happen as a result of accident or head trauma. This produces a serious life-threatening condition called hypotensive shock. Therefore, regulation and control of microcirculation is critically important.

Veins

As the blood leaves the capillary beds, it flows into small vessels called venules. The venules merge into larger and larger vessels, and ultimately into **veins**, like many small tributaries merging into a river. Veins are vessels that transport blood back to the heart (**FIGURE 12.9**).

By the time blood reaches the veins, it is moving slowly, and there is almost no blood pressure. Veins don't experience a shock wave of blood from the heartbeat, and they don't have a pulse. Veins lack the elastic fibers found in the arteries. They do, however, have thick, muscular walls. Since the effects of the heart's pumping don't reach the veins, the muscles in their walls contract to force blood along. Also, veins tend to run through larger muscles in your body, particularly in your arms and legs. As these muscles contract, they compress the veins and help to circulate blood.

Veins also have a number of one-way valves. When a vein is compressed, blood flows through the valve toward the heart, but the valve prevents the blood from flowing backward. Without these valves, blood movement in the veins would be controlled largely by gravity, and blood would eventually pool in your lower extremities. The veins leading from your head to your heart do not have valves. In this case, gravity works for you, moving venous blood in the proper direction. But if you lower your head or stand upside down, there are no

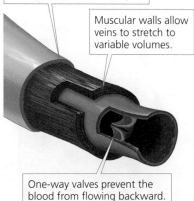

The outer layer is thick, although it is not as strong as the outer layer of an artery.

Muscular walls allow veins to stretch to variable volumes.

One-way valves prevent the blood from flowing backward.

FIGURE 12.9 Veins are blood vessels that transport blood back to the heart. By the time blood reaches the veins, there is almost no blood pressure, and so veins lack the elastic fibers found in the arteries.

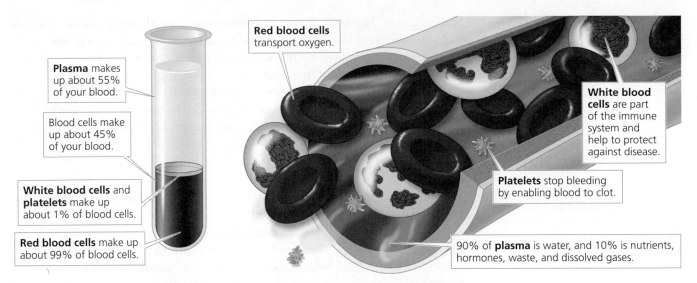

Plasma makes up about 55% of your blood.

Blood cells make up about 45% of your blood.

White blood cells and platelets make up about 1% of blood cells.

Red blood cells make up about 99% of blood cells.

Red blood cells transport oxygen.

White blood cells are part of the immune system and help to protect against disease.

Platelets stop bleeding by enabling blood to clot.

90% of plasma is water, and 10% is nutrients, hormones, waste, and dissolved gases.

FIGURE 12.10 Blood is composed of plasma, red blood cells, white blood cells, and platelets.

valves to prevent the backward motion of blood. Your blood quickly begins to pool in your head, which will lead to an increase in blood pressure.

Blood

The last component of the circulation system is the blood itself. Your blood has two components: a liquid part, called **plasma**, and a host of different types of cells (**FIGURE 12.10**).

Plasma makes up about 55% of your blood, and 90% of this is water. The remaining 10% of plasma consists of chemicals dissolved in your blood. This includes a little bit of everything: nutrients like proteins, amino acids, carbohydrates, and fats, as well as hormones, metabolic wastes, dissolved gases like oxygen, carbon dioxide, and nitrogen, chemicals from the immune system, and even measurable amounts of pollutants.

Blood cells make up 45% of your total blood volume. The vast majority of these (99%) are red blood cells, which transport oxygen. The remaining 1% is composed of platelets and various types of white blood cells. **Platelets** are small fragments of blood cells that are responsible for clotting blood to stop bleeding. **White blood cells** are part of your immune system and work in several different ways to protect you from disease, a process we will examine in Chapter 14.

Blood doping adds about an extra liter of red blood cells to your overall blood volume. This reverses the makeup of the blood so that it is about 45% plasma and 55% red blood cells. We mentioned earlier that doped blood is thicker and more viscous than regular blood. Here you can see why: by increasing the percentage of red blood cells, blood becomes more "solid" and requires more effort to pump.

plasma :: the liquid part of blood, 90% of which is water and 10% of which is chemicals dissolved in the blood

platelet :: small cell fragment that is responsible for clotting blood to stop bleeding

white blood cell :: part of the immune system that works to protect you from disease

12.3 How Is the Respiratory System Structured?

The respiratory system, like all other parts in all organisms, is the consequence of an evolutionary process. Like the circulatory system, the respiratory system illustrates the principle of branching as a means to increase surface area for gas exchange. However, in the respiratory system, all the branches lead to dead ends, and the same branches that bring fresh air in are used to send stale air out. In every branch, air moves in both directions, in and out. Imagine if our digestive system had this structure!

Nonetheless, our respiratory system does an excellent job of supplying us with oxygen and getting rid of carbon dioxide. In this section, we will consider how the structure of the respiratory system facilitates these functions (**FIGURE 12.11**).

Lungs

Your lungs consist of two large sacs: the left one, divided into two compartments (or "lobes"), and the right one, divided into three. These lobes are connected to a network of branched airways starting with your **trachea**, or windpipe, and leading deep into the lungs. The trachea connects with two tubes, called **bronchi**, one for each lung. Each bronchus branches into ever smaller and more numerous **bronchioles**. At the terminal end of the smallest bronchioles are tiny, thin-walled, dead-end sacs called **alveoli** (singular alveolus). The alveoli cluster like bunches of grapes at the tips of the bronchioles, and they are surrounded by capillary beds. Oxygen enters your bloodstream across the surface of the alveoli, and carbon dioxide leaves the same way.

This branching, sub-branching, and proliferation of tiny sacs has a dramatic effect on the surface area available for gas exchange. The total length of

trachea :: windpipe that leads deep into the lungs

bronchus :: one of the two tubes branching from the trachea

bronchiole :: a branch of the bronchus

alveolus :: thin-walled, dead-end sac at the terminal end of the smallest bronchioles, across the surface of which oxygen enters and carbon dioxide leaves the bloodstream

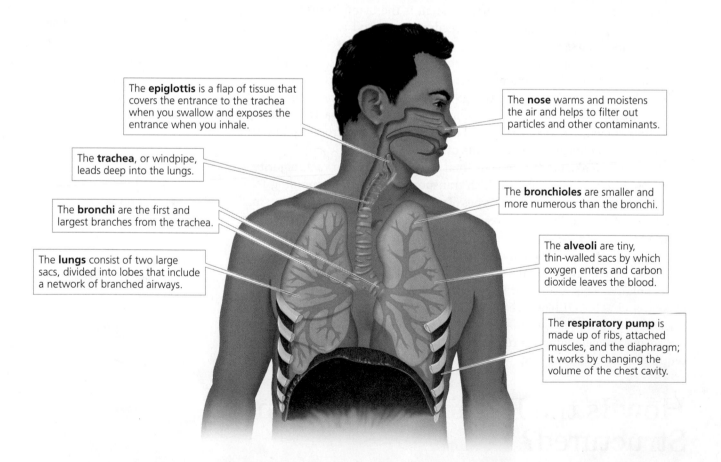

The **epiglottis** is a flap of tissue that covers the entrance to the trachea when you swallow and exposes the entrance when you inhale.

The **nose** warms and moistens the air and helps to filter out particles and other contaminants.

The **trachea**, or windpipe, leads deep into the lungs.

The **bronchi** are the first and largest branches from the trachea.

The **lungs** consist of two large sacs, divided into lobes that include a network of branched airways.

The **bronchioles** are smaller and more numerous than the bronchi.

The **alveoli** are tiny, thin-walled sacs by which oxygen enters and carbon dioxide leaves the blood.

The **respiratory pump** is made up of ribs, attached muscles, and the diaphragm; it works by changing the volume of the chest cavity.

FIGURE 12.11 The respiratory system supplies the body with oxygen and rids the body of carbon dioxide. It consists of lungs and a network of branched airways.

the airways in your lungs is an astonishing 1500 miles, and they contain from 300 to 500 million alveoli. The surface area of these alveoli is about 70 square meters, half the size of a volleyball court. That's a lot of material to pack into your lungs—which have a volume of only 6 liters, about the size of three 2-liter bottles of soda. This large surface provides ample opportunity for gas exchange, giving you all the oxygen you need even during heavy exertion.

Packing all this surface area into a relatively small volume is an impressive adaptation. Recall though that not all the oxygen in the lungs is actually picked up by the blood. The limiting factor in oxygen pickup is the amount of hemoglobin in the blood, which in turn is determined by the number of red blood cells. Blood doping allows athletes to tap into the extra oxygen in the lungs and increase oxygen delivery. A similar effect can be achieved with regular training to increase red blood cell count, and with much less risk.

The Respiratory Pump

Like the circulatory system, your respiratory system requires a pump. Your heart pumps blood in only one direction. But since the airways in your lungs are two-way streets, the pump for your lungs moves air in two directions, alternating between inhalation and exhalation.

The respiratory pump is made up of your ribs and the muscles attached to them, along with your **diaphragm**. This large circular sheet of muscle separates your chest from your lower abdomen. Taken together, your ribs and diaphragm form the chest cavity, a hollow compartment that contains your lungs (and heart). The respiratory pump works by changing the volume of the chest cavity. When the cavity increases in volume, your lungs expand to fill the extra space. This expansion draws air in. When the chest cavity compresses, the lungs compress as well, squeezing air out (**FIGURE 12.12**).

The muscles in your chest control the volume of the chest cavity. Generally, when your chest muscles contract, they expand the cavity, which produces an inhalation. When they relax, the cavity returns to its resting size, compressing the cavity and producing an exhalation.

The diaphragm plays an important role in this process. At rest, the diaphragm forms a dome that extends into the chest cavity. When it contracts,

diaphragm :: large circular sheet of muscles that separates your chest from your lower abdomen

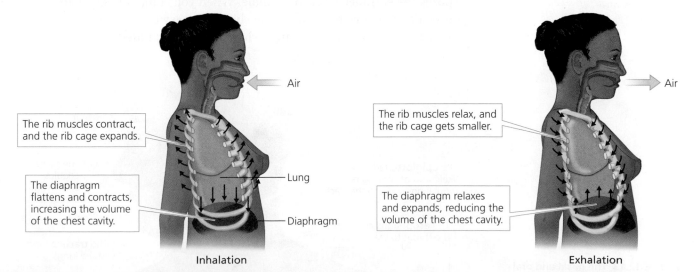

The rib muscles contract, and the rib cage expands.

The diaphragm flattens and contracts, increasing the volume of the chest cavity.

Air

Lung

Diaphragm

Inhalation

The rib muscles relax, and the rib cage gets smaller.

The diaphragm relaxes and expands, reducing the volume of the chest cavity.

Air

Exhalation

FIGURE 12.12 The ribs, muscle attached to the ribs, and diaphragm make a respiratory pump that brings air in and out of the body. The respiratory pump works by changing the volume of the chest cavity.

it flattens out, which increases the volume of the chest cavity. The muscles attached to your ribs rotate them upward and outward, further expanding the chest cavity and allowing even deeper inhalations. When the diaphragm relaxes, it expands back to its dome shape, reducing the empty space in the chest cavity. When the rib muscles relax, gravity and the natural elasticity of the ribs returns them to their resting position, further reducing the volume of the chest cavity and driving out more air for the exhalation. There are other muscles in the ribs that actively contract to add more power to the exhalation. These muscles typically are used when you are exerting yourself and need to breathe more deeply and at a faster rate.

Nose and Mouth

Your lungs have only one way in and one way out: your trachea, or windpipe. However, there are two ways to get to the trachea, through your nose or through your mouth.

Your nose is the passageway dedicated to your lungs. It warms and moistens the air you inhale and helps to filter out particles and other contaminants, so they don't enter your lungs. Your nose is also responsible for detecting odors in the air you breathe, your sense of smell.

You can breathe through your mouth as well, although you don't get the warming, moisturizing, and filtering that occur when you breathe through your nose. Because your nasal passages and oral passages are mostly separate, you can chew and breathe at the same time—a useful ability for mammals that can't stop breathing just to eat.

However, at the back of your mouth, the nasal passages and oral passages come together (**FIGURE 12.13**). There are advantages to this arrangement. If you have a cold and your nose is stuffed up, you can breathe through your mouth instead. There are disadvantages as well, mainly involving choking. Food or drink that should go down your esophagus to your stomach can occasionally go down your trachea instead. Violent spasms such as coughing act like powerful exhalations to expel the foreign material. Blocking the airway, even for a short time, can be fatal, and many people choke to death each year.

The epiglottis is a flap of tissue that reduces the possibility of choking. The epiglottis covers the entrance to the trachea when you swallow food, and it exposes the entrance when you inhale. When you compare how many times you swallow to how many times you choke, it's easy to see that the epiglottis does a good job of directing things where they are supposed to go.

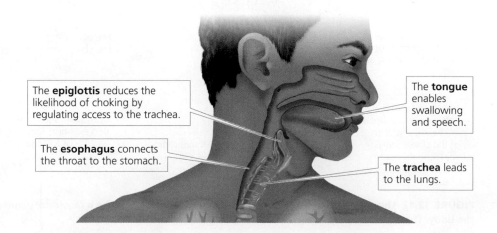

The **epiglottis** reduces the likelihood of choking by regulating access to the trachea.

The **tongue** enables swallowing and speech.

The **esophagus** connects the throat to the stomach.

The **trachea** leads to the lungs.

FIGURE 12.13 The nasal and oral passages come together at the back of the mouth.

12.4 How Do You Breathe?

Breathing involves not only taking air into your lungs, but also transporting oxygen from your lungs to every cell in your body. The carbon dioxide your body produces when your cells respire is removed as part of the breathing process as well. Overall, this process is referred to as **gas exchange**.

Getting Oxygen

We can start by examining what happens just after you finish an inhalation. Fresh air has moved into your lungs, down your bronchi and bronchioles, and into your alveoli. Fresh air is rich in oxygen (about 21%) and has a low concentration of carbon dioxide (0.04%). Recall that alveoli are closely associated with capillary beds and therefore close to blood.

The blood in these capillary beds comes from the right side of the heart, the side that gets "used" blood from the body. This blood has a very low concentration of oxygen and a relatively high concentration of carbon dioxide, picked up inside your body from actively metabolizing cells.

The concentrations of oxygen and carbon dioxide are therefore quite different on either side of the thin walls formed by the alveolus and capillary. This difference in concentration is referred to as a concentration gradient. Whenever a concentration gradient is set up across a thin cellular membrane, diffusion occurs (Chapter 11). Gases diffuse from areas of high concentration to areas of low concentration. As a result, oxygen diffuses into the blood, and carbon dioxide diffuses out of the blood into the alveolus. The oxygenated blood moves back to the left side of the heart, taking the oxygen with it. The carbon dioxide is expelled from the lungs when you exhale (**FIGURE 12.14**).

Carbon dioxide dissolves easily in water and blood and therefore moves across the alveoli with little difficulty. Recall that this is not the case with oxygen. If diffusion were the only principle operating, blood would become saturated with oxygen at a very low concentration and could not take in any more. But oxygen diffuses into the red blood cells, where it instantly binds to hemoglobin. As a result, oxygen is taken out of the liquid part of the blood, allowing more oxygen to enter from the lungs, again underscoring the important role of red blood cells and hemoglobin. It also explains the beneficial, though illegal, advantages of blood doping.

Transporting Oxygen

After oxygen-rich blood enters the left atrium, it is pumped into the left ventricle and out through the aorta as the heart beats. It then travels along the

gas exchange :: the process by which oxygen is taken into the lungs and transported to every cell in the body and carbon dioxide is removed from the cells

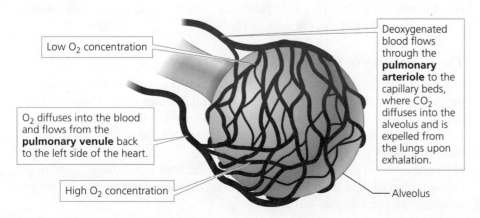

Low O₂ concentration

O₂ diffuses into the blood and flows from the **pulmonary venule** back to the left side of the heart.

High O₂ concentration

Deoxygenated blood flows through the **pulmonary arteriole** to the capillary beds, where CO₂ diffuses into the alveolus and is expelled from the lungs upon exhalation.

Alveolus

FIGURE 12.14 Oxygen enters the bloodstream across the surface of the alveoli, and carbon dioxide leaves the same way. A network of capillary beds surrounds the alveoli and facilitates the oxygen and carbon dioxide exchange.

aorta and throughout the systemic circuit, eventually reaching every capillary bed in your body. Some capillary beds are closed by capillary sphincters. The blood continues its circulation until it reaches an open capillary bed or is diverted into the venous part of the circuit.

Delivering Oxygen

All of the oxygen delivery occurs in the capillary beds. The cells around the capillaries typically are metabolizing, or undergoing cellular respiration. When a cell respires, it uses up oxygen and produces carbon dioxide. As a result, the concentration of oxygen in the cells is low, and the concentration of carbon dioxide is high.

The concentration of these gases in the blood is the opposite—high in oxygen and low in carbon dioxide. As in the alveolus, a concentration gradient is established across the capillary wall. In this case, oxygen diffuses out of the capillary and is delivered to the respiring cells. Carbon dioxide, a metabolic waste product, diffuses from the cells into the capillary, to be taken away with the blood.

All places in your body where materials move from one part to another face the same challenge. How is enough surface area provided to allow the exchange to occur? As we saw, your lungs achieved this by branching into smaller and more numerous airway tubes. In your circulatory system, the network of capillaries also greatly expands the available surface area in a similar way: by branching into millions of small tubes.

By most estimates, the capillaries in your body are over *60,000 miles long*. Laid out end to end, they would circle the world more than two times at the equator. All these small tubes provide a lot of surface area, between 800 and 1000 meters squared—about the size of the "red zone" on a football field, or almost the size of a hockey rink. The length of the capillaries allows them to reach virtually every cell in your body, and their surface area provides ample room for gas exchange to occur.

Capillary beds deliver more than oxygen. They deliver *everything* that cells need and use: nutrients, hormones, and other chemical signals. The capillary beds also pick up all the waste products cells produce, not only carbon dioxide.

Returning to the Lungs

As the blood leaves the capillary beds, it enters the venules, which merge with a series of larger and larger veins. Eventually this blood returns to the right side of the heart. After its passage through the capillary beds, this blood is low in oxygen and rich in carbon dioxide. This "used" blood enters the right atrium and moves through the right ventricle and into the pulmonary artery as the heart beats. The pulmonary artery delivers blood to the capillary beds around the alveoli, and the cycle repeats.

The Chemistry of Gas Exchange

We have mentioned several times that hemoglobin solves the problem of oxygen not dissolving readily in water or blood (**FIGURE 12.15**). Hemoglobin is composed of four protein molecules that come together to make the functional compound. Each of these four protein molecules contains an atom of iron. Iron combines easily with oxygen, forming iron oxide—what we know as rust. After oxygen molecules diffuse into the red blood cells, they immediately bind to the iron in hemoglobin. Red blood cells ferry this hemoglobin-bound oxygen around with them as they circulate.

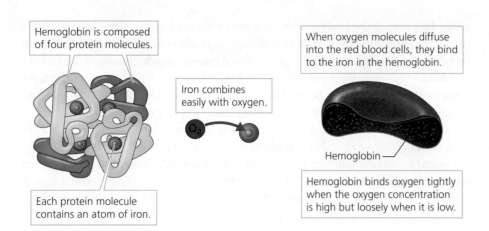

Hemoglobin is composed of four protein molecules.

Each protein molecule contains an atom of iron.

Iron combines easily with oxygen.

When oxygen molecules diffuse into the red blood cells, they bind to the iron in the hemoglobin.

Hemoglobin

Hemoglobin binds oxygen tightly when the oxygen concentration is high but loosely when it is low.

FIGURE 12.15 **Red blood cells carry most of the oxygen in the blood; each cell contains more than 200 million hemoglobin molecules.** Hemoglobin solves the problem of oxygen not dissolving readily in water or blood.

Hemoglobin has an important chemical property: it binds oxygen tightly when a lot of oxygen is present but binds it loosely when oxygen concentration is low. Let's consider how this property works in conjunction with the oxygen delivery system described earlier. In the lungs, the oxygen content is high, and very quickly every iron atom in the hemoglobin binds tightly to a molecule of oxygen. This oxygen stays with the hemoglobin as it passes through the heart, along the arteries, and into a capillary bed.

Cells around the capillary bed have been using oxygen for their own metabolism, so oxygen content here is low. Low oxygen changes the chemical properties of hemoglobin, and it is no longer able to hold on to the oxygen it carries. This oxygen is released, and it diffuses out of the red blood cell, out of the capillary bed, and eventually diffuses into the cells. One nice feature of this system is that cells that deplete oxygen the most (for example, muscle cells during exercise) are best at getting hemoglobin to give up its oxygen. As a result, active muscle cells become continually resupplied with oxygen. Blood doping enhances this effect because the blood has even more hemoglobin and therefore more oxygen to give up to muscle cells during exertion.

Carbon dioxide dissolves much more easily in water (and blood) than oxygen (**FIGURE 12.16**). Once dissolved, carbon dioxide quickly reacts with water to form carbonic acid, which gives carbonated beverages their fizzy, acidic

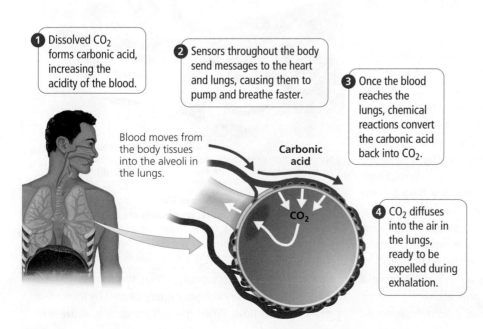

1 Dissolved CO_2 forms carbonic acid, increasing the acidity of the blood.

2 Sensors throughout the body send messages to the heart and lungs, causing them to pump and breathe faster.

3 Once the blood reaches the lungs, chemical reactions convert the carbonic acid back into CO_2.

Blood moves from the body tissues into the alveoli in the lungs.

Carbonic acid

CO_2

4 CO_2 diffuses into the air in the lungs, ready to be expelled during exhalation.

FIGURE 12.16 **Carbon dioxide dissolves more easily in blood than oxygen.**

taste. In areas of the body where cells are respiring heavily, not only is oxygen depleted, but a good deal of carbon dioxide is produced. As this carbon dioxide forms carbonic acid, the acidity of the blood increases, which lowers the blood's pH level. The declining pH tells your body that you need more oxygen. As sensors throughout your body detect an increase in blood acidity, they send messages to the heart and lungs, causing them to pump and breathe faster.

Carbon dioxide, in the form of carbonic acid or related chemicals, is carried both in the blood and in red blood cells. Once the blood reaches the lungs, chemical reactions convert these chemicals back into carbon dioxide. Since the concentration of carbon dioxide is higher in the blood than in the lungs, it diffuses out. This clears the blood of carbon dioxide and loads it into the air in the lungs, ready for you to exhale.

12.5 How Does Blood Pressure Vary Throughout Your Body?

blood pressure :: the force exerted by the blood on the arterial wall as it moves from the heart to the arteries

When blood moves from the heart into the arteries, it exerts a force on the arterial wall. This force is known as **blood pressure** (see *How Do We Know? Measuring Blood Pressure*). The arteries stretch in response to blood pressure, and pressure forces the blood farther down the arteries toward the capillary beds. Blood pressure varies during the day depending on what you are doing. At rest, your blood pressure is at its lowest, but when you are active, blood pressure increases to provide additional oxygen and remove excess carbon dioxide.

How Do We Know?

Measuring Blood Pressure

Blood pressure is one of the four vital signs—the routine measurements that health professionals take to monitor your body's physiological state. (The other three are pulse, respiration rate, and body temperature.) Both the heart and the major arteries contribute to blood pressure, which is why you have two numbers like 120 over 80.

When the heart contracts, it sends out a strong pulse of blood under high pressure, called the systolic pressure, and it is the first (and higher) of the two readings. But when the heart relaxes, the pressure does not drop down to zero, even though the heart exerts no pressure at this point. The arteries are stretched by the pulse, and when the heart relaxes, the arteries snap back to their normal size. This elastic recoil maintains blood pressure when the heart relaxes. This is called the diastolic pressure, the second and lower of the two readings.

Your blood pressure alternates smoothly between the systolic and diastolic pressures with every heartbeat. Medical personnel are able to measure your blood pressure by listening to the sounds in your artery. These sounds were first characterized in 1905 by the Russian physician Nikolai Korotkoff.

To measure your blood pressure, a medical professional places an inflatable cuff over your arm above the elbow. He or she inflates the cuff and allows it to deflate slowly,

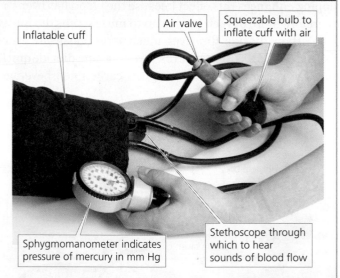

When the cuff pressure is above the systolic pressure or below the diastolic pressure, no sounds are heard. In between these two points, the restricted blood flow causes audible turbulence and Korotkoff sounds.

listening through a stethoscope. What the medical professional hears is turbulence in your artery caused by the cuff as it restricts blood flow. When the cuff pressure is above your systolic pressure, it cuts blood flow off completely, and no sounds are heard. As the cuff deflates, eventually the pressure falls below your diastolic pressure. At this point, the cuff no longer restricts blood flow, so again there is no turbulence and no sounds are heard.

In between these two points, the cuff is below systolic pressure, and so the heart can force blood through the arteries. But the cuff pressure is above diastolic pressure. This means the arterial pressure is not strong enough to force blood through, and the restricted blood flow causes audible turbulence. Within this pressure range, the turbulence causes Korotkoff sounds to be heard. These sounds start with a loud, repetitive snapping. These snaps become weaker and more drawn out and are eventually replaced by a murmuring sound. As the cuff deflates further, the murmurs become quieter and muted until eventually they disappear altogether.

When taking a blood pressure reading, then, two questions are asked. At what pressure are sounds first audible as the cuff deflates? This is the systolic pressure. At what pressure do the sounds go away? This is the diastolic pressure.

Pressure is measured in millimeters of mercury. A pressure of 120 represents enough pressure to hold up a column of mercury 120 mm high (just under 5 inches). If you imagine a device like that, you can see that as the pressure in the chamber rises, the mercury in the tube will rise. The height of the mercury column indicates the pressure in the chamber. ::

Recall that your heart is really two pumps, one for the pulmonary circuit (the lungs) and one for the systemic circuit (the body). These pumps don't run at the same pressure, which is one of the benefits of having two pumps. The pulmonary circuit is short. The delicate capillary beds that surround your lung's alveoli are not far from the heart. The right side of your heart, which pumps blood for the pulmonary circuit, does so at a relatively low pressure. This prevents high blood pressure from rupturing capillaries in the lungs.

The left side of your heart pumps blood throughout your entire body. This requires considerably more pressure, and the highest blood pressure occurs in the left ventricle and the first part of the aorta. High blood pressure can damage arterial walls or even tear their smooth inner lining. High blood pressure is a major consequence and cause of many circulatory system problems, such as atherosclerosis, or a narrowing of the arteries, and it is a direct cause of strokes, which we will discuss later.

Blood pressure drops significantly as blood moves through the systemic circuit (**FIGURE 12.17**), getting lower and lower as the arteries branch, until blood pressure is almost (but not quite) zero at the start of the capillary beds. Blood moves slowly through the capillary beds, and its pressure becomes further reduced, and by the time blood reaches the venules, pressure is nearly zero. Muscle contractions with valves that prevent backup are responsible for moving blood back to the heart.

Friction is one reason why blood pressure decreases along the arteries. Although arterial walls are smooth, friction still occurs between blood and the walls. Friction slows blood down, which also decreases its pressure. The main reason for the decrease in blood pressure involves volume, however. As the arteries branch and branch again, the vessels become smaller, but the total volume of the vessels gets larger. The capillaries are extremely small, but they are also numerous, and their total volume is as great as all the blood your body holds. Farther from the heart, the same volume of blood is moving through an increasing volume of vessels. When this happens, blood flow slows and pressure decreases (**FIGURE 12.18**).

You are probably familiar with this effect, though in reverse. When you put your thumb over the end of a hose, you know the water comes out faster and at higher pressure, so it squirts even farther. This happens because your

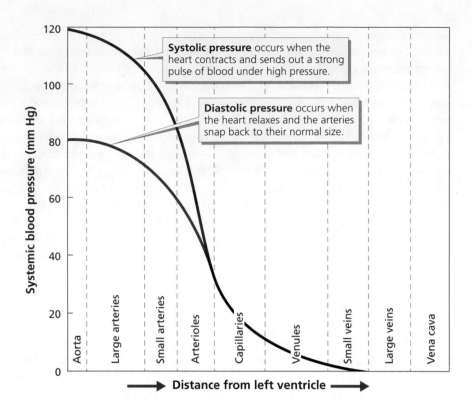

FIGURE 12.17 Blood pressure in the systemic circuit decreases the farther away from the heart that it moves.

thumb makes the hose opening smaller. The same volume of water escapes from the hose, but it has to go through a smaller opening. The only way it can do this is to go through the opening faster.

Now imagine running this process backward. Fast-moving water under high pressure *enters* the hose. The volume (diameter) of the hose is larger than

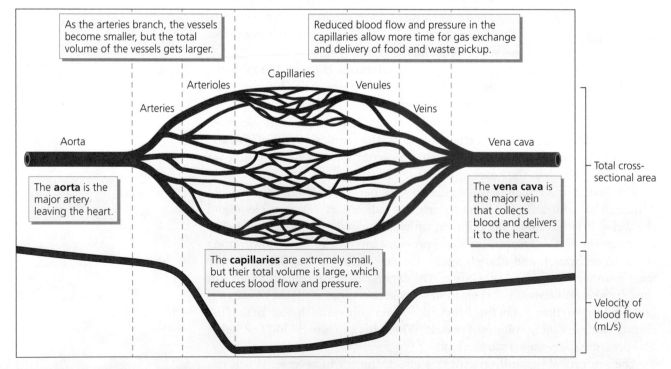

FIGURE 12.18 Blood does not flow at the same velocity in all blood vessels. Farther from the heart, the same volume of blood moves through an increasing volume of vessels, which slows the blood flow and pressure.

the opening into it, so the water slows down, and its pressure goes down as well. This is what happens in your body, and this is why blood pressure drops as blood reaches the capillaries.

As with the lung capillaries, the reduction in blood pressure and speed has important benefits in the systemic capillaries. It helps prevent rupture, for example. But mainly the slow speed allows more time for blood to move through the capillary bed. There is more time for gas exchange, and also more time to deliver food and other compounds and to pick up wastes.

12.6 How Are the Respiratory and Circulatory Systems Controlled?

You can hold your breath, or force yourself to breathe faster, and during exercise your respiration rate increases. Your heart rate slows down or speeds up as your need for oxygen varies. The respiratory and circulatory systems are coordinated, speeding up and slowing down together. But how does your body "know" when you need more oxygen? And when your body detects this need, what changes does it make to provide the additional oxygen?

Control of Respiration

The control of respiration rate is an example of homeostasis (Chapter 11). Breathing centers in your brain—in particular, two areas at the top of your spinal cord, the **medulla oblongata** and the **pons**—control your respiration rate. Nerves run between the medulla and the pons to the muscles in your chest, ribs, and diaphragm. The medulla and pons cause these nerves to fire in a cyclic pattern, and when they fire, it causes you to inhale. After each inhalation, and before the next round of nerve firing, these muscles relax and you exhale.

The medulla and pons produce a rhythm of nerve firings all on their own, which is why you don't need to control your breathing consciously. This rhythm corresponds to your resting rate of respiration. However, the medulla and pons have several ways of sensing your body's need for oxygen, and they adjust the breathing rate accordingly.

Two things occur when your body needs more oxygen. First, and most obviously, the oxygen levels go down. Second, carbon dioxide levels rise. In principle, either signal could be used to monitor oxygen need. Surprisingly, carbon dioxide is the signal your body actually uses. Recall that when carbon dioxide dissolves in blood, it produces carbonic acid, which lowers the pH level of the blood. The medulla and pons can detect changes in blood pH. If the pH of blood declines, there is too much carbon dioxide and the medulla and pons increase their firing rate, causing you to breathe more rapidly and more deeply. Conversely, when the pH of blood rises, the medulla and pons reduce their firing rate and breathing becomes shallower and slower.

Other pH sensors can be found in the aorta and in the carotid arteries, the two main arteries that supply blood to the head. These sensors send nerve impulses to the medulla and pons, informing them about changes in blood pH. The medulla and pons integrate this information with their own pH sensors and regulate breathing as just described (**FIGURE 12.19**).

How does blood doping affect the control of respiration? Recall that blood doping allows the delivery of more oxygen. But the brain's regulators don't

medulla oblongata :: a part of the most posterior portion of the brain; it controls breathing, heart rate, and digestion

pons :: a component of the brain that relays signals controlling respiration, swallowing, sleep, and bladder control

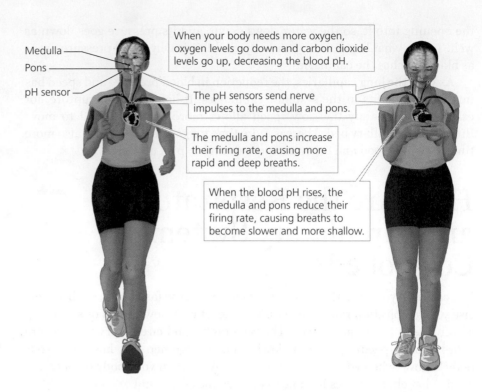

Medulla

Pons

pH sensor

When your body needs more oxygen, oxygen levels go down and carbon dioxide levels go up, decreasing the blood pH.

The pH sensors send nerve impulses to the medulla and pons.

The medulla and pons increase their firing rate, causing more rapid and deep breaths.

When the blood pH rises, the medulla and pons reduce their firing rate, causing breaths to become slower and more shallow.

FIGURE 12.19 The medulla and pons detect changes in blood pH to regulate respiration.

respond to the level of oxygen, so blood doping won't increase your respiration rate. The amount of air you inhale is still controlled by the amount of carbon dioxide you produce.

Control of Blood Pressure

Blood pressure increases when the heart contracts faster or more strongly, and it decreases when the contraction rate or strength decreases. This is simple enough, but a great many factors affect heart rate, and therefore blood pressure, and we don't know all the details of how they work. One of these factors involves pressure sensors (Chapter 11).

The main blood pressure sensors are found in the aorta and the carotid arteries (**FIGURE 12.20**). These pressure sensors are specialized nerve cells that respond to stretch in the artery when the heart beats. When the heart beats more quickly or contracts more strongly, blood pressure increases. This in turn causes the aorta and carotid arteries to stretch more. As a result, the nerves in the pressure sensors fire more rapidly. These nerve impulses are directed to the medulla oblongata and the pons. If these organs receive rapid firings from the pressure sensors, they send nerve impulses to the heart to slow it down. Conversely, if blood pressure is low, the pressure sensors fire more slowly. The medulla and pons detect this slowdown. They stop sending messages to the heart to slow down, so it speeds up.

These pressure sensors act like a thermostat to keep the blood pressure at roughly a constant level. High blood pressure sends a "negative" signal to the brain to slow the heart down. Low blood pressure sends a "positive" signal to the brain to speed up the heart. This is an example of negative feedback, the most common method of control in a homeostatic system.

Control of Heart Rate

As we just saw, negative feedback from blood pressure can regulate heart rate. But blood pressure is only one of many factors that affect how fast your heart beats.

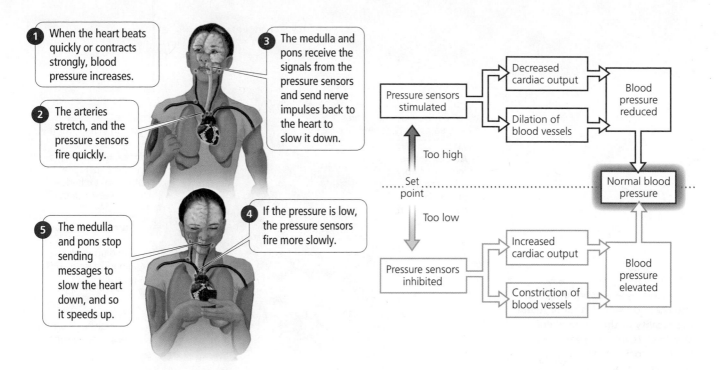

1. When the heart beats quickly or contracts strongly, blood pressure increases.

2. The arteries stretch, and the pressure sensors fire quickly.

3. The medulla and pons receive the signals from the pressure sensors and send nerve impulses back to the heart to slow it down.

5. The medulla and pons stop sending messages to slow the heart down, and so it speeds up.

4. If the pressure is low, the pressure sensors fire more slowly.

FIGURE 12.20 Blood pressure sensors located in the aorta and carotid arteries respond to the stretching of the arteries when the heart beats. They act like a thermostat to keep the blood pressure at roughly a constant level.

Your heart can beat all by itself with no regulation at all. The heart is composed of specialized muscle cells that naturally contract in a rhythmic pattern. Heart muscle cells do require a controller to make them contract *together* in a coordinated fashion.

This controller is called the pacemaker, and it is located in the right atrium. The pacemaker is different from the mechanical device of the same name that some patients need to keep their heart beating regularly. When the pacemaker cells contract, they send out a wave of electrical signals. These signals spread first to the left atrium and then travel down to the ventricles. As this electrical wave moves through the heart, it causes the heart muscles to contract (**FIGURE 12.21**). When the heart is functioning properly, the atria contract first, forcing blood into the ventricles. The ventricles

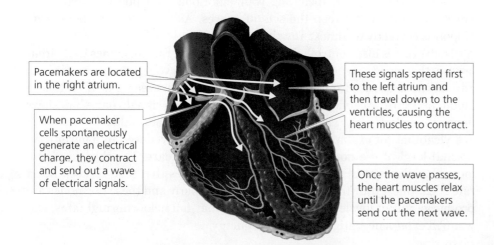

Pacemakers are located in the right atrium.

When pacemaker cells spontaneously generate an electrical charge, they contract and send out a wave of electrical signals.

These signals spread first to the left atrium and then travel down to the ventricles, causing the heart muscles to contract.

Once the wave passes, the heart muscles relax until the pacemakers send out the next wave.

FIGURE 12.21 Pacemakers located in the heart generate heart rhythm and control contraction.

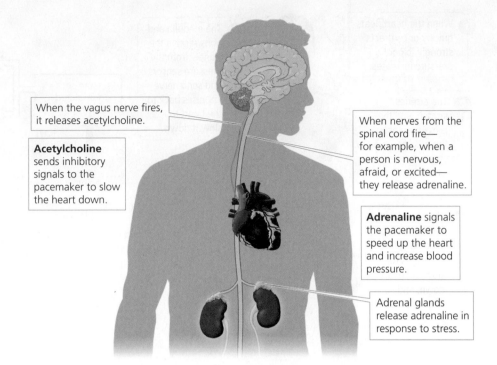

When the vagus nerve fires, it releases acetylcholine.

Acetylcholine sends inhibitory signals to the pacemaker to slow the heart down.

When nerves from the spinal cord fire— for example, when a person is nervous, afraid, or excited— they release adrenaline.

Adrenaline signals the pacemaker to speed up the heart and increase blood pressure.

Adrenal glands release adrenaline in response to stress.

FIGURE 12.22 **Heart rate is controlled by a number of nerves and chemical signals that regulate the activity of the pacemakers.** Many factors, including emotional and physiological state, affect the heart rate.

contract shortly thereafter, forcing blood into the arteries. Once the wave passes, the heart muscles relax for a short time before the pacemaker sends out the next wave.

If the pacemaker were the only thing controlling heart rate, your heart would beat about 100 times a minute. However, a number of signals from your body affect your heart, either speeding it up or slowing it down in a complex network of feedback loops (**FIGURE 12.22**).

The vagus nerve, a major nerve from your medulla, connects to the pacemaker, and when this nerve fires it slows your heart. A second group of nerves from your spinal cord act on the pacemaker to speed your heart up. Your actual heart rate at any given time depends on the balance of input from these two sets of nerves. For example, when the medulla detects the need for more oxygen, it decreases the input from the vagus nerve (the inhibitor), which shifts the balance in favor of the spinal nerves (the exciters), and heart rate speeds up. Many other factors, including your emotional and physiological state, affect the balance between the sets of nerves. As a result, your heart rate responds correctly to almost any situation.

Heart rate is also controlled by chemical signals, the hormones that circulate in the blood. For example, when you are nervous, afraid, or excited, adrenaline is released into your blood. Adrenaline increases heart rate and blood pressure. Acetylcholine is another chemical signal, one that slows down the heart rate. The vagus nerve slows the pacemaker by releasing acetylcholine, for example. Finally, heart rate can be reduced by a lower than normal level of the hormones that speed up the heart. The thyroid gland, located at the base of the neck, produces hormones that tend to keep heart rate high. But if your thyroid is not working properly and these hormones are produced only at low levels, your heart rate may fall below normal rates, and you will experience fatigue and loss of energy.

12.7 What Happens When the Circulatory and Respiratory Systems Malfunction?

Circulatory and respiratory diseases account for one-third to one-half of all deaths in the United States and other developed countries. Cells begin to die when deprived of oxygen for even a few minutes. If these dying cells are in the brain or heart, death of the body is often not far behind. Cardiovascular problems can appear quite suddenly and dramatically, as in heart attack or stroke, but often these sudden events are the result of many years of deterioration that went undetected or were ignored. Sometimes problems result from natural body processes that go awry. But often these problems arise from lifestyle choices we make, including the decision that some athletes make to engage in blood doping.

Atherosclerosis

Atherosclerosis is a disease of the arteries. It begins with inflammation in the arterial wall. The immune system may mistake this inflammation for a wound and begin what in other cases would be a normal healing process. White blood cells may invade the site, and platelets may also collect there, forming fibrous blood clots that infiltrate the tissue. Fats like cholesterol, which normally circulate in your blood, may accumulate in the site, and calcium deposits may form.

atherosclerosis :: a disease of the arteries

The resulting structure is called a **plaque** (**FIGURE 12.23**). As the plaque builds up, the inner part of the artery narrows. This narrowing has a number of effects, all of which are harmful: blood pressure increases upstream from the plaque, blood flow and oxygen delivery are reduced downstream, and there is a risk that the artery may become blocked.

plaque :: structure that forms from white blood cells, platelets, fats, and calcium deposits, narrowing the inner part of the artery as it grows

High blood pressure is the single most important indicator that you may have, or you may develop, problems with atherosclerosis. High blood pressure can cause damage to arterial walls, which results in plaque formation. Once plaques have formed, high blood pressure can indicate that your arteries are narrowing, a warning sign for advanced atherosclerosis. High levels of cholesterol circulating in your blood contribute to plaque formation, and therefore to atherosclerosis.

Atherosclerosis usually begins in the teenage years and slowly progresses without symptoms until problems are detected later in life. Scientists are making progress in understanding the causes of high blood pressure and atherosclerosis and also ways to treat and prevent it (see *Scientist Spotlight: Michael Ellis DeBakey*). Many risk factors are genetic. For example, you have many genes that control how cells produce fats like cholesterol and how they take it up from and release it into the bloodstream. These genes play a large role in determining how much cholesterol circulates in your blood, which directly affects plaque formation.

While genetics are important, life choices you make have a strong impact as well. Anything you do to lower blood pressure reduces your risk of atherosclerosis, as does anything that gives you a proper balance of cholesterol in your blood. If you smoke, quitting is one of the best things you can do to decrease your risk of heart attack and stroke: nicotine in cigarettes causes the muscles in your arteries to contract, narrowing them and raising blood

The artery narrows as a result of the **plaque**, which is composed of white blood cells, platelets, fats, and calcium deposits.

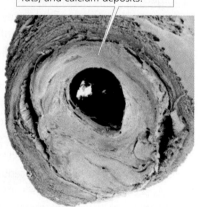

Plaque can lead to high blood pressure and reduced blood flow and oxygen delivery.

FIGURE 12.23 Atherosclerosis is a progressive disease of the arteries in which fatty deposits form plaques that can block blood vessels.

Scientist Spotlight

Michael Ellis DeBakey (1908–2008)

Have you seen lives saved on the television show *M*A*S*H*? Do you have a relative or a friend who has been treated at a veteran's hospital? Or do you know anyone who has had cardiac surgery to replace a damaged blood vessel or bypass the coronary arteries? If so, your life has been touched by Dr. Michael DeBakey.

DeBakey was born to Lebanese immigrants in Lake Charles, Louisiana, the oldest of five children. From childhood, DeBakey showed a keen intellect and wide-ranging interests. He read the entire *Encyclopedia Britannica* before college, participated in sports, played musical instruments, and learned to sew—a technique that later would help him when he became a surgeon.

In 1926, DeBakey entered Tulane University and his academic interests took center stage. In two years, he had earned enough credits to apply to medical school, but he worked out a deal to continue taking undergraduate credits while in medical school so he could complete both his bachelor's and medical degrees.

While still in medical school, DeBakey made his first major contribution to medical research. He invented a roller pump: a pump that gently pushes fluid through a tube compressed between two rolling pins. His pump was incorporated into cardiopulmonary bypass machines because it circulated blood more gently and caused less damage than other pumps available at the time.

After receiving his M.D. in 1932, DeBakey worked in many positions around the world. When World War II broke out, DeBakey volunteered for service and worked in the U.S. Surgeon General's office as a consultant. It was here that he developed his plan for mobile army surgical hospitals (MASH). By bringing hospitals to the front rather than transporting wounded soldiers, DeBakey saved countless lives. It also was at this time that DeBakey set up the organization that later would become the Veterans Administration.

After the war, DeBakey developed an impressive array of medical devices and surgical techniques. He developed several types of artificial hearts. On his wife's sewing machine, he made stretchy Dacron tubes that he used to replace damaged arteries. He also developed surgical procedures to implant these inventions, becoming one of the first surgeons to successfully perform cardiac bypass, artery replacement, and heart transplants.

DeBakey worked almost until the end of his life, contributing 75 years to medical research and practice. He authored

Dr. Michael Ellis DeBakey developed an impressive array of surgical techniques and medical devices, including several types of artificial hearts. He was one of the first surgeons to perform successfully cardiac bypass, artery replacement, and heart transplants.

numerous books and papers and received many honors, including the Presidential Medal of Freedom and the National Medal of Science. At the end of 2005, DeBakey suffered an aortic dissection, and at the age of 97 he had his aorta replaced with a Dacron graft, making him the oldest person to be saved by the pioneering technique he had invented almost fifty years previously. DeBakey died in 2008, a few months short of his 100th birthday. ::

pressure. Maintaining a healthy weight and participating in regular exercise also help control blood pressure. Finally, eating a sensible diet with reduced saturated and trans fats can help you maintain the proper cholesterol balance in your blood.

Heart Attack

The coronary arteries provide the heart with oxygen. These arteries branch directly from the aorta, and normally they are a guaranteed source of fresh, oxygenated blood for the heart. Atherosclerosis can affect the coronary arteries, just as it does other arteries. If these arteries become clogged with plaque, blood and oxygen flow diminish. The heart muscles may use up all of the oxygen, and when they run out, these cells begin to die; this is what happens during a heart attack (**FIGURE 12.24**).

Atherosclerosis causes heart attacks in other ways as well. When plaques form anywhere in the body, the blood clots in them can sometimes fragment and break away. When this happens, small, hard clot particles circulate throughout the blood vessels in your body. These clots may get stuck in the smaller vessels, partially or completely blocking them and shutting down blood flow. If a clot lodges in a coronary artery and shuts it down, a heart attack will follow. If the coronary arteries narrow or become blocked, cardiac bypass surgery may be needed to keep the patient alive (see *Technology Connection: Cardiopulmonary Bypass Machine*).

Here, the true risks of blood doping become evident. The increased viscosity of the blood and more labored beating of the heart can cause damage to the arterial lining. This in turn can cause inflammation that leads to atherosclerosis. Even more important, when blood is removed and stored, there is always the risk that tiny blood clots will be introduced. When the stored blood is re-infused, these tiny clots increase the risk of arterial blockage and heart attack. They also increase the risk of stroke.

Anyone who gets a transfusion runs similar risks. However, these risks are well worth taking when they are medically necessary, such as when you have lost a lot of blood from an operation or an accident. When the choice is death due to blood loss or accepting a higher risk of heart attack or stroke, the choice of a transfusion is easily made.

Stroke

Your brain occupies only about 2% of your body, but it requires about 20% of your oxygen. The brain is a very active organ, and like the heart it requires an uninterrupted supply of oxygen. All the problems that affect the heart can affect the brain as well (**FIGURE 12.25**, p. 371). If atherosclerosis narrows arteries in the brain or a blood clot shuts down the blood supply, oxygen-deprived brain cells begin to die. This is known as a stroke. Like a heart attack, a stroke can be fatal. However, the brain carries out many functions, while the heart has only one main function. The consequences of a stroke are more variable and depend on where in the brain it occurs (Chapter 13). Some "mini-strokes" may not have any detectable effect. Other strokes may result in the loss of speech or the inability to understand speech, loss of vision, and loss of the ability to move limbs.

Asthma

Asthma is an example of a respiratory disease that is due to a natural process going awry. It is an inappropriate allergic response that constricts the airways

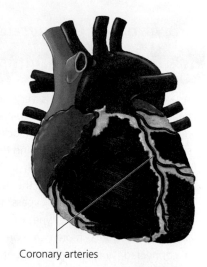

Coronary arteries

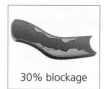
30% blockage

50% blockage

90% blockage

99% blockage

FIGURE 12.24 Heart attacks occur when the coronary arteries become clogged with plaque, which reduces blood and oxygen flow, and heart cells begin to die. Most people will feel symptoms of a heart attack with 50% or greater blockage.

Technology Connection

Cardiopulmonary Bypass Machine

As you might imagine, it is very difficult to operate on a beating heart. The contractions make delicate operations like replacing a valve or opening a blocked artery almost impossible. On the other hand, stopping the heart for long periods is not feasible since cells begin to die almost immediately, particularly in the brain and in the heart itself.

The cardiopulmonary bypass machine, or heart-lung machine, has made these types of operations possible, safer, and therefore much more common. Essentially this machine is connected "inline" with the patient's circulatory system and takes over the functions of both the heart and the lungs. With the machine in place, part of the circulation takes place outside the patient's body, where blood is oxygenated and pumped back into the body, bypassing the heart and lungs. The heart can then be chemically stopped while the operation takes place.

To see how this works, we can trace the path of blood outside the patient's body. First a tube is inserted into one of the body's large veins. This tube removes oxygen-poor blood from the body and directs it to a reservoir. The reservoir provides a continuous supply of blood to a pump, which directs the blood to an oxygenator. As the name implies, the oxygenator brings the blood into contact with oxygen, which diffuses into the blood just as it does in the lungs. Carbon dioxide diffuses out at the same time. The temperature of the oxygenator can be controlled to cool the blood, which lowers the patient's body temperature. This helps to lower metabolic rate and to minimize the need for oxygen while the operation proceeds. Oxygenated blood leaves the oxygenator and is pumped into a tube inserted into a major artery. Pressure from the pump gently forces blood throughout the body in the absence of the heartbeat.

The cardiopulmonary bypass machine contains a number of features that are not found in our circulatory system, such as a heat exchanger that warms or cools the blood. There are also a number of filters to remove tiny debris that

A cardiopulmonary bypass machine is used to maintain the body's vital functions while the heart is stopped temporarily during surgery.

may become trapped in the blood and block capillaries, and devices to prevent air bubbles from accumulating. However, most of the machine's features are there to simulate the adaptations that have evolved in our own circulatory and respiratory systems.

The best example of a design feature that simulates an adaptation is the oxygenator itself. The oxygenator allows blood to flow through a chamber with millions of tiny hollow rods. The oxygenator is attached to a pump, like the respiratory pump, which forces oxygen through the rods. These rods are analogous to the lung's alveoli: they are thin, extremely numerous, provide a very large surface area, and are filled with oxygen. Oxygen diffuses from these tubes into the blood, to be pumped back into the body. Carbon dioxide diffuses into the rods and is forced out the opposite end of the oxygenator. Although the surface area of the oxygenator is much smaller than that of the lungs, the rods are very thin and in close contact with the blood. These oxygenators are able to transfer over half a liter of oxygen per minute to the blood. ::

of the respiratory system, making it difficult to breathe. An asthma attack can be triggered by many things, specific to each person: pollen, animal fur or dander, cigarette smoke, cold air, shellfish, stress, and even exercise can bring on an asthma attack.

When the sensitive tissues of the airways are exposed to a particular irritant, the smooth muscles in the bronchioles contract, or spasm. Unlike the bronchi, bronchioles are not supported by rings of stiff cartilage. Consequently, the muscle spasms reduce or even close the airway. The problem is made even worse when the cells lining the respiratory tract release mucus.

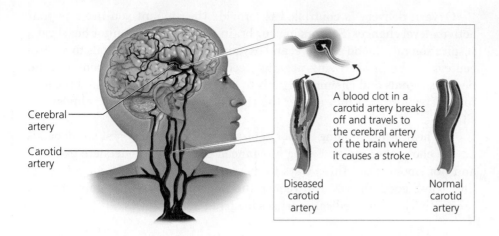

Cerebral artery

Carotid artery

A blood clot in a carotid artery breaks off and travels to the cerebral artery of the brain where it causes a stroke.

Diseased carotid artery

Normal carotid artery

FIGURE 12.25 Strokes occur when atherosclerosis narrows the arteries in the brain, or a blood clot shuts down the blood supply, and brain cells begin to die.

The inflammation, reduced diameter of airways, and mucus all make breathing difficult. Other physical consequences include wheezing, coughing, and a feeling of chest tightening.

Although asthma is incurable, it is treatable. First, a person with asthma learns to identify his or her triggers and reduce exposure to them. This behavior prevents many attacks from happening. Second, in the event of an asthma attack, an individual can use an inhaler to deliver drugs that can open constricted airways and restore breathing. Finally, there are also medications, some in inhaler form, that can be used to reduce tissue inflammation.

The incidence of asthma has skyrocketed in industrialized nations. According to the Centers for Disease Control and Prevention (CDC), the number of children with asthma more than doubled between 1980 and the mid-1990s. The reasons for this rise are not entirely understood, but several studies support the intriguing notion that *reduced* exposure to infection, due to improved hygiene, actually increases the chance of developing asthma.

For example, a 1999 study done in Austria showed that children from rural areas who had more exposure to animals were less likely to have asthma. Similarly, a 2000 study from Arizona revealed that children who attended day care or who had older brothers or sisters got sick from infections more often but were less likely to develop asthma. Some scientists hypothesize that asthma is the result of an "untrained" immune system. They argue that we need to reconsider the developmental role of routine childhood infections so that our immune systems do not overreact when exposed to potential irritants.

Biology in Perspective

What if your body doesn't get the oxygen it needs? As we have seen, if you are an athlete, you may be tempted to try blood doping to provide more oxygen, even though this practice is risky physiologically and may get you banned from your sport. If certain parts of the body are deprived of oxygen, atherosclerosis, blood clots, a heart attack, or a stroke may occur.

But under most conditions, your body does a good job of providing the oxygen you need. The respiratory and circulatory systems have evolved anatomical adaptations to ensure sufficient oxygen delivery.

These systems have biochemical adaptations as well. We have red blood cells packed full of hemoglobin that can carry 40 times more oxygen than the amount dissolved in blood.

Oxygen delivery is controlled to provide the amount you need as your activity level changes. Sensors in your brain continuously monitor heart rate, respiration rate, blood pressure, and oxygen levels and send signals to control centers in the brain. Using negative feedback, these control centers send nervous signals and chemical signals to adjust your circulatory and respiratory systems so that they deliver the right amount of oxygen for almost any situation.

The healthy functioning of our circulatory and respiratory systems depends on the choices we make. Eating a balanced diet, getting adequate exercise, and not smoking are three ways to increase our chances of a long, healthy life. These good choices also reduce the potential costs to society of work productivity lost or medical expenses incurred due to illness.

Chapter Summary

12.1 How Much Oxygen Do You Need?

Explain how oxygen needs vary depending on circumstances.

- Human adults use between 0.2 liters of oxygen (at rest) and 3 liters of oxygen (during heavy exercise) per minute.
- At rest, all of your 5 liters of blood circulate completely in about one minute.
- Hemoglobin in red blood cells increases the amount of oxygen blood can carry.
- When exercising, you use more oxygen than can be delivered.

12.2 How Is the Circulatory System Structured?

Describe the role of the different components of the circulatory system.

- Blood flows only one way in the circulatory system.
- The branching structure of the circulatory system increases surface area and provides for the transport and distribution of oxygen and other chemicals.
- The heart consists of two pumps lying side by side. The right pump receives used blood from the body and pumps it to the lungs. The left pump receives blood from the lungs and pumps it to the rest of the body.
- Arteries pump blood away from the heart and branch into smaller vessels, the smallest of which are arterioles.
- Arterioles break into capillaries, where oxygen, nutrients, and chemical signals are delivered to cells while carbon dioxide and wastes are picked up.
- Blood flows from capillary beds to venules, which merge into veins that transport blood back to the heart.
- Blood is composed of two parts, plasma and blood cells.

12.3 How Is the Respiratory System Structured?

Describe the role of the different components of the respiratory system.

- Lungs are two large sacs divided into lobes.
- A network of branched airways starts with the trachea, leading deep into the lungs, which branch into bronchi and then bronchioles, which end in tiny sacs called alveoli.
- Alveoli dramatically increase the surface area of the lungs.
- Oxygen enters and carbon dioxide leaves the bloodstream through the surface of the alveoli.
- Lungs pump air into and out of the lungs with muscles attached to your ribs and the diaphragm.
- At rest, the diaphragm forms a dome that extends into the chest. During inhalation, it contracts and flattens out, increasing the volume of the chest cavity.
- Air can enter the trachea either through the nose or through the mouth.

12.4 How Do You Breathe?

Explain how and where oxygen and carbon dioxide diffuse into and out of blood and the role of hemoglobin to facilitate oxygen transport.

- Breathing moves oxygen into your blood and carbon dioxide out.
- The right side of the heart gets the "used" blood and sends it to the alveoli of the lungs, where oxygen diffuses into and carbon dioxide diffuses out of the blood.
- Blood then goes into the left side of the heart, where it is pumped through the aorta to the rest of your body.

- Oxygen delivery occurs through capillary beds via diffusion.
- Oxygen binds tightly to the hemoglobin in the lungs and is released in the capillary beds.
- Once dissolved, carbon dioxide forms carbonic acid, which is converted into carbon dioxide in the lungs, where it diffuses out of the blood and is ultimately exhaled.

12.5 How Does Blood Pressure Vary Throughout Your Body?

Describe the differences in blood pressure in different positions in the circulatory system, as well as mechanisms that ensure blood flow.

- Blood pressure is lowest at rest and highest during exercise.
- The pulmonary circuit operates at a lower pressure than the systemic circuit. The highest blood pressure occurs in the left ventricle and the aorta.
- Blood pressure drops as blood moves through the systemic circuit.
- The venous side of the systematic circuit is not under pressure; muscle pressure is responsible for blood movement.
- Friction and the increasing volume of branched vessels decrease the pressure along the arteries.

12.6 How Are the Respiratory and Circulatory Systems Controlled?

Assess the effect that changes in the blood and nervous system have on respiration, heart rate, and blood pressure.

- The medulla oblongata and pons detect changes in blood pH and nerve firings to control breathing.

- Blood pressure sensors respond to stretch in the arteries when the heart beats.
- When the pressure is too high, the medulla oblongata and pons send signals to slow down the heart. When the pressure is too low, the medulla oblongata and pons send signals to speed up your heart rate.
- The pacemaker, the medulla oblongata, and your spinal cord working together determine your heart rate.
- Adrenaline increases your heart rate when you are nervous, excited, or frightened, and acetylcholine slows down the heart.

12.7 What Happens When the Circulatory and Respiratory Systems Malfunction?

Evaluate medical advice concerning the treatment and prevention of atherosclerosis, heart attacks, and asthma.

- Atherosclerosis occurs when an arterial wall is inflamed.
- Plaques increase the risk of a blocked artery.
- Genetics can play a role in atherosclerosis, as can lifestyle choices, including smoking, poor diet, low activity level, and drug use.
- During a heart attack, coronary arteries become clogged and kill heart cells.
- Strokes occur when brain cells are deprived of oxygen because arteries are narrowed or blocked.
- Asthma is an inappropriate allergic response that constricts the airways of the respiratory system, making it difficult to breathe.

Key Terms

alveolus 354
artery 350
atherosclerosis 367
atrioventricular valve 349
atrium 349
blood pressure 360
bronchiole 354
bronchus 354
capillary 350
circulatory system 344

diaphragm 355
gas exchange 357
hemoglobin 346
medulla oblongata 363
microcirculation 352
plaque 367
plasma 353
platelet 353
pons 363
pulmonary circuit 348

respiratory system 344
semilunar valve 349
sphincter 350
systemic circuit 348
trachea 354
vein 352
ventricle 349
white blood cell 353

Review Questions

1. What is blood doping? List one benefit and one risk of the procedure.

2. Make a table that lists how much oxygen is used by the following: a person at rest; a person during exercise; a world-class athlete during performance.

3. Does your blood absorb all of the oxygen you take into your lungs? Why or why not?

4. Which of the following is a benefit of the branching structure of the circulatory system?

   a. It dramatically increases the total surface area of blood vessels, ultimately leading to the delivery of more oxygen.

   b. It dramatically decreases the total surface area of blood vessels, ultimately leading to the delivery of more oxygen.

   c. It dramatically decreases the total surface area of blood vessels, ultimately leading to the delivery of more carbon dioxide.

   d. It allows blood vessels to get progressively larger.

   e. It ensures that the blood carries less carbon dioxide.

5. Briefly explain why the heart can be considered to be two pumps side by side. What is the function of each pump?

6. Define these terms: artery, arteriole, capillary bed, venule, and vein.

7. What is microcirculation?

   a. The process by which blood moves through the arteries.

   b. The process by which blood moves through veins

   c. The process by which oxygen moves from the heart to the lungs

   d. The process by which different capillary beds open and close to divert blood where it is needed

   e. The process by which bronchioles open and close to allow oxygen to enter the alveoli

8. How do capillary sphincters regulate microcirculation?

9. List the constituents of plasma. List the components of the cellular part of blood.

10. How does hemoglobin increase the amount of oxygen blood takes in from the lungs?

11. What is one reason why blood pressure drops as blood moves farther from the heart?

   a. The total volume of the blood vessels gets smaller as the arteries branch farther out.

   b. The total volume of the blood vessels gets larger as the arteries branch farther out.

   c. The hormone adrenaline is released into the blood.

   d. The hormone acetylcholine is released into the blood

   e. The concentration of carbon dioxide is higher the farther the blood moves away from the heart.

12. What happens when the medulla oblongata and the pons detect a decrease in blood pH? What does a decrease in blood pH say about the body's current need for oxygen?

13. What is atherosclerosis? Why does atherosclerosis increase the risk of heart attack or stroke?

14. Which of the following is NOT something that you should do to improve heart and lung health?

   a. Quit smoking, if you do.

   b. Maintain a healthy weight.

   c. Exercise regularly.

   d. Eat a reduced-fat diet.

   e. Consume a high amount of proteins like milk and cheese.

The Thinking Citizen

1. We have discussed two concentration gradients relevant to gas exchange: in the alveoli capillary beds and in the capillary beds that deliver oxygen to active cells. Explain why each gradient exists, and explain how each helps move oxygen in the proper direction.

2. Briefly discuss the need for a large surface area to maximize gas exchange (or any other situation in which materials move from one area to another across a membrane). Describe the ways in which the circulatory and respiratory systems have adapted to increase their gas exchange surface area.

3. Hemoglobin has two functions: to bind to oxygen and to release oxygen. How can one molecule do both? Explain how this property of hemoglobin allows it to have different functions in different parts of the body. Also explain how this property allows hemoglobin to deliver oxygen to the parts of the body where it is most needed.

4. When a medical professional draws blood from a patient, why is blood taken from a vein instead of an artery?

5. The way the body controls respiration rate is very similar to the way a thermostat controls temperature in a house. Describe the control of respiration rate in this context. The thermostat works by keeping temperature constant—what is the corresponding value in the control of respiration? How does negative feedback work in this process?

The Informed Citizen

1. Should blood doping be outlawed in sports? Some argue that there is nothing wrong with putting your own blood back in you and that doping simply speeds up the process of training. Others argue against it because of the risks involved and the fact that the enhanced performance has not been "earned" by the athlete. What is your position? Why?

2. Erythropoietin (EPO) is a hormone produced by the kidneys. When the kidneys sense low levels of oxygen in the body, they release EPO. EPO stimulates bone marrow to produce more red blood cells. Injecting an athlete with EPO is similar to blood doping, and in fact the two procedures are often used together. Why might EPO be considered a performance-enhancing drug? Why might it be considered a way to draw on the body's own natural processes to improve performance, like weight training to increase muscle mass?

3. According to the American Heart Association, the total cost for treatment, medication, and loss of productivity from heart disease was about $444 billion in 2010. Heart disease is the leading cause of mortality and disability in the United States, accounting for over 650,000 deaths (about 25% of all deaths). But many heart disease cases can be cured or minimized by simple lifestyle choices like quitting smoking, losing weight, maintaining a healthy diet, and exercising. Can we afford to rely on individual lifestyle choices to combat heart disease? Or will the government need to take a stronger role to improve public health and to reduce the costs of heart disease? Explain your answer.

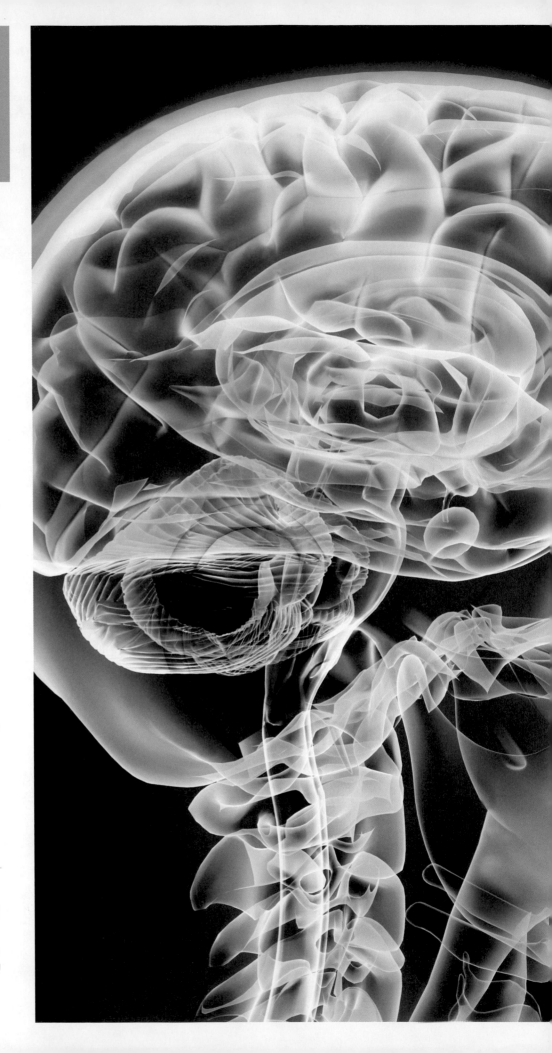

CHAPTER LEARNING OBJECTIVES

After reading this chapter, you should be able to answer the following questions:

The brain is the most obvious and important part of a nervous system that runs throughout the body. Understanding the biological basis of human thought, behavior, and emotion is probably the biggest challenge facing biologists in the 21st century.

CHAPTER 13

The Nervous System

Does Your Brain Determine Who You Are?

Becoming an Informed Citizen . . .

Our growing knowledge of the brain and the workings of the nervous system that runs throughout our bodies is exciting to many and unsettling for some. To what extent is our being human a product of brain and nervous system function? Neuroscientists are addressing some profound questions that in previous times were the exclusive domain of philosophy and religion.

e often say that we think with our brain and feel with our body. Yet that intuition is only part of the story of who we are. The brain is only the most obvious and important part of a nervous system that runs throughout the body, right down to our fingertips. Understanding the biological basis of human thought, behavior, and emotion is probably the biggest challenge facing biologists in the 21st century.

For example, an ever-growing number of people are being diagnosed with Alzheimer's disease and other types of memory loss disorders. Learning how the brain stores and retrieves memories may well lead to prevention of and effective treatments for these syndromes. Similarly, strokes, head injuries, and spinal injuries often cause irreparable damage to *cognitive* and *motor* functions: both thinking and moving are impaired. It is likely that new insights about the brain and nervous system will reveal ways to repair damage and restore function.

Research on the brain and nervous system is also likely to make us think more deeply about what it means to be human. Can we explain our feelings, desires, and emotions entirely by mapping out brain function? Do we make decisions freely, or are we "wired" to make certain choices? Are we responsible for our behaviors? What if that behavior depends on mind-altering substances, and how do those work? Does who we are as humans reside solely in the brain? Let's look at the astonishing experiences of Phineas P. Gage.

case study

"The Only Living Man with a Hole in His Head"

FIGURE 13.1 In the mid- to late 1840s, to clear a path for the Rutland-to-Burlington railroad in Vermont, workers had to blast through boulders and rock ledges—a dangerous task.

September 13, 1848, was a lovely autumn day in Cavendish, Vermont. Phineas Gage, a sturdy 25-year-old man, led a crew in the dangerous task of blasting through boulders and rock ledges to clear the path for the construction of the Rutland-to-Burlington railroad in Vermont (**FIGURE 13.1**).

The crew worked as a team. One man drilled holes in the bedrock at a specific angle and depth. Gage's assistant then placed coarse-grained gunpowder in each hole, one at a time. Gage used the tapered end of a tamping iron to stick a rope fuse into the powder. The assistant filled the rest of the hole with sand, and Gage pressed or tamped the sand down tightly with the blunt end of the tamping iron. The idea was to direct the explosion of the gunpowder down into the rock, blasting it to bits. Finally, men would come in and manually dig to remove the rubble.

Gage and his team had done these tasks thousands of times. They would set up for the blast, warn everyone, light the fuse, and run for cover. It sounds like a dangerous job, and it was, but Gage was so skilled that his men trusted him and followed his orders.

But on this day, something went terribly wrong. Somehow, sand was never poured down one of the holes. Gage turned his head to glance over his right shoulder, and the blunt end of his tamping iron slipped into the hole, sparking when it hit the rock. The spark lit the blasting powder and sent the tamping rod shooting into the air like a rocket. It landed with a loud clatter more than 100 feet away.

An examination of the rod revealed a gruesome detail about its flight—the rod was covered with blood and brain. Looking away from where the rod had landed and through a haze of smoke, the men saw Gage lying on the ground, his arms and legs twitching. The tamping iron had shot *through* Phineas Gage's head, up into the air, and back down to Earth, all in less than a second.

Everyone on the crew rushed over, and then an even more unexpected thing occurred. Gage sat up and began talking. When Gage's men brought him to town for medical care, he was able to walk, with minimal assistance, and to carry on normal conversation.

Gage's wounds were initially treated by Dr. John Harrow with the assistance of Dr. Edward H. Williams. The damage they observed was considerable. The tamping iron that shot through Gage's head was 3 feet, 7 inches long and weighed 13¼ lbs. Although the rod was 1½ inches in diameter, the exit hole through the top of his head was 3½ inches wide. The rod entered Gage's head just under his left cheekbone. It passed behind his left eye and through the front of his brain. It exited just above the hairline toward the middle of the forehead (**FIGURE 13.2**).

Harlow cared for Gage over the next two months. Harlow repositioned large pieces of Gage's skull and pulled the scalp over the hole to cover the unprotected brain. He kept the wound clean and drained infections as they occurred. Gage's strong constitution and good health enabled him to fight off infections that would have killed most men. Harlow declared Gage "healed" physically, and Gage eventually resumed his duties working for the railroad. However, Harlow and Gage's coworkers and friends soon suspected that something was not quite right.

Before the accident, Gage had been a responsible, hard-working, dependable man. He was well respected, a great favorite of his coworkers and considered by his employers their "most efficient and capable foreman." Now there was a "new" Gage who was unreliable and sometimes nasty. He was stubborn and foul-mouthed. He also couldn't make up his mind about what to do; he would make big plans but fail to follow through. He could no longer work effectively, and his employer had to let him go. Even his friends gave up on him, remarking that "Gage is no longer Gage."

What exactly happened to Phineas Gage? He survived his dreadful accident and retained his hearing and vision. He had no paralysis. He walked normally and could use his hands to do fine motor tasks. Gage also had no difficulty with speech or language. But his disposition, likes and dislikes, goals, and even dreams for his future had all changed dramatically. *Why?*

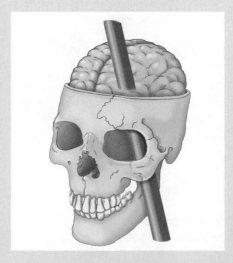
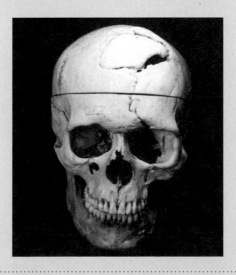

FIGURE 13.2 The tamping rod entered Phineas Gage's head just under his left cheekbone, passed behind his left eye, went through the front of his head, and exited just above the hairline toward the middle of his forehead.

13.1 What Are the Parts of the Human Brain, and What Do They Do?

When the newly injured Gage told people who had not witnessed his accident that an iron rod had shot through his head, he was not widely believed. In fact, Dr. Harlow's published report of the case and presentation at a medical conference were met with skepticism. It is easy to see why. Physicians, and others, thought it was unlikely that a brain could be so dramatically damaged without significant loss of function. Some thought that the whole brain operated together as a unit, and others argued that the brain was composed of "centers," each responsible for a particular task or behavior.

Both views are correct: the brain does operate as a whole, *and* there are activity centers. At the time of Harlow's report, however, many physicians thought that Gage's injury was evidence that the brain consisted of functional centers. They argued that the hole blown in his head apparently did not disrupt anything important. The real information about Gage's behavioral changes was not widely known until 1868, when details of the case were made fully public.

The average brain weighs around 3 pounds (**FIGURE 13.3**). Protected by the bones of the skull and the **meninges**, a durable connective tissue, the brain is organized in a modular manner, as a series of units. The various modules of the brain are themselves organized into smaller components and often further divided. Brain function sometimes comes from a subsection of the brain, but more usually from the communication and interactions from many parts. We will start with the top-level brain components: hindbrain, midbrain, and forebrain.

Hindbrain

The **hindbrain** is located at the base of the skull, right above the spinal cord (**FIGURE 13.4**). The hindbrain is responsible for the control of many basic bodily functions necessary for survival. It is composed of three subunits:

meninges :: a durable connective tissue that helps protect the bones of the skull

hindbrain :: located at the base of the skull, right above the spinal cord, and responsible for the control of many basic bodily functions necessary for survival; composed of the medulla oblongata, cerebellum, and pons

FIGURE 13.3 The average brain weighs about 3 pounds. Protected by the bones of the skull and durable connective tissue, known as meninges, the brain is organized in a modular manner as a series of units.

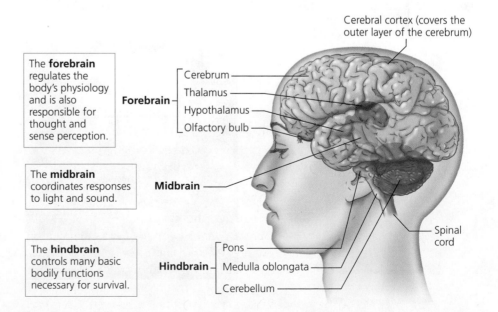

The **forebrain** regulates the body's physiology and is also responsible for thought and sense perception.

The **midbrain** coordinates responses to light and sound.

The **hindbrain** controls many basic bodily functions necessary for survival.

Cerebral cortex (covers the outer layer of the cerebrum)

Forebrain — Cerebrum
Thalamus
Hypothalamus
Olfactory bulb

Midbrain

Spinal cord

Hindbrain — Pons
Medulla oblongata
Cerebellum

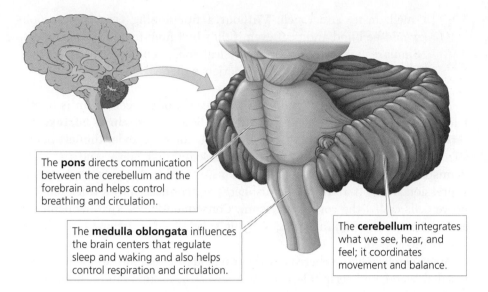

The **pons** directs communication between the cerebellum and the forebrain and helps control breathing and circulation.

The **medulla oblongata** influences the brain centers that regulate sleep and waking and also helps control respiration and circulation.

The **cerebellum** integrates what we see, hear, and feel; it coordinates movement and balance.

FIGURE 13.4 The hindbrain is located at the base of the skull, right above the spinal cord. It is responsible for the control of many basic bodily functions necessary for survival.

- *Medulla oblongata:* influences the brain centers that regulate sleep and waking. It also helps to control respiration and circulation (Chapter 12) and coordinates the motor responses that accompany certain reflexes like coughing or sneezing.
- *Cerebellum:* responsible for integrating what we see, hear, and feel. It coordinates movement and balance and also plays an important role in the control of manual dexterity and the use of language.
- *Pons:* directs the communication between the cerebellum and the forebrain. The pons also helps control breathing and circulation.

Midbrain

Along with the medulla oblongata and the pons, the **midbrain** forms the **brainstem** (**FIGURE 13.5**). All information moving to and from other brain regions passes through the brainstem, which selects what to send on. The midbrain itself plays an important role coordinating responses to light and sound.

Forebrain

By far the most highly developed and largest region of the brain, the **forebrain** is composed of the olfactory bulbs, thalamus, hypothalamus, and cerebrum (**FIGURE 13.6**):

- *Olfactory bulbs:* specialists in sensory information about smell.
- *Thalamus:* a relay switch. It sorts incoming and outgoing data.
- *Hypothalamus:* a master control center. It regulates body temperature, blood pressure, hunger, sex drive, emotions, thirst, and our physiological reactions to stress.
- *Cerebrum:* the largest and most sophisticated component of all. Accounting for 80–85% of brain mass in humans, the cerebrum is responsible for many of the characteristics that most people consider distinctly human: reasoning, mathematical ability, artistic ability, imagination, language, and personality. The cerebrum is responsible for creating the perceptions we gather with our senses: sight, hearing,

midbrain :: forms the brainstem along with the medulla oblongata and the pons; plays an important role in coordinating responses to light and sound

brainstem :: comprised of the medulla oblongata, the pons, and the midbrain; all information moving to and from other regions of the brain passes through this area

forebrain :: the most highly developed and largest region of the brain; composed of the olfactory bulbs, thalamus, hypothalamus, and cerebrum

cerebrum :: the largest and most sophisticated component of the forebrain, accounting for 80–85% of brain mass in humans; responsible for many of the characteristics that most people consider distinctly human: reasoning, mathematical ability, artistic ability, imagination, language, and personality

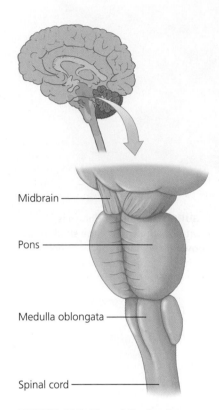

Midbrain

Pons

Medulla oblongata

Spinal cord

FIGURE 13.5 The midbrain plays an important role in coordinating responses to light and sound. Along with the medulla oblongata and the pons, it forms the brainstem, through which all information moving to and from other brain regions passes.

cerebral cortex :: the surface of the cerebrum that is highly folded and appears wrinkled

limbic system :: comprised of parts of the cerebrum, hippocampus, and amygdala; it is in charge of physical drive and instincts and is partly responsible for emotions, learning, and memory

smell, taste, and touch. Without a functioning cerebrum, people would be blind and deaf even if they had functioning eyes and ears. Similarly, tastes, odors, and physical sensations don't exist without the cerebrum enabling us to experience them.

A Closer Look at the Cerebrum The surface of the cerebrum is highly folded, producing a wrinkled structure called the **cerebral cortex**. The cerebrum has two hemispheres. In most but not all people, the left hemisphere has areas specializing in language, logic, and math, and the right hemisphere has areas devoted to emotions, intuitive thinking, and artistic expression. The left and right hemispheres are connected by a thick band of nerve cells called the corpus callosum. Consequently, the two hemispheres are in constant communication, and the specialized functions of each are deeply integrated.

Each cerebral hemisphere is further divided into four lobes—frontal, temporal, parietal, and occipital—each with specialized functions (**FIGURE 13.7**):

- *Frontal lobe:* speech; smell; problem solving; personality; and motor control.
- *Temporal lobe:* perception and processing of information from the eyes and ears; language comprehension; and ability to recognize patterns including faces.
- *Parietal lobe:* perception and processing of physical touch; and taking information from all senses and turning it into a single perception.
- *Occipital lobe:* perception and processing of information from the eyes.

Limbic System The forebrain is also home to the **limbic system** (**FIGURE 13.8**). Comprising parts of the cerebrum and two structures called the hippocampus and amygdala, the limbic system is in charge of physical drive and instincts. In addition, the limbic system is partly responsible for emotions, learning, and memory.

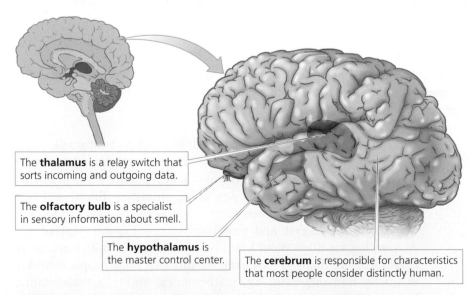

The **thalamus** is a relay switch that sorts incoming and outgoing data.

The **olfactory bulb** is a specialist in sensory information about smell.

The **hypothalamus** is the master control center.

The **cerebrum** is responsible for characteristics that most people consider distinctly human.

FIGURE 13.6 The forebrain is the most highly developed and largest region of the brain. It is responsible for thought and sense perception; it also regulates the body's physiology.

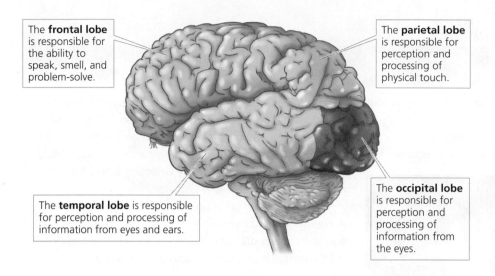

The **frontal lobe** is responsible for the ability to speak, smell, and problem-solve.

The **parietal lobe** is responsible for perception and processing of physical touch.

The **temporal lobe** is responsible for perception and processing of information from eyes and ears.

The **occipital lobe** is responsible for perception and processing of information from the eyes.

FIGURE 13.7 The cerebrum is divided into two hemispheres, and each hemisphere is further divided into four lobes. Generally, the left hemisphere specializes in language, logic, and math, while the right is devoted to emotions, intuitive thinking, and artistic expression.

What does this catalogue of parts mean, and what does it tell us about how we think, feel, emote, create, sense, learn, move, and remember? Is it even possible to understand the biological basis of these abilities? The goal of neuroscience research is to understand the human brain and nervous system so that it is possible to prevent or treat the thousands of disorders that impact millions of people in profound ways. For a discussion of how doctors and scientists use **functional magnetic resonance imaging (fMRI)** to measure brain activity, see *Technology Connection: Measuring Brain Activity*.

Ironically, most of what we know about the relationship between specific brain structures and functions comes from humans who have suffered brain trauma due to injury or illness. Physicians have published descriptions of patients with behavioral symptoms similar to those of Phineas Gage. Using modern techniques, they have been able to identify what part or parts of the brain are involved. Let's take another look at Gage to figure out exactly what happened to his brain.

functional magnetic resonance imaging (fMRI) :: a radiology technique that measures brain activity by detecting blood flow

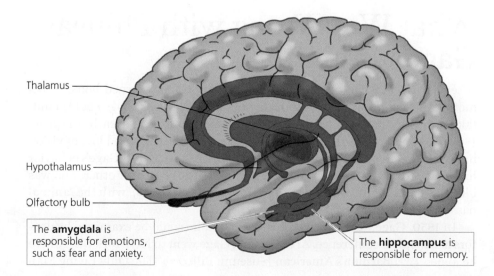

Thalamus

Hypothalamus

Olfactory bulb

The **amygdala** is responsible for emotions, such as fear and anxiety.

The **hippocampus** is responsible for memory.

FIGURE 13.8 The limbic system is comprised of the hypothalamus, hippocampus, and amygdala. It is involved in memory, emotion, physical drive, and instinct.

Measuring Brain Activity

Until the 1990s, there were only two ways to figure out which parts of the human brain were associated with a particular function or behavior. First, scientists could surgically disable or stimulate areas in animal brains, observe the

Auditory Decision Description Task

When the brain undergoes neural activity, blood flow to active brain regions increases dramatically. Functional magnetic resonance imaging can measure this flow.

responses, and attempt to relate the results to humans. Second, physicians could correlate functional changes and specific types of brain damage. Since the early 1990s, a technique called functional magnetic resonance imaging (fMRI), which measures brain activity by detecting blood flow, has revolutionized studies of brain activity.

When the brain undergoes neural activity, it needs a lot of energy. As a consequence, blood flow to active brain regions increases dramatically. fMRI can measure this flow, and the images it produces contain different colors for different levels of blood flow.

The initial studies done using fMRI confirmed that active brain areas were "lighting up." For example, scientists showed that Broca's area exhibited an increased blood flow when the test subject spoke.

Research using fMRI promises to answer many important questions about brain anatomy and function. A great deal has already been learned about neural circuits and systems involved in sensations, movement, and language. Practical applications of fMRI include planning for brain surgery and addressing chronic pain management.

Imaging brains also presents some bioethical challenges. It may soon be possible to identify brain activities that correlate with an increased likelihood of developing diseases like alcoholism or schizophrenia, or tendencies for behaviors like aggression or pedophilia. How sure are we about the reliability of this information? What should we do with this knowledge? Are there limits to personal privacy? ::

13.2 What Was Wrong with Phineas Gage's Brain?

After Dr. Harlow had pronounced him healed, Gage returned home to his mother in New Hampshire. Although still physically weak, Gage could count, take care of his personal needs, sing, understand, and speak clearly. In spring 1849, Gage returned to Cavendish, and except for the scars and loss of vision in his left eye, he seemed fit and ready for work. But he wasn't. Gage's behavior had changed. As Dr. Harlow described, he had little regard for others and was like "a child in his intellectual capacity and manifestations," with the "animal passions of a strong man."

In 1850, Gage was invited to Boston so that he could be examined by doctors at a medical conference. After Boston, Gage went to New York City, where he joined P. T. Barnum's American Museum. Billed as "The Only Living Man with a Hole in His Head," Gage and his tamping iron rod (which he always kept with him and which would eventually be buried with him on his death) were part of what was actually a "freak show" (**FIGURE 13.9**). He returned to New Hampshire in 1851 and got a job working with horses.

In 1852, Gage was hired by a man establishing a stagecoach line in Chile. Gage's skill with horses was well known, and so he was tasked with the Valparaiso–Santiago route. Driving a stagecoach is physically demanding, and it requires significant manual dexterity to control the horses. Gage drove seven hours a day on a regular schedule. He lived and worked in Chile for the next seven years.

It is obvious that Gage's injury had no effect on many of his abilities. Therefore, the regions of his brain responsible for those functions must have been untouched by the rod blasting through his head. It is also clear that Gage changed. Do the characteristics that were altered in Gage match up with the likely areas of his brain that were destroyed by the path of the rod?

In 1994, scientists used computer-imaging techniques and precise measurements of Phineas Gage's skull to calculate the tamping iron's most likely trajectory (**FIGURE 13.10**). The rod entered Gage's head below his left cheekbone and missed key areas on the sides and top of the forebrain, including the motor and sensory strips on top of the cerebrum. As a result, Gage was still able to maintain his balance, remember, and pay attention. The iron also missed a region in the left temporal lobe, **Broca's area**, which is responsible for producing language. If Broca's area had been damaged, it would have rendered Gage unable to speak. The main areas damaged were in the frontal lobe; as a result, the left hemisphere was affected more than the right.

Patients with damage to the same areas of the frontal lobes of the cerebrum display symptoms like Gage's. They do well in logic or math tests but have

FIGURE 13.9 Shown here is Phineas Gage. He posed with his tamping iron, which he always kept with him until he died.

Broca's area :: a region of the left temporal lobe responsible for producing language

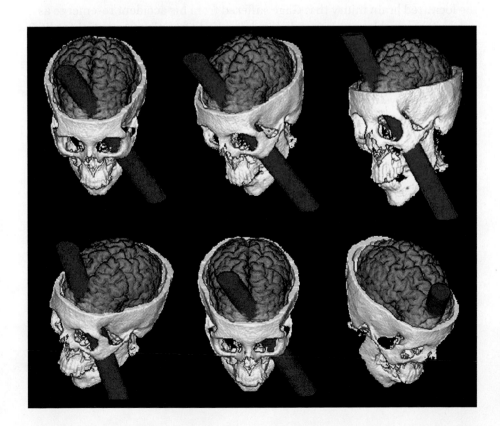

FIGURE 13.10 In 1994, scientists used computer-imaging techniques and precise measurements of Phineas Gage's skull to calculate the tamping iron's most likely trajectory.

trouble making decisions about personal and social things. They seem rather flat emotionally and exhibit little empathy. They are unable to "read" people and don't know how to interact socially. The change in Gage was evident to all who knew him. The areas of his brain damaged in the accident permanently altered his personality.

How is this possible? How could any organ, even the brain, be responsible for a person's identity? We really don't have a complete answer to this question, but let's tackle it by looking at how mental functions, simple and complex, are carried out by neural circuits in the brain and also how the entire nervous system engages different parts of the body.

13.3 How Is the Nervous System Organized, and What Does It Do?

In 1859, a tired and unhealthy Phineas Gage left Chile and made his way to California to find his family. Gage showed up at his sister's house too ill to take care of himself. Although he recovered after a few months and got work as a farm laborer, Gage continued to have difficulty working with people. In February 1860, Gage suffered an **epileptic seizure**, an electrical storm in the cells of his brain. Unlike his previous brain problems, a seizure is not localized. Large areas of the brain are involved, as is the entire body, which shakes and tenses its muscles.

This first seizure was followed by others. In fact, the seizures came in quick succession, with ever-shorter intervals between them until they were continuous. Finally, on May 21, 1860, Gage died; he was 38 years old. How did the localized brain injury that Gage suffered from his accident re-emerge as a more global problem—ultimately involving body functions essential for life? And how did misfiring in the brain lead to his death?

Overview of Organization

The nervous system is made up of specialized signaling cells called **neurons** (**FIGURE 13.11**) and various support cells that protect and insulate the neurons. The brain itself has more than 10 billion neurons, and the rest of the nervous system has even more. Each neuron is capable of communicating with at least 1000 other neurons. The potential for generating complex signals and information is staggering.

An analogy can give us a feel for how complex and variable the outputs can be that come from a communication network like the nervous system. A standard piano has 88 keys. Each key plays a specific note. The keys can be pressed in any combination and number to produce a seemingly unlimited variety of music. If put together in all possible combinations, 88 notes can make 2^{88}, or 100 trillion trillion, different "musical communications." Similarly, neurons communicating in all possible combinations can produce a virtually unlimited variety of outputs that enable us to think, walk, speak, and feel.

This seems almost like *too* much flexibility. How is this amazing potential for communication directed? And are the effects of neurological disease and injury to the nervous system due to the disruption of a well-coordinated communication network?

epileptic seizure :: an electrical storm in the cells of the brain; generally involves large areas of the brain and the entire body, which shakes and tenses its muscles

neuron :: a specialized signaling cell in the nervous system; the brain has more than 10 billion neurons, and the nervous system has even more

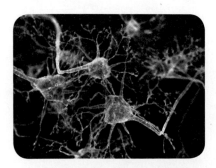

FIGURE 13.11 The brain has more than 10 billion specialized signaling cells called neurons, and the rest of the nervous system has even more. Each neuron is capable of communicating with at least 1000 other neurons.

The Central Nervous System

The brain is able to exert control over all of our functions and complex behaviors because it is connected, via a vast network of cells, to every part of the body. The brain and **spinal cord**—a bundle of nerve fibers in our backbone that serves as the communication pathway between the brain and the rest of the body—together comprise the **central nervous system (CNS)**, which controls the actions of the body (**FIGURE 13.12**). **Nerves** themselves are bundles of neurons wrapped together to form "cables."

The spinal cord carries sensory information from the body—touch, pain, and body position, for example—to the brain. The spinal cord also relays motor information from the brain to muscles, organs, and glands. A special population of cells called **interneurons** integrates the incoming and outgoing signals so that the body responds appropriately to stimuli. For example, if a person feels a mosquito land on his or her arm, this touch sensation will trigger movement to swat or brush the mosquito away, not movement to tap dance. The interneurons are located in the brain and spinal cord.

The spinal cord is also responsible for **reflexes**, movements in response to a stimulus (**FIGURE 13.13**). Reflexes can be simple or complex, but specific ones are always identical. They are also fast because they bypass the brain. For instance, most of us have experienced the "withdrawal reflex." If you unexpectedly touch a hot stove top, you will pull back your hand without even thinking. You will do it even before your brain has finished processing the information, "Oh, that's hot—ouch!"

The Peripheral Nervous System

The brain and the spinal cord need to communicate with the entire body, not just each other. A network of nerves radiating out from the CNS and throughout the body permits them to do so. This network, called the **peripheral nervous system (PNS)** (**FIGURE 13.14**), is intricately wired. It is physically connected to the CNS by cranial nerves that attach to the brain and spinal

spinal cord :: a bundle of nerve fibers in our backbone that serves as the communication pathway between the brain and the rest of the body

central nervous system (CNS) :: controls the actions of the body; comprised of the brain and spinal cord

nerve :: a bundle of neurons wrapped together to form "cables"

interneuron :: a special cell that integrates the incoming and outgoing signals so that the body responds appropriately to stimuli

reflex :: a movement in response to a stimulus

peripheral nervous system (PNS) :: a network of nerves radiating out from the CNS and throughout the body that enables the brain and spinal cord to communicate with the entire body

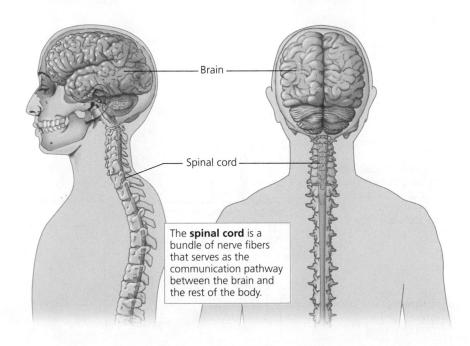

Brain

Spinal cord

The **spinal cord** is a bundle of nerve fibers that serves as the communication pathway between the brain and the rest of the body.

FIGURE 13.12 The central nervous system, which controls the actions of the body, is comprised of the brain and spinal cord.

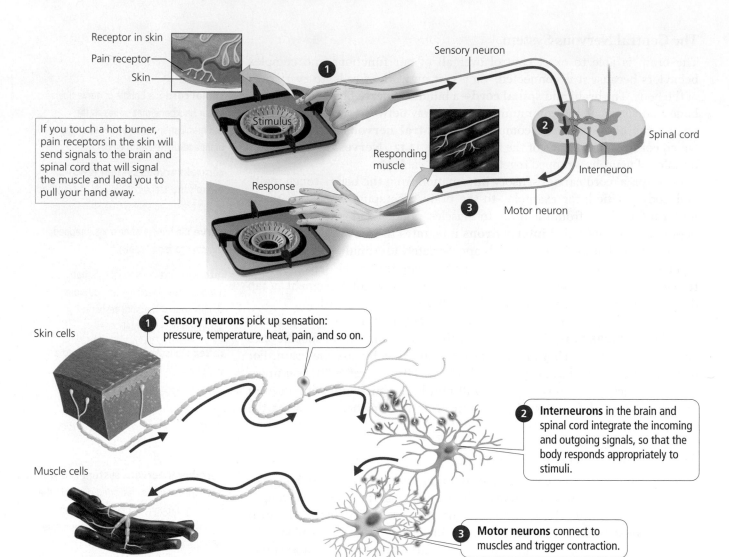

Receptor in skin
Pain receptor
Skin

Sensory neuron

1 Stimulus

If you touch a hot burner, pain receptors in the skin will send signals to the brain and spinal cord that will signal the muscle and lead you to pull your hand away.

Responding muscle

2

Spinal cord

Interneuron

Response

3

Motor neuron

1 **Sensory neurons** pick up sensation: pressure, temperature, heat, pain, and so on.

Skin cells

2 **Interneurons** in the brain and spinal cord integrate the incoming and outgoing signals, so that the body responds appropriately to stimuli.

Muscle cells

3 **Motor neurons** connect to muscles and trigger contraction.

FIGURE 13.13 Reflexes are simple or complex movements in response to a stimulus. They are fast, and they bypass conscious control by the brain.

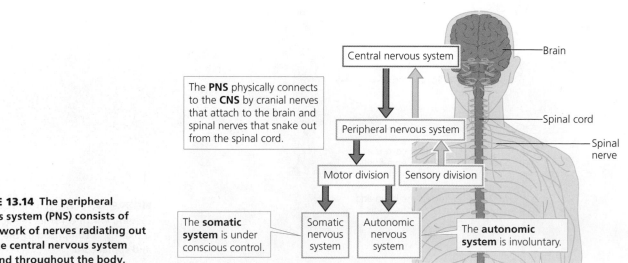

Central nervous system

Brain

The **PNS** physically connects to the **CNS** by cranial nerves that attach to the brain and spinal nerves that snake out from the spinal cord.

Peripheral nervous system

Spinal cord

Spinal nerve

Motor division

Sensory division

The **somatic system** is under conscious control.

Somatic nervous system

Autonomic nervous system

The **autonomic system** is involuntary.

FIGURE 13.14 The peripheral nervous system (PNS) consists of the network of nerves radiating out from the central nervous system (CNS) and throughout the body.

nerves that snake out from the spinal cord through the spaces between the bones of the spine. The cranial and spinal nerves branch out and connect to other nerves that connect to other nerves, and so on throughout the body.

Responsible for perceiving and carrying signals, the PNS is divided into two functional components: somatic and autonomic. The **somatic system** carries signals to and from the skeletal muscles, usually in response to an external stimulus. The somatic system is under conscious control. If you decide to move your foot, the somatic system controls the action. In contrast, the **autonomic system** is generally involuntary—or not under conscious control. It regulates the internal environment of the body by controlling the muscles of the heart and other internal organs as well as various glands.

The autonomic nervous system is further divided into the parasympathetic and sympathetic branches, which control organs in ways that balance function. The **parasympathetic branch** helps the body do things to gain and conserve energy. For example, it stimulates the digestive organs and decreases heart and breathing rates. The **sympathetic branch**, in contrast, prepares the body for intense action by increasing heart and breathing rates and diverting "nonessential" expenditures of energy. The sympathetic and parasympathetic branches of the autonomic system are in a state of constant interplay that coordinates the body's overall responses. When everything is calm, the parasympathetic branch dominates, but in the event of danger, excitement, or fear, the sympathetic branch rules (**FIGURE 13.15**). Phineas Gage never knew what hit him at the time of his accident. It does seem probable, however, that his coworkers experienced a jolt from their sympathetic branches as they ran to assist Gage and get help.

somatic system :: a functional component of the PNS that carries signals to and from the skeletal muscles, usually in response to an external stimulus; it is under conscious control

autonomic system :: a functional component of the PNS that regulates the internal environment of the body by controlling the muscles of the heart and other internal organs and glands; it is not under conscious control

parasympathetic branch :: part of the autonomic nervous system that helps the body do things to gain and conserve energy by stimulating digestive organs and decreasing heart rate and breathing rate

sympathetic branch :: part of the autonomic nervous system that prepares the body for intense action by diverting "nonessential" expenditures and increasing heart rate and breathing rate

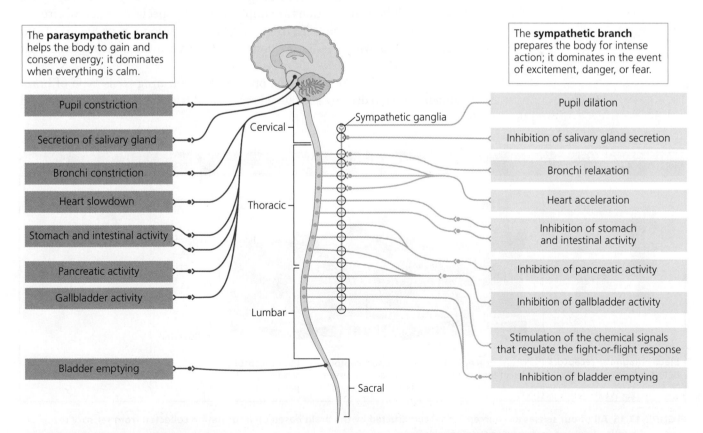

FIGURE 13.15 The autonomic nervous system is further divided into the parasympathetic and sympathetic branches.

Sensing

We don't see with our eyes, hear with our ears, or taste with our tongues. We don't even feel with our skin. All of our senses are *perceptions*—constructed by the brain based on information collected from sensory receptors on specific cells (**FIGURE 13.16**). When the brain malfunctions, as in hallucinations, it can create a vivid "real" world that does not actually exist. Similarly, if there is a flaw in the communication network, sensory information collected and sent to the brain may be misinterpreted or not noted at all.

Humans have five types of sensory receptors: pain receptors, thermoreceptors, mechanoreceptors, chemoreceptors, and electromagnetic receptors. As the name indicates, pain receptors register physical pain. Thermoreceptors detect temperature—hot, cold, and in between. Mechanoreceptors are sensitive to touch, pressure, motion, and sound. Chemoreceptors sense specific chemicals. Electromagnetic receptors detect light and electricity. These five types of receptors work in various combinations to send information to the brain for integration, thus producing our senses: sight, hearing, smell, taste, and touch. For a discussion of one scientist's investigations into chemoreceptors and **olfaction** (the sense of smell), see *Scientist Spotlight: Linda B. Buck.*

olfaction :: the sense of smell

Putting It All Together

The nervous system is too complex to depend on neurons always communicating independently with all other neurons. How is this problem resolved?

Neurons generally talk to a small subset of neighboring neurons organized into groups called **circuits**, which interconnect and communicate with each other. These circuits in turn form systems that communicate, which in turn form a "system of systems." The brain is actually a supersystem of systems. Within it, all mental functions are implemented by specialized neural circuits located in different regions. In this sense, brain function is both localized *and* interconnected with the entire body. It was this interconnectedness that probably killed Gage.

Gage's initial head injury was open and penetrating. This type of brain damage tends to destroy a localized region and with it specific functions and

circuit :: neurons organized into a group that interconnect and communicate with each other

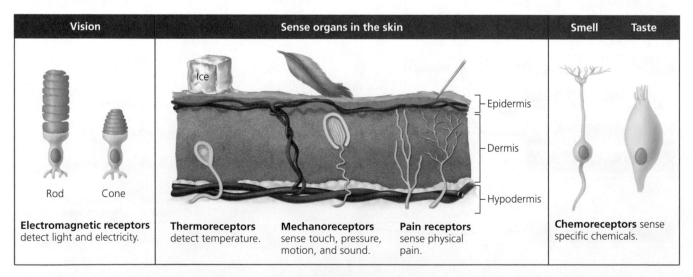

Vision	Sense organs in the skin	Smell Taste

Rod   Cone

Ice

Epidermis

Dermis

Hypodermis

Electromagnetic receptors detect light and electricity.

Thermoreceptors detect temperature.

Mechanoreceptors sense touch, pressure, motion, and sound.

Pain receptors sense physical pain.

Chemoreceptors sense specific chemicals.

FIGURE 13.16 **All of our senses are perceptions—constructed by the brain based on information collected from sensory receptors on specific cells.** Humans have five types of sensory receptors.

Scientist Spotlight

Linda B. Buck (1947–)

Even Nobel laureates can have trouble figuring out their futures when they are in college. Consider Linda Buck, winner of the 2004 Prize in Physiology or Medicine. She had first to be enchanted by the right question—the remarkable diversity of signals that we call our sense of smell.

Born in Seattle in 1947, Linda Buck enjoyed a pleasant, normal childhood with supportive parents who urged her to accomplish something worthwhile with her life. She went to college at the University of Washington, a few miles from home. Buck majored in psychology but found that college expanded her interests, making it difficult to choose a particular career path. For the next several years, Buck traveled, lived for a while on a nearby island, and took some classes. Her future finally became clearer when she took a course in immunology and decided that biology was for her. She completed her undergraduate degree in 1975; she was 28 years old.

Buck completed her Ph.D. in immunology in 1980 and went to Columbia University to learn molecular biology techniques. There, she was also introduced to the field of neuroscience and became intrigued by the diversity of signals and information that the brain processes. In 1985, Buck read a paper about olfaction and learned that humans and other mammals can detect more than 10,000 different odorous chemicals. Buck decided to study the diversity of neural signals and the perceptions they generate.

Often working 15-hour days for years, first at Columbia and then continuing at Harvard, Buck discovered more than 1000 genes that control the production of chemoreceptors that bind odorant molecules, and she showed that these receptors are found in a small area in the upper part of the nose. Buck and her colleagues also revealed that each cell in this area has only one type of odor receptor and that each receptor can detect a limited number of odors. As a result, each cell is specialized to detect only a few odors.

Once cells detect the odors, they send a signal to the olfactory bulb, an area of the brain that controls the sense of smell, and from there the information goes to other parts of the brain. Ultimately, information from several olfactory receptors is combined to produce a pattern we experience as a distinct odor.

Linda Buck's life shows how far a person can go once he or she becomes truly passionate about something. Before her, how our sense of smell worked was a mystery. She initiated the experimental work in 1988 and won the Nobel Prize just 16 years later. Pretty good for someone who had trouble deciding upon a major!

Linda Buck was the winner of the 2004 Nobel Prize in Physiology or Medicine. Buck discovered more than 1000 genes that control the production of chemoreceptors that bind odorant molecules.

behaviors. The change in Gage was evident and permanent. Why, then, did he take a dramatic turn for the worse in 1859?

There is no clear historical record, so we are left to speculate. Stagecoach driving over rough roads is not easy work. It is physically jarring. Perhaps years of bouncing and jerking around aggravated Gage's old injury. Or he may have hit his head on the carriage. Maybe he suffered a closed brain injury from which his brain swelled, causing damage elsewhere as the soft tissue squashed up against the hard skull. Although we will probably never learn the exact cause, we do know that when Gage came to California, he suffered increasingly serious seizures. Why did this kill him?

Seizures indicate that the brain is misfiring its signals and can't get things back into balance. The information then sent to the rest of the body from the brain instructs the muscles to contract. Consequently, Gage's body tensed and convulsed, twitching and moving uncontrollably. The body interpreted these muscle contractions as shivering and responded accordingly. Shivering is a beneficial behavior in cold temperatures, because it causes muscles to release heat. However, because Gage was trapped in continuous seizures, his muscles gave off so much heat that his body could not maintain its normal temperature. The seizures spent too much energy; Gage's body temperature actually began to drop.

Signals sent back to the brain warned that his body temperature was dropping. The brain sent back instructions for the circulatory system to shut down circulation to the hands and feet in an effort to keep essential organs warm. Gage kept losing heat from the convulsions. Eventually, circulation stopped in the internal organs and brain, and Gage died.

13.4 How Does Cell Communication Lead to Nervous System Function?

The nervous system controls and modulates an amazing array of functions and behaviors—everything from breathing and walking to creating art and music and falling in love. The modular organizations of the brain—with its functional systems and circuits, spinal cord, and PNS—are all built from the same basic unit: the neuron.

Neurons Are the Basic Building Blocks of the Nervous System

dendrite :: highly branched projection on a neuron that receives signals from other neurons

axon :: the long extension of the neuron cell body that "reaches" to the next neuron cell body

Neurons are highly specialized signaling cells able to receive communication and send information along to other cells (**FIGURE 13.17**). The structure of a neuron is ideal for carrying out its function. Long and skinny, a neuron has highly branched projections, or **dendrites**, on one end of its somewhat rounded cell body. Dendrites receive signals from other neurons. The other end of the cell body narrows into a long extension called the **axon** that "reaches"

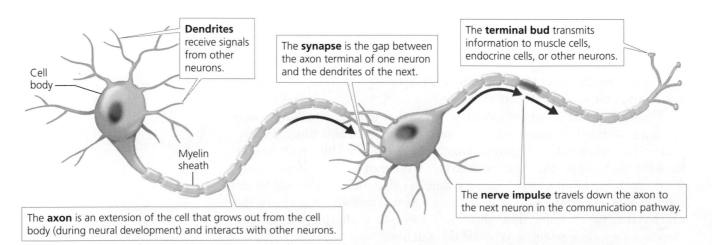

Dendrites receive signals from other neurons.

Cell body

Myelin sheath

The **axon** is an extension of the cell that grows out from the cell body (during neural development) and interacts with other neurons.

The **synapse** is the gap between the axon terminal of one neuron and the dendrites of the next.

The **terminal bud** transmits information to muscle cells, endocrine cells, or other neurons.

The **nerve impulse** travels down the axon to the next neuron in the communication pathway.

FIGURE 13.17 Neurons are highly specialized signaling cells able to receive communication and send information along to other cells.

to the next neuron cell body. The end of the axon divides into short branches, each of which ends with a **terminal bud**. These terminals transmit information to other cells. In most cases, there is a gap between the axon terminal of one neuron and the dendrites of the next one. This gap is called a **synapse**.

Messages travel as an electrical wave, called an **action potential**, through the axon. When this electricity reaches the axon terminal, the information is converted to a chemical signal so that it can bridge the synapse between the cells. The chemical signal is picked up by the receiving cells, which in turn convert it back to an electrical signal. This electrical signal travels down the axon, where it reaches the next terminal and is converted to a chemical signal, and so on as if the neurons were a bucket brigade. This process happens with lightning speed. How exactly do neurons signal one another to produce responses or behaviors? And how do complex outcomes like thought or emotion originate in cellular behaviors?

How Neurons Talk to Each Other . . .

The electrical signaling of neurons depends on two things. First, the membrane that encloses any cell is very selective about what can enter and leave cells. In fact, the membrane has specialized **channels**, or small tubes, made of protein that can open or close to regulate whether specific molecules can pass (**FIGURE 13.18**). Second, the fluid inside our bodies is salty. Think of how tears, sweat, and even blood taste. There are charged particles, or **ions**, that come from this salt. The charges can be negative or positive, like the poles of a battery. Some of these ions, sodium especially, are more abundant outside the cell than inside.

When a neuron is at rest, there is a difference in the ionic charge that is present on each side of the membrane. Because of the positively charged ions from the salt, the space outside the cell has a positive charge. The membrane keeps proteins and other relatively large molecules inside the cells.

terminal bud :: the end of the axon, which transmits information to other cells

synapse :: the gap between the axon terminal of one neuron and the dendrites of the next one

action potential :: an electrical wave that transports messages through the axon

channel :: specialized small tube made of protein that can open or close to regulate whether specific molecules can pass through a cell membrane

ion :: charged particle that comes from the salty fluid in our bodies; it may be negative or positive

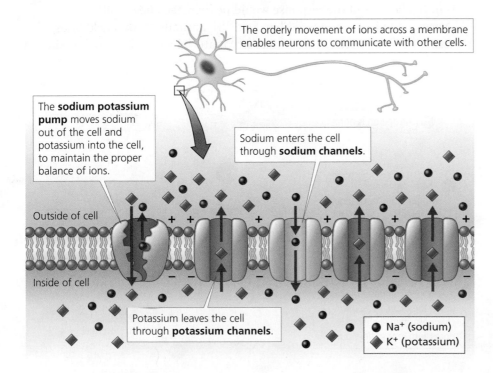

The orderly movement of ions across a membrane enables neurons to communicate with other cells.

The **sodium potassium pump** moves sodium out of the cell and potassium into the cell, to maintain the proper balance of ions.

Sodium enters the cell through **sodium channels**.

Outside of cell

Inside of cell

Potassium leaves the cell through **potassium channels**.

Na⁺ (sodium)
K⁺ (potassium)

FIGURE 13.18 The membrane that encloses any cell has specialized channels made of protein that can open or close to regulate whether specific molecules can pass.

These molecules are generally negatively charged. Therefore, a neuron has a positive charge built up on the outside and a negative charge within. The membrane keeps these charges separate until the neuron is excited.

Imagine being in Cavendish, Vermont, on that fateful day in 1848 and hearing a loud explosion, the whine of a projectile flying through the air, the clatter when it landed, and the screams of the people near Gage. Your nervous system would have kicked into high gear. How?

Once the sensory neurons in your eyes and ears had sent information to the CNS, the sympathetic branch of the autonomic component of the PNS would have signaled danger. The adrenal gland releases a chemical known as adrenaline. Adrenaline is a **neurotransmitter**. Neurotransmitters bind to specific receptor molecules on a neuron's membrane and initiate a **nerve impulse**. This impulse, or action potential, occurs because the stimulus, adrenaline in this case, triggers the openings of ion channels in the membrane where adrenaline binds (**FIGURE 13.19**). These openings let positive ions rush into the neuron, thus making the cell's interior shift from being negatively to positively charged. Opening up the membrane channels is analogous to switching on a battery-powered device, thus allowing the flow of current.

The electrical impulse in the neuron spreads down the axon because the flow of positive ions into the cells triggers the adjacent population of membrane channels to open, which now allow positive ions in. These ions trigger the next set of channels and so on—another bucket brigade. The electrical impulse goes in one direction down the axon toward the axon terminals because once activated, channels close up and go through a time period during which they are unable to respond to stimulus.

Once the nerve impulse travels all the way to the end of the axon, **vesicles**, or membrane-enclosed bubbles, containing adrenaline fuse with the cell membrane (**FIGURE 13.20**). In so doing, they squirt their contents, adrenaline, into the synaptic space. Adrenaline binds to receptors on the receiving cell membrane, and the whole scenario repeats until the signal finally reaches its destination, the **effector**. In this example, the effector would include muscle cells in the heart and the response would be increased heart rate.

Nerve impulses are incredibly fast. In animals with backbones, like humans, the axons are encased in a fatty material called the **myelin sheath**

neurotransmitter :: a chemical that binds to specific receptor molecules on a neuron's membrane and initiates a nerve impulse

nerve impulse :: an action potential that occurs as a result of a stimulus

vesicle :: membrane-enclosed bubble

effector :: the final destination of a signal

myelin sheath :: fatty material that encases the axons of neurons; acts like insulation on a wire by speeding up electrical impulses and preventing them from losing strength

FIGURE 13.19 A nerve impulse or action potential occurs when neurotransmitters bind to specific receptor molecules on a neuron's membrane, triggering the opening of ion channels.

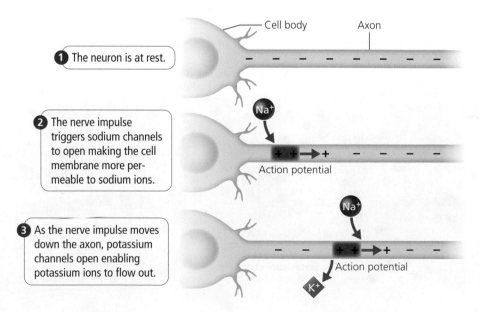

Cell body — Axon

❶ The neuron is at rest.

❷ The nerve impulse triggers sodium channels to open making the cell membrane more permeable to sodium ions.

Na⁺

Action potential

❸ As the nerve impulse moves down the axon, potassium channels open enabling potassium ions to flow out.

Na⁺

Action potential

K⁺

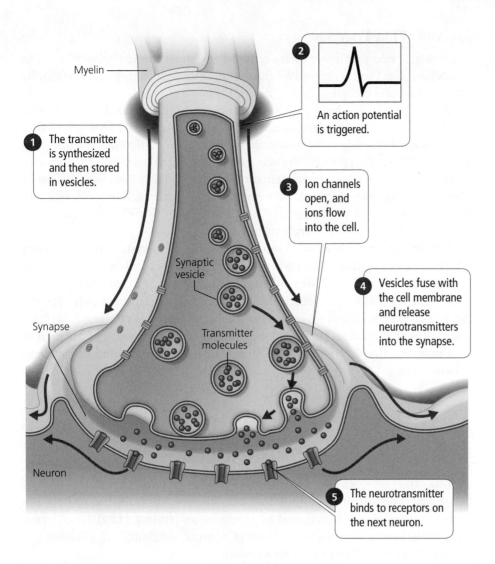

Myelin

1 The transmitter is synthesized and then stored in vesicles.

2 An action potential is triggered.

3 Ion channels open, and ions flow into the cell.

Synaptic vesicle

4 Vesicles fuse with the cell membrane and release neurotransmitters into the synapse.

Synapse

Transmitter molecules

Neuron

5 The neurotransmitter binds to receptors on the next neuron.

FIGURE 13.20 Once a nerve impulse travels all the way to the end of the axon, vesicles fuse with the cell membrane and release the neurotransmitter into the synaptic space.

(**FIGURE 13.21**). This sheath is composed of a chain of supporting cells that wrap themselves around the axon. The myelin sheath acts like insulation on a wire. It speeds up electrical impulses and prevents the impulse from losing strength. A nerve impulse can race down an axon as fast as 100 meters per second or greater than 200 miles per hour.

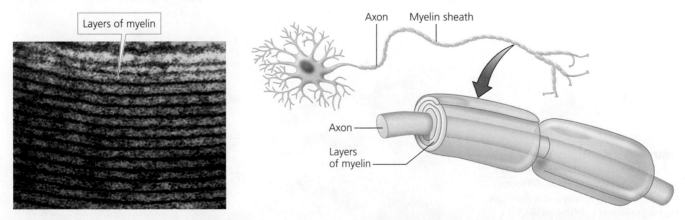

Layers of myelin

Axon   Myelin sheath

Axon

Layers of myelin

FIGURE 13.21 In animals with backbones, the axon is encased in a fatty material called a myelin sheath, which acts like insulation on a wire. The sheath speeds up electrical impulses and prevents the impulse from losing strength.

. . . And What They Say

A nerve gets many different forms of input and must "decide" what to do. The behaviors or actions triggered by the communication among neurons can be quite varied and complex.

First, there are at least 25 known neurotransmitters. Each is associated with particular characteristics, and generally none act alone. This means that neurons are bathed in neurotransmitter "cocktails." Different mixtures will produce different responses. Second, competing signals are summed up and integrated. For example, some neurotransmitters, like adrenaline, are excitatory; they endeavor to trigger a nerve impulse. Others, like serotonin, are inhibitory and work to prevent nerve impulses. Yet other chemicals magnify or dampen the effects of neurotransmitters.

We are bombarded with stimuli, and the nervous system must sort out which to pay attention to and which to ignore. If we monitored every sensation equally all of the time, we could not function. Imagine that one of Phineas Gage's coworkers had sore feet because he was breaking in a new pair of boots. All he could think about was the pain where his skin was rubbing raw. Then the explosion occurred. In an instant, he was no longer aware of his pain at that moment, thanks to nerve impulses, neurotransmitters, and the integration of a new set of information by his nervous system.

13.5 What If the Nervous System Is Injured or Diseased?

Because the nervous system is central to the control and regulation of all of the body's function and actions, injuries that impact the CNS and/or PNS can produce physical, emotional, and cognitive issues (**FIGURE 13.22**). In all cases, the problems are due to a breakdown of normal, efficient, and effective cell communication within the nervous system.

Injury

As you know from the case of Phineas Gage, localized injury of the brain can result in the loss of specific functions. In Gage's case, his personality was transformed. Other injuries, a blow to the head or a stroke, for example, can make people lose the ability to speak, walk, or even remain conscious. Sometimes a brain injury can cause a seizure disorder such as

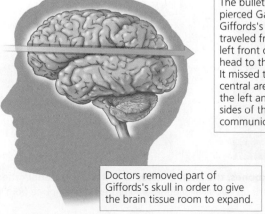

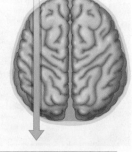

FIGURE 13.22 Critically injured by a gunshot wound to the head in 2011, congresswoman Gabrielle Giffords experienced profound brain damage that affected her ability to walk, talk, and think. After intensive rehabilitation, she recovered some but not all function.

The bullet that pierced Gabrielle Giffords's brain traveled from the left front of her head to the back. It missed the central area where the left and right sides of the brain communicate.

Doctors removed part of Giffords's skull in order to give the brain tissue room to expand.

Just 10 percent of individuals who have a bullet pierce their brain survive the injury.

Concussion

Imagine a hollow sphere with a ball of Jell-O inside taking up almost all of the available space. What do you suppose would happen to the Jell-O if you threw this sphere against a wall? Can you envision the Jell-O sloshing forward, hitting the inside of the sphere, and perhaps rebounding to hit the opposite surface? This is what happens when a person suffers a concussion.

A hard knock to the head can cause a concussion, or brain bruise. The force of the blow causes the brain to collide with the inner surface of the skull. In some cases, the collision between the brain and skull results in bleeding and the tearing of delicate nerve fibers. At its most extreme, the injury can cause dangerous brain swelling, which can produce widespread neurological damage and death.

Brain injuries are relatively common. In the United States, someone suffers a serious head injury every 21 seconds. The most common causes are falls, accidents, physical violence, and contact sports like football and hockey. Although protective headgear helps, it does not eliminate all risk of concussion. In fact, concussions can occur without direct head impact: simply shaking a person, especially a baby or small child, or sudden deceleration, like in a car accident, can cause brain injury.

If you receive a blow to the head that knocks you out or leaves you dazed, you are concussed. Other symptoms may include headache, nausea, amnesia, confusion, dizziness, and slurred speech. Your concussion may be obvious to those around you, or it may be more subtle. Symptoms can last for days to weeks and even longer. No matter how mild or serious the concussion, it is essential to minimize further brain trauma until the injury has healed.

In sports, there is sometimes a tendency to "shake it off" and get back into the game even after a hard hit in which the athlete "sees stars." This is a dangerous thing to do. The full extent of the injury from that initial hit might not be known for hours or perhaps days. Even after the severity of the concussion has been determined, the athlete should refrain from the sport until a medical professional can confirm that the concussion has healed. This caution is essential because once a person suffers a concussion, he or she is more likely to suffer another. Athletes, and others, who suffer multiple concussions over their lives experience cumulative neurological damage. No matter how important the event, brain damage goes far beyond "taking one for the team."

Facts About Brain Injury
- Every five minutes someone dies from a head injury.
- Every five minutes someone becomes permanently disabled due to a head injury.
- Over 50% of those who sustain a brain injury were intoxicated at the time of injury.
- Males between the ages of 14 and 24 have the highest rate of injury, and males are almost twice as likely to suffer serious brain injuries as females.
- Brain injuries kill more Americans under the age of 34 than all other diseases combined.

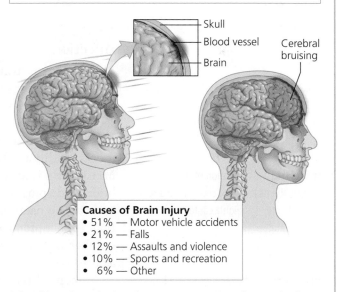

Skull
Blood vessel
Brain
Cerebral bruising

Causes of Brain Injury
- 51% — Motor vehicle accidents
- 21% — Falls
- 12% — Assaults and violence
- 10% — Sports and recreation
- 6% — Other

A hard knock to the head can cause a concussion, or brain bruise. The force of a blow to the head causes the brain to collide with the inner surface of the skull and may result in bleeding, dangerous brain swelling, and sometimes even neurological damage and death.

---

epilepsy. The seizures occur because the injury made the normal electrical activity in the brain become chaotic and uncontrolled. A violent blow to the head or neck can also produce a **concussion**, or brain bruise in which electrical activity in the brain is temporarily disordered (see *Life Application: Concussion*).

Spinal cord injury due to crushing or severing, usually in an accident, can have different outcomes depending on the precise location of the injury. If the damage is serious enough, there may be a loss of sensation and paralysis below the site of injury (**FIGURE 13.23**).

concussion :: a brain bruise in which electrical activity in the brain is temporarily disordered

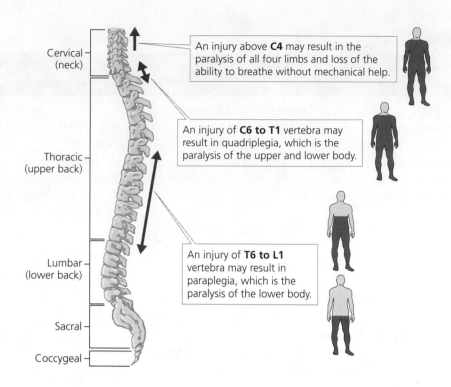

Cervical (neck)

An injury above **C4** may result in the paralysis of all four limbs and loss of the ability to breathe without mechanical help.

An injury of **C6 to T1** vertebra may result in quadriplegia, which is the paralysis of the upper and lower body.

Thoracic (upper back)

Lumbar (lower back)

An injury of **T6 to L1** vertebra may result in paraplegia, which is the paralysis of the lower body.

Sacral

Coccygeal

FIGURE 13.23 Spinal cord injury can have different outcomes depending on the precise location of the injury. If the damage is serious enough, paralysis below the site of the injury may result.

meningitis :: inflammation of the meninges, the tissues that cover the brain and spinal cord; it is highly contagious and often fatal

encephalitis :: inflammation of the brain due to infection; usually caused by a virus and includes symptoms such as fever, confusion, and seizures

Infection

Besides his survival from the accident itself, the other seemingly miraculous thing about Gage's injury was that he did not die from infection. Dr. Harlow certainly saved Gage's life by cleaning out the wound and then draining subsequent infections until they cleared up.

Even in these more modern times, with antibiotics and other treatments readily available, people suffer from infections of the nervous system. **Meningitis** is an inflammation of the meninges, the tissues that cover the brain and spinal cord. Meningitis is highly contagious and often fatal. It is caused by a bacterial or viral infection and less commonly by a fungal infection. Its symptoms include headache, stiff neck, and vomiting. An increasing number of colleges and universities require all students to be vaccinated against bacterial meningitis.

Encephalitis is a rare inflammation of the brain due to infection. Its cause is usually viral and its symptoms include fever, confusion, and seizures. Many survivors also experience serious brain damage. Concerns have increased in recent years because of the spread of mosquitoes that carry the encephalitis viruses. The mortality rate for humans is 33%.

Neurodegenerative Diseases

In addition to the potential external threats posed by injury and infection, the nervous system faces diseases and disorders that originate in the body itself. Neurodegenerative diseases are numerous in type and fairly common. There are more than 40 syndromes in which the delicate balance of structure and effective communication breaks down because neurons of the CNS deteriorate and die. In the United States, more than 6 million people suffer from such disorders at any given time. And because neurodegenerative diseases more often strike later in life, the number of individuals affected will certainly rise as the population ages. This means that as a society we face an increasingly difficult challenge: how can we care for an ever-growing number of people who need help to perform even routine functions?

Here is a sampling of the most prevalent of these disorders:

- *Alzheimer's disease (AD):* More than 5 million Americans suffer from AD. The first symptom is memory loss, which eventually becomes so severe that the person is rendered incapable of performing any type of daily activity. Although AD can occur before 50 years of age, it usually appears after age 65. AD patients have collections of an abnormal protein, called beta amyloid, that gum up axons. The regions of the brain most affected are the hippocampus and amygdala. AD patients also don't use the neurotransmitter acetylcholine properly. Drugs inhibiting the enzyme that breaks down acetylcholine can temporarily improve mental function in AD patients. There is no known cure for AD yet, nor is there a definitive way to prevent it.

- *Parkinson's disease (PD):* More than 1 million Americans have Parkinson's disease (**FIGURE 13.24**). PD is the second most common degenerative disease overall. PD usually begins between the ages of 50 and 60, although earlier onset is possible. PD patients experience a gradual but growing loss of motor control. The crux of the problem is neurons in the thalamus that release a chemical called dopamine. Dopamine is an inhibitory neurotransmitter, so when it is absent, excessive excitatory signals from motor regions of the brain trigger shaking, tremors, disordered muscle contractions, stiffness, and loss of coordination. Giving dopamine to patients temporarily helps to alleviate symptoms. There is no cure or prevention yet.

- *Multiple sclerosis (MS):* At least 400,000 Americans suffer from MS. It is a major cause of disability in younger adults. MS is an autoimmune disease, meaning that the body's immune system attacks itself. In MS, the myelin sheath is progressively destroyed. As a consequence, nerve impulses are slowed and a person with MS experiences muscle weakness, stiffness, difficulty walking, and slurred speech. MS can appear as early as one's 20s, although more commonly in one's 40s and 50s.

- *Amyotrophic lateral sclerosis (ALS):* Also known as Lou Gehrig's disease, ALS afflicts 30,000 Americans. ALS attacks motor neurons, and when muscles lose their nervous input, they shrink and atrophy. Eventually the neurons for breathing and swallowing are affected. Always fatal, ALS ultimately immobilizes all the muscles in the body; sensing and thinking remain completely functional, however, until death.

FIGURE 13.24 More than 1 million Americans have Parkinson's, which is a neurodegenerative disease that gradually results in loss of motor control. Diagnosed with young-onset Parkinson's in 1991, actor Michael J. Fox launched the Michael J. Fox Foundation for Parkinson's Research in 2000.

13.6 Can Mental Health Be Separated from Physical Health?

Because thought, behavior, emotions, and feelings all derive from brain function, distinguishing mental health as something separate from physical health does not really make sense. As was the case with Phineas Gage, his injury produced a permanent mental change that profoundly altered his personality. Indeed, many mental health problems can be explained in terms of nervous system structure and function. Mental health disorders can be due to anything from brain injury or defects to an imbalance in the mixture of neurotransmitters in which neurons bathe. Depending upon the specific cause and symptoms, some mental health problems can be addressed quite successfully. Two of the most common mental health issues diagnosed today are attention deficit hyperactivity disorder (ADHD) and depression.

Common Mental Health Issues

- *Attention deficit hyperactivity disorder (ADHD):* In the United States, 6% of all school-age boys and 2% of school-age girls take Ritalin to treat diagnosed ADHD. Symptoms of ADHD include forgetfulness, impulsiveness, distractibility, fidgeting, and impatience. Some, but not all, cases of ADHD can be attributed to low dopamine levels. Ritalin acts by increasing the release of dopamine, thus diminishing ADHD symptoms.

Many would argue that these ADHD symptoms are simply childhood behaviors and that ADHD is overdiagnosed. Others argue that when these symptoms cause problems for a child in at least two settings—school and home, for instance—the diagnosis of ADHD is appropriate. The questions for all of us to consider are these: Has there been a real increase in the number of children affected by the disorder? And, if so, what is the cause? Is there something wrong with our culture, or is it something in the environment? And, regardless of the cause, do all of these boys and girls really require medication to control their behavior?

- *Depression:* Depression affects 20 million adults in the United States, two-thirds of them women. Symptoms include sadness, loss of energy, changes in weight and sleep patterns (too much or too little), and a general loss of interest. In severe cases, a depressed person is unable to participate in daily life activities, may have intrusive thoughts about death, and in the most extreme case may commit suicide. The brain physiology of depression has been studied extensively. An imbalance of neurotransmitters is the culprit; in particular, low serotonin is associated with depression.

Antidepressant medications generally act to raise serotonin levels. However, although medication can be effective, as with other mental health issues, psychotherapy is also important. Sometimes used alone or in combination with medication, **psychotherapy** is the treatment of mental health disorders using psychological methods such as talking therapy. The ways people feel and act also influence brain chemistry. Often serious depression responds best to a combination of talking therapy and medication.

psychotherapy :: the treatment of mental health disorders using psychological methods such as talking therapy

Drug Use and Abuse

People put a lot of things in their bodies to feel good. Many of the substances that make us feel energized or happy or relaxed or sleepy do so by inserting themselves into some aspect of nervous system communication (**FIGURE 13.25**). Some drugs are too dangerous to use no matter how good they make you feel, because they can cause permanent physical harm or are addictive. Heroin and ecstasy are good examples. Other substances are relatively harmless (in

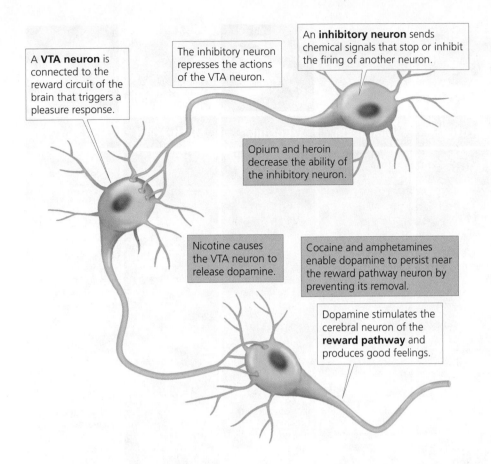

A **VTA neuron** is connected to the reward circuit of the brain that triggers a pleasure response.

The inhibitory neuron represses the actions of the VTA neuron.

An **inhibitory neuron** sends chemical signals that stop or inhibit the firing of another neuron.

Opium and heroin decrease the ability of the inhibitory neuron.

Nicotine causes the VTA neuron to release dopamine.

Cocaine and amphetamines enable dopamine to persist near the reward pathway neuron by preventing its removal.

Dopamine stimulates the cerebral neuron of the **reward pathway** and produces good feelings.

FIGURE 13.25 Many drugs affect some aspect of nervous system communication.

moderation) and are used so commonly we scarcely think of them as drugs. Caffeine is such a substance. No matter which drug a person uses, if he or she no longer has control, it means the line between use and abuse has been breached. The person now has a mental health problem.

Sometimes drug use is actually a symptom of an entirely different mental health problem. A person suffering from depression, for example, might use drugs in an effort to feel better. Sadly, such self-medication usually does not work, and the battle against depression is made worse by a substance abuse problem.

What are some of the more prevalent recreational drugs, and what do they do (**FIGURE 13.26**)?

- *Alcohol:* a depressant, which may act by increasing the effects of gamma-aminobutyric acid (GABA), an inhibitory neurotransmitter.
- *Caffeine:* a stimulant that acts by countering the effects of inhibitory neurotransmitters. Caffeine has an impact beyond cells of the nervous system; it increases the metabolic activities of all cells.
- *Nicotine:* a stimulant that acts by activating acetylcholine receptors. Nicotine also causes cells in the cerebral cortex to produce the neurotransmitters acetylcholine, epinephrine, and norepinephrine.
- *Marijuana:* not an easy drug to classify. Tetrahydrocannabinol (THC), the active drug in marijuana, binds to receptors in the brain normally used by neurotransmitters that play a role in pain, depression, appetite, and memory. It interferes with these functions, stimulating some and inhibiting others. THC also increases dopamine release by neurons.

	Alcohol	Nicotine	Marijuana	Cocaine	Mushrooms	Ecstasy
Type	Depressant	Stimulant	Cannabinoid	Stimulant	Hallucinogen	Stimulant/ Hallucinogen
Common method of administration	Swallowed	Smoked/ snorted/chewed	Smoked/ swallowed	Smoked/ snorted/injected	Swallowed	Swallowed
Mechanism of action	Increases the effect of gamma-aminobutyric acid (GABA), an inhibitory neurotransmitter	Activates acetylcholine receptors and causes cells in the cerebral cortex to produce the neurotransmitters acetylcholine, epinephrine, and norepinephrine	Tetrahydro-cannabinol (THC) binds to receptors in the brain that play a role in pain, depression, appetite, and memory; also increases dopamine release by neurons	Increases the release of norepinephrine and dopamine	Binds to serotonin receptors and stimulates the brain, especially the cerebral cortex	Causes the release of all serotonin, creating signal overload
Possible short-term side effects	Relaxation, lowered inhibitions, drowsiness, slurred speech, loss of coordination, and sexual dysfunction	Increased blood pressure and heart rate	Euphoria, relaxation, slowed reaction time, increased heart rate and appetite, and anxiety	Increased heart rate, blood pressure, body temperature, energy, and mental alertness; panic, anxiety, violent behavior, and paranoia	Hallucinations, nausea, nervousness, paranoia, and panic	Hallucinations, increased sensitivity, lowered inhibitions, anxiety, chills, muscle cramping, and seizures
Possible long-term health risks	Depression, neurologic deficits, hypertension, liver and heart damage, and fetal damage in pregnant women	Chronic lung disease, cardiovascular disease, stroke, multiple cancers, and adverse pregnancy outcomes	Cough, respiratory infections, and mental health decline	Psychosis, nasal damage from snorting, insomnia, cardiac complications, stroke, and seizures	Flashbacks, persisting perception disorder	Sleep disturbances, depression, and permanent memory problems and brain damage proportional to extent of use

FIGURE 13.26 Many drugs are addictive and can cause permanent physical harm.

- *Cocaine:* a stimulant that increases the release of norepinephrine and dopamine. Cocaine abuse can lead to symptoms similar to schizophrenia, a serious mental disorder.
- *Ecstasy:* a stimulant and hallucinogen. Ecstasy causes the release of all serotonin, creating signal overload. A high dose can produce panic, seizures, dangerously high blood pressure and heart rate, dramatically increased body temperature, organ failure, and death. Long-term consequences of ecstasy use include permanent memory problems and brain damage. The extent of the brain damage is proportional to the number of times the drug was used.
- *Lysergic acid diethylamide (LSD):* a hallucinogen that binds to serotonin receptors and increases the normal response to serotonin.

- *Amphetamines:* stimulants that work by mimicking the effects of the neurotransmitter norepinephrine. They also inhibit the enzyme that normally breaks down norepinephrine, which therefore increases the concentration of this signal molecule.
- *Tranquilizers:* depressants that work by activating receptors for GABA, an inhibitory neurotransmitter. Examples of tranquilizers include Valium and Xanax.
- *Opiates:* drugs such as codeine, morphine, and heroin, which bind to endorphin receptors reducing pain and producing euphoria or feelings of great pleasure. Endorphins are naturally produced molecules that decrease the perception of pain in the body.

Is all drug use bad? Should people be allowed to make their own decisions regarding drug use for pleasure? Or does society have a stake in each individual's behavior? These are questions that generate different answers depending on who you ask. Regardless, it is important to know the biology before deciding (see *How Do We Know? The Pleasure Centers of the Brain*).

13.7 Can Nervous System Injuries and Illnesses Be Cured?

Physical damage to the brain or spinal cord caused by injury or disease cannot be repaired. Neurons of the CNS do not regenerate. The type of injury suffered by Phineas Gage produced a permanent change in him. A person paralyzed due to a severed spinal cord will not recover from paralysis. ALS, PD, MS, and AD all produce irreversible deterioration of the nervous system.

These dire assertions are true today. But is there a chance that medical research will develop ways to fix these "untreatable" conditions? Scientists and physicians approach this question from several avenues.

Drugs

In addition to the medications already available to address the symptoms of AD, PD, MS, and other neurodegenerative diseases, drugs are being developed to get at the disease mechanisms themselves. More specifically, these drugs aim to correct defects in the genetic instructions and cellular processes that lead to the onset of the disease.

Surgery

Some studies have shown that surgical implantations of normal dopamine-releasing tissues have helped to alleviate the symptoms of PD. However, the fix is not permanent and does not work in all patients. Experimental surgeries have also been done on MS patients. Physicians transplanted Schwann cells, the cells that make the myelin sheath, from areas of the PNS into the brain to see whether the myelin sheath could be regenerated there. Other trials have involved gene therapy to treat AD. Surgical interventions show a lot of promise.

Regeneration and Stem Cells

Many people think that using stem cells to regenerate and replace damaged neural tissue shows the most promise for actually curing neurological disorders

How Do We Know?

The Pleasure Centers of the Brain

What if it were possible to experience happiness or even ecstasy at the flip of a switch? It turns out, it is.

In 1954, James Olds and Peter Milner, two postdoctoral scientists at McGill University, were studying the responses of rats to the electrical stimulation of specific brain regions. They planned to study a part of the brain that when stimulated caused the rats to run away.

Olds and Milner inserted electrodes into a rat's brain and placed the rat in a box with the corners labeled A, B, C, and D. If the rat went toward A, Olds and Milner administered a small electric current. Instead of running away from A, as expected, the rat returned to A. In fact, after repeated trials, the rat became quite attracted to A. The next day, the rat was even more interested in A. What was going on?

Once Olds and Milner got the rat into the habit of going to A, they attempted to lure it to B by giving an electrical current whenever the rat took a step in that direction. Within five minutes, Olds and Milner were able to train the rat to focus on B. They continued these studies and realized they could direct the rat anywhere by using electrical stimulation. Olds and Milner had accidentally discovered a brain region that when stimulated made the rats feel "good."

Next, Olds and Milner built an apparatus so that the rats could press a lever to deliver electrical stimuli to their own brains. Once the rats learned to use the lever, they stimulated their brains once every five seconds, with each jolt lasting a second or so. Some rats would self-stimulate 2000 times per hour, for 24 hours. Hungry rats would choose electrical stimulation over food, and thirsty ones would choose it over water. Male rats preferred to self-stimulate rather than to mate with receptive females. Self-stimulating mother rats ignored their newborn pups and would not nurse them.

Olds and Milner built an apparatus so that rats could press a lever to deliver electrical stimuli to their brains, which made them feel good. Once the rats learned to use the lever, they stimulated their brains once every five seconds.

Rats would repeatedly cross grids that delivered painful shocks to their feet to reach the self-stimulation lever.

Additional research showed that several regions deep in the brain form a pleasure, or "reward," circuit. The strength of the response varied depending on which specific spot in the brain the rats would self-stimulate. For some regions, rats would hit the lever 7000 times per hour.

Humans respond similarly to electrical stimulation of the pleasure circuit. Decades after Olds and Milner's experiment on rats, a human test subject was wired so that he could self-stimulate. He pressed his lever 1500 times during a three-hour session. The subject worked himself up into such a state that scientists disconnected him. He protested this decision vehemently. Maybe we aren't meant to be wired for "on demand" happiness.

---

and repairing CNS injuries. This research is still quite new; however, clinical outcomes may be years away. In addition, enthusiasm for this research is not shared by all. Some people still have concerns about whether the use of stem cells is appropriate, especially when they are collected from human embryos (**FIGURE 13.27**).

Minimizing Damage

Physicians and scientists are also developing techniques to minimize spinal cord damage resulting from an accident—usually a sports injury, fall, or car crash. Medications to reduce inflammation and realignment of the spine may both prevent an injury from becoming more serious. Rehabilitation involving muscle strengthening is also essential for restoring all possible function.

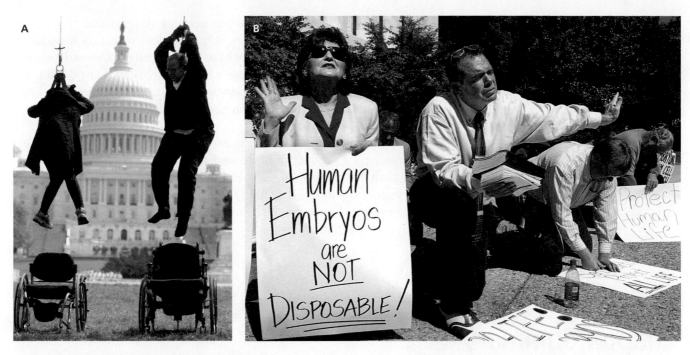

FIGURE 13.27 **Many people think that using stem cells to regenerate and replace damaged neural tissue shows the most promise for actually curing neurological disorders and repairing CNS injuries, but this enthusiasm is not shared by all.** Paraplegics stage a mock hanging to dramatize to senators the need for stem cell research (A), and protesters pray against stem cell research in front of the Senate Office Building (B).

A new experimental technique goes one step further to prevent additional nervous system damage after a spinal injury. Doctors lower the injured person's body temperature to 92°F (33.3°C) for a couple of days (normal body temperature is 98.6°F, or 37°C). Doing this slows down the body's metabolism and prevents swelling, therefore stopping the progressive worsening of a spinal injury.

Minimizing the damage of brain injuries once they occur is even more challenging. Scientists and physicians are focused on developing ways to reduce inflammation and swelling to protect the fragile tissues of the brain. Techniques include lowering the patient's body temperature and in some cases surgery to open the skull in order to relieve pressure.

Although there is reason to be optimistic about the development of treatments for individuals with CNS injury, prevention is by far the most sensible and effective action. Seatbelts, helmets for certain sports, and good decision making provide the best chances of keeping our brains and spines healthy. For Gage, putting sand in the hole would have made all the difference in the world.

Biology in Perspective

Our growing knowledge of the nervous system and its workings is exciting to many and unsettling for some. Compared to a generation ago, our ability to diagnose neurological disorders and in many cases treat them has advanced tremendously. Scientists are trying to figure out ways to predict disease onset at earlier stages both for more successful interventions and to find preventative measures. Techniques for dealing with injuries of the CNS are improving, and the future holds hope here too.

We also know how to manipulate our nervous systems and to measure activity. One challenge facing all of us is how to deal with this knowledge. Should we medicate individuals who behave eccentrically so they will be more "normal"? Should we evaluate everyone's brain activity to identify who has potentially harmful behavioral tendencies? What about screening brain activity to determine whether a person has the potential to develop a neurological disorder? Who decides?

And neuroscience opens the door to even bigger questions. To what extent is our humanness a product of brain and nervous system function? Or are we "more" than our bodies? Do we have conscious control over our decisions? In other words, do humans actually have free will? Neuroscientists are addressing profound questions that previously were the exclusive domain of philosophy and religion.

Chapter Summary

13.1 What Are the Parts of the Human Brain, and What Do They Do?

Identify the parts of the brain and predict the effect of damage to those regions.

- The hindbrain at the base of the skull controls many basic bodily functions. It contains three subunits: medulla oblongata, cerebellum, and pons.
  - o Medulla oblongata: regulates sleep, respiration, circulation, and reflexes like sneezing and coughing.
  - o Cerebellum: integrates sight and hearing, balance, manual dexterity, and language.
  - o Pons: directs communication between cerebellum and forebrain and helps control breathing and circulation.
- The midbrain and the hindbrain form the brainstem, a conduit for information from other regions, and also coordinate responses to light and sound.
- The forebrain is the most highly developed and largest part. It regulates the body's physiology and is responsible for thought and sense perception. It includes olfactory bulbs, thalamus, hypothalamus, cerebrum (the largest and most sophisticated part), and the limbic system.

13.2 What Was Wrong with Phineas Gage's Brain?

Assess how brain injury can alter personality using Phineas Gage as an example.

- The rod that went through Phineas Gage's skull missed key areas in the forebrain affecting balance, memory, and language.
- The main damage was on the left side of the frontal lobe: key areas affecting decision making, empathy, and social interactions.

13.3 How Is the Nervous System Organized, and What Does It Do?

Differentiate the roles of the central nervous system and the peripheral nervous system in controlling our actions, behaviors, and body functions.

- Neurons or nerve cells bundle together into nerves.
- The brain and spinal cord form the central nervous system (CNS), which controls actions of the body.
- The peripheral nervous system (PNS) consists of nerves radiating out of the CNS. It has two parts: the somatic system under conscious control and the autonomic system, which is generally involuntary and divided into parasympathetic and sympathetic systems.
- Five types of sensory receptors—pain receptors, thermoreceptors, mechanoreceptors, chemoreceptors, and electromagnetic receptors—send signals to the brain, constructing our perceptions of the world.
- Neurons are organized into interconnected groups called circuits, which form systems, all controlled by the brain.

13.4 How Does Cell Communication Lead to Nervous System Function?

Describe the process of a neuron firing and predict the impact of changes to neuron function.

- Neurons are signaling cells that receive and send information to other cells.
- Messages travel as an electrical wave called an action potential, controlled by the cell membrane.
- The nerve impulse travels neuron to neuron to the effector, where the response occurs.

13.5 What If the Nervous System Is Injured or Diseased?

Detail how injuries and disease can impact nervous system function.

- Seizures occur when the normal electrical activity in the brain becomes chaotic.
- Meningitis is an inflammation of the meninges; it causes headache, stiff neck, and vomiting.
- Encephalitis is an inflammation of the brain due to infection.
- Neurodegenerative diseases such as Alzheimer's disease, Parkinson's disease, multiple sclerosis, and amyotrophic lateral sclerosis occur when the neurons of the CNS deteriorate and die.

13.6 Can Mental Health Be Separated from Physical Health?

Evaluate the role of drugs on behavior, mental health, and overall human health.

- Some cases of ADHD are caused by low dopamine levels.
- Depression is caused by an imbalance of neurotransmitters.
- Some people take drugs to make themselves feel good. Generally, these drugs inhibit or enhance neurotransmitters or bind to receptors.

13.7 Can Nervous System Injuries and Illnesses Be Cured?

Assess the progress and potential of treatments for nervous system injury and disease.

- Physical damage to the brain or spinal cord cannot be repaired; CNS neurons do not regenerate.
- Drugs are being developed that address the mechanisms of disease by correcting genetic instructions and cellular processes that lead to disease onset.
- One type of surgical intervention involves the implantation of normal cells into damaged areas.
- Stem cells could be used to regenerate and replace damaged neural tissue.
- Minimizing damage from injuries is a continued focus of scientists, but prevention is still the most effective course of action.

Key Terms

action potential 393
autonomic system 389
axon 392
brainstem 381
Broca's area 385
central nervous system (CNS) 387
cerebral cortex 382
cerebrum 381
channel 393
circuit 390
concussion 397
dendrite 392
effector 394
encephalitis 398

epileptic seizure 386
forebrain 381
functional magnetic resonance imaging (fMRI) 383
hindbrain 380
interneuron 387
ion 393
limbic system 382
meninges 380
meningitis 398
midbrain 381
myelin sheath 394
nerve 387
nerve impulse 394

neuron 386
neurotransmitter 394
olfaction 390
parasympathetic branch 389
peripheral nervous system (PNS) 387
psychotherapy 400
reflex 387
somatic system 389
spinal cord 387
sympathetic branch 389
synapse 393
terminal bud 393
vesicle 394

Review Questions

1. What are the major components of the brain? What are the principal functions of each?

2. Which part of the brain is the largest and most highly developed?
   a. Thalamus
   b. Hindbrain
   c. Midbrain
   d. Forebrain
   e. Cerebellum

3. What are the principal functions of the frontal, temporal, parietal, and occipital lobes?

4. What are the main functions of the limbic system?

5. Which of the following was true of Phineas Gage after Dr. Harlow pronounced him healed?
   a. He slurred his speech but could take care of most of his personal needs.
   b. He seemed physically fit and ready for work, but his behavior and disposition had changed for the worse.
   c. He was unable to meet the rigorous physical demands of driving a stagecoach.
   d. He experienced many physical problems but his disposition improved.
   e. He went on to lead a long and happy life.

6. What are the basic cellular subunits of the nervous system?
   a. Neurons
   b. Nerves
   c. Interneurons
   d. Dendrites
   e. Axons

7. What are the components of the CNS? What is the function of each?

8. What are the main components of the peripheral nervous system?

9. What is the myelin sheath, and why does it speed up nerve transmission?

10. What are potential sources of injury for the central nervous system?

11. All of the following are potential symptoms of concussions EXCEPT which one?
    a. Amnesia
    b. Slurred speech
    c. Nausea
    d. Confusion
    e. Difficulty breathing

12. What are some ways to reduce the risk of brain injury?

13. What are the most common neurodegenerative diseases in the United States?

14. What are two common mental health issues?

15. Name three dangerous recreational drugs and explain what they do to the nervous system.

16. What are some avenues that scientists and physicians are exploring to prevent and treat nervous system injuries?

The Thinking Citizen

1. How does the organization of nervous system communication produce such a variety of outputs from the functioning of neurons?

2. How is it possible for reflex responses to be so fast?

3. How do the parasympathetic and sympathetic branches of the nervous system work together to balance organ function?

4. How could it be possible for a person to be deaf even if his or her ears function normally?

5. How does an action potential send information down an axon?

6. What happens when the nerve impulse reaches the axon terminal bud?

7. Why do you think closed head injuries might produce more brain damage than an open skull injury?

8. Our population is aging, and as a result, we will be facing an increase in the healthcare needs of many of our citizens. What do you think will be the societal repercussions of an ever-growing population of individuals who have neuro-degenerative diseases?

The Informed Citizen

1. As you have learned, serious brain injury can alter a person's behavior dramatically. If a person commits a terrible act after suffering brain damage, is he or she responsible for that act? Why or why not?

2. Why do you think so many children have been diagnosed with ADHD? Do you think we are simply better at diagnosis, so more cases are being identified? Or are more children developing ADHD than in previous generations? If ADHD is on the rise, why do you suppose this might be? Are there any actions we should take?

3. The number of individuals who will develop neurodegenerative disease is predicted to increase as the population ages. What steps should we take as a society to ensure the care of those in need without utilizing all healthcare resources? Who is responsible? Family? Society? Government?

4. We have the capacity to provide ourselves with physical pleasure through drug use or by electrically stimulating our brains. Should we have limits on these behaviors? Why or why not?

5. Can complex behaviors and emotions, like love, courage, loyalty, and humor, be explained entirely as the outcomes of nervous system function? Why or why not?

After reading this chapter, you should be able to answer the following questions:

14.1 What Causes Cholera?
Describe the history and biology of cholera.

14.2 What Invaders Do We Face?
Describe the array of potential pathogens humans are exposed to every day.

14.3 How Does the Immune System Protect Us?
Explain how the human immune system works and predict the effect of the removal of a portion of the system.

14.4 How Can We Harness the Immune System?
Assess the strengths and weaknesses of natural immunity, vaccination, and passive immunity.

14.5 What Can Help If Our Immune System Fails?
Assess the use and describe the production of antivenom, antibodies, and antibiotics.

14.6 How Can We Prevent the Spread of Infectious Disease?
Predict the impact of personal choices on your health and the health of those in your community.

Trillions of microbial cells live in and on us. The body has an immune system that identifies what does and does not belong and protects us from disease.

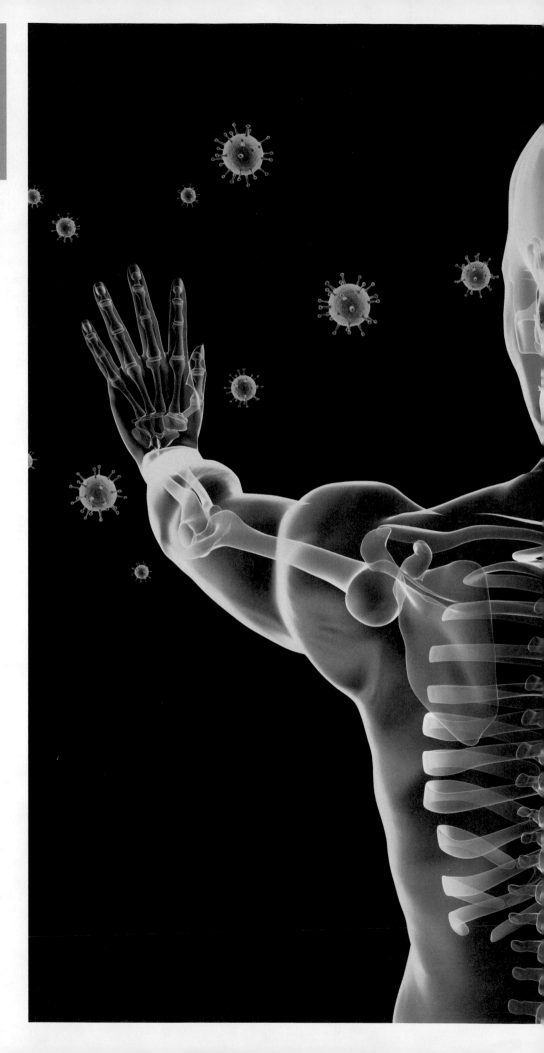

CHAPTER 14

Infectious Disease and the Immune System

How Are Invaders Repelled, Evaded, or Killed?

Becoming an Informed Citizen . . .

We have an immune system that protects us from disease, but it isn't invincible. Our personal behaviors influence our individual risks, and avoiding disease requires both individual and societal efforts.

e do not have our bodies completely to ourselves. We are each a society, a community of organisms—the cells that make us and the cells of tiny microorganisms, such as bacteria and fungi. Our bodies depend on communication, cooperation, and balance among all of our participants to function properly.

Most of the time everything runs fairly smoothly and we are healthy. Sometimes, however, the harmony is disrupted. An invasion comes from our surroundings. Maybe a large number of viruses sprayed during someone else's sneeze sneak in. Maybe we swallow a mouthful of water while swimming in a lake and some parasites enter. At other times, the invasion comes from within, as with cancer (Chapter 5). No matter the source, the body has an **immune system** to protect us from disease. This collection of cells, tissues, and organs has mechanisms to tell the body what does and does not belong.

In order to recognize "foreign," as opposed to "self," and then relentlessly eliminate the invaders, the immune system must balance its own behavior. An under-functioning immune system can't "see" or "hear" the invaders, thus leaving us vulnerable to infection. An overactive one "sees" and "hears" incorrectly, recognizes "self" as not belonging, and therefore attacks normal cells. Here, too, proper functioning depends on cooperation, communication, and balance.

When we become ill with an infectious disease, our immune system is in a race against time. If all goes well, the disease organisms will be beaten back, and we recover. Sometimes, however, the immune response is not strong enough or quick enough, and a person dies. Will the immune system save a life or die trying? To see how so many things influence the answer, consider just one disease, cholera, and the deadly Irish potato famine.

immune system :: the collection of organs, cells, and tissues that protect the human body from disease

case study

The *Summer of Sorrow*

The poor Irish farmers toiled on land they did not own, paying rent to landlords who lived in luxury in England. They had just one crop, potatoes, and in the early 1840s, a new disease, caused by a fungus, devastated the potato crop.

Then, in 1847, the entire potato crop of Ireland failed—and with it the only source of food for most tenant farmers (**FIGURE 14.1**). In a country of 8 million people, 2 million died of starvation, and more than 1 million made their way to the United States and Canada. The trip from Ireland to North America was difficult. The immigrant ships were overcrowded and dirty. Many people on board were malnourished and ill; disease spread easily. When people died during the trip, and many did, they were hastily wrapped up and tossed overboard like dead animals.

Few people welcomed the immigrants to North America. Some thought they brought disease and crime. In fact, immigrants were blamed for the spread of **cholera**, a serious intestinal infection, and one of the most dreaded diseases of the 19th century. Not only was cholera often fatal, but no one had any idea how it spread. Its principal symptom—uncontrollable diarrhea—was also discomforting and embarrassing.

cholera :: a serious infection of the intestines that leads to rapid dehydration from diarrhea and vomiting; left untreated, a person may die in as few as 24 hours

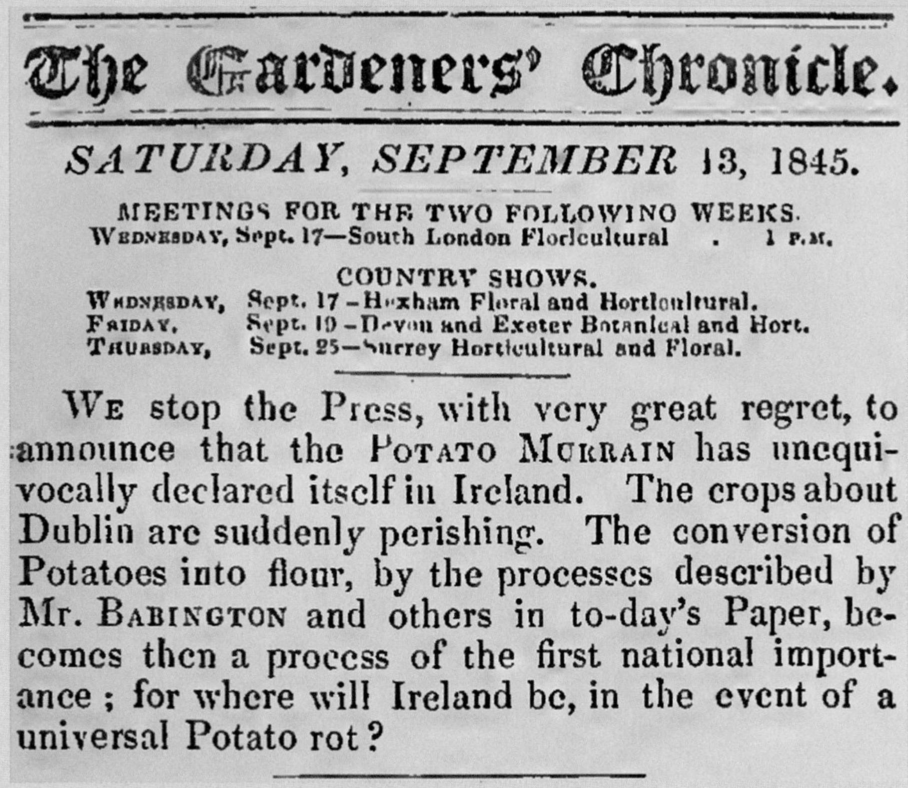

The Gardeners' Chronicle.

SATURDAY, SEPTEMBER 13, 1845.

MEETINGS FOR THE TWO FOLLOWING WEEKS.
WEDNESDAY, Sept. 17—South London Floricultural   . 1 P.M.

COUNTRY SHOWS.
WEDNESDAY,   Sept. 17 – Hexham Floral and Hortioultural.
FRIDAY,   Sept. 19 – Devon and Exeter Botanical and Hort.
THURSDAY,   Sept. 25 – Surrey Horticultural and Floral.

WE stop the Press, with very great regret, to announce that the POTATO MURRAIN has unequivocally declared itself in Ireland. The crops about Dublin are suddenly perishing. The conversion of Potatoes into flour, by the processes described by Mr. BABINGTON and others in to-day's Paper, becomes then a process of the first national importance ; for where will Ireland be, in the event of a universal Potato rot?

FIGURE 14.1 In 1847, the entire potato crop of Ireland failed, and 2 million people died of starvation. This excerpt from *The Gardeners' Chronicle* indicates how serious the problem had become two years prior, in 1845.

On May 15, 1847, the ship *Syria* landed on Grosse Île, an island in the St. Lawrence River, around 30 miles from Quebec, Canada. Grosse Île was used as a site to isolate immigrants before allowing them to enter mainland Canada (**FIGURE 14.2**). Of the 241 passengers who set sail, 9 died on the voyage. Within a week of landing, over 100 passengers were hospitalized, and 70 people died from cholera. *Syria*'s story was repeated over and over, as 398 ships came to Grosse Île during the summer of 1847. More than 5000 people died during their trips, and more than 5000 were buried in mass graves after dying on the island.

Grosse Île is a fairly rocky island, so by August the authorities had to import soil so that more bodies could be buried. Despite their efforts, rats came in from the ships to feed off the cadavers. A Celtic cross that sits atop a steep hill on Gross Île bears witness to that terrible summer of sorrow when more than 9000 people died from the cholera outbreak (**FIGURE 14.3**).

Did so many people have to die from cholera? Could anything have been done to stop the spread or to help the sick? Also, not everyone exposed to cholera caught it. Still others somehow survived and recovered. How was this possible? And is cholera still a risk today? This chapter looks at how our bodies protect us from infectious disease—and what we can learn to help ourselves stay healthy.

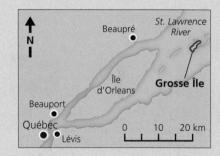

FIGURE 14.2 Grosse Île is an island about 30 miles from Quebec, Canada, which was used to isolate immigrants before allowing them to enter mainland Canada. During the summer of 1847, more than 5000 people died and were buried in mass graves on the island, after traveling on ships from Ireland.

FIGURE 14.3 The words carved into the Celtic cross that sits atop a steep hill on Grosse Île bear witness to the summer of 1847 when more than 9000 people died of cholera.

14.1 What Causes Cholera?

Cholera is a serious intestinal infection. An untreated person who catches it can dehydrate from diarrhea and vomiting, dying in as few as 24 hours (**FIGURE 14.4**). Fortunately, patients can be treated by rehydration. Unfortunately, this simple remedy was not realized over the course of many cholera outbreaks over many years. Sadly, treatment remains unavailable to some, particularly in underdeveloped nations.

Cholera has killed millions of people as it has repeatedly jumped from country to country and continent to continent. The disease originated in India and was localized there until 1816. The British troops stationed in India contracted cholera and brought it to their next ports of call. The global movement of cholera can be traced along the paths of armies and immigration.

In the 19th century, word of a cholera outbreak caused widespread panic. No one knew how it spread or why some people fell ill and others did not. And the death rates were terribly high. An 1831 outbreak in Cairo, Egypt, killed 13% of the population. If a disease caused similar mortality rates in a large city like New York today, more than 1 million people would die.

Cholera is not just a 19th-century disease, however. Cholera flairs up on a regular basis in parts of Africa, Asia, and Latin America. In the Dominican Republic, an outbreak that began in 2010 has infected over 26,000 people, with over 400 deaths suspected as of 2012. And, in Haiti, an outbreak that began as a result of an earthquake in 2010 has killed over 7500 people as of 2013.

Cholera is also what we call an opportunistic disease: it appears in the aftermath of natural disasters such as earthquakes and floods. Why is cholera still a threat? What is its cause, and how can we prevent or fight this disease? John Snow answered these questions one way, and Robert Koch in another, but both men were correct.

John Snow's Explanation for How Cholera Spreads

John Snow practiced medicine in the Soho district of London, where there was a cholera outbreak in 1848 and 1849. He treated his patients as effectively as he could, but he also wanted to figure out how the disease spread. He noticed that people who contracted cholera often got their drinking water from the same source.

In 1854, an even more terrible outbreak of cholera exploded in Soho. This time, Snow was prepared. He painstakingly traced the paths by which cholera spread. He identified what was different about households where people became ill and those that were untouched. Whatever caused cholera, it was present in a water pump at Broad Street (**FIGURE 14.5**). Sewage had leaked into water from this pump, contaminating it. Individuals contracted cholera because they were drinking water contaminated with feces from sick people.

Snow reported his conclusions to local officials. Although they did not believe him, they were so desperate to do something, anything, to stop the deaths that they agreed to his recommendation. The pump handle was removed, cutting off access to the Broad Street water. The cholera outbreak stopped.

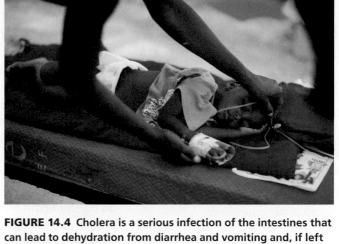

FIGURE 14.4 Cholera is a serious infection of the intestines that can lead to dehydration from diarrhea and vomiting and, if left untreated, ultimately death in as little as 24 hours. A little boy receives treatment following a cholera outbreak in Haiti that has killed thousands of Haitians since it began in 2010 following an earthquake.

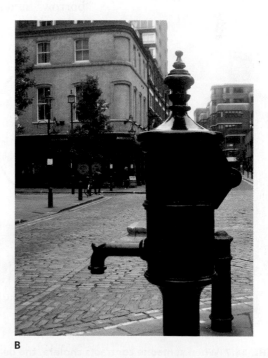

A   B

FIGURE 14.5 In 1854, John Snow created a map (A) tracing a cholera outbreak to a water pump at Broad Street in Soho, London (B). Sewage had leaked into the water, and individuals caught cholera because they were drinking water contaminated with feces from sick people.

FIGURE 14.6 In 1883, Robert Koch went to India to identify the type of bacteria that caused cholera. He microscopically examined the feces of patients and discovered a comma-shaped bacterium called *Vibrio cholerae.*

Vibrio cholerae :: a comma-shaped bacterium that causes cholera

electrolyte :: electrically charged substance necessary for muscle function and water balance

shock :: the body's response to an inadequate supply of blood for circulation; it usually results from blood loss

Robert Koch's Explanation for What Causes Cholera

Robert Koch looked at cholera's cause from another angle. He wanted to know what was *in* the water.

The germ theory of disease blossomed in the mid- to late 1800s. During that time, the bacteria responsible for diseases such as anthrax, tuberculosis, and diphtheria were discovered by scientists some have termed "microbe hunters." One of the stars of this group was Robert Koch.

In 1883, the government of Germany sent Koch to India to figure out what kind of bacteria was responsible for cholera. Koch examined the feces of cholera patients microscopically and discovered a comma-shaped bacterium, *Vibrio cholerae* (**FIGURE 14.6**). Ironically, Filipo Pacini had observed and described the same bacteria in the intestines of cholera victims in 1854. But no one had paid attention to Pacini's report; people had not yet been ready to accept that bacteria could cause disease.

Even with an understanding of how cholera is spread and what causes it, many questions remained for Snow and Koch. How can a tiny bacterium make a person *so* sick? Why doesn't everyone who swallows *Vibrio* develop cholera?

How Cholera Affects the Body

The migrating Irish farmers and families in the late 1840s probably felt as if they had been cursed. So many suffered from cholera either on the voyage to North America or when they arrived. The onset of the disease was swift. Individuals with severe cases of cholera could lose gallons of fluid within a day or so. Salts, or **electrolytes**, the electrically charged substances necessary for muscle function and water balance, also flooded out of their bodies. As people dehydrated, all of their organ systems struggled.

The dramatic loss of fluid sent people into **shock**, the body's response to an inadequate supply of blood for circulation. Usually shock results from blood loss. With cholera, the deluge of fluid from the intestines prompts the body to "borrow" fluid from the circulatory system (**FIGURE 14.7**). Unfortunately, this

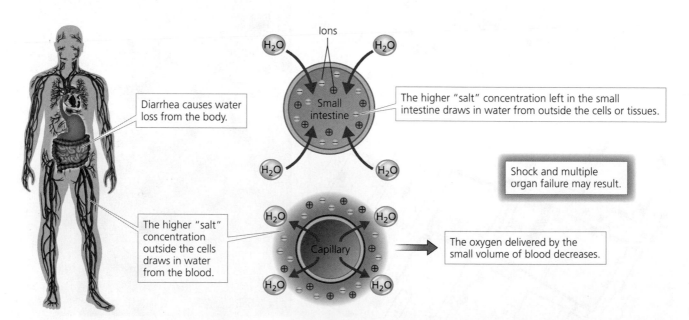

FIGURE 14.7 When someone contracts cholera, the deluge of fluid from the intestines prompts the body to "borrow" fluid from the circulatory system, which is also then lost. As a result, the person goes into shock, and, if untreated, multiple organ failure occurs.

fluid is lost too. People in shock are pale, faint, dizzy, and nauseous. They have a dangerously low blood pressure and high pulse—the heart tries to deliver oxygen to all cells and organs of the body. However, because the oxygen delivered by the small volume of blood is not adequate, eventually as many as 50–60% of individuals with untreated cholera experience multiple organ failure. And while all of this is happening, the sick suffer painful muscle contractions, and their skin turns blue as a result of blood leaking from ruptured capillaries. Finally, coma and death result.

But others who get sick recover. Some people surrounded by the sick and dying never become ill. Why?

How *Vibrio* Bacteria Survive and Spread Cholera

Trillions of bacteria live within and on our bodies. Most of them don't harm us at all. Even the potentially dangerous, or **pathogenic**, ones don't harm us 100% of the time. But when circumstances line up so that we are vulnerable, pathogenic bacteria can spring into action. In the case of cholera, the *Vibrio* bacteria are successful because they subvert some of the body's normal cellular mechanisms to help themselves. To do this (**FIGURE 14.8**), *Vibrio* must:

pathogenic :: capable of causing disease

- Enter the body
- Survive the trip through the stomach and into the intestines
- Evade the body's defenses
- Attach to cells and take up residence in the intestines
- Reproduce to make more bacteria
- Trigger the river of diarrhea that will deliver them to their next host

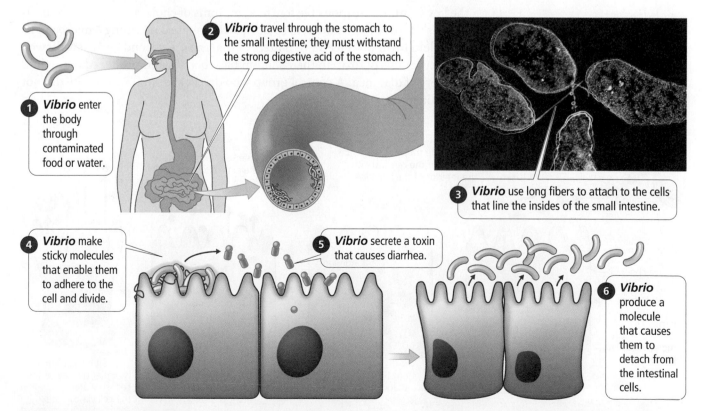

FIGURE 14.8 The *Vibrio* bacteria are successful because they subvert some of the body's normal cellular mechanisms to help themselves.

The strong digestive acid in the stomach is the first defense the *Vibrio* encounter. Normally, for *Vibrio* to cause cholera, at least 100 billion of the bacteria must be swallowed. When volunteers in a 1974 study decreased the acidity of their stomachs by drinking sodium bicarbonate (like the antacids available at the drugstore), it took much fewer *Vibrio*, as few as 10,000, to cause cholera.

Vibrio's next destination is the small intestine, where the *Vibrio* can trick the body into being a participant in its own troubles. First, *Vibrio* grab on to the cells that line the insides of the small intestine. The bacteria have long fibers that stick to specific molecules on the intestinal cell membranes. Once the *Vibrio* attach, they make more sticky molecules to strengthen their holds. Having gained a foothold, the *Vibrio* establish a constantly dividing population of cells. Next, *Vibrio* secrete a **toxin**, or poison. This substance binds to another molecule on the intestinal cell membrane. Once it does so, part of the toxin molecule enters the cell and short-circuits the cell communication pathway that controls water balance (**FIGURE 14.9**).

Under healthy conditions, certain hormones control whether the body secretes or retains water (Chapter 11). The cellular mechanism involved in this behavior involves a series of molecules that act like on/off switches. Cholera toxin bypasses the normal controls by binding permanently to a specific on/off molecule called **G protein**. The G protein gets stuck "on," and so does the pathway. The final component in this pathway is a molecule that transports chloride ions out of the cell and into the interior of the intestine. Because chloride rushes into the intestines, so do sodium and other electrolytes, so that electrical charges can be balanced. Chlorine has a negative charge, and sodium is positive. Together they make salt, and wherever salt goes, water follows.

Experiments with volunteers have shown unequivocally that the toxin is the key substance responsible for the symptoms of cholera. In one study, individuals who drank a solution of sodium bicarbonate containing 5 micrograms of cholera toxin produced 1 to 6 liters of diarrhea. A second group swallowed 25 micrograms of toxin in their sodium bicarbonate and produced more than 20 liters of diarrhea. A control group who drank sodium bicarbonate did not have diarrhea.

toxin :: poison

G protein :: a specific on/off molecule to which the cholera toxin binds

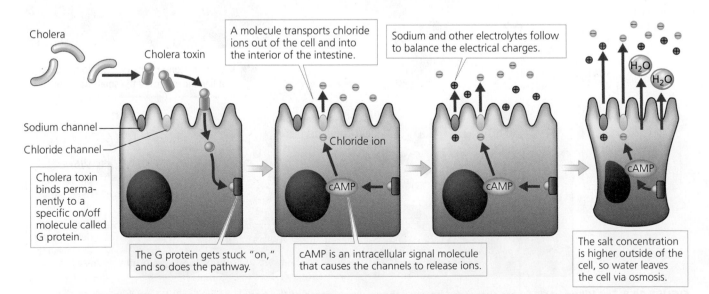

Cholera

Cholera toxin

A molecule transports chloride ions out of the cell and into the interior of the intestine.

Sodium and other electrolytes follow to balance the electrical charges.

H_2O
H_2O

Sodium channel

Chloride channel

Cholera toxin binds permanently to a specific on/off molecule called G protein.

The G protein gets stuck "on," and so does the pathway.

Chloride ion

cAMP

cAMP is an intracellular signal molecule that causes the channels to release ions.

cAMP

cAMP

The salt concentration is higher outside of the cell, so water leaves the cell via osmosis.

FIGURE 14.9 *Vibrio* **short-circuit the cell communication pathway that controls water balance.**

Finally, *Vibrio* need to leave the body to find a new host. The bacteria make yet another molecule that detaches them from the intestinal cells. Because the host pours out volumes of fluid, the *Vibrio* surf out. More than 100 million *Vibrio* are present in each milliliter of diarrhea. An individual with cholera can deliver trillions of *Vibrio* into their surroundings each day.

A person in crowded, dirty conditions surrounded by other people sick with cholera can easily ingest enough *Vibrio* to become ill. The immigrants on crowded ships from Ireland in the 19th century, or packed together in isolation centers like Grosse Île, were literally surrounded by *Vibrio*; they clearly had the odds stacked against them.

4.2 What Invaders Do We Face?

We are exposed to a practically infinite array of substances all day, every day. Not all of them pose a danger, however. Some are part of the community of organisms that make us the individuals we are. Others, like *Vibrio*, are part of the many potential problems that must be handled.

Bacteria

We are surrounded and covered by microorganisms that could make us sick, but most of the time they don't. Thousands of different species of bacteria live harmlessly on or in us. Bacteria are relatively simple **prokaryotic** cells, meaning they have no membrane-bound organelles. Their DNA is not packaged in a nucleus except in a small number of rare cases.

prokaryotic :: a simple cell with no membrane-bound organelles

Many bacteria are actually quite helpful to us. Some keep the growth of harmful bacteria in check. Others make it easier for us to digest foods we would have trouble handling alone. Nevertheless, some bacteria do cause serious and even fatal illnesses (**FIGURE 14.10**).

Besides cholera, other diseases caused by bacterial infection include tuberculosis, a respiratory disease; strep throat; gonorrhea and syphilis, two sexually transmitted diseases; dangerous "flesh-eating" staph infections; and even food poisoning. Bacteria are successful pathogens because they grow so fast that they can overwhelm a host's resources. One bacterial cell can divide every 20 minutes. A population with 1 cell can potentially reach 1 billion within 10 hours! Also, like *Vibrio*, many pathogenic bacteria make toxins that are responsible for the disease taking hold.

Protozoa and Other Eukaryotes

We can also be attacked by cells and organisms much more similar to us than bacteria (**FIGURE 14.11**). Protozoa are single-celled eukaryotic organisms. Like the cells in our bodies, protozoan cells have a nucleus and organelles. Examples of illnesses caused by protozoa include giardiasis, the most common waterborne diarrheal disease in the United States; amoebic dysentery, another diarrheal disease; sleeping sickness; and malaria.

Even multicellular eukaryotes can get into the act. In addition to destroying the potato crop and

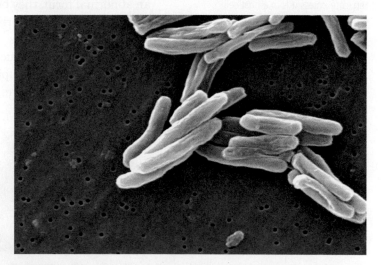

FIGURE 14.10 Bacteria are relatively simple prokaryotic cells that have no membrane-bound organelles. Some bacteria cause serious illnesses, such as *Mycobacteria*, which cause the respiratory illness tuberculosis.

FIGURE 14.11 Protozoa and eukaryotic cells may attack us.

Giardia is a protozoa that causes Giardiasis, a common waterborne diarrheal disease.

Athlete's foot is caused by a fungus (a multicellular eukaryote).

A Schistosome is a parasitic blood fluke that enters the body through the skin and can cause fever, fatigue, and diarrhea.

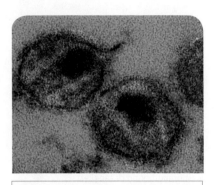

Viruses, such as the HIV virus responsible for AIDS, have a shell made of protein with genetic material inside.

FIGURE 14.12 Viruses are particles that bridge the gap between living and nonliving and cannot replicate unless they are in a host cell.

virus :: particle that bridges the gap between living and nonliving; cannot replicate unless it is in a host cell

prion :: an infectious protein; may lead to degenerative diseases of the nervous system

initiating a famine, different species of fungi also cause athlete's foot, jock itch, ringworm, and yeast infections.

People can also be infected with worms by eating contaminated fish or meat, or due to poor sanitation. Worms that take up residence in people include tapeworms, roundworms, blood flukes, and guinea worms.

Viruses and Prions

Finally, we can become infected by pathogens that are not even alive. **Viruses**, particles that bridge the gap between living and nonliving, cannot replicate unless they are in a host cell (**FIGURE 14.12**). Viruses have a shell or coat made of protein and inside it, genetic material. Viruses may contain DNA, like living cells do, *or* RNA. In order to replicate, the virus must get inside the host's cells. Once there, viruses pirate the host's metabolic machinery, turning it to the task of making virus proteins, nucleic acids, and ultimately viruses. Viruses are responsible for many diseases including AIDS, chicken pox, and even the common cold.

Prions are infectious proteins (**FIGURE 14.13**). Prion-like particles are found in the brain and have a normal function there, perhaps involving the storage of very long-term memories. When these normal proteins change their shapes to an abnormal form, they become pathogenic by forcing other prion-like particles to misfold into abnormal forms. The transformation from normal to "bad" spreads, and the prions make clumps in the brain, damaging delicate cell structures. Prions produce degenerative diseases of the nervous system like mad cow disease, scrapie (in sheep), and Creutzfeldt-Jakob disease (in humans). Animals or persons afflicted with these diseases lose control of their

Infectious prions cause normal prion proteins to fold into a new shape that is harmful to cells.

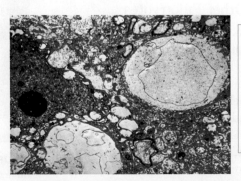

Prion-like protein particles in the brain lead to cell death and holes in the brain tissue (light green); this happens in neurodegenerative diseases such as Creutzfeldt-Jakob disease.

FIGURE 14.13 Prions are infectious proteins. They are the only pathogens invisible to the immune system, and the diseases they cause are untreatable at present.

nervous system and brain function. These diseases are extremely rare, which is fortunate because prions are virtually indestructible. The diseases they cause are untreatable at present. Prions are the only pathogens that are invisible to the immune system.

How Does the Immune System Protect Us?

The immune system has to protect us against a varied army of external and internal invaders. But it is not necessarily the body's first line of defense.

Not every Irish immigrant exposed to *Vibrio* contracted cholera. Not everyone who developed cholera died. Why? Some of the reasons might be quite simple. People who stayed away from the sick and who kept clean avoided swallowing *Vibrio*. In addition, individuals who drank liquids that were uncontaminated had a chance to remain healthy.

Other reasons are more complex and not fully understood. People's susceptibilities vary. A person with a very acidic stomach is less likely to be infected than someone with a less acidic one. Acid kills *Vibrio*. We also don't know why people with AB blood are extremely resistant to cholera, while those with type O blood are easily infected. Carriers of the gene associated with cystic fibrosis, a lung disease, are quite resistant too. This gene encodes for a faulty chloride transporter, the same transporter that operates at full blast in cholera patients. Perhaps the transporter's poor functioning protects the body from losing a large amount of water even if a person swallows *Vibrio*.

The most complex answer to the questions of why some people exposed to *Vibrio* remain healthy and why some who get sick recover can be found in the immune system. Our immune systems are constantly vigilant, checking out whether a particular cell, piece of dirt, or anything that has entered our bodies is friend or foe. Let's start with the first line of defense—before pathogens even encounter our immune system.

The First Line of Defense: Don't Let the Pathogens Get In

Imagine the people crammed into a crowded ship heading to Grosse Île and the dream of a better life. Food and clean water are in short supply. It is hard to sleep. The space is so tight that it is difficult to move around to get some exercise, fresh air, or even some sun. All around, cholera roared through the ship. How did some people remain healthy and others manage to recover from cholera?

Our bodies possess a battery of nonspecific defenses against infectious disease, as shown in **TABLE 14.1**. They work on any invader or potential pathogen. First, a person's general well-being influences whether or not he or she will become ill. Nutrition, age, climate (for instance, exposure to the elements, such as cold and rain), fatigue, and socioeconomic status all affect a person's well-being and therefore vulnerability to disease. The Irish immigrants were probably more or less likely to die from cholera based on these characteristics. Infants, children, and the elderly were more apt to die from cholera because they dehydrated faster than other age groups. Poor people were generally malnourished and did not have ready access to clean water for drinking.

Second, bacteria and other potential pathogens face both mechanical and chemical barriers that prevent disease. These are **nonspecific barriers**, meaning they don't target any particular pathogen. Skin blocks the direct

nonspecific barrier :: blocks bacteria and other potential pathogens without targeting specific ones; skin is a nonspecific barrier

Table 14.1:: Nonspecific Barriers That Prevent Infection

Barrier	How It Prevents Infection
General health and well-being	Lessens vulnerability to disease
Skin	Blocks the direct entry of bacteria into the body
Sweat	Contains chemicals that kill microbes
Saliva	Includes antimicrobial enzymes, which kill many bacteria that enter the mouth
Mucous membranes	Line the insides of the digestive, reproductive, respiratory, and urinary systems and trap microbes that enter the body; hairlike projections sweep mucus and any trapped microbes into the throat, where they are swallowed
Stomach acid	May destroy bacteria that enter the stomach
Urine	Keeps the urinary tract virtually free of bacteria
Tears	Contains antimicrobial chemicals
Helpful microbes	Keeps the growth of potentially harmful microbes in check

entry of bacteria into the body. Not only is skin too dry for bacteria to grow luxuriously, sweat contains chemicals that kill microbes. Also, skin cells are shed every day and with them any bacteria that are attached.

Third, mucous membranes that line the insides of the digestive, reproductive, respiratory, and urinary systems trap microbes that enter the body through these routes. If bacteria get into the mouth, many are killed there because of the antimicrobial enzymes in the saliva. If swallowed, bacteria are destroyed by acid in the stomach. The vagina is also an acidic environment thanks to helpful microbes called *lactobacilli* that secrete acid and prevent the growth of harmful bacteria. Hairlike projections on the cells lining the respiratory system sweep mucus and any trapped microbes into the throat, where it is all swallowed. And the regular passage of urine keeps the urinary tract virtually free of bacteria. Even the eyes have local protection: tears contain antimicrobial chemicals.

The Second Line of Defense: Don't Let the Pathogens Make a Home

Once pathogens break through the first line of defenses, they encounter the immune system's internal defenses. These are nonspecific and fall into three categories (**FIGURE 14.14**):

macrophage :: locates and kills microbes by engulfing and eating them

natural killer cell :: eliminates virus-invaded cells by penetrating their outer membranes and causing the infected cells to burst

interferon :: a chemical signal that enlists the help of noninfected cells to fight a virus

complement protein :: a protein that coats the surface of bacteria, making it easier for macrophages to eat them

inflammation :: the red, warm, and painful swelling that occurs with an infection

- *Defensive cells:* Two types of white blood cells attack and destroy invading pathogens. **Macrophages** locate and kill microbes by engulfing and eating them. **Natural killer cells** kill virus-invaded cells by penetrating their outer membranes, causing the infected cells to burst.
- *Defensive proteins:* Cells infected by viruses produce **interferons**, chemical signals that enlist the help of noninfected cells to fight the virus. **Complement proteins** coat the surfaces of bacteria, making it easier for macrophages to eat them. Complement proteins also poke holes in pathogens, causing them to burst.
- *Inflammatory responses:* The red, warm, and painful swelling that occurs with infection is called **inflammation**. Macrophages send out chemicals that increase blood flow and cause the blood vessels in the area to become leaky. Other macrophages then can get to the infection site more easily. Inflammation limits the infection to a local region.

Macrophages locate and kill microbes by engulfing and eating them.

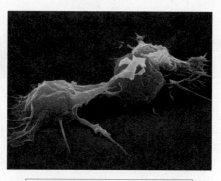

Natural killer cells penetrate the outer membranes of virus-invaded cells, causing them to burst.

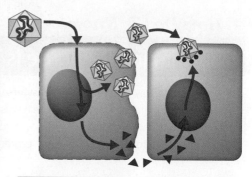

Cells attacked by viruses produce **interferons** that enter healthy cells and induce changes, making these cells more resistant to attack.

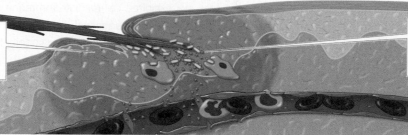

The red, warm, and painful swelling of **inflammation** limits the infection to a local region.

Complement proteins coat the surfaces of bacteria and also poke holes in pathogens.

FIGURE 14.14 The second line of defense against pathogens are the immune system's internal defenses: defensive cells, defensive proteins, and the inflammatory response.

Even when not fully successful, this second line of defense gives the body time to activate its final weapon: specific immunity.

The Third Line of Defense: Know Your Enemy

A person who swallows a big dose of *Vibrio* is likely to develop cholera. In the 19th century, as many as 50–60% of people with cholera died from the disease. Today, in areas of the world where medical care is available, less than 1% of individuals with cholera die from it. Why this improvement?

Physicians figured out that if they kept people with cholera hydrated and replaced electrolytes lost in diarrhea, most recovered. The cure, however, is not due directly to this treatment. It is the body's own specific immune response that beats back the *Vibrio*. The rehydration therapy simply provides the much-needed time for the body to marshal its own abilities to fight the pathogen. There are two types of specific immune responses: antibody mediated and cell mediated. Both rely upon the actions of white blood cells called **lymphocytes**. For a discussion of the discovery of these lymphocytes, see *How Do We Know? The Immune Response Is Two-Fisted*.

Antibody-Mediated Immunity: B Cells B lymphocytes, or **B cells**, are the main players in antibody-mediated immunity (**FIGURE 14.15**). This is the type of immunity that springs into action to fight cholera.

The body produces B cells continuously in the **bone marrow**, the spongy material inside the large bones of the body, and some other locations too. The cells have special receptor molecules on their membranes. These receptors recognize specific foreign molecules, or **antigens**. Each B cell recognizes only one type of antigen. When a person is infected with *Vibrio*, for example, the B cells that recognize its antigenic proteins will bind to them. This binding causes the B cell to divide, making many copies of itself. Each one of these

lymphocyte :: a type of white blood cell

B lymphocyte, or **B cell** :: produced by the bone marrow, it is the main player in antibody-mediated immunity

bone marrow :: the spongy material predominantly inside the large bones of the body; it produces B cells

antigen :: a foreign molecule

The Immune Response Is Two-Fisted

Research has a way of revealing the answers to questions that weren't even asked, at first anyway. Such was the case for the discovery that the immune system has two major components necessary for keeping the body protected: B cells and T cells.

Back in the early 1950s, Bruce Glick was a graduate student at Ohio State University. He was studying a structure called the bursa of Fabricius (BF) in chickens. He observed its development in embryos and noticed that it grew dramatically in the weeks after the chick hatched. Glick wanted to understand the function of the BF, so he removed it from hatchlings and monitored the birds.

Meanwhile, a graduate student friend of his, Timothy Chang, was studying the immune response of chickens to the bacterial pathogen *Salmonella*. He needed to inject birds with antigens from *Salmonella* so he could collect the antibodies the bird would make. Chang did not have any chickens and asked Glick if he could borrow a few.

Glick agreed, and Chang injected several young birds with *Salmonella* antigen. What happened next was quite unexpected. Several of the injected birds died, but *none* of the other injected birds produced antibodies. What was going on here?

Chang learned that the birds Glick had given him had no BFs. Could it be that non-BF birds can't make antibodies?

Glick and Chang tested this idea directly. They injected antigens into a population of chickens that had no BF (non-BF) and another population with the organ intact (BF). The results were astonishing. Almost 90% of non-BF birds failed to make antibodies, whereas only 13.7% of the BF birds failed to do so. We know now that the BF is the place where B cell development starts in embryos, although in mammals B cells are produced in the bone marrow throughout life.

What about cell-mediated immunity? The non-BF birds still exhibited cell-mediated immune responses. Some other observations suggested that the cells necessary for this component of the immune system originated in a gland called the thymus, which helps make white blood cells. Indeed, in experiments in which the thymus was removed from newly hatched chickens, cell-mediated immunity was lost. However, provided the bursa of Fabricius was present, the antibody-mediated immunity remained functional. Birds lacking both the BF and thymus had no specific immune response at all.

It is difficult to imagine most people getting very excited about basic research on the bursa of Fabricius. It may well be that Glick's family chuckled about his "odd project." If so, they laughed too soon, for it is just from these unpredictable origins that scientists make some of the most important discoveries with the potential to affect all of us. ::

Bursa of Fabricius

1 Glick noticed that the bursa of Fabricius (BF) grew dramatically in the embryos of chicks in the weeks after they hatched.

2 Glick removed the BF from the hatchlings and monitored the birds.

3 Chang borrowed some chickens (with the BF removed) from Glick.

4 Chang injected several young birds with *Salmonella* antigen.

Non-BF birds

Several of the injected birds died, but *none* of the other injected birds produced antibodies.

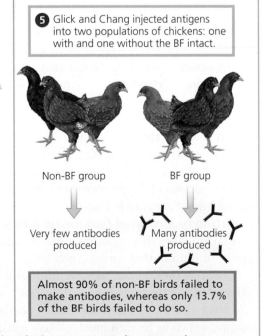

5 Glick and Chang injected antigens into two populations of chickens: one with and one without the BF intact.

Non-BF group | BF group

Very few antibodies produced | Many antibodies produced

Almost 90% of non-BF birds failed to make antibodies, whereas only 13.7% of the BF birds failed to do so.

In the early 1950s, Bruce Glick and Timothy Chang inadvertently discovered that the immune system has two major components necessary for keeping the body protected: B cells and T cells.

Unfortunately, not all diseases are preventable with vaccination. For example, pathogens that change their surface proteins over time, like the flu virus or malaria parasite, are poor candidates for vaccines that would offer permanent protection. Also, for some diseases a vaccine offers only partial protection. For cholera, the vaccines developed so far are only 25–50% effective. In fact, the World Health Organization (WHO) does not recommend the vaccine even if you travel to an area affected by cholera. Instead, they urge people to "boil it, cook it, peel it, or forget it": don't drink local water, don't eat raw food, and peel fruits yourself before eating. WHO's advice is excellent for avoiding any of the many infectious diarrheal diseases.

Passive Immunity

The memory cells produced after exposure to a pathogen or due to vaccination can last for years, even a lifetime for some diseases. One of the benefits of aging is that we become increasingly resistant to disease—at least up until the time we get so old that the immune system itself starts to falter. This means babies and children are especially vulnerable to infection.

Fortunately, we are not born without protection. Mothers provide **passive immunity** in the form of antibodies that circulate in the mother's blood and that cross the placenta and enter the fetal bloodstream (**FIGURE 14.18**). When a baby is born, he or she is outfitted with a collection of antibodies ready to fight off invaders. Also, breast milk, especially the first batch suckled by a newborn, has antibodies in it. These antibody gifts from the mother are a temporary protection lasting only several months. But they get a baby safely past the delicate newborn stage and provide time for his or her immune system to mature.

We also use vaccination programs to address the vulnerability of babies and children to infectious disease. Although the immune system does a masterful job fending off pathogens, some diseases—even if you are able to survive them—can damage the body terribly. For example, it is much better to avoid polio entirely than to risk its potentially horrible side effects such as paralysis.

passive immunity :: the temporary transfer of antibodies from one individual to another; for example, the antibodies that circulate in a mother's blood can cross the placenta and enter the fetal bloodstream

FIGURE 14.18 Passive immunity is the temporary transfer of antibodies from one individual to another. The antibodies that circulate in a mother's blood can cross the placenta and enter the fetal bloodstream; breast milk also contains antibodies.

What Can Help If Our Immune System Fails?

Sometimes a person is exposed to a fast-acting antigen for which they have no antibody. For example, suppose you are bitten by a poisonous snake for the first time. By the time your body mounts a good immune response, you will be dead. Fortunately, antivenom can treat a snake bite, while antibiotics can treat many diseases caused by bacteria.

Antivenom

Antibodies called **antivenom**, given by injection, can disable snake venom. Available in most hospitals, antivenoms are made by injecting goats or horses with increasing amounts of snake venom (**FIGURE 14.19**). The animals make antibodies to the venom, and eventually these antibodies reach high enough levels that they can be collected from the blood. These antibodies are concentrated and injected into the person bitten by the snake. As is true for the passive immunity enjoyed by newborns, this one is also temporary.

Antibodies

In some situations, a person can be treated with the antibodies isolated from another individual to help boost immunity. For example, a person in poor health who is exposed to measles can be treated with antibodies that fight the disease, thus preventing additional illness. In addition to isolating antibodies from other individuals, **monoclonal antibodies** (**MAbs**), which target only diseased cells, are another source (see *Technology Connection: Monoclonal Antibodies: Magic Bullets?*).

Antibiotics

Usually, we live in a balanced harmony with the microorganisms in and around us. When a pathogen manages to overcome our defenses, we can become ill so fast that we can't fight back quickly enough. People do die from infections. In fact, 25% of all deaths worldwide are caused by infection.

Microorganisms also live in competition with each other and produce chemicals to inhibit or kill other organisms that might harm them. These chemicals, or **antibiotics**, are used by us to treat bacterial infections (**FIGURE 14.20**).

WHO does not recommend antibiotics for most cases of cholera. Evidently, it is better to replace lost fluids and electrolytes and allow the body a chance to develop long-lasting immunity to cholera rather than to rely on antibiotics every time it reoccurs in the population.

FIGURE 14.19 Antivenom is an injection of antibodies that can disable snake venom. After the venom is extracted from a snake, it is injected into goats or horses that make antibodies, which are then collected.

antivenom :: antibodies given by injection that can disable snake venom

monoclonal antibody (MAb) :: referred to as a "magic bullet," a treatment that targets only a specific antigen

antibiotic :: a chemical produced by microorganisms to inhibit or kill other organisms that might harm them; this chemical is used by humans to treat bacterial infections

FIGURE 14.20 Microorganisms produce chemicals to inhibit or kill other organisms that might harm them. These chemicals, or antibiotics, are used by us to treat bacterial infections.

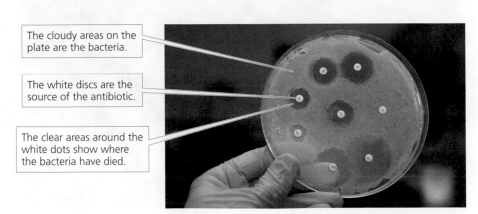

The cloudy areas on the plate are the bacteria.

The white discs are the source of the antibiotic.

The clear areas around the white dots show where the bacteria have died.

Technology Connection

Monoclonal Antibodies: Magic Bullets?

Sometimes the cure can seem worse than the disease. Many medicines, particularly some for serious problems like cancer, can produce awful side effects. Imagine having smart drugs, what Paul Ehrlich in 1900 called "magic bullets"—treatments that target only diseased cells while leaving healthy cells unharmed. In fact, Ehrlich's magic bullets are here—in the form of monoclonal antibodies (MAbs).

In 1975, Georges Köhler and César Milstein laid the groundwork for the development of therapies that harnessed the power of the immune system to fight disease. They realized that individual B cells make one, and only one, type of antibody. What if it were possible to isolate and grow a population of the B cell making a desired antibody? Unfortunately, B cells have a finite life span. Undeterred, Köhler and Milstein sought to make selected B cells immortal.

Köhler and Milstein injected an antigen into mice in order to elicit an immune response. They collected B cells from the mice and fused them with a type of blood cancer cell called a myeloma. The results were hybridomas, immortal cells that each produced one type of antibody. Köhler and Milstein grew large amounts of hybridoma cells that all made the same type of antibody. Because the antibodies came from one clone of B cells, they are called monoclonal antibodies.

Scientists and drug developers immediately saw the potential of MAbs for targeting illnesses. It took 10 years from Köhler and Milstein's breakthrough for the first MAb to be approved for sale. Today, more than 20 MAbs have won approval by the U.S. Food and Drug Administration (FDA). Several hundred more are making their way through the approval process.

MAbs are used for the detection of specific molecules. For example, pregnancy tests use MAbs to recognize human chorionic gonadotropin (Chapter 6). MAbs have also been developed to help prevent kidney transplant rejection; fight leukemia, lymphoma, and breast cancer; and prevent anthrax infection. They can also treat macular degeneration, which gradually destroys a part of the eye, the retina, causing vision loss.

The early generations of MAbs had some side effects of their own. Since the antibodies originated in mice, the human immune system sometimes reacted to them. Rashes, joint pain, nausea, kidney failure, and even death were possible.

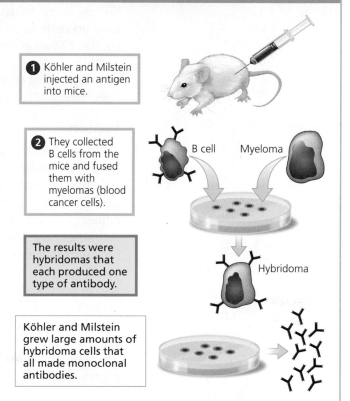

1 Köhler and Milstein injected an antigen into mice.

2 They collected B cells from the mice and fused them with myelomas (blood cancer cells).

B cell   Myeloma

The results were hybridomas that each produced one type of antibody.

Hybridoma

Köhler and Milstein grew large amounts of hybridoma cells that all made monoclonal antibodies.

In 1975, Georges Köhler and César Milstein laid the groundwork for the development of monoclonal antibodies (MAbs) that target only diseased cells while leaving healthy cells unharmed. Today, more than 20 MAbs have won approval by the U.S. FDA.

Scientists have since learned how to substitute human for mouse proteins in the MAbs. Research continues to make their biochemical structures fully human.

MAbs are likely to become increasingly common tools in 21st-century medicine. First of all, they work. Second, compared to most other drugs, the new generations of MAbs don't have toxicity problems. Finally, they are easier and quicker to develop than traditional drugs. Whereas on average it takes a company five years and $20 million to bring a drug to the patient testing phase, only two years and $2 million are needed to do the same for a MAb. It appears that Paul Ehrlich's vision from more than 100 years ago is becoming a reality. ::

We can probably generalize WHO's advice regarding antibiotic use beyond cholera. As described in Chapter 10, the overuse of antibiotics has created an entirely new problem—bacteria that are resistant to multiple types of antibiotics. There are now virtually indestructible "superbugs" that have the potential to be grave health threats to all of us.

14.6 How Can We Prevent the Spread of Infectious Disease?

We are in the midst of a worldwide cholera outbreak that has lasted more than 30 years so far. How is this possible, given our medical knowledge and abilities? More generally, how is any infectious disease spread, and can we do anything to stop it? *Scientist Spotlight: Anthony Fauci* describes one individual's ongoing efforts to address the biological, medical, political, and social components of infectious disease.

Wellness

Disease spreads more easily and quickly through a weak or diseased population. The immigrants on Grosse Île were malnourished and fatigued before they even got on the boat. Some had never had adequate nutritious food in their lives. They were overworked on their tenant farms and lived in the dark damp homes provided by their landlords. Few if any had reliable access to clean drinking water. Sewage treatment was nonexistent, and so people lived in close proximity to human and animal wastes.

The voyage made matters worse. People were so crowded that it was almost impossible to sleep enough. Food and clean water were still not reliably available. And conditions did not improve when they landed. With their immune responses and general physical functioning weakened, the immigrants were practically defenseless against the cholera that spread like wildfire. This relationship between a person's well-being and the likelihood of contracting an infectious disease is not specific to cholera alone.

Hygiene

Even when an outbreak is underway, people can influence whether they will contract the disease and spread it to others. Cholera is completely preventable—don't swallow *Vibrio*. Washing one's hands helps stop the spread, provided that infected water is decontaminated. So can washing the soiled clothing or bed linens of the sick. Being careful about what you eat and drink also helps: again, boil it, cook it, peel it, or forget it.

Other infectious diseases can also be contained if people follow some simple behaviors. Frequent hand washing is helpful for all infectious disease. For respiratory illnesses like influenza, avoid coughing or sneezing without covering your mouth and nose. Don't cough or sneeze into your hand; instead, use the crook of your elbow. In case of a highly infectious, dangerous respiratory pathogen, face masks are a good idea. The likelihood of catching infectious diseases spread by body fluids can be reduced by condom use (Chapter 6) and not sharing hypodermic needles or syringes.

Clean Water and Sewers

In North America, Japan, Australia, and Western Europe, cholera and other diarrheal diseases are no longer the serious concerns that they were in the 1800s. Cholera does, however, continue to plague parts of Africa, Latin America, and Asia in the 21st century. Why are populations in some parts of the world vulnerable, while others are not?

The simple answer is that countries with well-developed water and sewage treatment systems eliminate the biggest source of diarrheal disease pathogens. Once people no longer lived near piles of human waste that seeped into their drinking water, cholera and its kin disappeared. Places in the world that

Scientist Spotlight

Anthony S. Fauci (1940–)

"Incompetent idiot!" "Murderer!"

These were not the worst names hurled at Dr. Anthony Fauci by AIDS activists clamoring to gain access to experimental drugs for the disease. It was 1988 and Fauci, director of the National Institute of Allergy and Infectious Disease (NIAID), was the U.S. government's point person for the fight against HIV/AIDS.

Fauci was in his office as demonstrators stormed the NIAID campus. Instead of calling security, he took a good look at the protesters and saw only "people who were sick and scared." Fauci invited the leaders of the group into his office and listened to what they had to say. The result was the start of a dialogue that ultimately led to the participation of activists in many phases of planning. It also led to the idea that even gravely ill people can get access to experimental drugs. Such parallel tracking is now the model for testing treatments for other diseases, too.

Born in the Bensonhurst section of Brooklyn in 1940, Fauci is the grandson of Sicilian immigrants. Encouraged by his parents to work hard and aim high, he completed his degree in premed and classics at the College of the Holy Cross. After that, it was off to medical school at Cornell University. Upon completion of his residency, Fauci started as a research associate and has steadily climbed the ranks ever since.

In 1984, he was appointed director of NIAID. In this job, he oversees both basic and applied research in the prevention, diagnosis, and treatment of infectious diseases like HIV/AIDS, sexually transmitted diseases, influenza, tuberculosis, and malaria. NIAID also studies transplantation-related illnesses, autoimmune disease, immune-related illnesses, asthma, allergies, and potential bioterrorism. And Fauci continues to have his own research lab and sees patients. No wonder he works an 80-hour week.

Fauci is not simply a gifted leader. He has made some critical research breakthroughs. He developed therapies to successfully treat formerly fatal inflammatory and immune-related diseases. He has also helped reveal how the immune system works and how immunosuppressive drugs dial down the immune response. He has been at the forefront of basic research on how HIV destroys the body's defenses.

Anthony Fauci worked with AIDS activists to develop ways for gravely ill people to gain access to experimental drugs.

Voice of America has called Fauci a "tireless warrior." A man with a deep social consciousness, he is driven by the desire to discover new things and a dedication to public service. When asked what keeps him going on all burners, he explains, "It is an indescribable experience knowing that what you are doing will have an impact on the lives of tens, if not hundreds, of millions of people." ::

Source: Bill Snyder, "Unfinished Business: AIDS, Bioterrorism, and the Evolving Legacy of Anthony Fauci," *Lens* 2, no. 1 (Spring 2004): 16–21, http://www.mc.vanderbilt.edu/lens/download/vumcLENS_Spring04.pdf

remain underdeveloped are still in the 19th century as far as their public works are concerned and therefore still threatened by what many think of as 19th-century diseases.

Proper Antibiotic Use

Even though antibiotics play an important role in the fight against infectious disease, their improper use actually helps disease spread. Every time

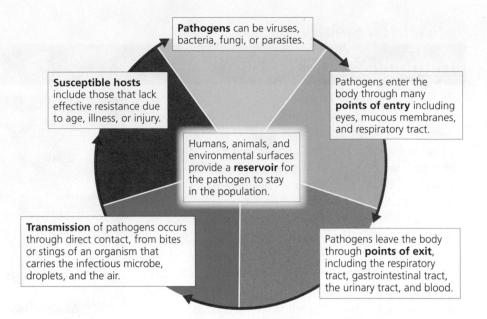

FIGURE 14.21 Pathogen reservoirs exist in the population, providing a continual source of infectious microbes. Improper antibody use is one factor that reinforces these reservoirs.

antibiotics—rather than the actions of the immune system—defeat an infection, another person remains a potential source of pathogen. In other words, they form a reservoir for the pathogen to stay in the population (**FIGURE 14.21**).

Imagine that 1000 people are infected with *Vibrio*. Let's say that 900 received no antibiotic and recovered after the disease had run its course, while 100 got better after antibiotic treatment. A second population of 1000 also contracted cholera, and all were treated with antibiotics and recovered. The next time cholera comes roaring through the first population, only 100 of the original 1000 are likely to get it again. In contrast, all 1000 of the second population are probably going to get sick gain. Clearly, the first population is less dangerous for others because it has only 100 people producing diarrhea with *Vibrio*. The second population will probably release at least 10 times as much *Vibrio* into the environment, forming a big reservoir or source of the pathogen.

Antibiotics, when not used properly, also enhance the spread of disease by selecting for pathogens that are resistant to treatment. When people demand antibiotics for illnesses that are caused by viruses, and therefore unresponsive to antibiotics, they are helping resistant bacteria take over the bacterial population. Also, failing to take an antibiotic as prescribed will increase the likelihood of resistant pathogens arising. Antibiotics used prudently are life-saving. Used carelessly, they put us all at risk.

Vaccines

For infectious diseases that are responsive to immunization, it seems obvious how a vaccine prevents the spread of disease. If you are vaccinated, you won't get sick. If you don't get the illness, you can't spread it to others.

True enough, but the power of vaccination is not solely about protecting the individual. The more important job of vaccination is to protect the population. If everyone is vaccinated to prevent a particular disease, the reservoir for the pathogen is eliminated. If only 50% of people receive a particular vaccine, then a huge pathogen reservoir remains possible.

Unfortunately, too, many vaccines are not 100% effective. Also, some individuals cannot tolerate vaccination because of other health issues like an immune system problem or cancer. Therefore, while vaccines certainly do protect the individual, they play a more critical role keeping the population healthy.

Biology in Perspective

Although the travails of the Grosse Île immigrants occurred more than 150 years ago, experiences similar to theirs continue to be suffered by millions of people today. Infectious disease is not confined to history.

Avoiding disease requires individual and societal efforts. Our immune systems are usually equal to the task of protecting us. However, personal behaviors influence our individual risks. Do we avoid contact with pathogens and generally take good care of our bodies? If we get sick, do we take steps to avoid spreading the pathogen? As a society, are we ensuring that everyone has access to clean water, adequate food, and decent shelter? Are there public works in place to treat waste? Do people have access to preventive medical care? Do we use antibiotics and vaccination responsibly?

The immune system works quite effectively, but it isn't magic. Its normal functioning is a necessary but insufficient condition for wellness. We need to provide the circumstances in which the immune system can succeed in its job of ridding our bodies of foes and leaving the friends in peace.

Chapter Summary

14.1 What Causes Cholera?

Describe the history and biology of cholera.

- Cholera is caused by an intestinal infection.
- John Snow stopped a cholera epidemic in London by identifying a water pump as the source of infection. Robert Koch determined that the bacterium *Vibrio cholera* caused cholera.
- Cholera makes people sick by causing them to lose gallons of fluid through diarrhea and vomiting. If left untreated, organ failure and ultimately coma and death may occur.
- To be successful, *Vibrio* needs to (1) be ingested, (2) survive stomach acid, (3) evade defenses, (4) attach to cells, (5) reproduce, and (6) trigger diarrhea to be transmitted to the next host.

14.2 What Invaders Do We Face?

Describe the array of potential pathogens humans are exposed to every day.

- Some prokaryotes live harmlessly on or in us, while others may cause diseases.
- Protozoa and multicellular eukaryotes, such as fungi and worms, can cause diseases.
- Viruses are particles that pirate host cell metabolic machinery to replicate.
- Prions are infectious proteins that can force other proteins to misfold into nonfunctional shapes.

14.3 How Does the Immune System Protect Us?

Explain how the human immune system works and predict the effect of the removal of a portion of the system.

- Nonspecific barriers—including mechanical and chemical boundaries—and helpful microbes work to prevent infection from getting into the body.
- Second-line, nonspecific barriers include defensive cells such as macrophages and natural killer cells; defensive proteins; and the inflammatory response.
- Antibody-mediated and cell-mediated specific immunity both rely on lymphocytes.
- Autoimmune diseases occur when B cells and T cells attack our own cells.
- Our body provides a weak defense when we are in poor health, depressed, or have an immunosuppressive disease like AIDS.
- B and T cells may not be sensitive to pathogens that change antigens quickly.

14.4 How Can We Harness the Immune System?

Assess the strengths and weaknesses of natural immunity, vaccination, and passive immunity.

- Ill individuals develop memory cells, so that when next confronted with the disease, they do not get as sick.

- Inoculation stimulates natural immune response through exposure to products, components, or weakened forms of pathogens.
- Babies and fetuses receive antibodies from their mother through the placenta or in breast milk; protection lasts only a few months.

14.5 What Can Help If Our Immune System Fails?

Assess the use and describe the production of antivenom, antibodies, and antibiotics.

- Antibodies produced by goats or horses in response to snake venom provide passive immunity.
- A concentrated form of antibodies from another individual or cell culture may be used to fight infection.
- Antibiotics inhibit or kill microorganisms; treated individuals may not develop natural immunity.

14.6 How Can We Prevent the Spread of Infectious Disease?

Predict the impact of personal choices on your health and the health of those in your community.

- People who are well nourished and healthy are less likely to get sick.
- Hand washing, drinking treated water, covering the mouth and nose when sneezing, and using condoms and clean hypodermic needles also prevent the spread of disease.
- Countries with well-developed water and sewage treatment systems eliminate the biggest source of diarrheal disease pathogens.
- Unnecessary antibiotic use decreases natural immunity and increases the selection of resistant pathogen strains.
- Inoculation can prevent the spread of disease by decreasing the number of individuals who can get sick and spread it to others.

Key Terms

antibiotic 430	immune system 412	passive immunity 429
antibody 425	immunosuppressive drug 426	pathogenic 417
antigen 423	inflammation 422	prion 420
antivenom 430	influenza 427	prokaryotic 419
B lymphocyte, or B cell 423	interferon 422	shock 416
bone marrow 423	lymphocyte 423	T lymphocyte, or T cell 425
cholera 412	macrophage 422	thymus 425
complement protein 422	memory cell 425	toxin 418
cytotoxic T cell 425	monoclonal antibody (MAb) 430	vaccination 428
electrolyte 416	natural immunity 428	*Vibrio cholerae* 416
G protein 418	natural killer cell 422	virus 420
helper T cell 425	nonspecific barrier 421	

Review Questions

1. What is cholera?

   a. A serious infection of the kidney

   b. A serious intestinal infection that leads to rapid dehydration from diarrhea and vomiting

   c. A serious disease that killed millions of people back in the 1840s but has since been eradicated because of vaccinations

   d. A serious intestinal infection that can be spread only through blood-to-blood contact

   e. A serious intestinal infection caused by a deadly virus

2. What is shock?

3. Define the function of each of the following types of cells: B cells, T cells, helper T cells, cytotoxic T cells, macrophages, natural killer cells, and memory cells.

4. What types of organisms cause illness?

   a. Only bacteria and viruses can cause illness.

   b. Only viruses and prions can cause illness.

   c. Only bacteria and protozoa can cause illness.

   d. Only viruses and protozoa can cause illness.

   e. Illness can be caused by bacteria, protozoa and other eukaryotes, viruses, and prions.

5. What is a prion, and how does it cause disease?

6. What is a virus, and how does it cause disease?

7. What is inflammation?

   a. The painful swelling that accompanies a bruise or contusion

   b. The red and painful swelling that occurs with infection and ensures that the infection spreads to other areas

c. The red, warm, and painful swelling that occurs with infection and limits the infection to a local region

d. The red and itchy swelling that occurs only as a result of an allergic reaction

e. The red, warm, and painful swelling that comes after an infection has cleared up and left the body

8. Explain how an infection with *Vibrio cholerae* can lead to shock.

9. What steps must *Vibrio* achieve to produce cholera in a person who swallowed the bacteria?

10. What are all the ways that the body keeps pathogens out?

11. What are the nonspecific immune responses used to fight pathogens?

12. Explain the B cell–mediated immune response.

13. Explain the T cell–mediated immune response.

14. How do babies acquire passive immunity?

a. Babies receive temporary protection from antibodies that circulate in the mother's blood and cross the placenta to enter the fetal bloodstream; antibodies are also present in breast milk.

b. Babies receive temporary protection solely from antibodies present in breast milk.

c. Babies receive permanent protection from antibodies that circulate in the mother's blood and cross the placenta to enter the fetal bloodstream.

d. Babies receive permanent protection solely from the antibodies present in breast milk.

e. Babies do not acquire passive immunity; they have natural immunity when they are born.

The Thinking Citizen

1. How does the overuse of antibiotics endanger everyone?

2. How does vaccination protect an individual from infection?

3. How does vaccination protect others in a population even if they themselves have not been vaccinated?

4. What are the relative advantages of preventing infectious disease compared to the relative advantages and disadvantages of treating infectious disease?

5. How would you demonstrate, experimentally, that *Vibrio cholerae*, and not some other bacterium, is the cause of cholera?

6. Why is there such variation in people's response to exposure to a pathogenic bacterium like *V. cholerae*? How do you explain that some individuals barely become ill, others are severely sickened, and some die?

7. If there were a cholera outbreak in your community, what steps would you take to limit the spread of the disease?

The Informed Citizen

1. Cholera is a completely preventable and treatable disease, yet many people in underdeveloped countries die from it. What specific solutions could solve this problem? How exactly could these solutions be implemented?

2. The shortage of organs for transplantation has produced a thriving black market. What are the ramifications of offering to pay for organs? Why do you think it is acceptable in some countries (such as Iran) but not in others (such as the United States)?

3. In the United States, people can check a box on their driver's license applications or carry organ donor cards to let it be known that in the event of their deaths, they wish to be organ donors. In Italy, Poland, and some other countries, the law states that people are to be considered organ donors unless they indicate on their license or other identification that they do not wish to be. Why do you think the United States requires active consent? Can you think of any potential dangers to automatically requiring organ donation, unless participants opt out?

4. Not everyone can be vaccinated. Individuals with poorly functioning immune systems and those who are undergoing cancer chemotherapy cannot, for example. Yet such people are vulnerable to infection and unlikely to be able to mount an adequate immune response to fight it. Should vaccination for contagious, infectious diseases be required by law? Why or why not?

5. Should patients be able to demand prescription antibiotics from their physicians? Why or why not? Why might some physicians give in to these demands even if they think the prescription is unnecessary?

CHAPTER LEARNING OBJECTIVES

After reading this chapter, you should be able to answer the following questions:

15.1 How Does What You Eat Influence Your Well-Being?
Identify and describe the components of a healthy diet.

15.2 How Does the Body Extract Nutrients from Food?
Trace the path that food travels from the mouth to the bloodstream and identify the key places where nutrients are extracted.

15.3 How Do Cells Extract Energy from Food?
Explain the process and chemistry of cellular respiration.

15.4 What Can a Person Do to Maintain a Healthy Weight?
Describe how a person's basal metabolic rate can be used to maintain a healthy weight.

15.5 What If Weight Becomes an Obsession?
Assess the various dangers associated with eating disorders that commonly afflict individuals.

15.6 How Do Muscles and Bones Keep Us Active?
Describe the components of the motor system and the ways muscles and bones work together to enable us to move.

15.7 How Can You Keep in Shape?
Discuss the safety and efficacy of various methods for getting and staying in shape.

15.8 How Do We Balance Looking Good with Being Healthy?
Distinguish between the value of looking good and being healthy.

The digestive system, which allows us to extract energy from food, and the motor system, including muscles and bones, are especially important to a healthy lifestyle. How can we use our knowledge of the body to live healthier lives?

CHAPTER 15

Nutrition, Activity, and Wellness

How Can We Live a Healthy Lifestyle?

Becoming an Informed Citizen . . .

Most people know what to do to be healthy: Eat properly, get enough sleep, and exercise. But fashioning a life that is balanced in the face of life's many competing demands is a challenge.

alance is the key to a healthy, satisfying life. In the last few chapters, we have looked at the systems that sustain our bodies through health and disease. Most often, in trying to maintain good health, we tend to focus on just one—particularly when something goes wrong. Some people might think of heart disease as their most important health challenge and direct their concern to the circulatory system. Others face deadly infections or the onset of Alzheimer's disease. We generally experience illnesses like these as a problem with a particular part or function: "He would be fine except for his heart." In reality, however, all of the body's systems function together as an intricate, coordinated whole.

Problems usually don't begin with the malfunction of one system: many contribute. Solutions must therefore address multiple systems, too—and so should prevention. Wellness, or good health, relies on the balanced functioning of all of our systems. If we don't eat properly, sleep enough, or get enough physical activity, everything can suffer: the heart, the brain, immunity, reproduction, and so on. That is why we started our look at physiology with the concept of homeostasis, or how our bodies balance their needs.

In a sense, the body is like a colorful woven tapestry in which all of the fibers come together to make a beautiful picture. This chapter introduces two more of those fibers—the digestive system, which allows us to extract energy from food, and the motor system, including muscles and bones. These are especially important to a healthy lifestyle. But if you pull on just a handful of fibers in the tapestry, the image can change in unexpected ways, and not always for the better.

How can we piece together our knowledge of the body to see the big picture? And how can we find ways to live in a healthy manner? For Heidi Guenther, who seemed a model of healthy living, from an active childhood to a career as a dancer, the question became a matter of life and death.

case study

Dying for the "Perfect Body"

Heidi Guenther seemed born to keep moving. She was a standout in gymnastics before even starting school, and she soon discovered that dancing was as natural to her as breathing. Her dance instructors recognized Heidi's gift, too, and her parents sent her to ballet school at age 12. Heidi's future was bright, and she was happy.

But things changed when Heidi went through puberty. By age 18, Heidi was built like a woman, not a child. She noticed that many of the girls in ballet school tried more than just dieting to stay slim: they used laxatives, starved themselves, and made themselves vomit. For them, being thin was as important to success as good dance technique. Heidi continued to be a standout, but the fears of other girls had started to become hers as well.

At age 19, Heidi joined the Boston Ballet Company (**FIGURE 15.1**). Her dream of becoming a professional ballerina was coming true. She had a real shot at being selected as prima ballerina—if her body did not betray her.

Female ballet dancers in top companies have a body form made popular in the 1950s by George Balanchine, choreographer at the New York City Ballet:

long, thin legs; narrow hips; short torso; small breasts; long necks; delicate arms; and arched feet (**FIGURE 15.2**). Even if a woman's body structure meets this physical ideal, however, it is very difficult to maintain.

When Heidi joined the Boston Ballet she was 5 feet 4 inches tall and weighed 110 pounds. But after injuring her foot, she had to curtail some of her physical activity; she put on 5 pounds as a result. The assistant artistic director of the ballet company told her to lose weight.

Distraught because her foot had not healed, Heidi convinced a doctor to prescribe diet pills and lost 10 pounds even while physically inactive. She also started smoking to control her weight and became gaunt. Other dancers thought she was purging—vomiting and using laxatives. Although Heidi denied it, she now weighed 93 pounds and still thought she was too big.

During summer break, on the way to her family's annual Disneyland visit, Heidi collapsed in the back seat of the car. Her upper body arched back, and her arms and legs stuck straight out. Her lips turned blue, and her throat made a gurgling sound. Her fingers clenched and toes curled under. Resuscitation efforts by paramedics at the scene and later in the ER were futile. Heidi died of cardiac arrest at age 22. She literally died trying to be thinner.

Taylor Hooton's "problem" was the opposite of Heidi's. He thought he was too small and wanted to be bigger, significantly bigger—like the other guys at his high school.

Taylor seemed to have everything going for him. At 6 feet 2 inches tall and 180 pounds, he had the agility and coordination to excel at baseball (**FIGURE 15.3**). He had a girlfriend, plenty of friends, and a loving, affluent family. But Taylor wanted to "look better."

Although only a high school junior, Taylor hoped to be the number one varsity pitcher. And he wanted results in a hurry. To help him get bigger more quickly, Taylor used drugs called steroids to bulk up. The steroids led him to uncontrollable behavior. Taylor's family realized that his violent outbursts—wall punching, phone throwing, and screaming rages—were not simply adolescent mood swings. They confronted him with their suspicions of drug

FIGURE 15.2 Female ballet dancers have a specific body type, with long arms and legs, a short torso, and narrow hips.

FIGURE 15.3 Taylor Hooton dramatically and quickly increased his size by using steroids.

use. After initially denying everything, Taylor eventually revealed that he was on the "juice" and promised to stop.

Abruptly stopping steroid use usually triggers serious depression. Anticipating this, Taylor's psychiatrist prescribed an antidepressant. Everyone thought that Taylor was on the road to recovery.

The entire family went on vacation to England for two weeks. On returning home, Taylor's parents discovered computer equipment in his suitcase that he presumably planned to sell in order to finance the steroid habit that he had not managed to quit. Taylor's parents grounded him. He went upstairs, buckled two belts together, wrapped one around his neck, attached the other to his bedroom door, and hung himself. He left a note that read, "I love you guys. I'm sorry about everything." He was 17 years old.

Heidi and Taylor were both trying to achieve the same goal: a body that was agile, athletic, strong, and attractive. Both were willing to do what they believed necessary to attain the perfect physique. How did things go so wrong, and who bears responsibility?

Heidi and Taylor are dramatic examples of individuals who suffered from body image problems. Like many people, they simply did not like how they looked and wanted to change. Is there a healthy way to slim down or bulk up? This chapter looks at how our bodies maintain our energy needs, as well as how our muscles and skeletal system enable us to keep active and in shape. We can then learn the important practices and behaviors for living a healthy life.

15.1 How Does What You Eat Influence Your Well-Being?

On New Year's Eve, roughly seven months before Heidi Guenther died, she was at a party and a strange thing happened. One of her hands clenched shut, and neither she nor her friends could uncurl her fingers. Then, not long before the day she died, Heidi told her mother that sometimes it felt like something was wrong with her heart and that it was racing.

What was happening to Heidi? To what extent had her eating habits caused these symptoms? What role does our diet play in supporting good health?

Food Quantity and Quality

Food plays an important cultural role in our lives. We mark celebrations like Thanksgiving or weddings by sharing elaborate meals. For many families, sitting down to dinner together is the way they connect and communicate. Food serves a biological function, too. It is the source of energy that drives everything the body does. Food molecules contain energy in the chemical bonds that hold atoms together (Chapter 1). When we eat and digest food, our bodies extract the energy from these molecules and convert it into a form that we can use. Our diets need to consist of *enough* food to supply energy for the body to perform all of its functions.

Not only is the quantity of food important, but so is the quality. No single food has everything we need, so we should eat a variety from each of the basic food groups: grains; vegetables and fruits; proteins, including dairy, meat, poultry, and/or beans; and small amounts of unsaturated fats. Unsaturated fats come from vegetables and are liquid at room temperature. The United States Department of Agriculture (USDA) establishes guidelines for a balanced diet, updated every five years to incorporate new research findings about nutrition (**FIGURE 15.4**).

Water

A varied diet includes all of the nutrients essential for health. We can easily overlook, though, the simplest and most essential component of all, water. Because water is necessary for all cellular activities, a person cannot survive for more than a few days without it. We need to take in around 3 liters of water per day to replace the fluid lost in sweat, urine, and feces. Most people get roughly half of that water from the food they eat. The other half must be replenished by drinking.

Adequate water is needed to maintain normal blood pressure, to eliminate solid wastes from the body, and to maintain body temperature. It is also required for the movement of nutrients, oxygen, and other materials through the bloodstream and into tissues and cells. Some of Heidi's physical symptoms—muscle cramps and increased heart rate, for example—were probably due to dehydration.

Carbohydrates, Proteins, and Fats

The major sources of energy for our cells come in three forms: carbohydrates, proteins, and fats. Carbohydrates, composed of sugars, are found in bread, cereal, rice, pasta, fruits, and vegetables (**FIGURE 15.5**). We will encounter some of the most important sugar molecules in the next section, when we see

FIGURE 15.4 No single food has everything we need, so we should eat a variety of grains, vegetables, fruits, and proteins. The USDA establishes guidelines for a healthy, balanced diet and updates them every five years.

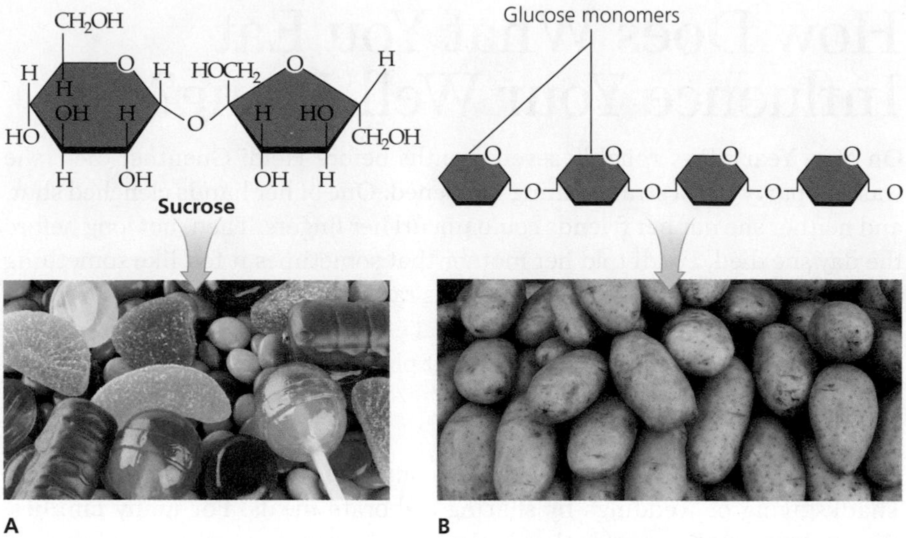

CH₂OH

Glucose monomers

Sucrose

FIGURE 15.5 Simple carbohydrates, such as those found in fruits, vegetables, and refined sugar products like cake and candy are composed of one or two sugars (A), while complex carbohydrates such as breads, pasta, and starchy vegetables like potatoes have more sugars and often a branching structure (B).

A

B

how food is digested. Simple carbohydrates are made from a single sugar, while in complex carbohydrates many sugars hook, or, more scientifically, bond, together to form branching chains. Complex carbohydrates are generally more beneficial to health because they are digested more slowly, thus releasing energy gradually.

Proteins are molecules made of subunits called amino acids (**FIGURE 15.6**). Twenty different types of amino acid are used in various numbers and combinations to make up all of the proteins in our bodies. We are able to synthesize 12 of them; the other eight, called the essential amino acids, must be obtained from our diets. Because the body does not store ingested proteins or amino acids for more than a day, it is important to take in essential amino acids every day. Fish, meat, eggs, grains, nuts, beans, and dairy products are all good sources.

Although proteins are a source of energy, they are also the only source of building materials to make new proteins. And the proteins we make are critical for building structures and for regulating chemical reactions in the body. With her focus on weight loss, Heidi's foot may have healed so slowly because she was unable to make the proteins required to replace injured cells.

Fats, a type of lipid, are molecules composed of subunits called fatty acids and glycerol. The breakdown of fat yields twice the energy that carbohydrates or proteins do (**FIGURE 15.7**). The average healthy person stores enough body fat to supply between four and five weeks of energy. Unlike many people in the modern, developed world, our ancestors faced uncertainty about the source of their next meal. People living in poverty in both developed and developing nations similarly lack consistent access to a balanced diet. For our ancestors, the ability to store fat was the difference between life and death, while in the modern world, for reasons that are not completely understood, being overweight or even obese is actually more prevalent in poor populations. As was the case for amino acids, our body can synthesize most of the fatty acids it requires. We must consume essential fatty acids in our diets because we can't make them. Meat, dairy products, nuts, and vegetable oils are all sources of dietary fat.

Fats have many important functions besides providing energy. They insulate the body from heat loss, act as "shock absorbers," and serve as building blocks for hormones and other signal molecules. It is likely that Heidi, like many ballerinas and elite female athletes, stopped menstruating because her level of body fat became too low to support the monthly cycle (Chapter 6).

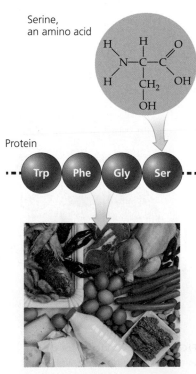

Serine, an amino acid

Protein

Trp   Phe   Gly   Ser

FIGURE 15.6 Proteins are composed of amino acids that regulate the chemical reactions in cells; many protein-rich foods have all the amino acids we need.

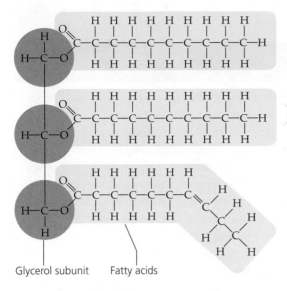

Glycerol subunit   Fatty acids

FIGURE 15.7 Fats, a type of lipid, are molecules composed of subunits called fatty acids and glycerol. The breakdown of fat yields twice the energy that carbohydrates or proteins do.

Vitamins, Minerals, and Dietary Fiber

Often referred to as micronutrients because only small amounts are needed in the diet, vitamins and minerals are not broken down for energy or destroyed by the body during use. **Vitamins** act as coenzymes, molecules that assist enzymes in the regulation of specific chemical pathways in the cell (**FIGURE 15.8**). Most vitamins can't be produced by the body and so must be included in the diet. See *Scientist Spotlight: Linus Pauling* for a

vitamin :: a micronutrient that is not broken down for energy or destroyed during use; it assists enzymes in the regulation of specific cellular chemical pathways

Scientist Spotlight

Linus Pauling (1901–1994)

Success can be due to many things: talent, opportunity, ambition, or good luck. Linus Pauling credited his accomplishments to his abilities to work hard, think deeply, and waste no time. They led him to two Nobel Prizes, one in chemistry and the other for peace.

Born in 1901 in Oregon, Pauling became fascinated with science as a boy. He started collecting butterflies and beetles and soon expanded his interest to rocks and minerals.

When Pauling was nine years old, his father died, and his mother struggled to support her three children. Pauling worked to help out, and by the time he was 16, he was earning a good salary from his job in a machine shop. However, Pauling realized that he had to go to college, and since tuition was free, he went to the Oregon Agricultural College (now Oregon State), where he studied chemical engineering. Later at Cal Tech, he immersed himself completely in the study of chemistry.

Pauling's career had four phases. First, his discoveries transformed our understanding of chemistry. Second, as a political activist, he championed nuclear disarmament and world peace. Third, he established the field of molecular medicine by showing that sickle cell disease is due to a faulty

gene. This interest in chemistry and health led to his fourth phase, which he called orthomolecular medicine.

Pauling hypothesized that optimum health occurs when all chemical reactions in the body run properly. When any

Linus Pauling's abilities to work hard, think deeply, and waste no time led him to two Nobel Prizes: one in chemistry and the other for peace. Pauling's discoveries transformed our understanding of chemistry and the field of molecular medicine, among many other accomplishments.

(Continued)

Scientist Spotlight

Linus Pauling (1901–1994) (*Continued*)

reactions are inhibited, disease then results. Pauling reasoned that the best way to achieve optimum health is to establish ideal reaction conditions—including the proper balance of chemicals. Pauling argued that we do not get an adequate amount of these "orthomolecular substances," which are normally present, required for life, and nontoxic even in high doses. He focused on vitamin C.

According to Pauling, the recommended daily amounts (RDA) of vitamins are much too low, because they are the minimum needed to prevent vitamin deficiencies. Dramatically higher quantities, he asserted, could improve health. He claimed that a balanced diet supplemented with vitamin C reduces the risk of cancer, heart disease, and infectious disease.

Was Pauling correct? It is not at all clear. His ideas about vitamin C have been met with enthusiastic support from some and skepticism from others. Some have shown that vitamin C reduces the duration of colds in humans. Other experiments have demonstrated that vitamin C suppresses tumor growth in mice. However, too much vitamin C has also been shown to produce side effects such as gastrointestinal distress and kidney stones. In fact, taking too much of any vitamin can lead to health problems, some of which can be serious. So, while it was relatively easy to determine the consequences of vitamin deficiencies and overdoses, it is much more difficult to figure out the complex interactions among food, vitamins, and overall human wellness. ::

Vitamin	RDI Males: age 19–50	RDI Females: age 19–50	Functions	Main sources	
A	900 mcg/d	700 mcg/d	Involved in immune system function, vision, reproduction, cellular communication; also plays a role in formation and maintenance of the heart, lungs, kidneys, and other organs.	Liver, sweet potatoes, carrots, butternut squash	
B6	1.3 mcg/d	1.3 mcg/d	Needed for more than 100 enzyme reactions involved in metabolism; also involved in brain development during pregnancy and infancy as well as immune function.	Rice and wheat bran, pistachios, fish, potatoes, and other starchy vegetables	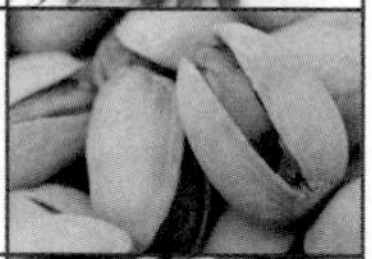
B12	2.4 mcg/d	2.4 mcg/d	Keeps the body's nerve and blood cells healthy and helps make DNA.	Clams, beef, liver, fish, poultry, eggs, milk, and other dairy products	
C	90 mcg/d	75 mcg/d	Acts as an antioxidant; also helps the body make collagen (protein involved in wound healing) and helps keep immune system working properly.	Citrus fruits (such as oranges and grapefruit) and their juices, as well as red and green pepper and kiwifruit	
D	5 mcg/d	5 mcg/d	Needed for health and to maintain strong bones; helps the body absorb calcium; also helps muscles move and nerves carry messages between the brain and every body part.	Fatty fish such as salmon, tuna, and mackerel; almost all of the U.S. milk supply is also fortified with vitamin D; sunlight	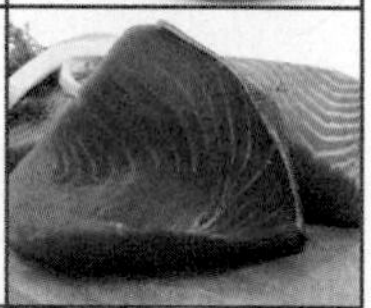
E	15 mcg/d	15 mcg/d	Acts as an antioxidant and boosts the immune system so it can fight off bacteria and viruses; also helps to widen blood vessels and keep blood from clotting.	Vegetable oils (like wheat germ, sunflower, and safflower); nuts (such as peanuts, hazelnuts, and almonds); and seeds (like sunflower seeds)	
K	120 mcg/d	90 mcg/d	Helps with blood clotting and increases bone density.	Kale, spinach, and collard greens	

mcg: microgram
RDI: recommended daily intake

FIGURE 15.8 Vitamins assist enzymes in the regulation of specific chemical pathways in the cell. Vitamin K is produced by bacteria that live in our gut. However, we must obtain all other vitamins, including the precursor molecules to vitamins A and D, and niacin, from our diet.

look at the important role vitamin C may play in the maintenance of good health.

Minerals are inorganic molecules, meaning they do not contain carbon. They are important for many functions—including bone and tooth development and maintenance, muscle contraction and relaxation, nerve impulses, and fluid balance in the body (**FIGURE 15.9**). Heidi's uncontrolled muscle spasms were probably due, at least in part, to inadequate mineral levels in her body.

Although not strictly a nutrient, **fiber** (or roughage) is an important component of our diet. Composed of complex carbohydrates that we can't digest fully, fiber helps to lower cholesterol and low-density lipoproteins (LDL, or "bad" lipids) in the blood. Also, because fiber in the diet keeps things "moving" through the digestive system, it ensures that solid wastes are eliminated regularly and efficiently. As a consequence, it greatly lowers the risk of developing cancers of the large intestine, or colon.

mineral :: inorganic molecule (does not contain carbon); helps with bone and tooth development, muscle contraction and relaxation, nerve impulses, and fluid balance

fiber :: roughage composed of complex carbohydrates that can't be fully digested; helps keep things moving through the digestive system

Mineral	RDI Males: age 19–50	RDI Females: age 19–50	Functions	Main sources	
Calcium	1000 mcg/d	1000 mcg/d	Maintains strong bones and enables muscles to move and nerves to carry messages between the brain and every body part.	Milk, yogurt, and cheese	
Phosphorus	700 mcg/d	700 mcg/d	Aids in the formation of bones and teeth; also crucial in the production of ATP.	Bran, meat, and milk	
Magnesium	400 mcg/d	310 mcg/d	Is needed for more than 300 biochemical reactions in the body; helps maintain normal muscle and nerve function, keeps heart rhythm steady, supports a healthy immune system, and keeps bones strong.	Green vegetables such as spinach, beans, nuts, seeds, and whole, unrefined grains	
Iron	8 mcg/d	18 mcg/d	Is an integral part of many proteins and enzymes that maintain good health; involved in oxygen transport; also essential for the regulation of cell growth and differentiation.	Red meats, fish, oysters, poultry, lentils, and beans	
Zinc	11 mcg/d	8 mcg/d	Helps the immune system fight off invading bacteria and viruses; aids in wound healing; also helps make proteins and DNA.	Oysters, red meat, poultry, seafood	
Selenium	55 mcg/d	55 mcg/d	Incorporated in proteins that make anti-oxidant enzymes.	Brazil nuts, plant foods, meat, fish, and poultry	
Copper	900 mcg/d	900 mcg/d	Helps in the formation of hemoglobin.	Liver (pâté), oysters and other shellfish, whole grains, beans, nuts, and potatoes	
Iodine	150 mcg/d	150 mcg/d	Needed for proper functioning thyroid gland; also needed for proper bone and brain development during pregnancy and infancy.	Seaweed, fish (such as cod and tuna), shrimp, and other seafood	
Chromium	35 mcg/d	25 mcg/d	Enhances the action of insulin; involved in metabolism of carbohydrates, proteins, and fats.	Broccoli, grape juice, whole grains, and beef	

mcg: microgram
RDI: recommended daily intake

FIGURE 15.9 Minerals are inorganic molecules that are important for bone and tooth development, muscle contraction and relaxation, nerve impulses, and fluid balance in the body.

15.2 How Does the Body Extract Nutrients from Food?

Heidi pursued her quest to be impossibly thin by restricting her eating and preventing her digestive system from doing its job. That includes breaking down whatever food she did ingest, absorbing the food molecules into her cells, and extracting the energy from the chemical bonds of these molecules. These steps are the job of the **digestive system**.

From the Mouth to the Stomach

The digestive tract is a 30-foot (10-meter) continuous tube that starts with the mouth and ends with the **anus**, the exterior, posterior opening (**FIGURE 15.10**). It runs through the entire body and consists of specialized regions that have distinctive structures and functions. Associated with the digestive tract are additional organs that also play essential roles in digestion—the mechanical and chemical breakdown of food.

Imagine eating a chicken taco. The first parts of the digestive system to start working are the lips, teeth, and tongue. The lips determine how far the mouth opens for the taco. They also help keep the bite of taco inside the mouth, so the

digestive system :: enables the extraction of energy from the chemical bonds of food molecules

anus :: the exterior, posterior opening of the digestive tract

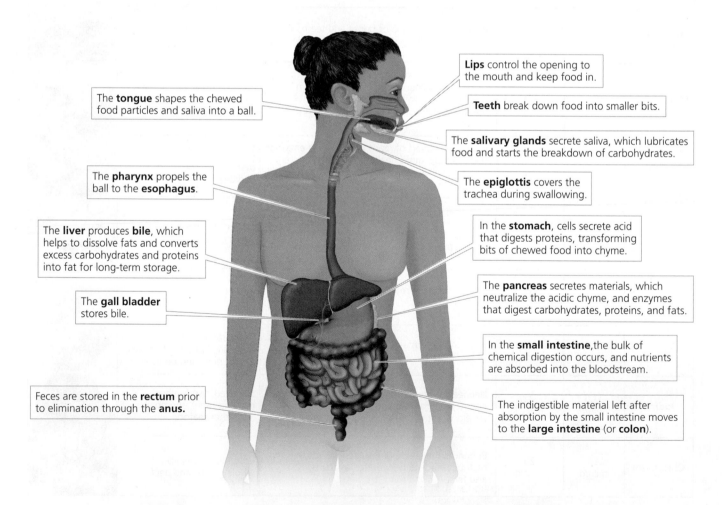

The **tongue** shapes the chewed food particles and saliva into a ball.

Lips control the opening to the mouth and keep food in.

Teeth break down food into smaller bits.

The **salivary glands** secrete saliva, which lubricates food and starts the breakdown of carbohydrates.

The **pharynx** propels the ball to the **esophagus**.

The **epiglottis** covers the trachea during swallowing.

The **liver** produces **bile**, which helps to dissolve fats and converts excess carbohydrates and proteins into fat for long-term storage.

In the **stomach**, cells secrete acid that digests proteins, transforming bits of chewed food into chyme.

The **gall bladder** stores bile.

The **pancreas** secretes materials, which neutralize the acidic chyme, and enzymes that digest carbohydrates, proteins, and fats.

In the **small intestine**, the bulk of chemical digestion occurs, and nutrients are absorbed into the bloodstream.

Feces are stored in the **rectum** prior to elimination through the **anus**.

The indigestible material left after absorption by the small intestine moves to the **large intestine** (or **colon**).

FIGURE 15.10 The digestive tract is a 30-foot (10-meter) continuous tube that starts with the mouth and ends with the anus, running through the entire body. It is responsible for the chemical breakdown of food and absorption of nutrients into the body.

teeth and tongue can do their tasks. Chewing allows teeth to break down the food into smaller bits. The tongue contains taste buds, which signal whether the food should be chewed and swallowed or spit out. Generally, the odor of spoiled food will be enough to prevent putting it in one's mouth.

The presence of food in the mouth stimulates the **salivary glands** to secrete saliva. Important for lubricating food as it is chewed, saliva also contains enzymes that begin the breakdown of carbohydrates into their simple sugar subunits. The tongue, a muscular organ, shapes the chewed food particles and saliva into a ball that is easy to swallow.

The process of swallowing moves food from the mouth to the **pharynx**, a chamber located at the back of the throat that connects the mouth to the digestive and respiratory tracts. More specifically, the pharynx leads to the **esophagus**, a muscular tube that leads to the stomach, and to the trachea, a pipe, as we saw in Chapter 12, that leads to the lungs. Everyone who has ever choked because "food went down the wrong way" knows that inhaled food can trigger choking or even block an airway. Fortunately, the action of swallowing causes a flap of tissue, called the **epiglottis**, to cover the trachea. This action prevents food, liquids, and saliva from entering the trachea.

The Stomach and Small Intestine

The esophagus propels swallowed food toward the stomach by means of **peristalsis**, coordinated waves of smooth muscle contraction (**FIGURE 15.11**). The esophagus is connected to the stomach, a muscular, *j*-shaped organ that can expand to hold more than 3 liters of material—a very large meal indeed. A ring of muscle, called a **sphincter**, regulates the opening between the esophagus and stomach. Normally, this sphincter prevents stomach contents from leaking "upstream" back into the esophagus. Heartburn, acid reflux, and acid indigestion are all names for the uncomfortable malady that occurs when this sphincter does not do its task properly.

Once in the stomach, significant chemical breakdown of food begins (**FIGURE 15.12**). The cells lining the inside of the stomach secrete acid that digests proteins and kills most bacteria that happen to be swallowed. The cells also secrete enzymes that take proteins apart into smaller subunits.

salivary gland :: secretes saliva at the presence of food in the mouth; saliva contains enzymes that begin the breakdown of carbohydrates into simple sugar subunits

pharynx :: chamber located at the back of the throat that connects the mouth to the digestive and respiratory tracts

esophagus :: muscular tube that leads to the stomach and to the trachea

epiglottis :: flap of tissue that covers the trachea when swallowing foods, liquids, and saliva

peristalsis :: coordinated waves of smooth muscle contraction that propel swallowed food toward the stomach and through the rest of the digestive system

sphincter :: ring of muscle that regulates the opening of a tube

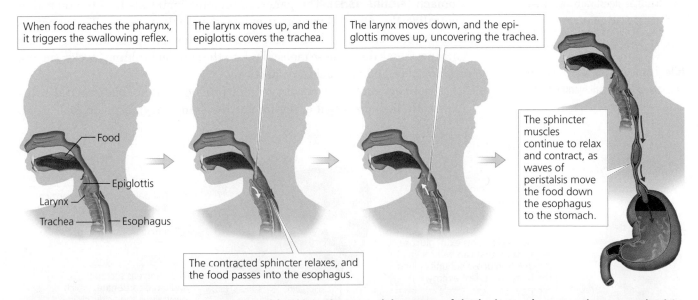

When food reaches the pharynx, it triggers the swallowing reflex.

The larynx moves up, and the epiglottis covers the trachea.

The larynx moves down, and the epiglottis moves up, uncovering the trachea.

The sphincter muscles continue to relax and contract, as waves of peristalsis move the food down the esophagus to the stomach.

Food

Epiglottis

Larynx

Trachea

Esophagus

The contracted sphincter relaxes, and the food passes into the esophagus.

FIGURE 15.11 The esophagus propels swallowed food into the stomach by means of rhythmic muscle contractions, or peristalsis.

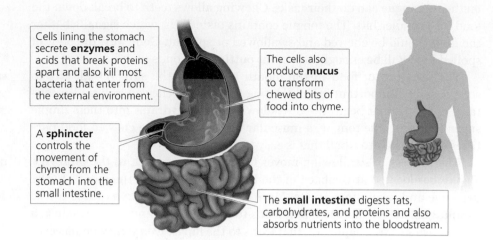

Cells lining the stomach secrete **enzymes** and acids that break proteins apart and also kill most bacteria that enter from the external environment.

The cells also produce **mucus** to transform chewed bits of food into chyme.

A **sphincter** controls the movement of chyme from the stomach into the small intestine.

The **small intestine** digests fats, carbohydrates, and proteins and also absorbs nutrients into the bloodstream.

FIGURE 15.12 Most of the chemical breakdown of food occurs in the stomach and small intestine.

In addition, the stomach cells produce mucus to lubricate and protect themselves from acid damage. What started off as bits of chewed food when they arrived in the stomach are transformed into a creamy, acidic, semiliquid mixture called chyme. Its destination is the small intestine.

Around 20 feet (6–7 meters) in length and approximately 1 inch in diameter, the small intestine is the place where most chemical digestion occurs. The small intestine is also the major organ for the absorption of nutrients into the bloodstream. As is the case for any structure involved in absorption or transport of materials across cell membranes, the surface area of the small intestine is enormous—approximately the size of a tennis court.

Just as a sphincter controls the movement of material from the esophagus into the stomach, another one guards the passage of chyme from the stomach into the small intestine. A few milliliters of chyme are squirted into the small intestine every 20 seconds or so.

The Pancreas and Liver

The **pancreas**, one of the accessory organs of the digestive system, is connected by a small duct to the small intestine very close to where it joins the stomach (**FIGURE 15.13**). The pancreas secretes materials into the duct that neutralize the acidic chyme as it enters the small intestine. The pancreas also secretes enzymes that digest carbohydrates, proteins, and fats. These enzymes break down these molecules into the smaller building blocks: sugars, amino acids, and fatty acids, respectively.

The **liver**, another accessory organ, produces **bile**. This material helps to dissolve fats, making it easier for fat-dissolving enzymes to do their work.

pancreas :: one of the accessory organs of the digestive system that secretes enzymes that digest carbohydrates, proteins, and fats

liver :: accessory organ that produces bile

bile :: material produced by the liver and stored in the gall bladder that helps dissolve fats

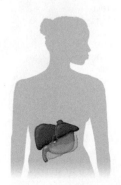
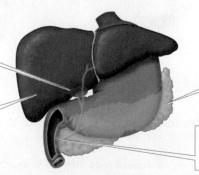

The liver produces bile that helps to dissolve fats.

The liver also stores extra glucose as glycogen that can be broken down into glucose subunits when the body needs fast energy.

The pancreas secretes enzymes that break down carbohydrates, proteins, and fats into sugars, amino acids, and fatty acids.

The pancreas secretes sodium bicarbonate, which neutralizes the acidic chyme.

FIGURE 15.13 The pancreas, liver, and gall bladder work with the small intestine to break down food molecules.

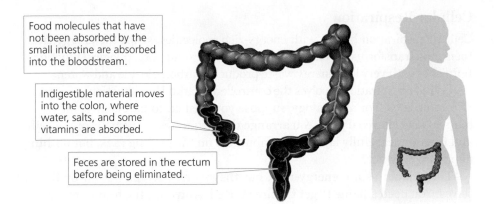

Food molecules that have not been absorbed by the small intestine are absorbed into the bloodstream.

Indigestible material moves into the colon, where water, salts, and some vitamins are absorbed.

Feces are stored in the rectum before being eliminated.

FIGURE 15.14 The large intestine (or colon) absorbs water from indigestible material, and the rectum stores feces until they are eliminated.

Bile is actually stored by the **gall bladder**, another organ. In addition to its role in fat digestion, the liver stores some vitamins and extra glucose as the multi-subunit molecule **glycogen**. Glucose is the most commonly used form of simple sugar that the body uses for energy. When the body needs fast energy, the liver will break down glycogen into its glucose subunits, providing an energy "boost." The liver is also capable of converting excess carbohydrates and proteins into fat for long-term storage in **adipose tissue**.

gall bladder :: organ that stores bile

glycogen :: multi-subunit molecule composed of glucose

adipose tissue :: stores excess carbohydrates and proteins that have been converted to fat by the liver

From the Intestine to the Bloodstream

Once food molecules are broken down into their simple subunits, nutrient absorption can take place. Molecules are transported into the cells lining the interior of the small intestine. From there, these molecules move into the fluid space that exists between cells and tissues. And from there, they finally move into the bloodstream. Once in the blood, nutrient molecules are delivered to cells and tissues, where the final stages of energy extraction will occur.

The indigestible material left after absorption by the small intestine moves into the **large intestine**, or **colon** (**FIGURE 15.14**). The colon, 3–6 feet long and 3 inches in diameter, absorbs water, salts, and some vitamins. The last six inches of the colon is the **rectum**. Feces are stored here until they can be eliminated.

large intestine, or **colon** :: coiled organ, 3–6 feet long and 3 inches in diameter, which absorbs water, salts, and some vitamins

rectum :: the last six inches of the colon; stores feces until they can be eliminated

Heidi's methods for preventing the absorption of nutrients likely included purging by vomiting and abuse of laxatives. The consequences were quite serious for her. Not only is it likely that purging dehydrated Heidi, it probably eliminated vitamins and especially minerals needed for normal heart function, resulting ultimately in her sudden death. Purging by vomiting also injures tissues in the esophagus by exposing them repeatedly to harsh stomach acids. Similarly, teeth and gums can be damaged terribly because of acid exposure. Many of these side effects can be permanent.

15.3 How Do Cells Extract Energy from Food?

After the digestive system has broken apart proteins, carbohydrates, and fats into their simpler subunits, the energy from the chemical bonds of these molecules is extracted. Food subunit molecules—sugars, amino acids, and fats—are rich in energy. However, we still have to transform this energy into energy that the body can use. That is the job of **cellular respiration**, the process by which cells extract energy from food molecules.

cellular respiration :: the process whereby cells extract energy from food molecules

Cellular Respiration

Cellular respiration begins with energy-rich molecules, strips the energy from them, and transforms the energy so it can be used to support biological functions. It finishes with very low-energy waste products: carbon dioxide and water.

Cellular respiration involves the controlled, gradual "combustion" of energy-rich molecules. For an analogy, suppose you wished to barbecue some hamburgers on a charcoal grill. You arrange the briquettes in the belly of the grill, and then you carefully layer the hamburgers on the grilling rack, but nothing happens. Why not?

The briquettes are energy-rich, but they won't become hot all by themselves. You need a flame to get the fire started. And once the briquettes catch, they will give off their energy as heat, a form good for cooking. Once the briquettes burn completely, you will have low-energy waste materials: carbon dioxide, water, and ash.

The combustion of food occurs in much the same way. Enzymes in the cell, not an elevation of temperature, prod high-energy molecules into action. The energy extracted is collected gradually and stored in other molecules, which can support the functions in cells and in the body overall. You can think of them as a cell's batteries.

The Chemistry of Cellular Respiration

Cellular respiration occurs inside cells and involves many chemical reactions. It is important to extract energy gradually in manageable doses so that cells are not damaged by a huge energy burst (like setting a briquette on fire).

The chemical reactions of cellular respiration are organized into three sets of steps: glycolysis, the Krebs cycle, and the electron transport chain (**FIGURE 15.15**).

glycolysis :: one of the chemical reactions of cellular respiration that occurs in the cytoplasm of cells; it begins with glucose and ends with pyruvic acid

- **Glycolysis** begins with glucose and ends with a less energy-rich molecule, pyruvic acid. This set of reactions occurs in the cytoplasm of cells. Enough energy is extracted from the glucose to produce two molecules

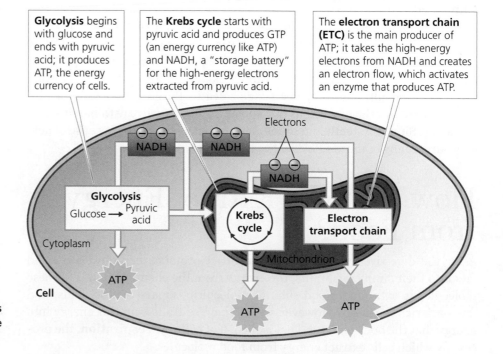

FIGURE 15.15 Cellular respiration, the process by which cells extract energy from food molecules, consists of glycolysis, the Krebs cycle, and the electron transport chain.

of **adenosine triphosphate** (**ATP**), the energy currency of the cell. ATP is used in practically all chemical actions that require an input of energy, from making proteins to moving muscle. ATP delivers just the right amount of energy to make other reactions go, but not so much that the cell is injured.

- The pyruvic acid that is produced by glycolysis is quickly converted to pyruvate; it still has some more energy that can be squeezed out during the reactions of the **Krebs cycle**. Taking place inside the mitochondria, the Krebs cycle produces GTP (an energy currency like ATP) and **NADH**, a "storage battery" for the high-energy electrons that have been collected during extractions of energy from pyruvic acid. NADH will be used in the last set of cellular respiration reactions, the electron transport chain.

- Also located in the mitochondria, the **electron transport chain** (**ETC**) is the main producer of ATP. The ETC operates by taking the high-energy electrons from the NADH storage batteries and creating an electron flow, which activates an enzyme that produces ATP.

adenosine triphosphate (ATP) :: the energy currency of the cell; used in practically all chemical actions that require an input of energy

Krebs cycle :: one of the chemical reactions of cellular respiration that takes place in the mitochondria of cells; produces GTP and NADH

NADH :: works like a storage battery for high-energy electrons collected during extractions of energy from pyruvic acid; used in the electron transport chain

electron transport chain (ETC) :: the main producer of ATP; located in the mitochondria

15.4 What Can a Person Do to Maintain a Healthy Weight?

When Heidi Guenther finished her first year at the Boston Ballet, the assistant artistic director told her that that she needed to lose 5 pounds. Was it appropriate for the artistic director to tell her to lose weight so she could be more aesthetically appealing for ballet? If so, was there a safer way for Heidi to lose weight rather than skipping meals, purging, and using laxatives?

Basal Metabolic Rate

Calories are the units of energy used to describe food. A calorie is the amount of energy needed to raise the temperature of one milliliter of water by one degree Celsius. The term Calorie is capitalized when it refers to a kilocalorie: one thousand calories are equal to one Calorie. To maintain health, the energy consumed in food Calories must balance the energy spent. If a person does not eat enough to supply the energy used, he or she will lose weight. And if an individual eats more than is needed, weight gain will result: excess Calories will be stored as fat.

Even if you are sedentary or completely inactive, your body still needs a basic level of Calorie intake to support life functions. This is known as the **basal metabolic rate**, and it is roughly 1300–1500 Calories per day for females and 1600–1800 Calories per day for males. The actual metabolic rate for any individual depends on sex, weight, age, genetics, and activity level. For example, metabolism slows with age, while exercise increases the Calorie requirement (**FIGURE 15.16**).

How many Calories should you consume each day? The answer is simple: take the weight you wish to maintain, and multiply by 11. Now add the number of Calories you expend during exercise. Consider the basic Calorie requirements for a dancer like Heidi:

Calorie :: the amount of energy needed to raise the temperature of one milliliter of water by one degree Celsius

basal metabolic rate :: the basic level of Calorie intake to support life functions

Desired weight:	110 pounds (as advised)
× 11	1210 Calories (if sedentary)
+ *exercise*	1600 Calories (dancing/practicing 5 hours per day)
Total Calories:	2810

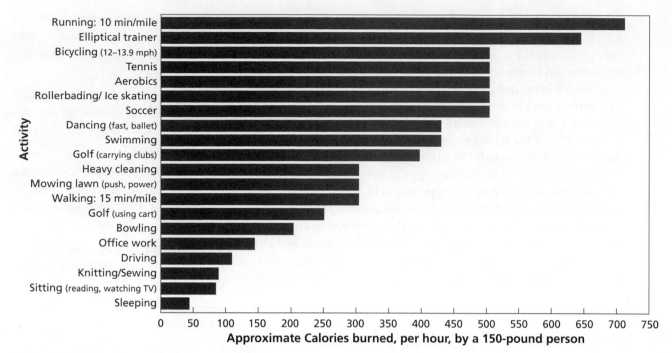

FIGURE 15.16 Physical activity and exercise of any type burns calories. The amount burned depends on the type of activity.

We can see that Calorie requirements are sensitive to the amount of exercise a person gets. Dancers and other athletes clearly need quite a lot of food to maintain their weight and remain healthy.

Understanding metabolic rate provides a guaranteed way to lose weight—decrease Calorie intake while still balancing nutrients, and increase exercise. Suppose you want to lose 10 pounds. A safe way to do so would be to reduce intake by 250 Calories a day and increase exercise to burn an additional 250 Calories a day. With this plan, you would lose 1 pound of fat each week, and it would take 10 weeks to reach your goal. This may seem like a long time, but slower weight loss is much safer and more likely to be permanent than quick, dramatic reductions.

Average Dimensions

Both Heidi and Taylor looked at their bodies and decided they were not the right weight. For both of them, the judgment was based entirely on how they perceived their appearance.

Healthy people actually come in a variety of shapes and sizes. The definition of "attractive" has changed over time. It also varies from culture to culture—and with what we see in movies, on television, in magazines, and online. The average American woman, for example, is 5 feet 4 inches and weighs 166 pounds, while the average American man is 5 feet 10 inches and weighs 195 pounds. By contrast, the average female runway model is 5 feet 10 inches tall and weighs between 100 and 120 pounds, while the average male runway model is 6 feet and weighs between 140 and 165 pounds (**FIGURE 15.17**). In reality, some athletes and models with "ripped muscles" probably have inadequate body fat for good health. A healthy woman should have approximately 22% body fat; below 12% or above 32% is not safe over the long term. A healthy man should have roughly 14% body fat and not less than 3% or more than 29%.

Obviously, we need to distinguish between the *cultural* and *biological* definitions of being overweight or underweight. How can we do this? Is there a way to tell whether you have a healthy weight?

A

B

FIGURE 15.17 The body dimensions of many female (A) and male (B) models do not match those of the average person, and they may sometimes also be unhealthy.

The Body/Mass Index

The **body/mass index** (**BMI**) is a tool for determining whether one's weight is in the healthy range (**FIGURE 15.18**). This number is calculated from a person's height and weight. A healthy BMI, based on the risk of illness and death, is generally 20–25.

body/mass index (BMI) :: tool for determining whether one's weight is in the healthy range; calculated from a person's height and weight

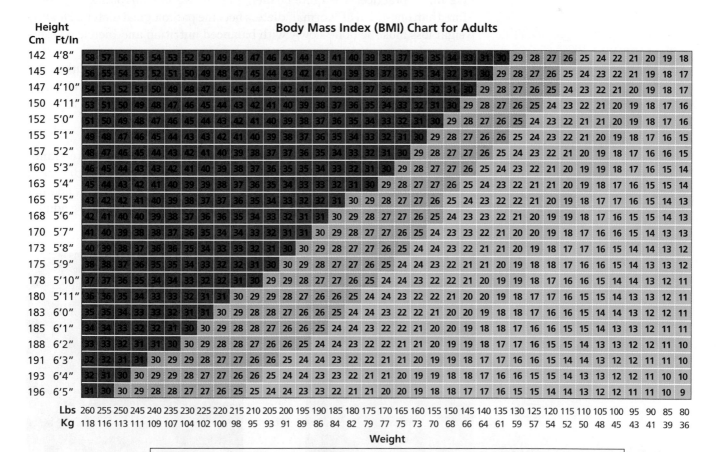

Body Mass Index (BMI) Chart for Adults

Height Cm	Ft/In																																					
142	4'8"	58	57	56	55	54	53	52	50	49	48	47	46	45	44	43	41	40	39	38	37	36	35	34	33	31	30	29	28	27	26	25	24	22	21	20	19	18
145	4'9"	56	55	54	53	52	51	50	49	48	47	45	44	43	42	41	40	39	38	37	36	35	34	32	31	30	29	28	27	26	25	24	23	22	21	19	18	17
147	4'10"	54	53	52	51	50	49	48	47	46	45	44	43	42	41	40	39	38	37	36	34	33	32	31	30	29	28	27	26	25	24	23	22	21	20	19	18	17
150	4'11"	53	51	50	49	48	47	46	45	44	43	42	41	40	39	38	37	36	35	34	33	32	31	30	29	28	27	26	25	24	23	22	21	20	19	18	17	16
152	5'0"	51	50	49	48	47	46	45	44	43	42	41	40	39	38	37	36	35	34	33	32	31	30	29	28	27	26	25	24	23	22	21	21	20	19	18	17	16
155	5'1"	49	48	47	46	45	44	43	43	42	41	40	39	38	37	36	35	34	33	32	31	30	29	28	27	26	26	25	24	23	22	21	20	19	18	17	16	15
157	5'2"	48	47	46	45	44	43	42	41	40	39	38	37	37	36	35	34	33	32	31	30	29	28	27	27	26	25	24	23	22	21	20	19	18	17	16	16	15
160	5'3"	46	45	44	43	43	42	41	40	39	38	37	36	35	34	33	32	31	30	29	28	27	27	26	25	24	23	22	21	20	19	19	18	17	16	15	15	14
163	5'4"	45	44	43	42	41	40	39	39	38	37	36	35	34	33	33	32	31	30	29	28	27	27	26	25	24	23	22	21	21	20	19	18	17	16	15	15	14
165	5'5"	43	42	42	41	40	39	38	37	37	36	35	34	33	32	32	31	30	29	28	27	27	26	25	24	23	22	22	21	20	19	18	17	17	16	15	14	13
168	5'6"	42	41	40	40	39	38	37	36	36	35	34	33	32	31	31	30	29	28	27	27	26	25	24	23	23	22	21	20	19	19	18	17	16	15	15	14	13
170	5'7"	41	40	39	38	38	37	36	35	34	34	33	32	31	31	30	29	28	27	27	26	25	24	23	23	22	21	20	20	19	18	17	16	16	15	14	13	13
173	5'8"	40	39	38	37	36	36	35	34	33	33	32	31	30	30	29	28	27	27	26	25	24	24	23	22	21	21	20	19	18	17	17	16	15	14	14	13	12
175	5'9"	38	38	37	36	35	35	34	33	32	32	31	30	30	29	28	27	27	26	25	24	24	23	22	21	21	20	19	18	17	17	16	16	15	14	13	13	12
178	5'10"	37	37	36	35	34	34	33	32	32	31	30	29	29	28	27	27	26	25	24	24	23	22	22	21	20	19	19	18	17	16	16	15	14	14	13	12	11
180	5'11"	36	36	35	34	33	33	32	31	31	30	29	29	28	27	26	26	25	24	24	23	22	22	21	20	20	19	18	17	17	16	15	15	14	13	13	12	11
183	6'0"	35	35	34	33	33	32	31	31	30	29	28	28	27	26	26	25	24	24	23	22	22	21	20	20	19	18	18	17	16	16	15	14	14	13	12	12	11
185	6'1"	34	34	33	32	32	31	30	30	29	28	28	27	26	26	25	24	24	23	22	22	21	20	20	19	18	18	17	16	16	15	15	14	13	13	12	11	11
188	6'2"	33	33	32	31	31	30	30	29	28	28	27	26	26	25	24	24	23	22	22	21	21	20	19	19	18	17	17	16	15	15	14	13	13	12	12	11	10
191	6'3"	32	32	31	31	30	29	29	28	27	27	26	26	25	24	24	23	22	22	21	21	20	19	19	18	17	17	16	16	15	14	14	13	12	12	11	11	10
193	6'4"	32	31	30	30	29	29	28	27	27	26	26	25	24	24	23	23	22	21	21	20	19	19	18	18	17	16	16	15	15	14	13	13	12	12	11	10	10
196	6'5"	31	30	30	29	28	28	27	27	26	25	25	24	24	23	23	22	21	21	20	20	19	18	18	17	17	16	15	15	14	13	13	12	12	11	11	10	9

Lbs 260 255 250 245 240 235 230 225 220 215 210 205 200 195 190 185 180 175 170 165 160 155 150 145 140 135 130 125 120 115 110 105 100 95 90 85 80
Kg 118 116 113 111 109 107 104 102 100 98 95 93 91 89 86 84 82 79 77 75 73 70 68 66 64 61 59 57 54 52 50 48 45 43 41 39 36

Weight

■ Obese (>30)   ■ Overweight (25–30)   ■ Normal (18.5–25)   ■ Underweight (<18.5)

FIGURE 15.18 The body/mass index is a tool to help determine whether a person's weight is in a healthy range.

BMI is not a perfect measure because it does not take into account sex, frame size, or muscle mass. Nevertheless, for most of us it is an excellent indicator. Consider Heidi's numbers. At 5 feet 4 inches, her BMI was 20 when she weighed 115 pounds and was told to lose weight. Her initial weight loss of 10 pounds brought her to 105 pounds and a BMI of 18. A BMI less than 18.5 signifies that a person is underweight. When she died, Heidi weighed 93 pounds, and her BMI was only 16, dangerously underweight.

15.5 What If Weight Becomes an Obsession?

It is a cruel irony that the abundance of readily available, tasty, Calorie-dense, un-nutritious foods has spawned two serious but very different problems: obesity and eating disorders. More than 60% of American adults are overweight or obese, according to their BMI values. Excess weight increases the risk of heart disease, stroke, high blood pressure, diabetes, joint problems, and certain cancers. For some, however, dieting substitutes one problem for another. Just as dangerous can be strategies like Taylor Hooton's for "bulking up."

Eating Disorders

Theoretically, losing weight is straightforward: burn more Calories than you take in. In practice, it is quite challenging, for we are surrounded by food. More than one-third of "normal" dieters become pathological dieters. Healthy weight loss must be done slowly with balanced nutrition and increased exercise as core components.

Millions of individuals in the United States suffer from an unhealthy relationship with food and an obsession with body weight and image (**FIGURE 15.19**). This obsession can lead to an **eating disorder**, a condition in which a person eats too much or too little to a degree that is detrimental to health:

eating disorder :: a condition in which a person eats too much or too little to the detriment of health

- *Anorexia nervosa:* Individuals with anorexia starve themselves continuously. They are always unsatisfied with their bodies and think they need to lose more weight. Besides starvation, excessive exercise is another symptom of anorexia. The peak of the onset is puberty to young adult.

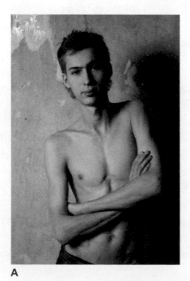

FIGURE 15.19 Body image issues can lead to undereating and dangerous weight loss, as in anorexia (A), or bulking up to an unhealthy degree, as in bigorexia (B).

A   B

- *Bulimia nervosa:* Individuals with bulimia consume large amounts of food in a short time (binge). This behavior triggers terrible guilt that leads them to balance the overeating by starving the rest of the day or purging by vomiting or using laxatives.
- *Binge-eating disorder:* Individuals suffering from binge eating gorge themselves on high-caloric food, often in secret, in an effort to fill a perceived emptiness. The binging often triggers feelings of embarrassment, but unlike bulimia, no purging occurs after the binge. Binge-eating disorder is associated with depression and feelings of low self-esteem.
- *Exercise disorder:* Individuals with exercise disorder exercise for several hours a day to lose weight or build muscles. In a sense, they are addicted to exercise. In some cases, excessive exercise is combined with anorexia and bulimia.
- *Bigorexia:* Generally more common in males, individuals suffering from bigorexia perceive themselves to be small, weak, and puny. They spend hours working out and may use steroids to bulk up.

See *Life Application: Drunkorexia* to see how body image issues, eating disorders, and alcohol use can combine to make a potentially dangerous situation.

Eating disorders like anorexia were once considered "female problems." In fact, while as many as 20% of college women suffer from the disorder, so do 5% of college men. Anorexia can affect children as young as five years, teenagers, young adults, and middle-aged males and females of any race or ethnicity. But compared to females, men seek help less and are more likely to be misdiagnosed.

Eating disorders like anorexia can produce dire consequences. The loss of minerals leads to kidney damage, muscle cramps, irregular heart rhythms, and even cardiac arrest. The depletion of body fat in women stops the production of hormones needed for the menstrual cycle and can result in permanent sterility. Because the sex hormone estrogen is necessary for healthy bones, eating disorders can also cause osteoporosis, a disease of thinning bones. Anorexia nervosa has the highest mortality rate of any mental illness.

A person with bulimia is at risk for all of the above *and* dental and gum problems. This is because of the continual exposure to acidic vomit, stomach rupture from repeated vomiting, and fatal dehydration.

The causes of eating disorders are complex, involving both genetic and environmental factors. Although anyone can be affected, athletes and performers in certain sports or activities are especially at risk compared to others. Many sports emphasize thinness or small size, including gymnastics, figure skating, dance, ballet, horse racing, ski jumping, crew, long-distance running, wrestling, modeling, acting, and synchronized swimming.

Eating disorders can be treated once a person recognizes he or she has a problem. Successful recovery depends on professional mental health counseling as well as support from family and friends. In Heidi's case, many people suspected the problem, but she denied it until her dying day. For Taylor Hooton, it is likely that bigorexia led to an obsession, drug abuse, and ultimately his death.

Steroid Abuse

Steroids are drugs that mimic the functions of normal sex hormones like testosterone and make muscles bigger. They can be taken orally, injected, or

steroid :: drug that mimics the functions of normal sex hormones such as testosterone and makes muscles bigger

Life Application

Drunkorexia

Weekends on most college campuses are often seen as a time to let off some steam. For many students, the fun is soaked in alcohol and "weekends" begin on Thursday. After all, if you work hard, shouldn't you play hard? Think again.

For many women the combination of wanting to be both "fun" and thin has produced a worrisome phenomenon called *drunkorexia*. Not actually a medical term, it means skipping meals to save Calories for alcohol consumption.

Drunkorexia is self-destructive and potentially dangerous. Even into their late 20s, many women will drink shots of hard liquor to avoid the Calories found in beer or mixed drinks. Dumping alcohol into an empty stomach produces a drunken individual very quickly. Even without starving themselves, women are generally more impaired by alcohol than are men due to physical differences.

Undesirable consequences follow. Blackouts from drinking are more common when food intake has been restricted. And alcohol on an empty stomach is more apt to produce irritation that triggers vomiting. The picture is not pretty—drunk, possibly vomiting, probably undernourished, and on the way to serious dehydration.

For some women, drunkorexia is a dangerous combination of an eating disorder and a substance abuse problem. More than 30% of college students meet the criteria for a diagnosis of alcohol abuse. Roughly 25% of anorexics and bulimics also have drug or alcohol issues. It is likely that serious drunkorexia is not uncommon, but studies are needed to pin down just how many are affected. Research is also needed to understand the relationships between eating disorders, substance abuse, and body image issues.

There is work to do to understand, prevent, and treat the problem. It won't be easy until attitudes change. People need to realize that drunkorexia is a real problem, not simply a lifestyle choice. ::

For many women, the combination of wanting to be both "fun" and thin has produced a worrisome phenomenon called *drunkorexia*. Not actually a medical term, it means skipping meals to save Calories for alcohol consumption.

rubbed on the skin as a gel or cream. They may be prescribed to help treat conditions in which the body fails to make enough testosterone or to help initiate puberty in cases in which it is delayed. However, although muscle size and strength increase dramatically in response to steroids, many other body systems are also involved.

Side effects of steroids include shrinkage of the testicles and reduced sperm counts in men, increased body hair and smaller breasts in women, and kidney damage, liver damage, heart disease, acne (especially on the back), and psychological problems in both men and women. Taylor Hooton experienced back acne, face and body bloating, serious violent outbursts, and depression.

Oftentimes, people abusing steroids will obtain them illegally, using them in much larger doses than would be prescribed by a physician to treat certain medical conditions. They may also use multiple drugs at the same time. As in Taylor's case, abuse of steroids may become addictive, dangerous, and even life-ending.

15.6

How Do Muscles and Bones Keep Us Active?

Would Taylor have obtained the results he wanted from a good exercise and nutrition program, or did he need steroids? How can we best increase muscle size and strength? To learn the answer, we must understand our body's motor system, including our muscles and bones.

Muscles and Bones

People can bulk up most effectively with exercise. Repeatedly using muscles to the edge of their capacity causes the production of new proteins, which in turn improves muscle performance. The movements of the body are governed by the actions of the **skeleton** and muscles. Composed of 206 bones, the skeleton supports the body, protects soft parts, and stores minerals, especially calcium. The bones also produce blood cells. The skull, spine, rib cage, and coccyx form the axial skeleton, while the appendicular skeleton includes hips, shoulders, and limbs (**FIGURE 15.20**).

Movement is made possible not only by muscles, but also by the hinges where bones connect to each other, held together by **ligaments** (**FIGURE 15.21**). Ball-and-socket joints, such as those found in the hips and shoulders, allow a three-dimensional range of motion. The hinge joints of the knees and elbows provide for two-dimensional motion. And a pivot joint makes it possible for the head to move side to side. All joints are lined with a firm but flexible tissue called **cartilage**, to protect the ends of bones, and with lubricating fluid.

Movement of the skeleton happens because skeletal muscles are attached to bones with **tendons**. These skeletal muscles can contract or shorten, and when they do, the bone moves. There are more than 700 skeletal muscles attached to bone. Approximately 40% of our weight comes from muscles. Athletes like Taylor Hooton target skeletal muscles in their quest for a larger physique.

Skeletal muscles are only one type of muscle. There are two more: cardiac and smooth. Cardiac muscle is located exclusively in the heart. Unlike skeletal muscle, cardiac muscle is not under our conscious control. The heart beats every day, all day, for a person's entire life. Its need for energy is tremendous, and the cells of this muscle contain more mitochondria, the organelle that produces ATP, than any other type of cell in the body.

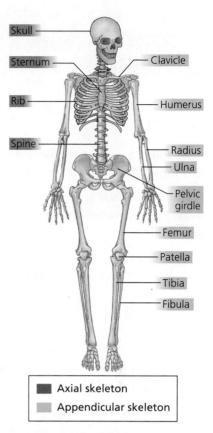

☐ Axial skeleton
☐ Appendicular skeleton

FIGURE 15.20 Composed of 206 bones, the skeleton supports the body, protects soft parts, and stores minerals. The skull, spine, rib cage, and coccyx form the axial skeleton, while the appendicular skeleton includes arms, legs, hips, and shoulders.

skeleton :: composed of 206 bones; the human skeleton supports the body, protects soft parts, and stores minerals

ligament :: band of tissue that connects two bones, cartilage, or joints together

cartilage :: flexible tissue that protects the end of bones

tendon :: cord of tissue that connects muscles to bones

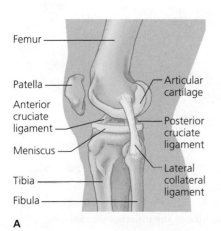

A

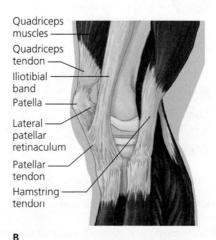

B

FIGURE 15.21 Movement is made possible because of the actions of ligaments (A), which connect bones together, and tendons (B), which connect bones to muscles.

Smooth muscle is found in layers built into blood vessels, bladder, intestines, and uterus. Smooth muscles produce slower, rhythmic, sustained contractions that can move blood, wastes, urine, and even babies along. Like cardiac muscle, smooth muscle is not under our conscious control.

Moving Muscles

All types of muscles contract thanks to their structure. Muscles are comprised of parallel arrays of filaments (**FIGURE 15.22**). Each of these filaments is made up of **muscle fibers**, in which each fiber is actually a single cell. Within each of these cells are long **myofibrils**, also aligned parallel to one another.

Every myofibril is composed of contraction units, called **sarcomeres**, which are attached end to end. The sarcomeres contain thin filaments made of the protein actin, and thick filaments, made of myosin, another protein. The thin filaments are anchored to z-discs, structures that form the ends of sarcomeres. Muscles contract simply by the shortening of individual sarcomeres.

Every muscle fiber is controlled by an individual neuron. When the signal to contract is received, the thin filaments slide over the myosin filaments, pulling the z-discs closer to one another. This sliding is caused by the actions of actin and myosin and is powered by ATP. Let's take a look at this sliding action, starting with the finish of the previous contraction (**FIGURE 15.23**).

Myosin is physically attached to the actin filament. Myosin is shaped like a golf club, a long filament with a hinged segment, or head, at one end. Unlike a golf club, the hinged end in myosin can move. ATP is required for actin and myosin to let go of each other. When ATP binds to myosin, the myosin lets go of the actin, and its "hinged" part bends in the direction farther down the actin filament. The myosin head, in its new location, binds to the actin filament. When this occurs, the myosin head snaps back into its initial orientation. When it does so, the actin filament is pulled along toward the center of the sarcomere. Because these events occur to many actin filaments on both sides of the sarcomere, the entire structure contracts. Every muscle cell has as many as 100,000 sarcomeres.

muscle fiber :: composed of a bundle of myofibrils aligned parallel to one another

myofibril :: composed of actin and myosin filaments arranged as sarcomeres attached end to end

sarcomere :: a contraction unit of muscle; made up of the proteins actin and myosin

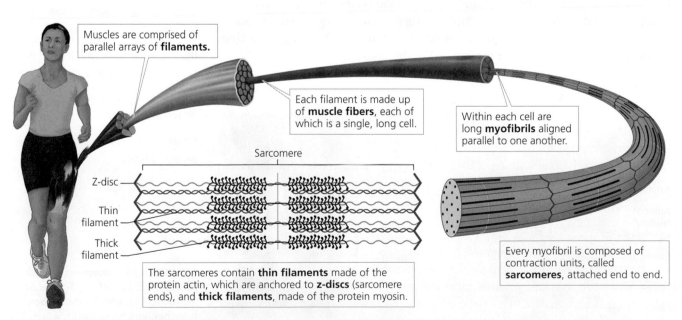

Muscles are comprised of parallel arrays of **filaments.**

Each filament is made up of **muscle fibers**, each of which is a single, long cell.

Within each cell are long **myofibrils** aligned parallel to one another.

Sarcomere

Z-disc

Thin filament

Thick filament

The sarcomeres contain **thin filaments** made of the protein actin, which are anchored to **z-discs** (sarcomere ends), and **thick filaments**, made of the protein myosin.

Every myofibril is composed of contraction units, called **sarcomeres**, attached end to end.

FIGURE 15.22 All types of muscles (cardiac, smooth, and striated) contract due to the structure of the muscle cells. Striated muscles are shown here.

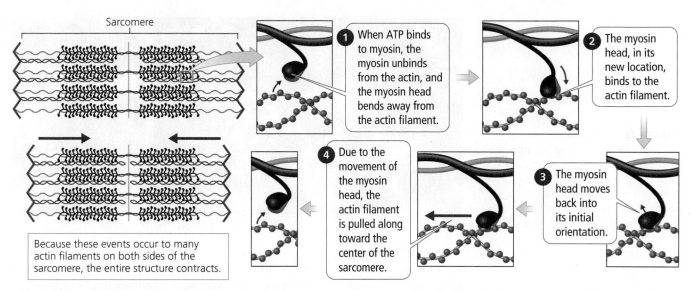

FIGURE 15.23 Muscle contraction occurs when sarcomeres shorten their lengths, resulting in a shortening of the entire muscle. ATP powers this muscle contraction cycle.

Skeletal muscles are arranged in antagonistic pairs, like the biceps and triceps. When one contracts because its sarcomeres shorten, the other relaxes when its sarcomeres lengthen (**FIGURE 15.24**).

15.7 How Can You Keep in Shape?

Taylor Hooton put on 30 pounds of muscle in six months. He went from struggling to do triceps extensions with 60 pound weights to easily doing two or three sets of 10–12 reps with 85 pound weights—all in just six weeks!

Was the problem that Taylor didn't use steroids safely? No, there is no safe level of steroid use. Steroids are a controlled drug and should be administered only by a doctor for specific health problems, never to enhance appearance or athletic performance. Could Taylor have enjoyed the same increases in size and strength with exercise alone? Probably not, and certainly not nearly as fast.

Exercise

Like everyone else, Taylor's muscles were composed of slow-twitch and fast-twitch fibers. Twitch refers to the time it takes to go from contraction to relaxation. Fast-twitch fibers, which contract and relax very quickly, often predominate in sprinters. Fast-twitch fibers do not have a lot of endurance: they fatigue quickly. Slow-twitch fibers, in contrast, work 10 times less quickly than fast-twitch, but they don't fatigue as readily. Slow-twitch fibers generally predominate in long-distance runners and other sports that require endurance.

Exercise can make muscles grow in size, increase in strength, or both (**FIGURE 15.25**). Weight lifters use sustained, long-term strength training with weights that are set at 80–90% of the maximum weight that they can move once. This type of training will increase muscle strength dramatically. It causes muscle cells to make more actin and myosin—and consequently more muscle fibrils. The increase in the number of muscle fibrils is directly responsible for the improved strength.

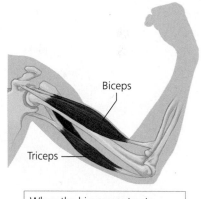

When the biceps contract (sarcomeres shorten), the triceps relax (sarcomeres lengthen).

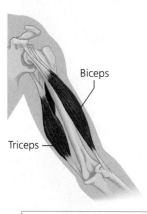

When the biceps relax (sarcomeres lenghten), the triceps contract (sarcomeres shorten).

FIGURE 15.24 Skeletal muscles are arranged in antagonistic pairs. Biceps and triceps work together opposing one another's actions; as one contracts, its antagonist relaxes.

A **B**

FIGURE 15.25 Different types of exercise can yield varied physical results. Weight lifters, such as Kendrick Farris of the United States, train principally to increase muscle strength (A), while bodybuilders, such as Bognar Gergo of Hungary, do so to increase muscle size (B).

Bodybuilders emphasize a different training routine. They work with weights at levels much lower than the maximum. The workout emphasizes multiple sets of repetitions—lifting four or more sets of 12–15 reps each. This type of exercise produces significant growth of muscle size by increasing the volume of a fluid-filled structure, which surrounds muscle fibers. Although they get bigger, the muscles do not become stronger.

What about someone like Taylor, who wanted more size *and* strength? The best approach is then to combine both approaches. Had Taylor done so, he may have eventually achieved his goal without steroids.

Taylor focused his efforts on strength training and increasing muscle mass; he paid less attention to cardiovascular, or aerobic, exercise. Jogging, swimming, running, and biking are all examples of exercises that elevate the heart rate. During exercise, the faster-beating heart pumps more blood, thus delivering more oxygen to all muscles, including the heart itself. Aerobic exercise strengthens the heart, making it function more efficiently even when a person is at rest. In fact, as the efficiency of the heart increases thanks to regular aerobic exercise, the heart rate during rest can actually decrease because the heart can deliver blood to the body more effectively. The Mayo Clinic recommends that adults exercise vigorously (e.g., running or aerobic dancing) for 75 minutes or moderately (e.g., brisk walking, mowing the lawn, or swimming) for 150 minutes per week. The exercise should be spread out over the week.

Exercise, of any type, must be approached in a balanced manner (see *Technology Connection: "Guaranteed" Six-Pack Abs*). Everyone can benefit from aerobic exercise, which strengthens the heart, and strength training, which improves muscle and bone health and endurance. A person with specific goals can get help from coaches, trainers, and other athletes regarding effective and safe ways to achieve them. Whatever your objectives for physical activity, the key to avoiding injury is taking things gradually and steadily, eating sensibly, and getting enough rest.

Supplements

Many athletes who have made the smart decision to reject steroids still seek faster, better results. Are there drugs or nutritional supplements that work? Are they safe? Let's look at a few:

- *Diuretics:* These drugs alter the balance of fluids and salts in the body. Some athletes use them to urinate in larger volumes and so decrease

"Guaranteed" Six-Pack Abs

You have probably seen products that promise the loss of "love handles." You can develop toned abdominal muscles, advertisers claim, with as little as 10 minutes a day of exercise. Do these machines work? Are there devices that can reduce fat in select areas of the body?

The product research team from *Popular Mechanics* rigorously tested seven types of ab machines over two months. They learned that the machines will strengthen abdominal muscles—if a person exercises vigorously for at least 15–20 minutes per day. Even then, unless the individual is already superthin, the toned muscles will not be visible. The *Popular Mechanics* team also learned that spot reduction is impossible. The only way to lose weight in one area is to lose weight everywhere.

According to Kurt Brungardt, a fitness expert and author of *The Complete Book of Abs*, and the American Council on Exercise, ab machines could be an effective way to tone abdominal muscles. However, focusing exclusively on those muscles doesn't make sense. In addition to regular muscle conditioning, aerobic workouts and a healthy diet are the keys to fitness.

It would be great if exercise machines could work the magic their manufacturers claim. But they are only a small piece in an effective fitness plan. ::

The product research team from *Popular Mechanics* rigorously tested seven types of ab machines. They learned that spot reduction is impossible, and the only way to lose weight in one area is to lose weight everywhere.

weight. Diuretics can have serious side effects, however, ranging from dehydration, muscle cramps, and fatigue to irregular heartbeat and even death.

- *Creatine:* This molecule, normally found in the body, can also be taken as a supplement. Creatine helps muscles make more ATP and delays muscle fatigue. Its side effects include muscle and stomach cramps, nausea, and diarrhea. It can also produce weight gain because of water retention.
- *Vanadium:* This strong-selling supplement is supposed to increase muscle mass, but there is no evidence that it does so. Vanadium does, however, cause diarrhea and stains the tongue green.
- *Boron:* This is a true nutritional supplement—but for plants, not for people.
- *Stimulants:* These can reduce fatigue and suppress appetite. Stimulants also improve alertness and sometimes make a person more aggressive. Side effects can include irritability, nervousness, insomnia, increased blood pressure, heart abnormalities, heart attack, and hallucinations.

Taking supplements is a big gamble. Not only do they have side effects, but their manufacture may not be quality controlled. Impurities can be present, and dosages vary from one bottle to the next. As alluring as the idea of supplements may be, there is no magic elixir. The only thing that works is the right combination of exercise, good nutrition, and adequate sleep—the same recipe as for good health.

15.8 How Do We Balance Looking Good with Being Healthy?

Heidi Guenther and Taylor Hooton both confused looking good with being healthy. Heidi became trapped in the quest to be thinner because she thought it would make her a better dancer. Taylor got hooked on steroids hoping to get bigger and to become a better pitcher. Except for the tragic endings, their stories are not so unusual. Many people in our culture focus on whether the "outside" of a person is attractive. This perspective can lead to unhealthy behaviors and consequences. Are there ways to look good *and* be healthy?

Weight and Body Image

Television, films, magazines, and the web bombard us with airbrushed images of superthin models, celebrities, and actors, as well as super-ripped athletes. These images skew what we all perceive as normal and invite, in not so subtle ways, the pursuit of a fantasy world where everyone looks perfect. One step toward good health is to reject those images as exemplars.

An obsession with weight and body image can have tragic consequences. While excess body weight is unhealthy, too little body weight can be dangerous as well. We need adequate body fat to regulate body temperature, reproduce, make hormones, insulate our bodies, protect our organs, and ensure adequate energy.

Most people know what is needed to maintain a healthy body. We thrive on a balanced, varied diet that emphasizes fruits, vegetables, and whole-grain food and that limits highly processed fatty foods and alcohol. We also benefit from regular physical activity, adequate sleep, and effective stress management.

Physical Activity

All of the systems of the body function together; none is isolated. If the body is healthy, then all body systems benefit. Regular physical activity is perhaps the most important key to wellness.

Regular physical activity prevents or improves many diseases, including coronary heart disease, heart attack, stroke, diabetes, high blood pressure, hip fractures, and obesity. Regular movement also benefits muscle strength, bone health, and joint structure and function. In addition, regular physical activity improves mental health, reduces the risk of depression, improves performance in school, and even bolsters self-esteem.

Together with a nutritious diet, physical activity is the linchpin to maintaining a good energy balance and healthy weight. Physical activity can take many forms: exercise, team sports, running, swimming, and yard work are all examples. What matters most is that the activity is regular: 30–60 minutes a day, at least five days a week.

How Do We Know?

Stress Weakens the Immune Response

Have you noticed that you, or people you know, get sick when work deadlines loom, or a romantic relationship ends, or the sight of a roommate makes you recoil even after living together for a semester? Is there a physiological connection between experiencing stress and becoming ill? There is. Stress, especially chronic stress, increases the susceptibility to infectious disease (and other types too).

More than 400 people (154 men and 266 women), aged 18–54 years and in good physical health, volunteered for a study in the 1990s at the Medical Research Council's Common Cold Unit in Salisbury, England. The subjects were given medical exams and evaluated for psychological stress.

One set of volunteers received nose drops containing cold viruses. A control group received saline nose drops. The volunteers were randomly assigned to the virus or saline groups; they did not know which type of drops they received.

All subjects were examined daily for two days before the exposure to virus or saline and for six days following. Signs of colds—sneezing, watery eyes, stuffed nose, runny nose, sore throat, sinus pain, cough, hoarse voice, and type of mucus produced—were monitored and recorded. The number of tissues each subject used was recorded, as was body temperature.

The types of stresses experienced by subjects were also categorized according to the following: current demands are overwhelming; several major stressful life events have been endured; or a current negative major life event is occurring (a death in the family, for example).

The results were startlingly clear. No matter what type of stress, exposure to virus produced colds in stressed individuals in much greater frequency than in low-stress subjects.

Can this increased susceptibility to colds be explained by the response of the immune system to stress? Probably. Another study showed that stressed people who became sick

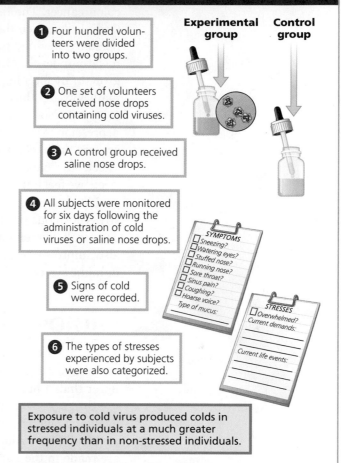

1 Four hundred volunteers were divided into two groups.

2 One set of volunteers received nose drops containing cold viruses.

3 A control group received saline nose drops.

4 All subjects were monitored for six days following the administration of cold viruses or saline nose drops.

5 Signs of cold were recorded.

6 The types of stresses experienced by subjects were also categorized.

SYMPTOMS
☐ Sneezing?
☐ Watering eyes?
☐ Stuffed nose?
☐ Running nose?
☐ Sore throat?
☐ Sinus pain?
☐ Coughing?
☐ Hoarse voice?
Type of mucus:

STRESSES
☐ Overwhelmed?
Current demands:

Current life events:

Exposure to cold virus produced colds in stressed individuals at a much greater frequency than in non-stressed individuals.

In the 1990s, more than 400 people volunteered for a study at the Medical Research Council's Common Cold Unit in Salisbury, England. Scientists wanted to determine whether there was a relationship between stress and illness.

from infections had a generally suppressed immune response. Here, then, is another reason to relax and keep a balanced perspective. Worrying really can make us sick. ::

Sleep, Laughter, and Stress Management

Every system of the body suffers when we don't sleep enough. If we sleep less than 6–7 hours a night, our risk of disease rises dramatically. Scientists and healthcare professionals recommend 7–9 hours of sleep each night.

One of the principal benefits of sleep is stress reduction (see *How Do We Know? Stress Weakens the Immune Response*). When we experience stress, our bodies go into "alert mode." This increases blood pressure and makes the body produce more stress hormones. High blood pressure increases the risk of stroke and heart attack. And stress hormones make it even harder to sleep, which in turn ups the stress hormones further. Stress hormones also increase the body's inflammatory response. As a consequence, sleep deprivation increases the risk of heart disease, diabetes, and cancer.

Besides keeping stress hormones lower, sleep may also prevent cancer in a different way. People who work the late shift have a higher risk of breast and colon cancer. The reason appears to reside in a molecule called melatonin. Ordinarily, when darkness falls, melatonin rises and makes us sleepy. If a person works nights, or stays awake at night, the exposure to light reduces melatonin levels. Melatonin, however, also has the ability to suppress tumor growth. By forgoing sleep, a person may be keeping melatonin levels abnormally low and thus short-circuiting a natural defense against cancer.

Sleep also bolsters memory, improves alertness, and reduces the risk of depression. And although it may seem counterintuitive, inadequate sleep increases the likelihood of being overweight or obese. Evidently a lack of sleep affects the balance of hormones that regulates appetite. Without enough sleep, we feel hungry and eat even when there is no metabolic reason to do so.

Laughter also plays a role in stress reduction. Besides lowering the level of stress hormones in the body, laughter increases the amount of beneficial brain chemicals like endorphins, which reduce pain and elevate mood. Laughter also improves immune system function by enhancing the effectiveness of immune system cells.

Biology in Perspective

The recipe for wellness is easy to understand: eat a varied, balanced diet; exercise or be physically active at least 30–60 minutes most days; sleep no fewer than 7 hours a night; don't drink excessively; don't use tobacco; don't use drugs for recreation; don't buy into media images of how we should all look; and laugh a lot.

Heidi and Taylor are examples of what can happen when poor decisions are made in the name of trying to meet unrealistic ideals for one's body. In Heidi's case, she could never be thin enough, and in Taylor's, he could never be big enough. They lost perspective about what was important for their well-being in the long term. They were, by themselves, unable to control their actions, which ultimately led to their tragic deaths. Heidi and Taylor might seem like dramatic examples, but their cases are not unique.

Knowing what to do to be healthy is not difficult. Fashioning a life that is balanced in a world with many competing demands is the challenge. There is no other pathway to good health, however. Your body's biology will respond to what you do and don't do no matter what. But you are in charge of decisions and so are ultimately the one in power.

Chapter Summary

15.1 How Does What You Eat Influence Your Well-Being?

Identify and describe the components of a healthy diet.

- Both the quantity and quality of food play a role in a healthy diet.
- Water is necessary for all cellular activities.
- Carbohydrates, proteins, and fats, are a major source of energy for our cells.
- Vitamins assist enzymes in regulating chemical reactions.
- Minerals are inorganic molecules essential for bone development, muscle movement, nerve impulses, and fluid balance.
- Fiber keeps things moving through the digestive tract.

15.2 How Does the Body Extract Nutrients from Food?

Trace the path that food travels from the mouth to the bloodstream and identify the key places where nutrients are extracted.

- Food breakdown begins in the mouth.
- The esophagus moves food to the stomach by peristalsis.
- In the stomach, cells secrete acid to digest protein and kill bacteria.
- Most chemical digestion occurs in the small intestine.
- The pancreas neutralizes acid and secretes enzymes to digest proteins, carbohydrates, and fats.
- The liver produces bile, which is stored in the gall bladder, and releases glycogen, the simple sugar most commonly used by the body for energy.

15.3 How Do Cells Extract Energy from Food?

Explain the process and chemistry of cellular respiration.

- ATP is the most commonly used energy source for cells.
- Cellular respiration produces ATP from food molecules in three steps:
  - During glycolysis, energy to produce two molecules of ATP is extracted from glucose.
  - The Krebs cycle produces GTP and NADH.
  - The electron transport chain produces most of the ATP.

15.4 What Can a Person Do to Maintain a Healthy Weight?

Describe how a person's basal metabolic rate can be used to maintain a healthy weight.

- Calories are the energy units in food.
- To maintain weight, Calories consumed must equal Calories spent.
- Basal metabolic rate is the minimal level of caloric intake necessary to support life.
  - It varies with age, weight, gender, activity level, and genetics.

  - The more active you are, the more energy (food) you need to maintain your weight.
- To lose weight, you must eat less and be more active.
- Healthy weight can be estimated using the body/mass index calculated from a person's height and weight.

15.5 What If Weight Becomes an Obsession?

Assess the various dangers associated with eating disorders that commonly afflict individuals.

- In an eating disorder, a person eats too much or too little to a degree that is detrimental to health.
  - Anorexia nervosa: Individuals starve themselves and/or exercise continuously.
  - Bulimia nervosa: Individuals binge on food, feel guilty, and so then starve or purge by vomiting or using laxatives.
  - Binge eating: Individuals gorge on high-Calorie food, often in secret; it is associated with depression and low self-esteem.
  - Exercise disorder: Individuals are essentially addicted to exercise.
  - Bigorexia: Individuals spend hours working out and/or taking steroids to get bigger.

15.6 How Do Muscles and Bones Keep Us Active?

Describe the components of the motor system and the ways muscles and bones work together to enable us to move.

- The skeleton supports the body, protects soft parts, stores minerals (especially calcium), and produces blood cells.
- Bones connect to each other with ligaments around joints; joints are cushioned and lubricated by cartilage.
- Skeletal muscles are attached to bones with tendons, and when muscles contract, bones move.
- Cardiac muscle, not under conscious control, causes the heart to beat.
- Smooth muscle, not under conscious control, is built into blood vessels, the bladder, intestines, and the uterus.
- Muscles are built of parallel filaments made of muscle fibers in which each fiber is a single cell that contains myofibrils composed of sarcomeres. When the sarcomeres shorten, using ATP for energy, the muscles contract.

15.7 How Can You Keep in Shape?

Discuss the safety and efficacy of various methods for getting and staying in shape.

- Steroids are not safe to use other than as prescribed by a doctor for a specific health problem.
- Exercise can increase size and strength of muscles.
- Aerobic exercise that strengthens the heart and improves muscle and bone health and endurance is good for everyone.

- Taking supplements is a gamble; some supplements do not work as advertised, and others have potentially dangerous side effects.

15.8 How Do We Balance Looking Good with Being Healthy?

Distinguish between the value of looking good and being healthy.

- Media images of "perfect" people are unrealistic: weighing too much is unhealthy, but weighing too little can be dangerous, too.
- To be healthy requires a balanced and varied diet, regular physical activity, and adequate sleep and laughter to reduce stress.

Key Terms

adipose tissue 451	esophagus 449	pancreas 450
adenosine triphosphate (ATP) 453	fiber 447	peristalsis 449
anus 448	gall bladder 451	pharynx 449
basal metabolic rate 453	glycogen 451	rectum 451
bile 450	glycolysis 452	salivary gland 449
body/mass index (BMI) 455	Krebs cycle 453	sarcomere 460
Calorie 453	large intestine, or colon 451	skeleton 459
cartilage 459	ligament 459	sphincter 449
cellular respiration 451	liver 450	steroid 457
digestive system 448	mineral 447	tendon 459
eating disorder 456	muscle fiber 460	vitamin 445
electron transport chain (ETC) 453	myofibril 460	
epiglottis 449	NADH 453	

Review Questions

1. What important role does each of the following play in the diet: water, protein, carbohydrates, fats, and fiber?

2. What roles do vitamins and mineral play in the diet?

3. Diagram the path that food takes during its digestion, starting in the mouth and ending at the anus.

4. What roles do the liver and gall bladder play in digestion?

5. What role does the pancreas play in digestion?

6. What are the three sets of reactions that comprise cellular respiration?

   a. Glycogen, ATP, and the Krebs cycle

   b. Glycogen, the Krebs cycle, and the electron transport chain

   c. Glycolysis, the Krebs cycle, and the electron transport chain

   d. Glycolysis, the Krebs cycle, and photosynthesis

   e. Glycolysis, ATP, and NADH

7. What are the functions of ATP and NADH?

8. What is the basal metabolic rate?

   a. The rate at which the body extracts energy from food

   b. The amount of Calories your body needs while you are at rest

   c. A tool for determining whether one's weight is in a healthy range

   d. The units of energy used to describe food

   e. The basic level of Calorie intake needed to support life functions

9. Describe in a few sentences each the following eating disorders: anorexia, bulimia, binge eating, bigorexia, and exercise compulsion.

10. Which of the following is NOT a common side effect of steroid use in males?

    a. Shrinkage of the testicles

    b. Smaller breasts

    c. Reduced sperm counts

    d. Kidney damage

    e. Acne

11. What are the potential benefits, if any, and the side effects of each of the following supplements: steroids, diuretics, creatine, stimulants, boron, and vanadium.

12. What is the difference between fast-twitch and slow-twitch fibers?

   a. Fast-twitch fibers contract and relax quickly, while slow-twitch fibers contract and relax slowly.

   b. Fast-twitch fibers are made of connective tissue, while slow-twitch fibers are made of bone.

   c. Fast-twitch fibers contract and relax slowly, while slow-twitch fibers contract and relax more quickly.

   d. Fast-twitch fibers contract and relax at a moderate speed, while slow-twitch fibers contract and relax slowly.

   e. There is no real difference between fast-twitch and slow-twitch fibers.

13. Diagram the process of sarcomere contraction.

The Thinking Citizen

1. The surface area of the small intestine is much greater than that of the esophagus, stomach, mouth, or large intestine. What is the functional significance of this difference?

2. Describe how the process of extracting energy from food molecules is similar to the combustion of charcoal. How is it different?

3. Determine your basal metabolic rate: the number of Calories you need per day to maintain your weight.

4. Determine your body/mass index (BMI). Do you think it places you accurately into a weight category? Why or why not?

5. Describe how a person could lose or gain weight in a healthy manner.

6. Explain why some types of exercise build muscle strength while others mainly increase muscle size.

7. How does inadequate sleep affect many body systems?

The Informed Citizen

1. The United States is suffering from an increasingly serious epidemic of childhood obesity. How do you think we can solve this problem and help children to maintain healthy weight?

2. Some dangerously obese children are not even three years old. What responsibility do parents play in providing a balanced diet for their children? If parents are unable to do so, how can they be helped? Is punishment for child abuse ever appropriate in these situations? Why or why not?

3. Many schools have removed "junk food" and soda machines from their buildings. Do you think this is an appropriate action to fight childhood obesity? Why or why not?

4. If an Olympic athlete is caught using a banned performance-enhancing substance (like steroids), he or she is suspended from the sport for two years. If an athlete is caught a second time, he or she is banned for life. Does this seem a reasonable standard for high school and college athletes too? Why or why not?

5. Should coaches be responsible and subject to suspension themselves if their athletes use banned substances? Why or why not?

6. Is it appropriate to screen middle school, high school, and college athletes for banned substances in order to be eligible to play? Why or why not?

7. Approximately 11% of all high school students in the United States have been diagnosed with an eating disorder; 90% of individuals with eating disorders are females between the ages of 12 and 25 years. What do you think are specific actions that should be taken to address this serious problem?

CHAPTER LEARNING OBJECTIVES

After reading this chapter, you should be able to answer the following questions:

16.1 How Do Species Adapt to Their Habitat?
Provide examples of the ways species adapt to the constraints of their habitats.

16.2 Why Do Species Compete?
Summarize how species can coexist without one outcompeting the other.

16.3 How Do Species Exploit One Another?
Describe the ways that species may exploit one another.

16.4 When Can Species Cooperate?
Identify circumstances of cooperation between species.

16.5 How Do Ecological Interactions Affect Us?
Summarize the beneficial and detrimental ecological interactions that we have with other species in the ecosystem.

16.6 What Does a Functioning Ecosystem Do?
Describe the movement of energy and nutrients through an ecosystem.

16.7 What Benefits Do We Get from a Functioning Ecosystem?
Enumerate the benefits we get from a functioning ecosystem.

Ecology is the study of interactions. As species compete, eat one another, and occasionally cooperate, they set up conditions that affect their survival and shape their evolution.

CHAPTER 16

Ecology

How Do We Benefit from a Functional Ecosystem?

Becoming an Informed Citizen . . .

All species, including our own, depend on the benefits ecosystems provide. These benefits may seem free of charge, but if we disrupt them, we must pay for them one way or another. Maintaining ecosystem services is an economic imperative as well as an ethical and cultural need.

s the English poet John Donne wrote, "No man is an island, entire of itself; every man is a piece of the continent, a part of the main" (Meditation XVII, 1624). Donne was reflecting on the social and spiritual interactions that connect people to one another. In reality, all organisms depend on others for survival. Biological interactions connect species to one another and to their habitat. To paraphrase Donne, no species is an island.

ecology :: the study of the interactions among species and their physical environments

Ecology is the study of interactions. As individuals of different species compete, eat one another, and occasionally cooperate, they set up conditions that affect their survival and shape their evolution. Populations also adapt to the challenges of their habitat, and this adaptation is another type of interaction. Interactions also allow an ecosystem to perform "work." Energy and nutrients are distributed throughout the ecosystem as organisms consume one another. Oxygen, carbon dioxide, and nitrogen recycle through the ecosystem, atmosphere, and soil as a result of living metabolism. All species, including our own, depend on the work ecosystems do for their survival.

In this chapter, we will examine how individuals of different species interact, and how their interactions drive an ecosystem's function. We will see how humans interact with other species, and how we have taken advantage of the adaptations produced when other species compete with, exploit, or cooperate with one another. Ultimately a species' ecology, its interactions with its habitat and with other species, determines whether it will survive or become extinct. Consider the fate of Kirtland's warbler.

case study

The Near-Extinction of Kirtland's Warbler

Warblers are small, insect-eating birds known for their colorful plumage and cheery songs. Most North American warblers are successful species and can be found nesting throughout wide regions of the continent. The Kirtland's warbler (scientific name *Dendroica kirtlandii*) (**FIGURE 16.1**) is an exception.

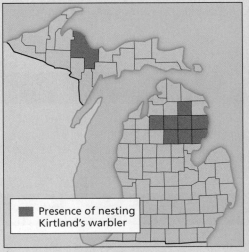

Presence of nesting Kirtland's warbler

FIGURE 16.1 Kirtland's warbler is extremely rare and nests in only a few sites in northern Michigan.

It is extremely rare and nests in only a few sites in northern Michigan. A thorough survey in 1973 detected only 200 singing males (400 birds if the males have mates) in the entire world. What is it about the ecology of these warblers that resulted in their precarious situation? What hypotheses could you develop to account for this near-extinction?

- The warbler's food source is rare.
- Some predator is killing off all the warblers.
- Some disease is killing off all the warblers.
- Human development is crowding the warblers out.
- Suitable habitat and nesting sites for the warblers are rare.

To distinguish among these five hypotheses, you would need to know how the warbler interacts with its habitat, and how it interacts with other species in the area. To make things easier, we'll describe some relevant discoveries about the Kirtland's warbler ecology.

Kirtland's warblers build their nests on the ground. A ground nest is vulnerable to predators and also to environmental problems like flooding during rainstorms. Kirtland's warblers address the predator problem by nesting under the lower branches of a young jack pine tree. These lower branches cover the nest and help to hide it from predators. Also, jack pines grow predominantly in dry sandy soil that drains well, reducing the risk of flooding.

In a young stand of jack pine, the trees are short and relatively spaced out. This allows lots of light into the stand, promoting the growth of grasses and shrubs between the trees. This grassy, shrubby area is home to many kinds of insects, which provide food for the warblers and their young. Young stands of jack pine are the ideal habitat for the Kirtland's warbler, providing nesting sites, shelter and protection, and food.

Jack pines are not rare. They are found in a broad swath across the northern United States and Canada. However, *young* stands of jack pine are rare. As a jack pine stand ages, the trees grow taller and block light from reaching the forest floor. The grassy, shrubby area between trees dies out due to lack of light, and with it goes the warbler's source of food. Tall jack pines also drop their lower branches, which leaves the warblers' nests exposed and vulnerable to predators.

Another ecological threat for the Kirtland's warbler comes from the brown-headed cowbird (scientific name *Molothrus ater*) (**FIGURE 16.2**). Cowbirds are nest parasites: they lay their eggs in the nests of other species—in this case, those of the Kirtland's warbler. The hapless warblers incubate the cowbird eggs along with their own. When the eggs hatch, they feed the young cowbirds along with their own young. Cowbird babies are demanding and aggressive and often succeed in getting so much of the warbler's attention and food that the warbler young starve.

We now have two clues about why Kirtland's warblers are rare: a limited number of young jack pine stands and nest parasitism from cowbirds. Let's now consider how young jack pine stands form. On dry, sandy soils, forest fires are a relatively common event. Jack pines are adapted to withstand fire; indeed, their reproduction depends on it. The cones on a jack pine stay closed, sealed with resin. Only when a fire comes through do the cones fall off the tree, and the fire melts the resin to release the seeds. The fire kills many adult trees, allowing light to reach the ground. Fire also burns away leaf litter and debris, producing a clean bed of soil that promotes seed germination. But fire is decidedly not good for human habitats near jack pines. Our practice of suppressing fire has reduced the number of young jack pine stands, and the warblers have fewer places to live as a result.

FIGURE 16.2 One ecological threat for the Kirtland's warbler comes from the brown-headed cowbird, which lays its eggs in the nests of the Kirtland's warbler. The warblers incubate and then feed the young cowbirds along with their own young; the cowbirds often get so much of the warbler's attention and food that the young warblers starve.

The dry sandy soil where jack pine forests grow also makes an ideal place for human homes (no flooded basements!). In addition, the soil is useful for agriculture: easy to work, and fertile with some addition of organic material. Jack pine forests are cleared to make way for houses and farms. While this change in habitat is detrimental to the Kirtland's warbler, it opens up opportunities for other species. Cowbirds, for example, get their name because they follow herds of bison and other large herbivores. Cowbirds eat soil insects stirred up by these herbivores' hooves. When the herds migrate, so do the cowbirds. This makes nesting difficult, and cowbirds have adapted by laying their eggs in the nests of other species rather than building their own nests. When agriculture opened up in the jack pine forests, perhaps 200 years ago, cowbirds invaded the traditional nesting habitat of Kirtland's warbler. Cowbird nest parasitism has reduced the size of the warbler population.

These ecological facts provide support for two of our hypotheses: *human development* and *limited nesting habitat*. Predators and disease do not seem to be significant, but another hypothesis that we did not consider earlier is *nest parasitism*. A number of other ecological factors play a role as well, including fire, soil conditions, and light penetration through the forest canopy. The Kirtland's warbler is part of a web of ecological interactions that constrain its distribution and abundance, nearly driving it extinct.

To help save Kirtland's warblers from extinction, Michigan has initiated a program of controlled burns and planting jack pines to increase suitable habitats. Cowbirds are being trapped to reduce nest parasitism. Educational programs inform people of the plight of Kirtland's warblers. These efforts appear to be successful. In 2010, there were 1747 breeding pairs, an almost nine-fold increase since 1973.

FIGURE 16.3 Kangaroo rats live in the deserts of North America. To minimize water loss, they forage for food at night and spend their days inside burrows that are humid (due to water evaporation from the kangaroo rats' lungs).

16.1 How Do Species Adapt to Their Habitat?

The *eco* in *ecology* comes from the Greek *oikos*, meaning "home." Ecology is the study of one's home—or, for nonhumans, the habitat. Many physical factors interact to determine whether a habitat is suitable for a particular species: the temperature range, water availability, and nutrients in the soil, for example. The habitat also provides shelter for survival and reproduction.

Species adapt in both their physiology and behavior to meet the challenges of their habitat. In this section, we will examine how species interact with and adapt to their habitat.

Adapting to Physical Conditions

Kangaroo rats (**FIGURE 16.3**) are a group of about 20 species living in the deserts of North America. They have powerful hind legs and hop about on them like tiny kangaroos. Like all animals, kangaroo rats must strike a physiological balance between water intake and water output. Living in a desert habitat makes achieving this balance very difficult.

Kangaroo rats lose water through three channels: urine, feces, and evaporation from respiratory surfaces of the lungs and nose. Compared to other mammals, however, they lose remarkably little water, particularly from their urine and feces. Kangaroo rats have extremely efficient kidneys. Kidneys filter toxic substances from the blood and flush them out in urine (Chapter 11). Kangaroo rat kidneys are able to make very concentrated solutions of these toxins and therefore require little water to get rid of them. Their kidneys use only one-quarter to one-sixth as much water as a human would to eliminate a given amount of waste. Their feces are dry as well. That's because the kangaroo rat colon is able to recover almost all the water as it leaves the digestive system.

The inner surface of the lungs must be kept moist in order for oxygen to be absorbed. Nasal passages must be moist to dissolve odors an organism smells. These moist surfaces are in contact with outside air, and water constantly evaporates from them. Kangaroo rats can't eliminate this water loss, but they have adaptations to minimize it.

One of these adaptations is that kangaroo rats are nocturnal: they spend the day inside their burrow. As water evaporates from their lungs, humidity in the burrow increases. This helps to reduce evaporative loss. At night, the rats forage for food, mainly dry seeds and grasses, which they take back to their burrow to eat and store. In the humid burrow, their food absorbs moisture. When the rats eat their food, they actually recover some of the moisture they lost previously from evaporation.

Water intake for kangaroo rats is a puzzle. They have no access to freestanding water, and in fact, in their natural habitat they do not drink. Instead, most of their water comes from metabolizing food. Animals digest food by oxidizing (essentially burning) it (Chapter 15). As part of this process, the oxygen an animal breathes is combined with hydrogen atoms in the food to produce water. Depending on the humidity, a kangaroo rat can get between 80% and 100% of its water through this metabolic process. Under conditions typically found in a desert burrow, a kangaroo rat takes in about 40% more water than it loses, without ever taking a drink or eating a succulent plant. This excess compensates for the water loss the rat experiences outside the burrow and enables it to live in the driest of deserts.

Adapting to Limited Shelter

Many species have specific needs for shelter and reproduction, such as suitable sites for burrows or nests. As we saw for the Kirtland's warbler, such sites may be rare finds throughout the habitat. There are several ways species respond to limitations on shelter and reproductive sites.

The yellow-bellied marmot (scientific name *Marmota flaviventris*) (**FIGURE 16.4**) lives high in the Rocky Mountains of Colorado. Marmots get shelter and reproduce in the larger cracks and crevices in the rocks, which they use as dens. Often there are not enough dens for each breeding female to have one of her own. A male marmot will claim and defend a suitable den, and two to five females will breed with this male and share the den. In this case, limitations on suitable sites for reproduction have resulted in a **polygynous** mating system, in which one male mates with several females. Since the sex ratio in marmots is 50:50, if one male monopolizes several females, some males will not have mates in any given year.

White-fronted bee-eaters (scientific name *Merops bullockoides*) (**FIGURE 16.5**) are birds that live in Central Africa and dig cavities in sandy riverbanks for

FIGURE 16.4 **Yellow-bellied marmots live high in the Rocky Mountains of Colorado.** They use cracks and crevices in the rocks as dens, but the limited number of suitable sites has resulted in a polygynous mating system, in which one male mates with several females.

FIGURE 16.5 **White-fronted bee-eaters are birds that live in Central Africa and dig cavities in sandy riverbanks for their nests.** Since there is often not enough room for all birds to nest, young bee-eaters frequently stay with their parents and help to raise another brood.

polygynous :: a mating system in which one male mates with several females

their nests. Since these sandy areas are limited in size, there is often not enough room for all birds to nest. Young bee-eaters frequently stay with their parents and help them raise another brood of birds by bringing food and helping with nest defense. These "helpers at the nest" provide a large benefit to their parents. Without them, almost half of a nest's hatchlings would die of starvation before they could fly. With them, survival rates can be three or four times higher.

16.2 Why Do Species Compete?

competition :: occurs when two or more species attempt to utilize the same resource; it has a negative effect on both species

exploitation :: an interaction between species in which one species benefits while the other is harmed

mutualism :: an interaction between species in which both species benefit

When individuals of different species come in contact, they interact in a variety of ways. **Competition** occurs when two or more species attempt to use the same resource. It has a negative effect on both species. In **exploitation**, one species benefits while the other is harmed. In **mutualism,** both species benefit. This is analogous to cooperation in humans: everybody wins.

TABLE 16.1 sums up the effects on each of the interacting species. In an ecological sense, benefit and harm are measured by their effects on abundance (population size) and distribution (where it can live). A species is harmed when the interaction reduces its population size or limits its geographic distribution. A species benefits when the interaction increases abundance or distribution.

We often speak of *competition, exploitation*, and *mutualism* quite apart from ecology. We think of competition as a sporting event, exploitation as a crime, and mutualism as active cooperation. The species involved in these interactions don't "think" of them in this way, however. Their interactions are simply the result of how each species makes its living.

The Competitive Exclusion Principle

When species compete, the resource must be in limited supply. All animals use oxygen, for example, but since usually there is enough to go around, this is not really competition. As Table 16.1 shows, competition is detrimental to both species since potentially it reduces the amount of the resource available to either species. With less of an important resource, it is more difficult for organisms to grow, survive, and reproduce, which can result in lower population size.

Since neither species benefits from competition, how can it be avoided? There are two ways this can happen: one competitor can win by driving the other species extinct, at least in the local area; or one or both species can change so that they no longer use the same resource. Competition tends to be an unstable interaction. In the long run, species tend to adapt in order to reduce or eliminate competition.

Table 16.1 :: The Effects of Species Interactions

	EFFECT ON:		
	Species 1	Species 2	Overall Effect
Competition	Negative	Negative	Competition has a negative effect on both species.
Exploitation	Negative	Positive	Exploitation benefits one species, while the other is harmed.
Mutualism	Positive	Positive	Mutualism benefits both species.

One of the first scientific studies of competition was performed by F. G. Gause in 1934. Gause worked in his laboratory with two microorganisms: *Paramecium aurelia* and *Paramecium caudatum* (**FIGURE 16.6**). Gause found that both species grew well when they were alone in their test tubes. However, whenever Gause grew them together, he found that *Paramecium caudatum* always died out. Gause reasoned that when grown together, the two species competed for food, and that *Paramecium aurelia* was the better competitor. Gause generalized his results, concluding that when species compete, the best competitor will always win and exclude the other species from the habitat. This conclusion is now known as the **competitive exclusion principle**, which says that two competitors cannot coexist.

The competitive exclusion principle holds up best under laboratory conditions. But even for microorganisms, a test tube is a relatively small world: there's only one food source, conditions in all parts of the tube are the same, and there really is no place to go. It's similar to throwing two fighters into the ring: they have no choice but to compete, and only one can win.

Conditions in nature are much more variable and dynamic, both in time and space. There are more species present, more food sources, and generally more ecological "freedom" for the species. If two species compete for the same food, it may be that one is the better competitor in warmer areas while the other is better in cold areas, for example. If both species were placed together in a constantly warm habitat, the competitive exclusion principle might hold, and one species would be driven extinct. In nature, however, the existence of variable temperatures may allow both species to coexist.

In nature, a species uses a whole suite of resources that promote growth, survival, and reproduction. The particular set of resources a species uses is called its **niche**. A niche includes biological factors like food sources and

competitive exclusion principle :: posits that two competitors cannot coexist

niche :: the particular set of resources a species uses to survive; includes biological factors like food sources and nesting sites and physical factors like temperature

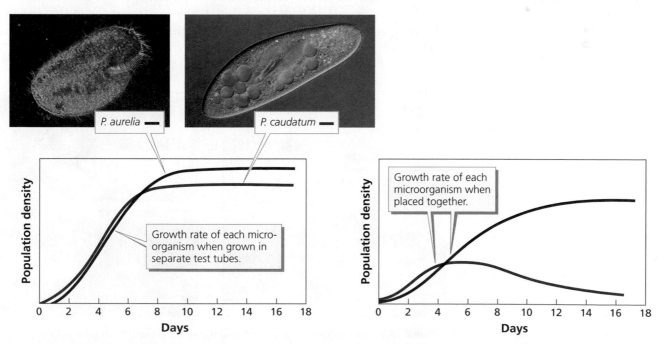

FIGURE 16.6 One of the first scientific studies of competition was F. G. Gause's 1934 research on two microorganisms, *Paramecium aurelia* and *Paramecium caudatum*. Gause discovered that both species grew well separately in test tubes, but when they were placed together, *P. caudatum* always died out.

nesting sites and physical factors like a minimum and maximum temperature. In short, a niche includes everything a species uses to survive. We can generalize the competitive exclusion principle to state that *two species cannot occupy the same niche*. However, they can vary aspects of their niche to reduce competition for some resources, even if they compete for others. In such cases, competitors may coexist. The following example shows how this process has worked on the Galapagos Islands.

Instances When Competitors May Coexist

The Galapagos Islands lie 600 miles off the west coast of Ecuador. They are home to 14 species of finch, made famous by Darwin and studied by many scientists up to the present day (see Chapter 9). Finches are primarily seed eaters, and the size of their beaks determines the size of seeds they can eat most easily. Finches with similar-sized beaks will compete for food.

Two similar species of Galapagos finch are the medium ground finch (*Geospiza fortis*) and the small ground finch (*Geospiza fuliginosa*) (**FIGURE 16.7**). The medium ground finch lives on Daphne Island, and the small ground finch lives on Los Hermanos. Scientists have measured the beak size of species on both islands, and they are about the same: 8–11 millimeters for the small ground finch, 8–12 millimeters for the medium. If these birds lived together, they would compete for the same seeds, but since they live on different islands, no competition occurs.

However, these two species do live together on Santa Cruz Island. Interestingly, on Santa Cruz both species have nonoverlapping beak sizes, which allow them to eat different seeds without competition. When species evolve a change in physical or behavioral characters that allows them to reduce competition, the process is known as **character displacement**. Generally, species tend to divide up resources in their habitat, reducing competition and allowing more species to coexist. See *Scientist Spotlight: Robert Helmer MacArthur*, for a closer look at a scientist who was interested in the problem of how species could coexist in a habitat without competitively excluding each other.

character displacement :: when species evolve a change in physical or behavioral characters that allows them to reduce competition

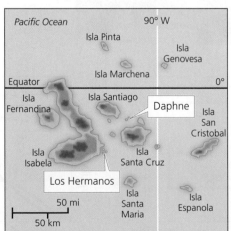

FIGURE 16.7 The medium ground finch and the small ground finch generally have similar beak sizes and live on different islands (Daphne Island and Los Hermanos, respectively), so they do not compete for food. However, on Santa Cruz Island, the finches live together and have developed different beak sizes so they can eat different seeds.

Scientist Spotlight

Robert Helmer MacArthur (1930–1972)

Most scientists find it a challenge to master their own field; it's the rare scientist that can master two. But sometimes it is precisely the blending of two separate disciplines that leads to new insight and new approaches that revolutionize a field. Robert Helmer MacArthur was one of these rare scientists, and his contributions transformed the field of ecology.

MacArthur was born in Toronto in 1930. His father was a professor of genetics who split his time between the University of Toronto and Marlboro College in Vermont. Like many ecologists, MacArthur's early passion was natural history. He particularly loved watching birds. In school, however, MacArthur studied mathematics. At 17, he moved to the United States, earning his bachelor's degree from Marlboro College and his master's in mathematics from Brown University. While pursuing his doctorate at Yale, MacArthur switched his field of study from mathematics to his earlier love, zoology. He earned his doctorate in 1957 and spent an additional year studying ornithology at Oxford University.

MacArthur was interested in the problem of how many species could "fit" into a habitat without competitively excluding one another. One of his earliest studies, now a classic, showed that several species of warblers could inhabit the same spruce tree by dividing up their foraging areas. One species foraged in the treetop, others in the inner branches, some in the outer branches, and others at the bottom.

MacArthur's main contribution, however, was to bring mathematical and experimental rigor to ecology. He developed mathematical models that showed the relationship between species number and habitat size that could be tested experimentally by other ecologists. Working with like-minded colleagues, MacArthur conducted research that provided a theoretical foundation for ecology. He published many papers involving both empirical and theoretical investigations of nature. In 1967, he and E. O. Wilson published *The Theory of Island Biogeography*, which addressed the question of why the number of species varied on islands of different size.

Robert Helmer MacArthur blended his expertise in mathematics and zoology to help revolutionize the field of ecology.

In 1971, MacArthur learned that he had cancer of the kidney and only a year or two left to live. He began working on his last book, *Geographical Ecology*, which summarized much of his life's work and identified open questions that ecologists are still studying. MacArthur died in November 1972. While his professional career lasted only 14 years, he helped to transform the science of ecology. In a memorial written by MacArthur's thesis advisor, G. E. Hutchinson, and colleague E. O. Wilson, they characterize a career cut short: "*Geographical Ecology* is both the reflective memoir of a senior scientist and the prospectus of a young man whose creative effort ended at the point of its steepest trajectory." ::

16.3 How Do Species Exploit One Another?

Recall that exploitation refers to an interaction in which one species benefits at the expense of another. Predation (in which a predator captures prey to eat), browsing and grazing (when organisms consume plants without killing them), parasitism (in which one organism lives off of another), and disease are all forms of exploitation. In competition, the best interests of both species are the same: each species benefits by avoiding the interaction altogether. In exploitation, the best interests of both species differ. The exploited species benefits from avoiding the interaction, while the exploiter depends on maintaining it.

Exploitation has two effects commonly observed in nature. First, the "population cycles" of both species can be linked: each population increases and decreases regularly over time. Second, an evolutionary "arms race" can develop: the exploited species develops better defenses, and the exploiter then adapts, too, with a better offense.

Population Cycles

Predator–prey systems provide good examples of population cycling, and it's easy to see why. Suppose both predator and prey start off with relatively small populations. In the absence of predators, the prey population can increase rapidly. But this provides more food for the predators, and their population size increases. As the number of predators increases, they begin to overexploit their prey, causing the prey population to decline. As prey decline, the predators begin to starve, and their population declines. This frees the prey to increase again (assuming not all are eaten), and the cycle continues.

One classic example involves the Canadian lynx (*Lynx canadensis*) and the snowshoe hare (*Lepus americanus*) (**FIGURE 16.8**). Both species live in northern North America, and both were extensively hunted for their pelts by fur traders of the Hudson Bay Company. For over a century, the Hudson Bay Company kept good records of how many pelts of each species it traded. Let's assume that the number of pelts taken in any given year is proportional to the size of the lynx and hare populations. We can then estimate the population size of both species over a 100-year period starting in 1845. The population cycles of both species are quite predictable: the lynx population rises just after that of the hare, and it declines just after the hare population begins to crash. These cycles have occurred every 10 years or so for over a century.

Many scientists have extended this example with more recent studies of their own, and as we learn more, the story becomes more complex. For one thing, the hare population cycle is driven not just by the lynx, but by food.

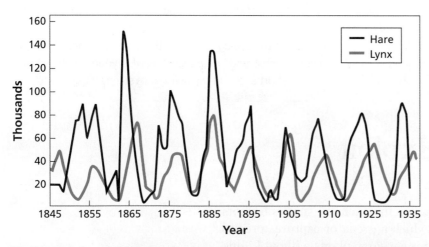

FIGURE 16.8 One classic example of population cycling involves the Canadian lynx and the snowshoe hare. Based on pelt trading records from the Hudson Bay Company, we know that the lynx population rises just after that of the hare and declines just after the hare population begins to crash; these cycles have occurred every 10 years or so for over a century.

As the hare population increases, hares can drastically overgraze the shrubs and twigs that provide winter food. That leads to another set of population cycles, with the *hare* as the exploiter. Also, other species besides the lynx eat snowshoe hares, and lynx have other prey as well. All these factors affect the lynx and hare populations.

An Evolutionary Arms Race

Exploitation can also create an evolutionary "arms race." Over time, the exploited species develops adaptations to prevent or avoid exploitation. The exploiter in turn develops adaptations that allow the exploitation to continue. This adaptation of one species to another is called **coevolution**.

Plants, for example, stay put, and they are loaded with vitamins and carbohydrates. That makes them an attractive target for exploitation by animals. Over long periods of time, however, plants coevolve in response. Many plants employ a chemical defense by making **secondary compounds**, chemicals that are usually poisonous and not needed by the plant for its metabolism. If an animal ingests these compounds, it could die, become sick, or become drugged and therefore vulnerable to its own predators. But some herbivores can adapt, too. Over time, they may develop resistance to the chemical. If other animals do not, they have the plant all to themselves. Of course, the plant may eventually develop a different secondary compound, and the arms race continues.

Milkweeds and monarch butterflies (**FIGURE 16.9**) are an example. Milkweeds are named for their thick, white sap. In many species of milkweed, this sap contains a variety of toxic chemicals known as cardiac glycosides. In high doses, cardiac glycosides speed up an animal's heart rate, sometimes to the point of cardiac arrest. As you might imagine, many animals avoid eating milkweed.

Monarch butterflies, on the other hand, eat nothing *but* milkweed, at least in their larval stage. Adult females lay their eggs only on milkweed plants. The larvae hatch into caterpillars that spend all their time living on—and eating—this milkweed plant. Monarch caterpillars have developed an interesting defense against the cardiac glycosides in their food. These chemicals are gathered up and stored in tiny compartments inside the caterpillar's cells. When the caterpillar metamorphoses into an adult butterfly, the cardiac glycosides remain there and cause no harm to it. However, a bird or other predator that eats a monarch caterpillar or butterfly is in for a rude surprise.

In this case, the monarch not only has won the arms race, it has coopted the milkweed's defensive "weapons" for its own defense. But no defense is perfect. The black-backed oriole of Mexico has developed a tolerance for cardiac glycosides and is one of the few predators that can enjoy a monarch snack.

A second example of the coevolutionary arms race involves nest parasitism, a behavior we introduced in the Kirtland's warbler case study. The European cuckoo is a common nest parasite in England. Unlike the Kirtland's warbler, some species recognize that the cuckoo's eggs are not their own and remove them from the nest. This response, known as **discrimination**, is a defensive adaptation to nest parasitism. The cuckoo, however, has developed a counter-adaptation. Female European cuckoos specialize in parasitizing the nest of only a single species. However, different females will specialize on different host species, so that a population of European cuckoos may parasitize half a dozen or more different hosts. Female cuckoos lay eggs that closely resemble, or mimic, the eggs of their host species (**FIGURE 16.10**). The close physical resemblance of an organism to another or to its surroundings is called **mimicry**. Egg mimicry is a coevolutionary response to discrimination.

FIGURE 16.9 Many plants, such as the milkweed plant, make secondary compounds that are poisonous to predators. The monarch butterfly, which eats only milkweeds, stores these compounds and uses them as a defense against other organisms.

coevolution :: the adaptation of one species to another

secondary compound :: chemical made by plants that is usually poisonous and that is not needed by the plant for its metabolism

discrimination :: when one species recognizes another as not its own

mimicry :: the close resemblance of an organism to another or to its surroundings

Cuckoo eggs

Host eggs

FIGURE 16.10 **The European cuckoo is a common nest parasite that lays eggs that generally resemble, or mimic, the eggs of its host species.** While the host will discriminate against cuckoo eggs, if the cuckoo's eggs are similar to its own, it may leave the eggs in its nest.

Although the host will discriminate against cuckoo eggs, if the cuckoo's eggs are similar to its own, it may sometimes mistakenly leave the cuckoo eggs in its nest.

Coevolution takes time. As mentioned previously, the Kirtland's warbler's lack of discrimination against cowbird eggs is probably due to the fact that the agricultural development that allowed the cowbird into the warbler's range is relatively recent. Kirtland's warbler may not have had time yet to evolve a discriminatory response.

16.4 When Can Species Cooperate?

Cooperation, or mutualism, is a win-win interaction: both species benefit. However, the species engage in mutualism just for that benefit, not to be nice to one another. We can look at mutualism as mutual exploitation. Over time, one species may even become a "cheater," keeping the benefits but no longer providing any to the other species.

Animals may act as mobile agents for plants, as pollinators, seed dispersers, and even defense against predators. Honeybees, for example, are legendary as hardworking pollinators. In return, the plant provides a renewable food source such as nectar, pollen, or fruit, or perhaps provides shelter.

Honeybees benefit plants by transporting pollen from one plant to another. This direct transport is efficient and allows the plant to reproduce without having to make massive amounts of pollen, saving the plant energy. The honeybee uses the plant's nectar and some of the pollen to make honey that feeds the hive's young.

Honeybees can pollinate many different species of plant, but some pollination systems are specific. Perhaps the ultimate example of specialized pollination involves fig trees and the fig wasps that pollinate them. There are 900 species of fig and 900 species of fig wasp. Each fig has its own wasp pollinator, and each wasp pollinates only one species of fig. Furthermore, the wasp reproduces only within the fig flower, and the fig flower is its only food. Neither species can reproduce without the other; the mutualism is obligatory. In these 900 cases, mutualism has become a dependency, and extinction of either pollinator or tree would almost certainly lead to extinction of the other species.

A mated female fig wasp will fly to the **inflorescence** (a plant structure containing many flowers) on her particular species of fig (**FIGURE 16.11**). As she crawls inside the inflorescence, she transfers pollen to the flowers, fertilizing them. She lays eggs on some of the flowers. These flowers develop a gall, a small compartment of plant tissue within which the larva hatches and on which it feeds. The other flowers develop normally and eventually produce

inflorescence :: a plant structure containing many flowers

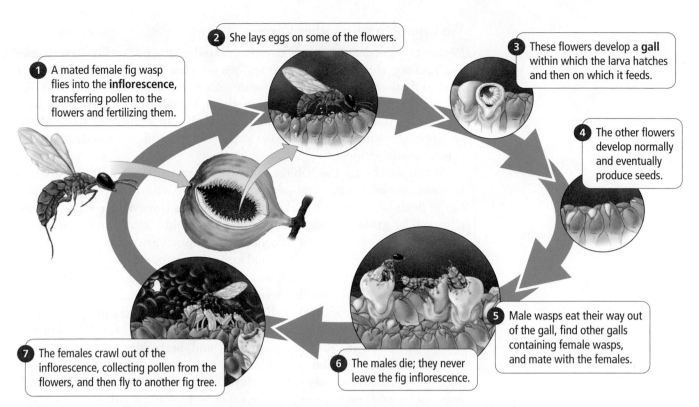

1 A mated female fig wasp flies into the **inflorescence**, transferring pollen to the flowers and fertilizing them.

2 She lays eggs on some of the flowers.

3 These flowers develop a **gall** within which the larva hatches and then on which it feeds.

4 The other flowers develop normally and eventually produce seeds.

5 Male wasps eat their way out of the gall, find other galls containing female wasps, and mate with the females.

6 The males die; they never leave the fig inflorescence.

7 The females crawl out of the inflorescence, collecting pollen from the flowers, and then fly to another fig tree.

FIGURE 16.11 The ultimate example of specialized pollination involves fig trees and fig wasps. There are 900 species of fig and 900 species of fig wasp; each fig has its own wasp pollinator.

seeds. Male wasps eat their way out of the gall, find other galls containing female wasps, and mate with them. Soon the males die; they never leave the fig inflorescence. The females crawl out of the inflorescence, collecting pollen from the flowers as they go. Then they fly to another fig tree of the correct species, and the cycle begins again.

There are a number of cases in which a plant–pollinator mutualism has broken down due to cheating. For example, robber bees chew holes at the base of a flower and extract the nectar without performing any pollination. Plants can be cheaters as well. Some species of orchid have flowers that resemble the females of particular species of bee or wasp (**FIGURE 16.12**). These flowers may also give off **pheromones**, chemical attractants similar to those used by females to attract males of their species. Males seek out these orchids and attempt to copulate with them. In so doing, they get covered with pollen. When they attempt to copulate with the next orchid, they transfer pollen and pick up some more. These orchids provide no nectar or other food for the pollinator, and clearly the "mating" will be unsuccessful as well. The wasp derives no benefit, but the orchid still benefits from pollination.

FIGURE 16.12 Some species of orchid, including _Ophrys speculum_ shown here, have flowers that resemble the females of particular species of wasp and may also give off pheromones that attract males. When the males attempt to copulate, they are unsuccessful, but the orchid benefits from pollination.

pheromone :: chemical attractant used by females to attract males of their species

16.5 How Do Ecological Interactions Affect Us?

Humans, of course, are part of the ecological community, too. We engage in interactions with other species, sometimes to our advantage and sometimes to our detriment. We also have learned to use the results of long-standing ecological interactions between other species for our own benefit.

We Compete with Other Species

We are very good competitors. Given our intelligence, our technological arsenal, and our ability to work together, we are able to compete effectively for land and other resources. When we develop large areas for our benefit—by cutting down a forest or draining a wetland, for example—we compete with the species that lived in these areas originally. In many cases like these, we apply the competitive exclusion principle: our competition drives many species to extinction.

When we change the habitat, we have a negative impact on the original resident species. But the new habitat opens up opportunities for other species. Large fields of cotton provide a haven for the boll weevil, for example, and silos full of grain provide good habitat for molds, mice, and seed-eating insects. The species we refer to collectively as pests are actually species that compete with us successfully for food and other resources.

We compete with weeds, insects, molds, birds, and mammals for the food we grow. Weeds are the fiercest competitors, though technically they compete with the crops rather than us, followed by insects and molds. The Environmental Protection Agency (EPA) estimates that $39.4 billion was spent worldwide in 2007 on pesticides and herbicides to combat these competitors—$11.8 billion in the United States alone (**FIGURE 16.13**). And this does not include the money spent each year on research to study crop pests. It does not include government programs that oversee and regulate research and development of pest management. It also does not include the cost of cleaning up pesticide pollution or the health costs related to exposure to pesticides.

In spite of our efforts, our competitors do very well. Estimates of food loss due to pests range from 20% to over 50% of the potential yield. The loss varies depending on the crop, where and how it is grown, and how the harvest is stored. Those of us in developed countries are largely buffered from these losses. We may experience them as an increase in price or decrease in selection at a supermarket. In a world where many people go hungry, however, losing even a small percentage of our food leads to loss of life.

We have historical evidence of another application of the competitive exclusion principle. There used to be several species of humans living at the same time in various parts of the world. Neanderthals (*Homo neanderthalensis*), for example, lived in Europe from about 200,000 years ago until about 28,000 years ago. Our species (*Homo sapiens*) initiated its migration from Africa approximately 80,000 years ago, arriving in Europe around 40,000 years ago. *H. sapiens* and *H. neanderthalensis* coexisted for 10,000 to

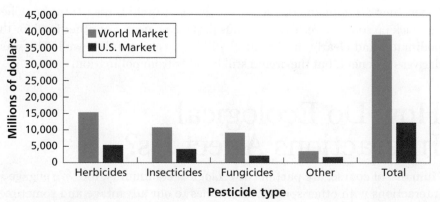

FIGURE 16.13 The Environmental Protection Agency (EPA) estimates that $39.4 billion was spent worldwide in 2007 on pesticides and herbicides—$11.8 billion in the United States alone.

Source: U.S. Environmental Protection Agency

15,000 years, but eventually *H. sapiens* was the only species remaining. There is good evidence that we were somewhat more flexible in our diet and tolerance to changing climatic conditions, and perhaps we were better hunters than the Neanderthals.

We Develop Useful Products and Ideas as a Result of Exploitation Interactions

In our prehistoric hunter-gatherer phase, we engaged directly in predator–prey interactions with many other species and may well have driven many prey species extinct. More recently, we have hunted species for food, pelts, fur, and sport, and we continue to hunt marine organisms such as fish and whales. However, today the bulk of our food comes from domesticated plants and animals. Although our exploitation interactions with natural populations continue, they have become less important to us as a source of food and material.

Exploitation interactions abound in nature, however. As we have seen, these interactions can lead to an evolutionary arms race and result in many adaptations. We have been exceptionally adept at taking advantage of these adaptations, both defensive and offensive, as evidenced by our use of plant secondary compounds.

The cardiac glycosides produced by milkweed to ward off herbivores can have a therapeutic function. They cause the heart to beat more forcefully, and in the proper dose they can be used to treat a weakening heart. Cardiac glycosides have been used to treat heart conditions for over 3000 years.

Young coffee trees produce high concentrations of caffeine. At these high concentrations, caffeine can be lethal to many insects and fungi that would eat parts of the seedling. Caffeine that diffuses into the soil can inhibit seed germination near the seedling, reducing competition from other plants. Caffeine is concentrated in coffee beans, the seed of the coffee plant, to deter herbivores from eating the next generation of coffee plants. Every time you drink your morning java, you reap the benefit of millions of years of defensive evolution in the coffee plant.

The compounds in spices such as pepper, nutmeg, mustard, cloves, cinnamon, mint, and wintergreen evolved as chemical defenses. Many spices have been used for medicinal purposes as well as flavoring. Scientists have isolated many of the chemical compounds responsible for the aroma or taste of spices and have found that they work to repel insects or kill bacteria. Intriguingly, spices are most heavily used in tropical and subtropical regions where food spoilage due to microorganisms is a significant threat.

The religious and recreational use of hallucinogenic drugs from plants and fungi is an ancient practice. For animals living in nature, an altered state of consciousness likely leads to a quick death from predators. This is an unlikely fate for humans, but the use of hallucinogenic and other plant drugs has its own set of physiological and legal consequences. Although developed as an herbivore defense, some of these drugs do have therapeutic value despite their dangers, such as the analgesic properties of morphine from the opium poppy.

Our understanding of exploitation interactions in nature can change our approach to pest control. Our pests have predators and parasites of their own. If we can successfully introduce these predators and parasites, they may take care of our pests for us, a procedure known as **biological control**. This is a promising approach, but it can be difficult to predict *all* of the possible consequences of introducing "pest-controlling" species into an environment. The coevolutionary arms race holds a lesson for us: it is a difficult, if not impossible, race to win.

biological control :: an approach to controlling pests by introducing predators of a pest into an area

Technology Connection

Biological Control of Schistosomiasis

When you think of medical technology, you probably don't think about crayfish. Nevertheless, crayfish may be part of the solution to a disease that affects over 200 million people annually: *schistosomiasis*.

Schistosomes are parasitic flatworms, or flukes, about a centimeter long. They burrow into human skin, find a blood vessel, and travel throughout the circulatory system. Males and females mate in the liver, and, depending on the species, a female can produce between 300 and 3000 eggs each day. Many of the eggs pass into the intestines and are eliminated with feces. But many eggs also get trapped in body tissues. The worms themselves cause little damage, but as the body's immune response attacks the eggs, the surrounding tissue can be damaged.

However, this is only half the story. Schistosomes have two hosts: one human and one a freshwater snail. When feces infected with schistosomes contaminate a pond or lake, the schistosomes hatch into swimming larvae that infect snails. Inside the snails, they reproduce asexually to make vast numbers of individuals called cercariae that break free of the snail. The cercariae swim and attach to the skin of humans who play or bathe in the infected water, completing the schistosome life cycle.

This is where the crayfish comes in. The Louisiana swamp red crayfish is a voracious predator of freshwater snails. If the snails are removed, schistosomes can't complete their life cycle, and humans will no longer get infected.

Bruce Hofkin has been studying the interaction of crayfish, snails, and schistosomes since the 1990s, and he has found that lakes and ponds that contain the crayfish rarely contain any snails. In an experiment, Hofkin compared infection rates at two schools. At the experimental school, crayfish were introduced into a nearby pond where students played. At the control school, no crayfish were introduced. Hofkin found that the crayfish established viable populations in the experimental pond. He also saw that the snail populations declined there, and students at the experimental school were significantly less likely to be infected with schistosomes.

Biological control can be a very effective technology, both medically and in agricultural systems to control pests.

A

B

Schistosomes (A) are parasitic flatworms that burrow into human skin and travel throughout the circulatory system. The Louisiana swamp red crayfish (B) is a voracious predator of freshwater snails, and if the snails are removed, Schistosomes can't complete their life cycle.

It is much cheaper than medical treatment or chemical control, and it reduces the need for drug treatment and the amount of pesticides introduced into the ecosystem. A good deal of research is needed before using biological control to make sure the control agents don't become worse than the pests or parasites they control. ::

The arms race between plants and herbivores has been going on for millions of years, and still neither side has won. By using pesticides and antibiotics, we have started an arms race of our own. Will we be able to defeat pests and pathogens this way, or will we simply encourage better defensive adaptations on their part? For another example of how biological control is being used, see *Technology Connection: Biological Control of Schistosomiasis.*

We Capitalize on Mutualisms

We have learned to capitalize on the results of mutualisms, just as we have on the results of the evolutionary arms race. A prime example is the mutualistic origin of fruit.

Technically, a fruit is the structure that contains a plant's seeds (Chapter 7). The fruit develops from the ovary of a flower, where the seeds form. The fruits we are most familiar with—for example, apples, oranges, and mangoes—are sweet and juicy. Fruits like these aid the plant in dispersing its seeds.

Mammals and birds eat the fruit, gaining nutrition and energy. The seeds are indigestible. They pass through the animal, and are deposited away from the parent plant with a dollop of fertilizer to help the seeds get started. Animal seed dispersal is an old plant–animal mutualism that results from the evolution of fruit. We have helped this process along by breeding larger and tastier fruits. But the fruits were already there, the result of a long-standing ecological interaction.

16.6 What Does a Functioning Ecosystem Do?

An ecosystem is like an engine, but how does it get its fuel? Interactions between species distribute the nutrients they need to survive. However, species also recycle the materials of planet Earth on a global scale to keep ecosystems in balance.

Exploitation Interactions Distribute Energy and Nutrients

Ultimately, all the energy in an ecosystem comes from the sun. Plants capture sunlight in the form of energy-rich chemicals, sugars, and starches (Chapter 7). Plants use these chemicals to drive their own metabolism. When an herbivore eats a plant, energy and nutrients are transferred, and when a carnivore eats an herbivore, they are transferred yet again. Ultimately, all living organisms die, and their bodies decompose, returning nutrients to the soil for plants to use. The ecological interaction of predation distributes energy and nutrients throughout the ecosystem, providing it with fuel.

If we make a diagram of "who eats whom (or what)," we get a multilinked structure called a **food web** (**FIGURE 16.14**). We can think of a food web as being divided into levels. At the first level are photosynthetic organisms. These are known as **primary producers** because they are the first link in capturing solar energy. Next we have herbivores, or **primary consumers**. Then several levels of carnivores can exist: these are known as **secondary consumers**, tertiary consumers, and so on. Finally we have **decomposers**, organisms that cause dead bodies to rot, which are the ultimate level of any food web. These various levels that organisms occupy on the food web are known as **trophic levels**. Trophic levels are helpful guidelines, not firm rules. Many animals eat at more than one level; for example, bears and humans eat both fish and berries.

Energy flow through a food web is not 100% efficient, and energy is lost every time one organism eats another. On average, an organism stores only about 10% of the energy content that it eats. Although this figure varies a lot, it means that about 90% of the energy in food is lost as indigestible material and heat. And this loss affects the entire food web: it means that the amount

food web :: a diagram showing a multilinked structure of "who eats whom (or what)"

primary producer :: photosynthetic organism that is the first link in the food web in terms of capturing solar energy

primary consumer :: herbivore

secondary consumer :: carnivore

decomposer :: organism that causes dead bodies to rot

trophic level :: level that organisms occupy on the food web

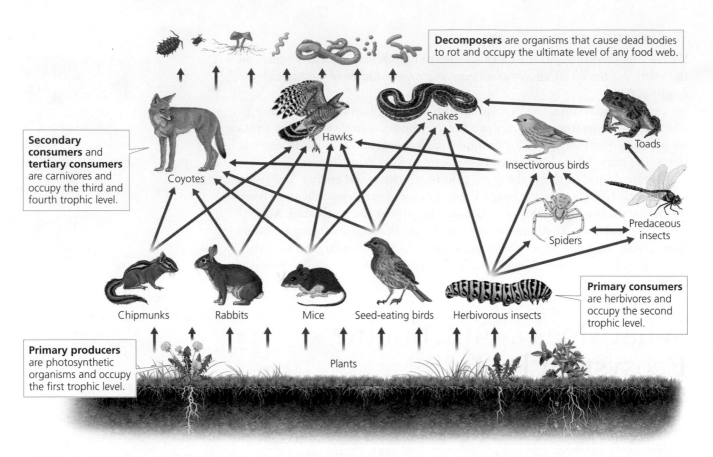

Decomposers are organisms that cause dead bodies to rot and occupy the ultimate level of any food web.

Snakes

Hawks

Toads

Insectivorous birds

Secondary consumers and **tertiary consumers** are carnivores and occupy the third and fourth trophic level.

Coyotes

Predaceous insects

Spiders

Chipmunks   Rabbits   Mice   Seed-eating birds   Herbivorous insects

Primary consumers are herbivores and occupy the second trophic level.

Primary producers are photosynthetic organisms and occupy the first trophic level.

Plants

FIGURE 16.14 A food web is a multilinked structure that identifies "who eats whom (or what)." Organisms occupy different trophic levels on the food web, but these are not firm, as many animals eat at more than one level.

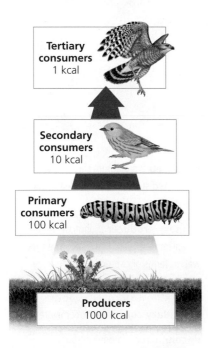

Tertiary consumers
1 kcal

Secondary consumers
10 kcal

Primary consumers
100 kcal

Producers
1000 kcal

FIGURE 16.15 An ecological pyramid shows the change in available energy at the different trophic levels. Population sizes decrease dramatically at higher trophic levels because less energy is available and fewer individuals can be supported.

of energy available decreases dramatically at higher trophic levels. Primary producers capture about 1% of the sunlight that hits their leaves, primary consumers get about 10% of that, secondary consumers get 10% of that, and so on up the trophic levels.

The availability of less energy at higher trophic levels means that fewer individuals can be supported. Population sizes decrease dramatically at higher tropic levels. We can picture the change from level to level with an **ecological pyramid** (**FIGURE 16.15**).

Ecosystems Recycle Material on a Global Scale

All of the matter on Earth is distributed among four reservoirs—the biosphere, atmosphere, hydrosphere, and lithosphere. The biosphere is the global ecosystem and encompasses all life on the planet. The remaining reservoirs are nonliving and refer to the gaseous, liquid, and solid parts of the Earth. **Biogeochemical cycles** are processes that move matter among these four reservoirs. The *bio* in *biogeochemical* tells us that the living organisms in ecosystems are major players in this recycling process. If recycling is disrupted, some things begin to build up while others become depleted, which can cause trouble on a global scale. Almost every chemical important to life is part of a biogeochemical cycle. In this section, we will discuss three of these cycles: carbon, oxygen, and nitrogen. See *How Do We Know? Long-Term Ecological Research* to learn how biogeochemical cycles are studied in nature.

The Carbon Cycle The carbon cycle refers to the movement of carbon throughout the ecosystem (**FIGURE 16.16**, p. 490). The drivers of the carbon

How Do We Know?

Long-Term Ecological Research

Ecology operates on a moderate time scale. It can take decades for changes to occur in an ecosystem, which makes studying ecosystems a challenge. In 1980, the National Science Foundation (NSF) established the Long Term Ecological Research (LTER) network, a collaboration of almost 2000 researchers at 21 sites throughout the United States and 2 in Antarctica, dedicated to studying ecosystem ecology on a large spatial and temporal scale.

One of the partners in the LTER network is the Hubbard Brook ecosystem study, located in the White Mountains of New Hampshire. The Hubbard Brook watershed is a valley that spans about 8 × 5 kilometers. Hubbard Brook flows east through the valley, and smaller tributaries flow into it from the south or north, carving out a number of smaller valleys that join up with the main valley.

Since the 1960s, researchers at Hubbard Brook have been studying nutrient and mineral cycling in the northeast hardwood forest ecosystem. They measure the input of water and dissolved minerals in rainfall, track their movement through the plants and trees in the forest, and measure the outflow of water and minerals in the forest streams. Because the forest sits on impermeable bedrock, any nutrients or minerals leaving the ecosystem do so in stream outflow. Researchers also have been monitoring the effects of acid rain, since acidified water and soils have a large effect on nutrient and mineral uptake by plants.

In a study designed to test the effect of plants on nutrient and mineral retention, Herbert Bormann of Yale and Gene Likens of Cornell worked with two adjacent valleys. Each valley had a single stream that flowed into Hubbard Brook. One valley was the control. In the experimental valley, they removed all the vegetation—every tree and every plant—and continuously removed all new plant growth during the three-year study, leaving them to decompose. In both valleys,

Hubbard Brook White Mountain National Forest in New Hampshire is the original site of the oldest ecological study in the United States and a model for research on watersheds.

they monitored the rainfall and the level of calcium, potassium, nitrogen, and other nutrients. They built a small spillway at the mouth of each stream and measured the volume of water leaving each valley as well as the concentrations of dissolved calcium, potassium, and nitrogen.

Their results were striking. In the control valley, plants took up about 40% of the rain. Virtually all of the calcium and potassium stayed in the forest, and nitrogen levels increased slightly. In the experimental valley, 100% of the rain left via the stream. Mineral and nutrient loss through the stream increased dramatically: a 10-fold increase for calcium, 20-fold increase for potassium, and a 60-fold increase for nitrogen. So much nitrogen was added to the stream that it encouraged the growth of algae, making the water undrinkable.

Long-term studies like the one at Hubbard Brook help to confirm the role forests play in maintaining soil fertility and reducing water pollution caused by runoff. ::

---

cycle are photosynthesis and respiration. Recall that photosynthetic organisms remove carbon dioxide from the atmosphere and convert it into energy-rich carbohydrates (Chapter 7). This is half of the cycle—movement of carbon from the nonliving reservoir of the atmosphere into the living reservoir. The other half of the cycle occurs through respiration. As organisms break down carbon compounds metabolically, they release carbon dioxide back into the atmosphere, completing the cycle (Chapter 15).

The Oxygen Cycle The oxygen cycle refers to the movement of oxygen throughout the ecosystem, and it is the mirror image of the carbon cycle (**FIGURE 16.17**). Oxygen is added to the atmosphere by photosynthesis and removed from the atmosphere by respiration. Burning also consumes

ecological pyramid :: a diagram of the changes in trophic level

biogeochemical cycle :: process that moves matter among the four reservoirs: the biosphere, atmosphere, hydrosphere, and lithosphere

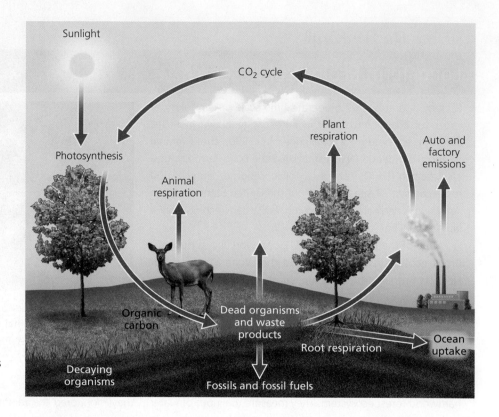

FIGURE 16.16 **The carbon cycle refers to the movement of carbon throughout the ecosystem.** The drivers of the carbon cycle are photosynthesis and respiration.

atmospheric oxygen, but since oxygen is so much more abundant in the atmosphere than carbon dioxide (oxygen makes up 21% of the atmosphere, carbon dioxide only 0.04%), fires and burning do not deplete oxygen appreciably.

The Nitrogen Cycle The nitrogen cycle refers to the movement of nitrogen throughout the ecosystem (**FIGURE 16.18**). Nitrogen's cycling is more complex than that of carbon or oxygen, and it involves a wider variety of species

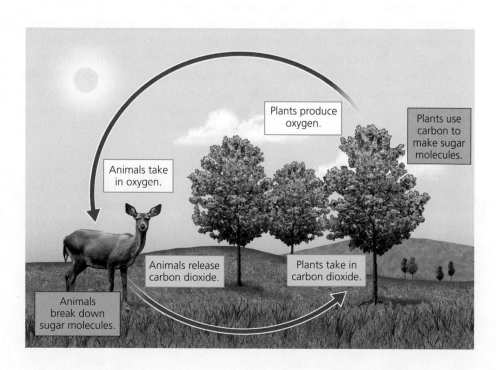

FIGURE 16.17 **The oxygen cycle refers to the movement of oxygen throughout the ecosystem.** It is the mirror image of the carbon cycle; oxygen is added to the atmosphere by photosynthesis and removed from the atmosphere by respiration.

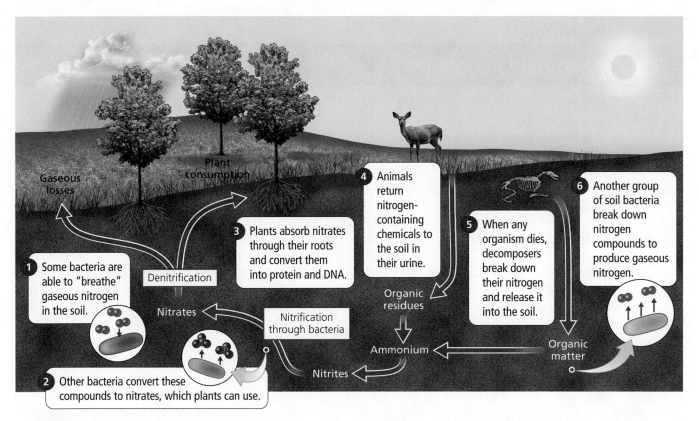

FIGURE 16.18 The nitrogen cycle refers to the movement of nitrogen throughout the ecosystem. Soil bacteria make nitrogen available in a two-step process.

working on a sort of biochemical assembly line. Nitrogen makes up 79% of the atmosphere, making it the most abundant gas. However, no plant or animal can use atmospheric nitrogen. This ability is found only in a number of species of bacteria that live in the soil. Since all living organisms require nitrogen to make protein and DNA, we are all dependent on the activity of these bacteria.

Soil bacteria make nitrogen available in a two-step process. First, some bacteria are able to "breathe" gaseous nitrogen in the soil and capture it in nongaseous chemicals similar to ammonia. Second, other bacteria convert these ammonia-like compounds to nitrates, which plants can use. Plants absorb nitrates through their roots and convert them into protein and DNA. Nitrogen compounds move throughout the food web, providing a usable source of nitrogen for all species. Animals return nitrogen-containing chemicals to the soil in their urine, and when any organism dies, decomposers break down their nitrogen and release it into the soil. To complete the cycle, another group of soil bacteria breaks down nitrogen compounds to produce gaseous nitrogen, which is returned to the atmosphere.

16.7 What Benefits Do We Get from a Functioning Ecosystem?

This is the most significant question for the study of ecology. The natural world provides us with numerous examples of ecology that are interesting to know about and study. Ultimately, however, we face the question of *preserving*

Life Application

Ecology and Human Conflict

The causes of human conflict are many and varied. There is probably no single cause for any major conflict, and seeking one inevitably leads to oversimplification. However, ecological pressure in conjunction with other factors can lead to conflict.

In modern times, it is difficult to find a more disturbing case of conflict than the genocide in Rwanda. This is a complex case, and the causes are many, but Jared Diamond in his book *Collapse* analyzes the ways that a failed ecology contributed to the conflict.

Between April and July 1994, approximately 800,000 Rwandans, about 10% of the population, were murdered, making it the third-largest genocide after the Holocaust. The genocide is usually portrayed as the result of ethnic hatred between the Hutu and the Tutsi, the two major ethnic groups in Rwanda, but that is not the whole story. Diamond reports on a study of the Kamana region of Rwanda, a region that consisted only of Hutu, but where genocide still occurred.

The population density of Rwanda in 1991 was 760 people per square mile, the highest of any African nation. In the Kamana region at the time, the population density was 2040 per square mile, one of the highest in the world. Rwanda has fertile volcanic soil, is at a sufficiently high elevation to receive significant rainfall, and is cool enough to be free of two of Africa's scourges: malaria and tsetse flies. Along with the introduction of modern medical techniques, the habitat supported rapid population growth, leading to the densities just mentioned. However, modern agricultural practices were not introduced. Rwandans continued to farm the way they always had. A combination of deforestation, overgrazing, and unsound techniques for tillage led to a nutrient depletion in the soil and loss of soil due to erosion.

As the land became poorer, the population increased and the average size of a family farm decreased proportionally. In Kamana in 1993, half of the farms spanned less than 0.75 acres, and these were split among family members so that a single family might have only one-tenth of that to farm.

With all the land taken, there was no chance for new families to start out with a farm of their own. Two-thirds of the young women and all of the young men stayed at home on the family farm, further stretching its resources and most likely generating a fair amount of tension and conflict. These small, poor-quality farms simply could not feed the families that owned them.

While most younger families in Kamana had small farms, a few older families had much larger ones. Farms owned by people in their 20s averaged only 0.4 acres, while farms owned by people in their 50s averaged 2 acres. Conflicts and disputes over land increased and became more violent up until the genocide started. When the killing began in Kamana in 1994, at least 5% of the population died. Researchers who studied Kamana concluded that the residents saw this as a chance to settle old scores and conflicts and to redistribute scarce farmlands among the survivors. While no one argues that ecological pressure was the only cause of the killing in Kamana, evidence suggests that it was an important contributing factor. ::

The causes of human conflict are many and varied, but ecological pressure in conjunction with other factors may contribute to conflicts, such as the one that occurred in Rwanda.

ecology. Interest alone does not provide sufficient motivation for the hard and costly work of preservation. In addition, we need to consider the economic benefits of a functioning ecosystem, as well as the costs of a dysfunctional ecosystem. These costs include the possibility of human conflict as we compete for ever-scarcer resources (see *Life Application: Ecology and Human Conflict*).

Ecosystem Services

We can think of the benefits we get from our ecosystem as **ecosystem services**: naturally occurring ecological processes that support our agriculture, technology, culture, and more. These services can be divided into four groups:

- *Provisioning:* these services provide us with food, fresh water, natural fibers, fuel, and biochemical compounds. With modern genetic techniques, even the genes of other organisms provide a useful resource.
- *Regulating:* these services regulate and stabilize the ecosystem, making it more reliable. They include regulating pest and disease control through the checks and balances discussed in this chapter: control of runoff and erosion, pollination, water and air purification, and waste decomposition.
- *Cultural:* these services relate to quality-of-life issues and are uniquely human. They include the recreational, aesthetic, and inspirational value of natural areas and the cultural, spiritual, and religious connections people have to nature. Cultural services also include the joy of learning about nature and the opportunities it provides for scientific discovery.
- *Supporting:* these are services on which all the others depend. They include the capture of solar energy by primary producers, biogeochemical cycling, and the production of soil and oxygen.

ecosystem service :: a naturally occurring ecological process that supports our agriculture, technology, culture, and more; can be divided into four groups: provisioning, regulating, cultural, and supporting

Scientists, conservationists, economists, and politicians are working together to measure the benefits we receive from ecosystem services. We already assign an economic value to provisioning services, since they provide products we can use or sell. Historically, though, we have placed little economic value on the other services, implicitly assuming that they are free and will always be there. We realize now that if we disrupt an ecosystem service, it will cost money to repair or to replace it. It may also cost lives. Honest, accurate bookkeeping requires that we take these costs into account in our development projects.

Disruptions to Ecosystem Services

Soil production is a vital ecosystem service, one on which all terrestrial life depends. New soil constantly forms at a rate controlled by the weathering of exposed rocks and the decomposition of dead bodies. Soil lost by erosion enters our streams and lakes as sediment, eventually reaching the ocean, where new rock is formed. The main defense against soil erosion is vegetation. The network of plant roots helps to hold soil in place. However, our agriculture, forestry, and development practices frequently reduce or remove vegetation cover. We have shifted the balance between soil loss and soil erosion dramatically: on a global scale, we lose about 7.5 to 10 tons of soil per acre annually. It is critical that development projects that may increase erosion take the cost of replacing soil into account. If not, the projects will appear to be more economically feasible than they really are.

Our exploitation interactions with other species can also disrupt ecosystem services and the food web. If we overexploit a predator species—by hunting, for example—the herbivores it normally eats can increase their population size. This in turn can have a negative effect on the plant populations the herbivores consume. Instability of the food web can lead to ecosystem collapse, which in turn reduces the ecosystem's ability to provide its services.

Finally, our activity can alter biogeochemical cycles. We can't do anything to change the total amount of carbon or nitrogen on the planet. But we can have a dramatic effect on where these compounds are stored. Burning forests and fossil fuels (which come from ancient forests) adds more carbon to the atmosphere than photosynthetic organisms can remove. This additional atmospheric carbon then becomes a cause of global climate change.

We also affect the nitrogen cycle. Often there is not enough nitrogen in the soil to support the high rate of plant growth our agriculture requires. Nitrogen is the main component of the fertilizers we apply to our crops. We have developed our own chemical process for capturing nitrogen from the atmosphere, and in fact the majority of nitrogen in our soil now comes from fertilizer rather than soil bacteria. Factories that make nitrogen fertilizer must burn fossil fuels to do so, which increases our impact on the carbon cycle. Excess nitrogen in the soil leaches into the watershed, promoting the growth of algae and altering the aquatic ecosystem.

Biodiversity

biodiversity :: the variety of species living in an ecosystem

There is a common denominator that relates to the health of an ecosystem, its ability to provide services, and the disruption human activity can cause: **biodiversity**. All of the ecological processes discussed in this chapter depend on having a diverse set of species present in the ecosystem. Reducing biodiversity is one of the most common results of development, and maintaining biodiversity may be our greatest ecological challenge. We will take a closer look at biodiversity in the next chapter.

 # Biology in Perspective

We benefit greatly from the ecological interactions among species. Ecological interactions account for many of the adaptations we observe in nature. These adaptations evolve when species develop tolerance for extreme environments. They evolve, too, when species compete, when they get better at catching prey or avoiding capture, and when they cooperate. Many of these adaptations are directly useful to us, providing food or medicine.

We also benefit greatly from the work that functioning ecosystems perform. Food webs fuel the ecosystem and mediate energy and nutrient flow. Biogeochemical cycles move important chemicals between living and nonliving reservoirs. Taken together, interactions among species enable an ecosystem to provide vital services that give us goods, maintain ecosystem health, and preserve the conditions necessary for life.

Our survival depends on the benefits the ecosystem provides. These benefits may seem free of charge, but if we disrupt them, we will pay for them one way or another. Maintaining these services intact is an economic imperative, as well as an ethical and cultural need.

Chapter Summary

16.1 How Do Species Adapt to Their Habitat?

Provide examples of the ways species adapt to the constraints of their habitats.

- Kangaroo rats live in North American deserts and have many physiological adaptations to conserve water, such as being nocturnal and having efficient kidneys.
- Marmots, large rodents in the Rocky Mountains, share dens to adapt to limited shelter.
- White-fronted bee-eaters are birds that nest in limited cavities; year-old offspring are often nest helpers for their parents.

16.2 Why Do Species Compete?

Summarize how species can coexist without one outcompeting the other.

- Competition occurs when two or more species attempt to use the same limited resource: in exploitation, one species benefits and one is harmed; in mutualism, both species benefit.
- The competitive exclusion principle holds that two species cannot share the exact same niche (complete set of physical and biological resources); one will always outcompete the other.
  - o Aspects of niches can be altered to reduce competition so competitors may coexist.
  - o Character displacement occurs when species evolve a change in physical or behavioral characteristics that allows them to reduce competition and coexist.

16.3 How Do Species Exploit One Another?

Describe the ways that species may exploit one another.

- In exploitation (predation, parasitism), one species benefits at the expense of another.
- Predator–prey systems are an example of population cycling: as prey increase, the predator population increases, but then as predators consume prey, the prey population declines; the lynx and snowshoe hare are an example.
- In an evolutionary arms race (coevolution), the exploited species develops adaptations to prevent or avoid exploitation and then the exploiter develops adaptations to counteract these; milkweed plants and cuckoo bird nest parasitism (mimicry) are examples.

16.4 When Can Species Cooperate?

Identify circumstances of cooperation between species.

- In cooperation or mutualism, both species benefit from the interaction.
- Honeybees transport pollen and in return receive nectar.

- Certain species of fig wasp have coevolved with certain species of figs, and neither can reproduce without the other.
- Sometimes species "cheat": for example, orchids attract bees with chemical pheromones, and bees collect their pollen but do not receive any nectar in return.

16.5 How Do Ecological Interactions Affect Us?

Summarize the beneficial and detrimental ecological interactions that we have with other species in the ecosystem.

- Humans are good competitors for land and other resources; we are driving many species to extinction.
- When we alter a habitat, we may drive the original inhabitants out but make it better suited to others.
- There were several species of humans living on Earth, but eventually *Homo sapiens* outcompeted them all.
- Humans don't directly exploit as many organisms now that most of our food is domesticated, but we outcompete many for habitat.
- We take advantage of many chemical adaptations of plants, including cardiac glycosides from milkweed to treat heart problems, caffeine from coffee plants, spices used for medicine and flavoring, and drugs such as opiates and hallucinogens.

16.6 What Does a Functioning Ecosystem Do?

Describe the movement of energy and nutrients through an ecosystem.

- Energy in an ecosystem comes from the sun.
- The food web based on the sun's energy is composed of trophic levels: primary producers or plants, primary consumers or herbivores, secondary consumers or carnivores, and decomposers.
- The ecological pyramid reflects the fact that 90% of energy is lost at each trophic level.
- Biogeochemical cycles recycle the materials in an ecosystem among four reservoirs: the biosphere, the atmosphere, the hydrosphere, and the lithosphere.
  - o During the carbon cycle, photosynthesis removes carbon dioxide from the atmosphere and respiration adds it.
  - o During the oxygen cycle, photosynthesis adds oxygen and respiration removes it from the atmosphere.
  - o During the nitrogen cycle, plants absorb nitrates from the soil, and nitrogen is released back into the soil through decomposition and converted back to gaseous form for release into the atmosphere.

16.7 What Benefits Do We Get from a Functioning Ecosystem?

Enumerate the benefits we get from a functioning ecosystem.

- Ecosystem services are naturally occurring ecological processes that support our agriculture, technology, and more; there are four types: provisioning, regulating, cultural, and supporting.
- We disrupt ecosystem services as we use land for our own purposes or alter biogeochemical cycles.
- Biodiversity is the common denominator that relates to the health of an ecosystem; all ecological processes depend on having a diverse set of species.

Key Terms

biodiversity 494
biogeochemical cycle 489
biological control 485
character displacement 478
coevolution 481
competition 476
competitive exclusion principle 477
decomposer 487
discrimination 481

ecological pyramid 489
ecology 472
ecosystem service 493
exploitation 476
food web 487
inflorescence 482
mimicry 481
mutualism 476
niche 477

pheromone 483
polygynous 475
primary consumer 487
primary producer 487
secondary compound 481
secondary consumer 487
trophic level 487

Review Questions

1. What are the ecological factors that make Kirtland's warbler a rare species?

2. What are the physiological and behavioral adaptations that allow kangaroo rats to live in the desert?

3. What is the competitive exclusion principle?

   a. A conclusion that says two competitors may peacefully coexist

   b. A conclusion that says two competitors cannot coexist

   c. A conclusion that says both species benefit

   d. A conclusion that says one species benefits while the other is harmed

   e. A conclusion that says that if two or more species attempt to utilize the same resource in a habitat, it will have a negative effect on the species that is newer to the habitat

4. What is a niche?

   a. The particular set of resources a species uses

   b. A mating system in which one male mates with several females

   c. An example of population cycling

   d. An evolutionary arms race

   e. A plant structure containing many flowers

5. Pick one of your favorite organisms and list some of the factors that determine its niche.

6. Why does exploitation (particularly predation) often lead to population cycles?

7. What causes a coevolutionary "arms race"?

8. Describe a mutualistic interaction between two (or more) species.

9. What is a food web?

   a. A food chain

   b. Helpful guidelines about which animals are most likely to be prey

   c. Processes that move matter among the biosphere, atmosphere, hydrosphere, and lithosphere

   d. A diagram of "who eats whom (or what)" in a multi-linked structure

   e. An ecological pyramid showing decreases in population size

10. What are trophic levels?

11. What is an ecological pyramid, and why does it narrow toward the top?

12. Draw a labeled diagram of the following: (a) the carbon cycle, (b) the oxygen cycle, and (c) the nitrogen cycle.

13. How is human activity altering the carbon cycle and the nitrogen cycle?

14. What is the original ecological function of spices found in plants like mint and cloves?

15. List the four types of ecosystem services.

    a. Biogeochemical, regulating, provisioning, and supporting

    b. Inflorescence, provisioning, cultural, and supporting

    c. Provisioning, regulating, cultural, and supporting

    d. Provisioning, biogeochemical, cultural, and supporting

    e. Biogeochemical, regulating, cultural, and supporting

The Thinking Citizen

1. Reread the discussion about the small and medium ground finches on the Galapagos Islands. If you assume that beak size controls the size of the seeds these birds eat, are there any islands on which these birds would eat the same seeds? If not, what is the evidence that these birds compete?

2. A typical food web has many pathways through which energy and nutrients flow, and a given species may be involved with more than one pathway. However, no pathway contains many steps. The number of species involved in the path from primary producer to primary consumer to secondary and higher consumers rarely exceeds four or five. What factors limit the length of a trophic pathway? Under what conditions or in what types of habitats would you expect to see the shortest pathways? The longest?

3. In some habitats with few species, the food web may become closer to a food *chain*: only a couple of herbivore species and even fewer carnivores. Compare the effects of losing one species in a food chain to losing one species in a more typical food web. Which is more stable (less likely to crash), a food chain or food web? Generally speaking, how are stability and number of species related, and why does this relationship exist?

4. Discuss the potential effects of the following on the reservoir of atmospheric carbon: (a) cutting down large tracts of rain forest and exporting the lumber; (b) burning fossil fuels on a global scale; and (c) warming of the surface waters of the oceans. (Hint: gases become less soluble in water as its temperature increases.) Think in terms of the input to and output from the atmospheric carbon reservoir: how might these activities change the balance?

5. Select one of the four types of ecosystem services. What ecological interactions among species allow the service to be delivered?

The Informed Citizen

1. Suppose you lived in a rural community that was considering reintroducing wolves into the ecosystem. What benefits and other consequences would you expect? What factors would be most important to you and your neighbors in reaching a decision? What other information would you need to know to make a good decision?

2. Agricultural pests compete with us successfully for our food crops. Pesticides are a major component of our competitive strategy, although they are not a complete or perfect solution. Review the pros and cons of pesticide use, considering biological, economic, and social factors. If your community were charged with designing a strategy for controlling agricultural pests, what role would pesticides play? What other control methods would you consider, and why?

3. There are several ways to assess the value of ecosystem services. One is *replacement cost*: how much would it cost to replace the naturally occurring service with one of our own design? Another is *avoidance cost*: how much money does the service save us by not having to provide it on our own? Select one category of ecosystem service and discuss some of the things we would have to do to replace it if it were disrupted. What are some of the things we *won't* have to do if the service remains intact?

CHAPTER LEARNING OBJECTIVES

After reading this chapter, you should be able to answer the following questions:

Biodiversity refers to the variety of species living in an area. It varies across the planet and has changed over time.

CHAPTER 17

Biodiversity and Human Affairs

How Is the Human Race Like a Meteorite?

Becoming an Informed Citizen . . .

Like all species, we use the resources available to us in our ecosystem for our own benefit. We are extremely good at exploiting our habitat, often at the expense of other species. Preserving biodiversity is critical to our own survival, but it is not easy or cheap, and it requires a number of changes in the way we live and work.

ou've never seen an ivory-billed woodpecker, a passenger pigeon, or a dodo bird. All three species have gone extinct within recorded human history. But have these extinctions made your life any less enjoyable or any less fulfilled? Why are these extinctions significant?

Saving endangered species is expensive and requires changes in the way we live, so it makes sense to understand why we should even try. In this chapter, we will discuss **biodiversity**, the variety of species living in an area. We will examine the impact humans have on biodiversity—and its impact on us. We will also address the biology behind biodiversity, such as the causes of its decline, how it varies across the globe, how it has changed over time, how it is monitored, and why preservation efforts make sense. What do you think our position on conserving biodiversity should be? How important *is* biodiversity? This chapter will help you decide.

biodiversity :: the variety of species living in an ecosystem

case study

The Discovery of America

FIGURE 17.1 Over 18,000 years ago, Beringia was a large low-lying region connecting continents that later would be named Asia and North America.

Undoubtedly the man had a name, but it was never recorded. If it had been, it would be listed today next to Leif Ericson and Christopher Columbus.

He stared out at the cold waters of the northern Pacific, watching his kin in their small boats catching fish or preparing to hunt seal and walrus. He stood on the subcontinent of Beringia, a large low-lying region connecting continents that later would be named Asia and North America (**FIGURE 17.1**). Legend held that the man's distant ancestors had come from the west, but glaciers a mile thick blocked travel to both the west and east. The relatively mild winds off the Pacific, however, created a pocket of habitable land in central Beringia where a community had prospered for thousands of years.

But it was getting warmer. As the glaciers slowly receded, they exposed habitable land along the coast to the east. Collecting their things, the man and other restless members of his village headed off in their boats in the direction of the rising sun. The weather grew warmer, and the coastal plains widened since the glaciers here began receding much earlier. Eventually, the group reached shore and surveyed their new home. The sea provided ample food, and the wide coastal plains provided a good place to live. The land looked to be a promising source of game. What they did not know is that they had discovered America.

This, or something very similar, took place starting about 18,000 years ago. A small group of humans, perhaps several thousand, was isolated in Beringia. A smaller group of perhaps several hundred colonized the Pacific coast of North America. Their population increased dramatically, and their descendants spread rapidly down the coast. By 15,000 years ago, humans had colonized the western coasts of both North and South America. As the ice age ended, glaciers continued to melt. Sea levels rose and the coastal plains were

FIGURE 17.2 Perhaps 12,000 years ago, humans began moving inland, and they found a continent that was home to many large mammals, including elephants, several species of camels, lions, deer, giant grizzlies, giant armadillos, and dire wolves, to name a few. All these animals are now extinct in North America.

submerged, along with much of the evidence for human habitation on the continents. Perhaps 12,000 years ago, humans began moving inland.

What they found in North America would astound you (**FIGURE 17.2**). The continent was home to many large mammals. There were elephants (mastodons and mammoths), several species of American camel, American lions like the saber-tooth cat, horses, many kinds of deer including four-horned antelopes, giant beavers the size of young grizzly bears, giant ground sloths larger than adult grizzlies, giant armadillos, rhinoceros, cheetahs, and dire wolves, to name but a few. All these animals are now extinct in North America (though the horse was reintroduced).

The extinctions began about 11,000 years ago and took place in a remarkably short period of time, perhaps 400 to 1000 years. Archaeological evidence suggests the extinctions occurred in a wave that moved west to east across the continent. This wave of extinction occurred at the same time as the wave of human expansion throughout North America (**FIGURE 17.3**). Was the timing of extinction and human expansion just a coincidence?

FIGURE 17.3 Archaeological evidence suggests that extinctions 11,000 years ago occurred at the same time as the wave of human expansion throughout North America. Here, an illustration depicts early humans called Hominids attacking a species of saber-toothed cat (Machairdontinae) using wooden branches and a spear.

17.1 What Are the Components of Biodiversity?

If you follow the news, you hear the terms *biodiversity* and *conservation* a lot. In most cases, the reports refer to species that are threatened with extinction. These bear on one important component of biodiversity: the diversity of species alive in an area, or **species richness**. Regions of high species richness allow for more varied interactions among species.

This species-level component is only part of the story, however. Another critical component of biodiversity is **genetic diversity**, the degree of genetic difference among individuals in a species. If a species is genetically diverse,

species richness :: the diversity of species alive in an area

genetic diversity :: the degree of genetic difference among individuals in a species

Life Application

The Importance of Genetic Diversity

How can a jellyfish help stop malaria? The jellyfish *Aequorea victoria* has a gene that makes green fluorescent protein (GFP), which glows green under the proper lighting conditions. Scientists studying a particular gene can insert the GFP gene alongside it, and when the native gene is expressed and makes its protein, GFP is made as well. When scientists observe the green glow, they know the gene they are studying has been turned on.

Scientists have inserted GFP into the larvae of the *Anopheles* mosquito—the mosquito that transmits malaria—next to a gene that is active only in the larval testes. Males have testes that glow green, and the females don't glow at all. This difference allows male and female larvae to be separated easily. In fact, a machine is available that uses laser light to separate 18,000 larvae per hour. Once the males have been isolated, they can be sterilized with gamma irradiation and released. Since female *Anopheles* mate only once, every female that mates with a sterile male will not produce any young. If enough sterile males are released, the number of malaria-carrying mosquitoes can be reduced significantly.

Osamu Shimomura, Martin Chalfie, and Roger Y. Tsien, the three scientists who first isolated the GFP gene from the jellyfish, were awarded the 2008 Nobel Prize in Chemistry for their work. Following their discovery, a whole host of

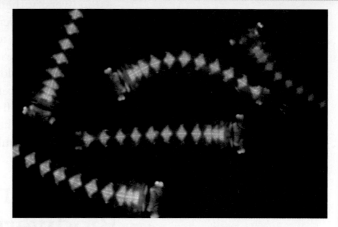

GFP has been inserted into the larvae of the *Anopheles* **mosquito that transmits malaria.** The GFP causes them to glow green, raising hopes that a gene could be introduced that would make mosquitoes unable to carry the protozoa that causes malaria.

fluorescent proteins have been found, allowing a veritable rainbow of color options for research. The important point for us is that if the jellyfish had gone extinct before this discovery, an entire subfield of molecular biology and chemistry—and maybe even a way to control malaria—would have been lost as well. Preserving biodiversity, even a single species, can be economically and medically valuable.

it is less vulnerable to extinction, which helps to conserve biodiversity. In a genetically diverse species, in which individuals vary in their traits, there is a better chance that some individuals will be resistant to new diseases, more tolerant of changing temperatures, or better able to face other challenges of a changing environment. Species with large populations and a wide geographic distribution are the most genetically diverse. (See *Life Application: The Importance of Genetic Diversity.*)

Genetic diversity is critically important to us. For example, there are only two species of cultivated rice, but we have bred over 120,000 genetically distinct varieties (**FIGURE 17.4**). Some varieties are drought tolerant and others resist disease or have high yield, while others cook more easily or taste better. Because we have bred a lot of genetic diversity into rice, we can grow it under a variety of conditions and plant varieties to meet specific needs. We can also cross varieties to produce hybrid strains that combine the best features of their parents. Nondomesticated species derive the same benefits from genetic diversity. A species spread out over a large area adapts genetically to a variety of different conditions. Some of these different forms may be able to thrive even if conditions change.

The final component of biodiversity is **habitat diversity**. Our planet is home to many different habitats, including coral reefs, rain forests, prairies, marshes, and more. Since different species are adapted to live in different habitats, it makes sense that more types of habitats result in greater species

FIGURE 17.4 If a species is genetically diverse, it is less vulnerable to extinction. While there are only two species of cultivated rice, we have bred over 120,000 genetically distinct varieties.

habitat diversity :: the range of different types of natural homes or environments in which organisms live

richness overall. The size of the habitat is important as well. A large forest supports more species than a small one. In fact, a single large forest supports more species than a whole collection of smaller ones, even if the total area is the same.

Biodiversity, then, includes species diversity, genetic diversity, and habitat diversity. Often when we speak of conserving biodiversity, we focus on species diversity alone. We are concerned with species that are on the endangered or threatened list, for example, or species that are close to extinction. Conservation requires that in addition to species diversity, we maintain genetic diversity and protect the diversity of the habitats where species live. Often a species is endangered in the first place because its habitat or genetic diversity has been reduced. Indeed, the small group of humans that moved out from Beringia expanded their numbers but left lower species diversity in their wake.

17.2 What Areas Have the Highest Biodiversity?

Regions of high geographic relief, such as mountains and their surrounding lowlands, often have higher biodiversity than flatter areas. That is because they provide many different types of habitat. A mountain may support a deciduous forest, or leaf-bearing trees, on its lower slopes, but conifers like pine trees higher up. It may become an alpine community above the treeline, where trees cannot grow. Protected valleys and exposed crags exist, along with streams, canyons, and caves. The windward side is likely to be moist, while the lee side is drier, and a grassland or desert may exist beyond the lee slope. These regions provide a diversity of habitats and physical conditions, and as a result many different species can live there.

The Latitudinal Gradient

As you travel north or south from the equator, however, a whole host of *other* physical features change. The temperature decreases, seasons become more pronounced, rainfall decreases, the length of daylight increases or decreases depending on the season, and the intensity of sunlight decreases, to name just a few changes. For a biologist, one of the most striking and consistent changes involves biodiversity, particularly species richness.

The number of species present in an area is highest near the equator and decreases as you travel north or south through the temperate zone into the polar regions. This pattern of species richness is known as the **latitudinal gradient** (**FIGURE 17.5**). The latitudinal gradient applies to almost every species. Compared to polar and temperate regions, tropical regions have more birds, more mammals, more snakes, more amphibians, more insects, more kinds of plants, more deadly disease pathogens—simply more of everything.

TABLE 17.1 compares species richness for several locations. Ecologists doing research in these areas collect data by counting the species they find. At this point, we have almost two centuries of data that confirm the latitudinal gradient. As Table 17.1 shows, in some cases, the tropics have over 100 times as many species as temperate and polar regions of the same size.

We have known about the latitudinal gradient for centuries. When European explorers began to sail across the globe, they brought back specimens

latitudinal gradient :: the decrease in the number of species as one moves away from the equator into the polar regions

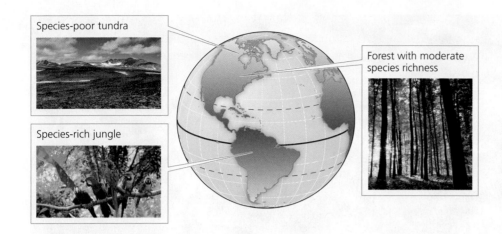

Species-poor tundra

Species-rich jungle

Forest with moderate species richness

FIGURE 17.5 The number of species present in an area is highest near the equator and decreases as you travel north or south through the temperate zone into the polar regions. This pattern of species richness is known as the latitudinal gradient.

of the plants and animals they found. These specimens found their way into universities, museums, zoos, and eventually into the public's imagination.

Why the Gradient Exists

Scientists have studied the gradient intensely at least since Darwin's time. We do not yet have a definitive answer as to why the gradient exists. However, a number of hypotheses provide partial answers.

- *The climate hypothesis:* According to this hypothesis, the number of species declines away from the equator because the climate becomes colder, harsher, more variable, and less predictable. Of course, polar bears, walruses, and reindeer do just fine in the north. If these species can adapt, why not others? The answer appears to hinge not on how many species can *adapt* to the ecology of high-latitude environments, but rather on how many species can *fit*. This idea gives rise to the second hypothesis.

Table 17.1 :: Latitudinal Variety of Species

Location	Number of Bird Species
Greenland	56
New York	195
Guatemala	469
Panama	1100
Columbia	1395
Location	**Number of Bat Species**
United States	40
Belize	84
Costa Rica	103
Location	**Number of Tree Species (per 100 m²)**
Eastern deciduous forest	15
Appalachian cove forest	30
Amazonian rain forest	283
Atlantic tropical rain forest	450

FIGURE 17.6 By far, the most diverse places on Earth are coral reefs.

- *The productivity hypothesis:* The tropics are characterized by warm, relatively constant temperatures, and a lot of rainfall and sunlight. These conditions are ideal for plants. And because vegetation captures solar energy, this translates into energy and food to support herbivores—which in turn support several layers of predators and parasites in a food web. Because a rich variety of food is so abundant in the tropics, many species can specialize on particular food items, allowing multiple species to live together because they eat different foods. Higher latitudes may require a species to be more of a generalist, taking advantage of any food source to get enough energy to survive. Under these conditions, competition for food may eliminate many species.

- *The stability hypothesis:* Over millions of years, ice ages come and go, and many species adapted to one climate go extinct when conditions change. It may take a long time for the ecological space to fill up with new species. The stability hypothesis notes that climate in the tropics has been more stable, and so species are not subjected to dramatic changes that could affect their survival. As a result, there may have been time for more species to evolve there.

Thousands of years ago, as humans made their way out of Beringia moving from west to east across the Americas, they encountered varied local environments that supported different levels of biodiversity. But by far the most diverse places on Earth are coral reefs (**FIGURE 17.6**). Interestingly, coral reefs combine the features of mountainous and tropical regions. Reefs stretch from the ocean floor to just below the surface, spanning a complete range of living conditions from cool, dark, deep water to warm, light surface water. Reefs have a complex three-dimensional structure with many tunnels and crevices to hide in and vast surfaces to grow on. And coral reefs are found only in the tropics, so the hypotheses just mentioned may help explain their astonishing diversity.

By some estimates, over 80% of the world's species live in the highly diverse tropics, including rain forests and reefs. These are precisely the regions that are most threatened by human population growth, development, and climate change.

17.3   What Can Islands Tell Us About Biodiversity?

One thing we learn from islands is that larger islands support greater species richness than smaller ones. This pattern has been known for centuries, but it wasn't until 1967 that R. H. MacArthur and E. O. Wilson proposed an explanation for it.

MacArthur and Wilson's **theory of island biogeography** is based on two facts of island life. First, every species present on an island is an immigrant. Each species flew, swam, floated, rafted, was ferried by a bird, or was blown there by a storm. Second, populations frequently go extinct on islands. The populations start off precariously small, perhaps just a single pregnant female. As populations grow on an island, their isolation causes them to reach resource limitations more quickly than a population on a mainland. Competition for these limited resources may be more intense than what occurs on the mainland.

Immigration will increase the number of species on an island, and extinction will reduce it. Large islands and islands closer to the mainland will have a high immigration rate because they are bigger targets. They are more likely to be found, particularly by species that arrive on the island by accident. Larger islands also provide many resources and habitats, which reduces competition and leads to lower extinction rates. With relatively high immigration rates and low extinction rates, large islands support a greater number of species. These conditions are reversed for smaller (or more distant) islands, where low immigration rates and high extinction rates lead to fewer species.

One important result from the theory of island biogeography is that there is a relationship between the area of an island and the number of species it has (**FIGURE 17.7**). This **species-area relationship** has been investigated by

theory of island biogeography :: a proposition that holds that the number of species found on an island is determined by the rates of immigration and colonization by new species and extinction of established species

species-area relationship :: the relationship between the area of an island, or any isolated area, and the number of species it has

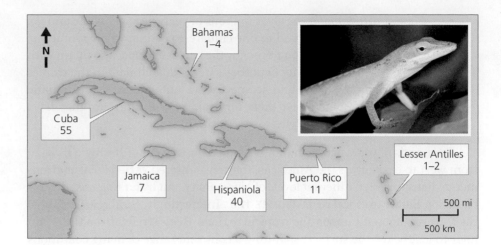

FIGURE 17.7 There is a relationship between the area of an island and the number of species it has. For example, there are about 138 species of *Anolis* lizards among the Caribbean Islands, and the number of species varies in proportion to island size.

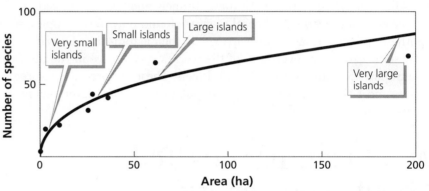

The hectare (ha) is a unit commonly used by ecologists; it equals 10,000 square meters.

FIGURE 17.8 Some ecologists have performed experiments to reveal the species-area relationship. A small island has many more species than a very small island, but a large island and a very large island have about the same number of species.

many ecologists, generally by counting the number of species of bird, lizard, or butterfly found on islands of different size. But some ecologists have also performed experiments that show us that the species-area relationship takes the shape of a curve (**FIGURE 17.8**). A small island has many more species than a very small island, but a large island and a very large island have about the same number of species. (See *How Do We Know? Experimental Island Zoogeography* for an account of Wilson and Simberloff's test of the theory.)

How Do We Know?

Experimental Island Zoogeography

Designing a good experiment is challenging under any conditions. Still, working in a climate-controlled lab offers numerous advantages compared to fieldwork. When ecologists experiment outdoors, they must be prepared for inclement weather, nosy observers (people and animals), blood-sucking bugs, and sunburn. Despite these difficulties, ecological theories need to be tested under natural conditions.

In 1967, R. H. MacArthur and E. O. Wilson developed their theory of island biogeography. They proposed that islands

have an equilibrium number of species, determined by the balance between immigration and extinction of species on the island. Their theory has been tested mainly by observation: counting species on various islands.

In 1969, Wilson and Daniel Simberloff designed an experimental test. They studied insects on six mangrove islands off the Florida Keys. These islands are made up of a few to many trees and range from 10 to 50 meters in diameter. They occur within a kilometer of the shore in waist-deep shallows.

Ants, beetles, butterflies, wasps, and many other insects live there.

Simberloff and Wilson hypothesized that if the theory of island biogeography were valid, the number of insect species on each island would be relatively constant. One way to test this theory is to count the insect species on each island, remove them all, and allow them to return. If the numbers of species before and after removal are similar, the theory would be supported.

Counting the species on an island is challenging all by itself. Simberloff and Wilson spent many hours in the Florida sun climbing through mangrove roots and sifting through leaves and hollow twigs to make sure they found them all. Removing all the insect species was even more difficult. Working in the water, they erected scaffolding 20–30 feet high in a cube around the island. They covered the scaffolding with a giant canvas bag and pumped in methyl bromide, the gas used to fumigate pest-infested homes. Once all the insects were dead, Simberloff and Wilson resurveyed the islands regularly for an entire year. They recorded when a species immigrated, when (or if) it went extinct, and how many species existed on each island.

Their results showed that as the islands were recolonized, the number of species on each leveled off at equilibrium.

In many cases, the number of species before and after fumigation was similar, though some variation is always expected in the real world. Considered overall, the findings of the study supported the theory of island biogeography. Interestingly, the identity of species varied quite a bit. Simberloff and Wilson typically found a different set of species before and after, even though the numbers were similar.

Table :: Species Counts Before and After Fumigation

	Number of Species	
Island Name	Before Fumigation	After Fumigation
E1	20	11
E2	36	29
E3	22	27
ST2	25	23
E7	23	26
E9	43	29

Source: Daniel S. Simberloff and Edward O. Wilson. "Experimental Zoogeography of Islands: The Colonization of Empty Islands." *Ecology* 50 (1969):278–296.

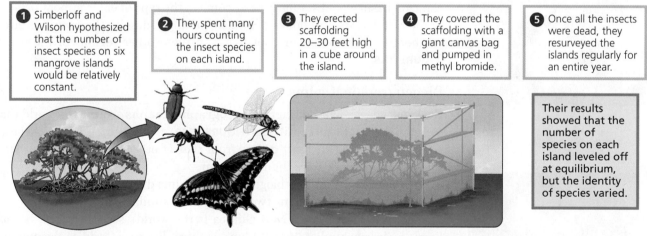

① Simberloff and Wilson hypothesized that the number of insect species on six mangrove islands would be relatively constant.

② They spent many hours counting the insect species on each island.

③ They erected scaffolding 20–30 feet high in a cube around the island.

④ They covered the scaffolding with a giant canvas bag and pumped in methyl bromide.

⑤ Once all the insects were dead, they resurveyed the islands regularly for an entire year.

Their results showed that the number of species on each island leveled off at equilibrium, but the identity of species varied.

In 1969, Wilson and Simberloff studied insects on six mangrove islands off the Florida Keys. They hypothesized that if the theory of island biogeography were valid, the number of insects on each island would be relatively constant.

Interestingly, the theory of island biogeography applies to any habitat that is like an island ecologically. Any area of good habitat surrounded by a "sea" of uninhabitable area is an ecological island. A marsh and a patch of forest surrounded by houses are good examples. This is important for biodiversity studies because it helps us understand the consequences of human activity. When we cut into a forested area to put up houses, the forest becomes fragmented into "islands." The theory of island biogeography tells us that these small patches will not be able to hold all the species present in the original forest. It is likely that the extinctions that coincided with the migration of humans across the Americas were more prevalent in environments that were ecologically similar to small islands.

Another thing we learn from islands is that many of the species on them are **endemic**—found nowhere else in the world. The finches and giant tortoises on the Galapagos Islands off the coast of Ecuador and the honeycreepers of Hawaii are examples. In fact, over 75% of the native species of Hawaii are endemics, found *only* on the island chain. That makes tropical islands a major source of diversity. It also means that extinction on the island means complete extinction, with no chance for recolonization from the mainland. Human colonization of islands has in fact led to the extinction of many island species, either from human activity or from the domesticated animals (and pests) that humans bring with them.

endemic :: species that are found in only one place in the world

biogeographic realm :: one of eight landmasses of the world with its own distinctive floras and faunas

17.4 Why Do Different Regions Have Different Species?

Rain forests, deserts, grasslands, tundra, freshwater lakes, and coastal intertidal zones—these ecosystems are found all over the globe. Each resident species, or species living in an area, is adapted to the conditions that occur in its ecosystem. It is probably true that a cactus from the Sonoran Desert of the southwestern United States could grow successfully in environmental conditions found in the Kalahari Desert of Africa, or that a monkey living in the Amazon basin would similarly be at home in the type of environment present in the rain forests of Indonesia. But, surprisingly, despite the favorable environments, these species are not found in these other regions of the world (**FIGURE 17.9**).

Biogeographic Realms

Ecologists have identified eight **biogeographic realms** around the world that correspond roughly to continents (**FIGURE 17.10**). Each realm is home to a large set of species, and each set differs from that of other realms, though many species are found in more than one realm.

A good example of how biogeographic realms differ is found in Southeast Asia. It is difficult to imagine two islands more similar than New Guinea and Borneo. At 786,000 km², New Guinea is the world's second-largest island. Borneo, close behind at 743,000 km², is third. Borneo straddles the equator and is in the Indo-Malay realm, and New Guinea is just south of it and in the Australasia realm. The islands are only about 800 miles apart. Both islands are mountainous and covered with rain forest, and like all tropical areas, their biodiversity is high.

The species found on each island differ dramatically, however. Borneo is home to plants and animals typical of Southeast Asia: tigers, orangutans, rhinoceros, monkeys, starlings, and woodpeckers, for example. New Guinea has none of these. Instead New Guinea is home to honey suckers and cockatoos, and in place of monkeys it has tree kangaroos! Most of the mammals on New Guinea are marsupials that care for their young in pouches like kangaroos and koalas do. Marsupials are absent from Borneo. All told, then, the total number of species in the region is much greater than it would be if they shared many species.

The barrel cactus (*Echinocactus grusonii*) is found in the deserts of the southwestern United States.

The African milk barrel (*Euphorbia horrida*) is from South Africa.

FIGURE 17.9 The barrel cactus and African milk barrel are not related, but they look similar and have developed similar adaptations to desert life, including thick, fleshy parts to store water and spines to deter herbivory. The fact that two species exist where one could suffice increases biodiversity.

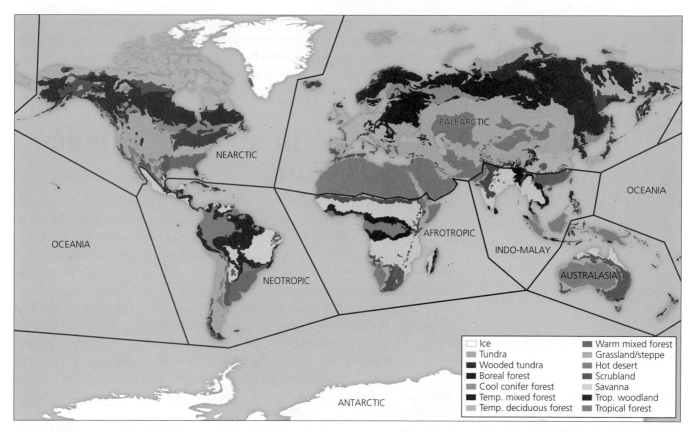

☐ Ice	■ Warm mixed forest
■ Tundra	■ Grassland/steppe
■ Wooded tundra	■ Hot desert
■ Boreal forest	■ Scrubland
■ Cool conifer forest	☐ Savanna
■ Temp. mixed forest	■ Trop. woodland
■ Temp. deciduous forest	■ Tropical forest

FIGURE 17.10 Ecologists have identified eight biogeographic realms around the world that correspond roughly to continents. Each realm is home to a large set of species, which increases global biodiversity.

Wallace's Line

Alfred Russell Wallace was the first to document the differences in species between the two islands. Wallace was a naturalist who in the 1850s traveled throughout the Malay archipelago, collecting specimens for British museums. He kept a map of which specimens were found on which islands in the archipelago. Wallace recognized that his specimens fell into two groups, Asiatic and Australian.

The line between the Asiatic and Australian groups—now known as Wallace's line—runs between the large islands of Borneo and Sulawesi (**FIGURE 17.11**). It continues through the narrow, deep-water strait that separates the small islands of Bali and Lombok. The islands west of this line all share the Asiatic fauna. East of the line the islands share the Australian fauna. It is astonishing that Bali and Lombok, separated by only 20 miles of ocean, could have such completely different resident species. The explanation for these differences can be found in the geological history of the region. The depth of the straits separating Bali and Lombok have always been deep with fast-running currents, hindering the migration of organisms between the islands. Prior to the melting of the glaciers after the last ice age (13,000 years ago), Bali was connected to the Asian mainland. When the glaciers melted, sea levels rose, and shallow water separated Bali from the mainland. Wallace's line is actually the eastern margin of the former Asian continent.

It is not surprising that biogeographic realms tend to be associated with continents. We know now that the continents move or "drift" slowly around the planet. When continents collide or are connected by land bridges, animals and plants can move from one continent to another. But drifting continents can be separated for long periods of time, as was the case with South America

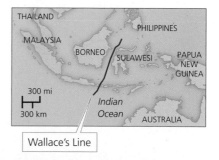

FIGURE 17.11 Alfred Russell Wallace was a naturalist who traveled throughout the Malay archipelago collecting specimens in the 1850s. He kept a map of which specimens were found on which islands and recognized that they fell into two groups: Asiatic and Australian; the line between the groups—now known as Wallace's line—runs between the large islands of Borneo and Sulawesi.

and Australia. When this happens, evolution occurs independently on the continent, resulting in a different set of species that make up its biogeographic realm. This helps to explain global patterns of biodiversity: for example, why Australian mammals are marsupials and why North American deserts are home to cactus, but deserts on other continents are not.

17.5 How Does Biodiversity Change Through Time?

Global biodiversity has varied dramatically throughout Earth's history as new species have arisen and others have gone extinct. But the pivotal time in the history of biodiversity was about 550 million years ago. Before that time, all species lived in the sea. Most species were unicellular and lived on the ocean floor, filtering bits of rotting organic matter from the sea for their food. Multicellular animals were small, soft, and slow.

All this changed 550 million years ago. There was a dramatic increase in biodiversity, unlike anything seen before or since. Within 20 million years, a geological blink of an eye, every major group of animals appeared in the fossil record. Ecologically complex communities formed containing vertebrates, arthropods, mollusks, sea stars, and their relatives. This event is known as the **Cambrian explosion**, named for the Cambrian period of geologic time, about 542–490 million years ago (**FIGURE 17.12**).

The cause of the Cambrian explosion is hotly debated. Some scientists argue that increasing oxygen levels allowed for larger and more active animals. Others argue that the explosion simply reflects the evolution of hard body parts, a new ecological "invention" that gave protection from predation. These hard parts fossilize better so we have more fossils of these organisms. Either way, the Cambrian period marks the beginning of global biodiversity.

Cambrian explosion :: the geological period of time from about 542–490 million years ago when every major group of animal appeared in the fossil record

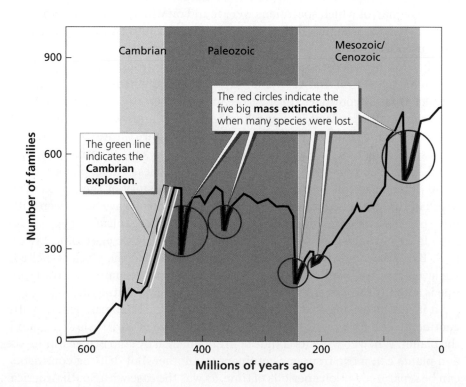

FIGURE 17.12 The Cambrian explosion occurred about 542–490 million years ago. During this explosion, every major group of animals appeared in the fossil record.

17.6 Why Is Biodiversity Needed for a Healthy Ecosystem?

Conservation efforts cost money and can be controversial. That is why it is important to understand the positive, even vital, reasons for preserving biodiversity. First, we look at the biological role of biodiversity, and then we return to the question of why biodiversity is important to us.

Productivity, Stability, and Ecosystem Health

Productivity and stability are two factors associated with ecosystem health. All living organisms require energy. Ultimately, an ecosystem's energy comes from sunlight, but sunlight must be converted into plant material before it can be used (Chapter 7). **Productivity** is an indicator of how effectively plants convert sunlight into food. Productivity is measured by the weight of plant material produced in the ecosystem. More plant material means higher productivity.

Another indicator of ecosystem health is **stability**. All ecosystems exist within a variable environment subject to disturbances: weather changes, drought, and fire are examples. Ecosystem stability refers to an ecosystem's ability to resist these disturbances and to return to its original configuration after them. Because biodiversity contributes to both productivity and stability, it is important to ecosystem health.

productivity :: an indicator of how effectively plants convert sunlight into food; measured by the weight of plant material produced in the ecosystem

stability :: the ability of an ecosystem to resist environmental disturbances and return to its original configuration after them

Why Biodiversity Increases Productivity

Ecologists have measured the productivity of grasslands, forests, farms, coral reefs, and many other ecosystems. One clear result that emerges from their work is that productivity is correlated with biodiversity. Biologically diverse grasslands produce more plant material than those that are less diverse, and this result holds for other types of ecosystems as well.

There are several reasons why biodiversity increases productivity. One reason is that a diverse collection of species can make more complete use of an ecosystem's resources. Ecosystems vary from place to place. Some areas are wetter, others have deeper soil or different concentrations of soil nutrients, some are shaded, and some are more protected from wind. A single species of plant is unlikely to grow equally well in all these conditions, and where it does grow, it may not use all the available resources. On the other hand, a diverse set of plant species, each adapted to different conditions, can make effective use of *all* the resources and mini-habitats available. More species in this case also means more individual plants, which results in higher productivity.

Some plants help other species to grow. For example, plants in the pea family (or legumes) work with bacteria in their roots to add nitrates to the soil (see Chapter 16). The presence of legumes thus helps species that require high levels of nitrogen. Another example occurs in ecosystems with harsh conditions, such as deserts. Under such conditions, it is hard for new seedlings to survive to adulthood. Woody shrub species provide shade, lower temperatures, and moisture so that seedlings of other species can germinate and grow. A more diverse ecosystem is likely to have more of these helper species, which allows even more species to occur, increasing biodiversity further.

Why Biodiversity Increases Stability

Biodiversity buffers an ecosystem from disruption, just as redundancies and backups protect computer systems. In a biologically diverse ecosystem, several different species can take on any given role. If disease or a parasite reduces the population size of one pollinator, for example, another species can step in. If some trees are killed by forest fire, other species, such as pitch pine and jack pine, have adaptations that allow their seeds to germinate and grow quickly.

In short, biodiversity helps ensure that at least some species will do well, whatever happens. This is the natural analogue to holding a diverse portfolio of stocks and bonds. An ecosystem with only a few types of species is a risky portfolio: if conditions become unfavorable, both may crash.

rivet hypothesis :: the idea that ecosystems are like airplanes with rivets; if too many species (or rivets) are lost, the ecosystem will crash quickly (just like an airplane)

Biodiversity promotes ecosystem health by increasing both productivity and stability. When biodiversity is reduced, the health of the ecosystem declines as well. In 1981, Paul and Anne Ehrlich described this process with the **rivet hypothesis**. Ecosystems are like airplanes, and species in the ecosystem are like the rivets that hold the plane together. An ecosystem can tolerate the loss of some species, just as an airplane can tolerate the loss of some rivets. However, if a particularly important species or too many species are lost, the ecosystem will crash quickly, just like a plane that loses too many rivets. Loss of biodiversity can have the same effect on an ecosystem.

How Biodiversity Keeps the Food Web Intact

Why is the world green? This seems like an odd question until you begin to think about it carefully. What this question really asks is why herbivores don't eat all the available vegetation (leaving us with a nongreen world) and then die out. One hypothesis is that plant defenses, like thorns and poisonous secondary compounds (Chapter 16), protect them from herbivores. Another hypothesis is that predators keep herbivores in check, so they don't eat all the plants.

Recently, John Terborgh devised an experiment to test these hypotheses. Terborgh worked in Lake Guri in central Venezuela. Lake Guri is an artificial lake, created in 1986 when a hydroelectric dam flooded 4300 square kilometers in Venezuela's Caroni Valley (**FIGURE 17.13**). What had previously been a valley floor with many rolling hills became a lake with hundreds of islands. Each island had its own set of species, depending on which species happened to be on the hill when the valley flooded. Terborgh and his colleagues monitored biodiversity on 14 sites in Lake Guri. Nine of these sites were free of predators, while five had a complete set of predators similar to that found on the mainland.

Between 1986 and 2002, Terborgh found that the predator-free islands had been virtually stripped of their vegetation. In contrast, the islands with predators had an ecosystem similar to the mainland.

Terborgh's results provide evidence that our world remains green because carnivores regulate the herbivore populations. More generally, biodiversity stabilizes the entire food web. In a diverse food web, if one species becomes too common, its predators will reduce the species' numbers. On the other hand, if a species becomes too rare, its predators can switch to other prey. A diverse food web provides a more diverse "menu," which helps keep population sizes in check.

An outcome of the species extinctions that occurred as humans moved into new habitats across the Americas may have been the collapse of the existing food web, which would have necessitated continued migration. Once gone, this food web would have been replaced with another, perhaps less biodiverse, one composed of different species.

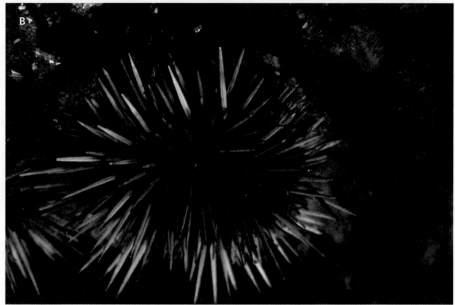

FIGURE 17.13 **Keystone species can have a disproportionate effect on an ecosystem's biodiversity.** When the sea otter population declined in the 1800s due to hunting (A), the sea urchin population (B) increased and overgrazed primary producers, sending ripples throughout the food web.

Sometimes a single species can have a disproportionate effect on an ecosystem's biodiversity. These are called **keystone species**. An example occurred off the Pacific Northwest coast. Throughout the 1800s, sea otters were hunted for their pelts, the densest fur known among mammals, and at one point the worldwide population of sea otters was as low as 2000.

Freed from predation by sea otters, sea urchin populations increased drastically in size. The sea urchins overgrazed kelp and seaweed beds, sending the populations of these primary producers into decline. The loss of primary producers sent ripples all through the food web, resulting in decreased biodiversity (Chapter 16).

keystone species :: a single species that has a disproportionate effect on an ecosystem's biodiversity

17.7 Why Should We Preserve Biodiversity?

Why, then, should we preserve biodiversity? To help us answer, let's look at an actual case that occurred not that long ago. We shall see that it tells us something about the fate of the ecosystem, and something about our fate as well.

FIGURE 17.14 In 1990, the northern spotted owl was placed on the U.S. List of Endangered and Threatened Wildlife. At the time there were an estimated 2000 pairs of spotted owls left.

indicator species :: a species whose presence indicates a healthy, functioning ecosystem

The Spotted Owl Controversy

In 1990, the northern spotted owl (**FIGURE 17.14**) was placed on the U.S. List of Endangered and Threatened Wildlife. At the time, there were an estimated 2000 pairs of spotted owls left. This move came about after years of negotiation and debate among environmental groups, timber companies, and the government.

The northern spotted owl lives exclusively in old-growth coniferous forests of the Pacific Northwest. These forests were over 300 years old, dominated by massive Douglas firs. Trees in the old-growth forest were worth about $4000 per acre, and logging accounted for about half of Oregon's economy and about a quarter of Washington's. To accommodate the spotted owl, over 10 million acres of old-growth forest had to be placed off-limits to logging. That's $40 billion of revenue lost in an economy that was already struggling. It is easy to imagine the controversy surrounding the spotted owl. Families, towns, and industries faced economic hardship to protect a bird most people had never seen.

However, framing the debate as "owls versus people" was a mistake that made it more difficult to understand the real issue. A more insightful way to frame the debate would have been the trade-off between the short-term and long-term benefits of using the forest's resources. Either way it is *people* that benefit from using the forest. It's simply a question of the timescale.

Long-Term Benefits of an Old-Growth Forest

The spotted owl is an **indicator species**. Its presence indicates a healthy, functioning old-growth forest. If we lose the spotted owl, it means we have lost the old-growth forest as well and most if not all of the species it contains. Other organisms may form a new ecosystem eventually, but it will not be as complex or as biodiverse as the old-growth forest it replaced.

The old-growth forests of the Pacific Northwest are a unique ecosystem (**FIGURE 17.15**): a temperate zone rain forest. These forests exhibit very high biodiversity, rivaling that of a tropical rain forest. As in a tropical rain forest, the upper branches of the canopy collect dust, dirt, and organic material and develop a layer of soil. This old-growth canopy supports many plants and animals, a hidden mini-ecosystem hundreds of feet above the forest floor. It takes about 300–1000 years for these forests to form a community of massive trees covered with life from top to bottom.

One valuable resource in these old-growth forests are the genes their species contain. For example, tamoxifen, a drug that was synthesized in the early 1960s and is used to treat breast cancer, originally comes from the Pacific yew tree, a resident of the old-growth forests. In addition to saving lives, tamoxifen provides economic benefit to pharmaceutical companies, accounting for over a billion dollars in sales. There is a good possibility that other useful products eventually can be discovered in this diverse ecosystem.

Many other benefits can be derived from old-growth forests. First, they hold soil well and prevent heavy rains from eroding soil into rivers and out to sea. In contrast, in areas where trees have been clear-cut, sediment from previously forested slopes clouds local streams and has a negative impact on salmon and other fisheries. Second, since water moves more slowly through forested soil, there is more time for the water to be filtered and cleansed of impurities, enhancing the supply of drinking water. Third, old-growth forests absorb a lot of carbon dioxide, helping to slow climate change. And fourth,

FIGURE 17.15 The old-growth forests of the Pacific Northwest are a unique ecosystem: a temperate zone rain forest. These forests exhibit high biodiversity, rivaling that of a tropical rain forest.

because they are unique and fascinating, old-growth forests support a strong tourism industry.

How Old-Growth Forests Provide Ecosystem Services

In Chapter 16, we introduced ecosystem services, the benefits we get from natural processes that occur in a healthy ecosystem. Let's take another look at these services and see how old-growth forests provide them.

- *Provisioning services:* Old-growth forests provide us with (some) lumber, and their genetic diversity has given us products like tamoxifen.
- *Regulating services*: Old-growth forests retain soil and purify water.
- *Cultural services:* Old-growth forests support a strong ecotourism industry and for many provide a spiritual connection with nature.
- *Supporting services:* Old-growth forests have tremendous productivity, create soil, and remove carbon dioxide from the atmosphere.

Let's return to the spotted owl controversy. In 1990, the rate of logging was so high that most of the old-growth forest would have been cut down in a decade or two, resulting in the same economic hardship. But the people would have been missing critical ecosystem services that the forest provides. The spotted owl legislation did not cause the economic problem; it only forced people to confront it earlier than they would have.

Choosing short-term benefits over long-term is like running up credit card debt. It appears to solve some immediate problems, but ultimately the bill must be paid. When the bill comes in the form of biodiversity loss and loss of ecosystem services, the people who run up the bill—and reap the profits—are not the ones who pay. These bills are paid by future generations. Another concern is that when biodiversity is lost and species go extinct, there is no way to bring them back. It may not be possible to "pay the bill."

The logging industry was likely aware of these issues and would have had plans to plant new, fast-growing trees to replace the ones they cut. This man-made ecosystem would have provided some ecosystem services. But it would have been much less mature and significantly less diverse. As a result, it would have been less productive and less stable, and it would have required our effort and money to maintain it. The most common consequence of our development projects is that we replace highly diverse natural ecosystems with artificial ones that require our care.

We began this section with a question: why preserve biodiversity? One answer is clear. We should preserve biodiversity to prevent long-term economic and ecological catastrophes. There is also an ethical or even spiritual side to preserving biodiversity. We are the only species with any concept of stewardship, and we are the only species capable of affecting biodiversity on a global scale. Those early humans did not understand the potential ecological impact of their actions, but we do. Given these circumstances, preserving biodiversity may be a moral obligation as well.

17.8 How Do We Keep Track of Biodiversity?

Scientists have identified about 1.3 million species. That's a lot, but every time a careful study is done of a new patch of rain forest, an inaccessible mountain peak, an isolated lake, or a deep sea community, many new species are found. Since scientists keep finding new species, it is clear that we have not come close to identifying them all. Using a variety of statistical methods, scientists estimate that our planet could be home to 30 million species or more, in which case we have identified only 4% of them. Most of the missing species are bacteria, other microscopic organisms, or animals like insects and worms. We have a fairly complete survey of mammals, birds, and plants.

The sheer number of species makes it almost impossible to characterize completely the biodiversity of any large area. Nevertheless, monitoring biodiversity allows us to understand how biodiversity changes in response to human activity and other causes.

Recognizing this need, scientists have devised a number of ways to measure and monitor biodiversity. The scale of the study area is important. Working with a smaller ecological community, scientists can be more thorough and more precise, while studying a large ecosystem requires making broader, and less precise, estimates. That includes the broadest ecosystem of all, the planet.

The Species Diversity Index

Recall that species richness refers to the diversity of species alive in an area. Many studies of biodiversity focus on a particular ecological community, such as a stream bed or an individual forest stand. For this type of study, two additional aspects of species diversity play a role. **Species abundance** refers to the number of individuals of each species present in the community.

species abundance :: the number of individuals of each species present in an ecological community

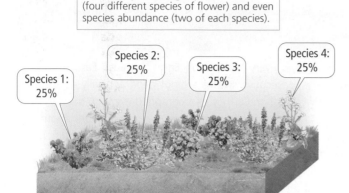

Community 1 has high species richness (four different species of flower) and even species abundance (two of each species).

Species 1: 25%
Species 2: 25%
Species 3: 25%
Species 4: 25%

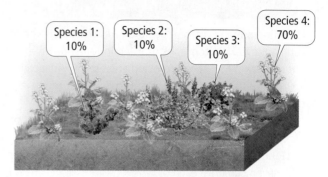

Community 2 has high species richness (four different species of flower) but low species abundance (dominated by one species).

Species 1: 10%
Species 2: 10%
Species 3: 10%
Species 4: 70%

FIGURE 17.16 Species diversity is highest when species richness is high and species abundance is relatively even (A); if a community is dominated by just a few very abundant species with a smattering of rare ones (B), we would not consider that community to be diverse.

Species evenness refers to the pattern of species abundance. Are most species rare or moderately common, for example?

Species diversity is highest when species richness is high and species abundance is relatively even. If a community is dominated by just a few very abundant species with a smattering of rare ones, we would not consider that community to be diverse. An extreme case of such a community is a cornfield. There may be many species of weeds here and there, but for the most part it's just corn—not very diverse. On the other hand, a prairie dotted with many species of wildflowers would be more diverse, even if its species richness were the same as the cornfield (**FIGURE 17.16**).

Ecologists who study biodiversity at the community level use a species **diversity index** to measure diversity. This mathematical formula takes into account *richness*, *abundance*, and *evenness*. If a community's species diversity index is lower than expected, this indicates adverse conditions that are preventing species from living there. Pollution or drought could eliminate the more sensitive species, for example. Or an invasive species may be outcompeting native species whose populations are limited by the natural checks and balances of ecological interactions. One of the consequences of the expansion of the human population and its migration across the Americas, for example, was a decrease in the species diversity index.

species evenness :: the pattern of species abundance

diversity index :: a mathematical formula that measures species diversity by taking into account richness, abundance, and evenness

Indicator Species and Satellite Images

Given the worldwide scope of human activity, it is important to monitor changes in biodiversity on a global scale. It is not possible to use diversity indexes on a global scale because it is impossible to collect and count all the species in a large geographic region. Instead, ecologists and other scientists use more approximate techniques to monitor biodiversity.

One technique is the use of indicator species, like the spotted owl. Recall that the presence of an indicator species implies that the ecosystem is healthy and will contain its regular set of species. In this case, we need to count only the indicator species. Scientists have learned that birds and flowering plants work well as indicator species of biodiversity, and they are relatively easy to survey. Scientists contribute much of the information, but interested amateurs are important as well. For example, around Christmas every year tens of thousands of bird-watching volunteers spend the day counting every bird they see or hear. The Christmas bird count has been going on since 1900 in the United States, Canada, and 19 other countries. By knitting together published

Technology Connection

Satellite Imagery

How fast is the rain forest being cut down? Satellite imagery can show you photographs of the rain forest taken at different points in time. The deforested areas stand out clearly, and you can use a computer program to compute their area and determine the rate of deforestation.

The U.S. Landsat program has been photographing global land cover since 1972. Landsat 7, which was launched in 1999, has an Enhanced Thematic Mapper (ETM+), a sensor with seven cameras, each of which captures light, but only in a particular band of wavelengths, or "color." Bands 1, 2, and 3 are the familiar blue, green, and red wavelengths of visible light. The remaining four bands are in the infrared zone, longer wavelengths of light that humans cannot see directly. Each camera records an image in its own particular wavelength. The seven resulting images can be combined in various ways to form composite, or "false color," images of the location.

Landsat 7 orbits Earth at an altitude of 705 kilometers (438 miles) and at a speed of 7.5 kilometers per second (16,776 mph). One complete orbit takes 99 minutes. The cameras in the ETM+ photograph a swath of Earth 185 kilometers (115 miles) wide. Each pixel in the camera corresponds to 15 × 15 square meters on Earth. Landsat can distinguish roads, clear-cuts, and larger buildings but not smaller houses, cars, or people.

Landsat 7 orbits from pole to pole, and as it orbits, Earth rotates below it. During each orbit, the satellite hovers over a different region of the globe. Landsat 7 is therefore able to photograph every location on Earth. Because Earth rotates quickly, consecutive swaths are not geographically adjacent. But every spot on Earth is photographed every 16 days.

As Landsat 7 continues to gather information, another orbiting satellite has been added to the sky for additional data collection: Landsat 8 was launched February 11, 2013.

The U.S. Landsat program has been photographing global land cover since 1972. A comparison of pictures taken in Mato Grosso, Brazil, reveals a dramatic loss of rain forest in just 25 years.

reports and encouraging these types of programs, a global system for monitoring the biodiversity of indicator species may emerge.

Satellite images are also used to monitor biodiversity at the ecosystem and global scale (See *Technology Connection: Satellite Imagery*). Landsat and other satellite systems have collected almost 40 years of data about ground cover. By comparing photographs in temporal sequence, scientists can see how glaciers have retreated, how forests have been cut down while deserts have expanded, and how urban centers have grown. If scientists can estimate how many species lived in the original habitat, they can estimate how many species have been lost when the size of that habitat changed.

17.9 Why Might We Be Facing the Sixth Mass Extinction?

About 65 million years ago, a meteorite smashed into the ocean off the coast of Yucatan near what is now the town of Chicxulub. The meteorite measured at least 10 kilometers in radius and was traveling at about 20 kilometers

Table 17.2 :: The Big Five Mass Extinctions

Geologic Period	Millions of Years Ago	Percentage of Species Extinct
Ordovician	439	85 ± 3
Devonian	367	83 ± 4
Permian	245	95 ± 2
Triassic	208	80 ± 4
Cretaceous	65	76 ± 5

Source: D. J. Futuyma. *Evolutionary Biology,* 3rd ed. (Sunderland, Mass.: Sinauer Associates, 1998), Table 25.1, p. 713.

per second. It was 2 million times more powerful than the largest atomic bomb humans have ever exploded. It burned its way through the ocean and deep into the sea floor, sending gigantic clouds of steam into the air and 1000-foot-high tsunamis throughout the Atlantic. The heat of its passage sent a shock wave of superheated air that leveled and burned forests in North and South America. The seabed erupted into a mountain higher than Everest and then quickly collapsed. Millions of tons of ash formed a blanket throughout the atmosphere, cutting out light. Carbon dioxide released by the blast produced a greenhouse effect, and sulfur released into the atmosphere caused global acid rain.

The meteorite completely collapsed global ecosystem services, resulting in the extinction of more than 75% of the species on Earth, including the dinosaurs. This was the Cretaceous extinction, the most recent—and mildest—of the "big five" **mass extinctions** (**TABLE 17.2**).

mass extinction :: a period when over 75% of Earth's species disappeared

The largest mass extinction so far, the Permian extinction, occurred 245 million years ago. Over 95% of all species on Earth vanished. But even so, hundreds of thousands of species survived. All life on Earth today descends from them.

The biological world that develops following a mass extinction differs greatly from the one that preceded it. Life will continue on Earth no matter what we do. But that doesn't mean we would like it, or even survive, in that future world.

The Blitzkrieg Hypothesis

Consider again the wave of human expansion into North America and its impact. Is it possible we could have hunted most of the large mammals on this new continent to extinction? Paul Martin thinks so. In 1973 Martin, then at the University of Arizona, wrote a paper whose title we borrowed for our case study, "The Discovery of America." Martin's position, that we are such a powerful force for extinction, has come to be known as the **blitzkrieg hypothesis**.

Donald Grayson at the University of Washington has argued against the hypothesis. Grayson cites limited evidence that humans actually hunted these large animals. Other hypotheses, too, could account for the extinction, such as effects of climate change after the ice age ended. He suggests that some scientists are using the blitzkrieg hypothesis to push a green political agenda that is not based on scientific evidence.

blitzkrieg hypothesis :: a theory proposed by Paul Martin that humans hunted most of the large mammals on the North American continent to the point of extinction

One point in favor of the blitzkrieg hypothesis is that the extinction of large mammals (and large birds) has occurred on *every* continent soon after humans arrived. South America, Europe, and Australia were affected almost as greatly as North America (**FIGURE 17.17**). Only Africa seems to have been immune, perhaps because the large mammals were more familiar with

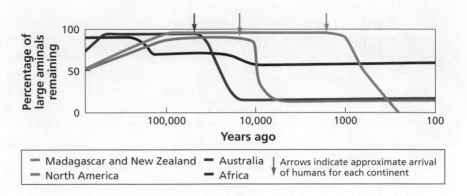

FIGURE 17.17 The blitzkrieg hypothesis proposed by Paul Martin holds that humans hunted most of the large mammals on the North American continent to the point of extinction. The extinction of large mammals (and large birds) has occurred on *every* continent soon after humans arrived, with the exception of Africa.

humans from having lived with them on the continent so long. Not all of these areas experienced climate change when humans arrived.

Humans may or may not have hunted the large mammals of North America to extinction, but there is no doubt that human activity has driven many species extinct—and threatens many more. We appear to be even more of a force for extinction than the meteorite that killed off the dinosaurs 65 million years ago.

Background Extinction

The fossil record shows us that extinction is a naturally occurring, ongoing process. Every species goes extinct sooner or later. The average life span of a species is between 1 million and 5 million years, though some species last much longer. Most of the species that have ever lived on Earth, around 99.9% by some estimates, are extinct.

The extinction rate is the percentage of species that go extinct in a given time period. This is a difficult value to measure. To compute a percentage of species, one needs to know the total number of species. Scientists are unsure of this number for living species. For species that lived in the past, the number is even more uncertain, since the fossil record has not revealed them all.

Paleontologists employ statistical techniques to make "best-guess" estimates of extinction rates. Marine invertebrates tend to last longer than terrestrial mammals, for example. Extinction rates also vary throughout time. A mass extinction is a period of very high extinction rate, whereas most of the time extinction rates are considerably lower. The **background extinction rate** is defined as the extinction rate during one of these "typical" times, not during a mass extinction. There is no exact value for the background extinction rate, but a commonly used estimate is one extinction per million species per year.

background extinction rate :: the extinction rate during "typical" times, not during a mass extinction

Current extinction rates are between 100 and 1000 times the typical rate of background extinction. This is in the same class as the "big five" extinctions in Earth's geologic history, and many have concluded that we are currently in the middle of the sixth mass extinction. According to the International Union for Conservation of Nature, 13% of all bird species are threatened with extinction. The threat is even more serious for other organisms: 25% of mammals, 40% of amphibians, 28% of reptiles, 27% of fishes, and 71% of flowering plant species are in danger of extinction. The causes of this ongoing mass extinction may not be as dramatic as a meteorite impact, but they are just as serious and inevitable. *If these extinction rates are accurate, 60% of all species alive today will be extinct within 100 years.*

Human Activity Threatens Biodiversity

Island species most vulnerable to extinction have already gone extinct. However, many of these extinctions happened long before our population began to increase significantly and before we developed technology. Initially, the population of humans that moved across the Americas was quite small. Yet their impact on biodiversity was dramatic. In the modern world, our burgeoning population requires space for farms, houses, and commercial ventures. We require resources and food that we harvest from the global ecosystem. We produce waste products and climate-changing gases. And piggybacking on our global commerce, many species are transported, intentionally or not, to new regions. All these activities have a negative impact on biodiversity.

Let's consider the ways that human activity causes biodiversity decline. In his book *The Diversity of Life*, E. O. Wilson compiled the percentages of endangered and threatened species from each activity (see *Scientist Spotlight: E. O. Wilson*):

Habitat destruction	73%
Displacement by invasive species	68%
Chemical pollutants	38%
Overharvesting	15%

Scientist Spotlight

E. O. Wilson (1929–)

What would a scientist be doing on a list that includes Oprah Winfrey, Jerry Seinfeld, Al Gore, Robert Redford, Sandra Day O'Connor, and Courtney Love? The list was *Time* magazine's 1995 list of the 25 most influential Americans. The scientist was Edward Osborne Wilson. Let's take a look at why Wilson made the list.

Wilson was born in 1929 in Birmingham, Alabama. He lived with his father and stepmother and moved around a lot as a child. Wilson loved nature and spent as much time as possible outdoors, hunting, fishing, and observing birds and insects. Insects, in particular ants, were to become his passion. As Wilson puts it, "Every boy goes through a bug stage. I just never grew out of it."

At age 13, Wilson discovered a colony of ants he'd never seen before down by the docks in Mobile, Alabama. This was the first colony of invasive, aggressive fire ants found in the United States. Wilson's first scientific paper, published in 1949 at the age of 20, documented the spread of fire ants throughout Alabama. He has studied ants ever since, publishing hundreds of papers and many books about them. His work has won recognition and many honors, including two Pulitzer Prizes, election to the National Academy of Sciences in 1969, and the National Medal of Science in 1976. Wilson earned his doctorate at Harvard and went on to join the Harvard faculty.

Impressive as these achievements are, they are not the reason Wilson was included on *Time*'s list of influential people. Building on a lifelong career in science, Wilson has become a philosopher, thinker, and writer in several areas.

In 1995, E. O. Wilson made *Time* magazine's list of the 25 most influential Americans. Wilson has been recognized as an important scientist and advocate of the conservation of biodiversity.

(Continued)

E. O. Wilson (1929–) (*Continued*)

He has also become passionate about the urgent need to pre-serve biodiversity. Wilson's work has taken him all over the globe and he has witnessed firsthand the effects of habitat destruction on biodiversity. He also understands the eco-nomic, cultural, and ethical benefits of preserving biodiversity. Wilson writes elegantly about the need to conserve and serves on many national and international organizations ded-icated to preserving biodiversity.

In 2006, Wilson published *The Creation: An Appeal to Save Life on Earth*, a book framed as an open letter to the religious community. It is an appeal to put aside differences over evolution and the origin of life so that science and religion can work together on the urgent issue of preserving biodiver-sity. Even if motivations differ, he argues, the preservation of nature is both a scientific and moral goal. "Science and reli-gion are two of the most potent forces on Earth and they should come together to save the creation." Despite the damage done so far, Wilson remains optimistic that biodiver-sity can be preserved.

Many species are threatened by more than one activity, so the total adds up to more than 100%. We will take a closer look at two of these: habitat destruction and overharvesting.

Habitat Destruction Habitat destruction is most obvious, and most harmful, in the deforestation of our planet's rain forests. Tropical forests account for about 7% of Earth's surface, but they are home to over 50% of Earth's species. Satellite images allow us to document the pattern of tropical deforestation (**FIGURE 17.18**). The process begins with roads into the forest, which provide access for logging operations. Selective logging occurs first, where trees of particular species and sizes are harvested for lumber. Villages grow up along the roads to support this industry, and the road system be-comes more extensive. Next farmers and ranchers move in and clear-cut the forest for agricultural purposes. The villages grow into towns or small cities as the area is settled.

About 1.1% of the rain forests in Southeast Asia and about 0.7% of the rain forests in Africa and South America are cut each year. In the Amazon basin, this translates into 25,000 square kilometers—about the size of Massachusetts—and the rate of deforestation is increasing. Even low rates of logging can clear the entire forest in a surprisingly short period of time. Wherever it occurs, habitat destruction attacks all three components of bio-diversity: species, genetic diversity, and habitat.

Overharvesting When individuals are taken from a population faster than they can grow back, this is known as overharvesting. It results in a negative

FIGURE 17.18 Satellite images are used to document the pattern of tropical deforestation. For instance, by comparing photographs in temporal sequence of Rondonia, Brazil, scientists can see how rain forests have been cut down.

June 1975—a single road is cut into the rainforest to provide access to loggers.

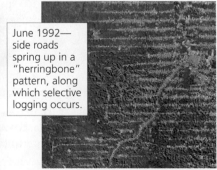

June 1992—side roads spring up in a "herringbone" pattern, along which selective logging occurs.

August 2011—over time, cleared patches fuse together as the rainforest is converted into farmland or cattle ranches.

growth rate that if continued will drive the population extinct. While habitat destruction targets all resident species, overharvesting targets particular species that are valuable to us. But as we have seen, removing even one species can destabilize an ecosystem and reduce the services it delivers, particularly if it is a keystone species.

Marine ecosystems have experienced a good deal of overharvesting. Whales, seals, sea otters, many types of fish, and crustaceans like crab and lobster have all experienced population declines due to overharvesting.

In marine ecosystems, it is not easy to monitor population size or to determine whether overharvesting is occurring in the early stages. The first symptoms may be a population crash, followed by the reduction in the catch and possibly the collapse of the harvesting industry. While many species are threatened by overharvesting, conservation programs have brought many of them back from the brink of extinction. Learning more about the population size, growth rates, and demography of the marine species we harvest will promote the development of more sustainable harvesting procedures.

7.10 How Can We Preserve Biodiversity?

There is no question that biodiversity is threatened on a global scale. The threat is a cause for concern, but not for pessimism. Preserving biodiversity is not easy or cheap, and it requires a number of changes in the way we live and work. But it can be done.

Become educated about the long-term benefits. Biodiversity enhances the productivity and stability of ecosystems. This in turn allows the ecosystem to provide more reliable services.

Develop economic policies that take advantage of long-term biodiversity benefits. Costa Rica has set aside over one-quarter of its land area for national parks and biological preserves. As a result, ecotourism in Costa Rica is a multibillion-dollar industry and has become one of the country's best performing economic sectors. Another way to capitalize on biodiversity is to "mine" diverse areas for new products, including medicines and genes.

Develop a cooperative approach. We have focused largely on the scientific issues surrounding biodiversity, but maintaining biodiversity requires the cooperation of scientists, politicians, economists, conservation organizations, organized religion, and of course informed citizens. International cooperation is required as well. Most of the regions of highest biodiversity are in developing countries. These countries will most likely require help if they are to forgo short-term exploitation of their biodiversity resources.

Curtail human population growth. As we will discuss in Chapter 18, global population growth shows evidence of slowing down, but it will still take decades before growth levels off. During this time, we will continue to require more space, more food, and more resources. And biodiversity will continue to decline as a result. It is clearly not possible to halt population growth quickly, but increased awareness of the need to reduce population growth can give us a long-term goal to work toward.

Design effective methods of conservation. Island biogeography has helped us understand the relationship between area and biodiversity. Continued research on particular species helps us understand their ecological needs.

This research can help us design parks and preserves more effectively and ensure that our conservation efforts help the species we are most concerned about.

None of these steps will be easy. But awareness of the benefits of biodiversity, and the costs of losing it, is becoming more widespread. Many governments, organizations, and individuals are working to curb biodiversity loss. There are reasons for cautious optimism, but we have a lot of work ahead of us.

Biology in Perspective

How is the human race like a meteorite? Like a meteorite, our activities are global in scale, and we have the ability to drive extinct a large percentage of Earth's species. Education about biodiversity and its benefits is the first step in developing approaches to preserve biodiversity.

Biodiversity comes from species diversity, genetic diversity, and habitat diversity. Biodiversity is highest in the tropics and in regions of high geographic relief. It has varied dramatically over time but has generally increased throughout Earth's history despite five episodes of mass extinction.

Like all species, we use the resources available to us in our ecosystem for our own benefit. With our intelligence and technology, we are extremely good at exploiting our habitat, often at the expense of other species.

Preserving biodiversity is critical to our own survival. We depend on the services that our ecosystem provides, and the ability of an ecosystem to deliver these services depends on its biodiversity. Life itself also has intrinsic value, and some argue that this is a sufficient reason to preserve species. Preserving biodiversity will not be easy and will require a number of political, economic, and social changes. In particular, preservation requires recognizing the long-term benefit of a biologically diverse ecosystem.

Chapter Summary

17.1 What Are the Components of Biodiversity?

Distinguish between species diversity, genetic diversity, and habitat diversity and describe how they relate to biodiversity.

- Biodiversity of a habitat depends on species richness (diversity of species alive in an area), genetic diversity (degree of genetic difference among individuals of one species), and habitat diversity (variety of types and sizes of habitats, each with different species).
- Conservation requires that we maintain all these factors.

17.2 What Areas Have the Highest Biodiversity?

Explain the relationships between geographic relief, latitude, and biodiversity and the reasons why the latitudinal gradient exists.

- Biodiversity depends on location.
- The latitudinal gradient is the pattern whereby species richness is highest at the equator and decreases as you move away; there are three hypotheses for why it exists: climate, productivity, and stability.

17.3 What Can Islands Tell Us About Biodiversity?

Describe the species-area relationship and relate it to the theory of island biogeography.

- Larger islands support greater species richness; the theory of island biogeography explains this:
  o Immigration increases the number of species on an island, while extinctions reduce them.

o Larger islands have higher immigration and more resources, while smaller or more distant islands have fewer immigrants and fewer resources.

- Species on islands are often endemic.

17.4 Why Do Different Regions Have Different Species?

Explain the relationship between Wallace's line and the distribution of species.

- Ecologists have identified eight biogeographic realms globally, roughly corresponding to the continents.
- The naturalist Alfred Russell Wallace described "Wallace's line" dividing neighboring South Pacific islands into either the Asiatic or Australian realm; continental movement explains why each realm has unique species.

17.5 How Does Biodiversity Change Through Time?

Describe, in general, the variation in global biodiversity over the course of Earth's history.

- Over Earth's history, new species arise and others go extinct; numbers vary dramatically.
- The Cambrian explosion, 550 million years ago, was a dramatic increase in species diversity, from unicellular or soft, simple multicellular organisms to every major group here on Earth now.

17.6 Why Is Biodiversity Needed for a Healthy Ecosystem?

Detail how biodiversity enhances the productivity and stability of an ecosystem.

- The overall health of an ecosystem depends on productivity and stability.
- The rivet hypothesis compares a healthy ecosystem with many species to an airplane with many rivets: a few rivets can be knocked out and the airplane can still fly, but if too many are lost, the plane will crash.
- If removal of a single species from an ecosystem causes its collapse, that species is the keystone species.

17.7 Why Should We Preserve Biodiversity?

Evaluate the importance of old-growth forests for providing ecosystem services.

- Northern spotted owls are an indicator species.
- In 1990, spotted owls were declared endangered, and over 10 million acres of old-growth forest were put off-limits for logging.

- Old-growth forests are a unique, highly diverse, and productive ecosystem that provide a number of ecosystem services, including provisioning in the form of lumber and medicines; regulating in the form of soil retention and purification; cultural in the form of ecotourism; and supporting in the form of carbon dioxide removal from the atmosphere.

17.8 How Do We Keep Track of Biodiversity?

Describe how biodiversity is monitored.

- The species diversity index allows us to monitor biodiversity.
- High richness and relatively even abundance equals the highest biodiversity.
- An unexpectedly low diversity index suggests adverse conditions.
- Monitoring indicator species or observing changes in ground cover from satellites are ways to monitor biodiversity globally over time.

17.9 Why Might We Be Facing the Sixth Mass Extinction?

Compare the background and current rates of extinction and explain how habitat destruction and overharvesting impact biodiversity.

- Over the history of life on Earth, there have been five mass extinction events, in which 75–95% of species living at the time disappeared.
- The blitzkrieg hypothesis holds that humans are a powerful force for extinction.
- Aside from mass extinctions, most species last between 1 million and 5 million years; the current extinction rate is 100–1000 times "normal."
- Habitat destruction by humans is responsible for about 73% of threatened and endangered species.
- Overharvesting is when humans remove individuals from populations faster than they can be replaced.

17.10 How Can We Preserve Biodiversity?

Examine the steps that can be taken to preserve biodiversity.

- Educate people about long-term benefits of stable ecosystems.
- Enact economic policies that take advantage of those long-term benefits.
- Develop a cooperative approach among scientists, politicians, economists, conservation and religious organizations, and informed citizens.
- Reduce human population growth.
- Develop effective methods of conservation.

Key Terms

background extinction rate 522
biodiversity 500
biogeographic realm 510
blitzkrieg hypothesis 521
Cambrian explosion 512
diversity index 519
endemic 510

genetic diversity 502
habitat diversity 503
indicator species 516
keystone species 515
latitudinal gradient 504
mass extinction 521
productivity 513

rivet hypothesis 514
species abundance 518
species evenness 519
species-area relationship 507
species richness 502
stability 513
theory of island biogeography 507

Review Questions

1. What is the "blitzkrieg hypothesis"? List one alternative hypothesis that could account for the same phenomenon.

2. Define the following terms: species richness, species abundance, and species evenness.

3. What are the three main components of biodiversity?

    a. Species diversity, species productivity, and species stability

    b. Species diversity, genetic diversity, and habitat diversity

    c. Species diversity, species-area relationship, and species evenness

    d. Genetic diversity, species abundance, and species evenness

    e. Species richness, species abundance, and species evenness

4. What is a biogeographic realm? List eight of them. Why do biogeographic realms increase biodiversity at the global level?

5. What was the Cambrian explosion, and approximately when did it occur?

6. What was the Permian extinction, and approximately when did it occur?

7. What three factors does a species diversity index take into account?

    a. Species abundance, species stability, and species richness

    b. Species productivity, species stability, and species richness

    c. Species richness, species diversity, and species productivity

    d. Species richness, species abundance, and species evenness

    e. Species evenness, species abundance, and species stability

8. How does an indicator species simplify the task of monitoring biodiversity?

9. What have ecologists learned about the relationship between biodiversity and productivity of an ecosystem?

10. What have ecologists learned about the relationship between biodiversity and the stability of an ecosystem?

11. Briefly describe one line of evidence that supports the hypothesis that predators prevent herbivores from eating all the plants, rather than the plants' defense systems keeping herbivores at bay.

12. What is one possible effect of removing a keystone species from an ecosystem?

    a. Increased species richness

    b. Increased biodiversity

    c. Decreased biodiversity

    d. Increased species stability

    e. Increased species evenness

13. What are the four main human activities that cause a loss of biodiversity?

    a. Preservation, habitat destruction, overharvesting, and use of chemical pollutants

    b. Conservation, preservation, use of chemical pollutants, and overharvesting

    c. Overharvesting, conservation, use of chemical pollutants, and invasive species displacement

    d. Habitat destruction, overharvesting, conservation, and use of chemical pollutants

    e. Habitat destruction, invasive species displacement, use of chemical pollutants, and overharvesting

The Thinking Citizen

1. How can you be sure that a species has gone extinct? This question refers to species that have been alive within human memory, not long-dead species like dinosaurs. What are some of the difficulties involved with assigning a cause to the extinction?

2. Consider the three hypotheses about why species diversity is high in the tropics. Probably all three play a role. But which of the three do you think offers the most compelling argument? Briefly explain your thinking.

3. The theory of island biogeography states that the number of species on an island is in equilibrium, reflecting the difference between the number of immigrant species and the number that have gone extinct. In the text, we point out that extinction rates are higher on smaller islands, which shifts the equilibrium toward fewer species. How does distance from the mainland affect an island's immigration rate? Consider a series of islands with the same area but at different distances from the mainland. How would distance affect the equilibrium number of species?

4. Many people have argued that losing one species here or one species there will not have a significant effect on anything important to humans. Briefly explain the Ehrlichs' "rivet theory." Does the rivet theory support this argument, refute it, or possibly support and refute it depending on circumstances? Explain your reasoning.

The Informed Citizen

1. Along with the spotted owl, the snail darter (a small freshwater fish) is the poster child for the conflict that frequently arises between conservation and development. Research the snail darter case, which began in 1973 and involved the Tellico Dam in Tennessee. What was the controversy? What were the arguments on both sides? How was the case ultimately decided? If you were in a position to make decisions about cases like this one, what set of guidelines would you use? Briefly explain your reasoning.

2. One of the biggest difficulties in arguing to preserve biodiversity is the trade-off between current and future benefits. A Brazilian farmer could choose to cut down his portion of rain forest now and use the land as a farm to feed his family. Or he could choose to preserve the rain forest because at some unspecified time in the future, a scientist may discover a plant that contains the next wonder cure for a disease. In most cases, the decision to farm would not be a difficult one. What are your suggestions for ways to preserve biodiversity in cases like these?

3. Another factor in the spotted owl controversy involved the barred owl. Over the past 30 years or so, the barred owl has expanded its range into the old-growth forests of the Pacific Northwest. The barred owl is larger and more aggressive than the spotted owl, and as its range expands, the spotted owl population continues to decline. If the barred owl drives the spotted owl extinct, did we waste millions of dollars to protect it? Were the lumber companies right in wanting to cut timber for profit and to provide jobs? Should we feel differently about this type of extinction than we do about extinction due to habitat destruction?

CHAPTER LEARNING OBJECTIVES

After reading this chapter, you should be able to answer the following questions:

18.1 How Can Populations Grow So Fast?

Distinguish between linear and exponential growth and explain how the growth rate is determined.

18.2 Why Don't Populations Grow Forever?

Describe the effects of population density and carrying capacity on the principle of logistic growth.

18.3 How Is Population Growth Influenced by Age and Sex?

Explain what age pyramids can tell us about a country's conditions and social challenges.

18.4 Why Do Developing and Developed Countries Grow Differently?

Compare and contrast the factors that contribute to growth in developing and developed countries.

18.5 How Do We Use Information About Population Growth?

Demonstrate how data on population growth can be used to plan for population shifts and resource depletion.

The continued growth of the human population is not sustainable. Fortunately, because we are capable of recognizing the problem, we are in a better position to find ways of solving it.

CHAPTER 18

Human Population Growth

How Many People Can a Single Planet Hold?

Becoming an Informed Citizen . . .

Our ability to reproduce successfully is programmed in our genes, the result of millions of years of evolution. However, the continued growth of the human population is not sustainable. To make the tough choices ahead, we need to first understand the principles behind population growth.

ne thing that all species are good at is reproduction. If they weren't, they would be extinct. Our ability to reproduce successfully is the result of millions of years of evolution. But will this ability be a source of our downfall? What happens if we reproduce without limit?

As a population gets larger, its habitat fills up, and overcrowding leads to death by starvation, disease, and predation. Advances in agriculture, medicine, engineering, and technology have supported healthy population growth for most of our history. However, the continued growth of the human population is not sustainable. As we become more numerous, we also deplete resources and produce more waste. Overcrowding can lead to political and economic instability as well.

We have an advantage over other animal populations: we can recognize the problem and discover ways to solve it. But we can do this only if we understand the principles behind population growth. In this chapter, we learn what affects population growth, and we compare patterns of growth in different regions of the globe. We begin by showing how fast and dramatic population growth can be—and how easy it is to underestimate.

case study

A Story About Bacteria

Bacteria have the simplest form of population growth possible. A bacterial cell divides in two, each of these divides to produce a total of four cells, then eight, and so forth. But even this simple form of growth can cause trouble. Let's consider the problems faced by one group of fast-dividing bacteria. Although we take some liberties with reality, the principles and consequences of our story are very real.

This particular species of bacteria divides once every minute (much faster than actual bacteria) and lives in test tubes, eating the culture medium in the tubes. If one bacterial cell is placed in a test tube at 11:00 a.m., at 11:01 there will be two cells, at 11:02 four cells, and this growth will continue until the tube is completely full and all the food is gone. This particular tube contained just enough culture medium so that the tube would be full, and the food all gone, at noon. When will the tube be *half* full?

Let's pretend these fictional bacteria, unlike real bacteria, are intelligent and capable of conscious thought and planning. They are fully aware of their finite, test tube world and their 12:00 deadline. They know they have to make plans for their long-term survival, but in the minutes just after 11:00 there are only a few bacteria and lots of food and room to go around. They decide to enjoy their current bounty and postpone making plans until their tube is half full.

As you probably expect, this is a bad decision. Their test tube becomes half full at 11:59, and because the bacterial population doubles every minute, it will take just one minute to be completely full. This gives the bacteria only one minute to plan for the complete ecological collapse of their habitat. Recognizing that they don't have enough time, intrepid bacterial explorers leave their tube and discover a remarkable resource: 15 more test tubes, each the size of their current home. Here's your second question: how long will it take

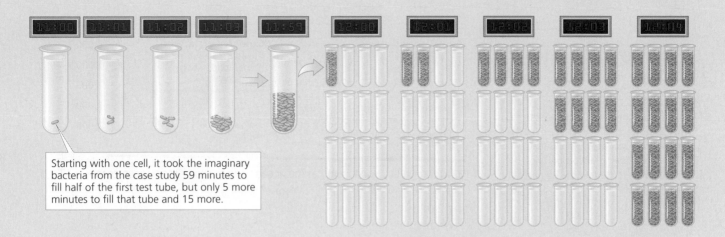

Starting with one cell, it took the imaginary bacteria from the case study 59 minutes to fill half of the first test tube, but only 5 more minutes to fill that tube and 15 more.

the bacteria to fill up these 15 tubes (assuming transportation to tubes is not a problem)?

At noon, the bacteria have 16 tubes (the original full tube plus the 15 new ones). You might think the extra tubes are a reprieve for the overcrowded bacteria, but they don't help much. If half of the bacteria in the first tube move to a second tube, both will be filled by 12:01. Then if half of the bacteria move from each of these tubes to two additional tubes, all four tubes will be filled by 12:02. Continuing this process, the bacteria will fill eight tubes by 12:03, and one minute later—12:04—they will have filled all 16 tubes. Starting with one cell, it took the bacteria 59 minutes to fill half of the first test tube, but only five more minutes to fill that tube and 15 more (**FIGURE 18.1**).

The parallels between us and our imaginary bacteria are clear. We, too, live in a finite world. We, too, need to be concerned about the finite nature of our resources and the Earth's ability to absorb waste and pollution. Underestimating population growth and how quickly we will use up natural resources leaves us vulnerable to the same mistake the bacteria made: not leaving enough time to prepare for its consequences. Understanding population growth requires a different type of reasoning than we use in our daily lives—the type of reasoning we explore in this chapter.

FIGURE 18.1 Even though bacteria have the simplest form of population growth possible, it is possible to draw parallels between their growth and human population growth. If we underestimate how quickly we will use up natural resources, we run the risk of not leaving enough time to prepare for its consequences.

18.1 How Can Populations Grow So Fast?

We begin by considering the pattern of human population growth. **FIGURE 18.2** shows our growth between 1000 AD and 1500 AD. If you lived in 1500, how would you project population growth throughout the rest of the millennium? Based on the data, you might decide to draw a straight line starting at 1500 AD and continuing through to 2010 AD. This is a logical approach. You would be assuming that the second 500 years would be like the first. Mathematically, you are assuming **linear growth**.

linear growth :: growth that proceeds at the same rate over a given time frame

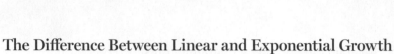

FIGURE 18.2 Linear growth proceeds at the same rate over time, while exponential growth proceeds at an increasing rate. Exponential growth is a fundamental property of all populations, including the human population.

exponential growth :: growth that proceeds at an increasing rate over a given time frame

compounding :: a principle that underlies exponential growth; occurs if new entities are themselves capable of creating new entities

growth rate :: how fast a population grows

The Difference Between Linear and Exponential Growth

But populations don't experience linear growth. The curve line in **FIGURE 18.2** shows that during the second half of the millennium, population growth actually followed a rapidly rising curve. This is known as **exponential growth**, and it is a fundamental property of all populations. Note that by the year 2000, linear growth predicts 500 million people, but the actual figure is 6 billion—the linear projection is off by 5.5 billion people. Starting in the year 1000, when there were fewer than a half billion people on Earth, it took us 960 years to reach 3 billion people, but only 40 more years to double to 6 billion.

The principle that underlies exponential growth is **compounding**. In any growth process, new entities are added to the population. Compounding occurs if the new entities are themselves capable of making still more new entities. Any compounding growth process can grow exponentially.

Without compounding, new entities contribute to growth just by being there, but they do not themselves make more new entities. For example, trees in the temperate zone add one new ring each year, but the rings themselves don't make new rings; the growth in the number of rings is linear. That is why the age of a tree equals the number of rings it has (**FIGURE 18.3**).

On the other hand, when a tree makes branches, each new branch can make several new branches. This is an example of compounding, and the number of branches on a tree grows exponentially (**FIGURE 18.4**). As we saw for bacteria, cell division is a compounding process, too, since each single cell can make two new cells.

Every biological population grows by adding new individuals. Since these new individuals can reproduce and make even more new individuals, this is a compound growth process as well. All biological populations therefore have the potential to grow exponentially.

Defining Growth Rate

Two questions are important for analyzing population growth. How fast does it grow? How big will it get? We'll consider the first question here and address the second in *How Do We Know? Modeling Population Growth*, p. 536.

A population grows when new members are added. Let's consider a small village of 100 people. Suppose five babies are born in the village and two people die during a particular year. Our village of 100 people grew larger by three people. This allows us to determine the **growth rate**: 3/100, or 3%.

FIGURE 18.3 **Trees in the temperate zone add one new ring each year, but the rings themselves don't make new rings; the growth in the number of rings is linear.** That is why the age of a tree equals the number of rings it has.

FIGURE 18.4 **When a tree makes branches, each new branch can make several new ones.** This is an example of compounding, and the number of branches on a tree grows exponentially.

The growth rate tells us how fast a population grows. It also tells us how many people are added to the population during a year (or other appropriate period of time). To figure out how many people are added, just multiply the population size by its growth rate. In this example, a population of 100 growing at 3% adds three people: $0.03 \times 100 = 3$.

These calculations illustrate the most important property of exponential growth: *the number of people added to the population is proportional to the number already present.* This means that when a population is small, few people are added and the population grows slowly. However, as the population gets larger, the number of people added each year increases as well. Population growth accelerates, starting off slowly but eventually increasing dramatically. When you graph exponential population growth over time, you get a characteristic *j*-shaped curve. Human population growth so far appears to follow an exponential curve, as do many animal populations in nature (**FIGURE 18.5**).

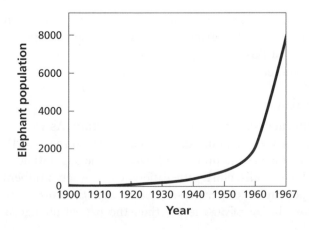

FIGURE 18.5 **Just as human population growth so far appears to follow an exponential curve, so do many other animal populations in nature, including that of elephants.** The graph is based on data from a 1900 to 1967 study of elephants conducted at South Africa's Kruger National Park.

Modeling Population Growth

Throughout this chapter, you will read statements like "The world's population will reach 9 billion by 2050." How do we make projections like this?

The first step is to figure out how fast a population is growing. We've already seen how to do this with exponential growth. The number added to the population is the growth rate multiplied by the population size. Mathematically we write it this way:

$$\text{number added} = rN$$

where r is the growth rate and N is the population size. To get the new population size, we make N bigger by the number of people that were added.

But this tells us only how much the population grew in a single year. One way to figure out how big the population will be in 50 years is to determine how many people are added in year 1, year 2, year 3, and so on for 50 years and then add them all together.

But this involves a lot of work, so mathematicians have devised a way to do all these additions in one step. To determine how big a population will be at some time in the future, you can use this equation:

$$\text{size} = N_0 e^{rt}$$

where N_0 is the starting population size, r is growth rate, and t is the number of years in the future you're considering; e is a mathematical constant, like π. Raising e to a power is called the *exponential function*, which is where *exponential growth* gets its name. You probably have a button labeled

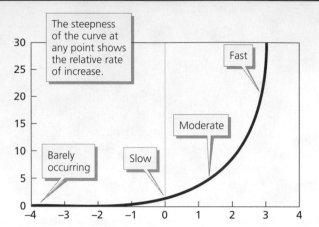

To predict population size, scientists calculate growth using *e*, the exponential function.

"e^x" on your calculator, which enables you to calculate this equation.

Let's work through an example. Suppose a population starts with 1000 individuals (N_0) and grows at 5% per year ($r = 0.05$). How big will the population be in 50 years (t)? Multiply r by t, raise e to this power, and multiply by N_0. The answer you get is 12,182 individuals, a 12-fold increase in 50 years.

This equation represents the exponential model of population growth, which works well for the early stages of a population's growth. But there are many other models that take into account factors that are missing in the exponential model. What happens when the population grows too large and starts to run out of food? How does a population's age structure affect its growth? Throughout this chapter, we will describe how these factors affect population growth.

The shape of the exponential growth curve helps to explain why we often fail to plan correctly for processes that grow exponentially. The early phase of slow growth lulls us into a false sense of security. Then we are caught unawares by the dramatic increase at the end.

Determining Growth Rate

When we speak of growth rate we really mean net growth rate. As we saw earlier, the number of babies born in a population doesn't determine growth rate by itself. You need to subtract the number of deaths in a population to arrive at the net change. We can also clarify the difference between numbers and rates. Rate refers to per capita (or per person) changes. If the *number* of individuals added to a population of 500 is 20, then the *rate* of change is 20/500: 0.04, or 4%.

In a population, birth rate is the number of babies born per capita, and death rate is the number of deaths per capita. The (net) growth rate is the birth rate minus the death rate. Consider our earlier example:

$$\text{growth rate} = (\text{birth rate} - \text{death rate})$$
$$= (5/100 - 2/100)$$
$$= 0.03$$

This value is sometimes referred to as the natural growth rate, since it is based on natural processes like birth and death. When ecologists study the population growth of animals and plants they often use this natural growth rate. For human populations, however, we need to introduce another consideration. Since we are a globally mobile species, migration can also affect a country's population growth. Like birth and death rates, migration breaks down into two components: immigration (those moving in) and emigration (those leaving). Like the birth rate, immigration adds to a population, and like the death rate, emigration subtracts from it. If we take migration into account, we get our final formula for the growth rate:

growth rate = (birth rate – death rate) + (immigration rate – emigration rate)

TABLE 18.1 lists these rates for the United States during the period 2010–2012. The country's growth was under 1%, and you can see how the growth rate was determined by the other four rates. When working with human populations, these rates are not expressed as percentages or fractions. As Table 18.1 shows, these values are expressed as number per 1000 individuals in a population (number per 100 would be a percentage). Although percentages are more familiar, we will adopt this standard for the remainder of the chapter and report rates as number per 1000.

Equilibrium

You can think of growth rate as having two positive components (birth and immigration) and two negative components (death and emigration). The growth rate is the "balance" among these four components. If the positive components outweigh the negative, the growth rate is positive and the population grows exponentially. However, if the negative components outweigh the positive, population growth is negative and the population gets smaller.

Table 18.1 :: U.S. Growth Rate Components: April 1, 2010–July 1, 2012

	Value	
	Number per 1000	**Per Capita Rate**
Birth rate	14.4	0.0144
Death rate	9.1	0.0091
Immigration rate	3.0	0.0030
Emigration rate	0	0
Growth rate	8.3	0.0083
Population size	308,745,538 (April 1, 2010)	

Source: United States Census Bureau, http://www.census.gov/popest/data/state/totals/2012/index.html.

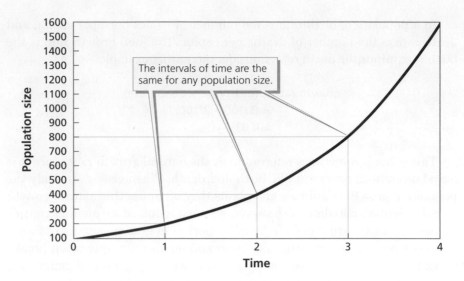

FIGURE 18.6 Doubling time is the amount of time required for a population to double in size. It depends only on the growth rate, not on the population size.

equilibrium :: a state when the population's size does not change

If birth and immigration exactly equal death and emigration, the growth rate is zero. When this occurs, the population is in **equilibrium** and its size does not change. As we will see later in the chapter, countries show a variety of growth rates. Many developing countries have large positive growth rates, while many European countries have negative growth rates.

Doubling Time

Exponential growth has an interesting and useful property: the amount of time required for a population to double in size is always the same (**FIGURE 18.6**). This **doubling time** depends only on the growth rate, not on the population size. To calculate doubling time, divide 0.69 by the growth rate (0.69 is the natural log of 2 and is a numerical constant used in calculations of exponential growth):

doubling time :: the amount of time required for a population to double in size

$$\text{doubling time} = 0.69/\text{growth rate}$$

Doubling time often gives a clearer and more intuitive interpretation of population growth than does the growth rate. For example, Table 18.1 shows that in 2010, the growth rate for the United States was 8.3 per 1000 individuals, or 0.0083—neither number is easily visualized. The doubling time makes more sense. At this rate of growth, the U.S. population would double in 0.69/0.0083, or 83 years. For comparison, Kuwait's population doubles in only 19 years. Doubling times can be meaningful for social planning purposes, such as deciding how many schools an area needs and how quickly they must be built.

 # Why Don't Populations Grow Forever?

Exponential growth as just described continues forever, with the population growing continuously larger and ever faster. If any natural population grew like this, we would surely notice. Before long, we'd be swamped with that species. The bacteria from our case study would blanket the entire planet in less than two hours. Clearly, exponential growth is at best temporary. But what makes it stop?

The Effects of Population Density

As a population grows larger, it begins to have a negative impact on its environment and on its members. The amount of food and space available to each individual decreases, and waste products and pollution increase. Diseases and parasites spread more easily. Close contact among individuals may lead to aggressive conflicts. Factors like these are called **density dependent**, because they grow worse with density, or crowding. Density-dependent factors make it harder for individuals to reproduce and less likely that individuals will survive as population size increases (**FIGURE 18.7**).

Is our population affected by density-dependent factors? The late 18th-century British scholar Thomas Malthus was among the first to argue that it is. He observed that the human population grows exponentially, but the human food supply grows only linearly: each newly cultivated acre produces only 1 acre of increased yield. Malthus predicted that human population growth would soon outstrip its food supply, and dramatically so. He feared that starvation and social collapse would surely follow.

Malthus's warnings have come true in many areas hit by famine, drought, or other causes of crop failure. Globally, food production has kept up with population growth because we have developed new varieties of high-yield crops and productive, technology-dependent agriculture. While some scientists believe that technological advances will allow the human population to grow without limit, most recognize that exponential growth is not possible if resources are finite. See *Scientist Spotlight: Donella Meadows* to learn about one of the first and most influential wake-up calls regarding sustainable living.

density dependent :: a factor responsive to population crowding

FIGURE 18.7 Density-dependent factors, such as limited food or space, grow worse as populations become larger and make it harder for individuals to survive and reproduce. Shown here, Pakistani women struggle as they try to order food outside of a subsidized food store on the outskirts of Rawalpindi, Pakistan.

Scientist Spotlight

Donella Meadows (1941–2001)

Donella Meadows was a true environmental visionary in the tradition of John Muir, Aldo Leopold, and Rachel Carson. Meadows shared their sensitivity to the environment, and she lived her personal life in a way that reflected her environmental ethic. She brought a particular skill to her professional work: a strong scientific background that allowed her to examine environmental problems analytically and quantitatively. Her work presented a formidable challenge to conventional wisdom about the environment and the economy, and the issue of population growth took center stage in her analysis.

Meadows grew up on the outskirts of Chicago. After earning her doctorate from Harvard, she became a research fellow at MIT, working with Jay Forrester. Forrester's research involved what he called "system dynamics," using a computer to simulate change in a complex system. They were able to ask how a whole host of factors like resource use, medical advances, and pollution affect population growth— and how population growth in turn affects them.

In 1972 Meadows, Forrester, and others published a seminal book, *The Limits to Growth*, in which they explain exponential and logistic growth. They simulated the growth of almost every important aspect of the world economy using system dynamics and reached a startling conclusion:

> If the present growth trends in world population, industrialization, pollution, food production, and resource depletion continue unchanged, the limits to growth on this planet will be reached sometime within the next hundred years. The most probable result will be a rather sudden and uncontrollable decline in both population and industrial capacity.

It is difficult to imagine the stir this caused. The very thought of limits collided with an economic system based entirely on growth. Meadows's work posed a severe threat to conventional thinking and legitimized the budding environmental movement.

Limits to Growth argued for sustainable living, rather than one based on growth. Meadows embraced this idea in her personal life. She taught environmental systems and ethics at Dartmouth for 29 years. In 1997, she founded the Sustainability Institute, which she referred to as a "think-do tank," rather than a think tank. For 16 years, Meadows wrote a syndicated column titled "The Global Citizen," in which she discussed many issues of environmental and social concern. She lived simply on an organic farm in New Hampshire. Meadows died in 2001 at the early age of 59, after a two-week battle with bacterial meningitis.

Donella Meadows was a true environmental visionary. She brought to the world's attention the fact that we would not survive unless we started living in a sustainable manner.

Logistic Growth

The reason why exponential growth assumes a population grows forever is that density-dependent factors are not taken into account. Leaving out density dependence makes sense in the early parts of a population's growth: when a population is small, there is plenty of food and space available, and for a while its growth *is* exponential.

But as density-dependent factors come into play, the population's growth rate slows down. Recall that growth rate is the difference between birth rate and death rate (we're ignoring migration here). Before density dependence comes into play, birth rate is high and death rate is low. But density-dependent factors make reproduction more difficult and dying more likely. As the population grows, birth rate declines and death rate increases.

One result of these changing rates is that the growth rate decreases. The population starts off growing exponentially, but eventually growth slows as density-dependent factors reduce the growth rate. A second result is that as birth rates go down and death rates go up, there must be a point at which they are equal. When the growth rate is zero, the population is in equilibrium and its size does not change. This unchanging population size has a name—the **carrying capacity**. You can think of carrying capacity as the maximum number of individuals the environment can "carry," or support.

When we take density dependence into account, we get a new pattern of population growth called **logistic growth**. Logistic growth starts off fast, almost exponentially, when the population is small. But growth slows down as the population increases. And growth stops altogether when the population reaches its carrying capacity. Logistic growth produces an *s*-shaped pattern of growth (**FIGURE 18.8**). Logistic growth is a more realistic description of how populations grow in nature, since density-dependent factors are a fact of life for all biological populations.

What happens when a population exceeds its carrying capacity? Individuals will starve and die until the population shrinks back to a sustainable level. The birth rate will decline, and the death rate will increase until the growth rate is negative.

carrying capacity :: the maximum number of individuals the environment can support

logistic growth :: a pattern of population growth that takes density dependence into account; starts off exponential and stops when the population reaches carrying capacity

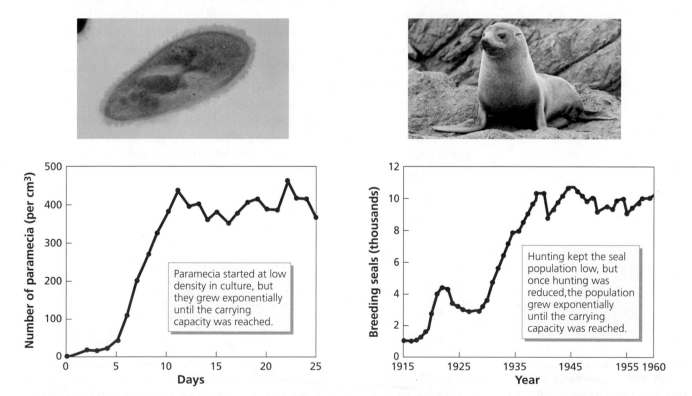

FIGURE 18.8 Logistic growth takes density-dependent factors into account: it starts off fast when the population is small, but then slows down, eventually leveling off, as the population gets larger. Logistic growth produces an *s*-shaped curve, as shown in population studies of paramecium and fur seals.

FIGURE 18.2 (p. 534) provides good evidence that our population is growing exponentially, at least for now—our population doubled between 1960 and 2000. If it continues to double every 40 years, we will reach 12 billion by 2040 and 24 billion by 2080. Most experts agree that our food production will not increase nearly as fast. Neither will our ability to provide clean, fresh water. In fact, according to a 2012 United Nations report, 783 million people do not have dependable access to water suitable for drinking. This represents 11% of the world population. Similarly, our abilities to remove waste and pollution, or to provide health care and education, lag behind our population's needs. If we are unable to provide basic services for the population, the impact on political and economic stability will be dire.

Fortunately, data show that we are approaching a logistic phase in our population growth. Many organizations—including the United Nations, the United States Census Bureau, and even the Central Intelligence Agency—monitor the global population. They have found that global population growth is slowing down. During the last half of the 20th century, global population growth was estimated at around 2% per year. Currently, growth is near 1.3% annually, and it is projected to decline to 0.5% by 2050. If these estimates are correct, our population size will reach 9 billion by 2050 and level off near 10 billion by 2200 (**FIGURE 18.9**).

Our Carrying Capacity

What is the carrying capacity for the human race? After more than 300 years of study, there still is no definitive answer. The carrying capacity depends on things that are harder to predict, from technological breakthroughs to subjective judgments regarding the quality of life. The global carrying capacity will be lower if every individual expects the quality of life we have in the United States, for example.

Despite the uncertainty, many scientists have taken a crack at assessing the carrying capacity. Estimates run as low as 2 or 3 billion, in which case we are already significantly above our carrying capacity. Most estimates, however, are in the range of 8–15 billion.

For now, the projected decline in growth rate is encouraging. Still, a 50% increase in population size by 2050 is significant. And the decline in growth rate is not universal. Population continues to increase in many developing countries, including regions that have difficulty sustaining the population they already have.

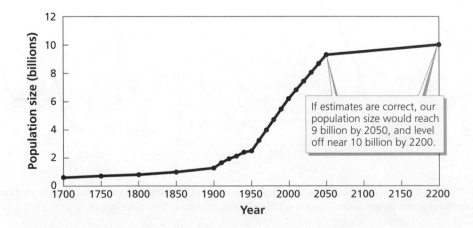

FIGURE 18.9 Many organizations that monitor population have found that global population growth is slowing down and we are approaching a logistic phase.

How Is Population Growth Influenced by Age and Sex?

Our story about the bacteria at the start of this chapter had a hidden assumption: that all bacterial cells in the population were the same. In other words, all individuals contributed equally to population growth. The exponential and logistic growth models alike make this assumption about populations. In reality, people do not contribute equally to population growth. Their contribution varies depending on their age and sex, which affect their fertility and their chances of survival.

Age, Sex, and Population Growth

Females have a greater impact on population growth than males. A population can't grow faster than the rate at which females produce children, so adding (or subtracting) females more directly affects how fast a population grows than does adding or subtracting males. In many animal populations, for example, only a few of the males actually mate—sometimes only one. Adding more males won't increase population growth, and removing males won't decrease it.

Not all females contribute equally to population growth, however, mainly due to age. Women go through three life-cycle stages: prepubescent, reproductive, and menopausal. A woman may not contribute at all to the birth rate for more than half her life. Even in their reproductive stage, women vary in their fertility. Typically, females in the middle of their reproductive years have more children than younger and older females.

Age also affects growth rate for another reason: the chances of survival. Diseases often affect people of different ages differently. A disease of the elderly has little effect on population growth. A disease that affects children has little effect on current population growth but may reduce population growth in the future. A disease that affects women at the peak of their reproductive years can have a dramatic effect on growth. In short, the age distribution of the population affects both reproduction and mortality.

Age Pyramids

Demography is the science that analyzes population growth by examining a population's age structure and how age affects fertility, survival, and population growth. An **age pyramid** (**FIGURE 18.10**) is a snapshot of the population at a given point in time. The age pyramid has separate sides for males and females, and each bar on the pyramid shows the number of individuals in a given five-year age class. Thus, an age pyramid shows both the sex ratio and age structure of a population.

Age pyramids fall into three shape categories. Typically, the age pyramid is widest at the bottom, since children usually are the largest age group, and narrows toward the top due to mortality. Other shapes are possible, however, including reasonably vertical sides for the non-elderly age classes, as well as a tapering bottom that is narrower than the top (**FIGURE 18.11**). Properly interpreted, the shape of an age pyramid tells us a great deal about conditions in a country and the social challenges it is likely to face.

From the end of World War II in 1945 until 1964, the United States experienced a growth spurt—the baby boom. In 1955, children from this growth

demography :: the science that analyzes population growth

age pyramid :: a snapshot of a population at a given point in time; displays the number of individuals in age categories

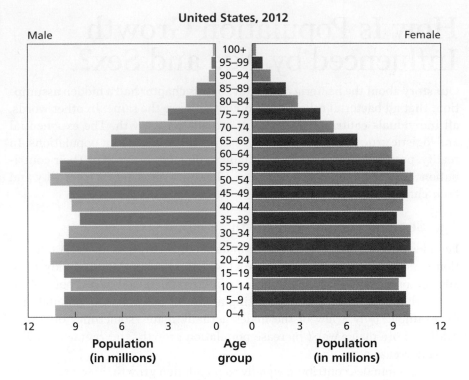

United States, 2012

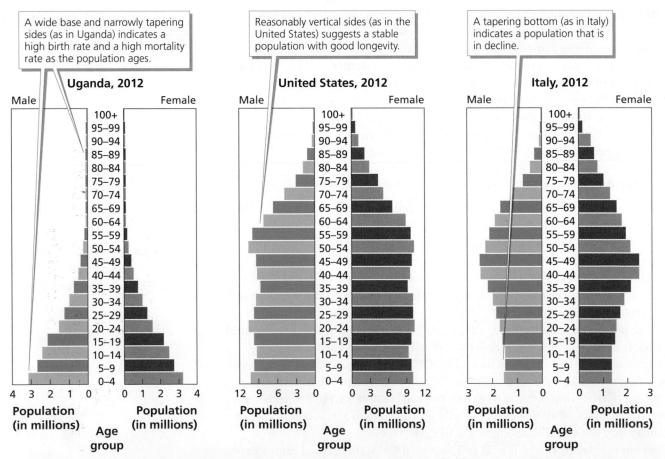

FIGURE 18.10 An age pyramid (such as the one shown for the United States in 2012) is a snapshot of the population at a given point in time. It shows both the sex ratio and age structure of a population.

A wide base and narrowly tapering sides (as in Uganda) indicates a high birth rate and a high mortality rate as the population ages.

Reasonably vertical sides (as in the United States) suggests a stable population with good longevity.

A tapering bottom (as in Italy) indicates a population that is in decline.

FIGURE 18.11 Age pyramids fall into three shape categories, but typically they are widest at the bottom and narrower toward the top. Properly interpreted, the shape of an age pyramid tells us a great deal about conditions in a country and the social challenges it is likely to face.

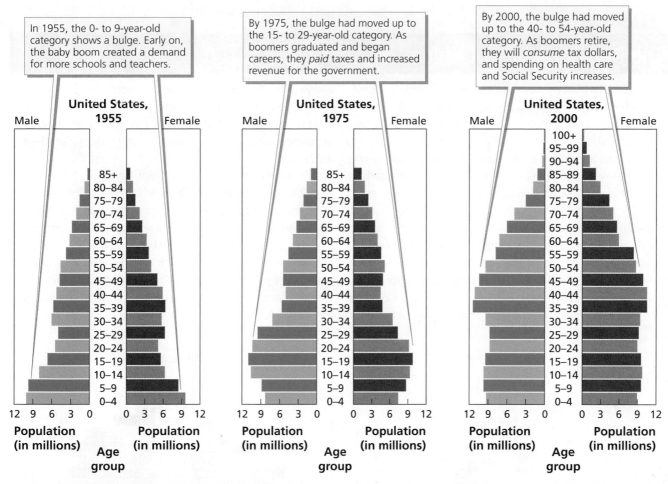

In 1955, the 0- to 9-year-old category shows a bulge. Early on, the baby boom created a demand for more schools and teachers.

By 1975, the bulge had moved up to the 15- to 29-year-old category. As boomers graduated and began careers, they *paid* taxes and increased revenue for the government.

By 2000, the bulge had moved up to the 40- to 54-year-old category. As boomers retire, they will *consume* tax dollars, and spending on health care and Social Security increases.

FIGURE 18.12 From the end of World War II in 1945 until 1960, the United States experienced a growth spurt known as the *baby boom*. The movement of the baby boom through the age pyramid has important consequences.

spurt were between 0 and 10 years old. The 1955 age pyramid (**FIGURE 18.12**) shows a bulge in these age classes. You can follow this bulge up the age pyramid as the baby boomers age. In 1975, for example, children born between 1945 and 1960 were between 15 and 30 years of age. In 2000, the boomers were between 40 and 55 years of age, and the bulge has moved up the pyramid accordingly.

The movement of the baby boom through the age pyramid has important consequences. Early on, the baby boom created a demand for more schools and teachers. As the boomers graduated and began careers, they paid taxes and increased revenue for the government. As boomers retire, however, they *consume* tax dollars, as spending on health care and Social Security will likely increase. At the same time, the number of individuals of working age will decline, along with the tax revenue they produce. Of course, policy decisions can compensate for the shifting age structure of the population. Tax rates and spending can both change. Nevertheless, it is easy to see how age pyramids help us decide what changes are needed.

The 2000 age pyramid shows a small population bulge about 25 years behind the baby boomers, the so-called baby boomlet (**FIGURE 18.12**). These are the children of the boomers, and you can predict that other and ever-smaller wavelets of growth may occur at 20- to 30-year intervals.

If you look closely at the numbers of baby boomers (the width of the bars), you will note that they don't change in size much. This is because survivorship

Life Application

The Demographics of China

No country has felt population pressure as strongly as China, which has almost a fifth of the world's total inhabitants. Its population control methods have been successful, but also controversial.

China's age pyramid for 1990 (the earliest data available on the Census Bureau site) shows the characteristic wide base and inward sloping sides of a growing population. By 1979, however, China had already announced its one-child policy: each couple was permitted only a single child, and financial penalties were imposed for violations. Children born between 1976 and 1980 would have been 10–14 in 1990. The "notch" in the age pyramid for 10- to 14-year-olds shows the result of the one-child policy. And as you might imagine, the rule wasn't enforced equally everywhere. For instance, it did not apply to rural areas, where children were needed for farm labor. Even so, one child per couple in principle results in a 50% decrease in population size, as each couple is only half replaced—a drastic step.

By some measures, China's efforts have been successful. Its population is predicted to peak at 1.46 billion around 2030 and decline thereafter. Perhaps as many as 300 million people will not have been born as a result. In 1990, China had 21 births per 1000 individuals, and couples had 2.2 children on average, still a positive growth rate. By 2050, the births may drop by 50%, and the aging population will actually decline. We can picture this prediction in an age pyramid for 2050.

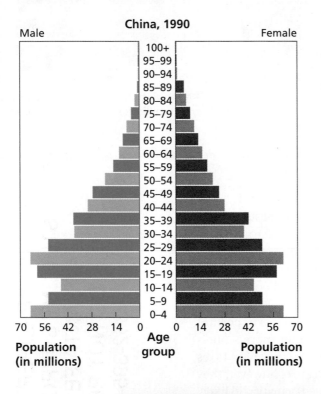

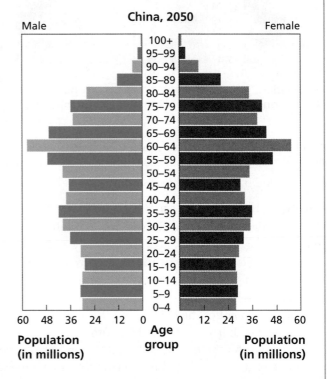

	1990	2050
Births/1000	21	10
Deaths/1000	7	13
Growth rate (%)	1.4	−0.3
Total fertility	2.2	1.7

China, home to nearly 20% of the world's population, has implemented dramatic population control measures. In 1990, China had 21 births per 1000 individuals, but by 2050 the births are expected to drop by 50%.

For all their success, government-imposed restrictions on reproduction have been very unpopular. Critics argue that the one-child policy has led to widespread abortion and even infanticide. It has definitely led to a preference for boys over girls, resulting as of 2009 in 1.13 male boys for every girl.

That means 17 million more boys than girls. Among adults there are 27 million more men than women.

All countries must eventually address population growth. Will that happen by choice, by natural causes, or by government regulation? It is an open question.

is high, and a relatively small fraction of the boomers die. The small increase in numbers is due to immigration of people in the same age classes as the baby boomers.

The age pyramids also show that longevity has increased. The 1955 and 1975 age pyramids show only 1–1.5 million over age 85. In 2000, there were over 4 million people in this age class. For younger people, the sex ratio is 50:50. But at 75 years and older, the number of males begins to decline more than the number of females. Men still live shorter lives than women, though both sexes live longer.

Finally, age pyramids are also useful for comparing growth of different countries. We will return to analyze age pyramids for Uganda, the United States, and Italy later in the chapter. *Life Application: The Demography of China* details the patterns of population growth and consequences of government measures instituted to control this growth.

18.4 Why Do Developing and Developed Countries Grow Differently?

Two additional demographic factors affect population growth: total fertility and the age of a woman at the time she has her first child. Developing and developed countries differ strikingly when it comes to these two factors.

Total Fertility and Age at First Reproduction

Total fertility is the number of children a woman *could* have, on average, if she lived at least until the end of her reproductive years. Birth rate is the average number of children a woman *does* have (**FIGURE 18.13**). Total fertility gives a higher estimate of growth than does birth rate, because some women will die before having all the children they could. On the other hand, total fertility gives a more accurate picture of a population's potential to grow if female longevity were not a factor.

As we have seen, population growth rates can be negative, zero, or positive. Total fertility provides similar information. If total fertility is 2, then each couple exactly replaces itself with children. We can call this zero population growth (ZPG). If total fertility is more than 2, each couple more than replaces itself, and population size increases. If total fertility is less than 2 the population size decreases.

total fertility :: the number of children a woman could have, on average, if she lived at least until the end of her reproductive years

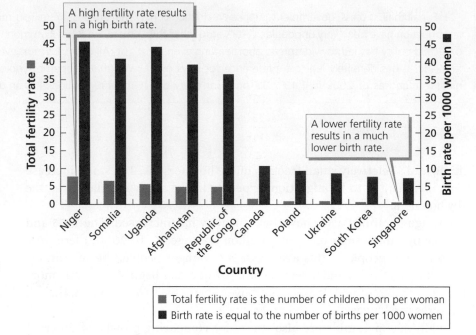

FIGURE 18.13 Total fertility is the number of children a woman *could* have, on average, if she lived at least until the end of her reproductive years. Birth rate is the average number of children a woman *does* have and is equal to the number of births per 1000.

A second important demographic factor is age at first reproduction (**FIGURE 18.14**). This refers to when women actually give birth to their first child, not the age of puberty. Women in developed countries tend to bear children later in life. This works to reduce population growth in several ways. First, a woman who postpones having children may not live to have as many. Second, postponing childbirth reduces the number of generations alive at any one time. If people live to see their great-grandchildren, then four generations are alive at once.

Fertility and Mortality Differences

Developing countries have a much higher total fertility rate than developed ones. **TABLE 18.2** shows that Uganda has a total fertility rate of 6.06 children per female, the fourth highest in the world. This compares to 2.06 and 1.41 for the United States and Italy, respectively. If we consider developing countries as a group, their total fertility is notably higher than that of developed countries.

Why is there such a big difference in total fertility? Clearly couples choose their family size based on a large number of economic, religious, and social considerations. Part of the explanation comes from infant mortality (**FIGURE 18.15**).

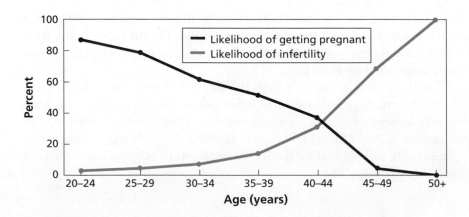

FIGURE 18.14 Age at first reproduction refers to when women actually bear their first child. Starting younger allows a woman to have more babies, as the likelihood of conception decreases as she ages.

Table 18.2 :: Demographic Data for Uganda,
United States, and Italy (2012)

Demographic Factor	Uganda	United States	Italy
Growth rate	3.3% (4th)	0.9% (124th)	0.38% (157th)
Birth rate/1000	45.8 (3rd)	13.7 (147th)	9.06 (208th)
Death rate/1000	11.6 (33rd)	8.4 (84th)	9.93 (54th)
Migration rate/1000	−0.02 (115th)	3.62 (26th)	4.67 (22nd)
Total fertility (births/female)	6.06 (4th)	2.06 (121st)	1.41 (203rd)
Infant mortality/1000	64.2 (22nd)	6.0 (173rd)	3.36 (215th)
Life expectancy (years)	53.45 (205th)	78.49 (51st)	81.86 (10th)

Source: CIA World Factbook, https://www.cia.gov/library/publications/the-world-factbook/.

In developing countries, about 48 of each 1000 infants die before they reach their first birthday, while in developed countries, 6 infants out of 1000 on average suffer this fate. In developing countries, bearing more children is a safety net to ensure that at least some of your offspring survive.

There are many causes of infant mortality in developing countries, including malnutrition, disease, lack of sanitation, and limited medical care. These causes are density dependent, which means they have a greater impact when population density is high. Overall, the population density of developed countries is about 24 individuals per square kilometer, while the density of developing countries is almost three times greater.

Family Planning Differences

The decision about when to start a family also varies greatly from individual to individual and nation to nation. Typically, women in developed countries begin having children later in life than women in developing countries.

FIGURE 18.15 In developing countries, about 48 of each 1000 infants die before they reach their first birthday, while in developed countries, 6 infants out of 1000 on average suffer this fate.

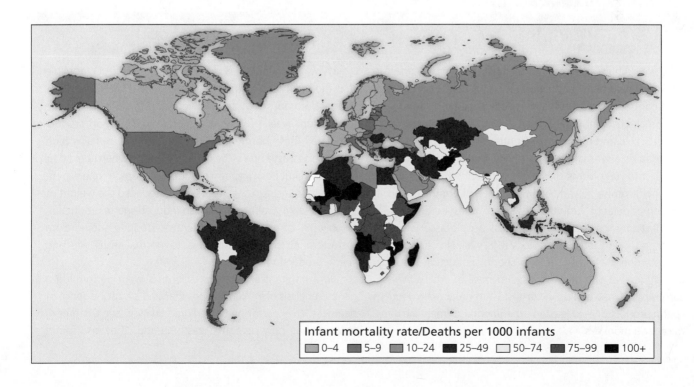

Infant mortality rate/Deaths per 1000 infants
□ 0–4 ■ 5–9 ■ 10–24 ■ 25–49 □ 50–74 ■ 75–99 ■ 100+

Biologically, many girls can become pregnant before they are 13. In some developing countries, cultural practices consider the onset of puberty to be the age of consent for marriage. Childbirth can follow quickly thereafter. In countries where life expectancy for parents is short, early marriage can provide a daughter with security for the future. This practice also results in fewer mouths for the parents to feed.

For example, girls in Uganda become sexually active at 16 on average, and in 2006 girls under 18 accounted for almost one-third of all pregnancies. Early marriage tends to increase total fertility and lower the age of first reproduction, both of which increase population growth. Sadly, growth is mitigated a bit because young girls can die of complications from pregnancy, as 6000 teenagers do annually in Uganda.

In contrast, women in developed countries often have greater access to family planning and may be more likely to postpone childbearing because of education or a career. In the United States, for example, about half of women marry by age 22, but the other half marry substantially later or not at all. In addition, in industrialized developed countries, children are not viewed as an important source of labor (as they are in agrarian developing countries), so there is less urgency to have children early. The result of these and other relevant factors is that total fertility is reduced, the age of first reproduction is raised, and there are fewer years devoted to childbearing, which reduces population growth. To learn more about research on the development of new methods of regulating reproduction, see *Technology Connection: Male Contraception*.

When it comes to population growth, perhaps the single most important difference between developing and developed countries is women's rights. In some countries, economic or religious conditions encourage women to marry young. For example, the average age for marriage is 20 or younger in Chad, India, Uganda, Bangladesh, and Niger. Some of these young women are denied an

Technology Connection

Male Contraception

Who would have guessed that scientists studying deafness in Iranians would discover a clue that could lead to an effective male contraceptive? But that is exactly what happened. Previously, males had only three options for contraception: condoms, vasectomies, and abstinence. Contraception has often been regarded as a woman's responsibility, but several lines of research into new forms of male contraception are beginning to show promise. Given the importance of contraception to limiting population growth, these discoveries are timely.

Testosterone is a hormone most people associate with males, although females produce it as well. Not surprisingly, testosterone plays a big role in sperm production, and low testosterone is one of the causes of male infertility. What is surprising is that adding *extra* testosterone actually *reduces* sperm production—at least in experimental animals like rams, mice, and monkeys.

Why this happens is not completely clear. It's possible that the brain uses testosterone level as its "fuel gauge" for several hormones that are involved with male fertility. If the testosterone level is too high, the brain may shut off production of *all* these hormones. Testosterone level remains high (because testosterone is being added), but when the other hormones shut down, sperm production drops. In one experiment with rams, external testosterone reduced the weight of the testis to 24% of normal. It also reduced sperm production to only 3% of normal. After scientists stopped adding extra testosterone, the rams returned to normal testis size and sperm production.

In a different experiment, scientists were performing a genetic analysis of Iranian families that had a high incidence of deafness. They performed a pedigree analysis (see Chapter 4) to determine how deafness was inherited. They also found that in two families, male infertility seemed to be inherited. The scientists analyzed DNA from these subjects and discovered that they had a mutation in a gene called *CATSPER1*. The *CATSPER1* gene makes a protein that is found on the flagellar "tail" of sperm cells. When a sperm cell encounters an egg, its tail begins to wiggle violently, a process known as hyperactivation. Hyperactivation is required for the sperm to get through the egg's outer membranes for fertilization. The *CATSPER1* mutation prevented the sperm cells from hyperactivating, so they could not fertilize the egg.

The *CATSPER1* mutation also suggests a method of contraception. If it were possible to knock out *CATSPER1* in males, they would not be able to father children. One way to do this is to inject antibodies that attack the protein, much in the same way that your immune system makes antibodies to attack the proteins in disease organisms. This approach is known as immunocontraception. The process would be reversible, just by stopping the injections. It would also be very specific. Unlike testosterone, which acts in many different ways throughout

the body, the protein produced by *CATSPER1* is found only in sperm cells, so these would be the only cells affected.

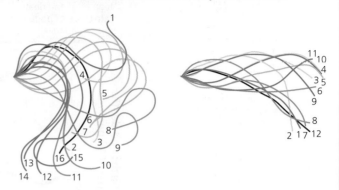

A Normal sperm   **B** Mutant sperm

Normal sperm exhibit hyperactivation swimming behavior (A), whereas CATSPER mutant sperm do not (B). The numbers refer to the positions of the sperm tail as it moved and then was captured in freeze-frame, at one beat per cycle and 16 frames in total.

education and the right to vote, are kept subservient to their husbands with little access to contraception, and have limited financial and social independence.

Education of women has a particularly powerful influence on reproduction. In Kenya, for example, adding one more year of schooling decreases by 10% the probability of giving birth while a teenager. Advancing women's rights, especially the right to an education, is a critical ethical goal, but it is also one of the most effective ways to control population growth.

Demographic Transition

Over time, developing countries become developed countries. Industrialization occurs, urbanization increases, and the number of people working on farms decreases. Education becomes more important as children become a less important part of the labor force. Demographic factors also change, a process known as **demographic transition**.

Demographic transition divides a country's history into several stages: pre-development, transitional, and developed (**FIGURE 18.16**). During the pre-development stage, the country experiences very high birth and death rates. These populations live close to their carrying capacity, and overall population growth is limited.

The transition begins when the country starts to develop due to improved agriculture and increased interaction with developed countries. This interaction can lead to better medical practices, more sanitation, a cleaner water supply, more technology, and better education, which increases the carrying capacity. The death rate drops rapidly, as more children survive and life expectancy increases. Since birth rates at first remain constant, population growth increases in an exponential fashion.

demographic transition :: the change in population growth patterns seen over time as developing countries become developed countries

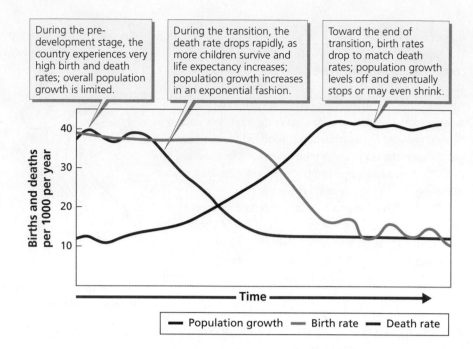

During the pre-development stage, the country experiences very high birth and death rates; overall population growth is limited.

During the transition, the death rate drops rapidly, as more children survive and life expectancy increases; population growth increases in an exponential fashion.

Toward the end of transition, birth rates drop to match death rates; population growth levels off and eventually stops or may even shrink.

FIGURE 18.16 Demographic transition refers to the changes in population growth patterns seen over time as developing countries become developed countries.

Toward the end of the transition, birth rates drop to match death rates. Urbanization increases, the status of women improves, access to contraception reduces the number of births, and women tend to postpone childbirth well past puberty. In time, population growth levels off and eventually stops. In some cases, the birth rate may drop below the death rate, causing the population to shrink.

Most of the world's developing countries are in the middle of their demographic transition. Death rates have dropped, but reproductive practices have not yet resulted in lowered birth rates. As a result, these countries are growing exponentially. They will account for the majority of population growth over the next century. We all stand to benefit if population growth slows, so economic, medical, and technological aid to developing countries is a worthwhile investment.

Demographic transition provides a uniquely human explanation for logistic growth. Ecologists studying animal and plant populations find that biological factors alone affect population growth. These species have been selected to maximize their lifetime reproduction, and their population growth is limited exclusively by starvation, disease, predation, bad weather, and similar factors. Over our evolutionary history, we have been selected in much the same way. But now we have intelligence, technology, and the ability to make choices about our reproduction. We should, therefore, be able to find better solutions to the challenges of population growth.

 18.5 # How Do We Use Information About Population Growth?

Every government keeps track of population data, and many nongovernment organizations do so as well. Insurance companies use population data to make predictions about life spans, and market researchers keep track of demographics when figuring out what to sell and to whom. The CIA keeps tabs on population data from every country, because it sees population pressure as a potential security threat.

Why is demographic information so important? It is an integral part of our representative government. Anticipating demographic shifts is vital to planning for almost all segments of society. And the demographic analysis of population growth also helps us analyze our use of critical resources.

The Constitution and the Census

The drafters of the U.S. Constitution recognized an important fact: a representative government requires detailed knowledge of how many people are being represented. A state's number of representatives in Congress and votes in the Electoral College depend on its population. Article 1, Section 2 of the Constitution, therefore, mandates a census at least every 10 years. The first national census was taken in 1790, and the Census Bureau was established as an official department in 1903.

Today, in addition to counting individuals, the Census Bureau keeps records of age, sex, ethnicity, family size, marriage, divorce, employment status, household income, and more. These data are used to allocate funds appropriated by Congress for public utilities, upkeep of roads and bridges, and school construction. About $300 billion is distributed annually, and accurate information on population size and growth is critical to making the distribution as equitable as possible.

Planning for Population Shifts

As we have seen, the shape of a country's age pyramid provides information about its future. By exploring a country's history, demographers can project what the country's pyramid will look like many years from now.

Recall the sharply narrowing sides of Uganda's age pyramid (Figure 18.11, p. 545), which reveal low rates of survival in all age classes, due in part to political upheaval. Following independence from British colonial rule in 1962, Uganda experienced the purges of a dictator (Idi Amin, 1971–1979) and the ravages of civil war (led by Milton Obote, 1980–1985). Up to half a million people were killed directly, and many more died from the loss of food, medical care, and other social services. Uganda was also hit early and hard by the AIDS pandemic. Ten years after the first case was diagnosed in 1983, the percentage of people infected reached 29% in the capital, Kampala, and 14% elsewhere.

The wide base of Uganda's age pyramid reflects a high birth rate and the early stages of demographic transition. Death rates are declining for several reasons. There is more political stability, and Uganda is often credited as being a model for sharply reducing the incidence of AIDS. The percentage of adults currently infected with HIV was reported to be 7.3% in 2012, a drop from 1993. Because births still exceed deaths, the population of Uganda is in an exponential phase of growth.

FIGURE 18.17 shows a projection of Uganda's age pyramid in 2050. At first glance, the age pyramids from 2012 and 2050 look similar. The sides of the 2050 pyramid narrow slightly less steeply, indicating improved survivorship. If you look at the horizontal axis, however, you'll see that the 2050 age pyramid predicts a much larger population size. During this 40-year period, Uganda is expected to double in size twice, from its current population of about 32 million to almost 138 million in 2050—half of them still children.

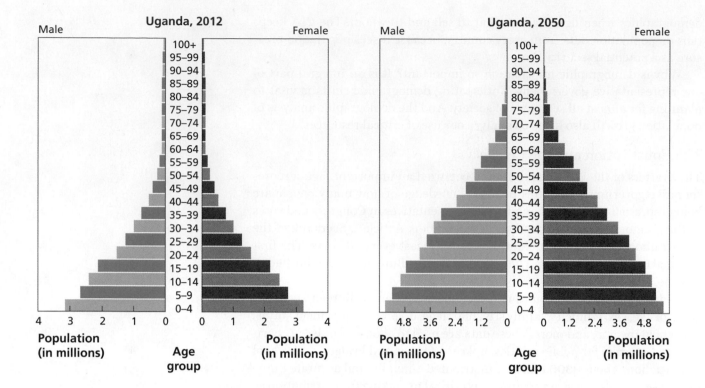

FIGURE 18.17 Uganda's age pyramids for 2012 and 2050 have a similar shape, with a wide base and sharply narrowing sides. However, the 2050 pyramid represents a population that is over four times larger.

Death rates in Uganda are likely to continue to drop. An infusion of agricultural techniques, improved medical care, and increased foreign aid will likely make higher survivorship possible. However, Uganda's projected age pyramid points to the challenges the country will have to address, especially the needs of childhood: improving nutrition and education, controlling childhood diseases, and eventually providing jobs.

While Uganda is a country of children, Italy is a country with an aging population, and its population is projected to decline. Italy's projected 2050 age pyramid (**FIGURE 18.18**) is similar in shape to that of 2012, though smaller and with the bulge of older individuals moved upward 40 years. As these individuals retire, the country's workforce will shrink. This could mean fewer workers to support the aging population. Italy's future needs are likely to be the opposite of Uganda's. Italy likely will be concerned with caring for the elderly: more focus on addressing cancer, heart disease, and palliative end-of-life care, and also on means of increasing worker productivity.

Resource Depletion

One lesson we learned from our story about bacteria is that population growth and resource depletion are two sides of the same coin: exponential growth of bacterial cells could also be viewed as exponential growth in the use of food or test tube space. The principles of growth are not limited to biology. The same ideas can help plan our use of finite resources, such as oil. Besides its obvious utility as a source of heat and combustion for the transportation of vehicles, oil is also used to manufacture a variety of other products that we use, including ink, floor wax, clothes, tires, car dashboards, deodorant, paint, soft contact lenses, and so on.

For the last 30 years, the United States has used about one-quarter of the world's annual oil production. **FIGURE 18.19** shows U.S. oil use in the period between 1860 and 1980. Despite some ups and downs, the data generally follow exponential growth.

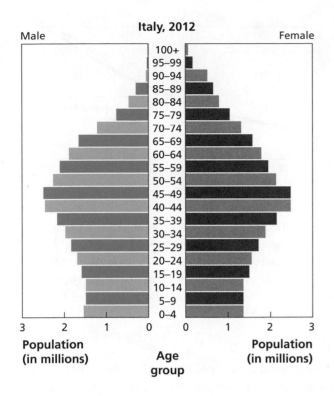

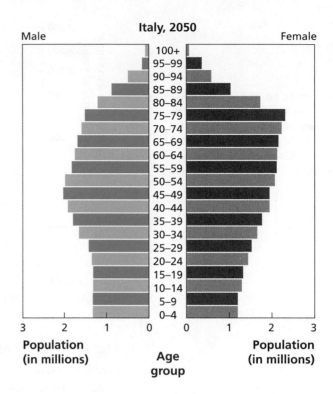

FIGURE 18.18 Italy's age pyramids for 2012 and 2050 show a declining population, with fewer workers and more retirees.

Developing countries currently consume relatively little oil. But as their development proceeds, they will consume more. In China between 1980 and 2009, the amount of oil consumed and the rates of consumption were strikingly similar to those in the United States between 1920 and 1960 (**FIGURE 18.20**). Currently, China uses only about 9% of the world's oil. But China's oil consumption is growing at 6% annually and is projected to double every 12 years, and other Asian countries are following a similar pattern.

There is a great deal of debate about how much oil is left, how long it will last, and when (or if) we will hit "peak oil": the point at which oil production begins to drop. It's a complicated debate, but we should not make the mistake of thinking about oil use as a linear process. It may have taken us 150 years to use about half our oil, but that doesn't mean the second half will last another 150 years. If our oil consumption continues to grow exponentially, once it is half gone, we will consume the second half in only one doubling period—about a decade or two.

Of course, we have another choice: reduce our oil consumption below the level of exponential growth. The cost of producing oil increases as some

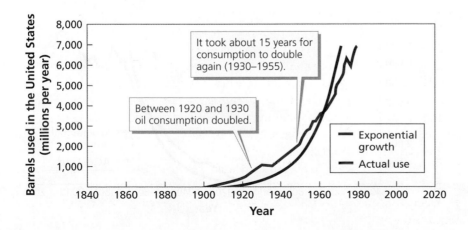

FIGURE 18.19 This graph shows that the use of petroleum in the United States from 1850 to 1980 increased exponentially. This curve has a growth rate of 7.4% and a doubling time of 9.4 years.

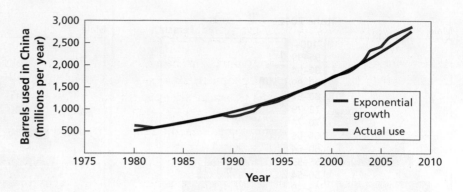

FIGURE 18.20 This graph shows that between 1980 and 2009, both the amount of oil consumed and the rates of oil consumption in China were strikingly similar to those in the United States between 1920 and 1960.

reserves run dry and more difficult reserves are tapped. As the price of oil goes up, consumption may come down. And when we understand that oil is finite, we can take steps to conserve it and to reduce consumption. Oil consumption in the United States declined during the oil crises of the 1970s. And since 1980, oil consumption has fallen way below the exponential increase that occurred previously (**FIGURE 18.21**).

The Limits to Growth

Population size, resource consumption, and pollution: all of these are growing, some linearly, some exponentially. But our planet is finite. There are experts who contend that our technology will overcome any ecological limitation and that growth can continue indefinitely. Most scientists argue, however, that there are limits to growth and the consequences of reaching those limits will be severe. Our future will depend on the choices we make now.

But since we do have choices, there is cause for optimism. Donella Meadows put it this way in her 1972 book *The Limits to Growth*:

> *If there is cause for deep concern, there is also cause for hope. Deliberately limiting growth would be difficult, but not impossible. The way to proceed is clear, and the necessary steps, though new ones for human society, are well within human capabilities. . . . The two missing ingredients are a realistic, long-term goal that can guide mankind to the equilibrium society and the human will to achieve that goal. Without such a goal and a commitment to it, short-term concerns will generate the exponential growth that drives the world system toward the limits of the earth and ultimate collapse. With that goal and commitment, mankind would be ready now to begin a controlled, orderly transition from growth to global equilibrium. (p. 183)*

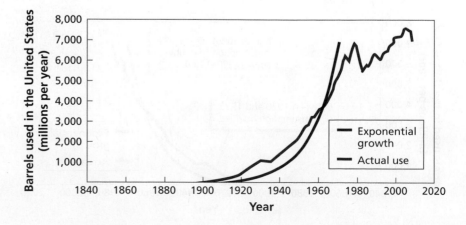

FIGURE 18.21 Since 1980, oil consumption in the United States has fallen below the exponential increase that occurred from 1850 to 1980.

Biology in Perspective

How many people can the Earth support? Our best current prediction is that our population will reach 10 billion people by 2200, so we can hope that it will hold at least that many. From the 1700s until the present, the human population has grown exponentially. We may be at a transition point where our population growth slows and becomes logistic, leading to the predicted leveling off around 2200.

To make the tough choices ahead, we need to understand population growth and how it differs in developing and developed countries. A country's age structure helps demographers make predictions about population trends that can guide future planning and resource allocation.

Demographic changes act as a brake on population growth in developing countries. Industrialization results in a decrease in birth and death rates. However, developed countries also deplete essential resources. If we don't take steps to limit our growth, density-dependent factors like disease pandemics, resource depletion, and insufficient food eventually may reduce our growth rate. If we acknowledge the problems of overpopulation, there is a chance that we can make choices that limit our growth in more acceptable ways.

Chapter Summary

18.1 How Can Populations Grow So Fast?

Distinguish between linear and exponential growth and explain how the growth rate is determined.

- Populations grow when new individuals are added; if this is the result of migration, the growth rate is linear, but if the individuals added reproduce, the growth is compounded, and the population grows exponentially.
- Growth rate addresses two questions: How fast will it grow? How big will it get?
- Population growth rate is calculated by adding births and immigrations and subtracting deaths and emigrations; a population is in equilibrium when the increase equals the decrease.
- Doubling time is the amount of time required for a population to double in size.

18.2 Why Don't Populations Grow Forever?

Describe the effects of population density and carrying capacity on the principle of logistic growth.

- Density-dependent factors—insufficient resources, waste production, disease, and aggression—eventually inhibit exponential growth.

- The density-dependent factors result in logistic growth for most populations; growth starts out fast, slows as the population becomes larger, and stops as it reaches carrying capacity (*s*-curve on a graph).

18.3 How Is Population Growth Influenced by Age and Sex?

Explain what age pyramids can tell us about a country's conditions and social challenges.

- Demography is the study of population growth based on its age structure.
- Females have a greater impact on population growth than males, and it varies based on their age.
- An age pyramid shows the demographics of a population, both the sex ratio and age structure.
- There are three possible shapes to age pyramids that describe conditions in a country and allow comparisons between countries:
  o Widest at the bottom shows that children are the largest group, and the number of individuals declines due to mortality associated with aging.

o A tapering bottom, narrower than the top, indicates an aging population with a low birth rate.

o Comparatively vertical sides suggest a stable growth rate and good longevity.

18.4 Why Do Developing and Developed Countries Grow Differently?

Compare and contrast the factors that contribute to growth in developing and developed countries.

- Total fertility is the number of children a woman could have, on average; birth rate is the average number she does have.
- A total fertility of two would just replace the parents and result in zero population growth (ZPG).
- Age at first reproduction is when women first reproduce, and it tends to be later in life in more developed countries.
- Developing countries tend to have both higher total fertility rates and lower ages of first reproduction, resulting in faster population growth.
- Countries undergo demographic transition as they develop.

18.5 How Do We Use Information About Population Growth?

Demonstrate how data on population growth can be used to plan for population shifts and resource depletion.

- Demographic information is necessary for anticipating shifts and planning government and resource allocation accordingly.
- The U.S. Census Bureau provides detailed information about the demographics and number of people living in the United States.
- The age pyramid at any given time allows predictions to be made about demographic shifts in the future.
  o Uganda has a lot of children, so its population will increase and the country will need to provide care for children and, eventually, jobs.
  o Italy's population is much older, so they will need to plan to support a retiring workforce with fewer younger workers.
- As countries develop, they increase their use of limited resources, so it's important to plan accordingly by developing new resources or minimizing consumption.

Key Terms

age pyramid 543
carrying capacity 541
compounding 534
demographic transition 551
demography 543

density dependent 539
doubling time 538
equilibrium 538
exponential growth 534
growth rate 534

linear growth 533
logistic growth 541
total fertility 547

Review Questions

1. Draw a graph with "population size" on the vertical axis and "time" on the horizontal axis. Sketch three curves that illustrate linear, exponential, and logistic growth.

2. What is compounding?
   a. A principle that underlines linear growth
   b. A principle that underlies exponential growth and that occurs when new entities are themselves capable of making more new entities
   c. A principle that underlies demographic transition
   d. A principle that underlies population density
   e. A principle that underlies logistic growth

3. Give an example of a process that exhibits compounding.

4. What are the four factors that determine a population's growth rate?
   a. Birth rate, death rate, immigration rate, and emigration rate
   b. Birth rate, death rate, compounding, and immigration rate
   c. Birth rate, death rate, doubling time, and compounding
   d. Birth rate, death rate, doubling time, and carrying capacity
   e. Birth rate, death rate, carrying capacity and emigration rate

5. If a population of 5000 is growing at a rate of 8% (0.08) per year, how long will it take for the population to double in size? Is it necessary to know how big the population is to determine the answer? Briefly explain.

6. What are density-dependent factors? Give examples of potential density-dependent factors for a top predator, like a wolf, and a hibernating creature, like a bear.

7. What is carrying capacity?
   a. Logistic growth that starts off fast but slows down as the population gets larger

b. Exponential growth

c. The maximum number of individuals an environment can support

d. Growth rate

e. The minimum number of individuals an environment can support

8. Why does population size level off as it approaches carrying capacity?

9. What information is shown in an age pyramid?

    a. Exponential growth and logistic growth

    b. Carrying capacity and sex ratio

    c. Logistic growth and age structure

    d. Demography and sex ratio of a population

    e. Sex ratio and age structure of a population

10. Briefly describe the overall shape of the age pyramid for a country whose population is: (a) increasing, (b) relatively stable, and (c) shrinking.

11. How do birth and death rates change during demographic transition?

12. How does a country's population size change in the early stages of demographic transition? How does it change toward the end of demographic transition?

13. List and briefly explain two reasons why a government would want an accurate census of its population.

The Thinking Citizen

1. Birth rates decrease eventually during demographic transition, and women tend to have their children later as the country develops and the transition proceeds. What are some of the reasons for postponing childbearing? What effect will this have on population growth in the country, and why?

2. When a population is at carrying capacity, its *growth rate* is zero but its *birth rate* is not zero. Does the population stay exactly at carrying capacity, or can it vary a bit? Imagine a deer population that is at carrying capacity, and that it is mating season. The deer, of course, don't know anything about carrying capacities, but they do know about mating. What will happen to the deer's population size over the next few years?

3. Normally we think about a country undergoing demographic transition. But is it possible that the entire global population does this, too? Review the changes that happen to birth and death rates during the three phases of demographic transition. Also, review global population growth. Do you think global population growth can be explained by the way birth and death rates change during the demographic transition of a single country?

4. The age pyramids of Uganda and Italy both change *size* between 2012 and 2050, but they do not change in terms of basic *shape*. What types of changes must occur in order for an age pyramid to change shape? What are the potential consequences of this shape change?

The Informed Citizen

1. Ultimately there are only three options for reducing a population's growth rate: decrease the birth rate, increase the death rate, or both. Which of these options do you find most acceptable? What are your suggestions for implementing your option? Do your suggestions work equally well in developing and developed countries, or do you need a different plan for each?

2. War is sometimes thought to have an ecological basis: an effort to reduce population pressure by acquiring more resources like gold, oil, or territory. Do you think war could be a density-dependent factor that limits the size of the human population? How does mortality from wars compare to mortality from disease? Do some online research to find mortality estimates from both causes. For diseases, you may want to research the Spanish flu epidemic of 1918 and the current AIDS pandemic.

3. Go online to the Census Bureau's International Database (http://www.census.gov/population/international/data/idb/informationGateway.php). Select a country of your choice and view its 1950, 2012, and (projected) 2050 age pyramids. For some countries, data from 1950 may not be available, so try for as early a year as possible. Analyze the pyramid's general shape to determine whether the country's population is increasing, relatively stable, or shrinking. What sorts of population-related problems do you predict this country will have in the future? Are there "bulges" or "pits" in the pyramid that indicate specific events in the country's history? Do some background research on the country to find out. You may also want to look at some of the other data and tables available for your country on the Census Bureau's website.

Answers to Selected Questions

Chapter 1

2. e
3. a
5. c
6. b

Chapter 2

1. c
3. a
5. d
10. e
12. b

Chapter 3

1. b
6. e
9. d
11. a

Chapter 4

3. c
4. a
5. d
6. b

Chapter 5

2. b
6. a
9. d
15. e

Chapter 6

4. c
5. a
13. d
15. b

Chapter 7

1. e
3. a
8. d

Chapter 8

1. b
2. a
5. e
8. c

Chapter 9

1. b
5. c
10. a
12. e

Chapter 10

1. c
2. a
9. e
11. b

Chapter 11

2. c
6. a
9. e
11. d

Chapter 12

4. a
7. d
11. b
14. e

Chapter 13

2. d
5. b
6. a
11. e

Chapter 14

1. b
4. e
7. c
14. a

Chapter 15

6. c
8. e
10. b
12. a

Chapter 16

3. b
4. a
9. d
15. c

Chapter 17

3. b
7. d
12. c
13. e

Chapter 18

2. b
4. a
7. c
9. e

Glossary

action potential an electrical wave that transports messages through the axon

active transport the movement of a molecule or ion against the concentration gradient from low concentration to high; requires energy

adaptation a change by a population to better align with specific environmental conditions

adenosine triphosphate (ATP) the energy currency of the cell; used in practically all chemical actions that require an input of energy

adipose tissue stores excess carbohydrates and proteins that have been converted to fat by the liver

adult stem cell a cell from an adult that can renew itself, but that is limited in its flexibility

age pyramid a snapshot of a population at a given point in time; displays the number of individuals in age categories

allele different version of the same gene

allele frequency the number of alleles present in a population

alveolus thin-walled, dead-end sac at the terminal end of the smallest bronchioles across the surface of which oxygen enters and carbon dioxide leaves the bloodstream

amino acid a subunit of protein

amniocentesis a method for examining fetal cells that can be done between 14 and 17 weeks of pregnancy

amnion sac in which the embryo is suspended

anaphase I in meiosis, the stage in the first round of cell division when paired chromosomes separate and move to opposite poles, leaving a haploid set of chromosomes on each side of the cell

anaphase II in meiosis, the stage in the second round of cell division when the two chromatids of each chromosome detach from each other and move to opposite ends of the cell; each half of the cell has a complete haploid set of chromosomes

androgen-insensitivity syndrome (AIS) a disorder of sexual development in which an individual who is XY does not develop as a male because the individual's cells lack receptors to bind to testosterone

anemia a disease in which a person is deficient in hemoglobin

angiogenesis a process by which successful tumors secrete chemical signals that cause blood vessels to sprout from the circulatory system

angiogenesis inhibitor a drug that stops the growth of blood vessels, inhibiting the ability of tumors to attach to the circulatory system and grow

antibiotic a chemical produced by microorganisms to inhibit or kill other organisms that might harm them; this chemical is used by humans to treat bacterial infections

antibody a molecule that binds specifically to an antigen and kills pathogens

anticodon a triplet nucleotide sequence at one end of tRNA

antidiuretic hormone (ADH) a water balance effector that reduces the volume of urine produced and thereby the amount of water lost in urine

antigen a foreign molecule

antivenom antibodies given by injection that can disable snake venom

anus the exterior, posterior opening of the digestive tract

APGAR scale an assessment of a newborn baby conducted at one and five minutes after birth to assess a baby's activity, pulse, grimace response, appearance, and respiration

apoptosis cell death

artery blood vessel that transports blood away from the heart

artificial insemination sperm from the male partner are placed in the female partner's vagina

asexual reproduction reproduction that involves a single parent and no sex

assisted reproductive technology (ART) reproductive procedure that is used to bring healthy sperm and eggs together so that fertilization is possible

atherosclerosis a disease of the arteries

atom the building block of every physical thing in existence; there is one type of atom for each element

atrioventricular valve keeps blood moving in the right direction; lies between the atrium and its corresponding ventricle

atrium an input, or reservoir, chamber of the heart

autonomic system a functional component of the PNS that regulates the internal environment of the body by controlling the muscles of the heart and other internal organs and glands; it is not under conscious control

axon the long extension of the neuron cell body that "reaches" to the next neuron cell body

B lymphocyte, or B cell produced by the bone marrow, it is the main player in antibody-mediated immunity

background extinction rate the extinction rate during "typical" times, not during a mass extinction

basal metabolic rate the number of Calories you need to maintain your body's processes when you are resting and fasting

benign harmless

bile material produced by the liver and stored in the gall bladder that helps dissolve fats

biodiversity the variety of species living in an ecosystem

biogeochemical cycle process that moves matter among the four reservoirs: the biosphere, atmosphere, hydrosphere, and lithosphere

biogeographic realm one of eight landmasses of the world with its own distinctive floras and faunas

biogeography the study of where different species live throughout the world

bioleaching the action of bacteria that have been genetically engineered to produce enzymes that dissolve metals from their ores, making the metal easier to extract

biological control an approach to controlling pests by introducing predators of a pest into an area

biological species concept it defines species as genetically distinct, independently evolving units

bioremediation the biological cleanup of pollution and contamination

blitzkrieg hypothesis a theory proposed by Paul Martin that humans hunted most of the large mammals on the North American continent to the point of extinction

blood pressure the force exerted by the blood on the arterial wall as it moves from the heart to the arteries

body/mass index (BMI) tool for determining whether one's weight is in the healthy range; calculated from a person's height and weight

bone marrow the spongy material predominantly inside the large bones of the body; it produces B cells

brainstem comprised of the medulla oblongata, the pons, and the midbrain; all information moving to and from other regions of the brain passes through this area

Broca's area a region of the left temporal lobe responsible for producing language

bronchiole a branch of the bronchus

bronchus one of the two tubes branching from the trachea

bulb short, underground stem with fleshy leaves that store nutrients and food

Calorie the amount of energy needed to raise the temperature of one milliliter of water by one degree Celsius

Calvin cycle the second stage of photosynthesis in which the solar energy captured from light reactions is used to make sugar

Cambrian explosion the geological period of time from about 542–490 million years ago when every major group of animal appeared in the fossil record

cancer a disease of uncontrolled cell proliferation

capillary an extremely tiny vessel through which red blood cells travel one at a time; located at the ends of arterioles

carbon fixation the attachment of carbon dioxide from the atmosphere to another carbon-containing compound

carpel the female part of a flower

carrying capacity the maximum number of individuals the environment can support

cartilage flexible tissue that protects the end of bones

cause and effect every event or outcome in nature has a source; if a scientist sets up the correct conditions, the results can be predicted in advance

cell the smallest and most basic unit of life

cell cycle four distinct phases of cell life: mitosis (the chromosomes separate into daughter cells), Gap 1 (the cell grows), DNA synthesis (the chromosomes replicate), and Gap 2 (the cell grows and prepares again for mitosis)

cell membrane outer boundary of the cell body, where the cell encounters its environment

cell theory the realization that all living organisms are made of cells; it was first proposed in 1839 by Matthias Schleiden and Theodor Schwann

cellular respiration the process whereby cells extract energy from food molecules

central nervous system (CNS) controls the actions of the body; comprised of the brain and spinal cord

cerebral cortex the surface of the cerebrum that is highly folded and appears wrinkled

cerebrum the largest and most sophisticated component of the forebrain, accounting for 80–85% of brain mass in humans; responsible for many of the characteristics that most people consider distinctly human: reasoning, mathematical ability, artistic ability, imagination, language, and personality

cervical cap a barrier contraceptive that keeps sperm out of the uterus

cervix the lower part of the uterus and the opening to the vagina

channel specialized small tube made of protein that can open or close to regulate whether specific molecules can pass through a cell membrane

character displacement when species evolve, a change in physical or behavioral characters that allows them to reduce competition

chemical bonding sharing of electrons in the outer shell of more than one atom

chemoprevention the action of natural or manufactured materials used to prevent or halt cancer development

chemotherapy a treatment method using drugs that impair the ability of cells to replicate; these medications may be taken orally or injected directly into the bloodstream

chlorophyll a green pigment used by plants in photosynthesis

chloroplast an organelle in a photosynthetic cell in which photosynthesis takes place

cholera a serious infection of the intestines that leads to rapid dehydration from diarrhea and vomiting; left untreated, a person may die in as few as 24 hours

chorionic villus sampling (CVS) procedure used to isolate fetal cells that can be done between 10 and 12 weeks of pregnancy

chromatid one of two copies of the DNA and proteins that make up a replicated chromosome; replicated chromosomes are composed of two chromatids

chromosome structure that contains most of the genetic information of the cell; housed in the nucleus

circuit neurons organized into a group that interconnect and communicate with each other

circulatory system carries the blood throughout the body

classification the assignment of every species to a genus, family, order, class, phylum, and kingdom

cleavage the start of embryogenesis when the cells divide into new cells

cloning the process of isolating and copying a gene

codon a specific mRNA sequence of three nucleotides

coevolution the adaptation of one species to another

colonoscopy a procedure used to examine the large intestine, rectum, and anus to screen men and women for colon or rectal cancer

commensalism an ecological interaction in which one species benefits and the other is neither helped nor harmed

competition occurs when two or more species attempt to utilize the same resource; it has a negative effect on both species

competitive exclusion principle posits that two competitors cannot coexist

complement protein a protein that coats the surface of bacteria, making it easier for macrophages to eat them

compounding a principle that underlies exponential growth; occurs if new entities are themselves capable of creating new entities

computerized tomography (CT) scan an advanced type of X-ray procedure that provides a more detailed look at internal structures by sending thin X-ray beams at different angles

concussion a brain bruise in which electrical activity in the brain is temporarily disordered

condom a barrier contraceptive that covers the penis and prevents sperm from entering the female reproductive tract

conduction heat transfer from a warmer solid object to a cooler solid object

conjoined twin identical twins whose bodies are physically attached at some location

conjugation two bacterial cells physically join together and transfer DNA from one cell to the other

contraceptive birth control

controlled experiment a test or manipulation in which a scientist keeps all variables (possible factors that could affect the outcome of the test) the same except for the one under investigation

controller the third part of the homeostatic system; it registers information from the sensors and sends signals to the appropriate effectors

convection heat transfer from a warmer object to a cooler, moving fluid, such as a liquid or gas, that surrounds it

cross mating

cyst a fluid-filled sac

cystic fibrosis a disease in which a thick, abnormal mucus accumulates in the lungs and the digestive system

cytokinesis the stage in cell division when the cells are pinched in two

cytoplasm in a cell, the cytosol and organelles together

cytoskeleton maintains the shape of the cell and the positions of the organelles

cytosol a viscous substance in a cell where small molecules are dissolved and organelles are embedded

cytotoxic T cell a type of T cell that specializes in killing cells infected by viruses, cancer cells, fungi, and worms

Darwin's finches the finches on the Galapagos Islands that occupied different habitats and developed different behaviors

decomposer organism that causes dead bodies to rot

demographic transition the change in population growth patterns seen over time as developing countries become developed countries

demography the science that analyzes population growth

dendrite highly branched projection on a neuron that receives signals from other neurons

density dependent a factor responsive to population crowding

descent with modification a phrase used by Darwin to describe the fact that species changed and that evolution occurs

diagnostic test precise evaluation of risks to an embryo or fetus

dialysis process in which blood is circulated outside the body through an artificial kidney to remove impurities

diaphragm large circular sheet of muscles that separates your chest from your lower abdomen; **Or** a barrier contraceptive placed in the vagina as close to the cervix as possible; it bars the way to the uterus

differentiation the change from cells that have unlimited potential to cells that are specialized

diffusion movement of molecules from a higher concentration to a lower concentration

digestive system enables the extraction of energy from the chemical bonds of food molecules

diploid cell cell with two complete sets of chromosomes

discrimination when one species recognizes another as not its own

disorder of sexual development (DSD) a condition sometimes noted at birth in which the genitalia are atypical

diversity the variety of life forms in the natural world

diversity index a mathematical formula that measures species diversity by taking into account richness, abundance, and evenness

DNA (deoxyribonucleic acid) a type of nucleic acid located in the cell nucleus; it is the physical material of which genes are made

dominant a trait that is seen even when one allele is present

doubling time the amount of time required for a population to double in size

eating disorder a condition in which a person eats too much or too little to the detriment of health

ecological pyramid a diagram of the changes in trophic level

ecology the study of the interactions among species and their physical environments

ecosystem service naturally occurring ecological process that supports our agriculture, technology, culture, and more; can be divided into four groups: provisioning, regulating, cultural, and supporting

ectopic pregnancy a pregnancy in which an embryo develops in the fallopian tube rather than in the uterus; ectopic pregnancies cannot develop successfully to term, and if left untreated can lead to rupture of the fallopian tubes

effector the second part of the homeostatic system; it has the ability to change the value of the system, typically by either increasing or decreasing it; **Or** the final destination of a signal

electrolyte electrically charged substance necessary for muscle function and water balance

electron negatively charged subatomic particle that orbits the nucleus of an atom

electron transport chain (ETC) the main producer of ATP; located in the mitochondria

electrophoresis a widely used method for separating mixtures of molecules, including proteins, DNA, and RNA

element a substance composed of only one type of atom

embryo the earliest stage of development

embryogenesis the development of cells into embryos and bodies that are able to perform the tasks necessary for life

embryonic stem cell cell that can produce any type of cell in the body

empirical evidence information that one gets from direct observation, from experience, or from the results of experiments and other tests of hypotheses

encephalitis inflammation of the brain due to infection; usually caused by a virus and includes symptoms such as fever, confusion, and seizures

endemic a species that is specific to a location and found nowhere else in the world

endomembrane system produces important molecules and delivers them to key locations within and outside the cell; consists of the endoplasmic reticulum, the Golgi apparatus, and lysosomes

endometriosis a painful condition in which uterine tissue grows elsewhere in the body, leading to inflammation and disruption of ovulation

endometrium the lining of the uterus produced during the menstrual cycle to support the development of an embryo if a pregnancy occurs

endoplasmic reticulum (ER) a series of membrane folds located near the nucleus; smooth endoplasmic reticulum handles the synthesis of fats and lipids, and rough endoplasmic reticulum produces proteins

endosperm cells that surround a plant embryo and form the bulk of the seed

enzyme protein that makes it easier for chemical reactions to occur

epidemic spread of a disease over a larger area and number of people than expected

epididymis coiled tube sitting atop each testis whose cells secrete chemical signals that help sperm to complete differentiation

epiglottis flap of tissue that covers the trachea when swallowing foods, liquids, and saliva

epileptic seizure an electrical storm in the cells of the brain; generally involves large areas of the brain and the entire body, which shakes and tenses its muscles

equilibrium a state when the population's size does not change

esophagus muscular tube that leads to the stomach and the trachea

estrogen a hormone secreted by the ovaries that promotes the development of the Müllerian ducts, oviducts, the uterus, and the upper end of the vagina

eugenics the notion that the way to improve human stock is to prevent biologically defective people from having children and encourage worthy people to have many

evaporation heat transfer by vaporization of a liquid, such as sweat, to a gas

evolution theory that explains how all living organisms are related and how existing populations adapt to their environments and new species arise

evolutionary tree diagram that takes into account the degrees of similarity among species based on their physical characteristics and genes

exploitation an interaction between species in which one species benefits while the other is harmed

exponential growth growth that proceeds at an increasing rate over a given time frame

fact something that has actual existence

factor a gene that possesses the information for the production of a specific protein

falsifiable able to be proved wrong

fertilization the process of a sperm cell from the father being fused with an egg cell from the mother

fetal alcohol spectrum disorder (FASD) encompasses a range of health problems experienced by children who were exposed to alcohol prenatally

fetal alcohol syndrome (FAS) the most severe FASD; symptoms may include poor growth, abnormal facial features, hyperactivity, poor reasoning skills, vision problems, hearing problems, and intellectual disabilities

fetus the stage of development that follows the embryo

fiber roughage composed of complex carbohydrates that can't be fully digested; helps keep things moving through the digestive system

filtration process that occurs in Bowman's capsule; solid materials are removed from a liquid

fitness the number of alleles, or forms of a gene, an organism passes on to the next generation

flower part of a plant that is used for sexual reproduction; it produces seeds and fruit

follicle a primary oocyte surrounded by a single layer of cells

food web a diagram showing a multilinked structure of "who eats whom (or what)"

forebrain the most highly developed and largest region of the brain; composed of the olfactory bulbs, thalamus, hypothalamus, and cerebrum

founder's effect a special case of genetic drift in which a few individuals of a population leave or are forced to leave and begin a new population

fruit any structure that develops from a flower's ovary

functional magnetic resonance imaging (fMRI) a radiology technique that measures organ activity by detecting blood flow

G protein a specific on/off molecule to which the cholera toxin binds

gall bladder organ that stores bile

gamete sex cell

gas exchange the process by which oxygen is taken into the lungs and transported to every cell in the body and carbon dioxide is removed from the cells

gastrulation a stage in embryonic development in which cells and tissues move to new locations, where they will grow into organs

gene hereditary unit consisting of DNA

gene expression the process during which the information encoded in genes is used

gene therapy the correction of a faulty gene

generality how widely a scientific investigation applies to situations other than the specific ones scientists tested

genetic code the information encoded in DNA; it is the language of nucleotides

genetic counseling an approach used to inform couples about their relative risk of passing on certain specific genes associated with a disease

genetic determinism the idea that our genes determine, direct, or cause everything about us

genetic diversity the degree of genetic difference among individuals in a species

genetic drift random changes in the frequency of an allele

genetic engineering the process of taking genes from one species and inserting them into another species

genetic essentialism the notion that being human means having a human genome

genetically modified organism (GMO) an organism whose genetic material has been altered

genital external sex organ

genome the term for all genes in an organism

germ-line gene therapy the correction of genetic problems in the eggs and sperm so that harmful mutations can never be passed on

glycogen multi-subunit molecule composed of glucose

glycolysis one of the chemical reactions of cellular respiration that occurs in the cytoplasm of cells; it begins with glucose and ends with pyruvic acid

glyphosate a broad-spectrum herbicide; it is the active ingredient in Roundup

Golgi apparatus the place where many proteins acquire their final structure

growth factor a signal molecule that enables normal cells to divide

growth inhibitor a signal molecule that prevents cell division

growth rate how fast a population grows

habitat diversity the range of different types of natural homes or environments in which organisms live

haploid cell a cell with just one set of chromosomes

HbA normal hemoglobin that carries oxygen molecules throughout the body

HbS abnormal hemoglobin; although it can carry oxygen, it forms long fibers once it has given up its oxygen, causing red blood cells to sickle

helper T cell a type of T cell that assists cytotoxic T cells by detecting foreign invaders, enhancing B cell responses, and alerting other T cells that there is an infection

hemoglobin a protein- and iron-containing compound found in red blood cells that binds to oxygen and transports oxygen throughout the body

hemophilia a disease in which an individual cannot form blood clots in the event of an injury

Herceptin a drug that is used to treat a certain type of breast cancer by binding to the *erbB-2* protein and stopping it from triggering cell division

hindbrain located at the base of the skull, right above the spinal cord, and responsible for the control of many basic bodily functions necessary for survival; composed of the medulla oblongata, cerebellum, and pons

homeostasis the ability of an organism to maintain constant internal conditions in the face of fluctuating external conditions

homologue one in a pair of chromosomes

horizontal gene transfer the transference of genes between individuals of the same generation, and often between individuals of different species

hormone chemical signal responsible for physiological or developmental response, including secondary sex characteristics

human development the process by which an individual grows and matures from a single cell embryo inside a mother's womb to a baby that can survive on its own

human embryonic germ cell a cell that comes from a six- to nine-week-old embryo and fetus; these cells ultimately form the cells that produce eggs and sperm

human embryonic stem cell a cell that comes from an inner cell mass and can be isolated from an embryo that is five to seven days old and grown in a lab

human papillomavirus (HPV) a sexually transmitted virus that is responsible for most of the world's cancers of the genitals and anus

hypothalamus a cone-shaped region at the base of the brain that is in charge of homeostasis for many of the body's systems; it secretes gonadotropin-releasing hormone (GnRH)

hypothesis possible cause or mechanism that could explain observations and facts

immune response the body's ability to fight off infection

immune system the collection of organs, cells, and tissues that protect the human body from disease

immunosuppressive drug medicine that keeps a transplanted organ safe but suppresses the immune system and so makes a patient more vulnerable to infection

immunotherapy treatment that directly targets a cancer or boosts the immune system to make other cancer therapies more effective

in vitro fertilization (IVF) immature oocytes are extracted and placed in a petri dish in a solution that permits them to mature, and they are combined with sperm from a male partner and then transferred into the uterus

indicator species a species whose presence indicates a healthy, functioning ecosystem

infertility failure to conceive and become pregnant after a year of regular, unprotected intercourse

inflammation the red, warm, and painful swelling that occurs with an infection

inflorescence a plant structure containing many flowers

influenza also known as the flu, a potentially serious respiratory disease caused by viruses

interferon chemical signal that enlists the help of noninfected cells to fight a virus

interneuron a special cell that integrates the incoming and outgoing signals so that the body responds appropriately to stimuli

interphase the stage in cell division when cells prepare materials that will be necessary for meiosis or mitosis to occur

intrauterine device (IUD) a contraceptive inserted into a woman's uterus to slow the development of an endometrium

intrauterine insemination sperm are placed directly into the uterus

ion charged particle that comes from the salty fluid in our bodies; it may be negative or positive

keystone species a single species that has a disproportionate effect on an ecosystem's biodiversity

kidney a fist-sized organ located near the spine and just above the waist; its main job is to make urine

kin selection selection for behaviors that help your relatives survive and reproduce

Krebs cycle one of the chemical reactions of cellular respiration that takes place in the mitochondria of cells; produces GTP and NADH

large intestine, or colon coiled organ, 3–6 feet long and 3 inches in diameter, which absorbs water, salts, and some vitamins

latitudinal gradient the decrease in the number of species as one moves away from the equator into the polar regions

leaf a broad or long, flat structure attached to the stem by a stalk; it is the place in the plant where photosynthesis occurs and food molecules are made

leukemia a type of blood cell cancer in which the tumor suppressor gene is mutated

ligament band of tissue that connects two bones, cartilage, or joints together

light reaction the first stage of photosynthesis in which solar energy is captured

limbic system comprised of parts of the cerebrum, hippocampus, and amygdala; it is in charge of physical drive and instincts and is partly responsible for emotions, learning, and memory

linear growth growth that proceeds at the same rate over a given time frame

liver accessory organ that produces bile

logistic growth a pattern of population growth that takes density dependence into account; starts off exponential and stops when the population reaches carrying capacity

lymphocyte a type of white blood cell

lysosome digests waste materials and worn-out organelles and recycles the molecules so they can be reused

macromolecule large molecule made up of many atoms

macrophage locates and kills microbes by engulfing and eating them

magnetic resonance imagery (MRI) scan a medical imaging technique that uses a magnetic field and radio waves to visualize the inside of a body

malaria a disease caused by a parasite carried by the Anopheles mosquito; it is often fatal

malignant cell growth that continues in an uncontrolled manner, resulting in cancer

mammogram a form of X-ray used to detect possible breast cancer

mass extinction a period when over 75% of Earth's species disappeared

materialism the idea that effects in the natural world all have natural causes, rather than supernatural ones

medulla oblongata a part of the most posterior portion of the brain; it controls breathing, heart rate, and digestion

meiosis a special type of cell division that produces eggs or sperm

melanoma an invasive and deadly skin cancer

memory cell a B cell that is produced in response to an infection but that does not produce antibodies; instead, it remains in the body to fight future infections of the same variety

meninges a durable connective tissue that helps protect the bones of the skull

meningitis inflammation of the meninges, the tissues that cover the brain and spinal cord; it is highly contagious and often fatal

menstruation the shedding of the lining of the uterus that supports fetal development

messenger RNA (mRNA) codons that identify precisely which amino acids should be joined together and in what order

metabolism all of the chemical reactions in cells that sustain life

metaphase I in meiosis, the stage in the first round of cell division when paired chromosomes line up in single file down the middle of the cell

metaphase II in meiosis, the stage in the second round of cell division when chromosomes line up in single file in the middle of the cell

metastasis a process by which individual cancer cells leave a tumor, enter the bloodstream, and produce secondary tumors at new sites

miasma harmful, toxic vapor supposedly exhaled by sick people or exuded by garbage or sewers into the air around them

microcirculation the process by which different capillary beds open and close to divert blood where it is needed throughout the body

microorganism a living thing that cannot be seen without the use of a microscope, such as bacteria

midbrain forms the brainstem along with the medulla oblongata and the pons; plays an important role in coordinating responses to light and sound

mimicry the close resemblance of an organism to another or to its surroundings

mineral inorganic molecule (does not contain carbon); helps with bone and tooth development, muscle contraction and relaxation, nerve impulses, and fluid balance

mitochondrion organelle responsible for extracting energy from food molecules

mitosis a type of cell division during which all of the chromosomes in a cell are copied or replicated and each new cell receives genetic information identical to the parent cell

modern synthesis the integration of genetics and evolutionary biology; one of its main architects was Sir Ronald Aylmer Fisher

molecular medicine a field of medicine that aims to understand the relationship between health and genes, proteins, and other molecules, in order to diagnose and treat disease

molecule formed through the chemical bonding of atoms

monoclonal antibody (MAb) referred to as a "magic bullet," a treatment that targets only a specific antigen

monohybrid a type of cross in which scientists pay attention to one trait at a time

muscle fiber composed of a bundle of myofibrils aligned parallel to one another

mutation a physical change in DNA; five examples of mutations are point, deletion, duplication, inversion, and translocation

mutualism an interaction between species in which both species benefit

mutualistic an ecological interaction in which both species benefit

myelin sheath fatty material that encases the axons of neurons; acts like insulation on a wire by speeding up electrical impulses and preventing them from losing strength

myofibril composed of actin and myosin filaments arranged as sarcomeres attached end to end

NADH works like a storage battery for high-energy electrons collected during extractions of energy from pyruvic acid; used in the electron transport chain

natural immunity immunity that occurs as a result of having recovered from a previous bout with an infectious disease

natural killer cell eliminates virus-invaded cells by penetrating their outer membranes and causing the infected cells to burst

natural selection the mechanism whereby individuals with certain heritable traits have an increased chance of surviving and producing offspring

necrosis tissue death

negative feedback the process of keeping a system at the set point; whichever way the system deviates, a force is applied in the opposite direction to bring it back in line

nephron a tubule in the kidney that works to make urine and to maintain water and salt balance in the body; each kidney has about a million such tubes

nerve a bundle of neurons wrapped together to form "cables"

nerve impulse an action potential that occurs as a result of a stimulus

neural tube the tissue that starts from a flat sheet of cells and rolls up, ultimately becoming the brain and spinal cord

neuron a specialized signaling cell in the nervous system; the brain has more than 10 billion neurons, and the nervous system has even more

neurotransmitter a chemical that binds to specific receptor molecules on a neuron's membrane and initiates a nerve impulse

neurulation formation of the earliest stages of the central nervous system

neutron neutral subatomic particle inside the nucleus of an atom

niche the particular set of resources a species uses to survive; includes biological factors like food sources and nesting and physical factors like temperature

nonspecific barrier blocks bacteria and other potential pathogens without targeting specific ones; skin is a nonspecific barrier

nucleotide a subunit of a DNA molecule

nucleus central sphere of an atom made of protons and neutrons, with electrons orbiting; **Or** the most prominent organelle inside the cell; it houses the chromosomes that contain the genetic information

observation what you can see, hear, smell, taste, or feel physically

olfaction the sense of smell

oncogene a gene that tells the cell to divide in the absence of normal instructions to do so

oogenesis the production of fertile eggs

orbital the outermost shell of an atom that can accommodate a specific number of electrons

organelle specialized structure in eukaryotic cells

osmosis the movement of water through a semipermeable membrane from an area where solutes are in low concentration to an area where they are in higher concentration

ovary female sex organ where eggs are produced

oviduct a small tube down which an egg travels when it is propelled from the ovary

ovulation the release of an egg from the ovary

pancreas one of the accessory organs of the digestive system that secretes enzymes that digest carbohydrates, proteins, and fats

pangenesis an idea that persisted for more than 2000 years and that held that each part of the body produced a characteristic seed that traveled to reproductive organs; this idea was debunked by August Weismann in the 1880s

Pap test a method used to detect cervical cancer by collecting and examining cervical cells

parasympathetic branch part of the autonomic nervous system that helps the body do things to gain and conserve energy by stimulating digestive organs and decreasing heart rate and breathing rate

passive immunity the temporary transfer of antibodies from one individual to another; for example, the antibodies that circulate in a mother's blood can cross the placenta and enter the fetal bloodstream

pathogen disease-causing organism, such as a bacterium, fungus, or worm, that invades the body and does damage

pathogenic capable of causing disease

pathologist a doctor who specializes in the study of disease

pedigree analysis a family history study; it shows a pattern of inheritance

pelvic inflammatory disease (PID) infertility caused by blocked oviducts that may be a consequence of a sexually transmitted disease or infection caused by an IUD

peripheral nervous system (PNS) a network of nerves radiating out from the CNS and throughout the body that enables the brain and spinal cord to communicate with the entire body

peristalsis coordinated waves of smooth muscle contraction that propel swallowed food toward the stomach and through the rest of the digestive system

permeable characteristic of a cell membrane through which molecules may pass

petal showiest part of the flower

pharynx chamber located at the back of the throat that connects the mouth to the digestive and respiratory tracts

pheromone chemical attractant used by females to attract males of their species

phloem part of the stem that carries food made in the leaves through the stems to the roots and other parts of the plant

photosynthesis a process that plants use to capture solar energy and then use it to make sugar

pituitary gland a pea-sized structure located in the brain connected to and just below the hypothalamus that produces many of the body's hormones, including antidiuretic hormone, follicle-stimulating hormone (FSH), and luteinizing hormone (LH)

placenta an organ that connects a developing fetus to the mother's uterine wall and provides for an exchange of nutrients and waste elimination

plaque structure that forms from white blood cells, platelets, fats, and calcium deposits, narrowing the inner part of the artery as it grows

plasma the liquid part of blood, 90% of which is water and 10% of which is chemicals dissolved in the blood

platelet small cell fragment that is responsible for clotting blood to stop bleeding

pluripotent a cell that is capable of producing any of the cells in the body but none of the support structures needed for the development of a baby

polygynous a mating system in which one male mates with several females

pons a component of the brain that relays signals controlling respiration, swallowing, sleep, and bladder control

prediction educated speculation about what an outcome will be

primary consumer herbivore

primary oocyte immature egg

primary producer photosynthetic organism that is the first link in the food web in terms of capturing solar energy

primary sex determination the step in sexual development that determines whether ovaries or testes form

prion an infectious protein; may lead to degenerative diseases of the nervous system

productivity an indicator of how effectively plants convert sunlight into food; measured by the weight of plant material produced in the ecosystem

progesterone a hormone secreted by the corpus luteum that is used in the development of the endometrium or for preventing the development of a second follicle during one menstrual cycle

prokaryotic a simple cell with no membrane-bound organelles

prophase I in meiosis, the stage in the first round of cell division when the chromosomes become shorter and thicker and line up with their homologues

prophase II in meiosis, the stage in the second round of cell division when chromosomes, still in their compact form, attach to the newly formed spindle

prostaglandin a chemical signal that causes muscle contractions in the female reproductive tract

prostate gland secretes a milky white fluid that neutralizes the acidity of the vagina

proton positively charged subatomic particle inside the nucleus of an atom

proto-oncogene a gene that encodes for a protein that regulates cell division

protozoan a simple, single-celled organism

pseudogene a gene carrying a mutation that makes its products nonfunctional

pseudohermaphroditism a disorder of sexual development in which an individual who is XY develops testes but is missing an enzyme that lets testosterone send the right signals until puberty; having started life as a girl, such an individual develops into a man

pseudoscience fake science

psychotherapy the treatment of mental health disorders using psychological methods such as talking therapy

pulmonary circuit the mechanism by which the right-hand pump of the heart receives used blood from the body and pumps it to the lungs

quackery promoting the use and/or purchase of remedies even when there is no scientific evidence or plausible rationale for their effectiveness

quorum a population of bacteria large enough to overcome the body's defenses

radiation energy transmitted as waves or subatomic particles; **Or** heat transfer through infrared rays

radiation therapy radiation delivered either externally or internally to kill cancer cells

reabsorption process in which compounds are removed from the filtrate and returned to the body

recessive a trait that is not seen when only one allele is present

rectum the last six inches of the colon; stores feces until they can be eliminated

red blood cell a cell that carries oxygen through the blood

reflex a movement in response to a stimulus

reinforcement natural selection that acts to reduce hybrid formation

replication the process by which the double helix of DNA unzips and new strands form

reproductive cloning creating an embryo with the intention of producing a baby

resistance the ability of an organism to survive exposure to a drug that previously had been able to kill or disable it

resistance gene gene that has mutated in such a way that the proteins it makes take on a different function or perform old functions in a new way

respiratory system enables you to breathe (take in oxygen and expel carbon dioxide)

ribosomal RNA (rRNA) forms ribosomes, clamp-like structures that hold mRNA and tRNA in the correct orientation so that the amino acids can be connected together

ribosome studs the surface of the rough endoplasmic reticulum (RER) and threads the protein into the interior of the RER as it is being made

rivet hypothesis the idea that ecosystems are like airplanes with rivets; if too many species (or rivets) are lost, the ecosystem will crash quickly (just like an airplane)

root an organ that anchors a plant to its surface and enables it to absorb water and nutrients from the soil

root hair tiny lateral extension of a root's outer cells that absorbs water and minerals from the soil

salivary gland secretes saliva at the presence of food in the mouth; saliva contains enzymes that begin the breakdown of carbohydrates into simple sugar subunits

sarcomere contraction unit of muscle; made up of the proteins actin and myosin

scientific method set of procedures scientists use in their investigations; includes four steps: observations and facts, hypotheses and predictions, testing, and evaluation and interpretation of results

scientific name the unique classification of each species from kingdom to species

screening test evaluation to see whether an embryo, fetus, baby, child, or adult is at risk for a particular problem

scrotum a pouch that holds the testes

secondary compound chemical made by plants that is usually poisonous and is not needed by the plant for its metabolism

secondary consumer carnivore

secondary sex determination the step in sexual development that governs the development of sex-related body characteristics such as external genitals, structure of pelvic bone, voice tone, and locations of body fat and hair

secretion compounds diffuse out of the blood and are moved by transport proteins into the filtrate

seed a plant embryo surrounded by a protective coating

seed coat a protective cover over the seed that keeps it from drying out

semen the final combination of sperm and secretions by the prostate gland, seminal vesicles, and bulbourethral glands

semilunar valve opens when the heart contracts to allow blood to leave the heart and closes when the ventricle relaxes to prevent blood from being sucked back in; lies between each ventricle and the vessel it uses to send blood from the heart

sensor the first part of the homeostatic system; it measures the property being regulated, such as temperature or water level in the blood

sepal the outer protective cover of a flower bud

set point the desired, or normal, value of a homeostatic system; for a human, the temperature set point is 37°C

sexual reproduction reproduction that occurs when an egg cell from a female fuses with a sperm cell from a male to form a new individual

shock the body's response to an inadequate supply of blood for circulation; it usually results from blood loss

sickle cell a type of abnormal red blood cell that is shaped like a sickle

sickle cell disease an inherited genetic disorder in which red blood cells sickle easily, leading to potentially serious physical consequences

signal transducer a molecule inside the cell that relays the information once a signal molecule binds to a receptor

skeleton composed of 206 bones; the human skeleton supports the body, protects soft parts, and stores minerals

somatic cell nuclear transfer (SCNT) therapeutic cloning in which a nucleus is taken from a healthy cell and placed into an egg that has had its own nucleus removed; the transplanted nucleus then directs the development of an embryo from which embryonic stem cells can be removed

somatic gene therapy the insertion of a normally functioning gene into a patient to cure a genetic disorder

somatic system a functional component of the PNS that carries signals to and from the skeletal muscles, usually in response to an external stimulus; it is under conscious control

speciation a process by which species branch off from existing species to form new species

species abundance the number of individuals of each species present in an ecological community

species evenness the pattern of species abundance

species richness the diversity of species alive in an area

species-area relationship the relationship between the area of an island, or any isolated area, and the number of species it has

spermicide drug that kills sperm

sphincter tiny band of muscles that forms a ring around the entrance to the capillary bed; it contracts and relaxes to control blood flow; **Or** ring of muscle that regulates the opening of a tube between the esophagus and stomach

spinal cord a bundle of nerve fibers in our backbone that serves as the communication pathway between the brain and the rest of the body

spindle formed from the microtubules of the cytoskeleton; this is a structure that helps to distribute chromosomes to new cells

stability the ability of an ecosystem to resist environmental disturbances and return to its original configuration after them

stamen the male part of a flower

stem connects a plant's roots to its leaves and flowers; it provides the structural support for the plant

stem cell a cell that can renew itself and give rise to other cells

steroid drug that mimics the functions of normal sex hormones such as testosterone and makes muscles bigger

surrogacy an embryo is placed in the uterus of a woman who is not the biological mother in order to establish a healthy pregnancy that results in the birth of a baby

sympathetic branch part of the autonomic nervous system that prepares the body for intense action by diverting "nonessential" expenditures and increasing heart rate and breathing rate

synapse the gap between the axon terminal of one neuron and the dendrites of the next one

systemic circuit the mechanism by which the left-hand pump of the heart receives fresh blood from the lungs and pumps it to the rest of the body

T lymphocyte, or **T cell** produced by the bone marrow, it is the main player in cell-mediated immunity

telophase I and cytokinesis in meiosis, the stage in the first round of cell division when the cell pinches and divides in two; each new cell has a complete set of organelles and chromosomes

telophase II in meiosis, the stage in the second round of cell division when the nuclear membrane reforms, enclosing the chromosomes

tendon cord of tissue that connects muscles to bones

terminal bud the end of the axon, which transmits information to other cells

test a procedure that sets up the conditions the predictions require

testability a procedure for determining the evidence in support of a hypothesis

testis male sex organ where sperm are produced

testosterone a hormone that promotes the development of the Wolffian ducts, the prostate gland, and the penis and regulates the descent of the testes into the scrotum

theory an idea, supported by evidence, that provides a bigger picture than a hypothesis of how some aspect of nature works; it may weave together supporting evidence from several scientific fields

theory of island biogeography the proposition that holds that the number of species found on an island is determined by the rates of immigration and colonization by new species and extinction of established species

therapeutic cloning producing embryos as sources of healthy embryonic stem cells for the medical treatment of individuals

thymus organ of the immune system that helps make white blood cells, especially in babies and children

total fertility the number of children a woman could have, on average, if she lived at least until the end of her reproductive years

totipotent a cell that is capable of producing any of the cells in a developing embryo

toxin poison

trachea windpipe that leads deep into the lungs

trait characteristic

transcription the process of copying or transcribing genetic information from DNA to RNA

transcription factor a molecule that regulates which genes are active in a cell

transduction the transfer of bacterial DNA from one bacterium to another by a virus

transfer RNA (tRNA) a carrier molecule; at one end is the anticodon, and at the other is a binding site to which a specific amino acid is attached

transformation the process by which a cell takes in and uses DNA from a foreign source

translation the process by which the information in RNA is converted to a new language: amino acid sequences

transmission the passage of a pathogen from one host to another

trophic level level that organisms occupy on the food web

tubal ligation a surgical procedure in which a woman's oviducts are cut and the ends are sealed so that the eggs cannot get into the uterus

tumor a lump that results from the production of extra cells

tumor suppressor a gene that encodes for a protein that tells a cell not to divide

ultrasound a device that relies on high-frequency sound waves to send back images of internal organs

urea nitrogen-containing compound found in urine

uterus the womb; the hollow, muscular organ located between the bladder and the rectum of females

vaccination the administration of material to provoke the development of immunity to a specific pathogen

variation traits that differ, often by geographic location

vasectomy a surgical procedure in which the vas deferens is snipped and tied off to prevent sperm from entering semen

vein a vessel that transports blood toward the heart

ventricle a muscular pump of the heart

vertical gene transfer the transference of genes from parents to offspring

vesicle membrane-enclosed bubble

Vibrio cholerae a comma-shaped bacterium that causes cholera

virulence a measure of a pathogen's ability to cause damage or death

virus particle that bridges the gap between living and nonliving; cannot replicate unless it is in a host cell

vitamin a micronutrient that is not broken down for energy or destroyed during use; it assists enzymes in the regulation of specific cellular chemical pathways

white blood cell part of the immune system that works to protect you from disease

Xeroderma pigmentosum (XP) a rare disorder in which individuals are extremely sensitive to ultraviolet (UV) light and more easily develop skin cancers if they are exposed to sunlight or other sources of UV light

X-ray a type of relatively high-energy radiation

xylem part of a stem that transports water and minerals from the roots to the leaves and flowers

zygote a one-celled embryo produced as a result of fertilization

Credits

CHAPTER 1

Chapter Opening Photo PHILIPPE HUGUEN/AFP/Getty Images; **1.1** ollyy / Shutterstock .com; **Technology Connection, p. 6** Tom McHugh / Science Source; Robert J. Erwin / Science Source; **1.2** AP Photo/Harry Cabluck; **1.3** Nastya Pirieva / Shutterstock.com; **1.4** Svetlana Valoueva / Shutterstock.com; cybervelvet / Shutterstock.com; **1.8** Artpose Adam Borkowski / Shutterstock.com; aodaodaodaod / Shutterstock.com; Elena Schweitzer / Shutterstock.com; **1.9** Eric V. Grave / Science Source; **Scientist Spotlight, p. 13** Chip Somodevilla/Getty Images; **1.11** Copyright © 2011, National Science Teachers Association (NSTA). Reprinted with permission from *The Science Teacher*, Vol. 78, No. 9, December 2011.; **Life Application, p. 18** Alfred Pasieka / Science Source; **1.15** Andrew Astbury / Shutterstock.com; **1.20** Elena Schweitzer / Shutterstock.com

CHAPTER 2

Chapter Opening Photo emin kulyev / Shutterstock.com; **2.1** Josef & Peter Schafer, {{PD-1996}}; **2.2** Universal History Archive/Getty Images; **2.3** University of Michigan Museum of Art, Collection of the University of Michigan Health System, Gift of Pfizer Inc., UMHS.26; **Life Application, p. 40** Public Health Image Library, Centers for Disease Control and Prevention; **Scientist Spotlight, p. 41** Science Source; PAUL J. RICHARDS/AFP/Getty Images; **2.7** Stefan Redel / Shutterstock.com; Sergey Panychev / Shutterstock.com; chinahbzyg / Shutterstock .com; Fotokostic / Shutterstock.com; Jan Martin Will / Shutterstock.com; VannPhotography / Shutterstock.com; Vladislav Gurfinkel / Shutterstock.com; **2.8** oksankash / Shutterstock.com; James Steidl / Shutterstock.com; **Technology Connection, p. 50 A and B** Public Health Image Library, Centers for Disease Control and Prevention; **Technology Connection, p. 50** Jarrod Erbe / Shutterstock.com; **2.9** Photosani / Shutterstock.com; Marco Brivio; Mike Tan C.T. / Shutterstock.com; Biophoto Associates / Science Source; Vlododymyr Goinyk / Shutterstock.com; Paul Maguire / Shutterstock.com; **2.10** Michael Richardson; **2.11** Art Lien; **2.12** Crepesoles / Shutterstock.com; iStockphoto/Juanmonino

CHAPTER 3

Chapter Opening Photo Denis Kukareko / Shutterstock.com; **3.1** Image courtesy of the Hensel Family and Figure 8 Films; **3.3** Science Source; **Scientist Spotlight, p. 63** Science Source; **3.5** Lebendkulturen.de / Shutterstock. com; Juan Gaertner / Shutterstock.com; **3.6** Michelangelus / Shutterstock.com; **3.7** CNRI / Science Source; **3.12** David M. Phillips / Science Source; **3.13** Michael Abbey / Science Source; Michael Abbey / Science Source; Michael Abbey / Science Source; Michael Abbey / Science Source; Michael Abbey / Science Source; **3.14** Claude Edelmann / Science Source; **Technology Connection, p. 79** Dmitry Kalinovsky / Shutterstock.com; **Life Application, p. 80** Rick's Photography / Shutterstock.com; **3.28** Blank Archives/Getty Images; **3.29** ASSOCIATED PRESS

CHAPTER 4

Chapter Opening Photo Eric Isselee / Shutterstock.com; **4.1** Henrik Larsson / Shutterstock .com; **4.2** Sebastian Kaulitzki / Shutterstock.com; **4.9** Tom Hollyman / Science Source; **4.12** © Bettmann/CORBIS; **4.13** image courtesy of the Mendel Museum and Abbot Lukáš Evžen Mastinec; **4.18** africa924 / Shutterstock.com; **4.19** Rachelle Burnside / Shutterstock .com; John Copland / Shutterstock.com; Kletr / Shutterstock.com; Migel / Shutterstock.com; **Scientist Spotlight, p. 110, A and B** Science Source; **4.31** Sebastian Kaulitzki / Shutterstock .com; **Life Application, p. 120** iStockphoto/ralcro

CHAPTER 5

Chapter Opening Photo Photo Researchers; **5.1** Science Source; Science Source; Scott Camazine / Science Source; Science Source; **5.4** Sakuoka / Shutterstock.com; CNRI / Science Source; **5.14** Nightman1965 / Shutterstock.com; **5.16** Dmitry Matrosov / Shutterstock.com; Michael Drager / Shutterstock.com; WM_idea / Shutterstock.com; Hemanta Kumar Raval / Shutterstock.com; OZaiachin / Shutterstock.com; Birgit Reitz-Hoffman / Shutterstock.com; 3DDock / Shutterstock.com; wonderisland / Shutterstock.com; **Scientist Spotlight, p. 142** Science Source; **5.20** Monkey Business Images / Shutterstock.com; **5.21 A** Levent Konuck / Shutterstock.com; **5.21 B** Simon Fraser / Science Source; **Technology Connection, p. 147 A** iStockphoto/yumiyum; **B** Nightman1965 / Shutterstock.com; **5.23** Science Sourcel; **5.24** Véronique Burger / Science Source; **Life Application, p. 150** Serg64 / Shutterstock.com

CHAPTER 6

Chapter Opening Photo Sergey Uryadnikov; **6.1** FRANCK FIFE/AFP/GettyImages; **6.3** Kletr / Shutterstock.com; **6.7** Carlo Allegri/Getty Images; **Life Application, p. 162** Bettmann/ Corbis / AP Images; **6.15**DUSAN ZIDAR / Shutterstock.com; Ikeskinen / Shutterstock.com; PHANIE / Photo Researchers, Inc.; Gary Parker / Science Source; Ray Ellis / Science Source; Scott Camazine / Science Source; GARO / PHANIE / Science Source; Phanie / Science Source; **How Do We Know?, p. 171** David M. Phillips / Science Source; **6.16** SPL / Science Source; Scimat / Science Source; SCIMAT / Science Source; Steve Gschmeissner / Science Source; **Technology Connection, p. 176** Tinydevil / Shutterstock.com; **6.19** Will & Deni McIntyre / Science Source; **6.20** Saturn Stills / Science Source; **6.21** ZouZou / Shutterstock.com; **Scientist Spotlight, p. 178** Science Source

CHAPTER 7

Chapter Opening Photo Monika Wisniewska / Shutterstock.com; **7.1** Eyerounds.org, University of Iowa; **7.2** iofoto / Shutterstock.com; Max Sudakov / Shutterstock.com; Drozdowski / Shutterstock.com; Lev Kropotov / Shutterstock.com; CoolR / Shutterstock.com; Sari ONeal / Shutterstock.com; **7.4** Dr. Jeremy Burgess / Science Source; **7.6** Martin Shields / Science Source; **7.7 A** Curioso / Shutterstock.com; **B** Mary Lane / Shutterstock.com; **C** Tamara Kulikova / Shutterstock.com; **7.11** Biehler Mitchell / Shutterstock.com; Alex James Bramwell / Shutterstock .com; Josef F. Steufer / Shutterstock.com; Paul Henjum; **7.15** Scott Camazine / Science Source; **7.16** Victor M. Vicente Selvas; **7.17** International Rice Research Institute; **Scientist Spotlight, p. 201** Vince Bucci/AFP/Getty Images; **Technology Connection, p. 203** Steve Jurvetson; **Life Application, p. 205** *Nicolle Rager Fuller, National Science Foundation* **7.21 A** LIU JIN/ AFP/Getty Images; **B** PO3 Patrick Kelley; **7.22** National Human Genome Research Institute; **7.23 A** Ted Kinsman / Science Source; **B** Blue Rose Man; **C** Nathan Shaner; **7.24** Associated Press

CHAPTER 8

Chapter Opening Photo Paul Cowan / Shutterstock.com; **8.1** M.E. Grenander Department of Special Collections & Archives, University at Albany, State University of New York; **8.2** M.E. Grenander Department of Special Collections & Archives, University at Albany, State University of New York; **8.3** American Philosophical Society; **8.4** University Hospitals of Cleveland, Case Western Reserve University, Dr. Guyuron; **8.5** sam100 / Shutterstock.com; **Scientist Spotlight, p. 223** Acey Harper//Time Life Pictures/Getty Images; **8.6** NYPL/Science Source; **8.8** Getty Images; **8.11** SARAH LEEN/National Geographic Creative; **8.12** Kara Jade Quan-Montgomery / Shutterstock.com; **8.14** Alex Wong/Getty Images; Chip Somodevilla/ Getty Images; **8.16** Simon Fraser / Science Source; Department of Health and Human Services; **Technology Connection, p. 238** Martin Shields / Science Source; **8.17** Jaren Jai Wicklund / Shutterstock.com; Vitalinka / Shutterstock.com; PUNIT PARANJPE/AFP/Getty Images

CHAPTER 9

Chapter Opening Photo Rita Kochmarjova / Shutterstock.com; **9.5** BIOPHOTO ASSOCIATES; **9.7** Ludmila Yilmaz / Shutterstock.com; Kerry L. Werry / Shutterstock.com; **9.9** Murray, Patti / Animals Animals; Pal Teravagimov / Shutterstock.com; worldswildlifewonders / Shutterstock .com; Christian Musat / Shutterstock.com; **9.10** H. Zell; **Scientist Spotlight, p. 257** A. Barrington Brown / Science Source; **9.17** K. Thalia Grant; **9.18** Stubblefield Photography / Shutterstock.com; **9.24** Rubberball/Mike Kemp; **9.27** Three Lions; **9.28** SPL / Science Source;

9.29 Rita Kochmarjova / Shutterstock.com; IrinaK / Shutterstock.com; Eric Isselee / Shutterstock .com; fivespots / Shutterstock.com; **9.30** Ardea/Boulton, Mark / Animals Animals; Mark Interrante / Shutterstock.com; Zhilitsov Alexandr / Shutterstock.com; **9.31** D.P. Wilson / FLPA / Science Source; **9.33** Martin Darley / Shutterstock.com; Jane Hamalainen / Shutterstock.com; Delmas Lehman / Shutterstock.com; **9.34** Stubblefield Photograph / Shutterstock.com; Martha Marks / Shutterstock.com; Gibson, Mickey / Animals Animals; Thomas & Pat Leeson / Science Source; Jason Patrick Ross / Shutterstock.com; **9.35** igor.stevanovic / Shutterstock.com; Matt Knoth / Shutterstock.com; Norman Bateman / Shutterstock.com; Ardea/Boulton, Mark / Animals Animals; Zhilitsov Alexandr / Shutterstock.com; **9.36** FPG/Archive Photos/Getty Images; **9.37** choosemyplate.gov

CHAPTER 10

Chapter Opening Photo Tischenko Irina / Shutterstock.com; **10.2 A** fusebulb / Shutterstock .com; **B** Sebastian Kaulitzki / Shutterstock.com; **C** Andrew Syred / Science Source; **How Do We Know?, p. 287** Joe Lertola, Bryan Christie Design; **10.6 A** Bryan Maudsley / Shutterstock.com; **B** Nigel Cattlin / Science Source; **Scientist Spotlight, p. 295** Associated Press; **10.11** Richard Harvey; **10.13** Science Source; **10.16 A** Bplanet; **B** Julia Lutgendorf; **C** O2creationz; **10.18 A** Kletr / Shutterstock.com; **B** Mark Bowler / Science Source; **10.19** Anti-Slavery International / Panos

CHAPTER 11

Chapter Opening Image jesadaphorn / Shutterstock.com; **11.1** Associated Press; **Life Application, p. 322** Getty Images; **11.9 A** Eric Isselee / Shutterstock.com; **B** Tyler Olson / Shutterstock .com; **11.12 A** Nancy Tripp / Shutterstock.com; **B** Johann Piber; **11.13** silver-john / Shutterstock .com; **Scientist Spotlight, p. 330** National Library of Medicine; **Technology Connection, p. 334** Picsfive / Shutterstock.com; **11.19** Sergii Figurnyi / Shutterstock.com; studioVin / Shutterstock.com; Peter Bernik / Shutterstock.com

CHAPTER 12

Chapter Opening Image Rich Carey / Shutterstock.com; **12.2** Alexander Gordeyev / Shutterstock .com; **How Do we Know?, p. 360** Leonid Shcheglov; **12.23** Science Source; **Scientist Spotlight, p. 368** Associated Press; **Technology Connection, p. 370** Antonia Reeve / Science Source

CHAPTER 13

Chapter Opening Image Digital Storm / Shutterstock.com; **13.1** Library of Congress Prints and Photographs Division; **13.2** Warren Anatomical Museum, Francis A. Countway Library of Medicine; **Technology Connection, p. 384** Living Art Enterprises / Science Source; **13.9** Author of underlying work unknown; **13.10** from H. Damasio, T. Grabowski, R. Frank, A.M. Galaburda & A.R. Damasio (1994), The Return of Phineas Gage: clues about the brain from a famous patient, *Science*, 264, 1102–1105. Dornsife Neuroscience Imaging Center and Brain and Creativity Institute, University of Southern California; **13.11** Juan Gaertner / Shutterstock .com; **Scientist Spotlight, p. 391** Associated Press; **13.21** Biophoto Associates / Science Source; **13.24** drserg / Shutterstock.com; **13.26** Alexey Lysenko / Shutterstock.com; kzww / Shutterstock .com; photopixel / Shutterstock.com; robtek / Shutterstock.com; Andrew Burns / Shutterstock .com; mikeledray / Shutterstock.com; **How Do We Know?, p. 404** Photo Researchers, Inc.; **13.27** Roll Call/Getty Images; AFP/Getty Images

CHAPTER 14

Chapter Opening Photo Sebastian Kaulitzki / Shutterstock.com; **14.1** *The Gardeners' Chronicle*; **14.3** Sébastien Champoux; **14.4** Julie Dermansky / Science Source; **14.5 A** John Snow; **B** Justin Cormack; **14.6** Knorre / Shutterstock.com; **14.8** Dr. Kari Lounatmaa / Science Source; **14.10** Science Source; **14.11** Dr. Stan Erlandsen, Dr. Dennis Feely; Biophoto Associates / Science Source; Science Source; **14.12** Dr. A. Harrison; Dr. P. Feorino; **14.13** PHANIE / Science Source; **14.14** SPL / Science Source; Eye of Science / Science Source; **14.18** Zurijeta / Shutterstock.com; **14.19** Millard H. Sharp / Science Source; **14.20** ggw1962 / Shutterstock.com; **Scientist Spotlight, p. 433** Michael Kovac/Getty Images for The Elton John AIDS Foundation

CHAPTER 15

Chapter Opening Photo mozakim / Shutterstock.com; **15.1** John Storey/Time & Life Pictures/ Getty Images; **15.2** Alexander.Yakovlev / Shutterstock.com; **15.3** Taylor Hooton Foundation (taylorhooton.org); **15.4** U.S. Department of Agriculture; **15.5 A** victoriaKh / Shutterstock.com; **B** aodaodaodaod / Shutterstock.com; **15.6** Elena Schweitzer / Shutterstock.com; **15.7** DUSAN ZIDAR / Shutterstock.com; **Scientist Spotlight, p. 445** Tom Hollyman / Science Source; **15.8** Dulce Rubia / Shutterstock.com; Artem Samokhvalov / Shutterstock.com; Anna Hoychuk / Shutterstock.com; Anna Kucherova / Shutterstock.com; Gregory Gerber / Shutterstock.com; Pakhnyushcha / Shutterstock.com; MARGRIT HIRSCH / Shutterstock.com; **15.9** tacar / Shutterstock.com; nito / Shutterstock.com; Dulce Rubia / Shutterstock.com; Joshua Resnick / Shutterstock.com; AVprophoto / Shutterstock.com; Leonid Shcheglov / Shutterstock.com; Lilyana Vynogradova / Shutterstock.com; Litvinenko Anastasia / Shutterstock.com; Oliver Hoffmann / Shutterstock.com; **15.17 A** crystalfoto / Shutterstock.com; **B** Dave Kotinsky / Shutterstock.com; **15.19 A** Elena Rostunova / Shutterstock.com; **B** Darren Hubley / Shutterstock .com; **Life Application, p. 458** Kzenon / Shutterstock.com; **15.25 A** Laurence Griffiths/ Getty Images; **B** Istvan Csak / Shutterstock.com; **Technology Connection, p. 463** Andresr / Shutterstock.com

CHAPTER 16

Chapter Opening Photo Henk Bentlage / Shutterstock.com; **16.1** B. & E. Boggs / Science Source; **16.2** gregg williams / Shutterstock.com; **16.3** bikeriderlondon / Shutterstock.com; **16.4** Sergey Yechikov / Shutterstock.com; **16.5** Albie Venter / Shutterstock.com; **16.6** Michael Abbey / Science Source; Lebendkulturen.de / Shutterstock.com; **16.7** Stubblefield Photography / Shutterstock.com; **Scientist Spotlight, p. 479** Seeley G. Mudd Manuscript Library, Princeton University; **16.8** Tom & Pat Leeson / Science Source; **16.9** Mark Herreid / Shutterstock.com; **16.10** © Natural History Museum, London; **16.12** Mauro Rodrigues / Shutterstock.com; **Technology Connection, p. 486 A** Science Source **B** Aleksey Stemmer / Shutterstock.com; **How Do We Know?, p. 489** Dr. Carleton Ray / Science Source; **Life Application, p. 492** Willem Tims / Shutterstock.com

CHAPTER 17

Chapter Opening Photo larus / Shutterstock.com; **17.2** Mark Hallett Paleoart / Science Source; **17.3** Mauricio Anton / Science Source; **Life Application, p. 503** Sinclair Stammers / Science Source; **17.4** AlenKadr / Shutterstock.com; **17.5** Nicram Sabod / Shutterstock.com; Juriah Mosin / Shutterstock.com; Julia Ivantsova / Shutterstock.com; **17.6** Tischenko Irina / Shutterstock .com; **17.7** John R. McNair / Shutterstock.com; **17.9** Jayne Chapman / Shutterstock.com; CT Johansson; **17.13** Mike Baird; Dr. Dwayne Meadows, NOAA/NMFS/OPR; **17.14** William F. Campbell//Time Life Pictures/Getty Images; **17.15** Sally Scott / Shutterstock.com; **Technology Connection, p. 520** United Nations Environment Program; **Scientist Spotlight, p. 523** Cindy Ord/Getty Images for World Science Festival; **17.18** U.S. Geological Survey

CHAPTER 18

Chapter Opening Photo jkirsh / Shutterstock.com; **18.3** Matthijs Wetterauw / Shutterstock .com; **18.4** Vadim Petrakov / Shutterstock.com; **18.5** Pal Teravagimov / Shutterstock.com; **18.7** AP Photo/Emilio Morenatti; **Scientist Spotlight, p. 540** Donella Meadows Institute; **18.8** Jubal Harshaw / Shutterstock.com; Nicram Sabod / Shutterstock.com

Index